DEFINITIONS

$$kW = kNm/s \qquad Pa = N/m^2 \qquad N = kg\ m/s^2 \qquad sl = lb/g$$

g = acceleration due to gravity at sea level

$$= 9.80665\ m/s^2 = 32.1740\ ft/s^2$$

UNITS

Force

$$1\ lb = 4.448\,221\ N \qquad 1\ N = 0.224\,808\,94\ lb$$

Length

$$1\ in. = 25.4\ mm \qquad 1\ micron = 10^{-6}\ m$$

Mass

$$1\ lbm = 0.453\,592\ kg \qquad 1\ kg = 0.068\,521\,76\ sl \qquad lbm = pound\ (mass)$$

$$1\ slug = 14.593\,903\ kg = 32.174\,05\ lbm \qquad sl = slug$$

Power

$$1\ hp = 550\ ft/s = 0.7457\ kW \qquad 1\ kW = 1.3410\ hp$$

$$hp = \pi Tn/198\,000 \qquad T = torque\ (in\text{-}lb) \qquad n = revolutions/minute$$

$$kW = \pi Tn/30\,000\,000 \qquad T = torque\ (Nmm)$$

Pressure and Stress

$$1\ bar = 100\,000\ Pa = 1.013\,25\ atm = 14.503\,8\ psi$$

$$1\ MPa = 145.037\,74\ psi \qquad 1\ psi = 6.894\,757\,2 \times 10^{-3}\ MPa$$

Volume

$$1\ gal. = 231\ in.^3 = 3.785\,411\,784\ liters \qquad 1\ liter = 0.001\ m^3 = 1000\ cm^3$$

Work

$$1\ in\text{-}lb = 112.9848\ Nmm \qquad 1\ Nm = 1\ J = 8.850\,746\ in\text{-}lb \qquad J = Joule$$

$$1\ Btu = 777.9805\ ft\text{-}lb = 251.996\ calories = 1055.056\ Nm\ (J)$$

$$1\ calorie = 4.186\,80\ Nm$$

MACHINE COMPONENT DESIGN

MACHINE COMPONENT DESIGN

WILLIAM C. ORTHWEIN

WEST PUBLISHING COMPANY

ST. PAUL · NEW YORK · LOS ANGELES · SAN FRANCISCO

Cover and text design: Adrianne Onderdonk Dudden
Copyediting: Virginia Dunn
Composition: Syntax International
Art: ANCO/Boston, Inc.
Cover Illustrations: *Top and middle:* Mechanical Drives Division,
  Zurn Industries, Inc. *Bottom:* Thomson-Saginaw Ball
  Screw Company.

Copyright © 1990   by WEST PUBLISHING COMPANY
                   50 W. Kellogg Boulevard
                   P.O. Box 64526
                   St. Paul, MN 55164-1003

Printed in the United States of America

96 95 94 93 92 91 90 89   8 7 6 5 4 3 2 1 0

Library of Congress Cataloging-in-Publication Data

Orthwein, William C., 1924–
    Machine component design / William C. Orthwein.
        p.   cm.
    Includes bibliographies and index.
    ISBN 0-314-24257-0
    1. Machinery—Design.   I. Title.
TJ230.0745 1989
621.8′15—dc19   88-28267
                                                 CIP

CONTENTS

CHAPTER THREE

STRESS AND DEFLECTION   91

CHAPTER TWELVE

BELT AND CHAIN DRIVES   709

CHAPTER THIRTEEN

SPUR GEARS   763

CHAPTER FOURTEEN

STRAIGHT BEVEL, HELICAL, AND WORM GEARS   837

CHAPTER FIFTEEN

MOTOR CHARACTERISTICS   889

CHAPTER SIXTEEN

HYDRAULICS   931

APPENDIX C

DESIGN EXAMPLE   C-1

PREFACE

The number of pages and the number of topics included in this book clearly indicate that it is not possible to study all of the material in this text in a one-semester course in machine design. This text is intended to fill one or more of the following functions:

1. To allow the professor a wide range for the selection of those topics that are important to the particular curriculum in the university department in which the course is being taught.

2. To allow the same text to be used in a two- or three-semester course sequence in machine design in those university departments that emphasize this aspect of mechanical engineering.

3. To provide the student with a text that may be retained for use in the practice of mechanical engineering after graduation.

Prerequisites are a knowledge of statics and dynamics, calculus through differentiation and integration, and vector analysis, and basic familiarity with numerical methods and computer programming. In most universities this is equivalent to stating that this text is intended for a junior or senior level machine design course.

The variety of topics and the depth of the coverage provided should enable the text to be used in one-, two, or three-semester machine design programs. In the case of a one-semester course, the instructor may select those topics suitable to the department's program and the requirements of the industries served by the university. Since all but Chapters 13 and 14 depend only upon Chapter 1 (for the interpolation and bisection routines), the order in which the components are studied may be selected to suit the instructor's preferences for a two- or three-semester program. It is expected that most one-semester courses would begin with Chapter 1 and all or part of Chapters 3 and 4 before moving on to component design and selection in Chapters 3 through 15. Chapter 16 differs from the preceding chapters in that a variety of components is considered in one chapter.

Because of additions to the theory and the availability of relatively inexpensive programmable calculators and personal computers, many of the chapters contain material that is new or is used in engineering in industry but is not included in other machine design texts at this time.

Chapter 1 includes listings in Microsoft Quick Basic and in the HP-41CX code for a bisection routine that may be used to solve many of the design relations in the following chapters that can only be solved numerically. The availability of numerical methods makes it practical to include many realistic design problems not found in other texts. Chapter 1 also includes a new interpolation formula that is especially suited to interpolating stress concentration factors in those regions where the concentration factors are not linearly dependent upon the parameters and where Lagrangian interpolation yields erroneous values. Visual interpolation of logarithmic scales, improvement of the precision in reading graphical data, and function reconstruction from logarithmic plots for use with programmable calculators or personal computers are also described.

The discussion of materials in Chapter 2 is more extensive than that in other machine design texts in that plastics and ceramics are also included. It is not, however, a substitute for texts on materials, or for the exhaustive data and discussion pertaining to metals found in the *Metals Handbook*, published by the American Society for Metals, Metals Park, Ohio.

Chapter 3 differs from earlier machine design texts by considering principal stresses in three dimensions and including a flowchart for a program for their determination. Use of that, or a similar, program eliminates the tedium usually associated with these calculations. With it we may find the significant principal stresses in many important machine components, such as shafts where gears, wheels, and similar machine components induce bending, torsion, and transverse shear stresses in the shaft and, perhaps, in the component itself. Although an explanation of the Mohr circle is given for those instances where stresses are two dimensional, use of the program is preferred because it specifically calculates shear stresses in all three directions, not just in the plane of the Mohr circle stresses. Finally, stress components are written in terms of the tensor notation, as employed in finite element analysis and advanced stress analysis.

Chapter 4 describes fatigue analysis, including recent criteria for three-dimensional fatigue analysis beyond the yield stress which provide improved agreement with experimental data and do not contain the internal contradictions that plagued earlier attempts to establish three-dimensional fatigue criteria. Chapter 4 also introduces the student to the Gerber-yield criterion for indefinite fatigue life and provides a detailed discussion of low-cycle fatigue that enables the reader to apply low-cycle, or strain, methods to the analysis of fatigue failure, as demonstrated in several problems. Analysis of finite fatigue life using the Palmgren-Miner and the Shanley methods is included along with relations to include the effect of mean stress.

The study of specific components begins in Chapter 5, which is devoted to helical tension, compression, and torsion springs, to Belleville springs, to shock mounts, and to pneumatic springs. Design formulas for Belleville and pneumatic springs are unique to this text at this time. The inclusion of diameter changes for torsional and compressive springs and a design formula to fit the MIL specifications for spring buckling also appear to be unique to this text, as is inclusion of stresses in the end loops for tension springs. Although the MIL specification formula (but not the curve) appears to be new, buckling and the other aspects are routinely considered in industry.

An iteration method is presented in Chapter 6 to faciliate the design of power screws by enabling the user to select the lead angle best suited for the power screw application at hand. This sets this chapter apart from texts that deal only with finding the characteristics of existing power screws, i.e., screws in which the lead angle and pitch are given. This chapter also contains the important design consideration of screw buckling according to the Euler and Johnson formulas for long and short screws, with an explicit criterion given to distinguish long from short screws. Ball and roller screws are also considered, and standard design formulas used in industry are used in the assigned problems for ball and roller screws. Inclusion of these screws and their selection formulas are also unique to this text.

Bolt/rivet, weld, and adhesive connections are discussed in Chapter 7 where vector formulas are used to calculate the bolt/rivet and weld stresses due to in-plane eccentric load. Examination of these formulas provides straightforward design criteria for the placement of bolts/rivets and welds, so that design problems may be assigned. These vectors formulas and presentation of formulas for adhesive joints is another distinction between this text and other machine design texts available in the United States at this time.

Power shaft design is described in Chapter 8, which also includes new material on key and keyway design and presents cam design without the unnecessary use of phasors. Although couplings are discussed in most of the existing machine design texts, the description of universal joints, flexible shafts, and retaining rings and their design formulas appears to be unique to this text. Shaft deflection is obtained by computer programs based upon the elastic energy method (or unit load method) introduced by van den Broek. This unique feature can eliminate much of the tedium associated with the design of realistic power shafts with numerous changes in diameter to position bearings and gears.

Chapter 9 deals with lubrication as applied to slider and journal bearings using the design curves introduced by Raimondi and Boyd based upon the Reynolds equation. The Reynolds equation is derived in the usual manner and it is also obtained from the Navier-Stokes equation to show the relation between the two and to introduce hydro-static and squeeze film lubrication. The chapter also includes sections on viscosity, additives, and solid lubricants. It closes with an interesting approximation routine for estimating the load, angle, and friction for full journal bearings.

Ball and roller bearings are described in Chapter 10, which begins with photographs and illustrations of all of the major types. Load–life relations follow for ball and roller bearing selection when subjected to steady and fluctuating loads. Unique to this text is the discussion, with load–life formulas, of opposed mounting of angular contact ball and tapered roller bearing and the integral formulation for the equivalent load. Bearing lubrication, mounting design, and seals and enclosures are also discussed.

Conventional friction brakes and clutches and newer nonfriction brakes and clutches, such as magnetic particle, hysteresis, and eddy current units, are described in Chapter 11. In addition to these features not found in other texts, dual shoe drum brakes are discussed, a flowchart for dual shoe brake design is presented, and a refer-ence is given to a specific program listing in the literature. Centrifugal and one-way, or overrunning, clutches are described along with examples of their application. Torque converters are introduced, their operating principles are given and an approximate formula for their torque is supplied. A tabular comparison of the major types of clutches and brakes closes this chapter.

Belt and chain drives are considered in Chapter 12 using formulas and criteria recommended by the belt and chain drive associations. Flowcharts are provided for programs to speed the design and to replace many, but not all, of the tables normally supplied for belt and chain drive design.

Chapters 13 and 14 are devoted to gear design and selection. All of the relations required to generate an involute tooth are given along with a flowchart for a program that may be used to calculate the involute profile and draw it on a screen or plotter.

Standard and nonstandard tooth proportions are considered. A unique feature of Chapter 13, which is devoted to spur gears, is a short program for recent methods of calculating the numbers of gear teeth required to achieve a specified speed ratio to any desired degree of accuracy. Bevel, helical, and worm gears are considered in Chapter 14.

An important aspect of machine design that has been omitted by most machine design texts has been that of selecting a driver for whatever machine is to be designed. This aspect is the subject of Chapter 15 where motor and engine torque–speed (or speed–torque) curves are given for common single-phase and three-phase ac motors, for dc motors, and for diesel and gasoline engines. Likewise, torque–speed curves are shown for a variety of driven machines and examples are given of selecting a suitable driver for a particular machine. Although this chapter is largely descriptive, formulas are given for the speed control of dc motors and for acceleration times to bring a load up to speed.

An important class of devices not considered by previous texts on machine design is that of hydraulic components. A drive past most construction sites, a visit to a farm, or a walk through a manufacturing plant will clearly demonstrate the prevalence of hydraulic equipment. Since hydraulic components are of primary importance in many machines throughout the world, it is difficult to justify their omission from a machine design text. They are, therefore, the topic of Chapter 16.

Chapter 16 is the longest chapter in the text because it covers all of the essentials in the design of hydraulic equipment from Bernoulli's theorem and a description of component such as cylinders, pumps, motors, and valves to accumulator sizing, line and component losses, pump and motor selection, and symbols used in hydraulic circuit schematics. Torque–speed curves for hydraulic motors are presented in this chapter rather than in Chapter 15 because of their close relation to other elements in the hydraulic system.

The material in these chapters has been written with the aim of providing the student or working engineer with a thorough understanding of the principles and major decisions that must be made in the design of those components listed in the table of contents. No attempt will be made at pontificating on the design process. It has been the author's experience that the appropriate design process depends upon the type of product being produced, the number of units to be produced, and the user's expected demands upon the item of equipment, its intended life, and so on. The key ingredient for a successful design in any situation is always an intelligent, well-educated design engineer who can also write and talk clearly.

Turning now to a different subject, the old slide-rule motivated habit of retaining only three and occasionally four significant digits has not been retained in this text. Since most pocket calculators used in engineering will accept up to nine digits, and since most personal computers will accept more, we shall use however many digits are either convenient or appropriate. We shall also err on the side of too many places rather than too few to minimize round-off and truncation errors in the intermediate calculations. While it is true that carrying nine digits, for example, in the elastic modulus, implies more accuracy than is statically justified in design problems, it is also true that by definition a competent engineer is one who learns enough about the machine to be designed, the components and materials to be used, and the machine's intended application to ascertain the required precision of the values used in a particular design calculation and to present the result to the appropriate number of places.

One of many examples in which retaining more than three or four digits is good practice is that of power shaft design, discussed in Chapter 8. In calculating minimum shaft diameters, maximum permissible stresses, or maximum safety factors it is important to round in the proper direction (not just based upon whether the next digit is greater or less than 5) to err on the side of safety. The easiest way to do this is to carry extra places and then round down in the case of calculated maximum stresses and safety factors and round up in the case of shaft diameters. Other examples can be found in the text and in industrial applications, such as fuel injector and turbine design.

Finally, first editions seem to always have a number of errors which go undetected until they appear in the final version. The author would appreciate being told of these errors and will acknowledge them by name and institution in subsequent editions. Errors, and their corrections, should be sent to the author at P.O. Box 3332, Carbondale, IL 62902-3332.

MACHINE COMPONENT DESIGN

CHAPTER ONE

PRELIMINARY CONSIDERATIONS

NOTATION

a	distance, or a coefficient (l), $(\ )$	m	mass (m)
b	distance, or a coefficient (l), $(\ )$	R	reactive force (ml/t^2)
F	force (ml/t^2)	r	radius (l)
$f(x)$	dependent variable, a function of x $(\ )$	X	dependent variable, orthogonal interpolation $(\ )$
g	acceleration due to gravity (l/t^2)	x	independent variable $(\ )$
L	load (force) (ml/t^2)	Y	dependent variable, orthogonal interpolation $(\ )$
M	moment (ml^2/t^2)		

This chapter provides background information that is used in most of the following chapters. In particular, the next two sections list and briefly discuss the units to be used and some of the organizations responsible for the standards used in the design of machine components. Several frequently used relations from statics are given to eliminate the task of repeatedly calculating reactions to statically determinant beam problems. We then turn our attention to interpolation formulas and function recovery. Finally, aids are given for estimating values from logarithmic plots and for improved accuracy in curve reading when grid lines are far apart.

1.1 UNITS

Both the Old English units of inches and pounds as well as the SI units of meters and kilograms will be used in this text. The term Old English, abbreviated OE, is used in preference to the term customary units because what is customary may change with time.

The abbreviations used in this text for the fundamental units of mass, force, length, and time are given in the following table.

| Quantity | OE | | SI | |
	Unit	Symbol	Unit	Symbol
mass	slug	sl[a]	kilogram	kg
force	pound	lb	newton	N
length	inch	in.	millimeter	mm
length	foot	ft	meter	m
time	second	s	second	s

[a] Symbol devised by the author (no standard symbol).

In each system the units of force and mass are often confused in commerce. Commercial scales that use OE units measure mass in units of force, pounds, so that in all engineering applications mass in units of force (i.e, weight) must be converted to correct units, slugs, according to

$$m = F/g \tag{1.1.1}$$

where g denotes the acceleration due to gravity at sea level, which is 32.174 ft/s^2 = 386.09 in./s^2 in OE units and 9.8066 5 m/s^2 in SI units. Whenever a mass is measured in terms of its weight at levels significantly different from sea level the corresponding g should be used in (1.1.1). Terminology is also an annoyance in the Old English system because the unit of mass, the slug, is defined only in units of lb-ft/s^2; there is no term for g when written in other convenient units, such as lb-in./s^2.

Although commercial scales in SI units measure mass correctly as kilograms, individuals seem to associate force with the mass that can be lifted by that force, so that in Europe force is often given in terms of kilograms rather than newtons. Hence, we must convert to correct units by using the relation

$$F = mg \qquad (1.1.2)$$

from which it follows that a newton is the force required to accelerate one kilogram one meter per second per second and that a pound is the force required to accelerate one slug 32.174 ft per second per second.

Following present practice, multiples and submultiples of SI units will be denoted by a series of prefixes:

Prefix	Symbol	Power of 10	Prefix	Symbol	Power of 10
tera	T	12	deci	d	-1
giga	G	9	centi	c	-2
mega	M	6	milli	m	-3
kilo	k	3	micro	μ	-6
hecto	h	2	nano	n	-9
deka	da	1	pico	p	-12

Although prefixes have been selected for exponents greater than 12, the more easily understood exponential notation will always be used in this text for numbers larger than 10^{12} or smaller than 10^{-12} and may be used for numbers with the listed prefix range when convenient.

Table 1.1.1 lists quantities other than mass, length, and time commonly used in machine design and their SI and OE units, along with the ratio of the SI to the OE units, which is, therefore, the conversion factor between them. All but the conversion factor between kilograms and pounds and between kg/cm^2 and psi are dimensionless. (For example, kilograms/slug has units of $\sim m/m = 1$, which is dimensionless.)

The ratio of kilograms to pounds is $0.4536 \ s^2/in.$ (mass/force) so that multiplication of the weight of an item in pounds by 0.4536 yields its mass in kilograms. The units of the conversion factor clearly indicate that the conversion is between physically different quantities, which, therefore, play different roles in engineering formulas.

Units of power are the kilowatt (kW) in the SI system and horsepower (hp) in the OE system. Units of pressure and of stress in the OE system are pounds/inch^2 (psi) and in the SI system they are either the pascal (Pa), which is one newton/meter^2, or the megapascal (MPa), which is one newton/millimeter^2. The kilogram/centimeter^2 is not fashionable in the United States because it does not have the proper units of force/area, but it is often encountered in Europe where it is physically understandable.

TABLE 1.1.1 COMMON ENGINEERING QUANTITIES, THEIR CONVERSION FACTORS AND UNITS

Quantity	SI units	OE units	Conversion factor[a]
Angle	radians	radians	1
	degrees	degrees	1
Moment of area	mm^4	$in.^4$	416 231
Linear velocity	m/s	ft/s	3.048
	mm/s	in./s	25.4
Angular velocity	rad/s	rad/s	1
	rpm	rpm	1
Moment of inertia (mass moment of inertia)	kg/mm^2	$sl\ in.^2$	0.0684 67
Force	N	lb	4.4483
Mass	kg	lb	$0.4536\ s^2/in.$
Moment	N m	lb f	1.3558
	N mm	lb in.	112.9848
Energy (work)	N m	ft lb	1.3558
	n mm	in. lb	112.9848
Power	kW	hp	1.342
Pressure and stress	Pa	psi	6.8948×10^{-3}
	kg/cm^2	psi	14.2231

[a] Conversion factor $= \dfrac{\text{SI magnitude}}{\text{OE magnitude}}$; example: 88 ft/s (3.048) = 268.224 m/s.

1.2 STANDARDS

Various organizations throughout the world have established standards for selected machine components, such as screws and bolts, to assure interchangeability and strength. To achieve this objective the standards may specify sizes, tolerances, and proportions. They may also specify the materials used, forming process heat treatment, proof strength, and other key steps in the production of a machine component.

By mutual agreement the organizations forming these standards do not overlap; no one item is subject to two standards for a particular feature, although it may be subject to more than one standard in its manufacture. For example, the dimensions of a screw thread are subject to dimensional standards established by ASME and to strength and marking standards established by SAE.

Some of the association standards have been endorsed by national organizations composed of the societies and associations whose members are the engineers who actually write the standards. The American National Standards Institute (ANSI) is the national organization in the United States. ANSI standards are the approved forms of standards written by member organizations, such as those listed below:

American Chain Association	ACA
American Gear Manufacturers Association	AGMA
American Iron and Steel Institute	AISI
American Society for Metals	ASM
American Society for Testing and Materials	ASTM
American Society for Mechanical Engineers	ASME
American Welding Society	AWS
Anti-Friction Bearing Manufacturers Association	ABMA
Mechanical Power Transmission Association	MPTA
National Electrical Manufacturers Association	NEMA
Rubber Manufacturers Association	RMA
Society of Automotive Engineers	SAE
Spring Manufacturers Institute	SMI
The Rubber Association of Canada	(acronym not used)

International cooperation on standards is handled by the International Organization for Standardization (ISO, not IOS) whose headquarters are in Geneva, Switzerland.

Not only are standards valuable in the standardization of components, such as belts, chains, and screw threads, but they also may provide accepted formulas for the design and selection of the component involved, as in the case of ANSI/AFBMA std. 9-1978, Load Ratings and Fatigue Life for Ball Bearings.

In countries where attorneys are eager to profit from product liability cases, as in the United States, these standards are valuable indications of accepted engineering practice as of the date of issue. Products designed according to these standards are protected from any liability claimed as due to engineering design if they conform to accepted engineering standards at the time of their design. This protection is justified by the practice of the engineering societies and associations involved in continually updating these standards as new techniques and knowledge become available.

1.3 STATICS

Most of the machine components analyzed in the following chapters may be studied under the assumption that they are in static equilibrium. Hence, the forces and reactions on the components may be determined from the following vector equations of static equilibrium:

$$\sum_{i=1}^{n} \mathbf{F}_i = 0 \tag{1.3.1}$$

and

$$\sum_{i=1}^{n} \mathbf{r}_i \times \mathbf{F}_i + \sum_{i=1}^{m} \mathbf{M}_i = 0 \tag{1.3.2}$$

In these equations vector forces $\mathbf{F}_i$ include applied loads and the reactions induced. Likewise, the vector moments $\mathbf{M}_i$ include those applied to the object and all moments induced at the supports.

Although (1.3.1) and (1.3.2) describe static equilibrium, the forces and moments under summation may include those caused by steady motion. An example is the viscous drag and hydrodynamic lift found in a journal bearing. Even though they are caused by motion, they enter into the static equilibrium equations that describe the steady state load on a journal bearing when the shaft turns at constant angular velocity. It is only during speed and load changes that the static equilibrium equations do not adequately describe the forces and moments involved. Steady radial acceleration affects a belt or a chain on a rapidly rotating sheave or sprocket as the centrifugal force acting about the center of rotation. Links in a chain and elements in a belt also experience a constant moment when moving in a circular trajectory about the sheave or sprocket.

Perhaps the most common statics problem in the design of machine components is to determine no more than two reactions in a statically determinant problem that involves any number of known forces in a single plane. A typical example is that shown in Figure 1.3.1 where the equilibrium equations (1.3.1) and (1.3.2) become

$$R_1 + R_2 = L_1 + L_2 \tag{1.3.3}$$

and

$$R_1 b_1 + R_2 b_2 = L_1 a_1 + L_2 a_2 \tag{1.3.4}$$

in which the reactions are positive upward and the loads are positive downward. Simultaneous solution of these equations yields

$$R_1 = L_1 \frac{b_2 - a_1}{b_2 - b_1} + L_2 \frac{b_2 - a_2}{b_2 - b_1} \tag{1.3.5}$$

$$R_2 = L_2 \frac{a_1 - b_1}{b_2 - b_1} + \cdots \tag{1.3.6}$$

Extension to n loads follows from the pattern observed in (1.3.5) and (1.3.6) to give

$$R_1 = \frac{1}{b_2 - b_1} \sum_{i=1}^{n} L_i(b_2 - a_i) \tag{1.3.7}$$

$$R_2 = \frac{1}{b_2 - b_1} \sum_{i=1}^{n} L_i(a_i - b_1) \tag{1.3.8}$$

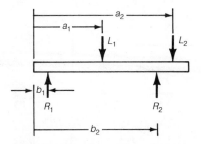

Figure 1.3.1 Static loads on a shaft.

where a_i is the distance of load L_i from the left end of the shaft or beam shown in Figure 1.3.1. It is convenient to write a program for a programmable calculator to solve for R_1 and R_2 as function of L_i. This may be accomplished by writing the program with two entry points: One is used when the program is first run to enter all parameters, and the second is used in succeeding runs where only new loads L_i and their a_i values are entered.

1.4 INTERPOLATION

When tabulated coefficients or the spacing between design curves varies linearly it is easy to estimate values between the tabulated or plotted quantities by linear interpolation. The formula is motivated by the geometry associated with the solid line assumed to lie through points $f(x_1)$ and $f(x_2)$ in Figure 1.4.1a.

From similar triangles it follows that

$$f(x) = \frac{x - x_1}{x_2 - x_1} [f(x_2) - f(x_1)] + f(x_1) \tag{1.4.1}$$

or simply, that

$$f(x) = (x - x_1) \frac{df}{dx} + f(x_1) \tag{1.4.2}$$

If differences between succeeding f values suggest an assumed curve as in Figure 1.4.1b, however, linear interpolation may be too inaccurate. In this circumstance we will consider two interpolation techniques that assume a curve between the data points.

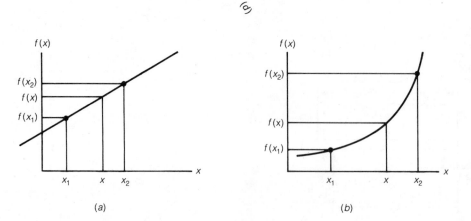

(a)   (b)

Figure 1.4.1 Assumed curves through discrete values of $f(x)$.

The first of these is Lagrangian interpolation, which assumes a polynomial curve through the discrete points. Although any order polynomial may be selected, one using functional values at three points (i.e., a second-order polynomial, or parabola) will prove to be the most practical in the following chapters because it may be easily programmed on many pocket calculators and may be used near the extreme values of tabulated data or set of curves. Accordingly, $f(x)$ is estimated from

$$f(x) \cong (x - x_0)(x - x_1)(x - x_2) \left[\frac{f(x_0)}{(x - x_0)(x_0 - x_1)(x_0 - x_2)} \right.$$

$$\left. + \frac{f(x_1)}{(x - x_1)(x_1 - x_0)(x_1 - x_2)} + \frac{f(x_2)}{(x - x_2)(x_2 - x_0)(x_2 - x_1)} \right] \quad (1.4.3)$$

Comparison of the performance of linear interpolation, according to (1.4.1), with that of Lagrangian interpolation, as shown in (1.4.3), may be made by using them to predict values of a known function. For instance, if the function chosen is $f(x) = e^x$ then interpolated values at $x = 2$ and $x = 4$ using Lagrangian interpolation based upon known values of x at $x = 0,1,3,5$, are as shown in Figure 1.4.2. In detail, linear interpolation at $x = 2$ used values at $x = 1$ and $x = 3$ and the interpolated value at $x = 4$

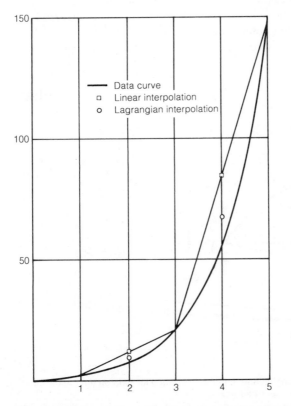

Figure 1.4.2 Comparison of linear and Lagrangian interpolation for $f(x) = e^x$.

used known values at $x = 3$ and $x = 5$. Lagrangian interpolation at $x = 2$ used values at $x = 0$, 1, and 3 and the interpolated value at $x = 4$ used known values at $x = 2$, 3, and 5. At both locations the errors are 54% for linear interpolation and 23% for Lagrangian interpolation.

Lagrangian interpolation using a second-order polynomial is not satisfactory in many engineering applications, however, because the assumed parabolic fit may be distinctly different from the actual curve through the data points for the interpolation. Suppose, for example, that the x and $f(x)$ values are taken from points 1, 2, and 3 on the curve described by short dashes in Figure 1.4.3. If Lagrangian interpolation is used to estimate the value of $f(x)$ for x between points 2 and 3 the $f(x)$ value will lie upon the solid curve, which may be almost 70% in error at its maximum deviation. Even though linear interpolation, represented by straight lines of long and short dashes, is a better approximation than Lagrangian interpolation between points 2 and 3, neither approximation is satisfactory in this and in many similar cases.

The second technique, orthogonal interpolation, does not use a polynomial approximation because it was designed for data that lie on curves that are monotonic, that is, curves whose first derivative either never increases, as in the case of the curve of short dashes in Figure 1.4.3, or never decreases. Stress concentration curves, which are introduced in Chapter 3, are examples of such monotonic curves.

Curves used in orthogonal interpolation occur in the forms shown in Figures 1.4.4 and 1.4.5 when the first and third data points lie one unit apart. As explained in Ref. 1 (which has several misprints) these curves may be represented by two equations of similar form:

$$Y = X^{\frac{b}{x+b}} \tag{1.4.4}$$

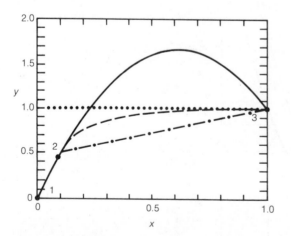

Figure 1.4.3 Comparison between Lagrangian and linear interpolation for data on the middle curve. Data taken from a stress concentration curve.

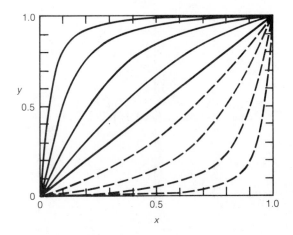

Figure 1.4.4 Representation interpolation curves for $f(x)$ values increasing to the right.

in which b is defined by

$$b = \frac{X_1}{\dfrac{\ln X_1}{\ln Y_1} - 1}$$

(1.4.5)

and

$$Y^* = X^{*\frac{b}{X_* + b}}$$

(1.4.6)

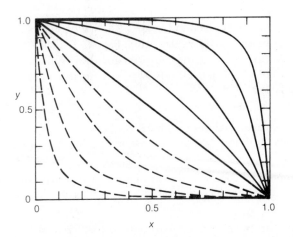

Figure 1.4.5 Representative interpolation curves for $f(x)$ values decreasing to the right.

in which b is defined by

$$b = \frac{X_1{}^*}{\dfrac{\ln X_1{}^*}{\ln Y_1{}^*} - 1} \tag{1.4.7}$$

The variables X, X_1, Y, Y_1, X*, $X_1{}^*$, Y*, and $Y_1{}^*$ are defined as follows:

$$X = \frac{x - x_0}{x_2 - x_0} \qquad Y = \frac{y - y_0}{y_2 - y_0}$$

$$X_1 = \frac{x_1 - x_0}{x_2 - x_0} \qquad Y_1 = \frac{y_1 - y_0}{y_2 - y_0} \tag{1.4.8}$$

and

$$X^* = \frac{x_2 - x}{x_2 - x_0} \qquad Y^* = \frac{y - y_2}{y_0 - y_2}$$

$$X_1{}^* = \frac{x_2 - x_1}{x_2 - x_0} \qquad Y_1{}^* = \frac{y_1 - y_2}{y_0 - y_2} \tag{1.4.9}$$

where $y = f(x)$ and where

$$1 - X = X^* \quad \text{and} \quad 1 - Y = Y^* \tag{1.4.10}$$

Equations 1.4.4 and 1.4.5 hold for curves on and above the diagonal in Figure 1.4.4 and on and below the diagonal in Figure 1.4.5. Equations 1.4.6 and 1.4.7 hold for curves on and below the diagonal in Figure 1.4.4 and on and above the diagonal in Figure 1.4.5. In all cases the diagonal is given by $b = \infty$ and axes $y = 0$ and $y = 1$ are given by $b = 0$ in the appropriate relations.

A flowchart for orthogonal interpolation is given in Figure 1.4.6. Since interpolation programs are especially suited for programmable calculators, Figures 1.4.7, 1.4.8, and 1.4.9 display the programs for linear, Lagrangian, and orthogonal interpolation as implemented on Hewlett-Packard HP-41CV and HP-41CX programmable pocket calculators.

Guided by the previous examples, we observe that:

1. Linear interpolation applies whenever curves or differences between tabulated values display a nearly linear relation between points in the interpolation interval.

2. Lagrangian interpolation applies whenever curves or differences between tabulated data indicate that points in the interpolation interval very nearly lie on a segment of a parabola.

3. Orthogonal interpolation applies whenever the curves or the differences between tabulated data indicate first derivatives that do not change algebraic sign in the interval of interpolation (i.e., the curves are monotonic).

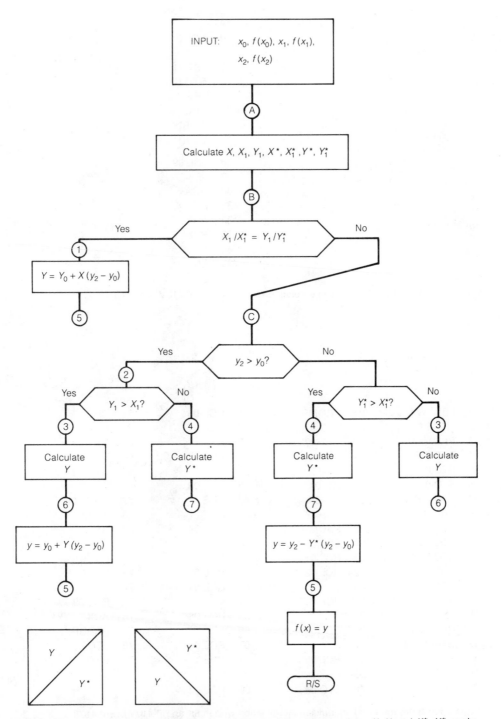

Figure 1.4.6 Flowchart for orthogonal interpolation and the regions where X, Y and X^*, Y^* apply.

01	LBL "LIN"		19	*
02	"F2=?"		20	RCL 04
03	PROMPT		21	+
04	"F1=?"		22	FS? 21
05	PROMPT		23	GTO 09
06	STO 04		24	"F="
07	−		25	ARCL X
08	"X2=?"		26	AVIEW
09	PROMPT		27	STOP
10	"X1=?"		28	LBL 09
11	PROMPT		29	"F="
12	STO 05		30	ACA
13	−		31	RCL X
14	/		32	ACX
15	"X=?"		33	PRBUF
16	PROMPT		34	STOP
17	RCL 05		35	END
18	−			

Figure 1.4.7 Program for linear interpolation written for an HP-41CV or HP-41CX.

01	LBL "W-L"	21	PROMPT	41	RCL 00	61	CHS
02	"X=?"	22	STO 07	42	RCL 02	62	/
03	PROMPT	23	"X2=?"	43	*	63	ST+09
04	STO 00	24	PROMPT	44	RCL 07	64	FS? 21
05	STO 01	25	ST−02	45	*	65	GTO 01
06	STO 02	26	ST−04	46	RCL 03	66	"F="
07	"XO=?"	27	ST−05	47	CHS	67	ARCL 09
08	PROMPT	28	"F2=?"	48	/	68	AVIEW
09	ST−00	29	PROMPT	49	RCL 04	69	STOP
10	STO 03	30	STO 08	50	/		
11	STO 05	31	RCL 01	51	ST+ 09	70	LBL 01
12	"F0=?"	32	RCL 02	52	RCL 00	71	"F="
13	PROMPT	33	*	53	RCL 01	72	ACA
14	STO 06	34	RCL 06	54	*	73	RCL 09
15	"X1=?"	35	*	55	RCL 08	74	ACX
16	PROMPT	36	RCL 03	56	*	75	PRBUF
17	ST−01	37	/	57	RCL 05	76	STOP
18	ST−03	38	RCL 05	58	CHS	77	END
19	STO 04	39	/	59	/		
20	"F1=?"	40	STO 09	60	RCL 04		

Figure 1.4.8 Program for Langrangian interpolation written for an HP-41CV or HP-41CX.

```
01◆LBL  "ORTHIN"      45  –           86   /           128◆LBL 06
   02  CLA            46  ST/01       87   j           129  RCL 08
   03  16             47  ST/02       88   –           130  RCL 07
   04  PSIZE          48  ST/04       89   RCL 02      131  –
   05  "X=?"          49  ST/05       90   X<>Y        132  RCL 10
   06  PROMPT         50  RCL 07      91   /           133  *
   07  STO 11         51  RCL 15      92   STO 09      134  RCL 07
   08  "XO=?"         52  –           93   RCL 01      135  +
   09  PROMPT         53  STO 03      94   RCL 09      136  STO 11
   10  STO 12         54  RCL 15      95   RCL 01      137  GTO 05
   11  "F0=?"         55  RCL 08      96   RCL 09
   12  PROMPT         56  –           97   +          138◆LBL 07
   13  STO 07         57  STO 06      98   /           139  RCL 08
   14  "X1=?"         58  RCL 07      99   Y↑X         140  RCL 08
   15  PROMPT         59  RCL 08      100  STO 10      141  RCL 07
   16  STO 13         60  –           101  GTO 06      142  –
   17  "F1=?"         61  ST/03                        143  RCL 10
   18  PROMPT         62  ST/06      102◆LBL 02        144  *
   19  STO 15                        103  RCL 02       145  –
   20  "X2=?"        63◆LBL B        104  RCL 03       146  STO 11
   21  PROMPT         64  RCL 02      105  X>Y?
   23  STO 14         65  RCL 05      106  GTO 03     147◆LBL 05
   24  "F2=?"         66  /                            148  FS? 21
   24  PROMPT         67  RCL 03     107◆LBL 04        149  GTO 08
   25  STO 08         68  RCL 06      108  RCL 05      150  "F="
                      69  /           109  LN          151  ARCL 11
                      70  X=Y?        110  RCL 06      152  AVIEW
  26◆LBL A            71  GTO 01      111  LN          153  STOP
   27  RCL 11                        112  /
   28  RCL 12                        113  1          154◆LBL 08
   29  –             72◆LBL C        114  –           155  "F="
   30  STO 01         73  RCL 07      115  RCL 05      156  ACA
   31  RCL 13         74  RCL 08      116  X<>Y        157  RCL 11
   32  RCL 12         75  X>Y?        117  /           158  ACX
   33  –              76  GTO 02      118  STO 09      159  PRBUF
   34  STO 02         77  RCL 05      119  RCL 04      160  STOP
   35  RCL 14         78  RCL 06      120  RCL 09
   36  RCL 11         79  X>Y?        121  RCL 04     161◆LBL 01
   37  –              80  GTO 04      122  RCL 09      162  RCL 08
   38  STO 04                         123  +          163  RCL 07
   39  RCL 14                         124  /           164  –
   40  RCL 13        81◆LBL 03        125  Y↑X         165  RCL 01
   41  –              82  RCL 02      126  STO 10      166  *
   42  STO 05         83  LN          127  GTO 07      167  RCL 07
   43  RCL 14         84  RCL 03                       168  +
   44  RCL 12         85  LN                           169  STO 11
                                                       170  END
```

Figure 1.4.9 Program for orthogonal interpolation written for an HP-41CV or HP-41CX.

1.5 FUNCTION RECOVERY

Rather than repeatedly interpolate between grid values to read data from graphs, it is often expedient to store the equation of the curve in a computer or pocket calculator and replace interpolation by direct calculation from the appropriate curve.

If the graphical data lie along a straight line when presented on linear paper, the equation is of the form

$$f(x) = ax + b \tag{1.5.1}$$

where a and b are found by evaluating equation 1.5.1 at x_0 and at x_1, as illustrated in Figure 1.5.1a. Consequently,

$$a = f(x_1) - \frac{b}{x_1}$$

$$b = \frac{x_1 f(x_0) - x_0 f(x_1)}{x_1 - x_0} \tag{1.5.2}$$

If the function is given as a straight line on semilog paper, as depicted in Figure 1.5.1b, then $\log f(x)$ replaces $f(x)$ in equation 1.5.1, so that*

$$\log f(x) = ax + b \tag{1.5.3}$$

which is equivalent to

$$f(x) = C \times 10^{ax} \tag{1.5.4}$$

where 10^b has been replaced by C, a constant. In terms of the quantities shown in Figure 1.5.1b

$$C = f(x_0)/(10^{ax_0})$$

$$a = \frac{1}{x_1} \log[f(x_1)/C] \tag{1.5.5}$$

Finally, a linear plot on log–log paper, as in Figure 1.5.1c, represents an equation of the form

$$\log f(x) = a \log x + \log b \tag{1.5.6}$$

which is equivalent to $f(x) = bx^a$, where

$$a = \frac{\log f(x_1) - \log b}{\log x_1} \tag{1.5.7}$$

---

*   In this text log denotes the logarithm to the base 10 and ln denotes the logarithm to the base e.

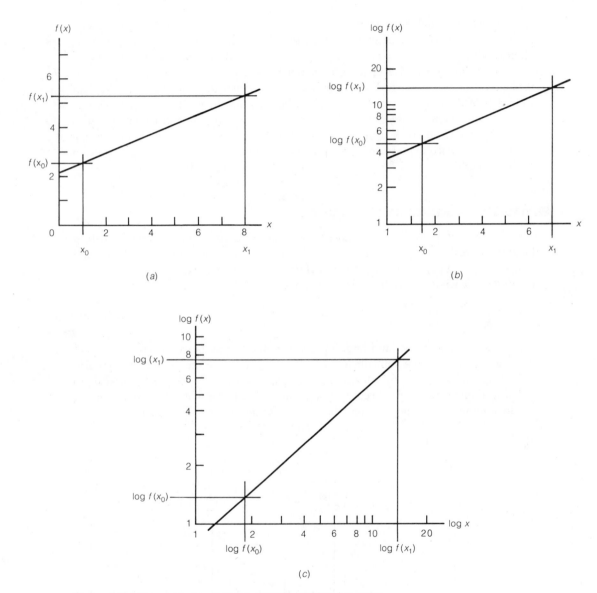

Figure 1.5.1 Linear graphs on linear, semilog, and log–log paper.

wherein

$$\log b = \frac{(\log x_1) \log f(x_0) - (\log x_0) \log f(x_1)}{\log (x_1/x_0)} \qquad (1.5.8)$$

Recovery of curves that are not linear when plotted on linear, semilog, or log–log paper is largely a matter of individual ingenuity in matching the curves to mathematical expressions. Although it is tempting to use the techniques of nonlinear regression analysis and polynomial curve fitting, as described in Chapter 6 of Ref. 2, these methods often produce curves that pass through some points and miss others by an excessive amount. As observed by Kopal (Ref. 3) polynomial curves in particular may pass through all of the prescribed points but have derivatives that are distinctly different from that of the desired curve almost everywhere within the interval of interpolation.

Reference 4, motivated by a photographic analysis of the cutting process in a large-diameter (20 ft) rock drill, provides several examples of individual curve fitting wherein the devised expressions agreed with the intended curves over a major portion of their lengths.

1.6 ESTIMATING LOGARITHMIC VALUES

Both experimental data and theoretical results are often presented on graphs with a logarithmic scale along the ordinate (semilog paper) or along both the ordinate and abscissa (log–log paper). The advantages of these plots are that exponential dependence upon the independent variable may be emphasized, since the relation

$$y = ax^b \tag{1.6.1}$$

appears as a straight line, and large ranges of x and y may be compactly plotted. A disadvantage is that it is difficult for the occasional user to interpolate between grid lines. This difficulty may be alleviated to some extent by comparing a linear scale to the logarithmic scale, as has been done in Figure 1.6.1. If the linear scale represents x, the logarithmic scale may be constructed by plotting 10^x opposite x. Hence, if the

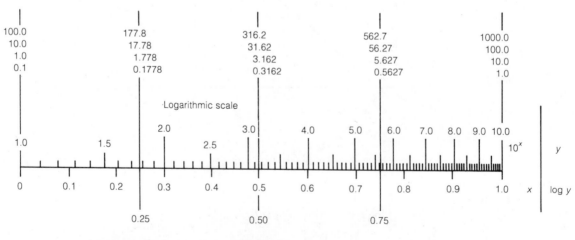

Figure 1.6.1 Comparison of linear and logarithmic scales.

logarithmic scale represents y, the linear scale represents log y. We may use this relation to produce log plots on linear paper.

By comparing the two scales in Figure 1.6.1 we find that the logarithmic value at the midpoint of the linear distance between 0.1 and 1.0 on a logarithmic scale is 0.3162. Likewise, at 0.25 and 0.75 of the linear distance between 0.1 and 1.0 the logarithmic values are 0.1778 and 0.5627. Visual interpolation along a logarithmic scale may be aided by recalling the relation between linear distances and logarithmic values as shown in Figure 1.6.1 and by using the above values as reference points. If a calculator is available it is easy to refine the estimate for y from a plot of log y by noting the linear position of the point where log y appears and then calculating

$$y = \beta \times 10^x \tag{1.6.2}$$

where β is the beginning value of the log cycle in which interpolation takes place. Thus, to interpolate for log y between 10^a and 10^{a+1} set $\beta = 10^a$. If log y is 0.6 of the distance from 10^a to 100^a then $y = 10^a(10^{0.6})$.

1.7 CURVE READING

The current practice in many engineering publications is to use large grid divisions in curve plots to emphasize curve visibility. This often makes it difficult to read the curves to the desired precision. To overcome this obstacle one may use a portion of graph paper with divisions of 0.10 in. or a machinist rule in which an inch is divided into 10, 50, or 100 equal divisions to provide a reference of length with decimal divisions. Place the reference length as shown in Figure 1.7.1 so that it intersects the curve at the

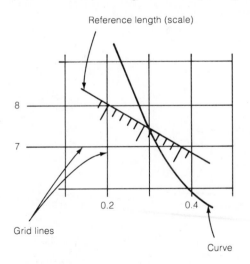

Figure 1.7.1 Interpolation between grid lines. Read 7.44 ± 0.01 at intersection of curve and 0.3 grid line. (For greater precision use scale with 100 divisions; *i.e.*, Pratt and Whitney No. C-2105R with 0.01-in. divisions.)

desired location and opposite ends of the reference length lie on adjacent grid lines. If the grid lines differ by other than 10 units it may be convenient to use a millimeter scale. For grid differences of 20 units the reference length would be 2 cm, and so on. Obviously the reference length should be equal to, or greater than, the perpendicular distance between grid lines.

1.8 BISECTION METHOD FOR FINDING SINGLE ROOTS

It has been shown that the bisection method is the fastest and simplest method yet proposed for finding a root x_1 of a function $f(x) = 0$ to within an acceptable error of

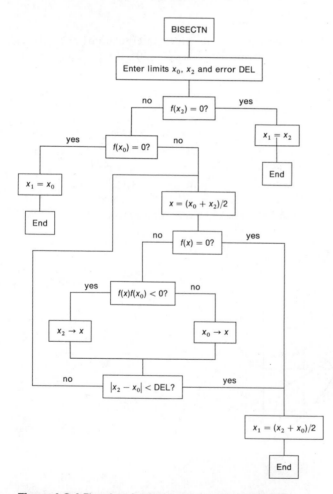

Figure 1.8.1 Flowchart for the bisection entitled BISECTN.

magnitude DEL when x_1 is known to be the only root between limits x_0 and x_2[10]. This method is useful in applying low cycle fatigue analysis, and in designing dual shoe brakes, belt and chain drives, and gears.

The procedure is to calculate $f(x_0)$ and $f(x_2)$. If the algebraic signs differ there is a root in the interval and we may find in which half it lies by calculating $f(x)$ for $x = (x_0 + x_2)/2$ and discarding that interval for which $f(x)$ does not change algebraic sign. Continuing in this manner until the length of the interval is less than DEL provides a value of x that differs from the exact root by an error less than DEL.

A flowchart for a computer routine to implement the bisection method is shown in Figure 1.8.1, a program for the HP-41CX/CV programmable calculator is displayed in Figure 1.8.2, and a routine for augmented BASIC which includes the IF . . . THEN . . . ELSE command is given in Figure 1.8.3. The function $f(x)$ is calculated by the subroutine whose name is stored in register 26 in the HP-41C/CX program listed in Figure 1.8.2 and by subroutine FUN in the BASIC program whose listing appears in Figure 1.8.3.

01 LBL BIS	19 RCL 24	37 RCL 24	55 LBL a
02 RCL 22	20 *	38 *	56 "OUT OF RANGE"
03 STO 20	21 X>0?	39 X<0?	57 PRA
04 XEQ IND 26	22 GTO 02	40 GTO 01	58 RTN
05 STO 24		41 RCL 20	
06 ABS	23 LBL B	42 STO 22	59 LBL 03
07 RCL 23	24 RCL 21	43 GTO B	60 RTN
08 X>Y?	25 RCL 22		61 END
09 GTO 03	26 +	44 LBL 01	
10 RCL 21	27 2	45 RCL 20	Registers used by BIS
11 STO 20	28 /	46 STO 21	20 x
12 XEQ IND 26	29 STO 20	47 GTO B	21 x_2
13 STO 25	30 XEQ IND 26		22 x_0
14 ABS	31 STO 25	48 LBL 02	23 Del
15 RCL 23	32 ABS	49 FS? 21	24 $ep(x_0)$
16 X>Y?	33 RCL 23	50 GTO a	25 $ep(x)$
17 GTO 03	34 X>Y?	51 "OUT OF RANGE"	26 function NAME
18 RCL 25	35 GTO 03	52 AVIEW	
	36 RCL 25	53 STOP	
		54 RTN	

Note: Function argument is taken from and left in register 20; *i.e.*, the main program must recall the solution obtained by BIS from register 20.

Figure 1.8.2 Listing of the bisection routine BIS for an HP-41CV/CX written for either display register or printer output.

```
            'Bisection subroutine written in Microsoft Quick BASIC
            CALL FUN(XO,EPO)
            IF EPO=0 THEN GOTO FIN1
            CALL FUN(X2,EP1)
            IF EP1=0 THEN GOTO FIN2
            IF EP1*EP2>0 THEN GOTO FIN0
ENT:    X=(X0+X2)/2
            CALL FUN(X,EP)
            IF EP=0 THEN
               GOTO FIN4
            ELSEIF EP*EP0<0 THEN
               X2=X
            ELSE
               X0=X
            ENDIF
            IF ABS(X2-X0)<DEL GOTO FIN3 ELSE GOTO ENT

            SUB FUN(X,EP) STATIC

            '. . . . . . . . . . . . .   (Write function formulas here)

            END SUB

FIN0:   PRINT "No root in the interval selected." : GOTO FINI
FIN1:   X=X0 : GOTO FIN4
FIN2:   X=X2 : GOTO FIN4
FIN3:   X=(X2+X0)/2
FIN4:   PRINT "Value of x is", X
FINI:   END
```

Figure 1.8.3 Sample program for the BISECTN routine written according to the rules for Microsoft Quick BASIC, version 3.0.

1.9 REFERENCES FOR OTHER MATHEMATICAL METHODS USED IN MACHINE DESIGN

A working knowledge of vector analysis is assumed, so that no references for a text on vector analysis are given here.

Numerical integration will involve integrands that may be given as analytical expressions or that may be defined by a series of discrete values for the dependent variable at discrete values of the independent variable. In either case the trapazoidal rule will be used because it is simple and sufficiently accurate when the subdivisions are small. Handling a large number of points is no obstacle when computers are available. Other methods, often originally motivated by hand calculation, may be found in Refs. 6 through 12.

Numerical solutions to ordinary and partial differential equations are not required in this text (although they would be helpful in the chapter on lubrication) because they are traditionally taught in other courses. They are often necessary when machine dynamics must be considered in the design of a machine to insure that it will operate as required. References 3 and 5 through 9 describe numerical methods for a variety of problems. Reference 2 outlines classical statistics and Ref. 11 describes the nonclassical Weibull distribution and its application to fatigue testing.

REFERENCES

1. Orthwein, W.C., Three-point interpolation formulas: A guide to selection, *Computers in Mechanical Engineering* 2 (6): 36–42 (1984).

2. Crow, E.L., Davis, F.A., and Maxfield, M.W., *Statistics Manual*, Dover, New York, 1960.

3. Kopal, Z., *Numerical Analysis*, 2d ed., Chapman & Hall, London, 1961.

4. Orthwein, W.C., Calculating point illumination using a programmable pocket calculator, *Journal of the Illuminating Engineering Society* 10 (3) 149–154 (April 1981).

5. Forsythe, G.E., Malcolm, M.E., and Moler, C.B., *Computer Methods for Mathematical Computations*, Prentice-Hall, Englewood Cliffs, NJ, 1977.

6. Kunz, K.S., *Numerical Analysis*, McGraw-Hill, New York, 1957.

7. Carnahan, B., Luther, H.A., and Wilkes, J.O., *Applied Numerical Methods*, Wiley, New York, 1969.

8. Forsythe, G.E., and Wasow, W.R., *Finite-Difference Methods for Partial Differential Equations*, Wiley, New York, 1960.

9. Davis, P.J., and Rabinowitz, P., *Methods of Numerical Integration*, Academic, New York, 1975.

10. Brent, R.P., *Algorithms for Minimization without Derivatives*, Prentice-Hall, Englewood Cliffs, NJ, 1973.

11. Johnson, L.G., *Statistical Treatment of Fatigue Experiments*, Elsevier, New York, 1964.

12. *Chains for Power Transmission and Material Handling, Design and Applications Handbook*. American Chain Association, Dekker, New York, 1982.

PROBLEMS

Section 1.3

1.1 Show that if a moment M acts upon a simply supported beam, the general expressions for the reactions R_1 and R_2 become

$$R_1 = [\sum L_i(b_2 - a_i) + M]/(b_2 - b_1)$$
$$R_2 = [\sum L_i(a_i - b_1) - M]/(b_2 - b_1)$$

1.2   Prepare a flowchart and a program to solve the equations in problem 1.1.

1.3   Use your program from problem 2 to find the reactions to the four loads shown in the figure. (dimensions are in mm).

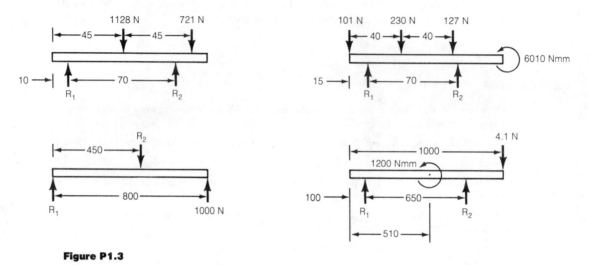

Figure P1.3

1.4   Use the rectangular, trapazoidal, and Simpson's rule to evaluate the area of the semicircle given by $49 = x^2 + y^2$ where $-7 \text{ mm} \leq x \leq 7 \text{ mm}$ and where $y \geq 0$ and compare with the analytically determined area $A = 76.9690 \text{ mm}^2$. Explain why one method is more accurate than the others. Under what conditions (for what boundary curves) would each give the more accurate answer. *Hint:* Evaluate $A = \int_{-7}^{7} (49 - x^2)^{1/2} \, dx$ where $da = y \, dx$ has been used.

[*Ans.* Area $= 76.358 \text{ mm}^2$ by Simpson's rule, 75.420 mm^2 by both the rectangular and trapazoidal rules. Simpson's rule is more accurate for curves resembling a parabola. The trapazoidal rule loses accuracy for curves that are all concave or convex, as does the rectangular rule. Symmetry improves the accuracy of the rectangular rule.]

Section 1.4

1.5   Find the stress concentration factor for the grooved shaft in tension shown in Figure P1.5 for $r/d = 0.10$ and $D/d = 1.60$. Use orthogonal interpolation and compare with linear interpolation.

1.6   Find the stress concentration factor for the grooved shaft in tension shown in the figure in problem 1.5 for $r/d = 0.16$ and $D/d = 1.40$ using linear and orthogonal interpolation. Compare with Lagrangian interpolation.

1.7   Find the stress concentration factor for the grooved shaft in tension shown in the figure in problem 1.5 for $r/d = 0.08$ and $D/d = 1.65$ using linear, Lagrangian, and orthogonal interpolation.

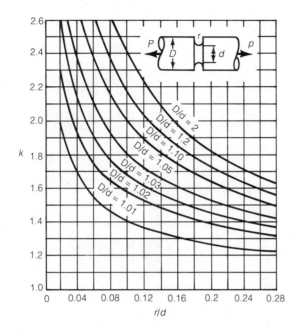

Figure P1.5

1.8 Estimate the stress concentration factor for the grooved shaft shown in the figure in problem 1.5 for $r/d = 0.113$ and $D/d = 1.71$ without aid of any calculations and compare that estimate with one obtained using linear interpolation between $r/d = 0.10$ and $r/d = 0.12$ to find the K value for $D/d = 1.1$, 1.2, and 2.0. Use these values with the orthogonal interpolation program to find K for $D/d = 1.71$.

1.9 A portion of Table 5-27, Horsepower Ratings for No. 100 Chain, 1.25-in.-Pitch Standard Single-Strand Roller Chain from Ref. 12 is reproduced on page 26. Use this table to compare the difference between horsepower ratings for a sprocket with 19 teeth for a No. 100 chain at 32 rpm as determined by linear and orthogonal interpolation. Use data at 10, 25, and 50 rpm. The small difference found indicates that the entries were carefully chosen to permit linear interpolation with a relatively small error.

   [*Ans.* Linear, 4.15 hp; orthogonal, 4.20 hp.]

1.10 A portion of the tabulated values for bearing Y factors used for bearing selection is reproduced from Table 2 of ANSI/AFBMA Std. 9-1978, Load Ratings and Fatigue Life for Ball Bearings (see page 27). Compare interpolated values of Y at $F_a/C_0 = 0.040$ found from linear, Lagrangian, and orthogonal interpolation.

1.11 Required fan input power was measured at the following set of input speeds to lower the cost of the measurement, with the intent that values between data points could be obtained by

HORSEPOWER RATINGS STANDARD SINGLE STRAND ROLLER CHAIN—NO. 100—1¼" PITCH

Revolutions per Minute—Small Sprocket

No. of Teeth Small Spkt.	10	25	50	100	150	200	300	400	500	600	700	800	900	1000	1100	1200	1300	1400	1600	1800	2000	2200	2400	2600	2700
9	0.65	1.49	2.78	5.19	7.47	9.68	13.9	18.1	22.1	26.0	29.6	24.2	20.3	17.4	15.0	13.2	11.7	10.5	8.57	7.19	6.13	5.32	4.67	4.14	0
10	0.73	1.67	3.11	5.81	8.37	10.8	15.6	20.2	24.7	29.2	33.5	28.4	23.8	20.3	17.6	15.5	13.7	12.3	10.0	8.42	7.19	6.23	5.47	4.85	0
11	0.81	1.85	3.45	6.44	9.28	12.0	17.3	22.4	27.4	32.3	37.1	32.8	27.5	23.4	20.3	17.8	15.8	14.2	11.6	9.71	8.29	7.19	6.31	1.29	0
12	0.89	2.03	3.79	7.08	10.2	13.2	19.0	24.6	30.1	35.5	40.8	37.3	31.3	26.7	23.2	20.3	18.0	16.1	13.2	11.1	9.45	8.19	7.19		
13	0.97	2.22	4.13	7.72	11.1	14.4	20.7	26.9	32.8	38.7	44.5	42.1	35.3	30.1	26.1	22.9	20.3	18.2	14.9	12.5	10.6	9.23	8.10		
14	1.05	2.40	4.48	8.36	12.0	15.6	22.5	29.1	35.6	41.9	48.2	47.0	39.4	33.7	29.2	25.6	22.7	20.3	16.6	13.9	11.9	10.3	9.05		
15	1.13	2.59	4.83	9.01	13.0	16.8	24.2	31.4	38.3	45.2	51.9	52.2	43.7	37.3	32.4	28.4	25.2	22.5	18.4	15.5	13.2	11.4	10.0		
16	1.22	2.77	5.17	9.66	13.9	18.0	26.0	33.6	41.1	48.4	55.6	57.5	48.2	41.1	35.7	31.3	27.7	24.8	20.3	17.0	14.5	12.6	11.1		
17	1.30	2.96	5.52	10.3	14.8	19.2	27.7	35.9	43.9	51.7	59.4	63.0	52.8	45.0	39.0	34.3	30.4	27.2	22.3	18.7	15.9	13.8	0.79		
18	1.38	3.15	5.88	11.0	15.8	20.5	29.5	38.2	46.7	55.0	63.2	68.6	57.5	49.1	42.5	37.3	33.1	29.6	24.2	20.3	17.4	15.0	0		
19	1.46	3.34	6.23	11.6	16.7	21.7	31.2	40.5	49.5	58.3	67.0	74.4	62.3	53.2	46.1	40.5	35.9	32.1	26.3	22.0	18.8	16.3	0		
20	1.55	3.53	6.58	12.3	17.7	22.9	33.0	42.8	52.3	61.6	70.8	79.8	67.3	57.5	49.8	43.7	38.8	34.7	28.4	23.8	20.3	17.6	0		
21	1.63	3.72	6.94	13.0	18.7	24.2	34.8	45.1	55.1	65.0	74.6	84.2	72.4	61.8	53.6	47.0	41.7	37.3	30.6	25.6	21.9	19.0	0		
22	1.71	3.91	7.30	13.6	19.6	25.4	36.6	47.4	58.0	68.3	78.5	88.5	77.7	66.3	57.5	50.4	44.7	40.0	32.8	27.5	23.4	20.3	0		
23	1.80	4.10	7.66	14.3	20.6	26.7	38.4	49.8	60.8	71.7	82.3	92.8	83.0	70.9	61.4	53.9	47.8	42.8	35.0	29.4	25.1	7.74	0		
24	1.88	4.30	8.02	15.0	21.5	27.9	40.2	52.1	63.7	75.0	86.2	97.2	88.5	75.6	65.5	57.5	51.0	45.6	37.3	31.3	26.7	0			
25	1.97	4.49	8.38	15.6	22.5	29.2	42.0	54.4	66.6	78.4	90.1	102	94.1	80.3	69.6	61.1	54.2	48.5	39.7	33.3	28.4	0			
26	2.05	4.68	8.74	16.3	23.5	30.4	43.8	56.8	69.4	81.8	94.0	106	99.8	85.2	73.8	64.8	57.5	51.4	42.1	35.3	30.1	0			
28	2.22	5.07	9.47	17.7	25.5	33.0	47.5	61.5	75.2	88.6	102	115	112	95.2	82.5	72.4	64.2	57.5	47.0	39.4	33.7	0			
30	2.40	5.47	10.2	19.0	27.4	35.5	51.2	66.3	81.0	95.5	110	124	124	106	91.5	80.3	71.2	63.7	52.2	43.7	10.0	0			
32	2.57	5.86	10.9	20.4	29.4	38.1	54.9	71.1	86.9	102	118	133	136	116	101	88.5	78.5	70.2	57.5	48.2	0				
35	2.83	6.46	12.0	22.5	32.4	42.0	60.4	78.3	95.7	113	130	146	156	133	115	101	89.8	80.3	65.8	55.1	0				
40	3.27	7.46	13.9	26.0	37.4	48.5	69.8	90.4	111	130	150	169	188	163	141	124	110	98.1	80.3	0					
45	3.71	8.47	15.8	29.5	42.5	55.0	79.3	103	126	148	170	192	213	194	168	148	131	117	45.3	0					

Type A   Type B   Type C

Figure P1.9

			Single Row Bearings		Single Row Bearings		Double Row Bearings		
Bearing Type			$\frac{F_a}{F_r} > e$		$\frac{F_a}{F_r} \leqq e$		$\frac{F_a}{F_r} > e$		e
			X	Y	X	Y	X	Y	

Radial Contact Groove Ball Bearings

$\frac{F_a}{C_0}$	$\frac{F_a}{iZD^2}$ Units Newtons, mm	$\frac{F_a}{iZD^2}$ Units lbs, in	X	Y	X	Y	X	Y	e
0.014	0.172	25		2.30				2.30	0.19
0.028	0.345	50		1.99				1.99	0.22
0.056	0.689	100		1.71				1.71	0.26
0.084	1.03	150	0.56	1.56	1	0	0.56	1.55	0.28
0.11	1.38	200		1.45				1.45	0.30
0.17	2.07	300		1.31				1.31	0.34
0.28	3.45	500		1.15				1.15	0.38
0.42	5.17	750		1.04				1.04	0.42
0.56	6.89	1000		1.00				1.00	0.44

Angular Contact Groove Ball Bearings with Contact Angle: 5°

For this type use X, Y and e values applicable to single row radial contact bearings

$\frac{F_a}{C_0}$	$\frac{F_a}{iZD^2}$ Units Newtons, mm	$\frac{F_a}{iZD^2}$ Units lbs, in	X	Y	X	Y	X	Y	e
0.014	0.172	25				2.78		3.74	0.23
0.028	0.345	50				2.40		3.23	0.26
0.056	0.689	100				2.07		2.78	0.30
0.085	1.03	150			1	1.87	0.78	2.52	0.34
0.11	1.38	200				1.75		2.36	0.36
0.17	2.07	300				1.58		2.13	0.40
0.28	3.45	500				1.39		1.87	0.45
0.42	5.17	750				1.26		1.69	0.50
0.56	6.89	1000				1.21		1.63	0.52

10°

$\frac{F_a}{C_0}$	Units Newtons, mm	Units lbs, in	X	Y	X	Y	X	Y	e
0.014	0.172	25		1.88		2.18		3.06	0.29
0.029	0.345	50		1.71		1.98		2.78	0.32
0.057	0.689	100		1.52		1.76		2.47	0.36
0.086	1.03	150		1.41		1.63		2.20	0.38
0.11	1.38	200	0.46	1.34	1	1.55	0.75	2.18	0.40
0.17	2.07	300		1.23		1.42		2.00	0.44
0.29	3.45	500		1.10		1.27		1.79	0.49
0.43	5.17	750		1.01		1.17		1.64	0.54
0.57	6.89	1000		1.00		1.16		1.63	0.54

15°

$\frac{F_a}{C_0}$	Units Newtons, mm	Units lbs, in	X	Y	X	Y	X	Y	e
0.015	0.172	25		1.47		1.65		2.39	0.38
0.029	0.345	50		1.40		1.57		2.28	0.40
0.058	0.689	100		1.30		1.46		2.11	0.43
0.087	1.03	150		1.23		1.38		2.00	0.46
0.12	1.38	200	0.44	1.19	1	1.34	0.72	1.93	0.47
0.17	2.07	300		1.12		1.26		1.82	0.50
0.29	3.45	500		1.02		1.14		1.66	0.55
0.44	5.17	750		1.00		1.12		1.63	0.56
0.58	6.89	1000		1.00		1.12		1.63	0.56

Bearing Type	X	Y	X	Y	X	Y	e
20°	0.43	1.00	1	1.09	0.70	1.63	0.57
25°	0.41	0.87	1	0.92	0.67	1.41	0.68
30°	0.39	0.76	1	0.78	0.63	1.24	0.80
35°	0.37	0.66	1	0.66	0.60	1.07	0.95
40°	0.35	0.57	1	0.55	0.57	0.98	1.14
Self-aligning Ball Bearings	0.40	$0.4 \cot\alpha$	1	$0.42 \cot\alpha$	0.65	$0.65 \cot\alpha$	$1.5 \tan\alpha$

Figure P1.10

interpolation. Compare linear, Lagrangian, and orthogonal interpolated values for power in kW at 250 and 450 rpm.

rpm	kW	rpm	kW	rpm	kW
10	0.774	300	14.062	500	110.703
100	1.903	400	38.219	600	300.211
200	6.172				

Sections 1.5 and 1.6

1.12 Guided by the relation between linear and logarithmic scales given in Figure 1.6.1, read the life in cycles for hot-rolled steel from Figure P1.12 for completely reversed stress magnitudes of 60 000, 50 000, and 40 000 psi for steel involved. Also read the life in cycles for the particular steel shown in the drawing for a ground surface at 60 000 and 40 000 psi. Finally, read the life in cycles for the as-forged surface at 60 000, 50 000, 30 000, 25 000, and 20 000 psi.

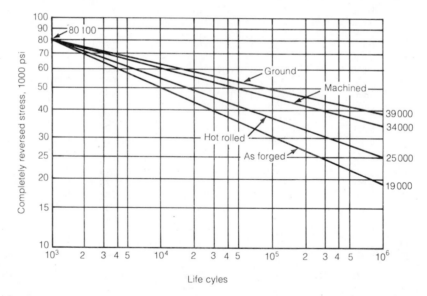

Figure P1.12

1.13 Figure P1-12 displays the S-N curves (stress cycles to failure) for steel having a Brinell hardness from 187 to 207 with ground, machined, hot-rolled, and as-forged surface finishes. Interpolation is uncertain to within the reader's ability to estimate the missing subdivisions on the logarithmic scale. Since numerical values are given at the ends of each of the lines in the figure, the formula for each curve may be obtained for improved accuracy. Find the values of a and b (from equations 1.5.7 and 1.5.8) and check the values read in problem 12 against those calculated from the resulting formulas.

1.14 Use the equation obtained in problem 1.13 to calculate the life in cycles of a steel link axially loaded with a completely reversed stress of 40 000 psi if the finish is as-forged. Find the increase in life if the finish is changed to hot rolled.

1.15 Using the bisection method, find the stresses corresponding to the strains $\varepsilon = 0.002$, 0.004, 0.006, and 0.008, where $E = 200$ GPa, $K' = 2310$ MPa, and $n' = 0.146$.

$$\varepsilon = \frac{\sigma}{E} + \left(\frac{\sigma}{K'}\right)^{1/n'} \qquad \begin{cases} \varepsilon \sim \text{strain} \\ \sigma \sim \text{stress} \end{cases}$$

This is the relation for the cyclic stress–strain curve in Chapter 4.

[*Ans.* For $\varepsilon = 0.002$, $\sigma = 398$ MPa.]

CHAPTER TWO

MATERIALS

Selected material characteristics of commonly used metals and plastics that affect their use in the design of machine components are enumerated in the following sections. Emphasis is on typical applications to machine components and, in the case of metals, on the classification systems used for the alloys available. Where possible, typical mechanical properties, such as elastic modulus and ultimate strength, are given. Crystal or molecular structures are usually introduced only as an aid to rationalizing variations in macroscopic properties or for correlation with similar materials.

In keeping with this application-oriented approach, no problems are found at the end of this chapter. Instead, problems related to this chapter will appear in those chapters where material considerations are entwined with the application at hand.

2.1 STRUCTURE OF IRON AND STEEL

Pure iron at room temperature is called *α-iron* or *ferrite* and is relatively soft and ductile. At commercial purity its tensile strength is less than 45 000 psi (pounds/square

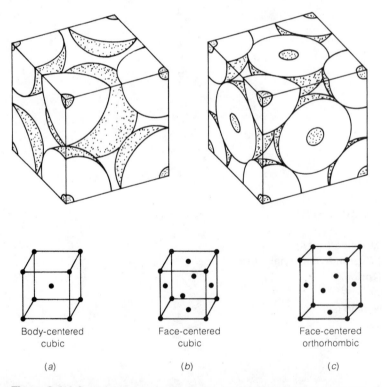

Body-centered cubic	Face-centered cubic	Face-centered orthorhombic
(a)	(b)	(c)

Figure 2.1.1 Crystal structure for (a) ferrite, (b) austenite, and (c) cementite. *From L.H. van Vlack, Material Science for Engineers, Addison-Wesley, Reading, MA, 1970. Reprinted with permission.*

inch). It has a body-centered-cubic structure; that is, the atoms are arranged as in Figure 2.1.1a according to a simple crystalline model. In this configuration there is little space between atoms; hence, the lattice cannot accept a carbon atom. At higher temperatures the structure changes to a face-centered cubic, Figure 2.1.1b, and can then accomodate carbon atoms with some lattice distortion. This is called *γ-iron* or *austenite*. It is still soft and ductile enough to be easily fabricated and it is stable at temperatures between 910 °C (1670 °F) and 1400 °C (2550 °F).

Carbon present in excess of the solubility limit forms a second phase known as *iron carbide*, Fe_3C. In this phase the crystal lattice contains three iron atoms and one carbon atom, for a total of 16 atoms in what is termed a face-centered orthorhombic lattice, shown in Figure 2.1.1c. Iron carbide is also known as *cementite*, properly suggesting that this phase is relatively hard and brittle.

The variety of steels that can exist at temperatures up to 1600 °C for a carbon content less than 7% by weight is shown in Figure 2.1.2.[1] It must be emphasized that this figure represents *equilibrium* conditions, which means that at 600 °C, for example, a steel with 0.50% carbon (known as SAE 1050 or UNS G10500 steel—see Tables 2.2.1

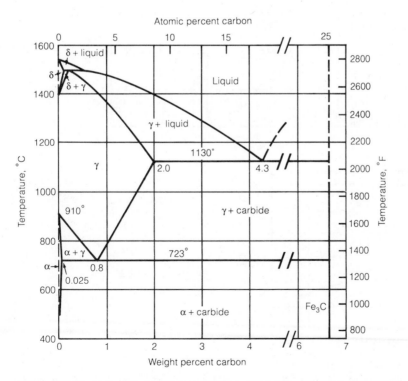

Figure 2.1.2 Iron–carbon equilibrium diagram. *From L.H. van Vlack*, Materials Science for Engineers, *Addison-Wesley, Reading, MA, 1970. Reprinted with permission.*

and 2.2.2) will be composed of ferrite and carbide. If the temperature is raised to 1000 °C (1832 °F) the ferrite (α) body-centered structure will transform into the austenite (γ) face-centered structure. This transformation takes time to accomplish because some of the atoms must migrate to new locations.

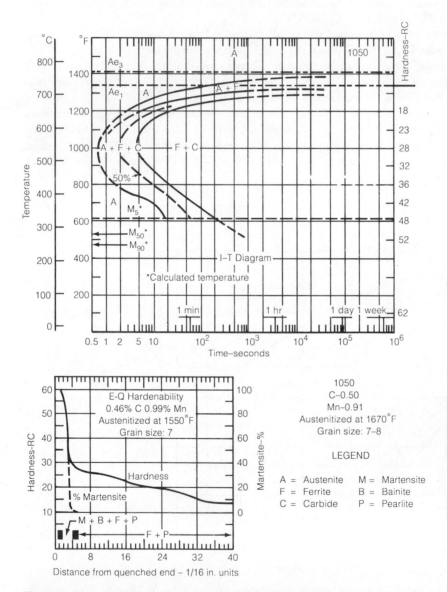

Figure 2.1.3 Isothermal transformation diagram for UNS G10500 steel. *From* Atlas of Isothermal Phase Diagrams, *United States Steel Corp., 1951. Courtesy United States Steel Corp., Pittsburgh, PA.*

The amount of time required for transformation from one phase to another is shown in an *isothermal transformation (IT) diagram*, as in Figure 2.1.3. This type of diagram is also known as a time–temperature transformation (TTT) diagram. The IT diagram for SAE 1050 steel shows that if the austenite could be cooled instantaneously

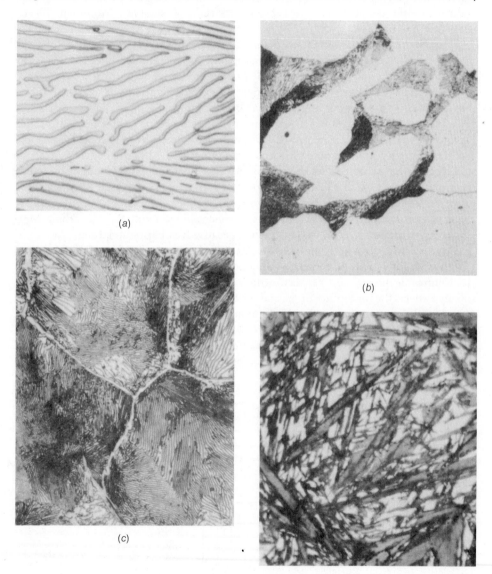

(a)

(b)

(c)

Figure 2.1.4 (a) Pearlite, (b) and (c) pearlite proportions with increased carbon, and (d) martensite. *Source: Metals Handbook, 8th ed., Vol. 7, American Society for Metals, Metals Park, OH, 1972, p. 152. Reprinted with permission.*

to 800 °F (417 °C), the first traces of ferrite would appear about 1.5 s later and the last of the austenite would vanish about 24 s after it reached 800 °C. Incidentally, the task of obtaining data for the isothermal diagram is very difficult, not only because of the temperatures themselves but also because any temperature change is relatively slow and because the temperature varies throughout the iron sample until an equilibrium temperature is reached.

The isothermal transformation diagram is the basis for *heat treatment* of steel, which generally consists of controlled heating and cooling and may be represented by a path in the isothermal transformation diagram. Only the end points of the isothermal transformation path may appear in the phase diagram, Figure 2.1.2, and then only if they are stable; that is, after any aging has been completed. For example, one cooling program can produce a lamellar structure of alternate ferrite and cementite, called *pearlite*, Figure 2.1.4*a*. Pearlite can be formed in many steels by particular heating and cooling programs, and the amount formed is influenced by the carbon content of the steel. If the steel is cooled fast enough to introduce a shear transformation (in which the molecules move in a shearlike fashion to form a different crystal structure[1]) the ferrite and cementite do not have time to form. A body-centered tetrahagonal phase called *martensite* is produced by this heat treatment in steel with 0.80% carbon. Martensite is important as a component of steel because it is hard and strong.

The effect of carbon on the yield strength of plain carbon steels (UNS G10100 to UNS G10900) is plotted in Figure 2.1.5 along with the tensile strengths obtained.

Control of the reaction between carbon and dissolved oxygen as the molten steel solidifies affects the uniformity of the end product. Immediate inhibition of the reaction, often by adding minute amounts of aluminum, produces *killed steel*. Alloy steels are made of killed steel, which has more uniform composition across its cross section than steels in which the reaction was allowed to continue for a longer time. *Rimmed steel* differs in chemical composition across its cross section because the carbon and oxygen

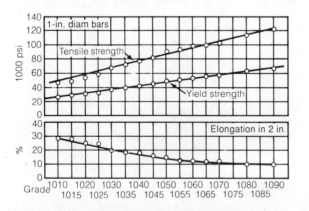

Figure 2.1.5 Effect of carbon content upon tensile strength and yield strength of low-carbon steel. *From Metals Handbook, 8th ed., Vol. 1, American Society for Metals, Metals Park, OH, 1961, p. 64. Reprinted with permission.*

process was allowed to continue. Intermediate between these extremes lie *semikilled steel* and *capped steel*. Capped steel begins as rimmed steel but the reaction is stopped shortly after pouring. Its surface of low-carbon steel is thinner than in rimmed steel and the core has the uniform properties of killed steel.

Further variation in the properties of steels may be obtained from heat treatment combined with alloying, adding elements other than carbon. The elaborate numbering systems discussed in the next section are indicative of this variety.

2.2 NUMBERING SYSTEMS FOR STEEL

At least seven different organizations supported their own numbering systems for steel. In 1975 a committee of users and producers of metals agreed upon a Unified Numbering System (UNS) for metals to replace the "often-confusing array of designation systems which had developed independently over the past 60 years."[2] Since the UNS system was intended only to unify the previously existing numbering systems, it did not set any new standards for any of the numbers. It is therefore necessary to retain an auxiliary list of previous numbering systems to fix the properties of any particular steel.

The numbering system of the Society of Automotive Engineers (SAE) and the American Iron and Steel Institute (AISI) is presented in Table 2.2.1 and assigned UNS numbers are shown in Table 2.2.2. Whenever possible the xxxxx digits were duplicates of earlier digital designations beginning with the left-most digit. Thus, UNS G10200 corresponds to AISI 1020, where G signifies what were AISI and SAE carbon and alloy steels. An extensive list of the association of UNS numbers with the earlier numbering systems is given in Ref. 2.

Mechanical properties of a representative set of steels are given in Table 2.2.3.

2.3 STEEL TYPES FOR VARIOUS APPLICATIONS

Engine blocks, heads, pistons, clutch plates, and similar parts may be made from gray cast iron, UNS F11401 to F13801. Corresponding designations by the American Society for Testing Materials (ASTM) are ASTM A-48, grades 20 to 55.[3] Gray iron has also been used for flywheels, gears, and machine frames, such as used in lathes, punch presses, and other factory equipment.[4] Axle housings, engine brackets, crankshafts, universal joint yokes, railroad brake parts, and some of the heavier parts in oil field and mining machines have been made of malleable cast iron, UNS F22200 to F22400.

Bolts, axles, shafts, and heavy duty, high-strength parts, such as heavy forgings and fears for power transmission, may be made from alloy steels represented by UNS G13400 (AISI 1340), UNS G40630 (AISI 4063), and UNS G43400 (AISI 4320). Automobile and truck chassis may use high-strength, low-alloy steel as ASTM A572.[5]

Stainless steel is frequently used where strength and moderately high temperature may cause corrosion and scale on other steels. Petroleum and chemical process equipment may use USN S31000 (AISI 310) stainless steel, while undersea equipment is often fabricated from UNS S31600 (AISI 316) steel because it resists corrosion in salt water.

TABLE 2.2.1   AISI–SAE SYSTEM OF DESIGNATIONS

Numerals and Digits[a]	Type of Steel and Nominal Alloy Content	Numerals and Digits[a]	Type of Steel and Nominal Alloy Content
Carbon Steels		**Nickel-Molybdenum Steels**	
10XX(a)	Plain carbon (Mn 1.00% max)	46XX	Ni 0.85 and 1.82; Mo 0.20 and 0.25
11XX	Resulfurized	48XX	Ni 3.50; Mo 0.25
12XX	Resulfurized and rephosphorized		
15XX	Plain carbon (max Mn range— 1.00 to 1.65%)	**Chromium Steels**	
		50XX	Cr 0.27, 0.40, 0.50 and 0.65
Manganese Steels		51XX	Cr 0.80, 0.87, 0.92, 0.95, 1.00 and 1.05
13XX	Mn 1.75		
		Chromium Steels	
Nickel Steels		50XXX	Cr 0.50
23XX	Ni 3.50	51XXX	Cr 1.02 } C 1.00 min
25XX	Ni 5.00	52XXX	Cr 1.45
Nickel-Chromium Steels		**Chromium-Vanadium Steels**	
31XX	Ni 1.25; Cr 0.65 and 0.80	61XX	Cr 0.06, 0.80 and 0.95; V 0.10 and 0.15 min
32XX	Ni 1.75; Cr 1.07		
33XX	Ni 3.50; Cr 1.50 and 1.57	**Tungsten-Chromium Steel**	
34XX	Ni 3.00; Cr 0.77	72XX	W 1.75; Cr 0.75
Molybdenum Steels		**Silicon-Manganese Steels**	
40XX	Mo 0.20 and 0.25	92XX	Si 1.40 and 2.00; Mn 0.65, 0.82 and 0.85; Cr 0.00 and 0.65
44XX	Mo 0.40 and 0.52		
		High-Strength Low-Alloy Steels	
Chromium-Molybdenum Steels		9XX	Various SAE grades
41XX	Cr 0.50, 0.80 and 0.95; Mo 0.12, 0.20, 0.25 and 0.30		
		Boron Steels	
Nickel-Chromium-Molybdenum Steels		XXBXX	B denotes boron steel
43XX	Ni 1.82; Cr 0.50 and 0.80; Mo 0.25		
43BVXX	Ni 1.82; Cr 0.50; Mo 0.12 and 0.25; V 0.03 min	**Leaded Steels**	
47XX	Ni 1.05; Cr 0.45; Mo 0.20 and 0.35	XXLXX	L denotes leaded steel
81XX	Ni 0.30; Cr 0.40; Mo 0.12		
86XX	Ni 0.55; Cr 0.50; Mo 0.20		
87XX	Ni 0.55; Cr 0.50; Mo 0.25		
88XX	Ni 0.55; Cr 0.50; Mo 0.35		
93XX	Ni 3.25; Cr 1.20; Mo 0.12		
94XX	Ni 0.45; Cr 0.40; Mo 0.12		
97XX	Ni 0.55; Cr 0.20; Mo 0.20		
98XX	Ni 1.00; Cr 0.80; Mo 0.25		

[a]   XX in the last two digits of these designations indicates that the carbon content (in hundredths of a percent) is to be inserted.

Source: Metals Handbook, 9th ed., Vol. 1, American Society for Metals Park, OH, 1978, p. 124. Reprinted with permission.

TABLE 2.2.2 UNIFIED NUMBERING SYSTEM FOR STEEL

Axxxxx	Aluminum and aluminum alloys
Cxxxxx	Copper and copper alloys
Exxxxx	Rare earth and similar metals and alloys
Fxxxxx	Cast irons
Gxxxxx	AISI and SAE carbon and alloy steels
Hxxxxx	AISI and SAE H-steels
Jxxxxx	Cast steels (except tool steels)
Kxxxxx	Miscellaneous steels and ferrous alloys
Lxxxxx	Low melting metals and alloys
Mxxxxx	Miscellaneous nonferrous metals and alloys
Nxxxxx	Nickel and nickel alloys
Pxxxxx	Precious metals and alloys
Rxxxxx	Reactive and refractory metals and alloys
Sxxxxx	Heat and corrosion resistant steels (including stainless), valve steels, and iron-base "superalloys"
Txxxxx	Tool steels, wrought and cast
Wxxxxx	Welding filler metals
Zxxxxx	Zinc and zinc alloys

Source: Metals & Alloys in the Unified Numbering System, 4th ed., © 1986 Society of Automotive Engineers, Warrendale, PA. Reprinted with permission.

TABLE 2.2.3 MECHANICAL PROPERTIES OF CARBON AND ALLOY STEELS[a]

UNS No. (AISI)	Treatment	Tensile Strength		Yield Strength	
		MPa	ksi	MPa	ksi
G10150	N	424.0	61.5	324.1	47.0
(1015)	A	386.1	56.0	284.4	41.3
G10200	N	441.3	64.0	346.5	50.3
(1020)	A	394.7	57.3	294.8	42.8
G10220	N	482.6	70.0	385.5	52.0
(1022)	A	429.2	62.3	317.2	46.0
G10300	N	520.6	75.5	344.7	50.0
(1030)	A	463.7	67.3	341.3	49.5
	205	848.0	123.0	648.0	94.0
G10400	N	589.5	85.5	374.0	54.3
(1040)	A	518.8	75.3	353.4	51.3
	205	896.0	130.0	662.0	96.0
G10500	N	748.1	108.5	427.5	61.0
(1050)	A	636.0	92.3	365.4	53.0
	205	1124	163	807	117
G10950	N	1013.5	147.0	499.9	72.5
(1095)	A	656.7	95.3	379.2	55.0
	205 w	1489.0	216.0	1048.0	152.0

(continued)

TABLE 2.2.3   (Continued)

UNS No. (AISI)	Treatment	Tensile Strength MPa	Tensile Strength ksi	Yield Strength MPa	Yield Strength ksi
G11370	N	668.8	97.0	396.4	57.5
(1137)	A	584.7	84.8	344.7	50.0
	205	1082.0	157.0	938.0	136.0
G13400	N	836.3	121.3	558.5	81.0
(1340)	A	703.3	102.0	436.4	63.3
	205	1806.0	262.0	1593.0	231.0
G41400	N	1020.4	148.0	655.0	95.0
(4140)	A	655.0	95.0	417.1	60.5
	205	1772.0	257.0	1641.0	238.0
G41500	N	1154.9	167.5	734.3	106.5
(4150)	A	729.5	105.8	379.2	55.0
	205	1931.0	280.0	1724.0	250.0
G43400	N	1279.0	185.5	861.8	125.0
(4340)	A	744.6	108.0	472.3	68.5
	205	1875.0	272.0	1675.0	243.0
G51400	N	792.9	115.0	472.3	68.5
(5140)	A	572.3	83.0	293.0	42.5
	205	1793.0	260.0	1641.0	238.0
G51600	N	957.0	138.8	530.9	77.0
(5160)	A	722.6	104.8	275.8	40.0
	205	2220.0	322.0	1793.0	260.0
G86500	N	1023.0	148.5	688.1	99.8
(8650)	A	715.7	103.8	386.1	56.0
	205	1937.0	281.0	1675.0	243.0
G92550	N	932.9	135.3	579.2	84.0
(9255)	A	774.3	112.3	486.1	70.5
	205	2103.0	305.0	2048.0	297.0
G12144	HR	390.0	57.0	230.0	34.0
(12L14)	CD	540.0	78.0	410.0	60.0

[a] N, normalized; A, annealed; 205, tempering at 205 °C; 205 w, tempering at 205 °C and water quenched; HR, hot rolled steel; CD, cold drawn steel.
Source: Abstracted from *ASM Metals Reference Book*, 2d ed., American Society for Metals, Metals Park, OH, 1983, pp. 211–216. Reprinted with permission.

Road salt and other highway chemical corrosion is resisted by UNS S43400 (AISI 434), so stainless steel can be used for automobile trim. High-temperature scale and corrosive effects may be reduced by using UNS S44600 (AISI 446) steel. Higher temperatures and hot gas abrasion may be resisted by the use of the so-called iron-base superalloys such as Incoloy Alloy 903 and A286.

Material choices are generally determined by availability, cost, ease of working and machining, weight, mechanical properties (tensile modulus, yield strength, endurance limit, etc.), corrosion resistance, lubrication requirements, hardness, and so on.

2.4 CAST IRON

Cast iron appears commercially in six varieties: pig iron, gray iron, white iron, chilled, malleable, and nodular iron. *Pig iron* is a high-carbon iron made by reduction of iron ore in a blast furnace. It therefore may be considered the raw product used in making the other types of cast iron.

Gray iron is an alloy of iron, carbon, and silicon with carbon in excess of that which can be held in solution in austinite at the eutectic temperature (the lowest temperature at which all components are liquid). Upon cooling, the excess carbon precipitates as graphite flakes.

Gray iron is the cheapest of all cast metals and has good wear resistance. Consequently, it has been widely used for guard rails, machine bases, valve housings, and lightly loaded gears. Its specifications are given in ASTM A48 in 9 classes, numbered 20, 25, . . . , 60. Relative strength of these classes is suggested by curves in Figure 2.4.1. Typical stress–strain curves are shown in Figure 2.4.2. Automotive uses are listed in Table 2.4.1 and associated specifications are found in SAE J431.[5] Mechanical properties of some of the more widely used gray iron grades are listed in Table 2.4.2.

SAE grades G1800 to G4000 (or UNS F10004 to F10008) typically span a minimum tensile strength range from 118 to 276 MPa, while ASTM classes 20 to 60 typically span a range from 152 to 431 MPa.

As the grades of gray cast iron increase from F10004 to F10008, or as the classes increase from 20 to 60, the tensile strength, the ability to produce a fine machine finish, the modulus of elasticity, and the wear resistance all increase. In contrast, the machinability (see Section 2.5), the resistance to thermal shock, the vibrational damping (often a valuable characteristic), and the ability to cast thin sections all decrease with an increase in class or grade. The machinability of gray cast iron is superior to that of most

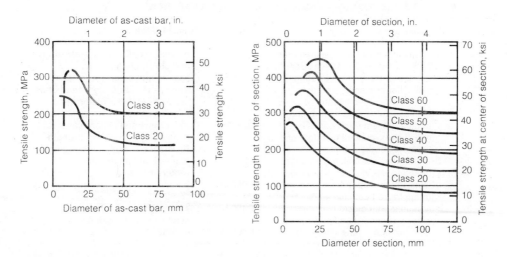

Figure 2.4.1 Relative strengths of gray cast iron as a function of bar diameter. *From* Metals Handbook, *9th ed., Vol. 1, American Society for Metals, Metals Park, OH, 1978, p. 14. Reprinted with permission.*

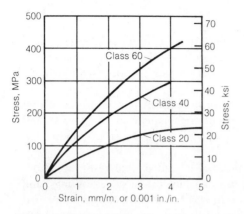

Figure 2.4.2 Typical stress–strain curves for gray cast iron. *From* Metals Handbook, *9th ed., Vol. 1, American Society for Metals, Metals Park, OH, 1978, p. 20. Reprinted with permission.*

steel because of the graphite. However, residual stresses may be present in the as-cast condition due to poor foundry control of cooling. They may be reduced by several stress-relieving procedures, but at added cost.

White iron is a mixture of pearlite and cementite and so is virtually free of graphite, which causes it to be extremely difficult to machine. Good wear characteristics justify its use for extrusion dies, ball mills, and as liners for containers of abrasive materials.

Chilled iron castings are made by casting against a metal or graphite chiller which produces an outer layer of white iron with a core of gray iron. Additives, such as manganese, sulfur, and phosphorus, may be used to control the chill depth while other ad-

TABLE 2.4.1   AUTOMOTIVE APPLICATIONS OF GRAY CAST IRON

Grade	Typical Uses	Grade	Typical Uses
G1800	Miscellaneous soft iron castings (as cast or annealed) in which strength is not a primary consideration.		pistons, medium duty brake drums and clutch plates.
G2500	Small cylinder blocks, cylinder heads, air-cooled cylinders, pistons, clutch plates, oil pump bodies, transmission cases, gear boxes, clutch housings, and light duty brake drums.	G3500	Diesel engine blocks, truck and tractor cylinder blocks and heads, heavy flywheels, tractor transmission cases, heavy gear boxes.
G2500a	Brake drums and clutch plates for moderate service requirements, where high carbon iron is desired to minimize heat checking.	G3500b	Brake drums and clutch plates for heavy duty service where both resistance to heat checking and higher strength are definite requirements.
G3000	Automobile and diesel cylinder blocks, cylinder heads, flywheels, differential carrier castings,	G3500c	Brake drums for extra heavy duty service
		G4000	Diesel engine castings, liners, cylinders, and pistons.
		G4000d	Camshafts.

Source: Metals Handbook, 9th ed., Vol. 1, American Society for Metals, Metals Park, OH, 1978, p. 19. Reprinted with permission.

TABLE 2.4.2 ELASTIC MODULI AND STRENGTHS FOR COMMONLY USED GRADES OF GRAY IRON

ASTM class	Tensile Modulus		Torsional Modulus		Tensile Strength		Torsional Shear Strength		Compressive Strength	
	GPa	10⁴ psi	GPa	10⁶ psi	MPa	ksi	MPa	ksi	MPa	ksi
20	66 to 97	9.6 to 14.0	27 to 39	3.9 to 5.6	152	22	179	26	572	83
25	79 to 102	11.5 to 14.8	32 to 41	4.6 to 6.0	179	26	220	32	669	97
30	90 to 113	13.0 to 16.4	36 to 45	5.2 to 6.6	214	31	276	40	752	109
35	100 to 119	14.5 to 17.2	40 to 48	5.8 to 6.9	252	36.5	334	48.5	855	124
40	110 to 138	16.0 to 20.0	44 to 54	6.4 to 7.8	293	42.5	393	57	965	140
50	130 to 157	18.8 to 22.8	50 to 55	7.2 to 8.0	362	52.5	503	73	1130	164
60	141 to 162	20.4 to 23.5	54 to 59	7.8 to 8.5	431	62.5	610	88.5	1293	187.5

Source: Metals Handbook, 9th ed., Vol. 1, American Society for Metals, Metals Park, OH, 1978, p. 19. Reprinted with permission.

ditives, such as chromium, may be added to improve wear resistance. Brinell hardness as a function of depth below the chilled surface is indicated in Figure 2.4.3.

Malleable iron refers to ferritic and to pearlitic iron, although in commercial practice the term alone usually denotes ferritic iron. Ferritic iron in turn includes two types: *blackhart* and *whitehart*. Only blackhart is produced in the United States, because of its superior malleability. Typical stress–strain curves for pearlitic and ferritic iron are shown in Figure 2.4.4 and uses for malleable iron are listed in Table 2.4.3.

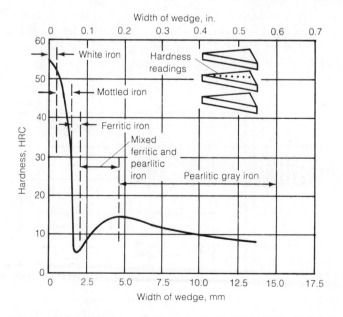

Figure 2.4.3 Chilled iron hardness and composition with depth below the surface (3.52 C, 2.55 Si, 1.01 Mn, 0.215 P, 0.086 S). *From* Metals Handbook, *9th ed., Vol. 1, American Society for Metals, Metals Park, OH, 1978, p. 14. Reprinted with permission.*

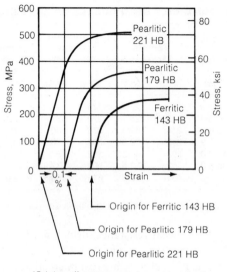

(Origins off-set to prevent curve overlap.)

Figure 2.4.4 Typical stress–strain curves for pearlitic and ferritic iron. Origins are offset to prevent curve overlap. *From* Metals Handbook, *9th ed., Vol. 1, American Society for Metals, Metals Park, OH, 1978, p. 65. Reprinted with permission.*

Nodular cast iron is also known as *ductile* iron and sometimes as *spherulitic iron* because of the tiny graphite spheres that are present. Stress–strain curves for ASTM A395-56T grade 60-45-15 are shown in Figure 2.4.5. Nodular cast iron is used for plow shares, crankshafts, rotors, and similar parts. It is often chosen for its easy machining.

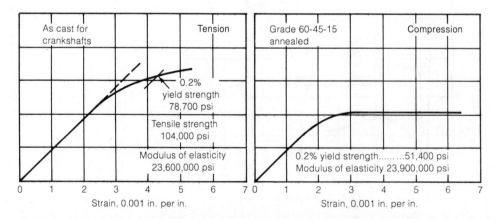

Figure 2.4.5 Stress–strain curves for ASTM A395-56T nodular iron. *From* Metals Handbook, *8th ed., Vol. 1, American Society for Metals, Metals Park, OH, 1961, p. 386. Reprinted with permission.*

TABLE 2.4.3   TYPICAL USES FOR MALLEABLE IRON CASTINGS

Specification No.	Class or Grade	Microstructure	Typical Applications
Ferritic			
ASTM A47; ANSI G48.1; FED QQ-I-666c	32510 35018	Temper carbon and ferrite	General engineering service at normal and elevated temperatures for good machinability and excellent shock resistance.
ASTM A338	32510 35018	Temper carbon and ferrite	Flanges, pipe fittings, and valve parts for railroad, marine, and other heavy duty service up to 345 °C (650 °F).
ASTM A197; ANSI G49.1	. . .	Free of primary graphite	Pipe fittings and valve parts for pressure service.
Pearlitic and Martensitic			
ASTM A220; ANSI G48.2; MIL-I-11444B	40010 45008 45006 50005 60004 70003 80002 90001	Temper carbon in necessary matrix without primary cementite or graphite	General engineering service at normal and elevated temperatures. Dimensional tolerance range for castings is stipulated.
Automotive			
ASTM A602; SAE J158	M3210	Ferritic	For low-stress parts requiring good machinability; steering gear housings, carriers, and mounting brackets.
	M4504	Ferrite and tempered pearlite(b)	Compressor crankshafts and hubs.
	M5003	Ferrite and tempered pearlite(b)	For selective hardening: planet carriers, transmission gears, differential cases.
	M5503	Tempered martensite	For machinability and improved response to induction hardening.
	M7002	Tempered martensite	For high-strength parts: connecting rods and universal joint yokes.
	M8501	Tempered martensite	For high strength plus good wear resistance: certain gears.

Source: Metals Handbook, 9th ed., Vol. 1, American Society for Metals, Metals Park, OH, 1978, p. 63. Reprinted with permission.

2.5 DATA SPREAD

Curves and tables displayed in previous sections for elastic properties such as the tensile stress at yield, the elastic modulus in tension (Young's modulus), the elastic modulus in shear, the ultimate strength (perhaps better termed the ultimate stress), and so on, have provided *typical* data. These data and those to follow may not exactly describe the commercially available material in the shop at the time of manufacture because of production variations. The observed spread of such data for selected steels, for example, is shown in Figure 2.5.1.

The data spread is due to the manufacturing tolerances set for standard metals by such organizations as the American Society for Metals (ASM) and for metals and other materials, such as plastics and ceramics, by the American Society for Testing and Materials (ASTM) and government agencies, such as the U.S. Department of Defense. This spread of data is recognized by some companies and manufacturers' associations in their recommended design procedures. For example, the Spring Manufacturers Institute lists only the *minimum tensile strength* for common spring materials and correlates this strength (which is in units of stress) with the heat treatment of the steel and the wire diameter after drawing it through a die. As used here, the term minimum tensile strength refers to the *minimum value* of the statistical spread of the *maximum tensile stress* for steels manufacturered to the given ASTM standards. According to this definition, the minimum tensile stress of the low-carbon steel whose test results are displayed in Figure 2.5.1 would be about 43.68 ksi.

In spring design it is convenient to define the maximum tensile strength as equivalent to the ultimate tensile stress as defined in the 1988 *SAE Handbook*[5], which is that the ultimate tensile stress is the maximum load divided by the original cross-sectional area.

2.6 HOT-FINISHED CARBON STEEL

Hot-finished carbon steel has been the cheapest steel per pound and it has many applications. It is supplied in two grades: merchant quality and standard quality. The two differ only in the wider chemical analysis limits set for acceptance of merchant quality steel.

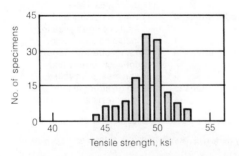

Figure 2.5.1 Representative mechanical properties of hot-rolled, low-carbon steel sheet. *From Metals Handbook, 9th ed., Vol. 1, American Society for Metals, Metals Park, OH, 1978, p. 158. Reprinted with permission.*

Machinability of hot-finished steel varies with the carbon content. That with low carbon content, such as AISI 1010 or 1015 (UNS G10100 or UNS G10200) may be soft and gummy in machinists' terms, whereas AISI 1025 (UNS G10250) to AISI 1045 (UNS G10450) may machine with little difficulty. Low-carbon steels are often used for stampings, structural sheets, kitchen utensils, appliance sheet metal, containers, and tanks.

Higher carbon steels, say UNS G10450 to UNS G10950 (AISI 1045 to AISI 1095) may have to be softened by annealing before machining. (Annealing involves heating to, and holding at, a particular temperature for the particular steel and then cooling at a particular rate to soften the steel.) Addition of manganese sulfide or lead improves the ease of machining in what are called free machining steels. The higher carbon steels are used for gears, shafts, and similar components.

2.7 COLD-FINISHED CARBON STEEL

Cold-finished steel is hot rolled to oversize dimensions, allowed to cool, and then reduced to the final size by grinding or drawing. Because of the earlier practice of rolling the cold bars to size, this steel is still often described as *cold rolled*. Effects of cold working on the tensile strength are shown in Figures 2.7.1*a*, and *b*. Residual stresses are relieved after cold finishing. Typical tensile strength of cold-drawn 1-in.-diameter bars is shown in Figure 7.1*b*. Cold drawing improves the machinability of the steels shown in Figure 2.1.5 by eliminating the gummy behavior at the low-carbon end of the range.

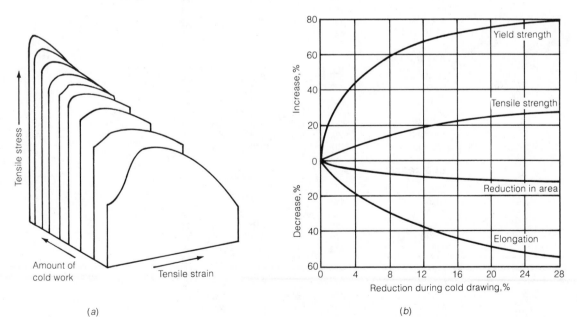

(a)   (b)

Figure 2.7.1 (a) Effect of cold work in tensile stress–strain relations for low-carbon steel and (b) the effect of cold drawing. *From* Metals Handbook, *9th ed., Vol. 1, American Society for Metals, Metals Park, OH, 1978, pp. 221–222. Reprinted with permission.*

Cold finished steel is used for axles, spark plugs, rifle barrels, hydraulic push rods, and similar products.

2.8 CAST STEEL

Cast steels usually have a total alloy content of less than 8%. Specifications for steel casting for several ASTM and SAE grades are listed in Table 2.8.1

Medium carbon steels (0.2 to 0.5%) represent the majority of cast steel production. They are always heat treated to relieve residual stresses and to improve ductility. Ten-

TABLE 2.8.1 TENSILE STRESS, YIELD STRESS, AND ELONGATION MINIMUM REQUIREMENTS

Class or Grade	Minimum Tensile Strength MPa	ksi	Minimum Yield Strength MPa	ksi	Minimum Elongation in 50 mm or 2 in., %
ASTM A27-77					
N-1	. . .	. . .	. . .	. . .	. . .
N-2	. . .	. . .	. . .	. . .	. . .
U60-30	415	60	205	30	22
60-30	415	60	205	30	24
65-35	450	65	240	35	24
70-36	485	70	250	36	22
70-40	485	70	275	40	22
ASTM A148-73					
80-40	552	80	276	40	18
80-50	552	80	345	50	22
90-60	621	90	414	60	20
105-85	724	105	586	85	17
120-95	827	120	655	95	14
150-125	1034	150	862	125	9
175-145	1207	175	1000	145	6
SAE J435c					
0022	. . .	. . .	. . .	. . .	. . .
0025	414	60	207	30	22
0030	448	65	241	35	24
0050A	586	85	310	45	16
0050B	690	100	483	70	10
080	552	80	345	50	22
090	621	90	414	60	20
0105	724	105	586	85	17
0120	827	120	655	95	14
0150	1034	150	862	125	9
0175	1207	175	1000	145	6
HA, HB, HC(d)	. . .	. . .	. . .	. . .	. . .

Source: Metals Handbook, 9th ed., Vol. 1, American Society for Metals, Metals Park, OH, 1978, p. 378. Reprinted with permission.

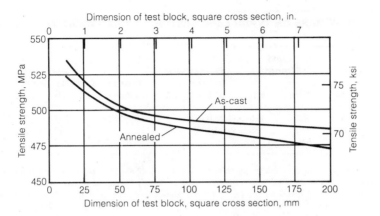

Figure 2.8.1 Decrease in tensile strength with increased casting size. *From* Metals Handbook, *9th ed., Vol. 1, American Society for Metals, Metals Park, OH, 1978, p. 392. Reprinted with permission.*

sile strength decreases as the casting size increases, as shown in Figure 2.8.1, because of different interior cooling.

Cast steel is used for aircraft landing gear components, for railroad locomotive frame components and railroad car suspension system parts, for valve bodies, for parts of steam turbines, and for frames of heavy machinery.

2.9 HIGH-STRENGTH, LOW-ALLOY STEEL

High-strength, low-alloy (HSLA) steel contains no more than 0.30% carbon, and usually less than 0.20%. Alloying elements may be nitrogen to aid precipitation hardening, manganese to strengthen ferrite components, silicon to improve density, phosphorus and chromium to resist corrosion, copper and nickel to resist corrosion and improve strength, molybdenum and niobium to enhance high temperature and yield strength, aluminum, vanadium, titanium, and zirconium for grain size and strength, and boron for hardenability. Minimum yield strength for these steels is commonly 50 000 psi (345 MPa) and tensile strength is commonly 70 000 psi (485 MPa), although for quenched and tempered HSLA steel plates from $\frac{3}{16}$ to $2\frac{1}{2}$ in. (5 to 64 mm) a yield stress of 100 000 psi (690 MPa) and a tensile stress of 115 000 psi (795 MPa) are typical. All of these steels, except A656, are used for bridges and buildings. A656 is used for truck frames, brackets, crane booms, rail cars, and support structure, but not for machine elements.

2.10 ULTRAHIGH-STRENGTH STEEL

Demand from the aircraft and defense industries motivated the development of ultrahigh-strength steels, all with a minimum yield strength of 200 000 psi (1380 MPa).

TABLE 2.10.1 EFFECTS OF MASS ON TYPICAL PROPERTIES OF HEAT-TREATED 4130 STEEL[a]

Bar Size		Tensile Strength		Yield Strength		Elongation in 50 mm or 2 in., %	Reduction in Area, %	Surface Hardness, HB
mm	in.	MPa	ksi	MPa	ksi			
25	1	1040	151	880	128	18.0	55.0	307
50	2	740	107	570	83	20.0	58.0	223
75	3	710	103	540	78	22.0	60.0	217

[a] Round bars oil quenched from 845 °C (1550 °F) and tempered at 540 °C (1000 °F). 12.83-mm (0.505-in.) diam tensile specimens were cut from center of 25-mm diam bar, and from mid-radius of 50- and 75-mm diam bars.
Source: Metals Handbook, 9th ed., Vol. 1, American Society for Metals, Metals Park, OH, 1978, p. 423. Reprinted with permission.

These steels include AISI 4130 and AISI 4140 (see Tables 2.10.1 and 2.10.2), which differ only in the higher carbon content of the latter. AISI 4130 has low to intermediate hardness and retains satisfactory tensile, fatigue, and impact properties to about 700 °F (370 °C). It is not plagued by temper embrittlement and it can be forged and nitrided.

TABLE 2.10.2 TYPICAL MECHANICAL PROPERTIES OF HEAT-TREATED 4140 STEEL[a]

Tempering Temperature		Tensile Strength		Yield Strength		Elongation in 50 mm or 2 in., %	Reduction in Area, %	Hardness, HB	Izod Impact Energy	
°C	°F	MPa	ksi	MPa	ksi				J	ft · lb
205	400	1965	285	1740	252	11.0	42	578	15	11
260	500	1860	270	1650	240	11.0	44	534	11	8
315	600	1720	250	1570	228	11.5	46	495	9	7
370	700	1590	231	1460	212	12.5	48	461	15	11
425	800	1450	210	1340	195	15.0	50	429	28	21
480	900	1300	188	1210	175	16.0	52	388	46	34
540	1000	1150	167	1050	152	17.5	55	341	65	48
595	1100	1020	148	910	132	19.0	58	311	93	69
650	1200	900	130	790	114	21.0	61	277	112	83
705	1300	810	117	690	100	23.0	65	235	136	100

EFFECTS OF MASS ON TYPICAL PROPERTIES OF HEAT-TREATED 4140 STEEL[b]

Diameter of Bar		Tensile Strength		Yield Strength		Elongation in 50 mm or 2 in., %	Reduction in Area, %	Surface Hardness, HB
mm	in.	MPa	ksi	MPa	ksi			
25	1	1140	165	985	143	15	50	335
50	2	920	133	750	109	18	55	202
75	3	860	125	655	95	19	55	293

[a] 12.7-mm (½-in.) diam round bars, oil quenched from 845 °C (1550 °F).
[b] Round bars oil quenched from 845 °C (1550 °F) and tempered at 540 °C (1000 °F). 12.83-mm (0.505-in.) diam tensile specimens were cut from center of 25-mm-diam bars, and from mid-radius of 50- and 75-mm diam bars.
Source: Metals Handbook, 9th ed., Vol. 1, American Society for Metals, Metals Park, OH, 1978, p. 423. Reprinted with permission.

TABLE 2.11.1 HEAT TREATMENTS AND TYPICAL MECHANICAL PROPERTIES OF STANDARD 18Ni MARAGING STEELS

Grade	Heat Treatment[a]	Tensile Strength		Yield Strength		Elongation in 50 mm or 2 in., %	Reduction in Area, %	Fracture Toughness	
		MPa	ksi	MPa	ksi			MPa $\overline{m}$	ksi $\overline{in}$.
18Ni (200)	A	1500	218	1400	203	10	60	155–200	140–220
18Ni (250)	A	1800	260	1700	247	8	55	120	110
18Ni (300)	A	2050	297	2000	290	7	40	80	73
18Ni (350)	B	2450	355	2400	348	6	25	35–50	32–45
18Ni (Cast)	C	1750	255	1650	240	8	35	105	95

[a] Treatment A: solution treat 1 h at 820 °C (1500 °F), then age 3 h at 480 °C (900 °F). Treatment B: solution treat 1 h at 820 °C (1500 °F), then age 12 h at 480 °C (900 °F). Treatment C: anneal 1 h at 1150 °C (2100 °F), age 1 h at 595 °C (1100 °F), solution treat 1 h at 820 °C (1500 °F) and age 3 h at 480 °C (900 °F).
Source: Metals Handbook, 9th ed., Vol. 1, American Society for Metals, Metals Park, OH, 1978, p. 448. Reprinted with permission.

Applications of AISI 4130 include automobile connecting rods, shafts, gears, engine mounts, and components for aircraft.

2.11 MARAGING STEEL

Maraging steels differ from the steels discussed thus far in that their hardening does not involve carbon. Their high strength, shown in Table 2.11.1 for commercial grades, along with machinability that is equal to or better than conventional steels of the same hardness (but lower tensile strength) makes them desirable for missile cases, aircraft forgings, Belleville springs, bearings, transmission shafts, and cannon recoil springs. They are also easily welded and possess good fatigue characteristics, as shown in Figure 2.11.1

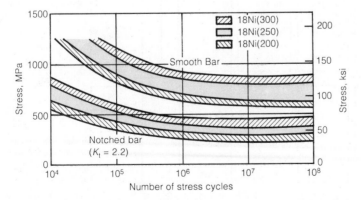

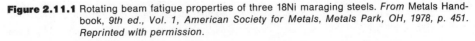

Figure 2.11.1 Rotating beam fatigue properties of three 18Ni maraging steels. *From* Metals Handbook, *9th ed., Vol. 1, American Society for Metals, Metals Park, OH, 1978, p. 451. Reprinted with permission.*

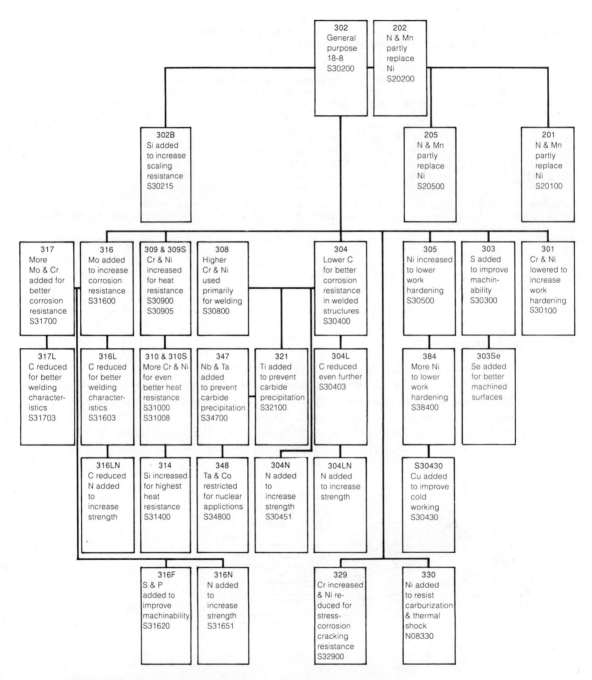

Figure 2.12.1 Family relationships for standard austenitic stainless steels. *From* Metals Handbook, *9th ed., Vol. 3, American Society for Metals, Metals Park, OH, 1980, p. 7. Reprinted with permission.*

2.12   STAINLESS STEEL

Stainless steel (UNS Sxxxxx) is used where resistance to corrosion and scale is important. Family relationships are shown in Figures 2.12.1, 2.12.2, and 2.12.3, in which their AISI designations are employed.[8] Representative strengths are listed in Table 2.12.1 for selected types of wrought stainless steel.[9] (Wrought steel is that in which the ingot can be formed to shape, usually while hot, without cracking.)

Proprietary stainless steels are not designated by the numbering system in Table 2.12.1. Examples of precipitation-hardened proprietary stainless steels are 15-5PH, 17-4PH, 17-7PH, and PH15-7Mo by Armco Steel Corporation, and AM 350 and 355 by Allegheny Ludlum Steel Corporation.

Examination of types 302, 304, 316, and 440B in Table 2.12.1 clearly shows the effects of tempering, size, and cold reduction upon tensile strength.

2.13   SUPERALLOYS

Superalloys are iron-, cobalt-, and nickel-base alloys which are stronger at high temperatures (800–900 °C or 1470–1650 °F) than conventional steels. They contain chromium to resist oxidation and hot corrosion as well as other elements to maintain

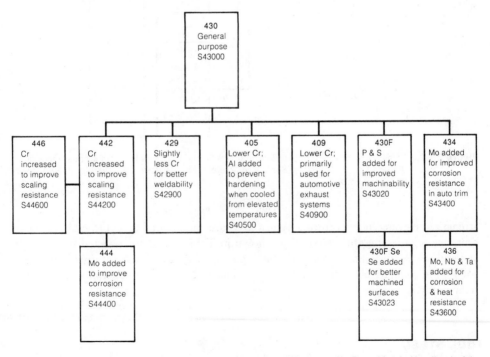

Figure 2.12.2 Family relationships for standard ferritic stainless steels. *From* Metals Handbook, *9th ed., Vol. 3, American Society for Metals, Metals Park, OH, 1980, p. 8. Reprinted with permission.*

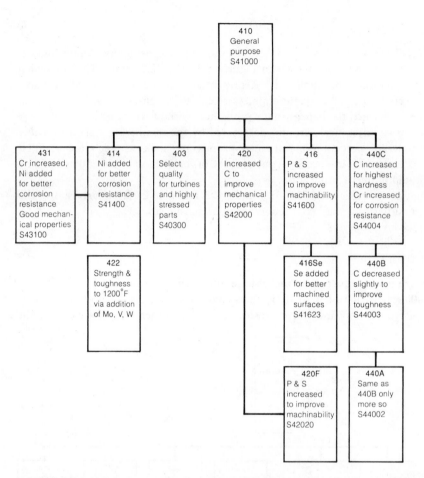

Figure 2.12.3 Family relationships for standard martensitic stainless steels. *From Metals Handbook, 9th ed., Vol. 3, American Society for Metals, Metals Park, OH, 1980, p. 8. Reprinted with permission.*

strength at elevated temperatures. Typical uses are enumerated in Table 2.13.1. Representative high-temperature strength is shown in Table 2.13.2 for selected proprietary superalloys. Rupture life for Udimet 700, Figure 2.13.1, at high temperature (850 °C or 1500 °F) is qualitatively representative of the loaded behavior of superalloys at similar temperatures.

2.14 TOOL STEEL

Tool steel is often a high-carbon, high-alloy steel that is very hard and resists abrasion, and in many cases retains these properties when heated during use. AISI and SAE classification is as in Table 2.14.1. Each class has several grades, with the grade indicated

TABLE 2.12.1 MECHANICAL PROPERTIES OF REPRESENTATIVE STAINLESS STEELS

UNS No. (AISI)	Condition	Tensile strength		Yield strength	
		MPa	ksi	MPa	ksi
S20100	Annealed	515	75	275	40
(201)	Full-hard	1280	185	965	140
S20200	Annealed	515	75	275	40
(202)	Full-hard	1030	150	760	110
S30100	Annealed	515	75	205	30
(301)	Full-hard	1280	185	965	140
S30200	Cold finished	620	90	205	30
(302)	and annealed				
	High tensile	2240	325	. . .	. . .
SS30400	Cold finished	620	90	310	45
(304)	and annealed				
	High tensile	2250	325	. . .	. . .
S30900	Cold finished	620	90	310	45
(309)	and annealed				
	Cold finished	620	90	310	45
S20910	Annealed	690	100	380	55
(Nitronic 50)					
S31600	Cold finished	620	90	310	45
(316)	and annealed				
	High tensile	1690	245	. . .	. . .
S44200	Annealed	550	80	310	45
(442)					

Source: Abstracted from *ASM Metals Reference Book*, American Society for Metals, Metals Park, OH, 1983, pp. 271–277.

TABLE 2.13.1 REPRESENTATIVE APPLICATIONS FOR HEAT-RESISTANT SUPERALLOYS

Aircraft Gas Turbines

Disks, bolts, shafts, cases, blades, vanes, burner cans, afterburners, thrust reversers

Steam-Turbine Power Plants

Bolts, blades, stack-gun reheaters

Reciprocating Engines

Turbochargers, exhaust valves, hot plugs, precombustion cups (Riccardo-type diesel), valve-seat inserts

Metal-working

Hot-work tools and dies, cast dies

Medical Applications

Dentistry, prosthetic devices

Space Vehicles

Aerodynamically heated skin, rocket engine parts

Heat Treating Equipment

Trays, fixtures, conveyor belts

Nuclear Power Systems

Control-rod drive mechanisms, valve stems, springs, ducting

Chemical and Petrochemical Industries

Bolts, valves, reaction vessels, piping, pumps

Source: Abstracted from *Metals Handbook*, 9th ed., Vol. 3, American Society for Metals, Metals Park, OH, 1980.

TABLE 2.13.2   TYPICAL MECHANICAL PROPERTIES FOR SELECTED COBALT-BASE AND NICKEL-BASE SUPERALLOYS

Temperature		Tensile Strength		Yield Strength		Elongation, %
°C	°F	MPa	ksi	MPa	ksi	
Cobalt-base alloys						
S-816, bar						
21	70	965	140	385	56	30
540	1000	840	122	310	45	27
650	1200	765	111	305	44	25
760	1400	650	94	285	41	21
870	1600	360	52	240	35	16
Nickel-base alloys						
Astroloy, bar						
21	70	1410	205	1050	152	16
540	1000	1240	180	965	140	16
650	1200	1310	190	965	140	18
760	1400	1160	168	910	132	21
870	1600	770	112	690	100	25
Inconel 706, bar						
21	70	1300	188	980	142	19
540	1000	1120	163	895	130	19
650	1200	1010	147	825	120	21
760	1400	690	100	675	98	32
René 95, bar						
21	70	1620	235	1310	190	15
540	1000	1540	224	1250	182	12
650	1200	1460	212	1220	177	14
760	1400	1170	170	1100	160	15
Udimet 700, bar						
21	70	1410	204	965	140	17
540	1000	1280	185	895	130	16
650	1200	1240	180	855	124	16
760	1400	1030	150	825	120	20
870	1600	690	100	635	92	27

Source: Metals Handbook, 9th ed., Vol. 3, American Society for Metals, Metals Park, OH, 1980, pp. 218–221. Reprinted with permission.

by a one- or two-digit suffix following the letter designation. Nominal properties of several L and S steels are listed in Table 2.14.2.[10]

Strength properties shown in Table 2.14.1 are used in the design of cutting tools in lathes and milling machines in which the cutting portion is essentially a cantilever beam loaded over a short edge length at the free end. Stress loads are cyclic bending, shear,

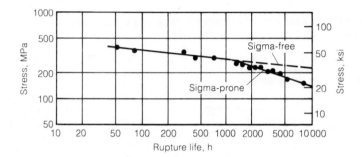

Figure 2.13.1 Log stress vs. log rupture life at 815 °C (1400 °F) for Udimet 700. *From* Metals Handbook, *9th ed., Vol. 3, American Society for Metals, Metals Park, OH, 1980, p. 277. Reprinted with permission.*

TABLE 2.14.1   CLASSIFICATION AND COMPOSITION OF PRINCIPAL TOOL STEELS

Designation	C	Mn	Si or Ni	Cr	V	W	Mo	Co
Water-Hardening Tool Steels[a]								
W1*	0.60 to 1.40[a]	. . . .	. . . .	. . . .	. . . .	. . . .	. . . .	. . . .
W2*	0.60 to 1.40[a]	. . . .	. . . .	. . . .	0.25	. . . .	. . . .	. . . .
W3	0.60 to 1.40[a]	. . . .	. . . .	. . . .	0.50	. . . .	. . . .	. . . .
W4	0.60 to 1.40[a]	. . . .	. . . .	0.25	. . . .	. . . .	. . . .	. . . .
W5	0.60 to 1.40[a]	. . . .	. . . .	0.50	. . . .	. . . .	. . . .	. . . .
W6	0.60 to 1.40[a]	. . . .	. . . .	0.25	0.25	. . . .	. . . .	. . . .
W7	0.60 to 1.40[a]	. . . .	. . . .	0.50	0.20	. . . .	. . . .	. . . .
Shock-Resisting Tool Steels								
S1*	0.50	. . . .	. . . .	1.50	. . . .	2.50	. . . .	. . . .
S2	0.50	. . . .	1.00 Si	. . . .	. . . .	. . . .	0.50	. . . .
S3	0.50	. . . .	. . . .	0.75	. . . .	1.00	. . . .	. . . .
S4	0.50	0.80	2.00 Si	. . . .	. . . .	. . . .	. . . .	. . . .
S5*	0.50	0.80	2.00 Si	. . . .	. . . .	. . . .	0.40	. . . .
Oil-Hardening Cold Work Tool Steels								
O1*	0.90	1.00	. . . .	0.50	. . . .	0.50	. . . .	. . . .
O2	0.90	1.60	. . . .	. . . .	. . . .	. . . .	. . . .	. . . .
O6	1.45	. . . .	1.00 Si	. . . .	. . . .	. . . .	0.25	. . . .
O7	1.20	. . . .	. . . .	0.75	. . . .	1.75	0.25 opt	. . . .
Air-Hardening Medium-Alloy Cold Work Tool Steels								
A2[b]	1.00	. . . .	. . . .	5.00	. . . .	. . . .	1.00	. . . .
A4	1.00	2.00	. . . .	1.00	. . . .	. . . .	1.00	. . . .
A5	1.00	3.00	. . . .	1.00	. . . .	. . . .	1.00	. . . .
A6	0.70	2.00	. . . .	1.00	. . . .	. . . .	1.00	. . . .
A7	2.25	. . . .	. . . .	5.25	4.50	. . . .	1.00	. . . .
High-Carbon High-Chromium Cold Work Steels								
D1	1.00	. . . .	. . . .	12.00	. . . .	. . . .	1.00	. . . .
D2*[b]	1.50	. . . .	. . . .	12.00	. . . .	. . . .	1.00	. . . .
D3*[b]	2.25	. . . .	. . . .	12.00	. . . .	. . . .	. . . .	. . . .

(continued)

TABLE 2.14.1   CLASSIFICATION AND COMPOSITION OF PRINCIPAL TOOL STEELS (Continued)

Designation	C	Mn	Si or Ni	Cr	V	W	Mo	Co
High-Carbon High-Chromium Cold Work Steels								
D4*	2.25	....	....	12.00	....	....	1.00	....
D5[b]	1.50	....	....	12.00	....	....	1.00	3.00
D6	2.25	....	1.00 Si	12.00	....	1.00	....	....
D7[b]	2.35	....	....	12.00	4.00	....	1.00	....
Chromium Hot Work Tool Steels								
H11	0.35	....	....	5.00	0.40	....	1.50	....
H12*	0.35	....	....	5.00	0.40	1.50	1.50	....
H13*[b]	0.35	....	....	5.00	1.00	....	1.50	....
H14	0.40	....	....	5.00	....	5.00	....	....
H15	0.40	....	....	5.00	....	....	5.00	....
H16	0.55	....	....	7.00	....	7.00	....	....
Tungsten Hot Work Tool Steels								
H20	0.35	....	....	2.00	....	9.00	....	....
H21*	0.35	....	....	3.50	....	9.50	....	....
H22	0.35	....	....	2.00	....	11.00	....	....
H23	0.30	....	....	12.00	....	12.00	....	....
H24	0.45	....	....	3.00	....	15.00	....	....
H25	0.25	....	....	4.00	....	15.00	....	....
H26	0.50	....	....	4.00	1.00	18.00	....	....
Molybdenum Hot Work Tool Steels								
H41	0.65	....	....	4.00	1.00	1.50	8.00	....
H42	0.60	....	....	4.00	2.00	6.00	5.00	....
H43	0.55	....	....	4.00	2.00	....	8.00	....
Tungsten High Speed Tool Steels								
T1*[b]	0.70	....	....	4.00	1.00	18.00	....	....
T2[b]	0.85	....	....	4.00	2.00	18.00	....	....
T3[b]	1.05	....	....	4.00	3.00	18.00	....	....
T4	0.75	....	....	4.00	1.00	18.00	....	5.00
T5	0.80	....	....	4.00	2.00	18.00	....	8.00
T7	0.75	....	....	4.00	2.00	14.00	....	....
T8	0.80	....	....	4.00	2.00	14.00	....	5.00
T15[b]	1.50	....	....	4.00	5.00	12.00	....	5.00
Molybdenum High Speed Tool Steels								
M1*[b]	0.80	....	....	4.00	1.00	1.50	8.50	....
M2*[b]	0.85	....	....	4.00	2.00	6.25	5.00	....
M3*[b][c]	1.00	....	....	4.00	2.40	6.00	5.00	....
M4	1.30	....	....	4.00	4.00	5.50	4.50	....
M6	0.80	....	....	4.00	1.50	4.00	5.00	12.00
M7	1.00	....	....	4.00	2.00	1.75	8.75	....
M10*[b]	0.85	....	....	4.00	2.00	....	8.00	....
M15	1.50	....	....	4.00	5.00	6.50	3.50	5.00
M30	0.80	....	....	4.00	1.25	2.00	8.00	5.00
M33	0.90	....	....	3.75	1.15	1.75	9.50	8.25
M34	0.90	....	....	4.00	2.00	2.00	8.00	8.00

TABLE 2.14.1 CLASSIFICATION AND COMPOSITION OF PRINCIPAL TOOL STEELS (Continued)

Designation	C	Mn	Si or Ni	Cr	V	W	Mo	Co
Molybdenum High Speed Tool Steels								
M35	0.80	....	....	4.00	2.00	6.00	5.00	5.00
M36	0.80	....	....	4.00	2.00	6.00	5.00	8.00
Low-Alloy Special-Purpose Tool Steels								
L1	1.00	....	....	1.25	....	....	....	....
L2	0.50 to 1.10[a]	....	....	1.00	0.20	....	....	....
L3	1.00	....	....	1.50	0.20	....	....	....
L4	1.00	0.60	....	1.50	0.20	....	....	....
L5	1.00	1.00	....	1.00	....	....	0.25	....
L6	0.70	....	1.50 Ni	0.75	....	....	0.25 opt	....
L7	1.00	0.35	....	1.40	....	....	0.40	....
Carbon-Tungsten Tool Steels								
F1	1.00	....	....	....	....	1.25	....	....
F2	1.25	....	....	....	....	3.50	....	....
F3	1.25	....	....	0.75	....	3.50	....	....
Low-Carbon Mold Steels								
P1	0.10 max	....	....	....	....	....	....	....
P2	0.07 max	....	0.50 Ni	1.25	....	....	0.20	....
P3	1.10 max	....	1.25 Ni	0.60	....	....	....	....
P4	0.07 max	....	....	5.00	....	....	....	....
P5	0.10 max	....	....	2.25	....	....	....	....
P6	0.10	....	3.50 Ni	1.50	0.20	....	....	....
P20	0.30	....	....	0.75	....	....	0.25	....
PPT	0.20	1.20 A1	4.00 Ni	....	....	....	....	....
Other Alloy Tool Steels[d]								
6G	0.55	0.80	0.25 Si	1.00	0.10	....	0.45	....
6F2	0.55	0.75	0.25 Si, 1.00 Ni	1.00	0.10 opt	....	0.30	....
6F3	0.55	0.60	0.85 Si, 1.80 Ni	1.00	0.10 opt	....	0.75	....
6F4	0.20	0.70	0.25 Si, 3.00 Ni	....	....	....	3.35	....
6F5	0.55	1.00	1.00 Si, 2.70 Ni	0.50	0.10	....	0.50	....
6F6	0.50	....	1.50 Si	1.50	....	....	0.20	....
6F7	0.40	0.35	4.25 Ni	1.50	....	....	0.75	....
6H1	0.55	....	....	4.00	0.85	....	0.45	....
6H2	0.55	0.40	1.10 Si	5.00	1.00	....	1.50	....

* Stocked in almost every warehousing district and made by the majority of tool steel producers.

[a] Various carbon contents are available in 0.10% ranges.

[b] Available as free-cutting grade.

[c] Available with vanadium contents of 2.40 or 3.00%.

[d] The designations of these steels are similar to those used in the 1948 Metals Handbook, except they were previously written with Roman numerals (VI F2, etc). Neither AISI nor SAE has assigned type numbers to these steels. *Source: Metals Handbook*, 8th ed., Vol. 1, American Society for Metals, Metals Park, OH, 1964. Reprinted with permission.

TABLE 2.14.2   NOMINAL ROOM-TEMPERATURE MECHANICAL PROPERTIES OF GROUP L AND GROUP S STEELS

Type	Condition	Tensile Strength		0.2% yield Strength		Hardness, HRC	Impact Energy	
		MPa	ksi	MPa	ksi		J	ft·lb
L2	Annealed	710	103	510	74	96 HRB	. . .	. . .
	Oil quenched from 855 °C (1575 °F) and single tempered at:							
	205 °C (400 °F)	2000	290	1790	260	54	28[b]	21[b]
	315 °C (600 °F)	1790	260	1655	240	52	19[b]	14[b]
	425 °C (800 °F)	1550	225	1380	200	47	26[b]	19[b]
	540 °C (1000 °F)	1275	185	1170	170	41	39[b]	29[b]
	650 °C (1200 °F)	930	135	760	110	30	125[b]	92[b]
L6	Annealed	655	95	380	55	93 HRB	. . .	. . .
	Oil quenched from 845 °C (1550 °F) and single tempered at:							
	315 °C (600 °F)	2000	290	1790	260	54	12[b]	9[b]
	425 °C (800 °F)	1585	230	1380	200	46	18[b]	13[b]
	540 °C (1000 °F)	1345	195	1100	160	42	23[b]	17[b]
	650 °C (1200 °F)	965	140	830	120	32	81[b]	60[b]
S1	Annealed	690	100	415	60	96 HRB	. . .	. . .
	Oil quenched from 930 °C (1700 °F) and single tempered at:							
	205 °C (400 °F)	2070	300	1895	275	57.5	249[c]	184[c]
	315 °C (600 °F)	2030	294	1860	270	54	233[c]	172[c]
	425 °C (800 °F)	1790	260	1690	245	50.5	203[c]	150[c]
	540 °C (1000 °F)	1680	244	1525	221	47.5	230[c]	170[c]
	650 °C (1200 °F)	1345	195	1240	180	42	. . .	. . .
S5	Annealed	725	105	440	64	96 HRB	. . .	. . .
	Oil quenched from 870 °C (1600 °F) and single tempered at:							
	205 °C (400 °F)	2345	340	1930	280	59	206[c]	152[c]
	315 °C (600 °F)	2240	325	1860	270	58	232[c]	171[c]
	425 °C (800 °F)	1895	275	1690	245	52	243[c]	179[c]
	540 °C (1000 °F)	1520	220	1380	200	48	188[c]	139[c]
	650 °C (1200 °F)	1035	150	1170	170	37	. . .	. . .
S7	Annealed	640	93	380	55	95 HRB	. . .	. . .
	Fan cooled from 940 °C (1725 °F) and single tempered at:							
	205 °C (400 °F)	2170	315	1450	210	58	244[c]	180[c]
	315 °C (600 °F)	1965	285	1585	230	55	309[c]	228[c]
	425 °C (800 °F)	1895	275	1410	205	53	243[c]	179[c]
	540 °C (1000 °F)	1820	264	1380	200	51	324[c]	239[c]
	650 °C (1200 °F)	1240	180	1035	150	39	358[c]	264[c]

[a] In 50 mm or 2 in. [b] Charpy V-notch. [c] Charpy unnotched

Source: Metals Handbooks, 9th ed., Vol. 3, American Society for Metals, Metals Park, OH, 1980, p. 431. Reprinted with permission.

and torsion, in the case of end mills, so that the cutting tool may vibrate unless it is massive enough to move its resonant frequency far from the exciting frequency. The high-speed tool steels listed in Table 2.14.1 were developed for high-speed machining where they remain strong and wear resistant at the high temperatures that are produced at the cutting edge. Because of these properties their use has now been extended to high-temperature springs, ultrahigh-strength fasteners, special purpose valves, and elevated temperature bearings.

2.15   HEAT TREATMENT AND HARDENING

Heat treatment of steel generally is a time- and temperature-controlled cycle to relieve stresses induced by manufacturing processes and/or to produce a particular metallic structure.

Stress relief heat treating involves heating the metal to a temperature below the transformation range, holding it at that temperature for a predetermined time, and then cooling at a *uniform* rate back to room temperature.

Austempering is an example of a process to produce a particular metallic structure. It depends entirely upon the path taken in the appropriate isothermal transformation diagram similar to that shown in Figure 2.1.3. Figure 2.15.1 indicates the difference between the path used for conventional quenching and tempering and that used for austempering. Postweld heat treatment is an example of a process designed to both relieve stress and produce a particular metallic structure.

Because hard, brittle, wear-resistant surfaces are extremely difficult to machine, it is common practice to form steel parts with the metal in a relatively soft state. When the metal forming is complete the part may then be hardened on the surface or throughout. Unfortunately, some distortion may accompany heat treatment, so a small amount of grinding or other forms of reforming may be necessary to bring the heat-treated

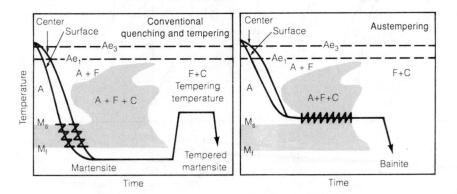

Figure 2.15.1 Schematic comparison of paths for conventional quenching and tempering and for austempering. *From Metals Handbook, 9th ed., Vol. 4, American Society for Metals, Metals Park, OH, 1981, p. 104. Reprinted with permission.*

part back to its proper shape and dimensions. Since the resulting strength properties of the steel usually are dependent upon the time–temperature schedule used, its selection and control are important in attempting to minimize distortion while obtaining the desired structure.

Surface hardening is often employed to improve wear and fatigue characteristics while retaining the toughness and flexibility of the core material. Improved fatigue resistance is due to compressive residual stresses being formed during the process; improved wear is due to the hardening itself. Common techniques for surface hardening are listed in the following paragraphs.

Carburizing with gases and easily vaporized hydrocarbon liquids is the main commercial process used. Natural gas and propane are preferred if they are available in adequate purity. Carburizing liquids, often proprietary, are usually fed as droplets to a hot plate in the furnace where they vaporize, dissociate, and provide a carburizing atmosphere. Solid carburizing compounds are used in pack carburizing to provide carbon monoxide at the steel surface, which may be at a temperature of 1550–2000 °F (850–1100 °C) in all of the carburizing methods.

Carbonitriding is accomplished using a carburizing carrier gas and ammonia in concentrations of 10% at 1300–1650 °F (700–900 °C). Nitrogen's effect upon case hardening ability is shown in Figure 2.15.2.

Nitriding produces a thin, hard surface by exposure of the material to nitrogen at temperatures of 950–1000 °F (500–550 °C). It is generally applied to steels with nitride-forming elements such as aluminum or chromium. Aluminum-content steels produce nitrided surfaces that are very hard and wear resistant, but have low ductility. Greater

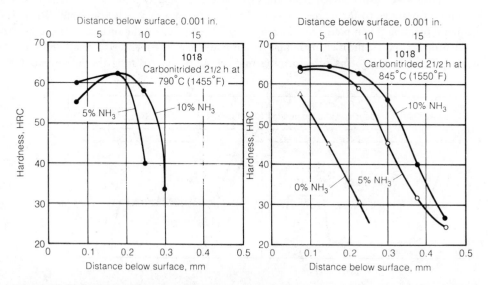

Figure 2.15.2 Effect of ammonia content of carbonitriding gas on the hardness gradient. *From* Metals Handbook, *American Society for Metals, Metals Park, OH, 1981, p. 182. Reprinted with permission.*

ductility, but lower hardness, results from low-alloy chromium steels. Nitriding also produces residual compressive surface stresses.

Gas nitriding introduces nitrogen at the surface from enveloping ammonia gas at a rate dependent upon gas flow rate and furnace temperature. Liquid nitriding uses a cyanide–cynate salt bath at 500–575 °C (950–1050 °F), and is effective on carbon steels. Similar case hardening results from proprietary gaseous nitrocarburizing using a mixture of ammonia and a proprietary carrier gas.

Ionitriding (glow-discharge nitriding) uses electric current to ionize low-pressure nitrogen gas and a potential difference to attract the ions to the part being nitrided. No additional heat is needed.

Flame hardening, of ductile iron, for example, is done by touching a flame to the surface for a few seconds to heat it to about 900 °C (1650 °F) and then quenching with a water spray. The maximum temperature and the cooling, or quenching, rates determine the depth and hardness of the treated region.

Induction hardening differs from flame hardening in that heat is supplied by an electrically induced oscillatory magnetic field. Temperature and depth of hardening may be controlled more precisely than in flame hardening by controlling a combination of the frequency, the strength of the magnetic field, and the exposure time.

Annealing is heating the metal to a certain temperature, holding it at that temperature for a particular time, and then cooling at a *particular* rate for purposes of reducing hardness, improving machinability, facilitating cold working, and so on.

Tempering is reheating either a quench-hardened or a normalized ferrous alloy to a temperature below the transformation range (a temperature range wherein austenite forms during heating and transforms during cooling) and then cooling at *any* convenient rate.

2.16 QUALITY GRADES OF STEEL

AISI has formally catagorized quality levels (accuracy of chemical composition, finish, freedom from inclusions, etc.) for carbon and alloy steels as in Table 2.16.1. These quality differences contribute in part to the variation of elastic constants exhibited in Figure 2.5.1.

2.17 ALUMINUM

Commercially pure aluminum is a face-centered-cubic metal which melts at about 1215 °F (655 °C) and its specific gravity is about one-third than of steel. Its coefficient of thermal expansion is about twice that of steel. Cast bars have a tensile strength of about 14 000 psi (97 MPa) with a modulus of elasticity that is approximately 10×10^6 psi (69 000 MPa). Although it is attacked by strong alkalies, aluminum is highly resistant to corrosion by many chemicals. It is high in the electrochemical series, and, therefore, is subject to galvanic attack by metals lower in the series, such as copper, when an electrolyte is present.

TABLE 2.16.1   FUNDAMENTAL QUALITY DESCRIPTION OF CARBON AND ALLOY STEELS

Carbon Steels

Semifinished for Forging

Forging Quality
  Special Hardenability
  Special Internal Soundness
  Nonmetallic Inclusion Requirement
  Special Surface

Carbon Steel Structural Sections

Structural Quality

Carbon Steel Plates

Regular Quality
Structural Quality
Cold Drawing Quality
Cold Pressing Quality
Cold Flanging Quality
Forging Quality
Pressure Vessel Quality
Marine Quality

Hot Rolled Carbon Steel Bars

Merchant Quality
Special Quality
  Special Hardenability
  Special Internal Soundness
  Nonmetallic Inclusion Requirement
  Special Surface
Scrapless Nut Quality
Axle Shaft Quality
Cold Extrusion Quality
Cold Heading and Cold Forging Quality

Cold Finished Carbon Steel Bars

Standard Quality
  Special Hardenability
  Special Internal Soundness
  Nonmetallic Inclusion Requirement
  Special Surface
Cold Heading and Cold Forging Quality
Cold Extrusion Quality

Hot Rolled Sheets

Commercial Quality
Drawing Quality
Drawing Quality Special Killed
Physical Quality

Cold Rolled Sheets

Commercial Quality
Drawing Quality
Drawing Quality Special Killed
Physical Quality

Porcelain Enameling Sheets

Commercial Quality
Drawing Quality

Long Terne Sheets

Commercial Quality
Drawing Quality
Drawing Quality Special Killed
Physical Quality

Galvanized Sheets

Commercial Quality
Drawing Quality
Drawing Quality Special Killed
Physical Quality
Lock Forming Quality

Electrolytic Zinc Coated Sheets

Commercial Quality
Drawing Quality
Drawing Quality Special Killed
Physical Quality

Hot Rolled Strip

Commercial Quality
Drawing Quality
Drawing Quality Special Killed
Physical Quality

Cold Rolled Strip
Specific quality descriptions are not
  provided in cold rolled strip,
  since this product is largely
  produced for specific end use.

Tin Mill Products

Specific quality descriptions are not
  applicable to tin mill products.

Carbon Steel Wire

Industrial Quality Wire
Cold Extrusion Wires
Heading, Forging and Roll Threading
  Wires
Mechanical Spring Wires
Upholstery Spring Construction Wires
Welding Wire

Carbon Steel Flat Wire

Stitching Wire
Stapling Wire

Carbon Steel Pipe

Structural Tubing

Line Pipe

Oil Country Tubular Goods

Steel Specialty Tubular Products

Pressure Tubing
Mechanical Tubing
Aircraft Tubing

Hot Rolled Carbon Steel Wire Rods

Industrial Quality
Rods for Manufacture of Wire Intended
  for Electric Welded Chain
Rods for Heading, Forging, and Roll
  Threading Wire
Rods for Lock Washer Wire
Rods for Scraples Nut Wire
Rods for Upholstery Spring Wire
Rods for Welding Wire

TABLE 2.16.1   FUNDAMENTAL QUALITY DESCRIPTION OF CARBON AND ALLOY STEELS (Continued)

Alloy Steels

Alloy Steel Plates

Regular Quality or Structural Quality
Drawing Quality
Pressure Vessel Quality
Structural Quality
Aircraft Quality
Aircraft Physical Quality

Hot Rolled Alloy Steel Bars

Regular Quality
Aircraft Quality or Steel Subjects to
   Magnetic Particle Inspection
Axle Shaft Quality
Bearing Quality
Cold Heading Quality
Special Cold Heading Quality
Rifle Barrel Quality, Gun Quality, Shell
   or A.P. Shot Quality

Alloy Steel Wire

Aircraft Quality
Bearing Quality
Special Surface Quality

Cold Finished Alloy Steel Bars

Regular Quality
Aircraft Quality or Steel Subject to
   Magnetic Particle Inspection
Axle Shaft Quality
Bearing Shaft Quality
Cold Heading Quality
Special Cold Heading Quality
Rifle Barrel Quality, Gun Quality,
   Shell or A.P. Shot Quality

Line Pipe

Oil Country Tubular Goods

Steel Specialty Tubular Goods

Pressure Tubing
Mechanical Tubing
Stainless and Heat Resisting Pipe,
Pressure Tubing, and Mechanical
   Tubing
Aircraft Tubing

Source: SAE Handbook, SAE J411 JUN81, Vol. 1, © 1988 Society of Automotive Engineers, Inc. Reprinted with permission.

Its strength is increased by alloying with copper, silicon, magnesium, manganese, zinc, and others. Aluminum and its alloys may be joined by fusion welding, resistance welding, soldering, and brazing. It is generally easy to machine if proper tools and techniques are used.

The SAE J993b designations for wrought aluminum and its alloys are found in Table 2.17.1 and for cast aluminum and its alloys in Table 2.17.2. The first digit of the designations in Table 2.17.1 indicates the alloy group, as shown. Zero as the second digit indicates the original alloy; digits 1 through 9 indicate modification to the original alloy. Digits three and four identify the alloy in groups 2xxx through 8xxx and the mini-

---

TABLE 2.17.1   DESIGNATION SYSTEM FOR WROUGHT ALUMINUM AND ALUMINUM ALLOY

Composition	Alloy No.
Aluminum, 99.0% min. and greater	1xxx
Aluminum alloys grouped by major alloying element	
Copper	2xxx
Manganese	3xxx
Silicon	4xxx
Magnesium	5xxx
Magnesium and silicon	6xxx
Zinc	7xxx
Other element	8xxx
Unused series	9xxx

Source: SAE J411 JUN81, SAE Handbook, Vol. 1, © 1988 Society of Automotive Engineers, Inc., p. 182. Reprinted with permission.

TABLE 2.17.2 DESIGNATION SYSTEM FOR CAST ALUMINUM AND ALUMINUM ALLOY

Composition	Alloy No.
Aluminum, 99.0%min. and greater	1xx.x
Aluminum alloy group by major alloying element	
Copper	2xx.x
Silicon, with added copper and/or magnesium	3xx.x
Silicon	4xx.x
Magnesium	5xx.x
Zinc	7xx.x
Tin	8xx.x
Other element	9xx.x
Unused series	6xx.x

Source: SAE J411 JUN81, SAE Handbook, Vol. 1, © 1988 Society of Automotive Engineers, Inc., p. 10.03. Reprinted with permission.

mum aluminum percentage in group 1xxx. Minimum aluminum percentage is indicated by the second and third digit in the 1xx.x group in Table 2.17.2 and the digit following the decimal point indicates a casting if it is 0 and an ingot if it is 1. The second and third digits in the 2xx.x through 9xx.x identify the alloy. Zero to the right of the decimal point indicates a casting; xxx.1 and xxx.2 designate ingots, but of different composition. Experimental alloys are designated by an X prefix until a number is assigned in accord with Tables 2.17.1 or 2.17.2.

TABLE 2.17.3 BASIC TEMPER DESIGNATIONS FOR ALUMINUM

F   As fabricated; applies to products of shaping processes
O   Annealed; applies to wrought products that are fully annealed
H   Strain hardened (wrought products only); followed by two or more digits:
      1x   Strain hardened only
      2x   Strain hardened and partially annealed
      3x   Strain hardened and stabilized by low temperature
      x   Indicates degree of strain hardening
W   Solution heat treated and aged at room temperature; it is specific only when followed by aging time: i.e., $w\frac{1}{2}h$
T   Thermally treated to give stable tempers other than F,O, and H; followed by one or more digits:
    1   Cooled from shaping at elevated temperature, naturally aged to substantially stable condition
    2   Annealed (cast products only)
    3   Solution heat treated then cold worked
    4   Solution heat treated, naturally aged
    5   Cooled from elevated temperature shaping, artificially aged
    6   Solution heat treated, artificially aged
    7   Solution heat treated, stabilized
    8   Solution heat treated, cold worked, artificially aged
    9   Solution heat treated, artificially aged, cold worked
   10   Cooled from an elevated shaping temperature, artificially aged, cold worked

Source: Abstracted from *SAE J993b, SAE Handbook*, Vol. 1, © 1988 Society of Automotive Engineers, Inc. Reprinted with permission.

Temper, when applied to nonferrous alloys, refers to the hardness and strength produced by either thermal treatment, mechanical treatment, or both, and is usually characterized by a certain structure or set of elastic properties. Heat treatments to increase strength and hardness in either cast or wrought aluminum employ precipitation hardening and aging. Temper designations for aluminum alloys that respond to heat treatment are given in Table 2.17.3 and the importance of time and aging temperature in determining tensile and yield strength is shown in Figure 2.17.1.

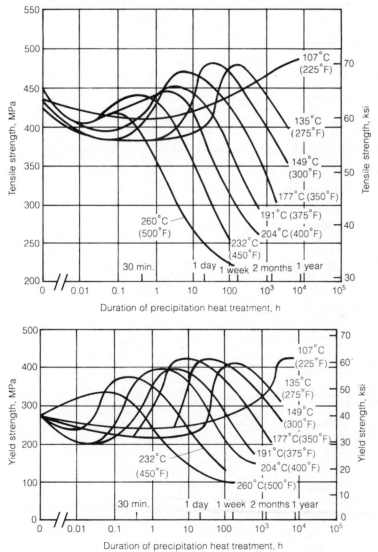

Figure 2.17.1 Typical artifical precipitation hardening aging curves for alloy 2014. *From* Metals Handbook, *9th ed., Vol. 2, American Society for Metals, Metals Park, OH, 1979, p. 42. Reprinted with permission.*

Aluminum tends to creep (i.e., to slowly elongate) at elevated temperatures, as exemplified by the behavior of alclad 2024 and 7075 shown in Figure 2.17.2. (Alclad is a "composite sheet produced by bonding either corrosion-resistant aluminum alloy or aluminum of high purity to base metal of structurally stronger aluminum alloy."[13]) Typical mechanical properties of selected alloys are listed in Table 2.17.4.

Alcad 2024-T3

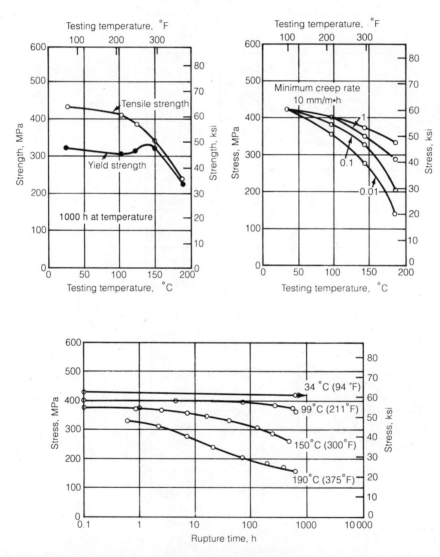

Figure 2.17.2 Effect of temperature on tensile properties of alclad alloys 2024 and 7075. *From* Metals Handbook, *9th ed., Vol. 2, American Society for Metals, Metals Park, OH, 1979, p. 57. Reprinted with permission.*

TABLE 2.17.4   TYPICAL   MECHANICAL   PROPERTIES
OF WROUGHT ALUMINUM ALLOYS

AISI[a]	Temper	Tensile Strength		Yield Strength	
		MPa	ksi	MPa	ksi
1050	0	76	11	28	4
	H18	160	23	145	21
1100	0	90	13	34	5
	H18	165	24	150	22
2014	0	185	27	97	14
	T6	485	70	415	60
2024	0	185	27	76	11
	T361	495	72	395	57
2124	T851	490	71	440	64
2218	T61	405	59	305	44
3105	0	115	17	55	8
	H18	215	31	195	28
5052	0	195	28	90	13
	H38	290	41	255	37
6005	T1	170	25	205	25
	T5	260	38	240	35
6262	T9	400	58	380	55
7050	6735	515	75	455	66
7175	T66	595	86	525	76

[a]   No specific correspondence with UNS has been published.
Source: Abstracted from *ASM Metals Reference Book*, American Society
for Metals, Metals Park, OH, 1983, pp. 235–238.

2.18   COPPER

Copper's important properties are its electrical conductivity and its resistance to corrosion along with strength and wearability. Alloying with silver, phosphorus, cadmium, chromium, lead, tellurium, beryllium, and other elements has produced a number of commercial metals that are widely used. Properties of several of these alloys and their uses in machine components are listed in Tables 2.18.1 and 2.18.2.

2.19   MAGNESIUM

Machine applications of magnesium alloys are often in parts that operate at high speed or experience large acceleration and deceleration where low inertia is desirable. Its good strength-to-weight ratio brings it into use wherever weight is important, from aircraft

TABLE 2.18.1 COPPER ALLOYS USED IN MACHINE DESIGN AND THEIR CHARACTERISTICS

Alloy	Typical Uses	Characteristics
C17000	Leaf springs, coil springs bellows, distributor breaker arm	These alloys and C17500 and C175000 can develop the highest mechanical properties with heat treatment.
C36000	Automatic screw machine parts, magneto parts, gears, locks	This is the standard free cutting brass often the standard for relating to other alloys.
C54400	Bearings bushings, gears thrust washers	Free cutting, good cold-working properties.
C61400	Gears, bolts, bushings threaded members	Good cold-working properties and corrosion resistance, high strength and ductility.
C62300	Spark plug inserts, gears, shifter forks, ball bearing components	Good hot-working properties, high strength at high temperature, acid and oxidation resistant.
C62400		
C63000	Retractable landing gear, propeller gears, high-pressure pump parts	Very high mechanical properties in heated condition, excellent corrosion resistance.
C67000	Diesel injector nozzles, cams, pistons	High strength and good wear resistance.
C67300	Forged water pump impellers, gears, bushings, bearings	Hot-forgeable, free-cutting alloy, good corrosion and wear resistance.
C67400	Connecting rods, shifter shoes, transmission synchronizing stop ring, differential idler pins, pump impellers	Hot-forgeable, high-strength, alloy with good wear and corrosion resistance.
C67500	Clutch disks, pump rods, shafts, valve stems	Strong, rigid, and abrasion resistant; adapted to hot forging and pressing, hot heading and upsetting.

Source: Abstracted from *SAE J461 SEP81, SAE Handbook,* © 1988 Society of Automotive Engineers, Inc.

TABLE 2.18.2 TYPICAL MECHANICAL PROPERTIES OF WROUGHT COPPER AND COPPER ALLOYS

Number	Form	Temper	Size, Over/Thru (mm)	Tensile strength (MPa)	Yield strength (MPa)
C16200	Rod	Drawn	–/25	415	—
			25/50	380	—
			50/75	345	—
C17000	Strip	Soft	All	415	—
		Hard	All	690	
		HT	All	1240	1070
C18400	Rod	Drawn	–/25	450	—
		or	25/50	415	—
		as forged	50/–	380	—
C23000	Strip	Hard	All	435	—
		Spring	All	540	—
		Extra hard	All	565	—
C51000	Rod	Spring Wire	–/0.65	1000	—
			0.65/1.6	930	—
			1.6/3.2	845	—
C67500	Bar	Soft	All	380	150
		Hard	–/25	550	385
			25/65	525	360
			65/–	505	330
	Shapes	Soft	All	470	310
C71500	Plate,	$\frac{1}{4}$ hard	—	400	—
	sheet,	$\frac{1}{2}$ hard	—	455	—
	Rolled	Hard	—	605	—
	bar	Extra hard	—	550	—

Source: Abstracted from *SAE J463 SEP81, SAE Handbook,* © 1988 Society of Automotive Engineers, Inc.

TABLE 2.19.1   STANDARD FOUR-PART ASTM SYSTEM OF ALLOY AND TEMPER DESIGNATIONS FOR MAGNESIUM ALLOYS

First Part	Second Part	Third Part	Fourth Part
Indicates the two principal alloying elements	Indicates the amounts of the two principal alloying elements	Distinguishes between different alloys with the same percentages of the two principal alloying elements	Indicates condition (temper)
Consists of two code letters representing the two main alloying elements arranged in order of decreasing percentage (or alphabetically if percentages are equal)	Consists of two numbers corresponding to rounded-off percentages of the two main alloying elements and arranged in same order as alloy designations in first part	Consists of a letter of the alphabet assigned in order as compositions become standard	Consists of a letter followed by a number (separated from the third part of the designation by a hyphen)
A—Aluminum E—Rare Earth H—Thorium K—Zirconium M—Manganese Q—Silver S—Silicon T—Tin Z—Zinc	Whole numbers	Letters of alphabet except I and O	F—As fabricated O—Annealed H10 and H11—Slightly strain hardened H23, H24 and H26—Strain hardened and partially annealed T4—Solution heat treated T5—Artificially aged only T6—Solution heat treated and artificially aged T8—Solution heat treated, cold worked and artificially aged

Source: *Metals Handbook*, 9th ed., Vol. 2, American Society for Metals, Metals Park, OH, 1979, p. 527. Reprinted with permission.

to extension ladders. Standard designations of alloy and temper are shown in Table 2.19.1 and typical strengths for several of these designations are shown in Table 2.19.2. Temperature effects upon several magnesium alloys are indicated in Table 2.19.3.

2.20   ZINC

Zinc has been used primarily as a die cast structural alloy because of its low cost and ease of casting at relatively low temperatures. Carburetor bodies, windshield wiper parts, and washing machine, oil burner, and typewriter housings are often made from zinc. Motor brackets, bearing supports, and similar housing and framing components of zinc have been used in recording machines, projectors, vending machines, garbage disposers, and fuel pumps. The mechanical properties of zinc, given in Table 2.20.1, are of interest in machine design because they indicate the behavior of frames that may support machine components.

TABLE 2.19.2 NOMINAL COMPOSITIONS AND TYPICAL ROOM-TEMPERATURE MECHANICAL PROPERTIES OF MAGNESIUM ALLOYS

Alloy	Composition						Tensile strength		Yield Strength — Tensile		Yield Strength — Compressive		Yield Strength — Bearing		Elongation in 50 mm or 2 in., %	Shear Strength		Hardness, HRB[b]
	Al	Mn[a]	Th	Zn	Zr	Others	MPa	ksi	MPa	ksi	MPa	ksi	MPa	ksi		MPa	ksi	
Sand and Permanent Mold Castings																		
AM100A-T61	10.0	0.1	...	...	...	...	275	40	150	22	150	22	...	...	1	...	...	69
AZ63A-T6	6.0	0.15	...	3.0	...	...	275	40	130	19	130	19	360	52	5	145	21	73
AZ81A-T4	7.6	0.13	...	0.7	...	...	275	40	83	12	83	12	305	44	15	125	18	55
AZ91C-T6	8.7	0.13	...	0.7	...	...	275	40	145	21	145	21	360	52	6	145	21	66
AZ92A-T6	9.0	0.10	...	2.0	...	...	275	40	150	22	150	22	450	65	3	150	22	84
EZ33A-T5	...	...	...	2.7	0.6	3.3 RE	160	23	110	16	110	16	275	40	2	145	21	50
HK31A-T6	...	...	3.3	...	0.7	...	220	32	105	15	105	15	275	40	8	145	21	55
HZ32A-T5	...	...	3.3	2.1	0.7	...	185	27	90	13	90	13	255	37	4	140	20	57
K1A-F	...	...	...	...	0.7	...	180	26	55	8	...	...	125	18	1	55	8	...
QE22A-T6	...	...	...	...	0.7	2.5 Ag, 2.1 Di	260	38	195	28	195	28	...	...	3	...	...	80
QH21A-T6	...	...	6.0	...	0.7	2.5 Ag, 1.0 Di	275	40	205	30	...	...	...	...	4	...	...	...
ZE41A-T5	...	...	...	4.2	0.7	1.2 RE	205	30	140	20	140	20	350	51	3.5	160	23	62
ZE63A-T6	...	...	...	5.8	0.7	2.6 RE	300	44	190	28	195	28	...	...	10	...	...	60–85
ZH62A-T5	...	...	1.8	5.7	0.7	...	240	35	170	25	170	25	340	49	4	165	24	70
ZK51A-T5	...	...	...	4.6	0.7	...	205	30	165	24	165	24	325	47	3.5	160	23	65
ZK61A-T5	...	...	...	6.0	0.7	...	310	45	185	27	185	27	...	...	...	170	25	68
ZK61A-T6	...	...	...	6.0	0.7	...	310	45	195	28	195	28	...	...	10	180	26	70
Die Castings																		
AM60A-F	6.0	0.13	...	...	...	...	205	30	115	17	115	17	...	...	6	...	...	...
AS41A-F[d]	4.3	0.35	...	...	...	1.0 Si	220	32	150	22	150	22	...	...	4	...	...	...
AZ91A and B-F[c]	9.0	0.13	...	0.7	...	...	230	33	150	22	165	24	...	...	3	140	20	63
Extruded Bars and Shapes																		
AZ10A-F	1.2	0.2	...	0.4	...	...	240	35	145	21	69	10	...	...	10	...	...	...
AZ21X1-F	1.8	0.02	...	1.2	...	...	...	...	...	...	...	...	...	...	...	...	...	...
AZ31 B and C-F[e]	3.0	...	...	1.0	...	...	260	38	200	29	97	14	230	33	15	130	19	49
AZ61A-F	6.5	...	...	1.0	...	...	310	45	230	33	130	19	285	41	16	140	20	60
AZ80A-T5	8.5	...	...	0.5	...	...	380	55	275	40	240	35	345	50	7	165	24	82
HM31A-F	...	1.2	3.0	...	...	...	290	42	230	33	185	27	195	28	10	150	22	44
M1A-F	...	1.2	...	...	...	...	255	37	180	26	83	12	...	...	12	125	18	...
ZK21A-F	...	...	...	2.3	0.45[a]	...	260	38	195	28	135	20	...	...	4	...	...	...
ZK40A-T5	...	...	...	4.0	0.45[a]	...	276	40	255	37	140	20	...	...	4	...	...	...
ZK60A-T5	...	...	...	5.5	0.45[a]	...	365	53	305	44	250	36	405	59	11	180	26	88
Sheet and Plate																		
AZ31B-H24	3.0	...	...	1.0	...	...	290	42	220	32	180	26	325	47	15	160	23	73
HK31A-H24	...	...	3.0	...	0.6	...	255	33	200	29	160	23	285	41	9	140	20	68
HM21A-T8	...	0.6	2.0	...	...	...	235	34	170	25	130	19	270	39	11	125	18	...
PE[f]	3.3	...	...	...	0.7	...	...	...	...	...	...	...	...	...	...	...	...	...

[a] Minimum.
[b] 500-kg load, 10-mm ball.
[c] A and B are identical except that 0.30% max residual Cu is allowable in AZ91B.
[d] For battery applications.
[e] Properties of B and C are identical, but AZ31C has 0.15 min Mn, 0.1 max Cu and 0.03 max Ni.
[f] Photoengraving grade.

Source: *Metals Handbook*, 9th ed., Vol. 2, American Society for Metals, Metals Park, OH, 1979, p. 528. Reprinted with permission.

TABLE 2.19.3 EFFECT OF ELEVATED TEMPERATURE ON TENSILE STRENGTH OF MAGNESIUM ALLOYS

Alloy	20°C MPa	(70°F) ksi	Tested at Exposure Temperature — Exposed 10 min at				Exposed 1000 h at				Tested at Room Temperature — Exposed 1000 h at			
			150°C MPa	(300°F) ksi	315°C MPa	(600°F) ksi	205°C MPa	(400°F) ksi	315°C MPa	(600°F) ksi	205°C MPa	(400°F) ksi	315°C MPa	(600°F) ksi
Castings														
AZ63A-T6	275	40	165	24	55	8	110	16	...	...	225	37	...	...
AZ92A-T6	275	40	195	28	55	8	115	17	...	...	270	39	...	...
EZ33A-T5	160	23	145	21	83	12	130	19	76	11	170	25	180	26
HK31A-T6	215	31	195	28	125	18	180	26	62	9	240	35	180	26
HZ32A-T5	200	29	145	21	83	12	115	17	76	11	220	32	235	34
ZH62A-T5	290	42	195	28	69	10	...	...	...	...	...	...	...	...
QH21A-T6	275	40	235	34	97	14	...	...	...	...	...	...	...	...
Extrusions														
AZ80A-T5	380	55	235	34	69	10	...	...	...	...	...	...	...	...
ZK60A-T5	365	53	180	26	41	6	...	...	...	...	315	46	315	46
HM31A-F	275	40	195[a]	28[a]	115	17	...	...	...	...	...	...	...	...
Sheet														
AZ31B-H24	285	41	145	21	48	7	90	13	62[a]	9[a]	255	37	260	38
HK31A-T6	255	37	180	26	115	17	...	...	55	8	255	37	215	31
HM21A-T8	235	34	140	20	97	14	...	...	...	...	...	...	...	...

[a] Tested at 260°C (500°F).

Source: Metals Handbook, 9th ed., Vol. 2, American Society for Metals, Metals Park, OH, 1979, p. 535. Reprinted with permission.

TABLE 2.20.1   AVERAGE PROPERTIES OF ZINC DIE CASTING ALLOYS

Properties	ASTM AG40A; SAE 903	ASTM AC41A; SAE 925	Alloy 7	ILZRO 16
Mechanical Properties				
Charpy impact strength, $\frac{1}{4}$-by-$\frac{1}{4}$-in. bar:				
As cast, J(ft·lb)	58 (43)	65 (48)	. . .	. . .
After aging indoors 10 yr, J(ft·lb)	56 (41)	54 (40)	275 (40)	. . .
Tensile strength:				
As cast, MPa (ksi)	285 (41)	330 (47.6)	285 (41)	230 to 235 (33 to 34)
After aging indoors 10 yr, MPa (ksi)	240 (35)	270 (39.3)	. . .	. . .
Elongation, % in 50 mm or 2 in.:				
As cast	10	7	14	5
After aging indoors 10 yr	16	13	. . .	. . .
Expansion, after aging indoors 10 yr at room temperature, μm/m	80	70	. . .	. . .
Other Properties and Constants of As-Cast Alloys				
Brinell hardness (HB)	82	91	76	75 to 77
Compressive strength, MPa (ksi)	415 (60)	600 (87)	. . .	. . .
Electrical conductivity, % IACS	27.5	26.5	. . .	. . .
Liquidus temperature, °C (°F)	387 (728)	386 (727)	. . .	417 (785)
Solidus temperature, °C (°F)	381 (717)	380 (716)	. . .	415 (780)
Modulus of rupture, MPa (ksi)	655 (95)	725 (105)	. . .	. . .
Shear strength, MPa (ksi)	215 (31)	260 (38)	. . .	. . .
Specific heat, J/kg (Btu/lb)	420 (0.10)	420 (0.10)	. . .	. . .
Thermal conductivity, W/m·K (Btu/ft·h·°F)[a]	113 (65.3)	109 (62.9)	. . .	. . .
Thermal expansion, μm/m·K (μin./in.·°F)	27.4 (15.2)	27.4 (15.2)	. . .	. . .
Transverse deflection, mm (in.)	6.9 (0.27)	4.1 (0.16)	. . .	. . .
Density, Mg/m^3 (lb/in.^{3})	6.6 (0.238)	6.7 (0.242)		

[a]   At 18 °C (64 °F).

Source: Metals Handbook, 9th ed., Vol. 2, American Society for Metals, Metals Park, OH, 1979, p. 633. Reprinted with permission.

2.21   MACHINABILITY

Machinability or the *machinability index* is usually intended to be a number that indicates how easily a material may be machined. It is difficult, however, if not impossible, to fulfill this intention with a single number because of the variety of factors and criteria involved. Of the criteria, the most common are (1) the life of the tool used in the machining operation, (2) the power required to perform the machining, and (3) the finish obtained. Some of the factors involved in each of these criteria are (1) the depth of cut, (2) the speed of cutting as a function of depth, (3) the particular tool steel used, (4) the angle of the tool's cutting edge, (5) the rigidity of the tool and its support, (6) the choice of coolant, and (7) temperature control of the cutter and of the material being cut.

The machinability index is generally lower for softer materials; melting of the chips at the edge of the cutter causes them to adhere to the cutting tool which results in a poor finish. Soft aluminum, pure iron, and low-carbon steels (with a carbon content of less than about 0.15%) are examples of materials with low machinability because of these effects. Machinability of steel increases with increased carbon up to 0.20%, but further increases in carbon decrease machinability because pearlitic grains become so numerous that they obstruct the cutting tool. Addition of lead, known as a free machining additive, improves machinability by providing lubrication to the cutting tool and by introducing inclusions in the steel which cause chips formed by the cutting tool to break into smaller sections. See Ref. 13 for additional details on machinability of specific metals by specific machining processes.

2.22 PLASTICS

Plastics are made up of long-chain molecules that may be intertwined with one another and bonded together in a variety of ways. Links in the chain are carbon atoms, which can have one or two common bonds (share one or two electron pairs) as well as bonding to other atoms. As an aid in describing this bonding structure, the carbon chain may be associated with a diagram as in Figure 2.22.1a. In it the carbon atoms are represented by their chemical symbols and their bonds by radial lines. Other elements, also represented by their chemical symbols, may be added to the open bonds to form a particular compound. In Figure 2.22.1b hydrogen has been added at each bond to make C_2H_6, ethane, while in Figure 2.22.1c only four hydrogen atoms have been added to make ethylene. Short chains such as these may be lengthened by adding carbon to some bonds to produce a chain as illustrated in Figure 2.22.2. As it lengthens the material becomes more dense and passes from a gas to a liquid and then to a waxy solid. Construction of long chains usually begins with a shorter chain material, such as ethylene,

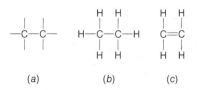

(a)   (b)   (c)

Figure 2.22.1 (a) Carbon chain with open bonds and compounds made by adding hydrogen to obtain (b) ethane and (c) ethylene.

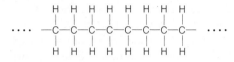

Figure 2.22.2 Section of a long carbon chain.

in which the carbon–carbon bond is opened so that the original short chains, or *monomers*, may be linked together to make a longer chain, known as a *polymer*. This linking together may continue until it is stopped by adding a terminal link, called a *terminator*, which satisfies the bonds at each end of the chain.

The simple chain just described is that of polyethylene, which is classed as a linear polymer because its chain diagram is a linear linkage of carbon atoms as in Figure 2.22.2. Mechanical properties of polyethylene appear to depend not only upon the nature of the chains but also upon how they intertwine in the solid.

Chains that are not linear may be formed from monomers of more complicated structure. For example, the basic amides, which have a bond arrangement like that sketched in Figure 2.22.3*a*, provide monomers such as acetamide and diamide (urea), shown in Figure 2.22.3*b* and *c*.

Polymers from the more elaborate monomers may produce plastics with distinctly different properties, not only because of the different monomers used, but because these long molecules may intertwine differently and may bond together at various points along their lengths. Plastics may be composed of polymers that are (1) interwound long-chain molecules, (2) interwound and branched long-chain molecules, or (3) interwound, branched, long-chain molecules that are bonded (cross-linked) together at various locations along the length of each chain.

Some of the stronger plastics are said to have a crystalline structure because the long-chain molecules are arranged in a repeating, regular, pattern. Such partial crystallization appears to be limited to long, regular, molecules without interfering side appendages.

Numerous plastics without sufficient strength from their molecular structure have been strengthened by adding fillers. Typical fillers are glass fibers, graphite, short-fiber cellulose, asbestos, and similar materials. The improvement gained is indicated by the curves in Figure 2.22.4 and the data for selected plastics are discussed in the following sections.

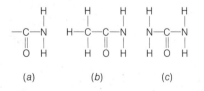

(a)   (b)   (c)

Figure 2.22.3 Characteristic bond diagram of (a) amides and monomers (b) acetamide and (c) diamide.

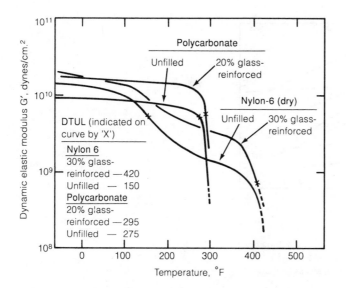

Figure 2.22.4 Dynamic shear (torsion) modulus vs. temperature for amorphous and semicrystalline glass-reinforced plastics: 264 psi load. *Reprinted with permission from* Modern Plastics Encyclopedia for 1984–1985, © *1984 McGraw-Hill Inc., New York. All rights reserved.*

Plastics may also be divided into two groups according to their behavior at elevated temperatures: *thermoplastic* and *thermoset*. Thermoplastic materials include linear polymers and are characterized by softening and eventually melting as the temperature increases. Thermoset plastics include cross-linked, tightly bound, long-chain molecules and are characterized by not softening and melting as the temperature increases. They tend to char instead.

Unfortunately not all plastics obediently fall into one or the other group. Thermoplastic polyimide, for example, becomes rubbery at 310 °C and can be compression molded at temperatures above 310 °C where it is held for at least 5 min to ensure good flow within the mold. Thermoset polyimide, however, exhibits heat resistance for a short time at 480 °C.

Perhaps the major practical distinction between metals and plastics is that many are *viscoelastic* at relatively low temperatures, below 120 °F (49 °C). In other words, they behave elastically under momentarily applied loads, but tend to flow as a viscous liquid under loads of long duration. Their behavior is time dependent. Elongation under load at various temperatures for selected plastics is illustrated in Figure 2.22.5. Typical stress–strain curves are shown in Figure 2.22.6 in which there may be no well-defined proportional limit. Moreover, a typical plastic may be characterized not by a unique stress–strain curve over a range of temperatures and strain rates, but rather by a continuum of curves that vary markedly with temperature, load, and the strain rate.

The flow under load behavior of plastics has been broadly classed as either ductile or nonductile, as in Figure 2.22.7. This behavior is distinctive in the region above the

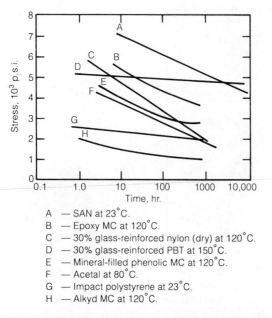

A — SAN at 23°C.
B — Epoxy MC at 120°C.
C — 30% glass-reinforced nylon (dry) at 120°C.
D — 30% glass-reinforced PBT at 150°C.
E — Mineral-filled phenolic MC at 120°C.
F — Acetal at 80°C.
G — Impact polystyrene at 23°C.
H — Alkyd MC at 120°C.

Figure 2.22.5 Creep rupture strength of plastics in tension and flexure. *Reprinted with permission from* Modern Plastics Encyclopedia for 1984–1985, © *1984 McGraw-Hill Inc., New York. All rights reserved.*

dashed creep rupture envelope where the ductile plastics exhibit a distinct strain increase for a comparatively modest stress increase.

Since machine components are useless if they change shape during normal load and operating temperatures, the selection of suitable plastics must be based upon both their room temperature characteristics (20–25 °C, 68–77 °F) and their characteristics at the

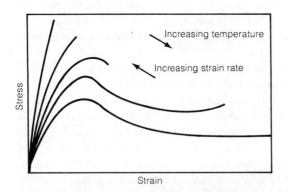

Figure 2.22.6 Typical tensile stress–strain curves of a ductile plastic, showing the effect of strain rate and temperature. *Reprinted with permission from* Modern Plastics Encyclopedia for 1984–1985, © *1984 McGraw-Hill Inc., New York. All rights reserved.*

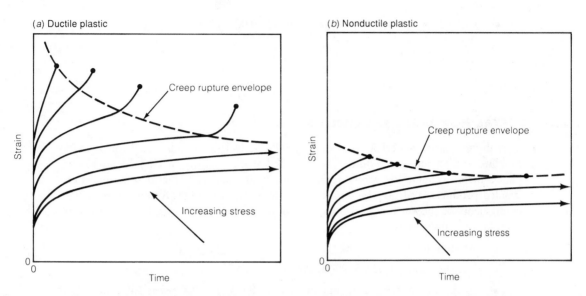

Figure 2.22.7 Representative creep behavior of ductile and nonductile plastic. *Reprinted with permission from* Modern Plastics Encyclopedia *for 1984–1985,* © *1984 McGraw-Hill Inc., New York. All rights reserved.*

highest expected operating and/or storage temperatures. If very low transportation or storage temperatures are expected the plastic behavior at these temperatures must also be examined.

Uses and mechanical properties of selected plastics that have been used in machines, appliances, and hand tools are found in the following sections. Because of time, load, and temperature effects on their mechanical properties, the ASTM test techniques used to measure the quantities listed are included. Properties listed are for preliminary comparison only; final selection should be made only after obtaining more complete data on strength, chemical resistance, time, and temperature effects from the manufacturer. An extensive list of plastics and their manufacturers has been provided in the annual editions of Refs. 11 and 12.

2.23 ABS

ABS is an acronym for the thermoplastic acrylonitrile-butadiene-styrene. It may be injection molded or extruded, and may be strengthened by inclusion of glass fibers. Automotive uses include wheel wells, head liners, fender extensions, and air scoops. Small tractor bodies and business machine shells have been made from ABS.

- Melting temperature: 80–100 °C
- Tensile strength (D638): 4000–7000 psi

- Compressive strength at rupture (D695): 1800–12 500 psi (This range holds for the medium-impact grade.)
- Tensile modulus (D638): 150–450 × 10^3 psi

2.24 ACETAL HOMOPOLYMER (DELRIN)

This thermoplastic material, sold commercially as Delrin, is well suited for injection molding and extrusion molding. The extruded stock is easily machined and has replaced metal in some low-stress applications. It has replaced zinc in some injection-molded products, and may be vibration, sonic, or spin welded. Applications include components of shower mixing valves, commode flushing apparatus, faucet cartridges, toys, and electric shavers.

- Melting temperature: 181 °C
- Tensile yield (D638): 9500–12 000 psi
- Compressive strength at rupture or yield (D638): 18 000 psi at 10% elongation
- Tensile modulus (D638): 520 × 10^3 psi.

2.25 ACETAL COPOLYMER (CELCON)

Screw-type injection-molding machines are recommended over the plunger type for forming this thermoplastic material. Celcon can also be extruded and processed by blow molding and roto-casting. Valve and pump parts, seat belt parts, gears and bearings for toys, and appliance parts such as gears, cams, and housings have been produced from acetal copolymer. It may also be used in some food-processing equipment, such as milk pumps.

- Tensile strength (D638): 8800 psi
- Compressive strength at rupture or yield (D695): 16 000 psi at 10% yield
- Tensile modulus (D638): 410 000 psi

2.26 FLUOROPLASTICS (TEFLON, HALON, FLUON)

The plastic commercially known as Teflon, Halon, or Fluon is also known as PTFE. It is crystalline thermoplastic that is a member of the fluoroplastics family, which contains plastics of paraffinic structure with some or all of the hydrogen replaced by fluorine. Fiberglass-reinforced PTFE has been employed in the manufacture of bushings, rider rings and seals in compressors, automotive power steering, and automatic transmissions.

- Tensile strength at break (D638): 4500–6000 psi
- Tensile modulus (D638): 150–300 × 10³ psi
- Melting temperature: 220 °C

2.27 PHENOLIC

Abrasive material in grinding wheels may be bonded together with this thermoset phenolic resin. Clutch faces and brake lining material are often made from filled phenolic, whereas pulleys, or sheaves, may be formed from high-strength, glass-reinforced phenolic. All of these uses stem in part from the relatively high heat and chemical resistance displayed by phenolics.

- Tensile strength at break (D638): 3.5–18.0 × 10³ psi (the upper yield strength is found in glass-reinforced material)
- Compressive strength at rupture or yield (D695): 20–35 × 10³ psi
- Tensile modulus (D638): 0.8–3.3 × 10³ psi

2.28 PHENYLENE OXIDE, MODIFIED (NORYL)

Pump and shower head parts and other plumbing items are made from this thermoplastic. Both rigid-foam and injection-molded parts appear in typewriters, high-speed printers, calculators, and copiers.

- Melting temperature: 100–135 °C
- Tensile strength at break (D638): 7800–18 500 psi (the upper strength pertains to 30% glass fiber-reinforced plastic)
- Compressive strength at rupture or yield (D695): 10 000–17 900 psi
- Tensile modulus (D638): 380–1200 × 10³ psi (again the larger value corresponds to 30% glass fiber-reinforced Noryl)

2.29 POLYAMIDE (NYLON, CAPRON NYLON, ZYTEL, FOSTA)

The plastics commonly known as nylon, Capron nylon, Zytel, or Fosta are members of the polyamide family, often generically known as nylon. Six types of nylon are produced in the United States: 6/6, 6, 6/9, 6/10, 6/12, and 11. They are distinguished by the chemical characteristics of their primary resins. Types 6/6 and 6 are the most commonly used, and so have become the cheapest. Selected physical, mechanical, thermal, and electrical properties of various types of nylon are listed in Table 2.29.1. Note the effect of moisture at 50% relative humidity.

TABLE 2.29.1 TYPICAL PROPERTIES OF NYLONS; DRY, AS MOLDED

ASTM or UL test	Property	Type 6/6	6	6/12	11	Castable
Physical						
D792	Specific gravity	1.14	1.13	1.06	1.04	1.15–1.17
D792	Specific volume (in.³/lb)	24.2	24.5	25.9	26.6	23.8
D570	Water absorption, 24 h. $\frac{1}{3}$-in. thick (%)	1.2	1.6	0.25	0.4	0.9
Mechanical						
D638	Tensile strength (psi)	12,000	11,800	8,800	8,500	11,000–14,000
D638	Elongation (%)	60	150	150	120	10–50
D638	Tensile modulus (10^5 psi)	4.2	3.8	2.9	1.8	3.5–4.5
D790	Flexural modulus (10^5 psi)	4.1	4.0	2.9	1.5	—
D256	Impact strength, izod (ft-lb/in. of notch)	1.0	1.0	1.0	3.3	0.9
D785	Hardness, Rockwell R	121	119	114	—	112–120
Thermal						
C177	Thermal conductivity (Btu-in./hr-ft²-°F)	1.7	1.7	1.5	—	1.7
D696	Coef of thermal expansion (10^{-5} in./in.-°F)	4.0	4.5	5.0	5.1	5.0
D648	Deflection temperature (°F)					
	At 264 psi	194	152	194	118	300–425
	At 66 psi	455	365	356	154	400–425
UL 94	Flammability rating, $\frac{1}{8}$ in.	V-2	V-2	V-2	—	—
Electrical						
D149	Dielectric strength (V/mil) Short time, $\frac{1}{8}$-in. thick	600	400	400	425	500–600[a]
D150	Dielectric constant At 1 kHz	3.9	3.7	4.0	3.3	3.7
D150	Dissipation factor At 1 kHz	0.02	0.02	0.02	0.03	0.02
D257	Volume resistivity (ohm-cm) At 73° F, 50% RH	10^{15}	10^{15}	10^{15}	2×10^{13}	—
D495	Arc resistance (s)	116	—	121	—	—

[a] 0.04 in. thick specimen.

Source: Reprinted from Machine Design, April 16, 1987. © Penton Publishing, Inc., Cleveland OH.

Gears, bearings, cams, conveyor rollers, and similar parts may be produced from injection-molding grades; power tool housings, automotive cooling fans, and other engine components may be made from glass-reinforced polyamide. High-impact, mineral-reinforced polyamide is used for housings for lawn and garden equipment. Glass mat laminate, which flows during stamping, has been used for automobile oil pans, rocker arm covers, and similar components requiring stiffness.

- Melting temperature: 216–275 °C

- Tensile strength at break (D638); 7000–28 000 psi (both type 6/6; 33% glass fiber-reinforced is the strongest)

- Compressive strength at rupture or yield (D695): 12 500–29 400 psi (the lower value is for antifriction molybdenum disulfide-filled, type 6/6, while the strongest is 33% glass fiber-reinforced type 6/6)

- Tensile modulus (D638); $100–1450 \times 10^3$ psi (the lower value corresponds to molding (type 6) and extrusion compound; the larger to 30–33% glass fiber-reinforced, type 6)

2.30 POLYAMIDE IMIDE (TORLON)

Injection molding of complex precision parts that are creep resistant and maintain a strength of about 7500 psi at 500 °F (260 °C) is possible with Torlon even though the material is thermoplastic. When compounded with graphite and fluoroplastics it forms a bearing material that has been used for valve seats. Polyamide imides in various grades have replaced some metal parts in engines, compressors, and in hydraulic and pneumatic devices. They are also used in transmissions and universal joints.

- Melting temperature: 275 °C

- Tensile strength at break (D1708): 18 100 psi (20% graphite powder, 3% PTFE); 29 800 psi (30% graphite fiber-reinforced, 1% PTFE)

- Compressive strength at yield or rupture (D695): 20 000 psi (graphite powder); 40 000 psi (unfilled)

- Tensile modulus (D638): 730×10^3 psi (unfilled, 73 °F)

2.31 UHMW POLYETHYLENE

The acronym denotes ultrahigh-molecular-weight polyethylene (UHMWPE), which has good abrasion resistance and impact strength. It is fabricated in plates, sheets, billets, and other shapes by either compression molding or ram extrusion. The end product is widely used in bulk material handling because of its combination of abrasion resistance and light weight when compared with steel. In machine applications it is used in parts for chemical pumps, gaskets, feed screws, or augers, guide rails, gears and bushings. It has also been used as bogy wheels, slide bars, and snowmobile sprockets. Textile industry uses include the picker block, gears, bushings, lug straps, and pickers.[11]

- Melting temperature: 125–135 °C

- Tensile yield strength (D638): 3100–4000 psi

- Tensile strength at break (D638): 5600 psi

- Tensile modulus (D638): $730–900 \times 10^3$ psi (30% glass reinforced)

2.32 THERMOPLASTIC POLYIMIDE

Although the material is said to become rubbery at molding temperatures, it apparently flows well enough to be compression molded and extruded. Molded and direct-formed parts include compressor piston rings and seals, while extruded rods and tubes serve as lubricated bearings and thrust washers. When filled with graphite, PTFE, or molybdenum disulfide, the extrusions may serve as self-lubricated bearings, thrust washers, and wear strips.

- Tensile strength at room temperature: 17.1×10^3 psi; at 288 °C: 4.4×10^3 psi
- Tensile modulus at room temperature: 188×10^3 psi; at 288 °C: 97×10^3 psi

2.33 THERMOSET POLYIMIDE

Fabrication techniques include powder metalurgy techniques (pressure and heat), injection molding, transfer molding, extrusion, and compression molding. Aircraft applications include blocker doors in a jet engine, bushings for aircraft turbine engines, and inlet guide vane bearings. It is also used for gears for speed reduction transmissions and for pistons in hydraulic equipment.

- Tensile strength at break (D638): 64 000 psi
- Compression strength at rupture or yield (D695): 34 000 psi (both data are for 50% glass-reinforced material)

2.34 POLYPHENYLENE SULFIDE (RYTON)

Parts are made by injection molding and compression molding of this thermoplastic material. They find machine applications in chemical processing equipment as submersible impeller vanes and as gears in gear pumps. Ryton has similar applications in the automobile industry and has been used for housings for small applicances, range components, and hair dryers.

- Tensile strength at break (D638):9500 psi–19 500 psi (injection-molding resin and 40% glass fiber-reinforced, respectively)
- Compression strength at rupture or yield (D695): 16 000–21 000 psi (injection-molding resin and 40% glass fiber-reinforced, respectively)
- Tensile modulus (D638): 480×10^3–1100×10^3 (injection molded and glass reinforced as above)

2.35 POLYURETHANE

Mechanical parts may be fabricated from unsaturated casting resin with properties represented by the data below. This resin is simply poured into a mold and allowed to cure. Curing and cross-linking can be accomplished by using a low-molecular-weight triol, such as trimethylolpropane (TMP) along with excess isocyanate and heating for several hours.

Applications include gears, gaskets, and a number of light-duty industrial machine components.

- Tensile strength at break (D638): $10–11 \times 10^3$ psi
- Compression strength at yield or rupture (D695): 20 000 psi
- Tensile modulus (D638): $10–100 \times 10^3$ psi

2.36 CERAMICS

There is no clearly defined boundary between ceramics and metals.[14] As with the metals and plastics already described, it is the atoms and the bonds between atoms that determine the characteristics of the material. Some are crystalline, such as graphite and alumina, and some are noncrystalline, such as glass. Most ceramics are brittle at room temperature, which means that their stress–strain curves resemble that shown in Figure 2.36.1. Consequently, the design engineer must carefully evaluate all stress concentrations (to be described and demonstrated in Chapter 3) and use a sufficiently large safety factor to assure that actual stresses during service conditions will not exceed the ultimate stress (ultimate strength) of the material.

Typical ceramic materials used for the manufacture of machine components are shown in Table 2.36.1, where they are identified by name and by their chemical formulas.

Many ceramic components are made from initially granular material, such as alumina and silicon carbide, which must be heated, pressed, and otherwise consolidated to

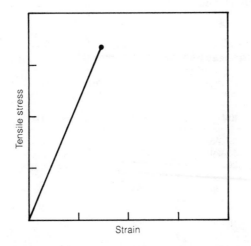

Figure 2.36.1 Typical stress–strain curve for a brittle material.

TABLE 2.36.1   COMMON CERAMICS AND REPRESENTATIVE APPLICATIONS

Ceramic	Application	Property
Alumina Al_2O_3	Sand-blast nozzles, wear plates, cutting tool nozzles, face seals against surfaces	Hardness, intermediate density
Berillia (BeO)	Nuclear reactors	Resists thermal shock up to 1500 °F/s
Boron carbine B_4C	Sand-blast nozzles, armor with Kelvar or fiberglass backing	High hardness, low density
Graphite C	Nuclear reactors, rocket nozzles, carbon fibers, face seals against alumina surfaces	Resists erosion of hot gases, high strength, high thermal conductivity, low thermal expansion
Silicon carbide SiC	Heat exchangers, shuttle reentry tiles, kiln furniture, gas turbine stators	High thermal conductivity, corrosion resistance
Silicon nitride Si_3N_4	Cylinder head valves in diesel engines, gas turbine stators	Thermal shock resistance during cooldown
Partially stabilized zirconia ZrO_2MgO $ZrO_2Y_2O_3$	Pumps, valves, impellers for abrasive liquids and slurries	Wear resistance

form the component. The tensile strength, thermal conductivity, and other mechanical properties are, therefore, a function of the size of the original granules and the manufacturing process. Hence, tabulated mechanical characteristics are imprecise, as indicated by the ranges shown in Table 2.36.2.

Ceramics also display a wider variety of responses to temperature than do either metals or plastics, as demonstrated by graphite in Figure 2.36.2. The increase in strength

TABLE 2.36.2   TYPICAL ROOM-TEMPERATURE MECHANICAL PROPERTIES

Ceramic	Tensile Stress		Average Elastic Modulus	
	MPa	ksi	GPa	psi $\times 10^{-6}$
Alumina Al_2O_3	200–310	30–45	380	55
Beryllia BeO	90–133	13–20	311	45
Stabilized zirconia, ZrO_2	138	20	138	20
Hot-pressed silicon nitride Si_3N_4 ($<1\%$ porosity)	350–580	50–80	304	44
Silicon nitride Si_3N_4 (15–20% porosity)	100–200	15–30	304	44
Silicon carbide SiC	—	—	414	60
Diamond C	—	—	1035	150

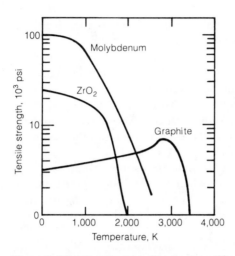

Figure 2.36.2 Tensile stress (tensile strength) as a function of absolute temperature for molybdenum, zirconia, and graphite. *From A.J. Rothman*, "Aerospace Nuclear Propulsion Technology," Ceramics for Advanced Technologies, *Howe and Riley, eds.,* © *1965 Wiley, New York.*

with increasingly higher temperatures is in distinct contrast to the behavior metals, plastics, and other common ceramics, such as zirconia.

Residual stresses due to cooling are much more important in some ceramics, principally zirconia, than in metals, because they produce immediate failure. Pure crystalline zirconia shatters when cooled below about 1000 °C because of a relatively large volume expansion (3.25%) due to a phase change from the tetragonal phase to the monoclinic phase. Addition of magnesium oxide, MgO, or ytterium oxide, Y_2O_3, produces a cubic crystal that is stable over the entire temperature range. Addition of MgO or Y_2O_3 is known as *stabilization*. *Stabilized* zirconia, however, is very sensitive to stress concentrations and impact. These problem characteristics may be alleviated by adding less MgO, or less Y_2O_3, and carefully controlling particle size and processing to produce a mixture of the stabilized cubic phase and the tougher monoclinic phase which is strong and impact resistant. This mixture is known as *partially stabilized* zirconia.

2.37 COMPOSITES

The term *composites* does not have a clear meaning when used in a discussion of materials. Concrete, wood, and some metals, such as nodular iron, may be considered to be composites since distinctly different materials with distinct boundaries may be found within the solid. In this text we use the term composites to denote solids in which fibers of a stronger material are embedded in a matrix of structurally weaker material in order to produce a solid with yield and ultimate stresses approaching those of the stronger material, at least in some directions.

By using composites we are able to produce custom-made materials that combine the desirable characteristics of different materials in a manner not available from one of them alone. Pneumatic tires are an early example of a composite in which the rubber, or elastomer, provided sealing, flexibility, and friction characteristics to the composite while the rayon, steel, or aramid fibers provide the strength. Newer applications include a carbon-fiber composite wing for the TAV-8B trainer version of the Harrier II V/STOL attack jet made by McDonnell-Douglas.

The advantages are obtained, however, at the expense of more complex design and testing. In specialized mechanical components where the directions of the principal stresses are known, the fibers are usually aligned along the corresponding principal directions. This composite material is not isotropic; its mechanical properties are not the same in all directions.

To locate the source of the complication added by using nonisotropic materials we may turn to the tensor formulation of the theory of elasticity where we find that for nonisotropic materials (materials that have different mechanical properties in different directions) up to 36 elastic constants may be required to relate stresses and strains instead of just the two required for an isotropic material; Young's modulus E and Poisson's ratio v. If the reinforcing fibers lie in only two mutually perpendicular directions the number of elastic constants is reduced to 6; if they lie in three mutually perpendicular directions the number of elastic constants is reduced to 3. Analysis of stress, strain, and displacement becomes much more complicated if the fibers are not mutually perpendicular, and the number of elastic constants may change from those given here.

Although composites were first used for structural applications, as mentioned at the beginning of this section, they are beginning to be used in mechanical components such as motor shafts, gears, and so on. Imperial Clevite in the United States has announced production of aluminum silicon pistons for internal combustion engines with ceramic fiber instead of cast iron inserts at the piston ring grooves and with ceramic fiber reinforcement at the surface of the combustion bowl to prevent cracking due to thermal fatigue.

Problems associated with the development of a new composite include forming a bond between the reinforcing fibers and the enveloping material, maintaining that bond in the working environment of the component, and maintaining proper positioning of the fibers during manufacturing.

REFERENCES

1.   van Vlack, L.H. *Materials Science for Engineers*, Addison-Wesley, Reading, MA, 1970.
2.   *Metals & Alloys in the Unified Numbering System*, 4th ed., a joint publication of the Society of Automotive Engineers, Inc. and the American Society for Testing and Materials, Society of Automotive Engineers, Warrendale, PA, 1986.

3. *Properties and Selection: Irons and Steels, Metals Handbook*, 9th ed., Vol. 1, American Society for Metals, Metals Park, OH, 1978.

4. *Properties and Selection of Metals, Metals Handbook*, 8th ed., Vol. 1, American Society for Metals, Metals Park, OH, 1964.

5. *SAE Handbook*, Vol. 1, Society of Automotive Engineers, Warrendale, PA, 1981.

6. Datsko, J., *Material Properties and Manufacturing Precesses*, Wiley, New York, 1966.

7. Horger, O.J., Ed., *Metals Engineering Design*, 2d ed., McGraw-Hill, New York, 1965.

8. *Properties and Selection: Stainless Steels, Tool Materials and Special Purpose Metals*, 9th ed., Vol. 3, American Society for Metals, Metals Park, OH, 1980.

9. *Ryerson Data Book*, Joseph T. Ryerson & Sons, Chicago, 1971.

10. *Properties and Selection: Nonferrous Alloys and Pure Metals*, 9th ed., Vol. 2, American Society for Metals, Metals Park, OH, 1979.

11. *Modern Plastics Encyclopedia, 1979–1980*, Vol. 56, Modern Plastics, Highstown, NJ, 1979.

12. *Modern Plastics Encyclopedia, 1981–1982*, Vol. 59, Modern Plastics, Highstown, NJ, 1981.

13. *Machinability, Metals Handbook*, 8th ed., Vol. 3, American Society for Metals, Metals Park, OH, 1961.

14. Richerson, D.W., *Modern Ceramic Engineering*, Dekker, New York, 1982.

15. Rothman, A.J., *Aerospace Nuclear Propulsion Technology*, Chap. 11, *Ceramics for Advanced Technologies*, J.E. Hove, and W.C. Riley, Eds., Wiley, New York, 1965.

CHAPTER THREE

STRESS AND DEFLECTION

A	Cross-section area, (l^2), constant		r_{cn}	Radii to locus of centroids (l)
a	Generic constant		T	Torque (ml^2t^{-2})
b	Generic constant, web width, width of contact area (l)		u_x, u_y, u_z	Displacements in the x, y, and z-directions, respectively (l)
B	Constant		V	Shear load (mlt^{-2})
C	Constant		w	Width (l)
D	Depth, diameter, or damage due to N cycles, determinant		x, y, z	Rectangular Cartesian coordinates (l)
da	Element of area (l^2)		z	Perpendicular distance from the neutral surface, beams (l)
E	Elastic modulus (Young's modulus) $(ml^{-1}t^{-2})$		α	Thermal coefficient of linear expansion (l/T)
e	Distance between neutral axis and locus of centroids (l)		α_i	Angles, $i = 1,2$ (l)
F	Force (mlt^{-2})		γ_i	Angles, $i = 1,2$ (1)
G	Shear modulus $(ml^{-1}t^{-2})$		δ	Beam deflection (l)
h	Thickness, distance (l)		$\varepsilon_{xx}, \dots$	Strain components (1)
I	Moment of area $(l)^4$		η	Safety factor (1)
$\hat{i}, \hat{j}, \hat{k}$	Unit vectors in x, y, z directions, respectively		θ	Generic angle (1)
J	Polar moment of area (l^4)		λ	Temperature (T)
K	Stress concentration factor (1)		v	Poisson's ratio (1)
K_D	Factor (1)		ρ	Radius to centroid, bent and initially curved beams (l)
k	Buckling parameter (1)		σ	Stress $(ml^{-1}t^{-2})$
l	Length (l)		$\sigma_I, \dots$	Principal stresses $(ml^{-1}t^{-2})$
M	Bending moment (ml^2t^{-2})		$\sigma_{xx}, \dots$	Stress components $(ml^{-1}t^{-2})$
$\hat{m}_i$	Unit normal vectors, $i = 1,2,3$ (1)		$\sigma'_{xx}, \dots$	Fictitious stress $(ml^{-1}t^{-2})$
n	Integer (1)		σ_u	Ultimate tensile stress $(ml^{-1}t^{-2})$
$\hat{n}_i$	Unit normal vectors, $i = 1,2,3$ or a,b,c (1)		σ_{yp}	Yield stress $(ml^{-1}t^{-2})$
n_x, n_y, n_z	Components of a unit normal vector (1)		$\sigma_{\eta\xi}$	Shear stress, Mohr circle $(ml^{-1}t^{-2})$
p	Axial buckling load, transverse load, (mlt^{-2}), generic point		$\sigma_{\xi\xi}$	Direct stress, Mohr circle $(ml^{-1}t^{-2})$
R_1, R_2	Radii (l)		τ	Shear stress $(ml^{-1}t^{-2})$
r	Radial distance (l)		$\tau_I, \dots$	Principal shear stresses $(ml^{-1}t^{-2})$
r_n	Radius to the neutral surface, curved beam (l)		ϕ	Function of θ (1)

Concern for the personal safety of those operating a machine and assurance by the manufacturer that it will perform its intended function both require that the engineer design each part so that the steady stresses expected during its operation not exceed those stresses that may cause it to permanently deform or break. Since the manufacturer has no control over the use of the machine once it is released to the buyer, the design engineer must also make a reasonable effort to design the entire machine so that if some component fails due to overload or misuse it will not lead to catastrophic failure of other parts of the machine, especially if their failure could cause injury to the operator.

Deformation must also be considered because severe deformation itself can be dangerous, as in extreme bending of the protective frame enclosing the operator in underground mining equipment, in agricultural (crop dusters) and military aircraft, and in trucks and automobiles.

This is the motivation for studying the material in this chapter. The first ten sections review formulas for calculating the dominant stress components and the associated deflections under common loading conditions. These dominant stress components are intimately related to the second area, that of estimating the maximum steady load that the part can sustain. If the loading is steady the maximum values calculated from the steady load stress criteria presented in the last three sections of this chapter are often used to determine component safety.

It is of central importance in the design of machine elements to recognize that the maximum tensile stress and the maximum shear stress are determined by all of the stress components displayed in Section 3.1. It is also important to recognize that the maximum tensile stress and the maximum shear stress can be correctly determined only by solving the three simultaneous equations displayed in 3.6.1. This elementary mathematical fact was often obscured in earlier texts by emphasis upon the Mohr circle method for finding two (not three) principal stresses when only stresses in a single plane were considered. This emphasis was justified in the older textbooks becase of the time-consuming and tedious calculations involved in finding the three roots, and hence the principal stresses, from the cubic eigenequation associated with equations 3.6.1. This justification no longer exists, however with easy access to pocket calculators that can be programmed to find the principal stresses and the associated principal directions. It is for this reason that Sections 3.6 and 3.7 are devoted to the solution of equations 3.6.1 and to a flowchart for a corresponding program. Once this program is coded into the calculator, or computer, not only is it easy to find the principal stress and principal directions, but the calculator, or computer, can output three principal directions and three principal stresses, thus assuring that one of them will not be forgotten.

3.1 STRESS NOTATION

Because stress is defined as force per unit area, the stress symbol should at least imply three quantities: stress magnitude, the direction of the force involved, and the orientation of the reference surface. All of these requirements are satisfied by the symbol σ_{xy}. Either the trio or the base letter itself may be thought of as representing the magnitude.

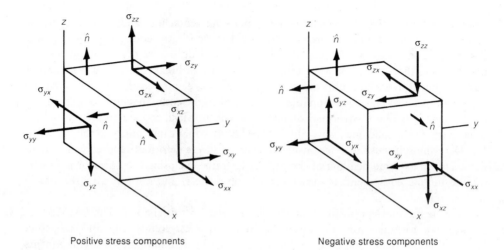

Positive stress components   Negative stress components

Figure 3.1.1 Positive and negative stress components in a rectangular Cartesian coordinate system.

One of the subscripts indicates the direction of the force component relative to an implied *xyz*-coordinate system and the other indicates the direction of the surface normal: a unit vector perpendicular to the surface. It is the symmetry of the stress tensor

$$\sigma_{xy} = \sigma_{yx}$$

that allows this interchange of significance.

The algebraic sign of a stress component is fixed by the rule:

$\sigma_{xx}, \sigma_{xy}, \sigma_{xz}, \sigma_{yy}, \sigma_{yz},$ and σ_{zz} are each positive if both the force component and the surface normal are in the positive or negative coordinate direction. A stress component is negative otherwise.

Positive and negative stress components are displayed in Figure 3.1.1. Note that tensile stresses are positive and compressive stresses are negative.

3.2 STRESSES IN BEAMS

By engineering convention a beam is a cylindrical object with any cross section (rectangular, circular, etc.) which supports loads and/or moments acting perpendicular to its longitudinal axis.

Distribution of its dominant stress, for lengths greater than three times the depth, is given by the assumptions that

1. plane cross sections initially perpendicular to the neutral axis remain plane during the deformation and are perpendicular to the deflected neutral axis,

2. the beam bends without twisting (i.e., that it is loaded and deflected in a longitudinal plane of symmetry), and

3. the elastic modulus (Young's modulus) in compression is equal to that in tension.

These assumptions also influence later deflection calculations.

Straight Beams

Consistent with these assumptions, the stress distribution within a straight beam is taken to be[1,2]

$$\sigma_{xx}(x,z) = -\frac{Mz}{I} \qquad \sigma_{xz}(x,z) = a(z)\frac{F}{A} \tag{3.2.1}$$

relative to the coordinate system shown in Figure 3.2.22 The form of $a(z)$ may be deduced by comparing with equation 3.2.22 which yields the explicit dependence of σ_{xz} upon z. Experiments have verified this stress distribution sufficiently far from the loads and reactions. Neglect of the complicated stress fields in the vicinity of loads and reactions may be justified by the St. Venant principle as stated by von Mises[3] and reinforced by Sternberg[4]:

> If the forces acting on an elastic body are confined to several distinct portions of its surface, each lying within a sphere of radius e, then the stresses and strains at a fixed interior point of the body are of a smaller order of magnitude in e as $e \rightarrow 0$ when the forces on each of the portions are in equilibrium than when they are not.

It is implied in this statement that all forces are to remain finite as $e \rightarrow 0$. Although the von Mises-Sternberg statement must be used if the principle is to be applied generally, the more common, but not always true, interpretation that

> if an actual distribution of forces is replaced by a statically equivalent system, the distribution of stress and strain throughout the body is altered only near the regions of load application

appears to describe stresses in beams and columns. Even though the stress distribution given by equations 3.2.1 may be assumed only in portions of the beam not near loads or supports, it is, in fact, common practice to assume it holds throughout. The error resulting from this assumption is usually negligible, especially if the deflection is very small.

Curved Beams

Stress within the elastic range in a beam in pure bending in its plane of symmetry, as shown in Figure 3.2.1, may be analyzed using the Bernoulli-Euler assumptions employed in the analysis of a straight beam; namely, that planes originally perpendicular to the undeformed neutral axis remain plane and rotate such as to be perpendicular to the deformed neutral axis. To do this select a polar coordinate system with its origin at the center of curvature at the location where the stress is to be calculated, again as

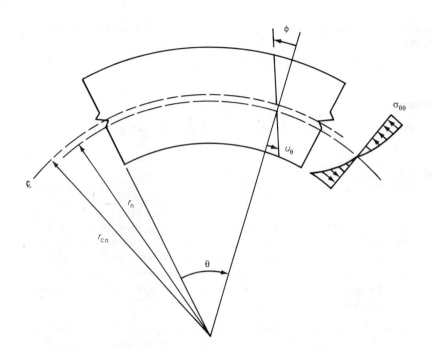

Figure 3.2.1 Curved beam geometry and quantities used in its stress analysis.

shown in Figure 3.2.1, so that the displacement in the circumferential direction may be expressed as

$$u_\theta = (r_n - r)\phi(\theta) \tag{3.2.2}$$

where $\phi(\theta)$ is as yet an unknown function of θ and where r_n is the radius to the neutral surface, which, incidentally, may not coincide with the locus of centroids of the cross-sections. Thus, from equation 3.10.1, to be introduced later, the strain may be written as

$$\varepsilon_{\theta\theta} = \frac{1}{r}\frac{\partial u_\theta}{\partial \theta} = \frac{r_n - r}{r}\frac{\partial \phi}{\partial \theta} \tag{3.2.3}$$

Since $\varepsilon_{\theta\theta}$ and $\sigma_{\theta\theta}$ are the only strain and stress components acting, it follows that if no force acts perpendicular to the cross section of the curved beam the summation of stress over the cross section must vanish. Hence,

$$F = \int_A \sigma_{\theta\theta}\, da = E \int_A \varepsilon_{\theta\theta}\, da = E \frac{\partial \phi}{\partial \theta} \int_A \frac{r_n - r}{r}\, da = 0$$

Since $\partial\phi/\partial\theta$ is independent of r, and since E is constant for a homogeneous beam, this equation can be zero only if

$$r_{\mathrm{n}} \int_A \frac{da}{r} - \int_A da = 0 \qquad (3.2.4)$$

which is to say that

$$r_{\mathrm{n}} = \frac{\int_A da}{\int_A \dfrac{da}{r}} \qquad (3.2.5)$$

Thus, equation 3.2.5 not only indicates that the neutral axis no longer coincides with the centroid, but it also gives its position.

From the summation of moments about the neutral axis it follows that

$$M = \int_A (r_{\mathrm{n}} - r)\sigma_{\theta\theta}\,da = E\frac{\partial\phi}{\partial\theta}\int_A \frac{(r_{\mathrm{n}} - r)^2}{r}\,da \qquad (3.2.6)$$

Upon solving for $E\,\partial\phi/\partial\theta$ from (3.2.3) after multiplying both sides by E to obtain

$$E\varepsilon_{\theta\theta} = E\frac{\partial\phi}{\partial\theta}\frac{r_{\mathrm{n}} - r}{r} = \sigma \qquad (3.2.7)$$

we find that since $\partial\phi/\partial\theta$ is constant over any cross section A

$$E\frac{\partial\phi}{\partial\theta} = \frac{\sigma r}{r_{\mathrm{n}} - r} = \text{Constant(over cross section } A) \qquad (3.2.8)$$

which implies that ϕ is a linear function of θ.

Substitution from (3.2.8) into equation 3.2.6 and expansion of the integrand in the manner shown leads to

$$\frac{\sigma r}{r_{\mathrm{n}} - r}\int_A \frac{(r_{\mathrm{n}} - r)^2}{r}\,da = \frac{\sigma r}{r_{\mathrm{n}} - r}\int_A \frac{r_{\mathrm{n}}^2 - 2rr_{\mathrm{n}} + r^2}{r}\,da$$

$$= \frac{\sigma r}{r_{\mathrm{n}} - r}\left\{r_{\mathrm{n}}\left[r_{\mathrm{n}}\int_A \frac{da}{r} - \int_A da\right] - r_{\mathrm{n}}\int_A da + \int_A r\,da\right\}$$

The term in square brackets vanishes because of equation 3.2.4. Since the centroid of the area is defined by

$$\int_A r\,da = r_{\mathrm{cn}}A \qquad (3.2.9)$$

it is evident that equation 3.2.6 reduces to

$$M = \frac{\sigma r}{r_n - r} A(r_{cn} - r_n) = F r_n \tag{3.2.10}$$

which in turn allows the stress $\sigma \equiv \sigma_{\theta\theta}$ to be calculated from

$$\sigma(r) = \frac{M}{Ae}\left(\frac{r_n}{r} - 1\right) \tag{3.2.11}$$

where $e = r_{cn} - r_n$. Observe that the maximum stress, whether tension or compression, always acts on the concave side of the curved beam and that the neutral axis migrates toward the center of curvature.

EXAMPLE 3.2.1

Calculate the maximum tensile stress in the transfer bracket shown in Figure 3.2.2a. Figure 3.2.2b shows that the cross section at B–B is composed of two circles of radius R_1 and R_2 connected by a web of thickness w. Forces F act through the center of curvature of the bracket, for which $r_0 = 3.00$ in., $R_1 = 0.80$ in., $R_2 = 0.50$ in., $w = 0.25$ in., $a = 1.5$ in., and $F = 1850$ lb.

The maximum tensile stress acts across plane B–B becuase it is perpendicular to the resultant of the applied forces. Since the force resultant does not act through the centroid of the cross section at B–B the force on the cross section may be represented by a force at the centroid and a moment acting about the centroid. This resolution of forces must be about the centroid, rather than about the neutral axis, in order that the stress distribution due to a tensile force will be uniform over the section whenever the moment is absent. This force system is illustrated in Figure 3.2.2c.

The simplest part of the stress calculation is that of finding the tensile stress due to tension only, namely, $\sigma_{\theta\theta} = P/A$. To find A turn to Figure 3.2.2d and note that the web will subtend an angle $2\gamma_1$ at its intersection with the circular cross section of radius R_1, where γ_1 is given by

$$2R_1 \sin \gamma_1 = w \tag{3.2.12}$$

Likewise, at its intersection with the circular cross section of radius R_2

$$2R_2 \sin \gamma_2 = w \tag{3.2.13}$$

The vertical distance of the web–circle intersection from the center of circle 1 is then

$$y_1 = R_1 \cos \gamma_1 \tag{3.2.14}$$

and the web–circle intersection with circle 2 is

$$y_2 = R_2 \cos \gamma_2 \tag{3.2.15}$$

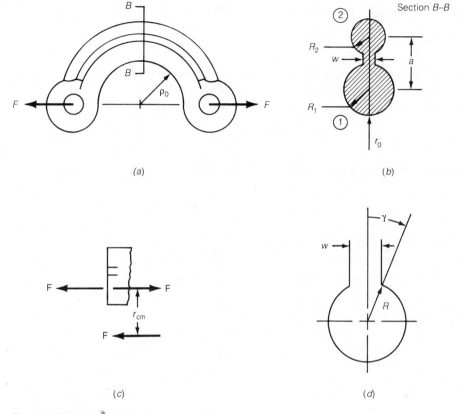

(a)

(b)

(c)

(d)

Figure 3.2.2 Curved link and its cross section.

from the center of circle 2. Consequently, the length of web is $a - y_1 - y_2$. The total area of the cross section is, therefore, given by

$$A = \left(1 - \frac{\alpha_1}{\pi}\right)\pi R_1{}^2 + \left(1 - \frac{\alpha_2}{\pi}\right)\pi R_2{}^2$$
$$+ w(a - R_1 \cos \alpha_1 - R_2 \cos \alpha_2) = 2.6889 \text{ in.}^2 \tag{3.2.16}$$

and the stress due to tension only is $1850/2.6889 = 688.01$ psi.

Evaluation of relation 3.2.11 next requires that both equations 3.2.5 and 3.2.9 be evaluated. Although the first of these may be integrated graphically, it appears simpler to handle both of them numerically because they have similar limits and closely related integrands. After recalling that the equation for a circle displaced a distance h along the y axis is

$$(y - h)^2 + x^2 = r^2$$

it is easy to write the equations for the circular boundaries as

$$x^2 + [r - (r_0 + R_1)]^2 = R_1{}^2 \tag{3.2.17}$$

for the region of radius R_1. In this equation the y coordinate used in writing the expression for the distance to the circular boundary in the y direction has been replaced by r to adapt the relation to the present situation. Thus, it follows that

$$x^2 + [r - (r_0 + R_1 + a)]^2 = R_2{}^2 \tag{3.2.18}$$

for the region of radius R_2. Hence, the radius to the centroid r_{cn} may be found from

$$\int_A r\,\frac{da}{r} = 2\int_{r_0}^{r_1} r[R_1{}^2 - (r - r_0 - R_1)^2]^{1/2}\,dr + w\int_{r_1}^{r_2} r\,dr$$

$$+ 2\int_{r_2}^{r_3} r[R_2{}^2 - (r - r_0 - R_1 - a)^2]^{1/2}\,dr = r_{cn}A \tag{3.2.19}$$

and the radius to the neutral surface r_n may be calculated from

$$\int_A \frac{da}{r} = 2\int_{r_0}^{r_1} [R_1{}^2 - (r - r_0 - R_1)^2]^{1/2}\,\frac{dr}{r}$$

$$+ w\int_{r_1}^{r_2} \frac{dr}{r} + 2\int_{r_2}^{r_3} [R_2{}^2 - (r - r_0 - R_1 - a)^2]^{1/2}\,\frac{dr}{r} = \frac{r_n}{A} \tag{3.2.20}$$

where

$$r_0 = r_0 = 3.00 \text{ in.}$$

$$r_1 = r_0 + R_1(1 + \cos \alpha_1) = 4.59017 \text{ in.}$$

$$r_2 = r_0 + R_1 + a - R_2 \cos \alpha_2 = 4.81588 \text{ in.} \tag{3.2.21}$$

$$r_3 = r_0 + a + R_1 + R_2 = 5.800 \text{ in.}$$

Numerical integration using Simpson's rule yields

$$r_{cn} = 4.4022 \text{ in.} \qquad r_n = 3.8714 \text{ in.}$$

so that $e = 0.5308$ in. Direct calculation of M yields

$$M = 1850\ r$$

Substitution of the appropriate values into equation 3.2.11 gives

$$\sigma_{\theta\theta} = 1129.176 \text{ psi}$$

for a maximum tension stress on plane $B-B$ of

$$1129.176 + 687.988 = 1817.164 \text{ psi}$$

Although four decimal places have been retained to prevent further degradation of the calculations due to additional rounding errors, the accuracy of Simpson's rule is such that the uncertainty may be of the order of 1.0 psi.

Shear Stress in Beams

Shear stress is also induced in a beam whenever it carries a transverse load. When no surface shear forces are applied, as is usually the case, the shear stress is zero at the surface and increases for points below the surface. Its distribution may be approximated at a distance z^* from the neutral surface by[2]

$$\sigma_{xz}(x,z^*) = \tau(z^*) = \frac{V}{Ib} \sum_{i=1}^{n} \bar{z}_i A_i \qquad (3.2.22)$$

where V is the shear force acting on the cross section and where $\bar{z}_i$ is the distance from the centroid of area A_i (Fig. 3.2.3) to the neutral surface of the beam. The summation in (3.2.22) is over all of the area between z^* and the farthest surface of the beam on the same side of the neutral axis. If, for example, we wish to calculate the shear on surface S in Figure 3.2.3, n would equal 2 and the area would be as shown.

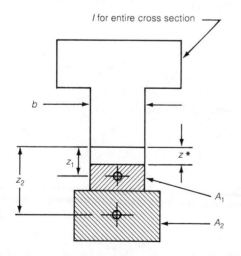

Figure 3.2.3 Quantities involved in equation 3.2.19. (Recall that the neutral surface passes through the centroid of the cross-sectional area if the beam is straight.)

The maximum value of σ_{xz} occurs on the neutral axis of a beam with a rectangular cross section and is equal to

$$\sigma_{xz} = \tau = 3V/(2A) \tag{3.2.23}$$

For beams with a circular cross section the maximum is given by

$$\sigma_{xz} = \tau = 4V/(3A) \tag{3.2.24}$$

where A is the total cross-sectional area for each beam.

EXAMPLE 3.2.2

Prove that equation 3.2.23 may be obtained from equation 3.2.22 for a beam with a rectangular cross section.

From Figure 3.2.3 we see that the cross section between z^* and the farthest surface of the beam is always that of a rectangle, so that the sum on the right side of equation 3.2.22 reduces to a single term. Thus

$$\sigma_{xz}(z^*) = Vz_1 A_1/(Ib) \tag{3.2.25}$$

where $I = bh^3/12$,

$$r_1 = \left[z^* + \frac{1}{2}\left(\frac{h}{2} - z^*\right) \right] = \frac{1}{2}\left(\frac{h}{2} + z^*\right)$$

and

$$A_1 = b\left(\frac{h}{2} - z^*\right)$$

Substitution of these expressions for r_1 and A_1 into relation 3.2.25 yields

$$\sigma_{xz} = \frac{12V}{b^2 h^3} \frac{b}{2}\left(\frac{h}{2} + z^*\right)\left(\frac{h}{2} - z^*\right) = \frac{6V}{bh^3}\left(\frac{h^2}{4} - z^{*2}\right)$$

Hence, the maximum value of σ_{xz} occurs at $z^* = 0$. Since $bh = A$, we have that

$$\sigma_{xz} = 3V/(2A)$$

in terms of the shear force V and the beam cross-sectional area A.

3.3 STRESSES IN COLUMNS—BUCKLING

By engineering convention a column is defined as a long cylindrical object (whose cross section is often not circular) supporting axial loads. Stresses on a cross section sufficiently far from any loads or supports in a centrally loaded column is simply

$$\sigma_{zz} = \frac{F}{A} \tag{3.3.1}$$

where F is the resultant of the central loads, and is assumed to act through the centroid of any cross-sectional area A. If the resultant axial load vector is not coincident with the longitudinal axis of symmetry stress σ_{zz} is given by

$$\sigma_{zz} = F\left(\frac{1}{A} - \frac{ey}{I}\right) \tag{3.3.2}$$

where I is the moment of area of the cross section and e is the load eccentricity, pictured in Figure 3.3.1.

The second term in equation 3.3.2 represents a bending moment of magnitude Fe which will cause the column to bend, so that in the bent configuration the axial force will cause an additional moment, as shown in Figure 3.3.1, whose magnitude at any particular point will depend upon the deflection at that point. If the deflection is small this additional moment will be small and the bent column will continue to support the axial load.

As the axial load increases, however, it will reach a critical value where regardless of how small e might be the slight deflection it causes will be sufficient for the additional moment due to the axial load to severely bend, and perhaps collapse, the column. This load is known as the *buckling load*.

Two primary formulas apply to calculation of the buckling load. If the column is long the classical Euler long-column buckling formula applies and if the column is short the Johnson formula applies. The Johnson formula is but one of a number of formulas that have been proposed. It will be used here because it agrees well with experimental data and, therefore, has been widely accepted.

A long column is one for which the *slenderness ratio* (l/r_g) obeys the inequality

$$\frac{l}{r_g} > \frac{\pi}{k}\left(\frac{2E}{\sigma_y}\right)^{1/2} \tag{3.3.3}$$

where l is the unconstrained length of the column and r_g is the *radius of gyration*, defined by

$$r_g^2 = I/A$$

in terms of the moment of area of the cross section (actually the second moment of area) and A is the cross-sectional area.

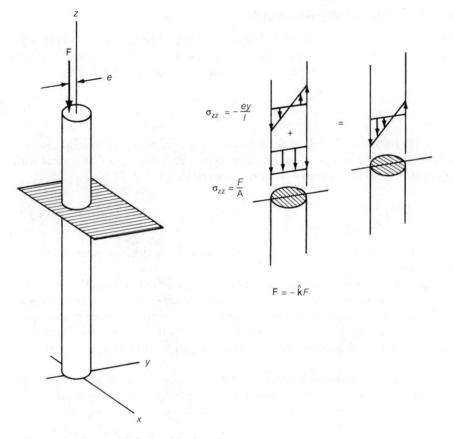

Figure 3.3.1 Column loading, eccentric resultant.

When relation 3.3.3 holds the maximum axial load that may be supported without buckling is given by

$$W = \frac{\pi^2 EI}{\zeta k^2 \ell^2} = \frac{EA}{\zeta \left(\dfrac{kl}{\pi r_g}\right)^2} \tag{3.3.4}$$

If the column is short, equation 3.3.4 must be replaced by

$$W = A \frac{\sigma_y}{\zeta}\left[1 - \frac{1}{2}\left(\frac{kl}{\pi r_g}\right)^2 \frac{\sigma_y}{2E}\right] \tag{3.3.5}$$

In the these equations ζ denotes the safety factor and σ_y in equation 3.3.5 denotes the yeild stress in tension. Since the yield stress in compression is often larger than that in tension, a cautious estimate of the buckling load W is had.

Factor k is determined by the end supports, also known as *fixities*. It takes the values

$$k = \tfrac{1}{2} \qquad \text{for fixed-fixed}$$
$$= 1/\sqrt{2} \qquad \text{fixed-pinned}$$
$$= 1 \qquad \text{pinned-pinned}$$
$$= 2 \qquad \text{fixed-free}$$

where these representations are represented by the bearing and end conditions shown in Figure 3.3.2. Chapter 9 on rolling element bearings gives further details on bearing mounts which closely approximate fixed or pinned support.

These values have been verified experimentally with very carefully controlled end support conditions. Safety factor ζ has been included the buckling load is very sensitive to the end condition and these conditions may not be duplicated in actual machine supports; that is, pin friction produces an end support that is between pinned and fixed.

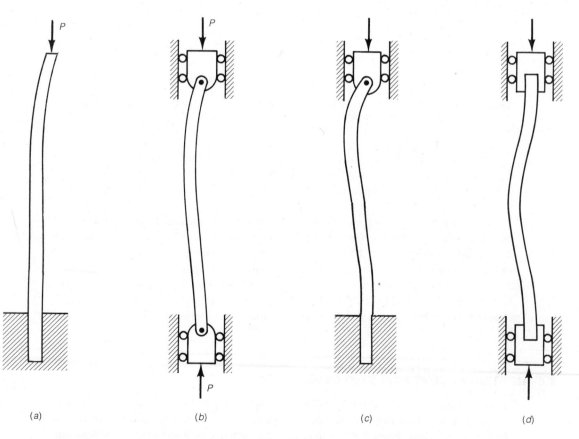

(a)   (b)   (c)   (d)

Figure 3.3.2 The four end conditions for (a) $k = 2$, (b) $k = 1$, (c) $k = 1/\sqrt{2}$, and (d) $k = 1/2$.

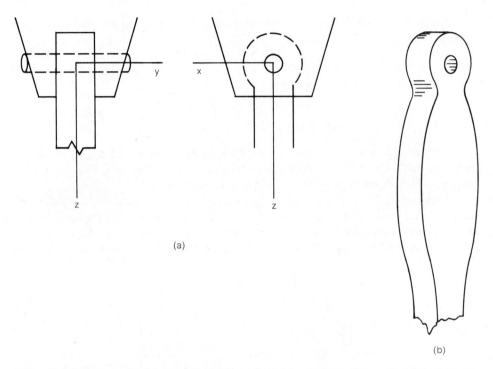

(a)

(b)

Figure 3.3.3 Example of a link and end condition where k is smaller in the xz plane but where I is smaller in the yz plane.

Two different end conditions may hold at one or both ends of a column in machine design applications. An example of this is shown in Figure 3.3.3*a* in which a link is pinned between two closely fitted blocks, so that the link is pinned in the x direction but fixed, or clamped, in the y direction.

If the body of the link away from the ends has a circular cross section it will buckle in the xz plane because the k value for bending in that plane is smaller than for bending in the yz plane. If, on the other hand, the body of the link away from the ends has a cross section whose area moment about an axis parallel the y axis is greater than about an axis parallel to the x axis, as found in Figure 3.3.3*b*, buckling may not be in the xz plane. Guided by this example, we see that it is the combination of k and I about various directions in the xy plane that ultimately determines the critical buckling load.

3.4 HERTZIAN (CONTACT) STRESSES

Hertzian stresses arise in the vicinity of the contact between two elastic objects whose geometry is such that they would make contact only at a point or along a line if they were perfectly rigid. Point contact would occur between two spheres or a sphere and

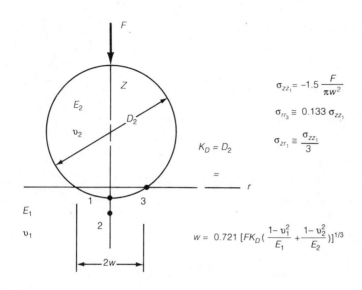

$$\sigma_{zz_1} = -1.5\,\frac{F}{\pi w^2}$$

$$\sigma_{rr_3} \cong 0.133\,\sigma_{zz_1}$$

$$\sigma_{zr_1} \cong \frac{\sigma_{zz_1}}{3}$$

$$K_D = D_2$$

$$=$$

$$w = 0.721\left[FK_D\left(\frac{1-v_1^2}{E_1} + \frac{1-v_2^2}{E_2}\right)\right]^{1/3}$$

(a)

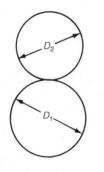

$$K_D = \frac{D_1\,D_2}{D_1 + D_2}$$

F   pounds

E, p   psi

$D_1,\ D_2,\ w.$ inches

(b)

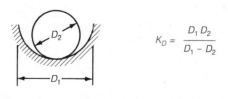

$$K_D = \frac{D_1\,D_2}{D_1 - D_2}$$

(c)

Figure 3.4.1 Hertzian stresses for a sphere in contact with (a) a plane, (b) another sphere, and (c) a spherical cavity. *Source: R.C. Roark and W.C. Young,* Formulas for Stress and Strain, *5th ed., McGraw-Hill, New York, 1975.*

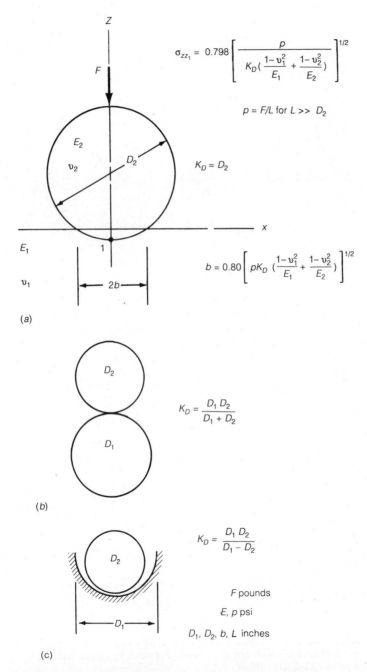

$$\sigma_{zz_1} = 0.798 \left[\frac{p}{K_D \left(\dfrac{1-v_1^2}{E_1} + \dfrac{1-v_2^2}{E_2} \right)} \right]^{1/2}$$

$$p = F/L \text{ for } L \gg D_2$$

$$K_D = D_2$$

$$b = 0.80 \left[pK_D \left(\frac{1-v_1^2}{E_1} + \frac{1-v_2^2}{E_2} \right) \right]^{1/2}$$

(a)

$$K_D = \frac{D_1 D_2}{D_1 + D_2}$$

(b)

$$K_D = \frac{D_1 D_2}{D_1 - D_2}$$

F pounds

E, p psi

D_1, D_2, b, L inches

(c)

Figure 3.4.2 Hertzian stresses for a circular cylinder of length L in contact with (a) a plane, (b) another circular cylinder, and (c) a circular channel. *Source: R.J. Roark and W.C. Young, Formulas for Stress and Strain, 5th ed., McGraw-Hill, New York, 1975.*

a plane, and line contact would occur between a circular cylinder and a plane. Actual point and line contact cannot exist because any small force between the contacting objects would cause infinite stresses at the point or line of contact.

Hertz, using analytical techniques beyond the scope of this text, calculated the stresses to be expected between a number of such objects after an initial deformation and published the results in 1881. Some of the results pertinent to the design of machine components are listed in Figures 3.4.1 and 3.4.2 as taken from Chapter 13 of Ref. 1.

3.5 THE MOHR CIRCLE AND ITS STRESS NOTATION: PRINCIPAL STRESSES IN TWO DIMENSIONS

The Mohr circle is an aid in transforming stresses from one set of two-dimensional coordinates through a point P to another set through the same point. It is necessary because stresses are tensors and, therefore, do not transform as vectors.

The Mohr circle may be applied to a point on either a flat or curved surface because it represents a stress transformation at a point. This is justified by a theorem from differential geometry that it is always possible to describe a tensor relation in the vicinity of a point in terms of a local Cartesian coordinate system.*

Recall from the derivation of the Mohr circle that it is drawn in an abstract orthogonal stress space in which one coordinate axis corresponds to a normal stress (tension or compression), represented by σ, and the other corresponds to a shear stress, represented by τ.

A convenient first step is the labeling of two perpendicular sides of the element about P as side a and side b as shown in Figure 3.5.1a before laying out axes and τ as in that figure. Tensile stresses are taken as positive, compressive stresses as negative, and shear stresses are positive if they tend to rotate the element in a clockwise direction. This rule is one of the disadvantages of the Mohr circle: It uses its own sign convention for shear stresses.

Next recall that angles between directions in stress space are twice those in the real world. Thus, the stresses acting on sides a and b, whose normals are 90° apart in the real world, Figure 3.5.1a, plot as the ends of a diameter of the Mohr circle to be 180° apart in stress space, as indicated on the Mohr circle in Figure 3.5.1b. With a known diameter it is a simple matter to draw a circle centered at the midpoint of the diameter and find the principal stresses as the intersection of this circle with the σ axis, shown as σ_I and σ_{II} in Figure 3.5.1b, and the maximum shear stresses lying in the plane through P as the radius of the Mohr circle. Thus, points e and f represent the maximum and minimum values of the in-plane shear stress in Figure 3.5.1b. Mohr circle users often forget that there are two other shear stresses as well, $\sigma_I/2$ and $\sigma_{II}/2$, as we shall see from equations 3.6.4.

---

* J.L. Synge and A. Schild, *Tensor Calculus*, University of Toronto Press, Toronto, 1966, Section 2.6.

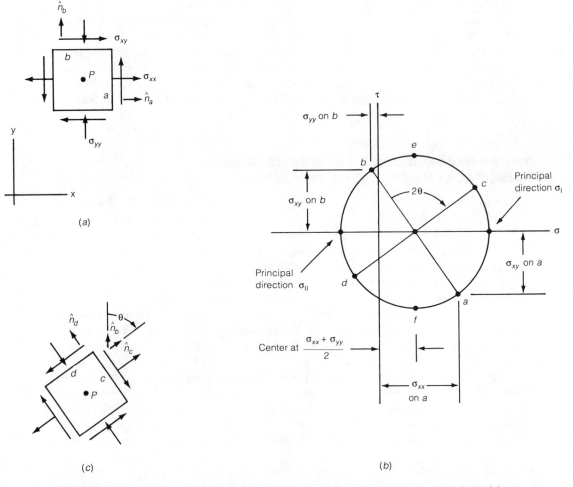

Figure 3.5.1 Mohr circle (*b*) constructed from stresses in (*a*) used to find stresses on planes (*c*).

The Mohr circle may also be used to find the shear and direct stresses acting across a plane making an angle θ with one of the sides of the element in Figure 3.5.1*a*, such as on the element about P shown in Figure 3.5.1*c*. If we are interested in the stress components acting across plane *c*, Figure 3.5.1*c*, whose normal makes an angle θ in the clockwise direction from the normal to plane *b*, Figure 3.5.1*a*, we simply move in a clockwise direction around the circle point *b* an amount 2θ to locate a corresponding *c* on the circle. Its coordinates are the direct and shear stresses acting across plane *c* through point P in the real material. The stress acting across plane *d*, perpendicular to plane *c* in the real world, may be read from the coordinates of a point on the Mohr circle 180° from *c*, namely, point *d* in Figure 3.5.1*b*.

EXAMPLE 3.5.1

Find the principal stresses and principal directions for a part on which strain gage measurements indicate $\sigma_{yy} = 200$ MPa, $\sigma_{yz} = -110$ MPa, and $\sigma_{zz} = -80$ MPa. Also find the maximum shear stress and the planes across which it acts.

Draw the stresses on an element centered at P using the standard sign convention as shown in Figure 3.5.2a.

Guided by the geometry of the circle in Figure 3.5.1 we find that the center lies at $(\sigma_{yy} + \sigma_{zz})/2 = (200 - 80)/2 = 60$ MPa. Also from the geometry we see that the radius is given by $[(\sigma_{yy} - \sigma_{xx})^2/4 + \sigma_{yz}^2]^{1/2}$. Substitution yields $[(200 + 80)^2/4 + 110^2]^{1/2} = 178.04$ MPa as the magnitude of the radius.

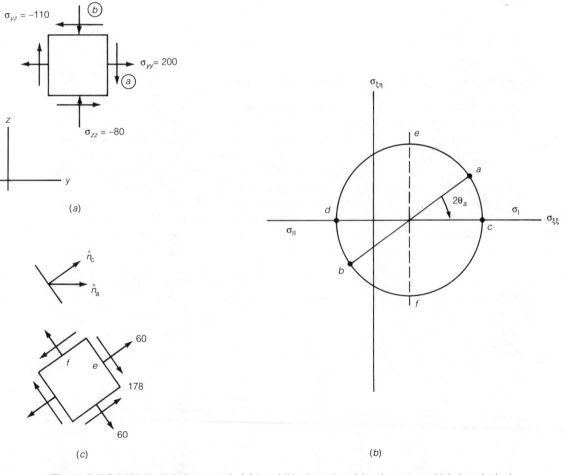

(a)

(b)

(c)

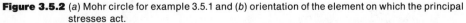

Figure 3.5.2 (a) Mohr circle for example 3.5.1 and (b) orientation of the element on which the principal stresses act.

Hence, according to Figure 3.5.2b

$$\sigma_{\mathrm{I}} = 60 + 178.04 = 238.04 \text{ MPa},$$

$$\sigma_{\mathrm{II}} = 60 - 178.04 = -118.04 \text{ MPa}.$$

$$\tau = 178.04 \text{ MPa}.$$

From Figure 3.5.2b we see that the angle between point a and σ_1 on the Mohr circle may be found from $\phi = \tan^{-1} 110/140 = 38.157°$, so that angle between points a and e on the Mohr circle is $\pi/2 - \phi = 51.843°$. Since e is counterclockwise from a on the Mohr circle, the normal to the plane across which the shear acts will be $51.843/2 = 25.922°$ counterclockwise from the normal to a in the real material. The element on which the shear stresses corresponding to points e and f on the Mohr circle act is shown in Figure 3.5.2c.

3.6 PRINCIPAL STRESSES IN THREE DIMENSIONS

In most machine applications the loads applied to the mechanical components give rise to stresses with components in three directions, so that the Mohr circle analysis is inappropriate. For example, whenever a shaft provides torque to a gear, sheave, sprocket, or traction wheel it experiences a radial load at the gear, sheave, or sprocket, which gives rise to stress component σ_{rr}. This radial load in turn causes a bending stress σ_{zz}. To provide torque the shaft experiences a twist which gives rise to stress component $\sigma_{z\theta}$. Depending upon the means of fastening the gear, sheave, or sprocket to the shaft, there may be either a $\sigma_{\theta\theta}$ stress component (from a keyway wall), a $\sigma_{r\theta}$ stress component (from an interference fit), or both, added to transfer torque from the shaft. Additional forces and moments are added whenever bevel or single helical gears are used.

Principal stresses, maximum shear stresses, and the orientation of the planes across which these stresses act may be found from the matrix equation

$$\begin{pmatrix} \sigma_{xx} - \sigma & \sigma_{xy} & \sigma_{xz} \\ \sigma_{yx} & \sigma_{yy} - \sigma & \sigma_{yz} \\ \sigma_{zx} & \sigma_{zy} & \sigma_{zz} - \sigma \end{pmatrix} \begin{pmatrix} n_x \\ n_y \\ n_z \end{pmatrix} = 0 \tag{3.6.1}$$

and the condition that the normal vectors are of unit length; namely, that

$$\hat{\mathbf{n}} \cdot \hat{\mathbf{n}} = 1 \tag{3.6.2}$$

where

$$\hat{\mathbf{n}} = \hat{\mathbf{i}} n_x + \hat{\mathbf{j}} n_y + \hat{\mathbf{k}} n_z$$

The roots of the eigenequation of the coefficient matrix in (3.6.1)

$$
\sigma^3 - (\sigma_{xx} + \sigma_{yy} + \sigma_{zz})\sigma^2 - (\sigma_{xy}^2 + \sigma_{yz}^2 + \sigma_{zx}^2
$$
$$
- \sigma_{xx}\sigma_{yy} - \sigma_{yy}\sigma_{zz} - \sigma_{zz}\sigma_{xx})\sigma - (\sigma_{xx}\sigma_{yy}\sigma_{zz}
$$
$$
+ 2\sigma_{xy}\sigma_{yz}\sigma_{zx} - \sigma_{xx}\sigma_{yz}^2 - \sigma_{yy}\sigma_{xz}^2 - \sigma_{zz}\sigma_{xy}^2) = 0 \qquad (3.6.3)
$$

give the three principal stresses, which will be represented by σ_1, σ_2, and σ_3. The principal directions for each of the principal stresses may be found by substituting each of these roots into equations 3.6.1 and 3.6.2. The values of n_x, n_y, and n_z which are found are the direction cosines of the principal direction for that principal stress which was substituted into (3.6.3). The θ and ϕ values required to draw each of the unit vectors lying along a principal direction as shown in Figure 3.6.1 may be found from the relations

$$
\phi = \tan^{-1} n_y/n_x \qquad \theta = \cos^{-1} n_z \qquad (3.6.4)
$$

relative to the coordinate system used to define the stress components themselves

The magnitudes of the maximum shear stresses are given by the relations

$$
\tau_1 = |\sigma_2 - \sigma_3|/2 \qquad \tau_2 = |\sigma_1 - \sigma_3|/2
$$
$$
\tau_3 = |\sigma_1 - \sigma_2|/2 \qquad\qquad (3.6.5)
$$

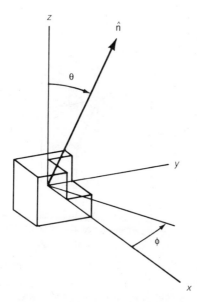

Figure 3.6.1 Coordinate angles θ and ϕ describing unit vector $\hat{n}$.

They act across those planes that bisect the angles between the planes whose direction cosines are given by n_x, n_y, and n_z for each of the principal stresses involved. If we introduce another subscript on the direction cosines to associate them with their principal stresses, such that n_{1x}, n_{1y}, and n_{1z} are the direction cosines of the principal stress σ_1 (i.e., components of the normal vector $\mathbf{n}_1$) and so on, we may find the unit vectors to the planes across which τ_1, τ_2, and τ_3 act from

$$\hat{n}^*_1 = (\hat{n}_2 + \hat{n}_3)/\sqrt{2} \qquad \hat{n}^*_2 = (\hat{n}_3 + \hat{n}_1)/\sqrt{2}$$
$$\hat{n}^*_3 = (\hat{n}_1 + \hat{n}_2)/\sqrt{2}$$

(3.6.6)

respectively. The values of θ and ϕ for each of these unit normals may be found from relations similar to those given in (3.6.4). These planes are shown in Figure 3.6.2.

A flowchart for a program to calculate the principal stresses and their principal directions is given in Appendix B. This program was used in Ref. 6 along with a graphics subroutine to display the physical components of the principal stress tensors as viewed from a point along each of the positive coordinate axes used to describe the stress components involved in calculating the principal stresses. It also allows the trio of principal stresses to be viewed from any direction defined by θ and ϕ. (Unfortunately not all of the principal directions were shown in all of the figures in Ref. 6. Also, the values given for coefficents a and b in equation (8) should be those given for b and c, respectively, with $a = 1$.)

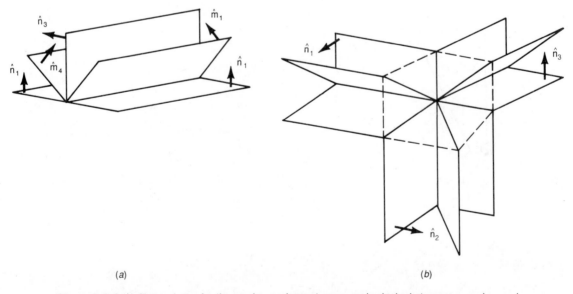

(a) (b)

Figure 3.6.2 (a) Shear planes for the maximum shear stress τ_1 and principal stresses σ_1 and σ_{11} and (b) planes for the principal stresses and three of the six planes for shear stress.

EXAMPLE 3.6.1 (from Ref. 6)

Calculate the principle stresses and find the principal directions for the stresses in the vicinity of a keyway in a motor shaft fitted with a sprocket for a chain drive when the key and keyway are on the side of the shaft farthest from the driven machine.

The stresses in the shaft near the keyway are shown in Figure 3.6.3 at a point removed from the stress concentration near the end of the key. Stress $\sigma_{z\theta}$ is the shear stress component due to transmitted torque, and $\sigma_{r\theta}$ is the shear stress component caused by the key force against the wall of the keyway as it attempts to peel off the outer section of the shaft. Its presence is evidenced by shaft failures initiated by circumferential cracks. Stress $\sigma_{\theta\theta}$ is due to compression at the keyway wall and tensile stress component σ_{zz} is due to shaft bending because of chain tension. Component σ_{zr} is the shear stress associated with chain tension and σ_{rr} is due to the radial force generated by the key as it tends to roll in the keyway because of the force of the sprocket against that part of the key in the sprocket's keyway.

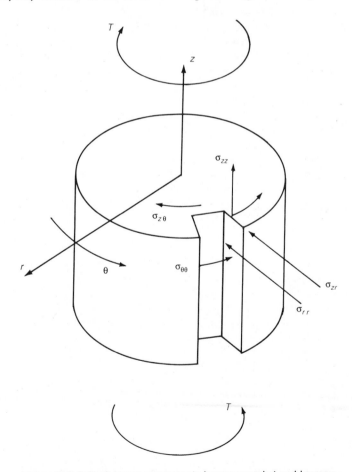

Figure 3.6.3 Shaft stress components hear a sprocket and keyway.

A radial compressive stress would also be generated if an interference fit were used between shaft and sprocket.

Representative values of these stresses may be

$$\sigma_{rr} = -120 \text{ MPa} \qquad \sigma_{r\theta} = \sigma_{\theta r} = 350 \text{ MPa}$$

$$\sigma_{\theta\theta} = -300 \text{ MPa} \qquad \sigma_{\theta z} = \sigma_{z\theta} = -485 \text{ MPa}$$

$$\sigma_{zz} = 110 \text{ MPa} \qquad \sigma_{zr} = \sigma_{rz} = 80 \text{ MPa}$$

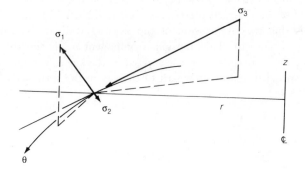

(a)

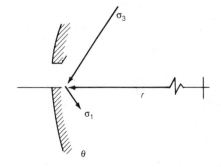

(b)

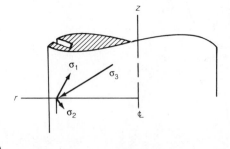

(c)

Figure 3.6.4 Principal stresses and directions (a) as viewed in three dimensions, (b) as viewed parallel to the axis, and (c) as viewed from the side of the shaft.

To demonstrate the consequences of omitting one or more of these stress components, suppose we ignore all but the shear stress acting across the shaft cross section $\sigma_{z\theta}$. The calculated principal stresses are then ± 485 MPa. If we improve our principal stress calculation by including $\sigma_{zr} = 80$ MPa the principal stresses increase to ± 491.6 MPa. They still lie in a plane, but the plane is no longer perpendicular to the x_1 axis. Inclusion of $\sigma_{zz} = 110$ MPa produces $\sigma_1 = 549.6$ MPa, $\sigma_2 = -439.6$ MPa, and $\sigma_3 = 0.0001$ MPa. Inclusion of the remainder of the stress components yields $\sigma_1 = 464.9$ MPa, $\sigma_2 = 17.7$ MPa, and $\sigma_3 = -792.6$ MPa. It is important to observe that although the maximum tensile stress is less than that estimated from the unrealistically simple assumption that only $\sigma_{\theta z}$ need be considered, the maximum shear stress has increased from 485.0 to 628.8 MPa, an increase of 29.6%.

The directions and magnitudes of these principal stresses are pictured in Figure 3.6.4 where a three-dimensional view from a point within the cylinder is shown in Figure 3.6.4a. When viewed from the end of the shaft in the positive z direction the principal stresses appear as in Figure 3.6.4b; when viewed from outside of the cylinder they appear as in Figure 3.6.4c. Principal stress σ_2 is slightly exaggerated to make at least the arrowhead visible.

After studying the figure long enough to visualize the principal directions in three dimensions it becomes evident that the shear plane is tangent to an augerlike surface at the point where the principal stresses are displayed.

3.7   STRESS CONCENTRATION FACTORS, SAFETY FACTORS

Before discussing stress concentrations it is necessary to define *nominal stress* as the stress that acts across a surface at a point on that surface as calculated by elementary methods. Unfortunately, this definition is not entirely satisfactory because there is no uniform rule for selecting the so-called elementary method to be used for a particular object, i.e., beam, column, shaft, or more complex mechanical part. Examples of what are normally accepted as elementary methods are

$\sigma = Mc/I$ stress at distance c from the neutral axis due to bending,

$\tau = Tr/J$ stress at radius r in a shaft due to torsion, and

$\sigma = F/A$ stress due to pure tension or compression.

With this definition we can define the stress at a point in a mechanical part as the product of the nominal stress and the *stress concentration factor* at that point. Stated differently, the stress concentration factor is defined by

$$K = \frac{\text{actual stress at a point } P}{\text{nominal stress at a point } P}$$

Although stress concentration factors can also be caused by abrupt load discontinuities,[7] in the remainder of this section will be concerned only with geometrically induced stress concentrations.

Analytical methods from theoretical elasticity may be used to find stress concentration factors for simple geometries, such as that due to a small transverse hole in a plate whose lateral dimensions are large compared with the hole diameter. More complicated geometries require use of experimental methods involving strain gages, Moiré patterns, or photoelastic measurements, depending upon the size and location of the affected region. Finite element methods are not always reliable for finding stress concentration factors because of their dependence upon the size of the mesh selected and upon the details of the load distribution.

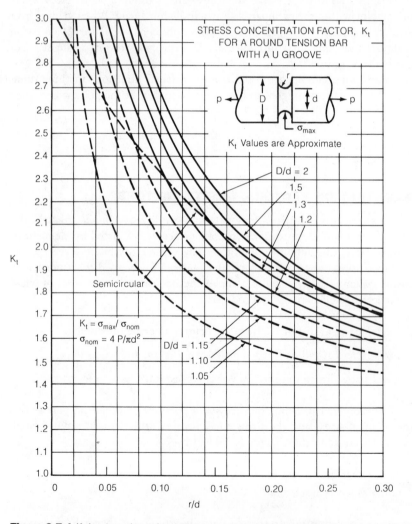

Figure 3.7.1 *K* due to a circumferential groove in a round bar in tension. *Source:* Stress Concentration Factors, *R.E. Peterson, copyright © 1974 John Wiley & Sons, Inc.*

Curves for stress concentration factors for 15 geometrical changes often encountered in machine design are given in Appendix A to this chapter.

EXAMPLE 3.7.1

Find the stress at a U-shaped groove in a tension–torsion bar whose outside diameter is 1.230 in. Groove depth is 0.115 in. and the bottom radius is 0.094 in. The axial load is 10 180 lb and the torque is 814 in.-lb.

If δ denotes the groove depth, then d in Figure A.8.3 is given by $d = D - 2\delta = 1.230 - 0.230 = 1.000$ in. The nominal tension and torsion stresses, therefore, are

$$\sigma_{nom} = F/(\pi r^2) = 10\,180/0.25\pi = 12\,961.58 \text{ psi}$$

$$\tau_{nom} = Tr/J = 2T/(\pi r^3) = 2(814)/(0.53\pi) = 4145.67 \text{ psi}$$

The actual tensile stress may be found from Figure 3.7.1, which is a copy of Figure A.3.8. Enter at $r/d = 0.094$, read the following D/d and K values,

D/d	K
1.15	2.250
1.20	2.395
1.30	2.532

and use orthogonal interpolation to obtain $K = 2.458$ at $D/d = 1.23$ as the stress concentration factor for tension. The expected tensile stress at the groove is given by

$$\sigma = 2.458(12\,961.58) = 31\,859.56 \text{ psi}$$

Torsional stress may be found by entering Figure 3.7.2, a copy of Figure 3.A.5, at $D/d = 1.23$ and reading the following r/d and K values

D/d	K
0.05	1.840
0.10	1.530
0.20	1.325

Upon using orthogonal interpolation we find that $K = 1.559$ at $r/d = 0.094$ so that the torsional stress becomes

$$\tau = 1.559(4145.67) = 6463.04 \text{ psi}$$

Safety Factors

Seldom can the design engineer be assured that the design loads will not be exceeded (that the machine will not be abused or misused), that all materials will be within specifications and free of hidden defects, and that all stress concentration factors are

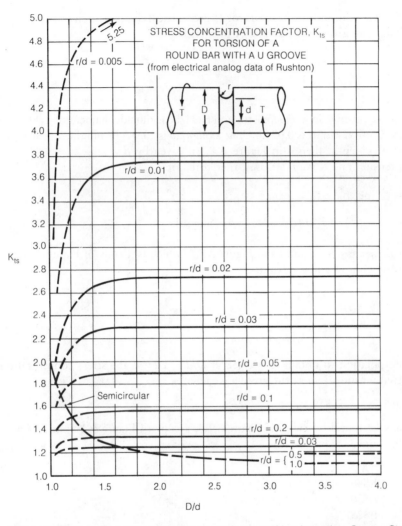

Figure 3.7.2 *K* due to a circumferential groove in a round bar in torsion. *Source:* Stress Concentration Factors, *R.E. Peterson, copyright © 1974 John Wiley & Sons, New York. Reprinted by permission of John Wiley & Sons, Inc.*

exactly correct. Multiplicative safety factors are introduced to attempt to artificially increase the design loads to compensate for these uncertainties. Since we are dealing with uncertainties there are obviously no formulas for selecting safety factors; they are based upon failure histories for those machine parts that have been in production on previous models of similar equipment and they are arbitrarily selected by the design engineer for newly designed equipment that has no service history. Typical guidelines may be as follows.

Safety Factor	Circumferences
8	Stress concentrations present and human life endangered by component failure.
6	Stress concentrations present, life not endangered, component failure causes expensive repairs and down time is also expensive.
5	Small or stress concentration factors present, human life may be endangered and/or repairs are expensive.
4	Small or stress concentrations, life not endangered, but there may be expensive repairs and costly down time.
1.5–3	Small or stress concentrations, life not endangered, repairs troublesome but not abnormally expensive.

These safety factors may vary from company to company and from industry to industry. It is important to emphasize that they are guidelines to be used only in the absence of more specific guidelines for the particular machines and industries. More closely selected safety factors, such as 3.5, are common in many cases. Integer values serve as general boundaries.

It is also important to decide whether the safety factors are selected relative to yield stress, as in the case of ductile materials and machine components, or relative to ultimate stress (ultimate strength). Yield stress usually is used where component deformation could cause a malfunction that could result in damage or loss of life.

Inclusion of a safety factor in the formulas for σ and τ in Example 3.7.1 yields for a safety factor $\zeta = 3.0$.

$$\sigma = K_t \zeta_t \frac{F}{\pi r^2} \qquad \tau = K_s \zeta_s \frac{Tr}{J}$$

$$\sigma = 95\,578.68 \text{ psi} \qquad \tau = 19\,389.12 \text{ psi}$$

where ζ_t and ζ_s denote the safety factors in tension and shear, respectively, and K_t and K_s denote the stress concentrations in tension and shear in that order. Usually the safety factors in tension and shear are equal, as in the above example. They may differ whenever the uncertainties in the respective stress concentration factors are different, if loading uncertainties differ, or if the consequences of the two failures differ.

3.8 STRAIN NOTATION

In this text we will use the dual index strain notation in which one index refers to the direction of the displacement involved and the other refers to the coordinate axis involved. In rectangular Cartesian coordinates strain components ε_{xx}, ε_{yy}, ε_{zz}, ε_{xy}, ε_{yz}, and ε_{zx} are given by

$$\varepsilon_{xx} = \frac{\partial u_x}{\partial x} \qquad \varepsilon_{yy} = \frac{\partial u_y}{\partial y} \qquad \varepsilon_{zz} = \frac{\partial u_z}{\partial z}$$

$$\varepsilon_{xy} = \frac{1}{2}\left(\frac{\partial u_x}{\partial y} + \frac{\partial u_y}{\partial x}\right) \qquad \varepsilon_{yz} = \frac{1}{2}\left(\frac{\partial u_y}{\partial z} + \frac{\partial u_z}{\partial y}\right) \qquad \varepsilon_{zx} = \frac{1}{2}\left(\frac{\partial u_z}{\partial x} + \frac{\partial u_x}{\partial z}\right)$$

(3.8.1)

where u_x, u_y, and u_z represent displacements in the x, y, and z directions, respectively. Derivation and physical justification of these relations will not be given here under the assumptions that were included in an earlier course in the strength of materials or an introduction to elasticity theory. In circular cylindrical coordinates these relations take on the form

$$\varepsilon_{rr} = \frac{\partial u_r}{\partial r} \qquad \varepsilon_{\theta\theta} = \frac{1}{r}\frac{\partial u_\theta}{\partial \theta} + \frac{u_r}{r} \qquad \varepsilon_{zz} = \frac{\partial u_z}{\partial z}$$

$$\varepsilon_{r\theta} = \frac{1}{2}\left(\frac{1}{r}\frac{\partial u_r}{\partial \theta} + \frac{\partial u_\theta}{\partial r} - \frac{u_\theta}{r}\right) \qquad \varepsilon_{\theta z} = \frac{1}{2}\left(\frac{\partial u_\theta}{\partial z} + \frac{1}{r}\frac{\partial u_z}{\partial \theta}\right) \qquad (3.8.2)$$

$$\varepsilon_{zr} = \frac{1}{2}\left(\frac{\partial u_z}{\partial r} + \frac{\partial u_r}{\partial z}\right)$$

As implied by their subscripts, u_r, u_θ, and u_z represent displacements in the r, θ, and z directions of a circular cylindrical coordinate system as shown in Figure 3.8.1. Recall that the phrase "displacement in the θ direction" means displacement in a direction tangent to the θ coordinate at the point of evaluation. Large deflections are not encountered in machine design because they may result in permanent deformation, thus rendering the machine useless in most cases. We therefore will not display the nonlinear, large deformation, strain–displacement relations.

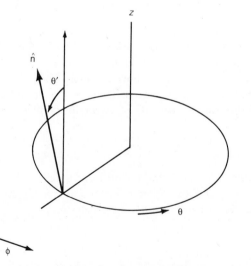

Figure 3.8.1 Angles θ' and ϕ relative to circular cylindrical coordinates. *Note:* Symbol θ is conventionally used for the polar angle in circular cylindrical coordinates and for the declination angle. Declination will be replaced by θ' when circular cylindrical coordinates are used.

3.9   STRESS–STRAIN RELATIONS

The stress–strain relations take on the same form for either rectangular Cartesian or circular cylindrical coordinates, and are displayed here for convenience, again under the assumption that they have been derived and discussed in an earlier course. The linear form of these relations in rectangular Cartesian coordinates is

$$\varepsilon_{xx} = \frac{\sigma_{xx}}{E} - \frac{v}{E}(\sigma_{yy} + \sigma_{zz}) \qquad \varepsilon_{yy} = \frac{\sigma_{yy}}{E} - \frac{v}{E}(\sigma_{zz} + \sigma_{xx})$$

$$\varepsilon_{zz} = \frac{\sigma_{zz}}{E} - \frac{v}{E}(\sigma_{xx} + \sigma_{yy}) \tag{3.9.1}$$

and

$$\varepsilon_{xy} = \frac{1+v}{E}\sigma_{xy} \qquad \varepsilon_{yz} = \frac{1+v}{E}\sigma_{yz} \qquad \varepsilon_{zx} = \frac{1+v}{E}\sigma_{yz} \tag{3.9.2}$$

The *shear modulus* G is defined by

$$G = \frac{E}{2(1+v)} \tag{3.9.3}$$

so that (3.9.2) may be rewritten in terms of G as

$$\varepsilon_{xy} = \sigma_{xy}/(2G) \qquad \varepsilon_{yz} = \sigma_{yz}/(2G) \qquad \varepsilon_{zx} = \sigma_{zx}/(2G) \tag{3.9.4}$$

Young's modulus is represented by E in the above equations and Poisson's ratio by v.

The stress–strain relations for circular cylindrical coordinates may be obtained from (3.9.1) and (3.9.2) by replacing x, y, and z by r, θ, and z, respectively.

Many older texts employ definitions of strain that omit the factor $\frac{1}{2}$ in equations 3.8.1 and 3.8.2 and compensate for this omission by omitting the 2 from equations 3.9.4. The above form has been adopted to conform to the form of the physical components of the strain tensor used in theoretical elasticity and continuum mechanics. Obviously, G was defined when the older strain–displacement relations were widely accepted in order to achieve similarity between 3.9.4 and 3.9.2.

Perhaps because of the practical difficulties in measuring lateral distortion before the 1970s, the terms involving Poisson's ratio were usually neglected in laboratory measurements of axial stress versus axial strain; that is, lateral thinning and necking were largely ignored. Consequently, the resulting data appeared as in Figure 3.9.1*a* for low-carbon steel and gave the erroneous implication, due to the negative slope between points 2 and 3, that the specimen lengthened and failed with decreased load. When this defect was recognized and the change in cross section was included the resulting plot of stress versus strain appeared as in Figure 3.9.1*b*. Not all plots show the cusp

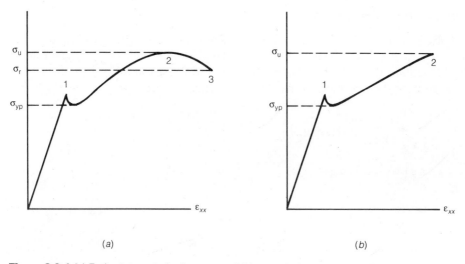

Figure 3.9.1 (a) Early stress–strain diagram, and (b) corrected stress–strain diagram. σ_u = ultimate strength, σ_r = rupture strength, σ_{yp} = yield limit.

at point 1 in both figures because it is difficult to observe without careful, computer controlled data collection.

The stress calculated by including the lateral reduction in area was known as the "true stress" and the earlier stress was known as "engineering stress." It is the author's opinion that such terminology should be discouraged because it implies that engineering is synonymous with "sloppy" or "unthinking."

When tensile failure finally occurs in ductile materials that neck (reduce their cross section) before failure, the stress field in the necked region is no longer simple tension because of the changed geometry. Precise measurement of the internal stresses in that portion of the specimen is difficult because the shape changes with the load. An X-ray photograph of a necked region taken by MacGregor[11] and reproduced in Figure 3.9.2 shows tension failure initiated internally and that final failure is due to shear in the outer ring surrounding the internal void. The significance of this failure mechanism is that the terminal portions of the curves in Figures 3.9.1a and b probably do not correctly represent the behavior of the material in pure tension.

SAE Information Report J1099 distinguishes between these two curves by defining the *ultimate strength*, which will be designated the *ultimate stress* in the remainder of this text, as the maximum load divided by the original cross-sectional area, and the *true fracture strength* (which is a stress) as the load at failure divided by the minimum cross-sectional area after failure with correction for triaxial stress due to necking. A general correction method is not defined.[9]

Aluminum, copper, and other materials display a stress–strain curve similar to that shown in Figure 3.9.3a. In this type of curve the stress at the upper limit of the initially linear portion is termed the *proportional limit*. A slightly larger stress, by some arbitrary amount, may be termed the *elastic limit*. This amount is often taken to be

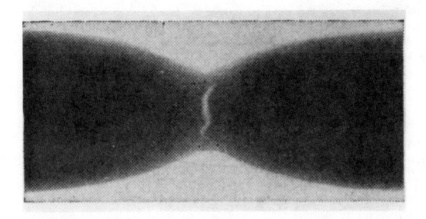

Figure 3.9.2 First internal crack in a round aluminum bar. *Source: Reprinted with permission from "Relations between stress and reduction in area for tensile tests of metals" by C.W. MacGregor, The Metallurgical Society, 420 Commonwealth Drive, Warrendale, Pennsylvania 15086, 1937.*

the stress corresponding to a strain of 0.002 or 0.0002, depending upon the reference cited. The stress at this point may be known as the *yield stress*. These values obviously coincide in materials whose stress–strain curve is similar to that shown in Figure 3.10.1, where they are usually defined at the base of the cusp instead of at its tip.

Design approximations have been made in the past by replacing the curve in Figure 3.9.3*a* by that shown in Figure 3.9.3*b*. In this simplified form the stress at point 1 is known as the yield stress and is denoted by σ_{yp}. That portion of the stress–strain

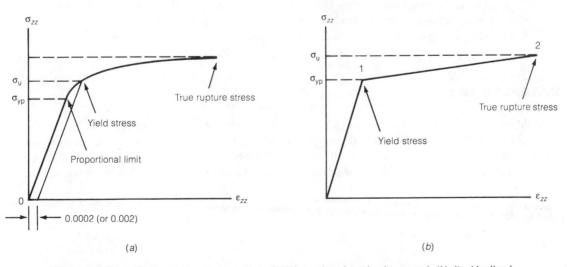

(a)   (b)

Figure 3.9.3 (a) Stress–strain diagram in uniaxial tension for aluminum and (b) its idealized representation.

profile between points 1 and 2 is sometimes termed the *work-hardening* region if the slope is positive, as shown.* If the slope is not positive that portion is known as the *yield region.* Although ultimate failure is assured for loads that produce stresses in the yield region, the stress–strain curve is incapable of indicating the rate of flow and the elapsed time to failure. This is the subject of courses in viscoelasticity and in non-linear elasticity which are beyond the scope of this text.

3.10 THERMAL STRESSES

Temperature effects may be added to equations 3.9.1 to obtain

$$\varepsilon_{xx} = \frac{\sigma_{xx}}{E} - \frac{v}{E}\left(\sigma_{yy} + \sigma_{zz}\right) + \alpha(\lambda - \lambda_0)$$

$$\varepsilon_{yy} = \frac{\sigma_{yy}}{E} - \frac{v}{E}\left(\sigma_{zz} + \sigma_{xx}\right) + \alpha(\lambda - \lambda_0) \qquad (3.10.1)$$

$$\varepsilon_{zz} = \frac{\sigma_{zz}}{E} - \frac{v}{E}\left(\sigma_{xx} + \sigma_{yy}\right) + \alpha(\lambda - \lambda_0)$$

in which α is the *linear coefficient of thermal expansion*, often called the *coefficient of thermal expansion* (the linear coefficient of thermal expansion is a different number from the volumetric coefficient of thermal expansion). Present temperature is represented by λ and the reference temperature by λ_0. No additions are made to equations 3.9.2 for shear strains.

Thermal stresses occur in a variety of situations, such as when a portion of a part is at a temperature different from the remainder of the part, or when different parts of a machine have different coefficients of expansion.

EXAMPLE 3.10.1

Find the stress in a 1-in.-diameter shaft for a roller for hot material on a conveyor. It has zero axial clearance at 70 °F. Assume the frame to be rigid, and that the temperature distribution is as shown in Figure 3.10.1. $E = 30 \times 10^6$ psi, $\alpha = 8 \times 10^{-6}/°F$, $v = 0.30$. Neglect the lateral constraints of the bearings

From 3.10.1 written in circular cylindrical coordinates and the condition that the frame is rigid

$$\varepsilon_{zz} = \frac{\sigma_{zz}}{E} - \frac{v}{E}\left(\sigma_{rr} + \sigma_{\theta\theta}\right) + \alpha(\lambda - \lambda_0) = 0.$$

---

* Prager, W., and Hodge, P.G., *Theory of Perfectly Plastic Solids*, John Wiley & Sons, Inc., New York 951, p. 2.

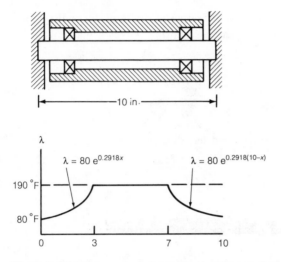

Figure 3.10.1 Temperature distribution for Example 3.10.1 for roller shown.

Because there are no lateral constraints, $\sigma_{rr} = \sigma_{\theta\theta} = 0$, hence

$$u_z = \int_0^{10} \varepsilon_{zz}\, dz = \int_0^{10} \left[\frac{\varepsilon_{zz}}{E} + \alpha(\lambda - \lambda_0) \right] dz = 0 \qquad (3.10.2)$$

so that integration of (3.10.2) with respect to z yields

$$10\frac{\sigma_{zz}}{E} = 10\alpha\lambda_0 - 80\alpha \int_0^3 e^{0.2918z}\, dz - \alpha(190 - 70)4$$
$$- 80\alpha \int_7^{10} e^{0.2918(10-z)}\, dz$$

where $\lambda_0 = 70\,°F$ because this is the reference temperature at which there is zero clearance. Thus

$$10\frac{\sigma_{zz}}{E} = 10(8 \times 10^{-6})70 - \frac{80(8 \times 10^{-6})}{0.2918} e^{0.2918z} \Big|_0^3$$
$$- 480(8 \times 10^{-6}) + e^{2.918}\frac{80(8 \times 10^{-6})}{0.2918} e^{-0.2918z} \Big|_7^{10}$$

Upon evaluation and substitution for $E = 30 \times 10^6$, the result is that

$$\sigma_{zz} = -13141 \text{ psi}$$

where the negative sign indicates compression. This is indicative of the large thermal stresses that can occur in machinery that is not properly designed for the temperature ranges it may encounter.

3.11 BEAM DEFLECTION

Deflection of straight beams whose unsupported length/depth ratio is greater than 8 may be found from the Bernoulli-Euler relation

$$\frac{1}{\rho} = \frac{M}{EI}$$

in which ρ is the radius of curvature of the bent neutral axis. Replacement of $1/\rho$ according to

$$\frac{1}{\rho} = \frac{d^2y/dx^2}{[1 + (dy/dx)^2]}$$

and use of the approximation that

$$\frac{1}{\rho} \cong \frac{d^2y}{dx^2} \qquad \text{when} \quad \left(\frac{dy}{dx}\right) \ll 1$$

produces

$$\frac{d^2y}{dx^2} = -\frac{M(x)}{EI} \tag{3.11.1}$$

In this equation $y(x)$ is positive downward, as illustrated in Figure 3.11.1, I represents the area moment of inertia, and E denotes the modulus of elasticity. $M(x)$ is the moment that would be required to act on an imaginary cross section at x to maintain equilibrium if the beam were cut along that transverse plane, and all loads and reactions remained unchanged. The algebraic sign of M is dependent upon the positive directions chosen for y and x relative to the applied loads. If they are selected as in Figure 3.11.1, $M(x)$ is positive when it tends to bend the segment on which it acts into a curve that is concave downward.

Alternatively, the algebraic sign for $M(x)$ may be determined by first defining a right-hand coordinate system in three dimensions from the x and y axes implied by equation (3.11.1). In it, $\hat{i} \times \hat{j} = \hat{k}$ and the sign of **M** must be selected to satisfy the vector relation

$$(\hat{\mathbf{M}} \cdot \hat{\mathbf{k}})(\hat{\mathbf{n}} \cdot \hat{\mathbf{i}}) = -1, \qquad \hat{\mathbf{M}} = \mathbf{M}/|\mathbf{M}|$$

where $M(x)$ is determined by the right-hand rule and n is the outward normal to the

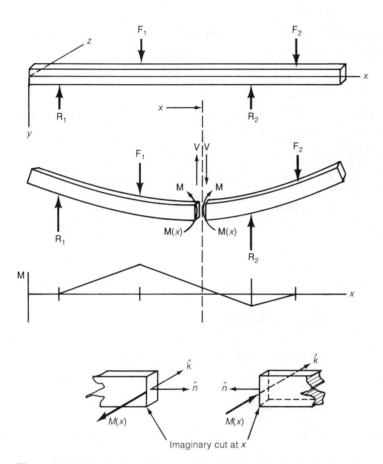

Figure 3.11.1 Illustration of the sign convention used to determine the algebraic sign of the bending moment $M(x)$ acting at x to maintain equilibrium.

cross section upon which $M(x)$ acts. (According to this convention the moment on the left end of the beam segment shown at the lower left of Figure 3.11.1 is positive relative to the right-hand xyz-coordinate system shown at the upper left of that figure.) Shear force V is positive in the direction that will satisfy the relation $V = dM/dx$.

Although (3.11.1) may be solved numerically, it is usually easier to solve it by traditional methods, such as the unit load method[10] or the conjugate beam method. Either method agrees satisfactorily with actual deflection for elastic beams with length/depth ratio equal to or greater than 8 (greater than 15 for an I beam if the web is relatively thin).

In the remainder of this text we will employ only the unit load method because of the ease with which it can be used with a numerical integration routine. It will be further described and demonstrated in Chapter 8, where it will be used to find shaft deflections.

Deflection of initially curved beams may be found from the Bernoulli-Euler relation by replacing ρ by $\rho = l/\theta$ to obtain

$$\theta = Ml/El \tag{3.11.2}$$

As implied by this equation, the deflection is measured by the change in angle between tangents to the neutral axis.

Restricting the application of the formulas collected in panels 1 through 14 of Figure 3.11.2 to the span/depth ratio ranges listed above is motivated by the observation that shear deflection, which becomes important for smaller span/depth ratios, has been neglected in panels 1 through 14.

Beam deflection due to shear stresses may be included by adding the shear deflection to that calculated due to bending to obtain the total deflection y_t as given by

$$y_t = \left[1 + C\frac{1 + v}{\left(\dfrac{l}{h}\right)^2} \right] y \tag{3.11.3}$$

where y represents the deflection due to bending, l the length of the beam, h the depth, and v Poisson's ratio. Constant C is determined by the cross section of the beam and its support conditional. For example, for an end-loaded cantilever

$$
\begin{aligned}
C &= 0.600 && \text{rectangular cross section} \\
&= 0.417 && \text{circular cross section}
\end{aligned}
$$

and for a simple supported beam with a concentrated load at the midpoint,

$$
\begin{aligned}
C &= 2.400 && \text{rectangular cross section} \\
&= 1.667 && \text{circular cross section}
\end{aligned}
$$

With this these values we may extend the displacements given in panels 5 and 6 to stub beams, with the caveat that further correction may be needed for span/depth ratios less than 1.00.

Deflection due to shear for other cross sections and other end conditions may be estimated from the unit load method for shear, which is that

$$\delta = F \int_0^l \frac{Vv}{AG}\, dx \tag{3.11.4}$$

where $F = \frac{6}{5}$ for rectangular cross sections and $\frac{10}{9}$ for circular cross sections. Parameter v is the shear due to a unit load at the location where the deflection is desired and V is the shear due to the applied loads and their reactions. As indicated by the above notation, integration is over the length of the beam.

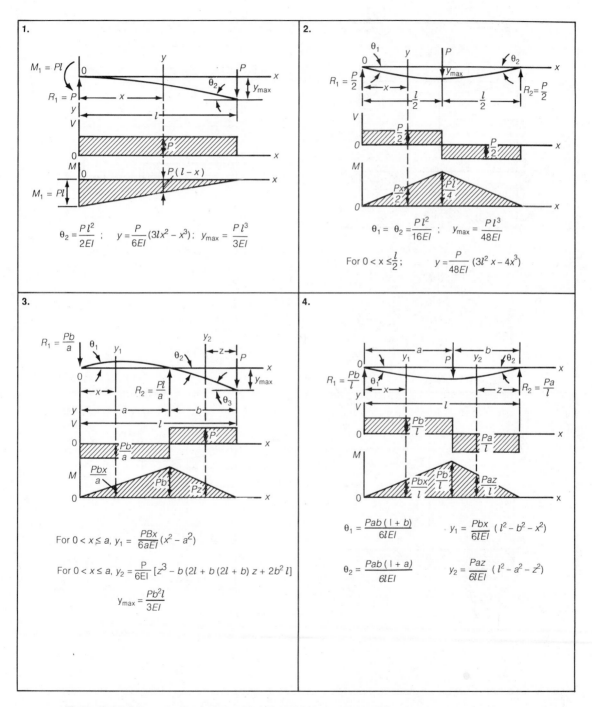

Figure 3.11.2 Formulas for beam deflections and slopes, Panels 1–4.

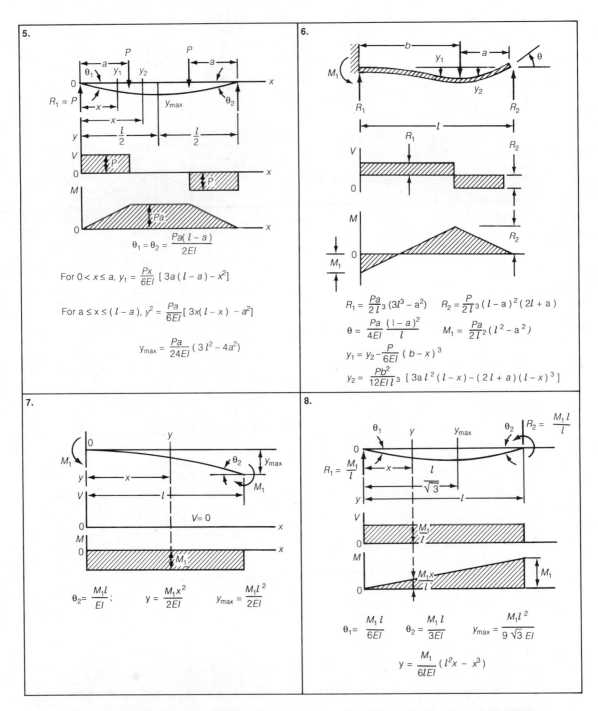

5.

$$\theta_1 = \theta_2 = \frac{Pa(\,l-a\,)}{2EI}$$

For $0 < x \le a$, $y_1 = \frac{Px}{6EI}\,[\,3a\,(\,l-a\,)-x^2\,]$

For $a \le x \le (\,l-a\,)$, $y_2 = \frac{Pa}{6EI}[\,3x(\,l-x\,)\,-a^2\,]$

$$y_{max} = \frac{Pa}{24EI}\,(\,3\,l^2-4a^2\,)$$

6.

$$R_1 = \frac{Pa}{2\,l^3}\,(3\,l^3-a^2)\qquad R_2 = \frac{P}{2\,l^3}\,(\,l-a\,)^2\,(\,2\,l+a\,)$$

$$\theta = \frac{Pa}{4EI}\,\frac{(\,l-a\,)^2}{l}\qquad M_1 = \frac{Pa}{2\,l^2}\,(\,l^2-a^2\,)$$

$$y_1 = y_2 - \frac{P}{6EI}\,(\,b-x\,)^3$$

$$y_2 = \frac{Pb^2}{12EIl^3}\,[\,3a\,l^2\,(\,l-x\,)-(\,2\,l+a\,)\,(\,l-x\,)^3\,]$$

7.

$$\theta_2 = \frac{M_1\,l}{EI};\qquad y = \frac{M_1 x^2}{2EI}\qquad y_{max} = \frac{M_1 l^2}{2EI}$$

8.

$$R_2 = \frac{M_1\,l}{l}$$

$$\theta_1 = \frac{M_1\,l}{6EI}\qquad \theta_2 = \frac{M_1\,l}{3EI}\qquad y_{max} = \frac{M_1 l^2}{9\sqrt{3}\,EI}$$

$$y = \frac{M_1}{6lEI}\,(\,l^2 x - x^3\,)$$

Figure 3.11.2 *Continued*, Panels 5–8.

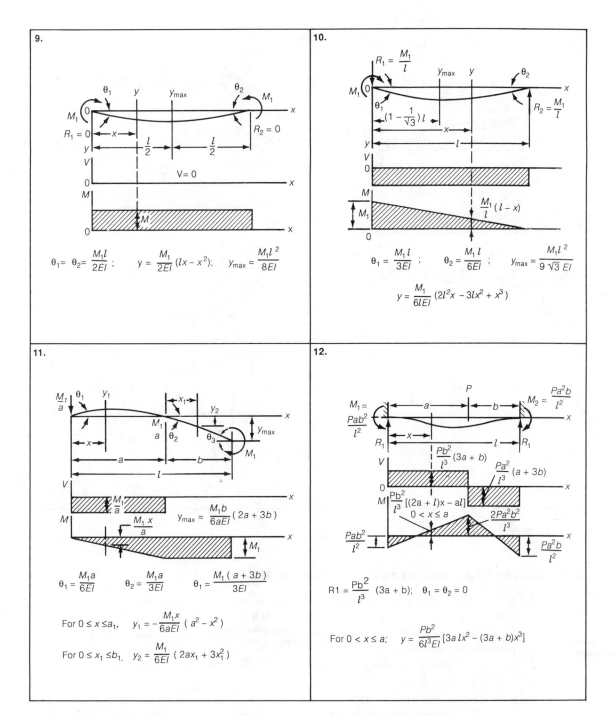

9.

$$\theta_1 = \theta_2 = \frac{M_1 l}{2EI} \; ; \qquad y = \frac{M_1}{2EI}(lx - x^2); \qquad y_{max} = \frac{M_1 l^2}{8EI}$$

10.

$$\theta_1 = \frac{M_1 l}{3EI} \; ; \qquad \theta_2 = \frac{M_1 l}{6EI} \; ; \qquad y_{max} = \frac{M_1 l^2}{9\sqrt{3}\,EI}$$

$$y = \frac{M_1}{6lEI}(2l^2 x - 3lx^2 + x^3)$$

11.

$$\theta_1 = \frac{M_1 a}{6EI} \qquad \theta_2 = \frac{M_1 a}{3EI} \qquad \theta_1 = \frac{M_1(a + 3b)}{3EI}$$

$$y_{max} = \frac{M_1 b}{6aEI}(2a + 3b)$$

For $0 \le x \le a_1, \quad y_1 = -\frac{M_1 x}{6aEI}(a^2 - x^2)$

For $0 \le x_1 \le b_1, \quad y_2 = \frac{M_1}{6EI}(2ax_1 + 3x_1^2)$

12.

$$R1 = \frac{Pb^2}{l^3}(3a + b); \quad \theta_1 = \theta_2 = 0$$

For $0 < x \le a; \quad y = \frac{Pb^2}{6l^3 EI}[3a\,lx^2 - (3a + b)x^3]$

Figure 3.11.2 *Continued,* Panels 9–12.

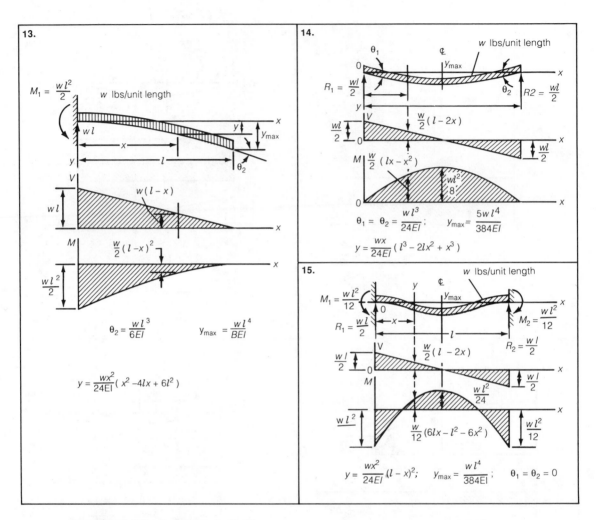

Figure 3.11.2 *Continued*, Panels 13–15.

3.12 DEFLECTIONS AND ELASTIC ENERGY METHODS

Deflections within the elastic range may be used as an aid in calculating the forces involved in simple indeterminate structures (only one indeterminant support) because the principal of superposition is valid within this range. Suppose, for example, that we wish to find the force R_0 on a rigid support at A when the shaft is rigidly supported at each end and loaded as shown in Figure 3.12.1a. This may be accomplished in a two-step process.

Step 1. Remove the support at A and calculate the deflection at A due to F_1 where R'_1 and R'_2 are selected to hold the beam in equilibrium. Call this deflection δ_1, Figure 3.12.1b.

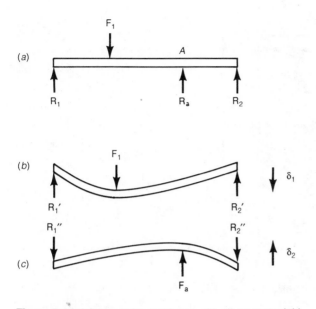

Figure 3.12.1 Singly redundant beam with all supports rigid.

Step 2. Remove all loads and calculate the deflection at A due to force F_a at point A where R_1'' and R_2'' are again selected to hold the beam in equilibrium. Call this deflection δ_2, Figure 3.12.1c.

Step 3. Calculate the required force R_a at A from

$$R_a = (\delta_1/\delta_2)F_a \tag{3.12.1}$$

The reasoning behind this process is that the final deflection of an elastic body at any point within that body is the sum of the deflections caused at that point by each of the individual loads and the reaction at the redundant support, regardless of their order of application. Hence, we can consider the deflection of the support point as the sum of the deflection due to the loads and the deflection due to the supporting force. If the support is rigid this sum must be equal to the deflection allowed by the support, which is zero in the case shown in Figure 3.12.1. One multiplication step may be saved by letting F_a be a unit force.

The process is somewhat more complicated if the support is flexible enough for its deflection to be included in the calculation. Although we will again use the principle of superposition in this situation, we will add the notion of a *spring constant k*, where

$$k = \frac{\text{applied force}}{\text{deflection}}$$

Let us again consider a beam with a single load and three supports, as illustrated in Figure 3.12.2a. With the support at A removed the deflection of the beam under its

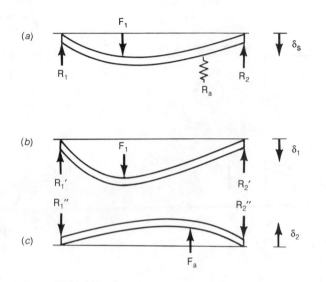

Figure 3.12.2 Singly redundant beam with a flexible redundant support.

applied load may be that shown in Figure 3.12.2b, where the deflection at A is again represented by δ_1. Again remove all loads from the shaft, apply F_a at point A, calculate the deflection of point A, and again let it be represented δ_2, as shown in Figure 13.2.2c. Since the supporting structure is elastic it will resist deflection at A with a force proportional to its actual deflection δ_s. Hence, the support force is $k\delta_s$.

Following the same reasoning as for the rigid support, the force required to raise the shaft at A from deflection δ_1 to δ_s may be found to be $F_a(\delta_1 - \delta_s)/\delta_2$. Hence,

$$F_a \frac{\delta_1 - \delta_s}{\delta_2} = k\delta_s \qquad (3.12.2)$$

which may be solved for δ_s to obtain

$$\delta_s\left(k + \frac{F_a}{\delta_2}\right) = F_a \frac{\delta_1}{\delta_2} \qquad (3.12.3)$$

Observe that equation 3.12.2 and hence equation 3.12.3 reduce to equation 3.12.1 if we let k increase and δ_s decrease such that $k\delta_s = R_a$ is always true. Under these conditions (3.12.2) approaches (3.12.1) in the limit as δ_s goes to zero.

More complicated situations may be analyzed by the *elastic energy* method, which is also known as the *unit load* method. It appears to have been first described in the United States by van den Broek in Ref. 9 in which he showed it to include and be more general than Castigliano's method, although both are derived from the energy of deformation.

According to the elastic energy method the lateral deflection at an arbitrary P along the length of a beam whose neutral axis is coincident with the x axis is given by

$$\delta = \int \frac{mM}{EI}\, dx \qquad\qquad (3.12.4)$$

where m denotes the moment due to a unit force at P in the direction of δ, M the moment due to all of the actual forces and reactions acting on the beam, and E and I the elastic modulus and the cross-sectional moment of area, respectively. As we find from the following example, this method is superior to the moment area method because deflection is not measured from a tangent.

To demonstrate the elastic energy method, consider the deflection of a simply supported beam at the point of application of a force F acting at distance a from the left end of the beam as in Figure 3.12.3a. Imagine force f acting at A as shown in Figure 3.12.3b with both R_1' and R_2' chosen accordingly. The moment due to f will be

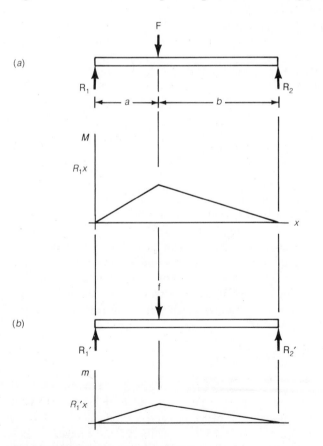

Figure 3.12.3 Simply supported beam.

given by

$$m(x) = \left(1 - \frac{a}{L}\right)fx \qquad \text{for } 0 \leq x \leq a$$

$$= \left(1 - \frac{x}{L}\right)fa \qquad \text{for } a \leq x \leq L$$

Since F also acts at a distance a from the left end we may easily write that the moment due to actual force F is given by

$$M(x) = \left(1 - \frac{a}{L}\right)Fx \qquad \text{for } 0 \leq x \leq a$$

$$= \left(1 - \frac{a}{L}\right)Fa \qquad \text{for } a \leq x \leq L$$

Substitution into equation 3.12.4 and setting $f = 1$ yields

$$\frac{EI}{F}\delta = \int_0^a \left(1 - \frac{a}{L}\right)^2 x^2\, dx + \int_a^L \frac{a^2}{L^2}(L - x)^2\, dx$$

Integration yields

$$\delta = \frac{Fa^2}{3EI}\frac{(L - a)^2}{L}$$

which agrees with the earlier formula in panel 7 of Figure 3.11.2.

Although the method may also be used to find redundant forces in statically indeterminant structures, such as shown in Figures 3.12.1 and 3.12.2, it is not now practical in these load arrangements because formulas for them have already been derived.

It is, however, an excellent method for finding forces and stresses in more complicated structures, such as a truss reinforced beam as shown in Figure 3.12.4, and it may be straightforwardly programmed. We will use this truss-beam combination to demonstrate the method in the following example, which was taken from Ref. 10.

EXAMPLE 3.12.1

Find the maximum uniformly distributed load w between points A and B of Figure 3.12.4a that may be supported by a beam having a cross-sectional area of 84 in.^{2} and a moment of area of 1372 in.^{4}. Truss members 1, 4, and 5 have a cross-sectional area of 0.25 in.^{2}, members 2 and 3 have a cross-sectional area of 2.0 in.^{2}. Young's modulus for the beam is 1 500 000 psi, and its maximum stress in tension is

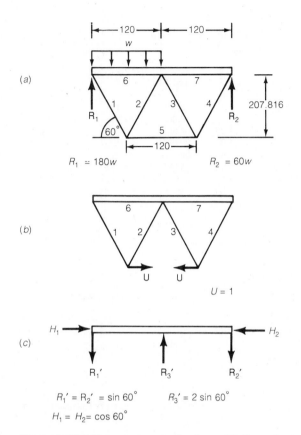

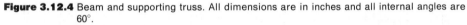

Figure 3.12.4 Beam and supporting truss. All dimensions are in inches and all internal angles are 60°.

4500 psi and in compression is 13 000 psi. Young's modulus for all of the rods is 30 000 000 psi and their yield stress is 60 000 psi. Use a safety factor of 1.5. Other dimensions are shown in Figure 3.12.4.

According to the elastic energy method, the elastic energy for the rods is given by CfS for each rod, where $C = L/(AE)$ in terms of the rod length L, rod cross-sectional area A, and Young's modulus E (elastic modulus). The force in each member due to a unit force in either it or another member is denoted by f and the force in that member due to the actual load is denoted by S. Since the beam must support axial compressive loads its elastic energy includes a contribution from both the axial loads as well as the bending loads.

By replacing one of the bars with two equal but opposite unit loads acting as shown in Figure 3.12.4*b* the displacement will be zero. Hence,

$$\sum_{i=1}^{7} C_i f_i S_i + \int \frac{mM}{EI}\, dx = 0 \tag{3.12.5}$$

From the figure and problem statement we may calculate the values shown in Table 3.12.1.

TABLE 3.12.1   VALUES OF CALCULATED QUANTITIES FOR EXAMPLE 3.12.1

Bar number	f (lb)	A (in.^{2})	S	c_i 10^{-6}	$c_i f_i S_i$ 10^{-7}
1	1.0	0.25	$1.0S_e$	32.000	320.000
2	−1.0	2.00	$-1.0S_e$	4.000	40.000
3	−1.0	2.00	$-1.0S_e$	4.000	40.000
4	1.0	0.25	$1.0S_e$	32.000	320.000
5	1.0	0.25	$1.0S_e$	32.000	320.000
6	−0.5	15.38	$-0.5S_e$	1.040	26.006
7	−0.5	15.38	$-0.5S_e$	1.040	26.006

$E = 30 \times 10^6$ psi for steel truss members

$\quad = 1.5 \times 10^6$ psi for the beam

All truss members are 240 in. long and pin connected.

Turning again to Figure 3.12.4 we may write the moment acting on the beam due to the actual loads in terms of S_e as

$$M(x) = (180w - 0.86603S_e)x - wx^2/2 \qquad \text{for } 0 \leq x \leq 240 \text{ in.}$$

$$M(x) = 28\,800w - 415.692S_e + (0.86603S_e - 60w)x \qquad \text{for } 240 \leq x \leq 480 \text{ in.}$$

The moment induced by the unit loads acting in place for bar 5 when the actual loads are removed are given by

$$m(x) = -0.86603x \qquad \text{for } 0 \leq x \leq 240 \text{ in.}$$
$$\quad\;\; = -207.846 + 0.86603x \qquad \text{for } 240 \leq x \leq 480 \text{ in.}$$

Substitution into the integral in (3.12.5) yields

$$\int \frac{mM}{EI}\,dx = -a \int_0^{240} (bx - cx^2)\,dx + d \int_{240}^{480} (-e + fx)(g + hx)\,dx$$

$$(3.12.6)$$

where

$$a = 0.86603/EI \qquad\qquad e = 415.692$$

$$b = 180w - 0.86603S_e \qquad f = 0.86603$$

$$c = w/2 \qquad\qquad\qquad g = 28\,800w - 415.692S_e$$

$$d = 1/EI \qquad\qquad\qquad h = 0.86603S_e - 60w$$

After performing the indicated integration and substitution into (3.12.5) we find from

$$0.1176w = 0.0015815S_e$$

that

$$S_e = 74.347w$$

Recall that the maximum moment occurs where the shear vanishes, so that at upon substituting for S_e in

$$180w - 0.86603S_e - wx = 0$$

we find that

$$(180 - 64.213 - x)w = 0$$

which is satisfied by

$$x = 115.787 \text{ in.}$$

Substitution of this value for x into the expression for $M(x)$ between 0 and 240 in. yields

$$M_{max} = 6703.311w$$

Since the pipe is in axial compression due to the reactions at the truss connections, the maximum tensile stress is given by

$$\sigma = -\frac{P}{A} + \frac{Mc}{I} = -\frac{0.5(74.347)}{15.381}w + \frac{6703.311)4(7.376)}{\pi(7.376^4 - 5.901)^4}w = 33.610w$$

Using the safety factor of 1.5 to reduce the maximum tensile stress from 4500 to 3000 psi we find that

$$w = 89.258 \text{ psi}$$

For this loading the maximum stress in the truss will be in those members with a cross section of 0.25 in.^{2} where

$$\sigma = 74.347w/0.25 = 26\,544 \text{ psi}$$

which is well below the maximum stress of $60\,000/1.5 = 40\,000$ psi.

APPENDIX A
Stress Concentration Factor Graphs

The following 16 graphs present the stress concentration factors associated with the cross-sectional area changes shown by a small sketch within each graph. All of these graphs, with the exception of Figure 3.A.7, have been reproduced from Ref. 11, which was compiled while its author, Dr. Peterson, was employed at the Westinghouse Research Laboratory. As mentioned in the section on stress concentration factors, these factors were generally determined experimentally for all but those geometries that could be described analytically. Experimental results are still a major contributor to a knowledge of stress concentration factors, although finite element methods may also be used to find these factors if skillfully applied, if all load distributions are realistic, and if a sufficiently small mesh is used in the vicinity of the stress concentration.

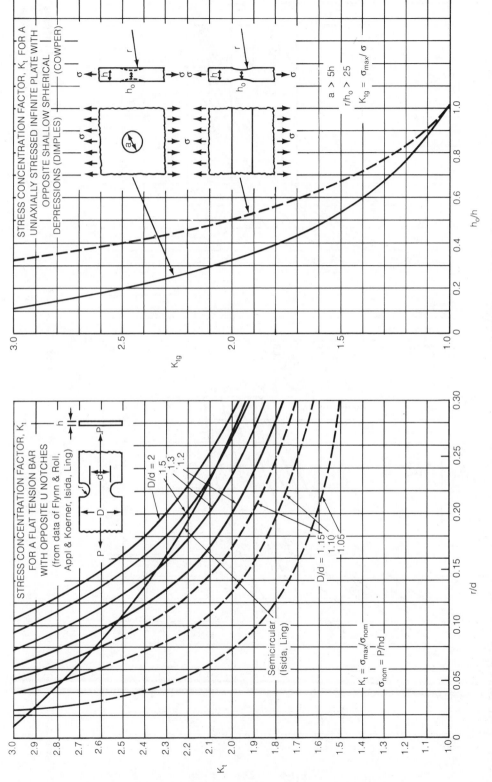

Figure 3.A.1 K for a flat tension bar with opposite U-shaped notches. *Source: Stress Concentration Factors, R. E. Peterson, copyright © 1974 John Wiley & Sons, New York. Reprinted by permission of John Wiley & Sons, Inc.*

Figure 3.A.2 K for a flat plate in tension with spherical depressions and troughs of circular section. *Source: Stress Concentration Factors, R. E. Peterson, copyright © 1974 John Wiley & Sons, New York. Reprinted by permission of John Wiley & Sons, Inc.*

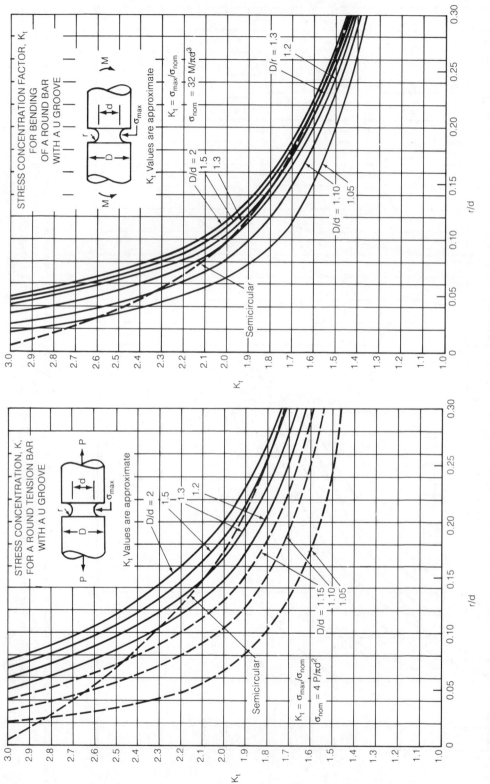

Figure 3.A.3 K due to a circumferential groove in a round bar in tension. *Source: Stress Concentration Factors, R.E. Peterson, copyright © 1974 John Wiley & Sons, New York. Reprinted by permission of John Wiley & Sons, Inc.*

STRESS CONCENTRATION, K_t FOR A ROUND TENSION BAR WITH A U GROOVE

K_t Values are approximate

$K_t = \sigma_{max}/\sigma_{nom}$

$\sigma_{nom} = 4P/\pi d^2$

Semicircular

D/d = 2
1.5
1.3
1.2

D/d = 1.15
1.10
1.05

Figure 3.A.4 K due to a circumferential groove in a rounded bar in bending *Source: Stress Concentration Factors, R.E. Peterson, copyright © 1974 John Wiley & Sons, New York. Reprinted by permission of John Wiley & Sons, Inc.*

STRESS CONCENTRATION FACTOR, K_t FOR BENDING OF A ROUND BAR WITH A U GROOVE

K_t Values are approximate

$K_t = \sigma_{max}/\sigma_{nom}$

$\sigma_{nom} = 32M/\pi d^3$

D/d = 2
1.5
1.3

D/r = 1.3
1.2

D/d = 1.10
1.05

Semicircular

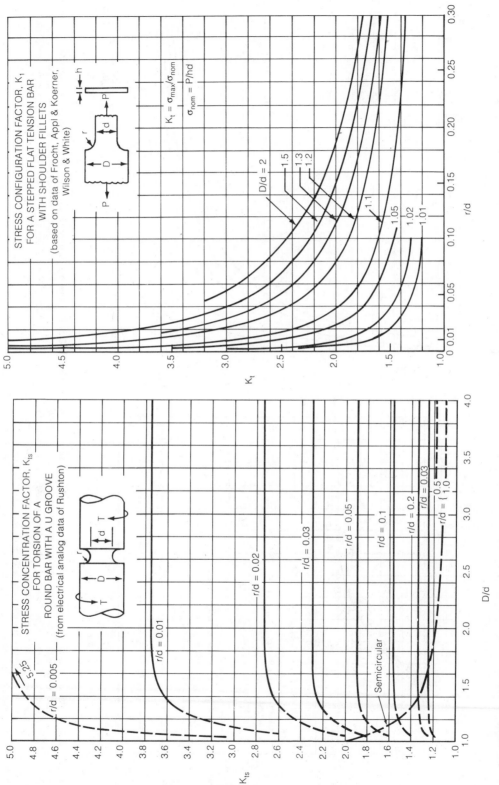

Figure 3.A.5 K due to a circumferential groove in a round bar in torsion. *Source: Stress Concentration Factors, R. E. Peterson, copyright © 1974 John Wiley & Sons, New York. Reprinted by permission of John Wiley & Sons, Inc.*

Figure 3.A.6 K due to a step with shoulder fillets in a flat bar in tension. *Source: Stress Concentration Factors, R. E. Peterson, copyright © 1974 John Wiley & Sons, New York. Reprinted by permission of John Wiley & Sons, Inc.*

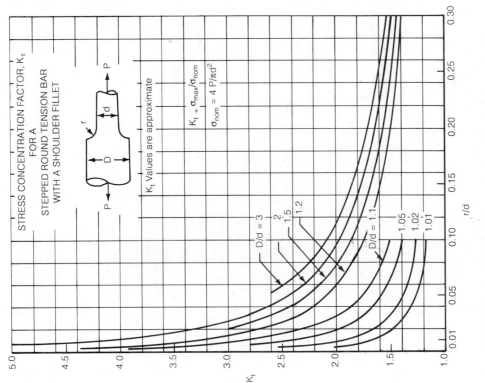

Figure 3.A.8 K due to a step with shoulder fillets in a round tension bar. *Source: Stress Concentration Factors, R.E. Peterson, copyright © 1974 John Wiley & Sons, New York. Reprinted by permission of John Wiley & Sons, Inc.*

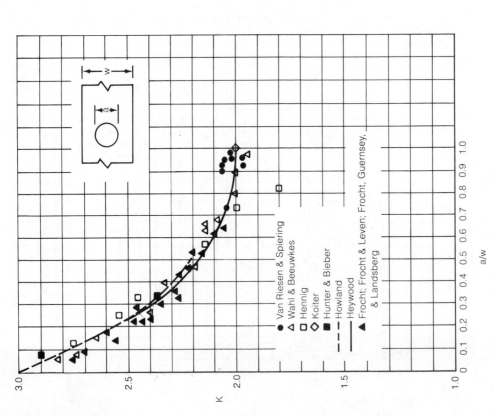

Figure 3.A.7 K for a circular hole in a plate in tension. Note that K approaches 2.00 as hole diameter approaches plate width. *Source: W.J. Bell and P.P. Benhan, The Effect of Mean Stress on Fatigue Strength of Plain and Notched Stainless Steel Sheet in the Range from 10 to 10⁷ Cycles. American Society for Testing and Materials, STP 338, 25–46, 1963. Copyright ASTM. Reprinted with permission.*

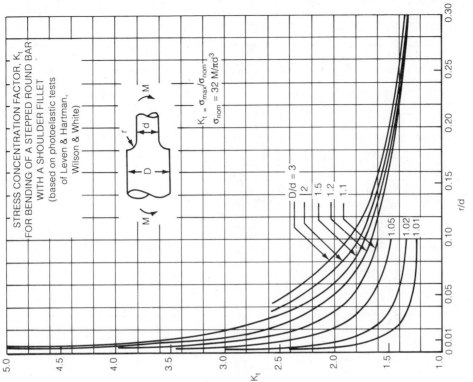

Figure 3.A.9 K due to a step with shoulder fillets in a flat bar in bending. $\sigma_{nom} = 6M/(hd^2)$. *Source: Stress Concentration Factors, R.E. Peterson, copyright © 1974 John Wiley & Sons, New York. Reprinted by permission of John Wiley & Sons, Inc.*

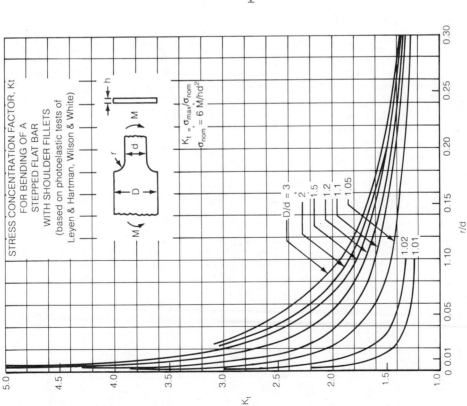

Figure 3.A.10 K due to a step with shoulder fillet in a round bar in bending. *Source: Stress Concentration Factors, R. E. Peterson, copyright © 1974 John Wiley & Sons, New York. Reprinted by permission of John Wiley & Sons, Inc.*

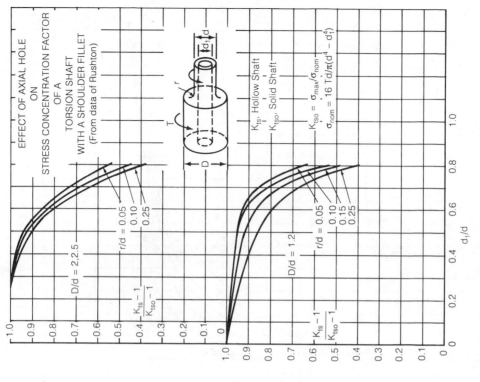

Figure 3.A.11 K at a shoulder with fillet in a round bar in tension. *Source:* Stress Concentration Factors, R. E. Peterson, copyright © 1974 John Wiley & Sons, New York. Reprinted by permission of John Wiley & Sons, Inc.

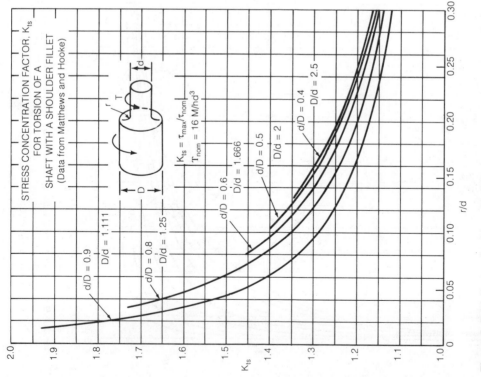

Figure 3.A.12 Effect of an axial hole on K from Figure 3.19. *Source:* Stress Concentration Factors, R. E. Peterson, copyright © 1974 John Wiley & Sons, New York. Reprinted by permission of John Wiley & Sons, Inc.

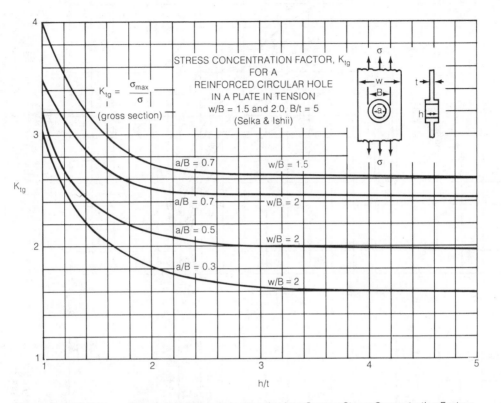

Figure 3.A.13 *K* for a reinforced circular hole in a tension bar. *Source:* Stress Concentration Factors, *R.E. Peterson, copyright © 1974 John Wiley & Sons, New York. Reprinted by permission of John Wiley & Sons, Inc.*

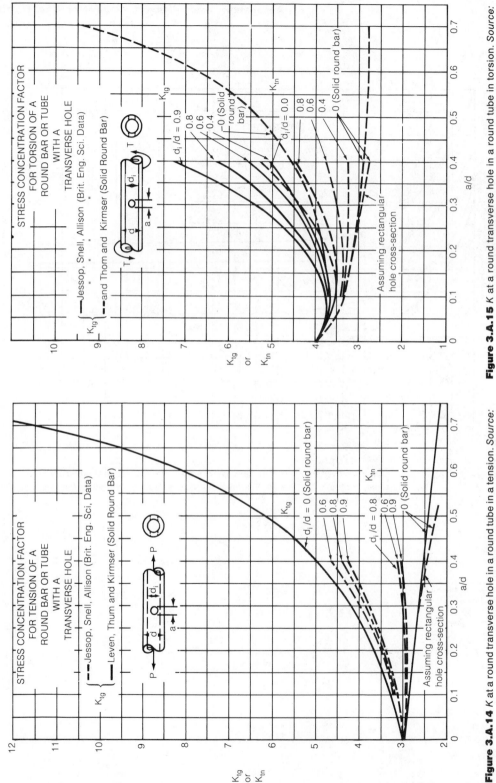

Figure 3.A.14 *K* at a round transverse hole in a round tube in a tension. *Source: Stress Concentration Factors, R. E. Peterson, copyright © 1974 John Wiley & Sons, New York. Reprinted by permission of John Wiley & Sons, Inc.*

Figure 3.A.15 *K* at a round transverse hole in a round tube in torsion. *Source: Stress Concentration Factors, R. E. Peterson, copyright © 1974 John Wiley & Sons, New York. Reprinted by permission of John Wiley & Sons, Inc.*

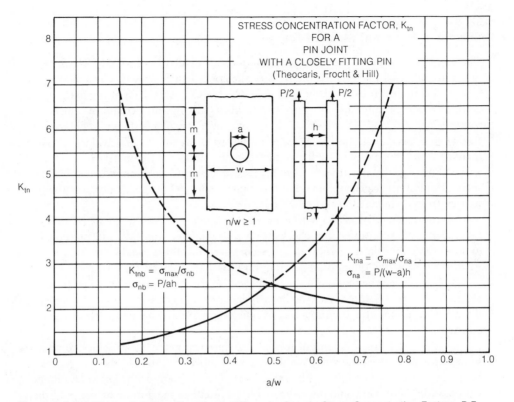

Figure 3.A.16 *K* for a pin joint with a closely fitting pin. *Source:* Stress Concentration Factors, *R.E. Peterson, copyright © 1974 John Wiley & Sons, New York. Reprinted by permission of John Wiley & Sons, Inc.*

APPENDIX B
Flowcharts for a Principal Stress Program

The program consists of the main program, whose flowchart is entitled PRINS, shown in Figure 3.B.1, which tests for the rank of the coefficient matrix by checking for zero rows and columns. It calls subroutines YZ, ZX, or XY respectively if the first, second, or third rows (or columns) are zero and notes if all rows and columns vanish. Subroutines YZ, ZX, and XY are similar to subroutine YZ which is shown in Figure 3.B.2. Rather than executing another check for rank, it solves for the eigenvalues of the examining 2×2 matrix which will give the principle stresses if the matrix is nonsingular and will be zero if the matrix is singular. Principal stresses are found using QUAD to solve the quadratic eigenequation. No flowchart is shown for QUAD because it may be written from a standard, well-known relation.

Subroutine XYZ is called whenever the 3×3 matrix is nonsingular. As shown in Figure 3.B.3 the subroutine first checks for a zero diagonal and calls subroutine CUBE to solve the cubic eigenequation in the event that not all diagonal elements are zero. As in the case of the quadratic equation, no flowchart is given for CUBE.

Because the eigenequation of the coefficient matrix does vanish, it is helpful to call SELECT, Figure 3.B.4, to select the linearly independent equations used in solving for the direction cosines for the three sets of unit vectors which define the three principal direction.

OUTPUT contains the output formatting statements and the graphics commands suitable for presenting the data. No flowchart is shown because of the differences between computers and computer software. A program written for the Digital Equipment Corporation's Pro 350 computer is available from either the author or the publisher.

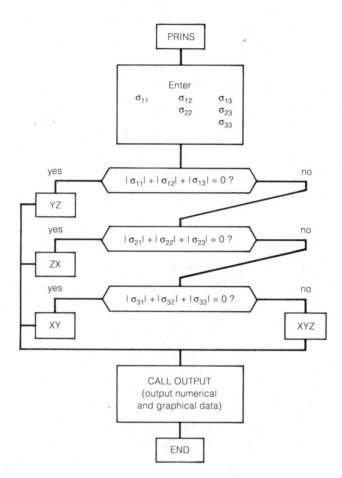

Figure 3.B.1 Flowchart for main program PRINS.

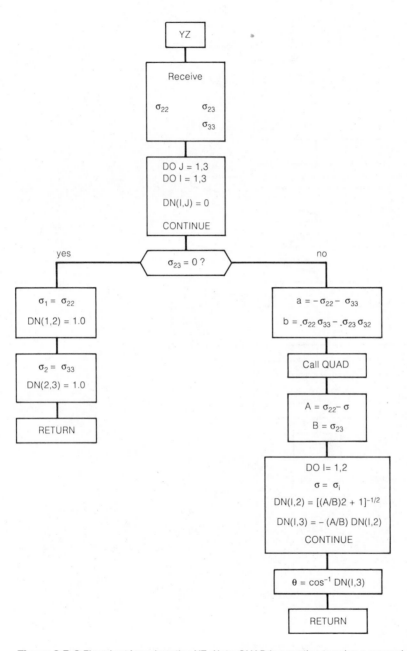

Figure 3.B.2 Flowchart for subroutine YZ. *Note*: QUAD is a routine to solve a general quadratic equation with real coefficients, i.e., $x^2 + ax + b = 0$

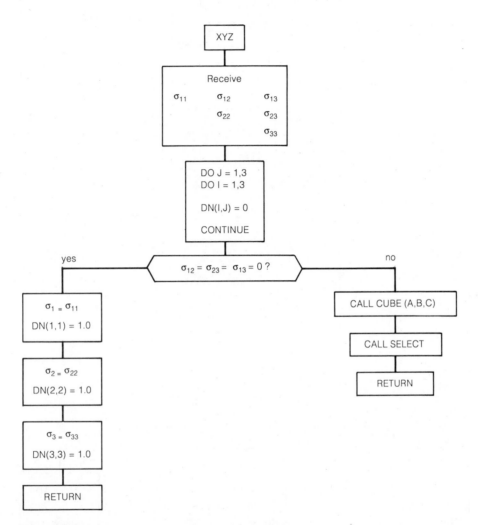

Figure 3.B.3 Flowchart for subroutine XYZ. *Note:* CUBE is a routine to solve a cubic equation with real coefficients, i.e., $x^3 + ax^2 + bx + c = 0$

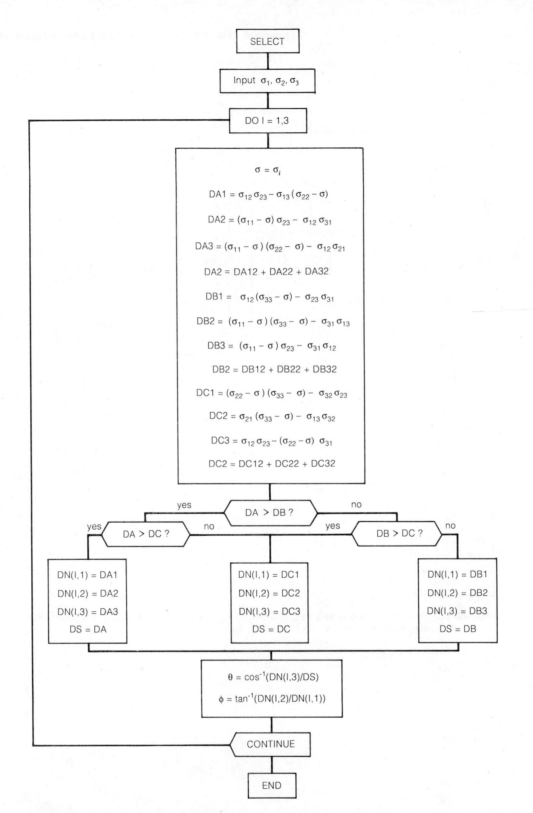

Figure 3.B.4 Flowchart for subroutine SELECT.

REFERENCES

1.   Roark, R.J., and Young, W.C., *Formulas for Stress and Strain*, 5th ed., McGraw-Hill, New York, 1975, pp. 526, 184–186.

2.   Popov, E.P., *Mechanics of Materials*, 2d ed., Prentice-Hall, Englewood Cliffs, NJ, 1975, pp. 125, 152.

3.   von Mises, R., On Saint-Venant's principle, *Bulletin of the American Mathematics Society*, 51, 555 (1945).

4.   Sternberg, E., On Saint-Venant's principle, *Quarterly of Applied Mathematics*, 11, 392–402 (1954).

5.   Olsen, G.A., *Elements of the Mechanics of Material*, 3d ed., Prentice-Hall, Englewood Cliffs, NJ, 1974, pp. 418–419.

6.   Orthwein, W.C., Finding principal stresses in three dimensions, *Computers in Mechanical Engineering*, 4 (4), 51–58, (1986).

7.   Orthwein, W.C., Keyway stresses when torsional loading is applied by the keys, *Experimental Mechanics*, 15, 245–248, (June 1975).

8.   MacGregor, AIME Technical Publication 805 (1937).

9.   1988 *SAE Engineering Handbook*, Society of Automotive Engineers, Warrendale, PA, 1988.

10.   van den Broek, J.A., *Elastic Energy Theory*, Wiley, New York, 1942, Chapter 5.

11.   Peterson, R.E., *Stress Concentration Factors*, Wiley-Interscience, New York, 1974.

12.   Orthwein, W.C., Adding ribs for maximum strength, *Machine Design*, 53 (4), 113–116 (1981).

PROBLEMS

3.1   Draw and label the stresses on an element in a grinding wheel
  a   at $r = 0.8R$ at point A away from any tangential forces (consider centrifugal forces only)
  b   at point B, at $r = R$, where the wheel contacts the work being ground.

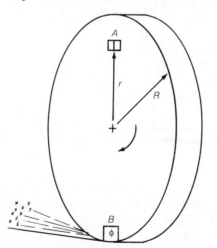

3.2   Draw and label the stress components acting on an element in the sidewall of an inflated tire on the load bearing side.

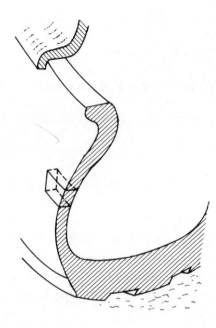

3.3   Draw and label stresses acting on elements centered at points *A*, *B*, and *C* of a beam when loaded as shown in the drawing.

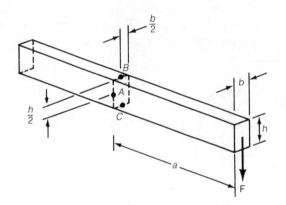

3.4   Draw and label stress components acting on elements at points *A*, *B*, and *C* on the surface of a drive shaft as shown in the drawing. Neglect the rigidity of the sheave (pulley) and the bearing friction. Give your answer in terms of T_1, T_2, a, r, and L.

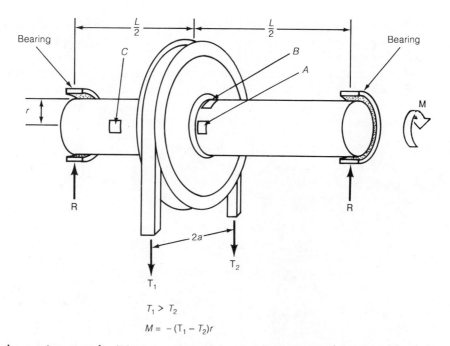

$$T_1 > T_2$$
$$M = -(T_1 - T_2)r$$

3.5   Find the maximum and minimum compressive stress in a rectangular support loaded as shown in the drawing. Load is 4000 lb and support thickness is 1 in. Load is uniformly distributed across the thickness.

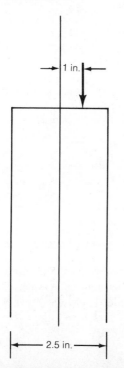

3.6   Find the principal stresses and principal directions at points A and B in problem 3.4.

3.7   Find the principal stresses and principal directions at A and B in problem 3.1.

3.8   Find the principal stresses and directions at the base of the cutting edge of the threads on a tap (a tool to cut internal screw threads) whose force components are shown below. The induced stresses are $\sigma_{rr} = 50$ MPa, $\sigma_{r\theta} = 130$ MPa, $\sigma_{rz} = 15$ MPa, and $\sigma_{zz} = -5$ MPa.

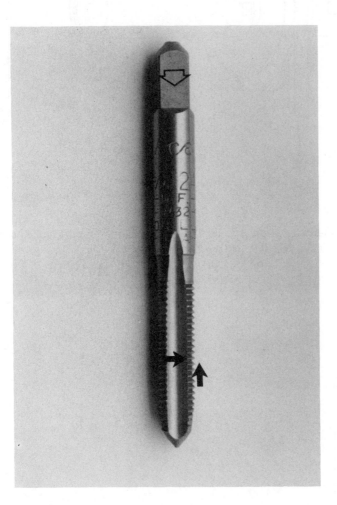

3.9   Calculate the principal stresses and principal directions at a point 0.5 in. above the centerline of the 1.25 in. diameter rear axle and 4.0 in. inboard of the plane of symmetry of the left-hand wheel. The right-hand wheel does not resist lateral motion. Each wheel provides a torque of 1100 ft-lb.

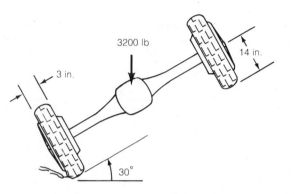

3200 lb

3 in.

14 in.

30°

3.10   Write an expression for moment M and shear force V as a function of angle θ for $0 \le \theta \le \frac{3}{2}\pi$ for the beam shown in the figure. Use the convention that positive M causes the radius to increase.

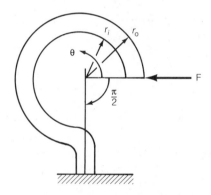

r_i r_o

θ

$\frac{\pi}{2}$

F

3.11   Find r_n and the maximum permissible F for the curved beam in problem 3.10 when $r_i = 4$ in. and the beam cross section in the curved portion is square, 0.2 in. on each side, and if the maximum working stress is 30 000 psi.

3.12   Rework problem 3.11 for a circular cross section with diameter 0.2 in.

3.13   Draw the stress distribution and indicate its maximum and minimum values on the cross section of the beam in problem 3.11 at $\theta = \pi/2$ and at $\theta = \pi$.

3.14   Draw the stress distribution and indicate maximum and minimum values on cross sections A and B in the circular shaft shown in the drawing. All dimensions are in inches.

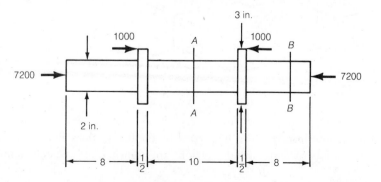

3 in.

1000   A   1000   B

7200   7200

2 in.

A   B

8        $\frac{1}{2}$ 10 $\frac{1}{2}$ 8

3.15 Write an expression for each of the stress components acting across plane *A* in the column shown in the drawing. The cross section is circular and loads act at 1.5 in. from the center line.

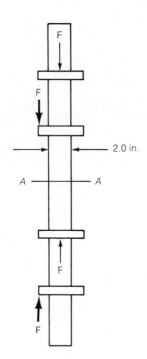

3.16 A circular trough in a tension bar is used to receive a holding clamp in an automatic freight car positioning device. Find the maximum tensile stress for a load of 328 500 lb. All dimensions are in inches.

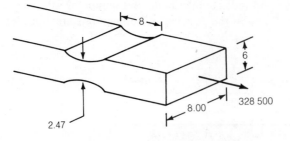

3.17 What is the maximum load that can be carried by the 0.5 in. thick link shown if the working stress is not to exceed 36 000 psi? All dimensions are in inches and the hole diameter is 0.75 in.

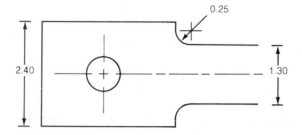

3.18 Would you recommend hole reinforcement to prevent it from causing failure of the link in problem 3.17 if the fillet radii are increased to 0.50 in. and D is increased to 2.9 in.? Why?

3.19 What is the percentage change in cross section of a circular bar if $v = 0.20$ and $\sigma_{zz}/E = 1 \times 10^{-3}$? Does the percentage change if the cross section is rectangular with width/depth $= 100$?

3.20 Find the thermal stress at 130 °C in a steel frame around a copper radiator. Equivalent cross-sectional areas are $A_c = 700$ mm² for the radiator and $A_s = 580$ mm² for the steel frame. At 21 °C there is no force between the top of the radiator and the steel frame. Thermal coefficients of linear thermal expansion are

$$\alpha_c = 1.73 \times 10^{-5}/°C \qquad \alpha_s = 1.17 \times 10^{-5}/°C$$

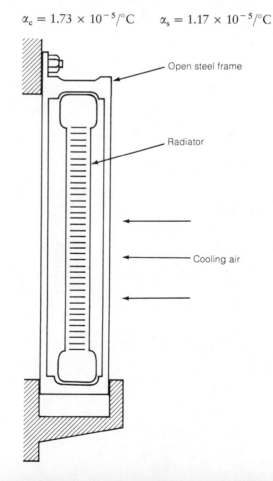

the elastic moduli are

$$E_c = 124 \times 10^3 \text{ MPa} \qquad E_s = 207 \times 10^3 \text{ MPa}$$

and the two values of Poisson's ratio are

$$v_c = 0.29 \qquad v_s = 0.27.$$

3.21   Find the stress at point 0 on the outer surface of a plate as shown in part *a* of the drawing and compare it with the stress at point 1 on the outer surface of a rib attached to the plate, part *b*. (*Hint:* Compare stresses in a representative section of width $w + b$, where w is the width of the rib and b is the space between ribs, as shown.)

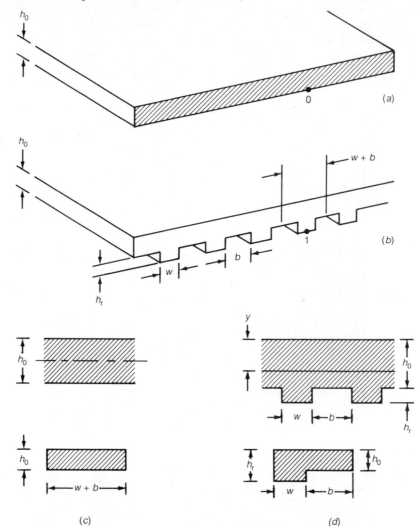

3.22 Plot σ_1/σ_2 as given in problem 3.21 for $w/b = 0.2$ for h_r/h_0 from 0.0 to 1.0 and compare with the corresponding curve in the drawing below. Use the curves in the drawing to decide whether to approve adding ribs to the floor plate of a motor-generator housing for use on a deep sea fishing boat. The original floor plate was 10 mm thick and was occasionally subjected to its maximum working stress. The replacement plate is to have ribs 2 mm high, 6 mm wide, and spaced 26 mm between centers.

[Ans. Do not approve (w/b = 6/20 = 0.3, h$_r$/h$_0$ = 2/10 = 0.2, and $w + b$ = 26 mm)]

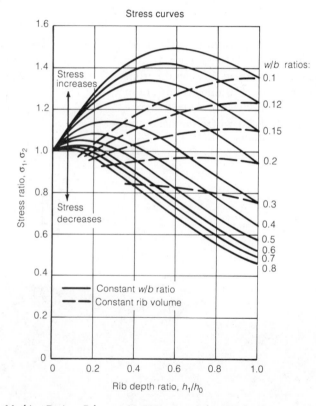

Source: Machine Design, *February 26, 1981. Copyright 1981 by Penton Publishing Co., Cleveland, OH.*

3.23 The drawing shows the cross section of the front axle of a riding mower (*a*) which pivots about its center, as shown in (*b*). (All dimensions are in millimeters, and are not those of a commercial lawn mower.) Find the shear stress at the neutral surface. Force = 2500 N.

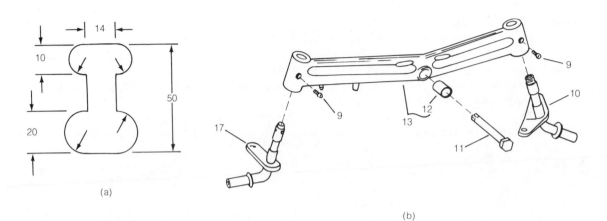

(a)

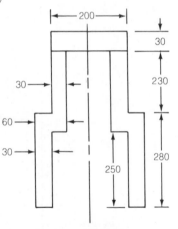

(b)

3.24 The drawing shows a simplified cross section of the head, cylinder block, and crankcase walls of a farm tractor (*a*) which also serves as the main frame between the front and rear wheels, as shown in (*b*). (All dimensions are in millimeters.) Find the shear stress at the neutral surface and compare it with that calculated where the head and cylinder block meet, that is, at the head bolt shear plane. $V = 7200$ N.

(a)

(b)

3.25   Find the buckling load for a column having a circular cross section 15 mm in diameter and 215 mm long using a safety factor of 3.10. Each end is welded to a rigid support. $E = 205$ GPa and $\sigma_{yp} = 99.3$ MPa. Justify your use of either the long- or short-column buckling formulas.

3.26   After 3 years of service a circumferential crack extending over 280° was found in the weld. The equipment operator has requested the approval of the manufacturer to continue using the equipment but with the safety factor reduced to 1.2, with the manufacturer still maintaining the warranty and being liable for an accidents. Would you, as the field engineer, forward this request to the home office? Justify your answer.

3.27   A push rod 2.1 m long is pivoted at each end to allow rotation about the x axis but not about the y axis. Find the buckling load if its cross sectional area is 40 mm^2, its moment of area (area moment of inertia) about the x axis is 1066.7 mm^4, and its moment of area about the y axis is 416.7 mm^4. Use $\sigma_y = 762$ MPa, $E = 207$ GPa, and $v = 0.3$.

3.28   A 450 000-N force on a cylindrical shaft which is supported by two self-aligning bearings is shown in the drawing (all dimensions are in millimeters). Calculate the vertical position of the center line of the shaft directly under the load relative to a plane through the center of each bearing, which acts as a simple supports. Justify inclusion or neglect of shear deformation if the calculated displacement is to be accurate to within 0.01 mm. $E = 207$ GPa.

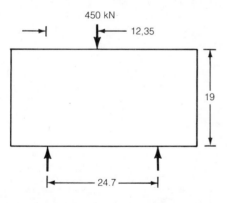

3.29   Find the force at rigid support C for the simply supported shaft with the loads shown in the figure (all dimensions are in millimeters). The shaft diameter is 50 mm and $E = 207$ Gpa.

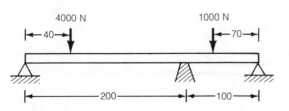

3.30   Find the force at C in the figure for problem 3.29 where the rigid support at C is 0.006 mm below the undeflected shaft.

3.31 Find the shaft deflection and the force at C in the figure for problem 3.29 if the shaft is supported by a cantilever 100 mm long (from the fixed end of the cantilever to the point where it contacts the shaft), 50 mm deep, and 25 mm wide. The shaft and support are in contact with zero force between them before the shaft is loaded. The cantilever cross-section is rectangular.

3.32 Find the reaction at C for the support described in problem 3.31 when the initial clearance between shaft and support is 0.006 mm, that is, before the loads are applied to the shaft.

CHAPTER FOUR

STATIC FAILURE AND FATIGUE (DYNAMIC FAILURE) CRITERIA

NOTATION

a	Exponent in formula for *S-N* (σ-*N*) line *(1)*		t	Time *(t)*
b	Coefficient in formula for *S-N* line or exponent in strain formula for low-cycle fatigue $(ml^{-1}t^{-2})$, *(1)*		α, α_i	Fraction of a duty cycle or of total life *(1)*
c	Exponent in strain formula *(1)*		ε_f'	Fatigue ductility coefficient *(1)*
BHN	Brinell hardness number *(1)*		$\Delta\varepsilon$	Total strain, low-cycle fatigue, local strain approach *(1)*
D, d	Dimensions (diameter, width) *(l)*			
D, D_i	Damage after various stress levels *(1)*		$\Delta\varepsilon_e$	Elastic strain, low-cycle fatigue, local strain approach *(1)*
E	Elastic (Young's) modulus $(ml^{-1}t^{-2})$			
K	Stress concentration factor, steady load *(1)*		$\Delta\varepsilon_p$	Plastic strain, low-cycle fatigue, local strain approach *(1)*
K'	Cyclic strength (stress) coefficient $(ml^{-1}t^{-2})$		ζ	Safety factor *(1)*
K_f	Stress reduction factor, infinite life *(1)*		η	Cycles/second (t^{-1})
K_f'	Stress reduction factor, finite life *(1)*		σ_a	Alternating stress $(ml^{-1}t^{-2})$
k	Exponent in the Shanley relation *(1)*		σ_{a_e}	Effective alternating stress $(ml^{-1}t^{-2})$
N_e	Number of cycles to failure, various stresses *(1)*		σ_e	Endurance limit (endurance stress) $(ml^{-1}t^{-2})$
N_f	Number of stress reversals *(1)*		$\sigma_I, \sigma_{II}, \sigma_{III}$	Principal stresses $(ml^{-1}t^{-2})$
N_i	Number of cycles to failure for stress σ_i *(1)*		σ_m	Mean stress $(ml^{-1}t^{-2})$
			$\sigma_{rr}, \sigma_{r\theta},$ $\dots, \sigma_{zz}$	Stress components in circular cylindrical coordinates $(ml^{-1}t^{-2})$
n_i	Number of cycles at stress σ_i in total life *(1)*			
n_i	Number of cycles at strain ε_i in a block *(1)*		σ_f'	Fatigue strength (stress) coefficient $(ml^{-1}t^{-2})$
P	Probability that a machine will last a specified time *(1)*		σ_u, σ_{u_t}	Ultimate strength (ultimate stress), tension $(ml^{-1}t^{-2})$
P_i	Probability that component i will last a specified time *(1)*		$\sigma_{u_t}, \sigma_{u_c}$	Ultimate stress in tension, compression $(ml^{-1}t^{-2})$
q	Notch sensitivity factor or number of loads *(1)*		σ_y, σ_{y_t}	Yield stress in tension $(ml^{-1}t^{-2})$
			$\sigma_{xx}, \sigma_{xy},$ $\dots \sigma_{zz}$	Stress components in rectangular Cartesian coordinates $(ml^{-1}t^{-2})$
R	$\sigma_{min}/\sigma_{max}$ *(1)*			
r	Fillet radius *(l)*		$\sigma_{\xi\xi}, \sigma_{\eta\xi}$	Direct and shear stress in Mohr circle coordinates $(ml^{-1}t^{-2})$
S	Stress (used only in relation to *S-N* curves) $(ml^{-1}t^{-2})$		σ_1, σ_2	Maximum and minimum direct stress components $(ml^{-1}t^{-2})$
			τ	Shear stress $(ml^{-1}t^{-2})$
			τ_I	Maximum shear stress $(ml^{-1}t^{-2})$

This chapter deals with topics that are important to the design engineer, but that are still only poorly understood in many instances. Conseqently, the design procedures and criteria discussed are those that have been accepted because of their agreement with a selected number of laboratory tests which are believed to simulate common operational conditions fairly well. When the predicted strength and /or life is compared with that found from a machine component in actual service the agreement may not always be as satisfactory as it was for the laboratory data. It is, therefore, necessary to remember that these estimates are intended primarily for use in the preliminary design of a component

After a new mechanical component is produced a number of specimens should undergo strength and life tests in actual service to determine the mean and, if possible, the standard deviation of its service strength and its life. Even though these test results may generally be susceptible to a statistical analysis using standard methods, so that no new statistical theory need be developed, the data collection and analysis programs in themselves are, with few exceptions, expensive. This is often one of several reasons why manufacturers are reluctant to modify designs that have proven to be satisfactory in the past.

The table of contents for this chapter has been divided into three sections to emphasize the distinctions between criteria for failure under a steady load as opposed to failure under repeated or fluctuating loads, also known as cyclic loading.

Section 1 relates the reliability of a machine to the probability of failure of each of its components.

Single load criteria are grouped in Sections 2 through 5, which describe several criteria for preventing failure due to excessive stress from a steady load. Section 5 differs from Sections 2 through 4 in that it displays relations that are now seldom used: Recent data have indicated dependence upon so many variables that no generally accepted criteria for failure of brittle materials presently exist which agree with most of the data.

Service failures and laboratory tests together have demonstrated that mechanical components may fail when exposed to cyclic loads of magnitude less than that required to cause failure under a single loading; that is, they fail even though the stresses never reach the ultimate stress of the material. Failures of this sort are known as fatigue failures.

Sections 6 through 11 are devoted to criteria that have been formulated to prevent failures that are due to a combination of steady and fluctuating loads. They include the older Soderberg formula for estimating indefinite life under cyclic loading as well as the more satisfactory Goodman-yield (modified Goodman) and Gerber-yield criteria. A recent extension of the last of these to multiaxial loading is given in Section 11.

Some important and widely used machine elements, such as ball and roller bearings, chain components, and gear teeth, require cyclic contact over an extremely small area, they are subjected to unavoidably large stresses over these regions. Since these stresses often exceed the maximum specified by the criteria given in Sections 4 through 11, we need formulas for estimating the number of cycles to failure in order to make the best of a bad situation. These formulas are given in Sections 12 through 17, and include methods for accounting for stress concentrations, mean stresses, and continuously varying loads.

4.1 RELIABILITY

The estimated life of a machine is often an important factor in its design because of its role in determining the price and the size of the market for the machine. According to probability theory, if the life of a machine depends upon the life of a certain number n of its individual components, then the probability of machine failure is given by

$$P(t) = \prod_{i=1}^{n} p_i(t) \tag{4.1.1}$$

where $P(t)$ is the probability that the machine will fail at time t and $p_i(t)$ is the probability that component i will fail at time t. If we assume that one or more of these n components will fail if given enough time (such as ball and roller bearings) it follows that

$$\int_0^\infty P(t)\, dt = \int_0^\infty \prod_{i=1}^{n} p_i(t)\, dt = 1 \tag{4.1.2}$$

Probabilities $P(t)$ and $p_i(t)$ are termed *density functions* by statisticians. Most mechanical components satisfy density functions $p(t)$, representative of $p_i(t)$, which are included in the family of *Weibull distributions* and which may be written as

$$p(t) = \frac{\beta(t-\delta)^{\beta-1}}{(\theta-\delta)^{\beta}} \exp\left[-\left(\frac{t-\delta}{\theta-\delta}\right)^{\beta}\right] \tag{4.1.3}$$

where β is known as the shape factor, δ the minimum life, and θ the characteristic life.

Since the integral of $P(t)$ from time zero, when the machine was first put into service, to the present is the probability that it will have failed, it follows that the reliability $R(t)$ should be defined by

$$R(t) = 1 - \int_0^t P(\lambda)\, d\lambda \tag{4.1.4}$$

which is to say that $R(t)$ is the probability that the machine will not fail from startup until time t; it is exactly what is meant by reliability. The integral in (4.1.4) is known as the *distribution function*.

Perhaps a more useful expression for the reliability of a machine whose operation depends upon n components is that

$$R(t) = \prod_{i=1}^{n} [1 - p_i(t)] \tag{4.1.5}$$

where $p_i(t)$ is has been redefined to denote the probability that component i will fail *at or before* time t.

SINGLE-LOAD FAILURE CRITERIA

4.2 MAXIMUM TENSION CRITERION

Failure prediction has been a troublesome problem throughout the history of the development of the theory of elasticity. No single theory has yet been developed that is able to accurately predict failure for all materials under a given loading, or to predict failure in any one material for all stress distributions that can lead to ultimate failure. Some materials, such as cast iron, when subjected to simple tension (i.e., $\sigma_{zz} > 0$ is the only nonzero stress component) fail across a plane perpendicular to the direction of the applied load. Necking is negligible and the entire failure surface is perpendicular to the applied force—there is no central failure in tension surrounded by a ring that failed in shear, as in ductile materials.

This suggests that for other loading conditions the material may fracture over a plane perpendicular to the largest principal tensile stress whenever it becomes equal to the ultimate strength (ultimate tensile stress) of the material. In other words, if a specimen of the material fails in simple tension at a stress σ_u, then it is postulated that under loading that gives rise to components such as $\sigma_{xx}, \sigma_{xy}, \ldots, \sigma_{zz}$, the material will fail whenever one of the principal stresses, say σ_I, increases to σ_u.

Implicit in this failure criterion is the restriction that the loading be such that the Mohr circle, for a two-dimensional stress field, be centered at the origin of the stress coordinates. This may be easily seen from Figure 4.2.1, where shear and tensile failure stress limits are shown as dashed lines. If a material of this nature is loaded in shear only, then the corresponding Mohr circles will increase in diameter from circle 1 to circle 2 as the shear stress increases. As shear stresses continue to increase the principal stresses increase as well until the Mohr circle contacts the tensile limit. At this point $\sigma_I = \sigma_u$ and rupture occurs.

This type of behavior may be easily demonstrated by placing a piece of chalk or a cast iron rod in simple torsion. The chalk example may be demonstrated by hand if one is careful to avoid any bending moments as the chalk is twisted, with the results shown in Figure 4.2.2. From Figure 4.2.1 it is clear that the principal direction for tension is $45°$ away from a generator of the circular cylindrical surface of the chalk, so that a tension failure will occur along a helix, as displayed in Figure 4.2.2. After one revolution of the helix the body of the rod has failed and the center of rotation

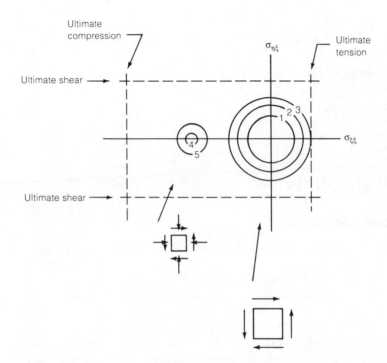

Figure 4.2.1 Mohr circle with tension, torsion, and compression limits shown.

has moved toward the line joining the ends of the helix so that the final tension failure is over a small plane containing the *z*-axis.

Different behavior may be encountered if the loading corresponds to Mohr circle 4 in Figure 4.2.1. As the load increases and the circles enlarge through 5 and larger until a circle encounters either the ultimate strength in shear or in compression and the material fails accordingly.[1]

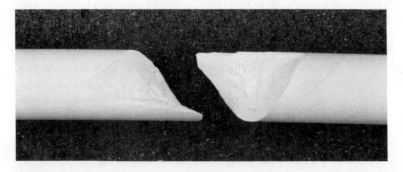

Figure 4.2.2 Tension failure in chalk placed in torsion. Similar failure may be demonstrated in the laboratory with cast iron rods in torsion.

The maximum normal stress criterion for failure may be expressed as

$$\sigma_I = \frac{\sigma_{u_t}}{\zeta} \qquad (4.2.1)$$

From a practical point of view, a machine component has failed once it is distorted from its original shape. It does not have to break to be unusable. Hence, equation 4.2.1 may be replaced by

$$\sigma_I = \frac{\sigma_{y_t}}{\zeta} \qquad (4.2.2)$$

for a material that yields before failing in tension. In this relation σ_{y_t} is the yield stress in tension and ζ is the factor of safety.

4.3 MAXIMUM SHEAR CRITERION

The maximum shear theory is often applied to ductile materials under the assumption that they will fail in shear for both shear and tensile loads. This theory may be verified experimentally for shear loads, but only in part for tensile loads. Final failure of steel rods of circular cross-section loaded in tension is, in fact, in shear. As was noted earlier in Chapter 3, however, and as verified by X-ray photograph reproduced in Figure 3.9.2, the initial internal failure in the necked-down region is in tension. It is only after formation of the internal cavity that final failure in shear occurs in the wall between the internal cavity and the outside surface. It is important to note that because of the modified geometry at the time of shear failure the stress field at the fracture is no longer that of simple tension.

In Mohr circle coordinates a material loaded in shear only (i.e., simple shear) may be represented by circle 1 in Figure 4.3.1 at a load less than the failure load. As the

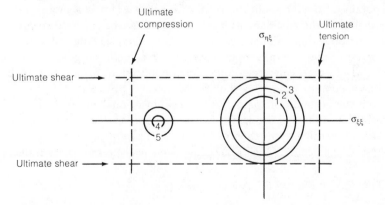

Figure 4.3.1 Mohr circle coordinates with tension, compression, and shear ultimate stresses superimposed.

shear forces increase the shear stresses will increase, so the circle will grow from circle 1 through circle 2 and may eventually reach circle 3. Circle 3 is tangent to the failure stress in shear, as shown in Figure 4.3.1, shear failure at this point terminates the loading.

It is commonly assumed that ultimate strength in shear is half that in tension, so the shear criterion is often written as

$$\tau_I = \frac{1}{2}\frac{\sigma_{u_t}}{\zeta} \tag{4.3.1}$$

where τ_I is the maximum shear stress associated with the principal directions.

As in the previous section, equation 4.3.1 is often replaced by

$$\tau_I = \frac{1}{2}\frac{\sigma_{y_t}}{\zeta} \tag{4.3.2}$$

as a practical limit to the shear stress.

According to Figure 4.3.1 equation 4.3.2 may agree with experiment even in the presence of small in-plane tensile or compressive stress. If either are so large that the tensile or compressive limits contact the Mohr circle before it reaches the shear limit, failure will not be in shear. Laboratory tests of shear in the presence of in-plane compression are difficult to interpret because the failure is gradual, rather than sudden, as in tension failure.[1] With this admonition, however, it appears that under large compressive stress the ultimate shear stress may effectively increase.[1]

4.4 HUBER-VON MISES-HENCKY CRITERION

Comparison of the principal stresses obtained from a three-dimensional stress distribution with those obtained from the application of the Mohr circle clearly demonstrates that except for special cases the two-dimensional method cannot be expected to provide correct results for three-dimensional stress fields. For example, failure due to a maximum normal stress or due to a maximum shear stress may depend upon a principal stress not included in a particular two-dimensional application of the associated maximum normal or maximum shear stress failure theories.

One attempt to better account for some of the three-dimensional effects is to consider one or more of the invariants of the deviatoric stress tensor (see texts on advanced plasticity theory) since, according to the theory of tensor analysis, these quantities may have particular physical significance. One of these invariants leads to the criterion that

$$2\sigma_{y_t}{}^2 = (\sigma_I - \sigma_{II})^2 + (\sigma_{II} - \sigma_{III})^2 + (\sigma_{III} - \sigma_I)^2 \tag{4.4.1}$$

where σ_I, σ_{II}, and σ_{III} represent the principal stresses at the point under study. Condition 4.4.1 is known as the Huber-von Mises-Hencky yield criterion or as the maximum distortion [sic] energy theory, even though it is not a theory but a yield condition from plasticity theory. In earlier texts it was referred to as the von Mises-Hencky, Mises-Hencky, or often just the Hencky yield criterion. In the remainder of this text it will be denoted as the HMH criterion. In terms of the stress components from which the principal stresses were computed, the HMH criterion may be written as

$$\left(\frac{\sigma_{y_t}}{\zeta}\right)^2 = \sigma_{xx}^2 + \sigma_{yy}^2 + \sigma_{zz}^2 - \sigma_{xx}\sigma_{yy} - \sigma_{yy}\sigma_{zz} - \sigma_{zz}\sigma_{xx} \\ + 3(\sigma_{xy}^2 + \sigma_{yz}^2 + \sigma_{zx}^2) \tag{4.4.2}$$

in terms of an xyz-coordinate system; in terms of a circular cylindrical $r\theta z$ system the criterion appears as

$$\left(\frac{\sigma_{y_t}}{\zeta}\right)^2 = \sigma_{rr}^2 + \sigma_{\theta\theta}^2 + \sigma_{zz}^2 - \sigma_{rr}\sigma_{\theta\theta} - \sigma_{\theta\theta}\sigma_{zz} - \sigma_{rr}\sigma_{zz} \\ + 3(\sigma_{r\theta}^2 + \sigma_{\theta z}^2 + \sigma_{zr}^2) \tag{4.4.3}$$

EXAMPLE 4.4.1

Find the factor of safety according to the normal stress criterion, the shear stress criterion, and the HMH criterion for a material in which

   a $\sigma_{xx} = 80$ MPa

   b $\sigma_{xx} = 80$ MPa, $\sigma_{yy} = -20$ MPa

   c $\sigma_{xx} = 80$ MPa, $\sigma_{xy} = 20$ MPa, $\sigma_{xz} = 20$ MPa   and $\sigma_{yy} = -20$ MPa

Yield stress in tension is 200 MPa and note that a factor of safety less than unity implies failure.

   a   From 4.2.1 for the normal stress criterion

$$\zeta = \sigma_{yp}/\sigma_1 = 200/80 = 2.5$$

while from 4.3.2 and the Mohr circle

$$\tau = \sigma_{xx}/2 = 40$$

so

$$\zeta = \sigma_{yp}/2\tau = 200/80 = 2.5$$

for the shear stress criterion. From equation 4.3.2 the HMH criterion yields

$$(200/\zeta)^2 = (80)^2 \quad \text{or} \quad \zeta = 2.5$$

Thus, all three agree for simple tension.

b From the Mohr circle $\tau = 50$ MPa, so according to 4.2.2

$\zeta = 200/80 = 2.5$ from the normal stress criterion

however,

$\zeta = 200/100 = 2.0$ from the shear stress criterion

finally

$$\left(\frac{200}{\zeta}\right)^2 = (80)^2 + (20)^2 + 20(80) = (91.65)^2$$

so that

$\zeta = 2.18$ from the HMH criterion

In the absence of experimental data the most restrictive factor of safety is usually used; $\zeta = 2.0$

c Using the method of Section 3.6 the principal stresses are found to be

$\sigma_I = 88.23$ MPa $\tau_I = 56.38$ MPa

$\sigma_{II} = -3.70$ MPa $\tau_{II} = 45.96$ MPa

$\sigma_{III} = -24.53$ MPa $\tau_{III} = 10.42$ MPa.

Thus, the normal stress criterion, equation 4.2.2, yields

$\zeta = \sigma_{yp}/\sigma_I = 200/88 = 2.27$

while the shear stress criterion, equation 4.3.2, yields

$\zeta = \sigma_{yp}/(2\tau_I) = 200/[2(56.38)] = 1.77$

Finally, the HMH criterion, equation 4.4.1, yields

$$\left(\frac{200}{\zeta}\right)^2 = 80^2 + 20^2 + 20(80) + 3(20^2 + 20^2) = (103.92)^2$$

so that $\zeta = 1.92$.

As the example indicates, the factor of safety given by the HMH criterion is never less than both factors given by the normal stress and the shear stress criteria. In general, it has been used instead of the shear and normal stress criteria when experiments indicate that its factors of safety are realistic.

4.5 BRITTLE MATERIAL

Under the assumption that failure is due to fracture when the principal stresses exceed the ultimate strength of the material either in tension or compression, the failure criterion for steady stress becomes

$$\frac{\sigma_I'}{\sigma_{u_t}} + \frac{\sigma_{III}'}{\sigma_{u_c}} = \frac{1}{\zeta} \tag{4.5.1}$$

where

$$
\begin{aligned}
\sigma_I' &= \sigma_I && \text{if } \sigma_I > 0 \\
&= 0 && \text{otherwise}
\end{aligned} \tag{4.5.2}
$$

$$
\begin{aligned}
\sigma_{III}' &= \sigma_{III} && \text{if } \sigma_{III} < 0 \\
&= 0 && \text{otherwise}
\end{aligned} \tag{4.5.3}
$$

In equation 4.5.1 σ_{u_t} represents the ultimate strength in tension and σ_{u_c} represents the ultimate stress in compression; it is negative according to the standard sign convention. When applied to two-dimensional stress distributions the smaller principal stress, normally denoted by σ_{II}, replaces σ_{III} in equation 4.5.1. It is subject to the conditions (4.5.3).

This criterion, introduced in 1950,[3] is seldom repeated in the newer texts because of the considerable spread in failure data from brittle materials. At low temperatures steel may exhibit brittle fracture for reasons that are as yet largely unknown, as Parker[4] observed in 1957. In spite of data collected since then, Parker's observation is still true.

FATIGUE (CYCLIC LOADS)—CRITERIA FOR INDEFINITE LIFE

4.6 FATIGUE LIFE: *S-N* CURVES (σ-*N* Curves)

Failure of cyclically loaded iron and steel parts after many load cycles, where the maximum stress applied during each cycle was within the safe limits for static loading, was observed many years ago and first verified experimentally by Wohler in Germany in 1858. Data obtained by Wohler and subsequent investigators produced stress–number of cycles (*S-N*) curves of various steel alloys similar to those shown in Figure 4.6.1. Although other materials may display *S-N* curves having somewhat different characteristics, we will restrict our attention to those for steel alloys to display the methods involved in their application to mechanical design.

The fluctuating stress components shown in Figure 4.6.1 may be written as

$$
\begin{aligned}
S &= \frac{\sigma_{max} + \sigma_{min}}{2} + \frac{\sigma_{max} - \sigma_{min}}{2} \sin 2\pi\eta t \\
&= \frac{\sigma_{max}}{2} [1 + R + (1 - R) \sin 2\pi\eta t]
\end{aligned} \tag{4.6.1}
$$

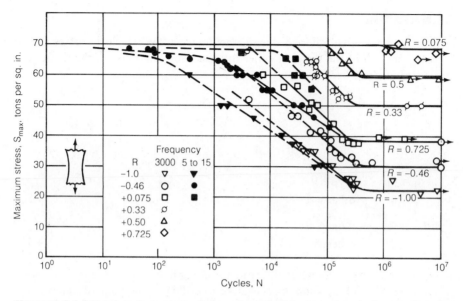

Figure 4.6.1 *S-N* curves for various stress ratios, unnotched specimens. *Source: W.J. Bell and P.P. Benham,* The Effect of Mean Stress on Fatigue Strength of Plain and Notched Stainless Steel Sheet in the Range from 10 to 10^7 Cycles, American Society for Testing and Materials, STP 338, 1963, pp. 25–46.

in which

$$R = \frac{\sigma_{min}}{\sigma_{max}} \qquad (4.6.2)$$

is the ratio referenced in Figure 4.6.1. Its physical significance is illustrated in Figure 4.6.2.

Plotting the *S-N* curves on log–log paper more clearly displays the demarcation between cyclic stress levels that induce ultimate failure and those that do not. In fact, the *S-N* data for parts forged from SAE 1035 and SAE 4063 steel which appear as in Figure 4.6.3[5] are typical of the *S-N* data for most steel alloys. Solid squares and circles represent data points for parts A and B as labeled and are indicative of the variation found in other steel alloys. The sharp changes in slope of the lines that bound this data distribution occur at what is obviously a physically significant stress level, which is now known as the *endurance limit*, or *endurance stress*.

The endurance limit, which will be considered in the next section, is of special importance in the design of machine components because a machine part subjected to cyclic tensile stresses that are below the endurance limit will not fail because of these stresses. In particular, these curves suggest that stress levels that can be endured for 10^7 cycles can be endured indefinitely.

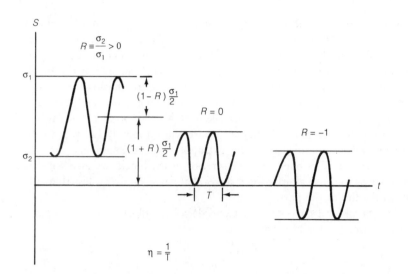

Figure 4.6.2 Relation of *R* used in Figure 4.6.1 to the mean stress and the stress amplitude.

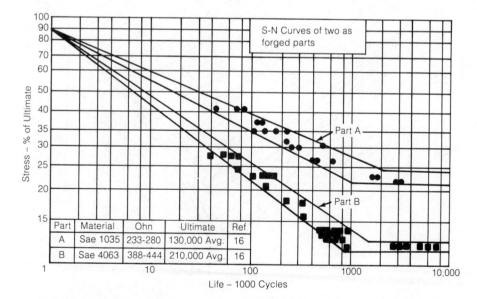

Figure 4.6.3 Relation between endurance limit and tensile strength. Unnotched specimens in reversed bending. *Source: G.C. Noll and M.A. Erickson,* Allowable Stresses for Steel Members of Finite Life, Proceedings of the Society for Experimental Stress Analysis, *Vol. 5, Society for Experimental Mechanics, Brookfield, CT, 1948, pp. 132–143.*

Separation of the effects of the mean stress amplitude, the $(\sigma_1 + \sigma_2)/2$ term in (4.6.1), from the alternating stress level, the $(\sigma_1 - \sigma_2)/2$ term in (4.6.1), may be accomplished by cross-plotting the data as a function of the mean stress for a family of curves similar to those shown in Figure 4.6.4. Those curves on the side of the block portray the effect of cyclic loading when the mean stress is zero; the load fluctuates between equal values of tension and compression. This is known as *completely reversed loading*. Those curves at the end of the block, taken at $N = 10^7$ cycles, portray the effect of the mean stress in determining whether a part will ultimately fail when subjected to a superimposed alternating stress. By extrapolating the imaginary surfaces implied by these intersection lines back into the interior of the block, we also see that addition of a mean stress to those alternating stresses that cause failure at less than 10^7 cycles will lenghten the life of the part.

These important families of curves on the two sides of the block form the foundation for widely used formulas for life estimation.

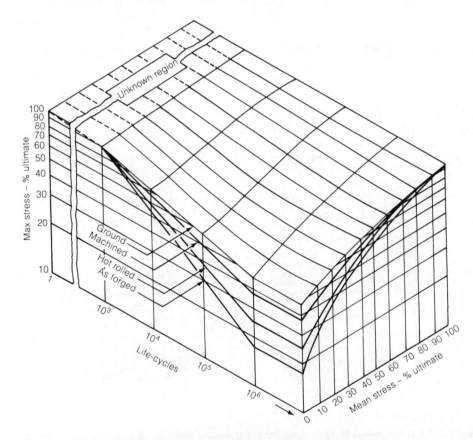

Figure 4.6.4 Fatigue strength of steel as a function of mean stress, alternating stress, and cycles to failure. *Source: G.C. Noll and M.A. Erickson,* Allowable Stresses for Steel Members of Finite Life, Proceedings of the Society for Experimental Stress Analysis, Vol. 5, *Society for Experimental Mechanics, Brookfield, CT, 1948, pp. 132–143.*

4.7   THE ENDURANCE LIMIT (ENDURANCE STRESS)

The endurance limit, often denoted by σ_e, associated with the *S-N* curves, is seldom listed among the cataloged properties of a metal, perhaps because there appears to be good correlation between the Brinell hardness and the endurance limit, as shown in Figure 4.7.1 for given conditions of surface finish and tensile strength.

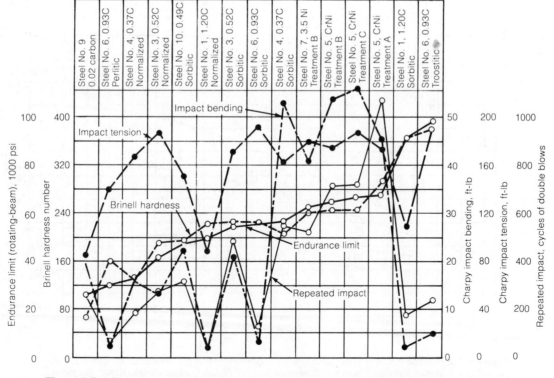

Figure 4.7.1 Comparison of endurance limits of steels with other mechanical properties. *Source: T.J. Dolan,* Mechanical Properties, ASME Handbook, Sec. 7.4, Metals Engineering De-sign, 2d ed., *American Society of Mechanical Engineers, New York, 1965.*

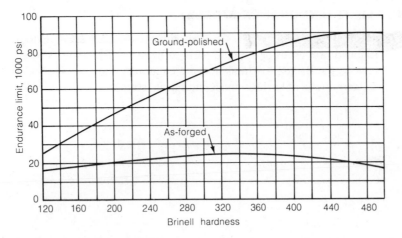

Figure 4.7.2 Relation between endurance limit of steel and Brinell hardness. *Source: T.J. Dolan, Mechanical Properties, ASME Handbook, Sec. 7.4, Metals Engineering Design, 2d ed., American Society of Mechanical Engineers, New York, 1965.*

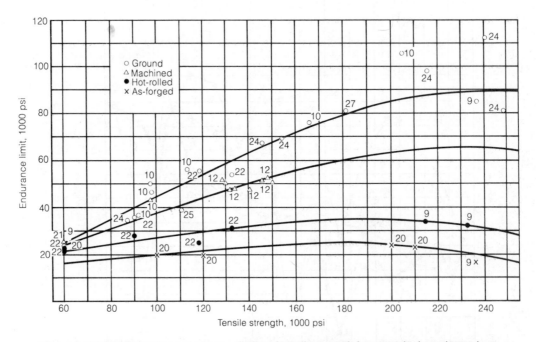

Figure 4.7.3 Relation between endurance limit and tensile strength for unnotched specimens in reversed bending. *Source: B.C. Hanley and T.J. Dolan*, Surface Finish, ASME Handbook, Sec. 7.5, Metals Engineering Design, 2d ed., *American Society of Mechanical Engineers, New York, 1985.*

Effects of surface finish are shown in Figure 4.7.2, which may be used as a means of estimating the endurance limit as a function of the Brinell hardness for the two finishes shown. Similar correlation has also been observed between the tensile strength and the surface finish, as illustrated in Figure 4.7.3, as may be expected from the good correlation already found between Brinell hardness and tensile strength for a given surface finish. This set of curves is more useful to the designer than those in Figure 4.7.2 because it explicitly shows the effect of machined and hot-rolled surface conditions. The relation between the endurance limit in reversed bending (tension and compression) and the endurance limit in reversed torsion (reversed shear) is shown in Figure 4.7.4. Note that some nonferrous materials may obey similar relations, as indicated by the McAdam data points. As implied by these figures, most of the fatigue studies to date have been concerned with ferrous materials, hence, our discussion has emphasized steel alloys. Other metals, as indicated by Figure 4.7.4, appear to have qualitatively similar behavior. Plastic behavior appears to have much more variety and often to be more dependent upon cyclic rate, temperature, and the chemical environment.

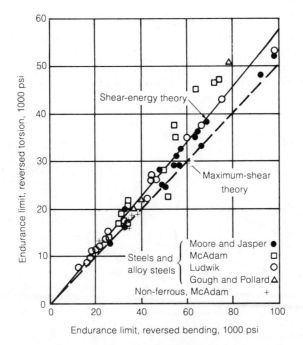

Figure 4.7.4 Relation between endurance limits in bending and in torsion. *Source: T.J. Dolan*, Mechanical Properties, ASME Handbook, Sec. 7.4, Metals Engineering Design, 2d ed., *American Society of Mechanical Engineers, New York, 1965.*

4.8 NOTCH SENSITIVITY

Not only does cyclic loading appear to reduce the strength of a material by causing it to fail at stresses less than its ultimate strength (ultimate stress) under a steady load, it also appears to reduce the effect of stress concentrations in many materials. This effect is known as the *notch sensitivity* of the material.

Notch sensitivity is related to, but different from, the stress concentration factor at a notch. The stress concentration factor at a point is a multiplication factor which gives the actual stress at that point when the applied load is static; the notch sensitivity, on the other hand, indicates the effect of the notch upon the fatigue strength of a machine component when it is subjected to cyclic loading.

The notch sensitivity factor is defined as[6,7]

$$q = \frac{K_f - 1}{K - 1} \tag{4.8.1}$$

in which the stress concentration factor is denoted by K and the fatigue strength reduction factor K_f is defined by

$$K_f = \frac{\text{fatigue strength of unnotched specimens at } N \text{ cycles}}{\text{fatigue strength of notched specimens at } N \text{ cycles}} \tag{4.8.2}$$

Brief examination of equation 4.8.1 reveals that it ranges from 0 for no notch effect ($K_f = 1$) to 1.0 for the full notch effect ($K_f = K$) where the stress concentration factor and the fatigue strength reduction factors are equal.

A representative relation between the notch sensitivity, the fatigue strength reduction factor, and the stress concentration factor for some steels is shown in Figure 4.8.1.[7]

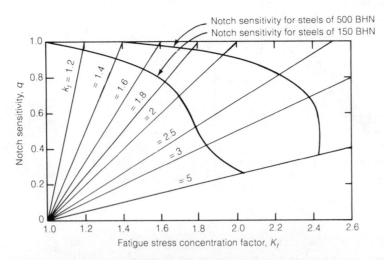

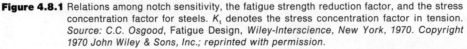

Figure 4.8.1 Relations among notch sensitivity, the fatigue strength reduction factor, and the stress concentration factor for steels. K_t denotes the stress concentration factor in tension. *Source: C.C. Osgood, Fatigue Design, Wiley-Interscience, New York, 1970. Copyright 1970 John Wiley & Sons, Inc.; reprinted with permission.*

In this figure the radial lines are plots of

$$K_f = q(K - 1) + 1$$

which begin at the origin and terminate at values of K along the upper edge of the graph: that is, on the $q = 1$ line. The two curves are indicative of the q values that hold for the two steels shown for some particular heat treatment and for some particular manufacturing process.

To demonstrate the use of these curves, suppose that we have a machine part with a groove that produces a stress concentration factor of 1.6 when the part is placed in tension under a steady load. If that part was produced from that steel having a Brinell hardness of 150 (i.e., 150 BHN) which was used to obtain the data shown in Figure 4.8.1 we may use that figure to find K_f by projecting a line vertically downward from the intersection of the $K = 1.6$ line with the q curve for 150 BHN to read $K_f = 1.5$, approximately, on the $q = 0$ line at the bottom of the graph. Reading horizontally from the intersection point yields $q = 0.5/0.6 = 0.833$.

Because the curves in Figure 4.8.1 hold for a steel of unknown nature, except for its hardness, it should not be used for the design of actual machine components: the data for the steel actually used must enter calculation. For example, AISI 1020 steel can have a range of mechanical properties depending upon the grade of AISI 1020 and upon the manufacturer of the steel.

Juvinal stated in 1967 that there was no universal agreement upon the effect of K_f upon the mean and average stress in their determination of the fatigue life of the material.[6] Since that controversy has yet to be resolved, we will arbitrarily select the opinion that K_f affects only the alternating stress.

4.9 DESIGN APPROXIMATIONS: SODERBERG, GOODMAN, AND GERBER DIAGRAMS

Extraction of design criteria from the preceding experimental observations has been accomplished by replacing the data scatter by curves and lines that represent the minimum performance of the material in order to obtain a cautious estimate of its behavior. As indicated by the curves on the side and end of the block in Figure 4.6.4, two types of curves are of interest: those that assure infinite life and those that predict when failure will occur when infinite life is impossible. This section will be concerned with the first of these, criteria for infinite, cautiously termed *indefinite*, life.

By observing that the alternating stress is zero along a diagonal line from the lower left corner to the upper corner of the end of the block in Figure 4.6.4, we can replot the curves on the end of that block in a rectangular coordinate system as shown in Figure 4.9.1. Since the as-forged curve permits the smallest alternating stress, it represents the minimum performance of the material and is, therefore, often used as the design curve for all surface finishes.

The simplest mathematically and the most cautious of the commonly accepted design approximations is the *Soderberg line* which is a straight line from the endurance

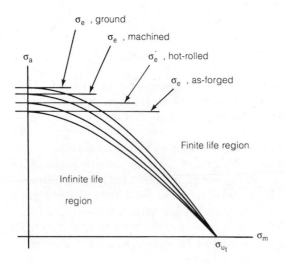

Figure 4.9.1 Infinite life boundary curves in σ_a, σ_m coordinates.

stress (endurance limit) on the alternating stress axis to the yield point (yield stress) on the mean stress axis, as shown in Figure 4.9.2. According to the *Soderberg criterion* any combination of alternating and mean stresses that form the coordinates for a point lying in the triangular region bounded by Soderberg line and the σ_a and σ_m axes will allow infinite life.

The *Goodman line* differs from the Soderberg line only in that it extends from the endurance stress σ_e on the alternating stress axis to the ultimate stress (ultimate strength) σ_u instead of the yield stress on the mean stress axis. The *Goodman criterion* is accord-

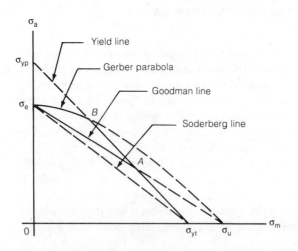

Figure 4.9.2 Approximations to curves in Figure 4.7.2 and the yield line.

ingly defined relative to this line and the σ_a and σ_m axes. An improved criterion for machine design is that obtained by adding the *yield line* which extends from the yield stress on the alternating stress axis to the yield stress on the mean stress axis, as in Figure 4.9.2. Any combination of alternating and mean stresses corresponding to a point that lies in the region bounded by the Goodman line from σ_e to point A, the yield line from point A to σ_{yp}, and by the σ_a and σ_m axes will not only assure indefinite life, but will also assure freedom from yielding. Points within this region are said to satisfy the *Goodman-yield* (modified Goodman) criterion.

The *Gerber-yield* criterion is a better approximation to the bounding curves shown in Figure 4.9.1, but it is also requires more calculation. It is constructed by passing a parabola that is symmetric about the σ_a axis through the endurance stress σ_e and the ultimate stress σ_u. This is the *Gerber parabola*, and together with the yield line it forms the Gerber-yield criterion. Any combination of alternating and mean stresses that forms the coordinates of a point lying within the region bounded by the Gerber parabola between σ_e and point B, by the yield line from B to σ_{y_t}, by the σ_m axis between the origin and σ_{y_t}, and by the σ_a axis from the origin to σ_e in Figure 4.9.2 will, according to this criterion, assure indefinite life without yielding.

The Soderberg Equation

In the case of the Soderberg criterion we can write the equation of a straight line from σ_e to σ_{y_t} in the σ_a, σ_m coordinate system as

$$\sigma_a = a\sigma_m + b \tag{4.9.1}$$

where constants a and b may be determined from the conditions that

$$\begin{aligned} \sigma_a &= \sigma_e & \text{when } \sigma_m = 0 \\ \sigma_m &= \sigma_y & \text{when } \sigma_a = 0 \end{aligned} \tag{4.9.2}$$

Thus,

$$\sigma_a = \sigma_e - \sigma_e \frac{\sigma_m}{\sigma_y} \tag{4.9.3}$$

From Figure 4.9.2 we observe that moving any of these curves toward the origin corresponds to the introduction of a safety factor. Thus, if we denote the safety factor by ζ we may write the equation of a shifted Soderberg line by replacing σ_e and σ_{yp} by σ_e/ζ and σ_{yp}/ζ in (4.9.3) to obtain

$$\frac{\sigma_a}{\sigma_e} + \frac{\sigma_m}{\sigma_y} = \frac{1}{\zeta} \tag{4.9.4}$$

If we include the effect of the fatigue strength reduction factor K_f and assume that it affects only the response of the material to alternating stresses, as assumed in the last section, equation 4.9.4 becomes

$$\frac{\sigma_m}{\sigma_y} + K_f \frac{\sigma_a}{\sigma_e} = \frac{1}{\zeta} \tag{4.9.5}$$

which is known as the Soderberg equation.

The Goodman Equation

Since the Goodman line differs from the Soderberg line only in that it intercepts the means stress axis at the ultimate stress, we may obtain the Goodman equation from the Soderberg equation merely by replacing σ_{yp} by σ_u. Thus,

$$\frac{\sigma_m}{\sigma_u} + K_f \frac{\sigma_a}{\sigma_e} = \frac{1}{\zeta} \tag{4.9.6}$$

represents the Goodman equation.

The Goodman-Yield Equations

This relation may be formed by adding the equation for the yield line to that of the Goodman line, equation 4.9.6. Upon returning to (4.9.1) and using the conditions that

$$\begin{aligned} K_f \sigma_a &= \sigma_y \quad \text{when } \sigma_m = 0 \\ \sigma_m &= \sigma_y \quad \text{when } \sigma_a = 0 \end{aligned} \tag{4.9.7}$$

we find that the yield line may be written as

$$K_f \sigma_a = \frac{\sigma_y}{\zeta} - \sigma_m \tag{4.9.8}$$

in which ζ represent the safety factor. The Goodman-yield criterion, is therefore, represented by equations 4.9.6 and 4.9.8.

Although we could solve (4.9.6) and (4.9.8) simultaneously to find the coordinates of the intersection point, that information does not simplify the application of the Goodman-yield criterion. It is, in fact, simpler to program a computer calculator to solve both (4.9.6) and (4.9.8) for σ_a, σ_m, or ζ and accept the smaller of the two values for any one of these than it is to evaluate the expression for the intersection point, select the pertinent equation, and solve it.

Regardless of the program details, the numerical solution is superior to graphical methods both in accuracy and speed. The time required to install the program listed in

Ref. 8 may be less than that required for one careful plot, and it is certainly less than that required for two plots.

The Gerber-Yield Equations

We may obtain the Gerber-yield equations by replacing equation 4.9.6 by the equation for a parabola passing through the endurance stress on the alternating stress axis and the ultimate stress on the mean stress axis. The equation for a parabola in these co-ordinates is

$$\sigma_a = \sigma_e - a\sigma_m{}^2 \tag{4.9.9}$$

where constant a may be determined from the condition that

$$\sigma_a = 0 \quad \text{at} \quad \sigma_m = \sigma_u \tag{4.9.10}$$

Thus, (4.9.9) becomes

$$\sigma_a = \sigma_e - \sigma_e \left(\frac{\sigma_m}{\sigma_u}\right)^2 \tag{4.9.11}$$

Upon dividing by σ_a, including safety factor ζ, and introducing K_f as before, this relation becomes

$$K_f\zeta\frac{\sigma_a}{\sigma_e} + \left(\zeta\frac{\sigma_m}{\sigma_u}\right)^2 = 1 \tag{4.9.12}$$

which is the equation for the Gerber parabola. The Gerber-yield criterion is, therefore, represented by equations 4.9.8 and 4.9.12.

Extension of the Soderberg, Goodman-yield, and Gerber-yield conditions to alternating and mean shear stress requires only that each stress term be replaced by the corresponding shear term. When experimental data for the endurance stress in shear and the ultimate stress in shear are unavailable it is common practice to assume either that $\tau_u = \sigma_u/2$ and $\tau_e = \sigma_e/2$, based upon the Mohr Circle for uniaxial tension, or that $\tau_e = \sigma_e/\sqrt{3}$, based upon the Huber-von Mises-Hencky relation and as justified experimentally by the data shown in Figure 4.7.4, and that $\tau_u = \sigma_u/\sqrt{3}$, based upon an extension of the Huber-von Mises-Hencky relation.[10]

Application of Goodman-yield and Gerber-yield is greatly simplified by writing computer programs to calculate the mean stress, the alternating stress, and the safety factor for each. Reference 8 gives an explicit form for each of these relations along with a program written for the HP-41CV/CX programmable calculators. Note that in Ref. 8 there is a misprint in Figure 2 which may be corrected by comparing with formula 7 in that reference. A misprint in line 174 of the programmable calculator program may be corrected by replacing that line with the command GTO 01.

EXAMPLE 4.9.1

Find the maximum completely reversed (tension and compression) load according to the Soderberg criterion that can be superimposed with a factor of safety of 2.0 on a notched plate which is designed to support a steady load of 18 000 lb. Dimensions of the steel link, which has BHN = 150, are as shown in Figure 4.9.3, $\sigma_{yp} = 50\,000$ psi and $\sigma_e = 20\,000$ psi.

We can find the stress concentration factor from Figure 3.A.1 and the ratios

$$r/d = 0.35/1.25 = 0.28$$

$$D/d = 1.4125/1.25 = 1.130$$

However, since the D/d ratio is not explicitly shown in the figure, enter the graph at $r/d = 0.28$ and read from the curves shown (the third place is estimated)

D/d	K
1.05	1.517
1.10	1.648
1.15	1.732

in order to use orthogonal interpolation to find $K = 1.705$. Turn next to Figure 4.8.1 to read $K_f = 1.43$, approximately.

The nominal stress is obtained from $\sigma = P/A = 18\,000/[1.25(0.80)]$ to give $\sigma = 18\,000$ psi. According to the assumption that K_f applies only to the fluctuating stress, write

$$\sigma_a = \frac{\sigma_e}{K_f}\left(\frac{1}{\zeta} - \frac{\sigma_m}{\sigma_{yp}}\right) = \frac{20}{1.43}\left(\frac{1}{2} - \frac{18\,000}{50\,000}\right) \text{psi}$$

$$= 1958 \text{ psi}$$

Since the cross-sectional area is 1.0 in.², the alternating force that can be sustained is 1806 lb.

It should be emphasized that although we have calculated the stress and load to four significant places, the formulas used include a number of approximations that qualitatively represent the behavior deduced from scattered data, so that an error of 20% about this value may not be unexpected.

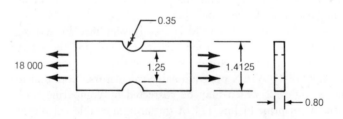

Figure 4.9.3 Tension link for Example 3.9.

EXAMPLE 4.9.2

Find the maximum fluctuating tensile load for the link described in Example 4.9.1 according to the Gerber-yield condition.

Upon solving the Gerber relation for σ_a we obtain

$$\sigma_a = \frac{\sigma_e}{K_f \zeta}\left[1 - \left(\zeta\frac{\sigma_m}{\sigma_u}\right)^2\right]$$

$$= \frac{20\,000}{1.43(2.0)}\left[1 - \left(2 - \frac{18\,000}{50\,000}\right)^2\right] = 5739 \text{ psi}$$

From the yield condition, written as

$$K_f\sigma_a = \frac{\sigma_{yp}}{\zeta} - \sigma_m$$

we find that

$$\sigma_a = [\tfrac{1}{2}(50\,000) - 18\,000]/1.43 = 4895 \text{ psi}$$

The smaller of these two values is the one that satisfies the Gerber-yield conditions, so that the estimated load is 4895 lb.

This large difference between the Gerber-yield and the Soderberg criteria is due to the unreasonably cautious estimate produced by the Soderberg criterion, which has caused some authors to suggest that it never be used.[6]

4.10 COMPRESSIVE LOADS AND STRESS LIMITS

It has been tacitly assumed, and sometimes explicitly stated, that the Goodman equation and its mirror image about the σ_a axis may be used to bound the safe region in both tension and compression. The resulting life criterion is, however, especially cautious over part of the compressive region, as may be seen from Figure 4.10.1, because the data are not symmetric about the σ_a axis.[1]

If the compressive yield stress is taken as negative σ_{yp} the yield lines become symmetric about the σ_a axis. Thus, the boundary of the region for infinite life with no yielding is defined by straight lines CD and DE in Figure 4.10.2 when the mean stress is compressive. The composite region according to the Gerber-yield condition is then as shown in Figure 4.10.2.

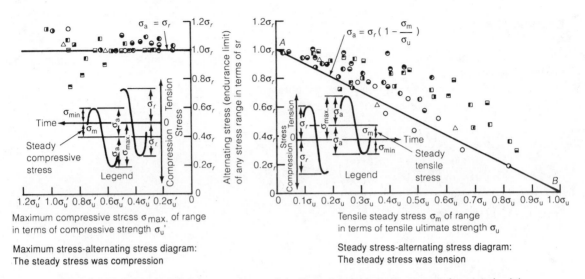

Maximum stress-alternating stress diagram:
The steady stress was compression

Steady stress-alternating stress diagram:
The steady stress was tension

Figure 4.10.1 Data for notch-free specimens of ductile metals subjected to ranges of repeated axial stress. *Source: T.J. Dolan*, Stress Range, ASME Handbook, Sec. 7.2, Metals Engineering Design, 2d ed., *American Society of Mechanical Engineers, New York, 1965.*

Analytical criteria for the life of a machine part may be formulated by reference to the boundaries that correspond to the equations shown for each segment. From Figure 4.10.2 we find that these criteria are

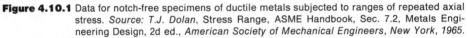

$$\zeta = \text{minimum of} \begin{cases} \dfrac{\sigma_e}{K_f \sigma_a} \\[2ex] \dfrac{\sigma_{yp}}{K_a - \sigma_m} \end{cases} \tag{4.10.1}$$

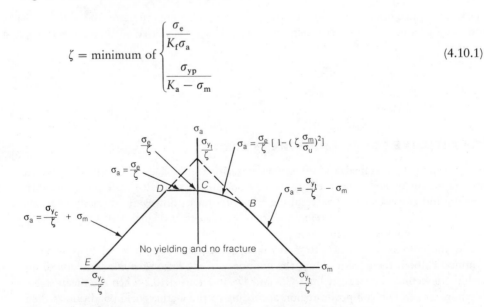

Figure 4.10.2 Region free of yielding and fracture and its bounding curves.

for $\sigma_m \leqq 0$ and that according to the Gerber-yield criterion

$$\zeta = \text{minimum of} \begin{cases} \dfrac{\sigma_{yp}}{K_f \sigma_a + \sigma_m} \\[4mm] \dfrac{1}{2}\left(\dfrac{\sigma_u}{\sigma_m}\right)^2 \left\{\left[\left(K_f \dfrac{\sigma_a}{\sigma_e}\right)^2 + 4\left(\dfrac{\sigma_m}{\sigma_u}\right)^2\right]^{1/2} - \left(K_f \dfrac{\sigma_a}{\sigma_e}\right)\right\} \end{cases} \tag{4.10.2}$$

for $\sigma_m \geqq 0$. Mean and alternating stresses that will allow infinite life are those that produce a safety factor ζ equal to or less than unity when substituted into equations 4.10.1 and 4.10.2.

Torsional fatigue does not distinguish between positive and negative shear stress, so that the region of infinite life in shear is similar to that portion of Figure 4.10.2 where the mean stress is positive, but with direct stresses replaced by shear stresses.

4.11 COMBINED STRESS CRITERIA

In this section we consider two criteria for a machine part to have indefinite life under multiaxial loading.

The older of the two criteria was obtained from the HMH criterion after replacing the individual stress components by their equivalent static stresses as determined by the Soderberg relation. The reasoning behind this process is that since the HMH criterion for the initiation of plastic yielding relates individual stress components to the yield stress in tension divided by the safety factor σ_y/ζ we may extend that criterion to cyclic loads by replacing each of the individual stress components by an equivalent static stress component $\sigma_{uv}'' = (\sigma_y/\zeta)_{uv}$ as determined from the Soderberg relation according to

$$\sigma_{uv}'' = \sigma_{uv_m} + K_{f_c} \frac{\sigma_{y_c}}{\sigma_{e_c}} \sigma_{uv_a} \tag{4.11.1}$$

where $c = \text{t}$ when uv take on values xx, yy, or zz to indicate that values of K_f and σ_y for tension/compression loading pertain and where $c = \text{s}$ when uv take on values xy, yz, or zx to indicate that values of K_f and σ_y for shear loading apply.

According to this criterion a machine component will have an indefinite fatigue life with a safety factor of ζ when the stresses are equal to or less than those that satisfy the equation

$$\left(\frac{\sigma_{y_t}}{\zeta}\right)^2 = \sigma_{xx}''^2 + \sigma_{yy}''^2 + \sigma_{zz}''^2 - \sigma_{xx}''\sigma_{yy}'' - \sigma_{yy}''\sigma_{zz}'' - \sigma_{zz}''\sigma_{xx}''$$

$$+ 3(\sigma_{xy}''^2 + \sigma_{yz}''^2 + \sigma_{zx}''^2) \tag{4.11.2}$$

The second criterion for indefinite fatigue life involves four equations: a modified yield condition for fatigue loading, an expression reminiscent of, but different from, equation 4.11.2, and two equations defining the terms contained in these first two equations.

The modified yield condition for one-, two-, and three-dimensional stress fields is formed from the HMH criterion and written as

$$\left(\frac{\sigma_{y_t}}{\zeta}\right)^2 = \sigma_{xx}'^2 + \sigma_{yy}'^2 + \sigma_{zz}'^2 - \sigma_{xx}'\sigma_{yy}' - \sigma_{yy}'\sigma_{zz}' - \sigma_{zz}'\sigma_{xx}'$$
$$+ 3(\sigma_{xy}'^2 + \sigma_{yz}'^2 + \sigma_{zx}'^2) \tag{4.11.3}$$

where

$$\sigma_{uv}' = K_f \sigma_{uv_a} + \sigma_{uv_m} \tag{4.11.4}$$

in which σ_{uv}' is the equivalent static yield stress. The material will remain elastic if none of the stress components exceed those that satisfy equations 4.11.3 and 4.11.4.

The condition for indefinite fatigue life is written as

$$\left(\frac{\sigma_{u_t}}{\zeta}\right)^2 = \sigma_{xx}^{*2} + \sigma_{yy}^{*2} + \sigma_{zz}^{*2} - \sigma_{xx}^*\sigma_{yy}^* - \sigma_{yy}^*\sigma_{zz}^* - \sigma_{zz}^*\sigma_{xx}^*$$
$$+ 3(\sigma_{xy}^{*2} + \sigma_{yz}^{*2} + \sigma_{zx}^{*2}) \tag{4.11.5}$$

where

$$\sigma_{uv}^* = \frac{1}{2}\left(K_{f_c}\sigma_{u_c}\frac{\sigma_{uv_a}}{\sigma_{e_c}}\right) + \frac{1}{2}\left[\left(K_{f_c}\sigma_{u_c}\frac{\sigma_{uv_a}}{\sigma_{e_c}}\right)^2 + 4\sigma_{uv_m}^{\ 2}\right]^{1/2} \tag{4.11.6}$$

and where c, u, and v play the same roles in equations 4.11.4 and 4.11.6 as they do in equation 4.11.1. Equation 4.11.6 was derived from the Gerber condition, and, therefore, equations 4.11.5 and 4.11.6 reduce to the Gerber condition when only one stress component acts.[10]

Together equations 4.11.3 and 4.11.5 form an extension of the Gerber-yield criterion for indefinite life to those cases where the stresses and loads may be in one, two, or three dimensions.

As in the Gerber-yield criterion for one dimension, equations 4.11.3 and 4.11.5 define the maximum stresses and the maximum safety factor combination which can assure no plastic yielding, equation 4.11.3, and indefinite life, equation 4.11.4. Hence, we once again solve both equations for the particular stress component or safety factor required. The smaller of these two solutions is clearly the maximum value of that stress component or the safety factor that will allow infinite life and only elastic deformation.

Formal calculation of equivalent static stress components from the Goodman relation and substitution into the HMH condition to form a combined stress criterion has not been discussed because it produces a self-contradictory result. Substitution of an equivalent static stress calculated from the Goodman relation into equation 4.11.5 yields a relation that is internally consistent but that does not agree as well with experimental data.[10]

Equations 4.11.3 through 4.11.6 may be easily programmed on either a personal computer or a programmable pocket calculator, such as the Hewlett-Packard HP-4ICX, outlined in Ref. 22, to easily solve for the alternating stress σ_{uv_a}, the mean stress σ_{uv_m}, or the safety factor ζ.

EXAMPLE 4.11.1

Find the largest alternating torsional stress that may be applied to a cylindrical shaft while it is subjected to an alternating cyclic tensile stress of 129 MPa and a mean tensile stress of 200 MPa. For this steel

Normal   *Shear*

$$\sigma_u = 1120 \text{ MPa} \qquad \sigma_u = 750 \text{ MPa} \qquad q = 0.81$$

$$\sigma_y = 900 \text{ MPa} \qquad \sigma_y = 600 \text{ MPa} \qquad \zeta = 1.40$$

$$\sigma_e = 760 \text{ MPa} \qquad \sigma_e = 440 \text{ MPa}$$

$$K_t = 2.14 \qquad\qquad K_s = 1.84$$

Begin by calculating the fatigue strength reduction factors in tension and torsion from equation (4.8.1)

$$K_f = 0.81(2.14 - 1) + 1 = 1.923$$

$$K_f = 0.81(1.84 - 1) + 1 = 1.680$$

Substitution into (4.11.6) yields

$$\sigma_{zz}^* = \frac{1}{2}\left[1.923(1120)\frac{120}{760}\right] + \frac{1}{2}[(340.067)^2 + 4(200)^2]^{1/2} = 432.543 \text{ MPa}$$

and

$$\sigma_{r\theta}^* = 1.68(750)\frac{\sigma_{r\theta}}{440} = 2.864\sigma_{r\theta}$$

Substitution of these values into relation (4.11.5) yields

$$\left(\frac{1120}{1.4}\right)^2 = (432.543)^2 + 3(2.864\sigma_{r\theta})^2$$

so that

$$\sigma_{r\theta} = 135.666 \text{ MPa}$$

From (4.11.4)

$$\sigma_{zz}' = 1.932(120) + 200 = 430.760 \text{ MPa}$$

$$\sigma_{r\theta}' = 1.68\sigma_{r\theta}$$

Substitution into (4.11.3) yields

$$\left(\frac{900}{1.4}\right)^2 = (430.76)^2 + 3(1.68\sigma_{r\theta})^2$$

so that

$$\sigma_{r\theta} = 163.992 \text{ MPa}$$

In this instance it is the Gerber criterion that produces the smaller limit for the maximum fluctuating shear stress due to torsion: 135.666 MPa.

EXAMPLE 4.11.2

Find the largest fluctuating torsional stress for the conditions of Example 4.11.1 if the mean tensile stress is increased to 350 MPa.

Only the equivalent stress components σ_{zz}^* and σ_{zz}' are altered. They become

$$\sigma_{zz}^* = \frac{1}{2} 340.067 + \frac{1}{2}[(340.067)^2 + 4(350)^2]^{1/2} = 559.150 \text{ MPa}$$

and

$$\sigma_{zz}' = 1.923(120) + 350 = 580.760 \text{ MPa}$$

Substitution of these values along with the previously calculated expressions involving $\sigma_{r\theta}^*$ and $\sigma_{r\theta}'$ into equations 4.11.3 and 4.11.5 and solving for the cyclic torsional stress yields

$$\sigma_{r\theta} = 115.338 \text{ MPa}$$

from (4.11.5) and

$$\sigma_{r\theta} = 94.730 \text{ MPa}$$

from (4.11.3). Thus, we find that when the axial mean stress is increased to 350 MPa the yield condition provides the limiting value for the cyclic torsional stress.

As implied by the use of stress components in circular cylindrical coordinates r, θ, and z, equations 4.11.1 through 4.11.6 apply to stress components in any orthogonal Cartesian coordinate system. (For the mathematically inclined, this is because equations 4.11.3 and 4.11.6 were derived from tensor invariants. Quantities referred to as stress components in this discussion should be described as the physical components of the stress tensor. This was not done in deference to the common, but less accurate, terminology used in traditional machine design texts.)

FATIGUE (CYCLIC LOADS)—ESTIMATES OF CYCLES TO FAILURE

4.12 CUMULATIVE DAMAGE: HIGH-CYCLE FATIGUE FAILURE

We now turn our attention to the case of a single stress component, say σ_{xx}, where the maximum stress is larger than the endurance stress, so that eventual failure is guaranteed. This condition is unavoidable, for example, in and near the contact regions between ball bearings and their races and between crowned gear teeth in contact.

Two estimation formulas will be considered: the Palmgren-Miner and the Shanley, both of which apply to cyclic tension and cyclic torsion. Both of these formulas rely upon S-N data as displayed on the side of the three-dimensional plot shown in Figure 4.6.4. The data for that plot and other S-N plots for different hardness ranges are found in Table 4.12.1.

Rather than plot all of this data on a series of graphs similar to that in Figure 4.12.1, as was done by Noll and Erickson,[5] it is now easier to use a computer or pocket calculator to calculate these values from the formulas implied by these curves.

We may utilize the function recovery formulas given in Chapter 1 to find the formulas used in constructing these S-N graphs by making the following associations

$$x \to N \qquad x_1 \to 1000 \qquad x_1 \to 1\,000\,000$$

$$f(x) \to \sigma \qquad f(x_0) \to 0.9\sigma_u \qquad f(x_1) \to \sigma_e$$

TABLE 4.12.1 Stress at 1000 cycles is taken as 0.90 of the tensile strength in drawing the *S-N* curves and in the corresponding formulas

Brinell hardness	Tensile strength (psi)	Yield point (psi)	Endurance limit (psi)			
			Ground surface	Machined surface	Hot-rolled surface	As-forged surface
160–187	77 000	48 000	33 000	30 000	24 000	18 000
187–207	89 000	60 000	39 000	34 000	25 000	19 000
207–217	99 000	69 000	44 000	37 000	27 000	20 000
217–229	103 000	74 000	46 000	38 000	27 000	20 000
229–241	109 000	80 000	49 000	40 000	28 000	20 000
241–255	114 000	86 000	51 000	42 000	29 000	21 000
255–269	121 000	94 000	55 000	44 000	30 000	21 000
269–285	127 000	101 000	57 000	46 000	30 000	22 000
285–302	135 000	110 000	61 000	49 000	31 000	22 000
302–321	142 000	120 000	64 000	51 000	32 000	23 000
321–352	151 000	130 000	68 000	53 000	33 000	23 000
352–375	166 000	147 000	74 000	57 000	34 000	24 000
375–401	176 000	158 000	78 000	59 000	35 000	25 000
401–429	188 000	171 000	82 000	62 000	35 000	24 000
429–461	202 000	183 000	86 000	64 000	35 000	24 000
461–495	217 000	196 000	88 000	65 000	34 000	22 000
495–514	233 000	210 000	89 000	66 000	32 000	20 000
514–555	241 000	217 000	89 000	65 000	31 000	19 000

Source: G.C. Noll and M.A. Erickson, *Allowable Stresses for Steel Members of Finite Life, Proceedings of the Society for Experimental Stress Analysis*, Vol. 5, Society for Experimental Mechanics, Brookfield, CT, 1948, pp. 132–143.

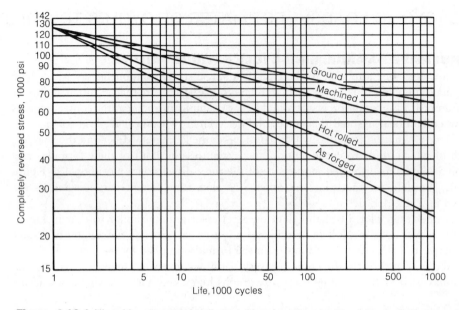

Figure 4.12.1 Allowable stresses for limited life of parts whose minimum Brinell hardness specification lies in the range of 302 to 321. *Source: G.C. Noll and M.A. Erickson, Allowable Stresses for Steel Members of Finite Life, Proceedings of the Society for Experimental Stress Analysis, Vol. 5, Society for Experimental Mechanics, Brookfield, CT, 1948, pp. 132–143.*

After simplifying the resulting expressions we have that

$$\sigma_a = \frac{b}{N^a} \quad \text{or} \quad N = \left(\frac{b}{\sigma_a}\right)^{1/a} \qquad \text{for } \sigma > \sigma_e \tag{4.12.1}$$

where

$$a = \frac{1}{3} \log \frac{0.9\sigma_u}{\sigma_e} \tag{4.12.2}$$

and

$$b = \frac{(0.9\sigma_u)^2}{\sigma_e} \tag{4.12.3}$$

Using these relations and the data from Table 4.12.1 for Brinell hardness from 302 to 321 we find, for example, that for the as-forged surface

$$\sigma_a = \frac{710\,123.475}{N^{0.248\,266\,673}}$$

and

$$N = (710\,123.478/\sigma_a)^{4.027\,910\,640}$$

describe the corresponding line in Figure 4.12.1. Plotted data as in Figure 4.12.1 are difficult to read unless one has the logarithmic scale memorized in order to locate the missing subdivisions. This problem is eliminated by the use of the two equations shown in display (4.12.2) along with the definitions in (4.12.3).

Although Figure 4.6.3 indicated that an uncertainty in N of the order of 15% is to be expected, we will continue to calculate life to the nearest cycle as an aid to following the calculations and to carefully locate the mean of the implied distribution of cycles and stresses before rounding to the nearest several hundred or thousand cycles.

4.13 EFFECTS OF STRESS CONCENTRATION AND MEAN STRESS ON LIFE

Noll and Erickson[5] observed that the effects of both a mean stress and a fatigue strength reduction factor may be accounted for by associating them with an *equivalent alternating stress* σ_{a_e} that is defined by

$$\sigma_{a_e} = \frac{K_f \sigma_a}{1 - \left(\dfrac{\sigma_m}{\sigma_u}\right)^v} \tag{4.13.1}$$

in which

$v = 2$ whenever the Gerber criterion is used

$= 1$ whenever the Goodman criterion is used

Noll and Erickson considered only the Goodman criterion in which $v = 1$.

The general expression for the relation between stress and fatigue life that we shall use henceforth is therefore

$$\sigma_{a_e} = \frac{b}{N^a} \quad \text{and} \quad N = \left(\frac{b}{\sigma_{a_e}}\right) \tag{4.13.2}$$

where quantities a and b are still defined by equations 4.12.2 and 4.12.3. Note that $\sigma_{a_e} = \sigma_a$ whenever $\sigma_m = 0$ and $K_f = 1$ in equation (4.13.1).

EXAMPLE 4.13.1

Calculate the life of an as-forged steel part whose ultimate stress is 105 000 psi and whose Brinell hardness is 310 when subjected to an alternating stress whose maximum value is 68 000 psi and whose minimum value is 32 000 psi. Use the Goodman line, $v = 1$.

First calculate the mean stress from

$$\sigma_m = (68\,000 + 32\,000)/2 = 50\,000 \text{ psi}$$

and then substitute into (4.15.1) to find that the equivalent alternating stress is

$$\sigma_{a_e} = 105\,000\,\frac{68 - 50}{105 - 50} = 34\,364 \text{ psi}$$

From (4.12.3)

$$N = \left(\frac{701\,260}{34\,364}\right)^{4.042\,613\,230} = 197\,260 \text{ cycles}$$

is the estimated life of the part in cycles.

EXAMPLE 4.13.2

Calculate the equivalent alternating stress using the Gerber parabola rather than the Goodman line—use $v = 2$—for Example 4.13.1.

According to (4.13.2)

$$\sigma_{a_e} = \frac{18\,000}{1 - \left(\dfrac{50}{105}\right)^2} = 23\,279 \text{ psi}$$

so that substitution into equation 4.12.3 yields

$$N = \left(\frac{701\,260}{23\,279}\right)^{4.042\,613\,230} = 952\,487 \text{ cycles}$$

which clearly indicates the reason for turning to the Gerber curve. It predicts longer life with no increase in failure probability simply because it uses a more realistic fit to experimental data.

4.14 PALMGREN-MINER AND SHANLEY FAILURE PREDICTIONS

Palmgren-Miner Theory

This theory is concerned with loading schedules (or histories) as represented in Figure 4.14.1 and is based upon the assumption that damage to the material from cyclic loading is dependent only upon the number of cycles experienced at a particular stress. It

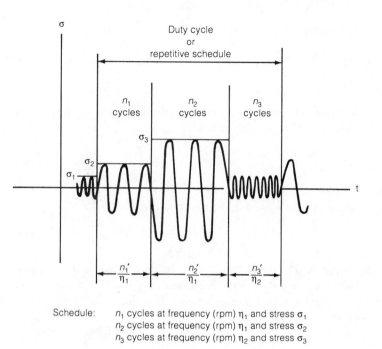

Schedule: n_1 cycles at frequency (rpm) η_1 and stress σ_1
                   n_2 cycles at frequency (rpm) η_1 and stress σ_2
                   n_3 cycles at frequency (rpm) η_2 and stress σ_3

Figure 4.14.1 One period of a repetitive loading schedule of n'_1, n'_2, and n'_3 cycles at stresses σ_1, σ_2, and σ_3.

is also assumed that the order of application of the various stress levels does not affect the life of the component. Consistent with these assumptions, let

$$N_i \equiv N_i(\sigma_i)$$

denote the total life in cycles of the mechanical component when subjected to a completely reversed cyclic loading with amplitude σ_i and let

$$n_i \equiv n_i(\sigma_i)$$

denote the number of completely reversed cycles accumulated at stress σ_i throughout its life under a load schedule that includes other amplitudes of cyclic stress.

Using these parameters, the Palmgren-Miner theory postulates that the fraction of the component's life consumed by n_i cycles at σ_i is given by

$$\frac{D_i}{D} = \frac{n_i}{N_i} \tag{4.14.1}$$

where D_i is the life consumed by loading at stress σ_i and D is the entire life of the mechanical component. Whenever the component is exposed to several different stress amplitudes, failure is expected when

$$\sum_{i=1}^{q} \frac{D_i}{D} = \sum_{i=1}^{q} \frac{n_i}{N_i} \qquad (4.14.2)$$

where q is the number of different stress amplitudes involved. If these loads cause the machine part to fail after N_e cycles, it must be true that

$$N_e = \sum_{i=1}^{q} n_i \qquad (4.14.3)$$

The proportion α_i of a component's life used in n_i cycles is given by

$$\alpha_i = n_i/N_e \qquad (4.14.4)$$

where α_i may also be interpreted as the proportion of the duty schedule during which stress σ_i acts if only one duty schedule acts throughout the life of the component. Substitution for n_i from equation 4.14.4 into equation 4.14.3 shows, as expected, that the sum of the proportions must equal unity:

$$\sum_{i=1}^{q} \alpha_i = 1 \qquad (4.14.5)$$

Next, substitution for n_i from equation 4.14.4 into the right-hand equality in display (4.14.2) yields the Palmgren-Miner relation, namely,

$$\frac{1}{N_e} = \sum_{i=1}^{q} \frac{\alpha_i}{N_i} \qquad (4.14.6)$$

EXAMPLE 4.14.1

Estimate the life in hours according to the Palmgren-Miner criterion of a piston rod made from UNS G10450 (AISI 1045) steel and used in a reciprocating pump that operates at 30 rpm. The rod is ground and has a Brinnel hardness of 310. The expected load schedule calls for

- 20 min at 68 000 psi

- 30 min at 83 000 psi

- 10 min at 65 000 psi

Although (4.14.6) is written in terms of N_e and N_i, it may be rewritten in terms of times t_i and t_e by using the relations

$$N_i = nt_i \qquad N_e = nt_e$$

whenever a constant frequency n holds throughout the life of the mechanical component. Thus, from (4.14.6)

$$\frac{1}{nt_e} = \frac{1}{N_e} = \sum_{i=1}^{q} \frac{\alpha_i}{N_i} = \sum_{i=1}^{q} \frac{\alpha_i}{nt_i}$$

or

$$\frac{1}{t_e} = \sum_{i=1}^{q} \frac{\alpha_i}{t_i} \tag{4.14.7}$$

From formula 4.12.1 for N as a function of σ_a for a ground surface find ($a = 0.100\,116\,960$, $b = 255\,200.62$) with $\sigma_u = 142\,000$ psi, $\sigma_y = 120\,000$ psi, and $\sigma_e = 64\,000$ psi for a ground surface, that has the following criteria:

Time	α_i	N_i
20 min	0.333	545 781 cycles
30 min	0.500	74 531 cycles
10 min	0.167	856 530 cycles

where we have chosen to work in cycles because the S-N curve is in cycles.

Substitution into the Palmgren-Miner equation when solved for N_e yields

$$N_e = \frac{1}{\sum_{i=1}^{q} \frac{\alpha_i}{N_i}} = \frac{1}{\dfrac{0.333}{545\,781} + \dfrac{0.500}{74\,531} + \dfrac{0.167}{856\,530}} = 133\,079 \text{ cycles}$$

$$= 73.93 \text{ h}$$

EXAMPLE 4.14.2

Find the maximum stress that can be applied during the 30-min interval in Example 4.14.1 if the life is to be increased to 120 h.

A life of 120 h corresponds to

$$120 \times 60 \times 30 = 216\,000 \text{ cycles}$$

From

$$\frac{1}{N_e} = \frac{\alpha_1}{N_1} + \frac{\alpha_2}{N_2} + \frac{\alpha_3}{N_3}$$

we find that

$$N_2 = \cfrac{\alpha_2}{\cfrac{1}{N_e} - \cfrac{\alpha_1}{N_1} - \cfrac{\alpha_3}{N_3}}$$

$$= \cfrac{0.500}{\cfrac{1}{216\,000} - \cfrac{0.333}{545\,758} - \cfrac{0.167}{856\,530}} = 130\,754 \text{ cycles}$$

Substitution of this value into formula 4.12.1 yields

$$\sigma_a = 255\,200.62(130\,754)^{-0.100\,116\,960}$$
$$= 78\,458 \text{ psi}$$

EXAMPLE 4.14.3

Estimate the expected life of the pump in example 4.14.1 if it is operated according to the following schedule:

- 20 min at 20 rpm at a stress of 68 000 psi
- 30 min at 15 rpm at a stress of 79 100 psi
- 10 min at 36 rpm at a stress of 67 300 psi

Calculation of α_i begins with finding the number of cycles for each time interval. Thus,

20 min at 20 rpm	400 cycles
30 min at 15 rpm	450
10 min at 36 rpm	360
	1210 cycles/(60min)

from which it follows that

α_i	N_i
400/1210 = 0.3306	545 781
450/1210 = 0.3719	120 534
360/1210 = 0.2975	605 207

where the life for each time period was found from the *S-N* curve data as before. Thus,

$$N_e = \frac{1}{\dfrac{0.3306}{545\,781} + \dfrac{0.3719}{120\,435} + \dfrac{0.2975}{605\,207}} = 239\,078 \text{ cycles}$$

and the time in hours is found by dividing the number of cycles by the cycles per hour to get

$$\text{life} = \frac{239\,078}{1210} = 197.58 \text{ h}$$

Shanley Theory

Various other cumulative damage formulas have been proposed, motivated by scattered evidence that in some cases the Palmgren-Miner relation may predict a fatigue life longer than that observed experimentally. Of these, the Shanley formula is one of the more readily applicable because it is based upon the *S-N* already considered.[11] It differs from the Palmgren-Miner relation in that it assumes that the portion of the component's life consumed during its exposure to stress σ_i is given by the damage ratio

$$\frac{D_i}{D} = \frac{n_i}{N_e}\left(\frac{N_e}{N_i}\right)^k \tag{4.14.8}$$

The purpose of the $(N_e/N_i)^k$ term is to magnify the significance of loads that produce large stresses as indicated by small values of N_i associated with these stresses.

Proceeding as in the derivation of the Palmgren-Miner relation, we find that

$$N_e = \left[\sum_{i=1}^{q} \frac{\alpha_i}{(N_i)^k}\right]^{-1/k} \tag{4.14.9}$$

In design applications where experimental data are not yet available it has been suggested that $k = 2$.[9]

EXAMPLE 4.14.4

Compare the Palmgren-Miner estimate of the life of the piston in Example 4.14.1 with that estimated by the Shanley method for $k = 1.5$, 2.0, and 2.5.

Substitution into equation 4.14.9 yields

$$N_e = \left(\frac{0.333}{54\,781^k} + \frac{0.500}{74\,531^k} + \frac{0.167}{1\,905\,284^k}\right)^{-1/k}$$

so that for $k = 1.5$ we find $N_e = 115\,537$ cycles, or 64.187 h; for $k = 2.0$ we find $N_e = 104\,727$ cycles, or 58.182 h; and for $k = 2.5$ we find $N_e = 98\,160$ cycles, or 54.534 h.

EXAMPLE 4.14.5

Compare the maximum stress that can be applied during the 30-min interval in Example 4.14.1 as calculated by the Palmgren-Miner relation with the stress estimate calculated from the Shanley relation. Rewrite equation 4.14.9 as

$$\frac{1}{N_e{}^k} = \frac{\alpha_1}{N_1{}^k} + \frac{\alpha_2}{N_2{}^k} + \frac{\alpha_3}{N_3{}^k}$$

and solve for N_2 to obtain

$$\frac{1}{N_2} = \left[\frac{1}{\alpha_2} \left(\frac{1}{N_e{}^k} - \frac{\alpha_1}{N_1{}^k} - \frac{\alpha_2}{N_3{}^k} \right) \right]^{1/k}$$

The corresponding stress is, therefore, given by

$$\sigma = \frac{b}{N^a} = b \left[\frac{1}{a_2} \left(\frac{1}{N_e{}^k} - \frac{\alpha_1}{N_1{}^k} - \frac{\alpha_3}{N_3{}^k} \right) \right]^{a/k}$$

where $a = 0.100\,169\,20$ and $b = 255\,200.62$ psi. Upon substitution for k we find the following combinations

k	σ (psi)
1.15	77 649
2.0	77 031
2.5	76 608

See problems 4.45 through 4.47 for additional properties of the Palmgren-Miner criterion and for the Shanley relations.

4.15 CONTINUOUSLY VARYING LOADS

The summation over a discrete number of loads becomes the integral over the number of cycles when the load changes from cycle to cycle. In terms of the Shanley criterion, which includes the Palmgren-Miner relation as a special case when $k = 1.0$, the life N_e of a machine component may be calculated from

$$\frac{1}{N_e} = \left(\frac{1}{N_0} \int_0^{N_0} \frac{dn}{N^k} \right)^{1/k} \tag{4.15.1}$$

where $d\alpha_i = dn/N_0$. From (4.12.2) we may express N in terms of σ according to

$$N_i = (b/\sigma_{ae})^{1/a} \tag{4.15.2}$$

Upon substituting for N in (4.15.1) from (4.15.2) and recalling that b is a constant, we find that (4.15.1) may be written as

$$\frac{1}{N_e} = \left(\frac{1}{N_0 b^{k/a}} \int_0^{N_0} \sigma_{ae}^{k/a} \, dn \right)^{1/k} \tag{4.15.3}$$

Integration of either (4.15.1) or (4.15.3) usually must be performed numerically because the exponent k/a is generally not an integer. Moreover, most realistic loads do not conform to those functions listed in integral tables.

EXAMPLE 4.15.1

Compare the expected life calculated from the Palmgren-Miner form of equation 4.15.3 with that calculated from the Shanley form for a completely reversed cyclic load whose stress amplitude increases linearly from 35 000 to 55 000 psi over 100 cycles and then repeats. The surface of the part is machined and its hardness is BHN 200.

First observe that the stress induced by the load may be written as

$$\sigma_a = 35\,000 + 200n$$

where n is the number of cycles. From Table 4.12.1 find that $\sigma_u = 89\,000$ psi and $\sigma_{yp} = 34\,000$ psi. Substitution into relations 4.12.3 produces $a = 0.124\,051\,20$ and $b = 188\,706.18$. Inserting these values into (4.15.3) yields

$$\frac{1}{N_e} = \left[1.565\,715 \times 10^{-81} \int_0^{100} (35\,000 + 200n)^{15.316\,257} \, dn \right]^{1/1.9}$$

to give

$$N_e = 4684 \text{ cycles}$$

from the Shanley condition. To obtain the Palmgren-Miner value, set k equal to 1.0 to get

$$\frac{1}{N_e} = 2.957\,341 \times 10^{-45} \int_0^{100} (35\,000 + 200n)^{8.061188}\, dn$$

so that

$$N_e = 69\,389 \text{ cycles}$$

4.16 CUMULATIVE DAMAGE: LOW-CYCLE FATIGUE FAILURE

Low-cycle fatigue life prediction, also known as the local strain approach, was developed more recently than high-cycle fatigue life prediction methods and is presently limited to one-dimensional tensile and compressive loading. Although the experimental curves used for obtaining the coefficients and exponents involved are often plotted to a million cycles, the theory itself is generally applied where failure is expected before from one to ten thousand cycles, depending upon the material involved.

Since the life predictions are based upon strain, stress–concentration factors need not be explicitly considered because the strain is by definition a function of the stress and stress concentrations at the point where the strain appears.

Low-cycle fatigue prediction is also known as the local strain approach because is based upon at least some small, or local, region of the machine component being strained beyond the elastic limit, after which any continued strain is said to be *plastic* strain, as illustrated in Figure 4.16.1*a*. Both elastic and plastic strain may coexist in test specimens similar to those shown in Figure 4.16.1*b*.

The stress–strain history of the component after a number load cycles consisting of alternating tension and compression may form a number of closed loops, known as *hysteresis* loops, as illustrated in Figure 4.16.2*a*. As shown in Figure 4.16.2*b*, the stress amplitude of a loop is conventionally designated as $\Delta\sigma$, the associated total strain by $\Delta\varepsilon$, and its elastic and plastic strain components by $\Delta\varepsilon_e$ and $\Delta\varepsilon_p$, respectively.

To emphasize that these stresses are to be calculated on the basis of the locally changing cross section of the component as the strains become plastic, and that these strains are based upon a changing reference length, the strains are termed "true" strains to distinguish them from the approximate strain e (also unfortunately known as engineering strain) given by $\varepsilon = (\Delta L)/L$ and often used in calculating strain from strain gage data, where L is the original gage length.

"True" strain is calculated for a changing reference length L from

$$\varepsilon = \int_L^{L+\Delta L} du/u = \ln[(L + \Delta L)/L] = \ln(1 + e) \tag{4.16.1}$$

"True" is placed in quotation marks because ε is also an approximation to the actual strain. (See Ref. 12 for the theoretically correct expression.)

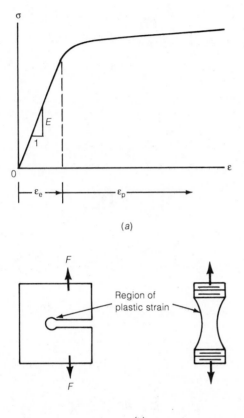

(a)

(b)

Figure 4.16.1 Elastic ε_e and plastic ε_p strain domains in a stress–strain curve in which stress is calculated from the minimum cross section at each strain in (a). Typical test specimens are shown in (b) indicating regions where plastic strain begins.

Experimental data indicate that the relation between cyclic total strain and the number of cycles to failure for many metals, with the exception of gray cast iron and annealed austenitic stainless steel, is of the form shown in Figure 4.16.3. Since the elastic curve may be fitted by

$$\frac{\Delta\varepsilon_e}{2} = \frac{\sigma_f'}{E}\,(2N_f)^b \tag{4.16.2}$$

and since the plastic curve may be fitted by

$$\frac{\Delta\varepsilon_p}{2} = \varepsilon_p'(2N_f)^c \tag{4.16.3}$$

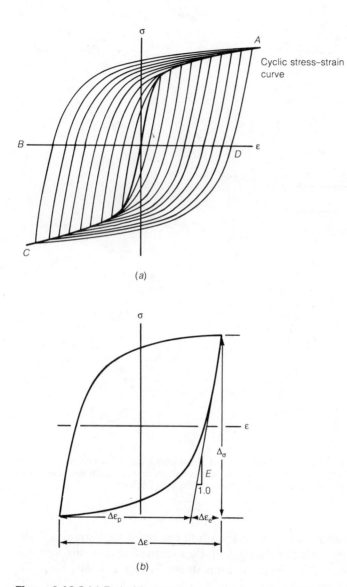

Figure 4.16.2 (a) Typical hysteresis loops and the cyclic stress–strain curve formed by their tips. (b) Definition of $\Delta\varepsilon_e$, $\Delta\varepsilon_p$, and $\Delta\sigma$ for a hysteresis loop.

it follows that the exponents and coefficients may be found from Figure 4.16.3. Thus, the total strain amplitude $\Delta\varepsilon = \Delta\varepsilon_e + \Delta\varepsilon_p$ may be found from

$$\frac{\Delta\varepsilon}{2} = \frac{\sigma'_f}{E}(2N_f)^b + \varepsilon'_p(2N_f)^c \tag{4.16.4}$$

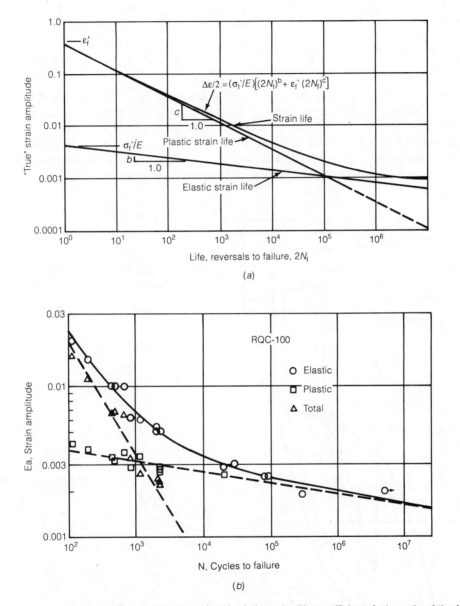

(a)

(b)

Figure 4.16.3 (a) Plot used to determine the fatigue ductility coefficient ε_f', the ratio of the fatigue strength (stress) coefficient divided by the elastic modulus σ_f'/E, the fatigue ductility exponent c, and the fatigue strength (stress) exponent b. (b) Plot showing experimental points for RQC-100 steel produced by Bethlehem Steel. *Sources: (a) SAE J1099, Technical Report of Fatigue Properties, 1988 SAE Handbook. ©1988 Society of Automotive Engineers, Inc. Reprinted with permission. (b) N.E. Dowling, W.R. Brose, and W.K. Wilson,* Advances in Engineering, *Vol. 6, Notched Member Fatigue Life Predictions by the Local Strain Approach, Fatigue under Complex Loading, Society of Automotive Engineers, Inc., 1977. Reprinted with permission.*

in the absence of any mean stress and by

$$\frac{\Delta\varepsilon}{2} = \frac{\sigma_f' - \sigma_0}{E}(2N_f)^b + \varepsilon_f'(2N_f)^c \tag{4.16.5}$$

whenever mean stress σ_0 is not zero.[14,15] In this expression N_f denotes the number of cycles to failure and $2N_f$ represents the number of associated strain reversals to failure.[16]

Equation 4.14.5 was written in terms of $2N_f$ to indicate that low-cycle fatigue failure appears to depend only upon complete cycles in most situations. We must, therefore, be able to count complete loading cycles either from automatically recorded

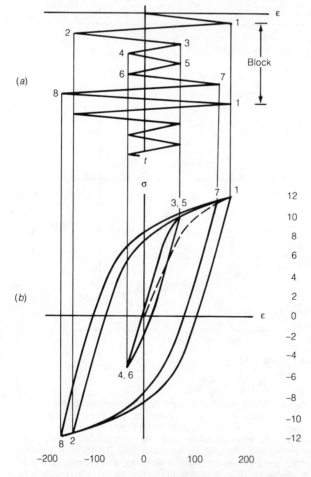

Figure 4.16.4 (a) Initial plot of a repeating load block, strain vs. time (b) hysteresis loops corresponding to one loading block, stress vs. strain.

pen plots (or digital records) of strain or from anticipated strain schedules whenever (4.16.5) applies. If the plot or strain schedule is composed of short blocks that are repeated until failure, the associated hysteresis loops may be easily reconstructed as shown in Figure 4.16.4. It is done by recognizing that points 1 to 8 on the strain vs. time plot in this figure correspond to reversals 1 to 8 on the associated hysteresis loops. The dashed curve represents the initial tensile load and is usually not included in the cycle count.

In particular, the first block begins by increasing the strain with time along the dashed line from the origin to point 1 in Figure 4.16.4*a*, corresponding to the dashed curve from the origin to point 1 in the stress–strain plot in Figure 4.16.4*b*. As the strain is reduced from point 1 to point 2 in (*a*) the lower portion of a hysteresis curve from 1 to 2 in (*b*) is traced. As the strain is increased along a straight line to point 3 in (*a*) the upper portion of a potential hysteresis curve is traced in (*b*). The constant amplitude fluctuation from point 3 to 4 to 5 corresponds to a loop, or load cycle, about a mean stress σ_0. A second cycle is completed as the strain decreases from point 5 to point 6 and back to point 5 on the way to point 7. A third cycle is completed as the strain is reduced from that at point 7 to that at point 2 on the way to point 8. Finally, a fourth cycle is closed as the strain returns to point 1 to begin the second block of cyclic loads. These blocks will continue until the component fails.

Manual counting of complete cycles in-blocks that contain many strain reversals or in loads displaying no periodicity may be done by breaking the strain history into sections that terminate with a complete cycle. Alternative methods for manually counting complete cycles are briefly discussed in the appendix to Ref. 17 where the rain flow method is introduced.

A computer program may be written to automate the counting in which the cyclic stress–strain curve shown in Figure 4.16.5 is represented by

$$\frac{\Delta\varepsilon}{2} = \frac{\Delta\sigma}{2E} + \left(\frac{\Delta\sigma}{K'}\right)^{1/n'} \tag{4.16.6}$$

This expression also describes the initial loading path, shown by the dashed curve in Figure 4.16.4.[14,15] The first term on the right side of equation 4.16.6 represents the elastic strain, the second represents the plastic strain, and K', the cyclic strength coefficient, appearing in the second term may be found from experimental data plotted as in Figure 4.16.5.

Upper hysteresis paths, such as path *ABC* in Figure 4.16.2, may be represented by

$$\frac{\varepsilon - \varepsilon_r}{2} = \frac{\sigma - \sigma_r}{2E} + \left(\frac{\sigma - \sigma_r}{2K'}\right)^{1/n'} \tag{4.16.7}$$

in which ε_r and σ_r represent the coordinates of point *C*, (each of which is negative) while lower hysteresis paths, such as *CDA*, may be represented by

$$\frac{\varepsilon_r - \varepsilon}{2} = \frac{\sigma_r - \sigma}{2E} + \left(\frac{\sigma_r - \sigma}{2K'}\right)^{1/n'} \tag{4.16.8}$$

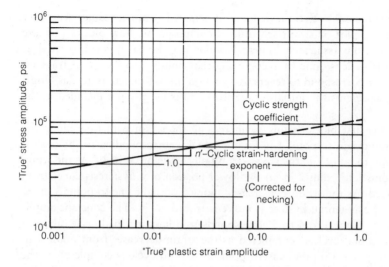

Figure 4.16.5 Stress amplitude vs. plastic strain amplitude plot used to determine the cyclic strain
hardening exponent and the cyclic strength (stress) coefficient. *Source: SAE J1099,
Technical Report of Fatigue Properties, 1988 SAE Handbook. ©1988 Society of Automotive Engineers, Inc. Reprinted with permission.*

Individual reversals may be important whenever failure occurs after a few reversals
or when it is caused by several major half-cycles associated with large plastic strains.[17]
In this event half-cycles must be included (each half-cycle is terminated by a strain
reversal in the strain–time plot) and $2N_f$ in equation 4.16.5 is replaced by N_r, the
number of strain reversals.

Once the N_f have been found for each strain amplitude from equation 4.16.5 using
a numerical method, such as the bisection routine, we may find the life of the machine
component for a particular loading schedule by again calling upon the Palmgren-Miner
criterion, equation 4.13.6. Upon multiplying equation 4.13.6 by N_e and recalling that
$\alpha_i N_e = n_i$ we have that

$$1 = \sum_{i=1}^{q} \frac{n_i}{N_i} \qquad (4.16.9)$$

where n_i is the number of cycles at strain ε_i (linearly related to σ_i for an elastic material) for which the fatigue life is $N_i = N_{fi}$. Whenever the loading is periodic each
period corresponds to a block of β cycles so that if B is the life in blocks then the total
life N_e is given by $N_e = \beta B$ and (4.16.9) may be written as

$$1 = B \sum_{i=1}^{q} \frac{n_i}{N_i} \qquad (4.16.10)$$

where n_i denotes the number of cycles at strain ε_i during any one block.

EXAMPLE 4.16.1

Find the expected life for a machine component made from UNS G10450 steel (quenched and tempered, BHN = 410) when it is subjected to alternating tension and compression as given by the strain–time plot similar to that shown in Figure 4.16.5. The strains at points 1 through 8 are as follows

Point	Strain	Point	Strain
1	0.0170	5	0.0066
2	−0.0141	6	−0.0018
3	0.0066	7	0.0141
4	−0.0018	8	−0.0170

and a mean stress of 250 MPa acts during the cycles between points 3 and 4. From Tables 1 and 2 of Ref. 14 we find that for the SAE 1045 steel used

strain hardening exponent	$n' = 0.146$
elastic modulus	$E = 200$ GPa
fatigue strength (stress) coefficient	$\sigma'_f = 1862$ MPa
fatigue ductility coefficient	$\varepsilon'_f = 0.60$
fatigue strength (stress) exponent	$b = -0.073$
fatigue ductility exponent	$c = -0.70$

According to Figure 4.16.4 the largest strain occurs at points 1 and 8 so that upon recalling the notation shown in Figure 4.16.2b we have that $\varepsilon = 0.0340$. Substitution of this value and the above constants into equation 4.16.5 yields

$$\frac{0.0340}{2} = \frac{1862}{2 \times 10^5} (2N_f)^{-0.073} + 0.60(2N_f)^{-0.70}$$

After calling upon a bisection routine with an initial estimate of the fatigue life between 5 and 2000 cycles we find that $N_{f1} = 153.92$ cycles. Likewise with $\varepsilon = 0.0282$ for the loop with vertices at 2 and 7 we find a fatigue life $N_{f3} = N_f = 232.19$ cycles. Inclusion of the mean stress of 250 MPa acting when the strain cycles between points 3, 4, 5, 6, and 7 leads to

$$\frac{0.0084}{2} = \frac{1862 - 250}{2 \times 105} (2N_f)^{-0.073} + 0.60(2N_f)^{-0.70}$$

and to $N_{f2} = N_f = 16\,027.26$ cycles. Substitution of these values into equation (4.16.10) and solving for 1/B yields

$$\frac{1}{B} = \frac{1}{153.92} + \frac{2}{16\,027.26} + \frac{1}{232.19} = 0.010\,928$$

from which we find that

$$B = 91.504 \text{ blocks} \qquad N_e = 4B = 366 \text{ cycles}$$

Life N_e is given by $4B$ because each block contained four cycles.

Although there is no accepted formula for the uncertainty of this calculation, an error of 20%, or about 70 cycles, may not be unexpected.

Use of two decimal places in the expression for lives N_{fi} implies much more accuracy than the methods justify. They have been retained to avoid the practice of rounding all calculations to one or two places, which can lead to serious errors when dealing with small differences of large numbers or with exponents. In fact, if b and c are changed by 0.005 to -0.068 and -0.695, respectively, the predicted life becomes 392 cycles, for an additional error of slightly over 7%.

The test specimens shown in Figure 4.16.1 were chosen with forethought. The one with the hourglass shape was selected for ease of experimental measurements while the one with the notch was selected to represent a typical machine application where plastic strain is limited to a very small region. This latter consideration is very important in simulating machine elements because practically all machines are designed to maintain their shape and tolerances throughout their lifetime. Because of the hysteresis effect any large-scale plastic deformation would result in a deformed mechanical part. The larger diameter hole at the end of the notch was a concession to the difficulty of measuring strain at the end of a notch with a small radius.

The alternate designation of the low-cycle fatigue method as the local strain approach recognizes that this method has been used for failures beyond 10^4 cycles when the high stress region is very small relative to the cross section supporting the load. Analysis of such regions is important because a crack initiated there can produce a large stress concentration at its tip, which may then initiate crack growth across the entire cross-section.

No clearly defined, absolute guidelines have yet been established to indicate whether high-cycle or low-cycle prediction methods should be applied to a particular situation. In broad terms, however, it appears that high-cycle fatigue prediction should be used where any plastic deformation is constrained by a surrounding elastic region, and that low-cycle, or local strain, prediction should be used where plastic deformation may cause a crack to spread. This choice has been buttressed by a recent paper by Harris and Ioannides which indicates that an endurance limit exists for rolling element bearings even though they undergo plastic strain in the contact region between the rolling element and the race.[18] This is consistent with the use of a high-cycled fatigue method for predicting the life of rolling element bearings. Crack growth in this case is inhibited, perhaps, by the increased cross-section in the direction of crack growth and by the particular load distribution.

4.17 OTHER FACTORS

Other factors that may affect the fatigue life of a machine component are the size of the component, its temperature, the loading rate, and its surface conditions, including corrosion. Exposure to nuclear radiation also may have an effect.

The size effect becomes important when extrapolating from small test specimens to large mechanical parts. It has been stated that the basic size effect is an inverse relation between size and fatigue resistance, with the degree of dependence determined in large part by the working of the metal in producing the machine element.[7] We will find this relation to hold when we study spring design in a later chapter and it is indicated in Figure 4.17.1 for a series of undefined light alloys for the completely reversed loading conditions noted.

Increasing temperature is often associated with decreased fatigue life. As indicated by the following graph, however, this depends upon the material involved and the temperature range. From Figure 4.17.2 we find that while increased temperature beyond a certain value ultimately reduces the fatigue resistance of a material, an increase over some ranges for some materials may be beneficial. In the case of mild steel an increase in the range from 100 to 330 °C may improve fatigue resistance. A less pronounced change holds for cast iron from about 100 to 400 °C.[19] It is because of this temperature dependence of fatigue properties that predicting fatigue life in the presence of rapid thermal cycling becomes particularly difficult. Rapid thermal cyclic may occur with machines operating between conventional room temperatures and furnace or cryogenic temperatures.

Loading rates may become important in instances of acoustic fatigue, as arises on aircraft wings behind jet engines.[7]

Since residual stresses may be considered as mean stress in the region in which they act, it follows that surface conditions are important in that machining, cladding

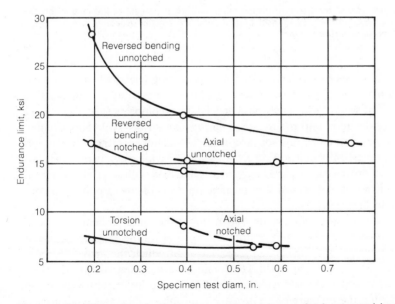

Figure 4.17.1 Size effect for light alloys subjected to completely reversed loading. *Source: C.C. Osgood,* Fatigue Design, *Wiley-Interscience, New York, 1970. Copyright 1970 John Wiley & Sons, Inc.; reprinted with permission.*

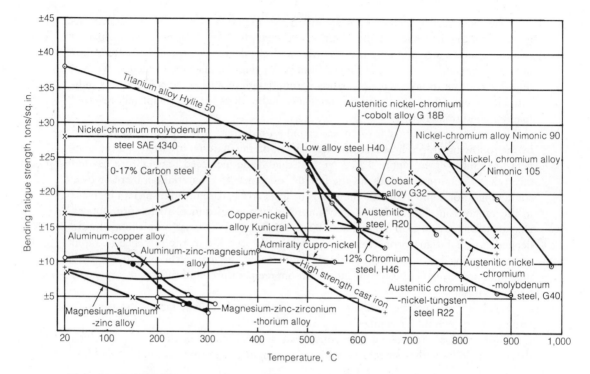

Figure 4.17.2 The influence of temperature on the fatigue strength (stress) of the metals labeled. *Source: P.G. Forrest, Fatigue of Metals, Addison-Wesley, Reading, MA. Copyright 1962 Pergamon Books, Ltd.*

(bonding outer metal sheets to a different inner metal), plating (electrochemical deposition of a different metal on the surface), quenching, and so on, often produces a state of residual stress on the surface which may be beneficial if it is compressive. Cold rolling is commonly used on axles and shafts to provide a compressive surface layer. Shot peening, where small metal or glass balls, or shot, are propelled against a target surface with sufficient velocity to plastically indent it, is often used to provide a compressive surface layer in springs and other such an irregular shape that cold rolling is not practical. The comparative effects of plating and shot peening on a steel plate of undefined composition is shown in Figure 4.17.3.[20]

Fatigue effects due to corrosion are dependent upon the material being corroded, the corrodant, the pressure, and the temperature. An example of the effect of the chemical activity of ambient gases is found in Figure 4.17.4 in which the growth rates of fatigue cracks at constant stress for 2014-T6 aluminum are compared for specimens immersed in mercury, which actively exposed aluminum, and in air, nitrogen, argon, and helium. Increasing the pressure of a corrosive atmosphere increases its density at thus shortens the fatigue life of the material being attacked, as is shown in Figure 4.17.5.[21]

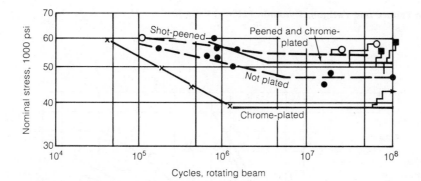

Figure 4.17.3 Effect of chrome plating and of shot peening on fatigue life. *Source: H.O. Fuchs and E.R. Hutchinson, Shot peening, Machine Design, 30 (3), 116–125. © 1958 Penton Publishing Inc.*

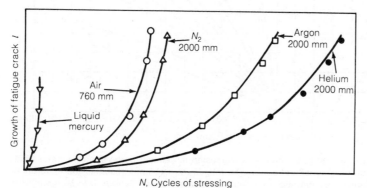

Figure 4.17.4 Crack growth in 2014-T6 aluminum in the media shown. *Source: C.C. Osgood, Fatigue Design, Wiley-Interscience, New York, 1970. Copyright 1970 John Wiley & Sons, Inc.; reprinted with permission.*

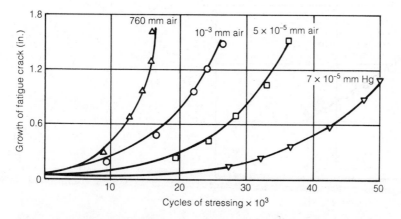
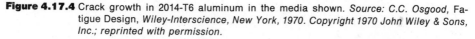

Figure 4.17.5 Effect of reduced air pressure (less than 20 h) on the fatigue life under uniaxial testing of an aluminum alloy. *Source: R.H. Christensen and R.J. Bellinfante, Some Considerations in the Fatigue Design of Launch and Space Structures, NASA CR-242, June 1965. Courtesy National Aeronautics and Space Administration.*

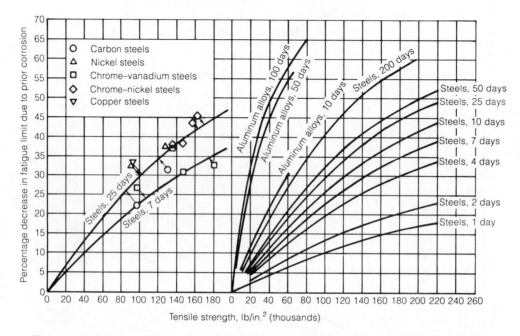

Figure 4.17.6 Decrease in the fatigue limit due to corrosion in calcium carbonate and water as a function of initial tensile strength. The fatigue (stress) limit becomes the endurance (stress) limit when the specimen surface is uncorroded. *Source: D.J. McAdam, Jr. and R.W. Clyne, Influence of chemically and mechanically formed notches on fatigue of metals, Journal of Research of the National Bureau of Standards, 13, 527–572 (Oct. 1934). Courtesy of the National Bureau of Standards.*

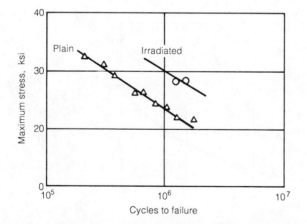

Figure 4.17.7 Effect of prior neutron irradiation (nvt. $\sim 2 \times 10^{18}$ fast neutrons/cm^2) on the fatigue life of 7075-T6 aluminum alloy. *Source: R.H. Christensen and R.J. Bellinfante, Some Considerations in the Fatigue Design of Launch and Space Structures, NASA CR-242, June 1965. Courtesy National Aeronautics and Space Administration.*

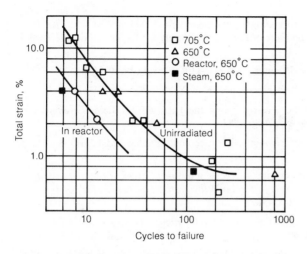

Figure 4.17.8 Fatigue life of Type 304 stainless steel tubing with and without the effect of neutron radiation. *Source: C.C. Osgood,* Fatigue Design, *Wiley-Interscience, New York, 1970. Copyright 1970 John Wiley & Sons, Inc.; reprinted with permission.*

Surface corrosion prior to fatigue loading can reduce the fatigue life of a mechanical part by causing pitting on the surface which induces stress concentrations which in turn may cause yielding and crack formation. This conclusion is based upon Figure 4.17.6 along with the observation that the corrosion caused a negligible reduction is cross section of specimens under test.[19]

Fretting corrosion is that situation in which two surfaces rub together, in the presence of a corrosive agent, so that any fragments produced by making and breaking cold welds are chemically attacked. If the corrosion product is harder than the parent metal they may produce pits that exacerbate the already roughened surface and increase the probability of a pit causing a stress concentration that can result in a surface crack.[7]

Tests indicate that nuclear radiation may be either beneficial or harmful. A beneficial effect of neutron radiation upon 7075-T6 aluminum is shown in Figure 4.17.7 while a harmful effect upon type 304 stainless steel is shown in Figure 4.17.8.[7,21]

REFERENCES

1. Bridgman, P.W., *Studies in Large Plastic Flow and Fracture*, McGraw-Hill, New York, 1952, pp. 75, 255–257, 264–268.

2. Bisplinghoff, R.L., Mar, J.W., and Pian, H.H., *Statics of Deformable Solids*, Addison-Wesley, Reading, MA, 1965, pp. 153–159.

3. *Handbook of Experimental Stress Analysis*, Hetenyi, M. (ed.), Wiley, New York, 1950, pp. 448–450.

4. Parker, E.R., *Brittle Behavior of Engineering Structures*, Wiley, New York, 1957, p. 57.

5. Noll, G.C., and Erickson, M.A., *Proceedings of the Society for Experimental Stress Analysis*, Vol. 5, *Allowable Stresses for Steel Members of Finite Life*, Society for Experimental Mechanics, Brookfield, CT, 1948, pp. 132–143.

6. Juvinall, R.C., *Stress, Strain, and Strength*, McGraw-Hill, New York, 1967, pp. 250–276

7. Osgood, C.C., *Fatigue Design*, Wiley-Interscience, New York, 1970, pp. 103–106

8. Orthwein, W.C., Estimating Cyclic Stresses below the Fatigue Threshold, *Computers in Mechanical Engineering*, 1 (1), pp. 80–83, Aug. 1982.

9. Spotts, M.F., *Design of Machine Elements*, Prentice-Hall, Englewood Cliffs, NJ, 1985, Chapter 2.

10. Orthwein, W.C., Estimating fatigue due to cyclic multiaxial stress, *Transactions of the American Society of Mechanical Engineers, Journal of Vibration, Acoustics, Stress Analysis and Reliability in Design*, 109 (1), pp. 97–102, 1987.

11. Leve, H.L., in *Cumulative Damage Theories, Metal Fatigue, Theory and Design* (A.G. Madayag, Ed.), Wiley, New York, 1969.

12. Doyle, T.C., and Erickson, J.L., *Advances in Applied Mechanics*, Vol. 4, *Nonlinear Elasticity*, Academic, New York, 1956, pp. 53–115.

13. Dowling, N.E., Brose, W.R., and Wilson, W.K., *Advances in Engineering*, Vol. 6, *Notched Member Fatigue Life Predictions by the Local Strain Approach, Fatigue under Complex Loading*, Society of Automotive Engineers, Warrendale, PA, 1977, pp. 55–84.

14. Technical Report on Fatigue Properties—SAE J1099, *SAE Handbook*, Vol. 1, Society of Automotive Engineers, Warrendale, PA, 1987.

15. *Fatigue Design Handbook*, Advances in Engineering Design, AE-4, Society of Automotive Engineers, Warrendale, PA, 1968, p. 27.

16. Raske, D.T., and Morrow, J., Mechanics of Materials in Low Cycle Fatigue Testing, *Manual on Low Cycle Fatigue Testing*, ASTM STP 465, American Society for Testing and Materials, 1969, pp. 1–25.

17. Dowling, N.E., Fatigue predictions for complicated stress–strain histories, *Journal of Materials*, 7 (1), pp. 71–87, Mar. 1972.

18. Harris, T.A., and Ioannides, E., A new fatigue life model for rolling bearings, *Ball Bearing Journal*, No. 224, SKF, The Netherlands, 1987.

19. Forrest, P.G., *Fatigue of Metals*, Addison-Wesley, Reading, MA, 1962.

20. Fuchs, H.O., and Hutchinson, E.R., Shot peening, *Machine Design*, 30 (3), pp. 116–125 (6 Feb. 1958).

21. Christensen, H.R., and Bellinfante, R.J., Some Considerations in the Fatigue Design of Launch and Spacecraft Structures, NASA CR-242, June 1965.

PROBLEMS

Section 4.1

4.1 What is the reliability of a machine rated at 10 000 h if the probability of failure at 10 000 h is 0.15 for the motor, 0.20 for the ball bearings, and 0.25 for the drive chain?

Section 4.2

4.2   Find the factor of safety based upon the maximum tensile stress for a small shaft having a diameter of 4.2 mm if the yield point for the steel used is 600 MPa and if tensile and torsional loading is as pictured in the figure.

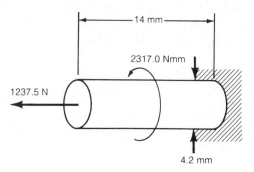

Section 4.3

4.3   Find the factor of safety for the shaft and loading described in problem 4.2 based upon the maximum shear stress if the yield stress in shear is 336 MPa.

Section 4.4

The problems for this section may use the program for principal stresses given in Chapter 3.

4.4   A raise-drill mining machine's shaft supports an axial stress of $\sigma_{zz} = 114$ MPa. What are the maximum shear stress components $\sigma_{z\theta}$ and σ_{zr} that can be added if $\zeta = 2.20$ and if $\sigma_{z\theta} = \sigma_{zr}$? Use the HMH criterion and a yield stress of $\sigma_{yp} = 271$ MPa.

4.5   Based upon the yield stress, select a steel for a replacement shaft in Problem 4.4 if the new shaft is to be hollow, if the safety factor is to be 2.50, and if it is to be pressurized such that $\sigma_{rr} = \sigma_{\theta\theta} = 80$ MPa. Use the HMH criterion, obtain steel data from Chapter 2, and let σ_{zz} remain at 114 MPa.

4.6   A pressurized pipe (inside radius r_i and outside radius r_o) to a cutting head in a mining prototype is to support an axial load of 2378 lb, a torque of 200 ft-lb, and an internal pressure of 1500 psi. Find the factor of safety according to

    a   the maximum normal stress criterion,
    b   the maximum shear stress criterion, and
    c   the HMH criterion.

The yield stress in tension is 90 000 psi. Circumferential and radial stresses are given by

$$\sigma_{\theta\theta} = p\,\frac{r_o{}^2 + r_i{}^2}{r_o{}^2 - r_i{}^2} \qquad \sigma_{rr} = -p$$

where p represents the internal pressure (gage) to calculate the stresses in the circumferential

(θ) and radial (r) directions in the pipe. Its inside diameter is 1.463 in. and its outside diameter is 1.500 in. Assume the shear stress $\sigma_{r\theta}$ is uniformly distributed over the pipe cross section.

4.7   If the pressure is lost in the pipe in problem 4.6, find the factors of safety listed in that problem.

Section 4.6

4.8   Estimate the percentage uncertainty in cycles N and stress σ_e for parts A and B in Figure 4.6.3 relative to the average values of N and σ_e, respectively. Use Figure 1.6.1 as an aid in determining the values for the unlabeled grid lines for various values of N.

Section 4.7

4.9   From Figure 4.7.4 derive an expression for the endurance limit (endurance stress, σ_e) in shear as a function of the endurance limit (endurance stress) in tension using the shear energy theory line. At $\sigma_{e_t} = 100\,000$ psi the method of Chapter 1, Section 1.7, may be used to read $\sigma_{e_s} = 57\,600$ psi.

Section 4.8

4.10   Find the fatigue strength reduction factor K_f for materials for which $q = 0.7, 0.84$, and 1.0 for a component having $K = 1.21$.

4.11   Check Figure 4.8.1 for BHN 500 for $K = K_t = 2.0, 2.5$, and 3.0 and for BHN 150 at $K = K_t = 1.4, 1.8$, and 2.0.

Section 4.4–4.9

4.12   A hanger bracket for the track of an overhead crane is subjected to a minimum tension of 12 000 psi when the crane is farthest from the bracket and to a maximum stress of 32 000 psi when the crane is at the bracket and carrying its rated load. The stress concentration factor is 1.41 and the notch sensitivity factor is 0.88. Find the factor of safety according to the Gerber-yield and Goodman-yield criteria for $\sigma_e = 18\,000$ psi, $\sigma_{yp} = 48\,000$ psi, $\sigma_u = 77\,000$ psi.

4.13   Find the maximum steady load that can be carried by a machine cut bolt of annealed UNS S31600 (AISI 316) stainless steel whose root diameter is 1.8466 in. (nominal 2.0-in. bolt) when the bolt is subjected to a completely reversed alternating load of 11 130 lb. A stress concentration factor of 6.0, based upon the root diameter, exists at the nut–thread junction. The notch sensitivity factor is 0.83. Use a safety factor of 1.5 and base the design upon the Gerber-yield criterion. See Chapter 2 for the mechanical properties of stainless steel and Figure 4.7.3 for σ_e.

4.14   Repeat problem 4.13 using the Goodman-yield criterion as the basis for the design.

4.15   Repeat problem 4.13 using the obsolete Soderberg criterion and find the percent load reduction penalty relative to the Gerber-yield criterion.

4.16   Find the maximum alternating stress according to the Gerber-yield condition that can be applied to a leaf spring that has a mean stress of 50 MPa without risk of failure. The spring is made from AISI 1095 (UNS G10950) high-carbon steel, tempered at 205 °C and water quenched. Use a fatigue strength reduction factor of 1.7. See Table 2.2.3 and Figure 4.7.3. Use the hot-rolled surface.

4.17   A vertically oscillating mixer is to be made as shown in the drawing with the spring made from UNS G10950 steel, treated as in Problem 4.16 and with a hot-rolled surface. The motor mass is 5 kg and the eccentric mass of 3 kg (for a total mass of 8 kg) which rotates at 100 rpm has its center of gravity 90 mm from the axis of rotation. Select the spring dimensions b and h such that it will last indefinitely and that the maximum deflection will be 200 mm due to the weight plus the centrifugal force. Design according to the Gerber-yield condition with a safety factor of 1.5. (*Hint:* Find bh^2 from the Gerber-yield condition and find bh^3 from deflection. Then $h = bh^3/(bh^2)$ and $b = bh^2/h^2$.)

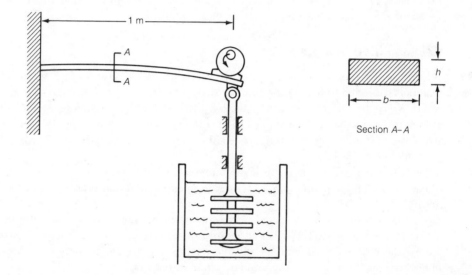

Section A–A

4.18   Find the maximum load, with a safety factor of 1.4, that can be applied to the rotary embosser and shaft shown in the drawing. The stress concentration factor at the fillet is 2.40 in bending and the notch sensitivity factor is 0.80. The shaft diameter is 1.00 in. and is machined from UNS G1020 cold-drawn steel, for which $\sigma_u = 78\,000$ psi, $\sigma_y = 66\,000$ psi, and $\sigma_e = 31\,500$ psi. Use the Gerber-yield condition.

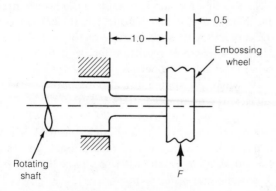

4.19 A mechanical part subjected to a tensile stress of 421 MPa has a stress concentration factor of 2.29 at the point of maximum stress and is made from a material for which $q = 0.90$. If $\sigma_u = 2590$ MPa, $\sigma_y = 1460$ MPa, and $\sigma_e = 448$ MPa, find the maximum alternating stress that may be added to assure indefinite life with a safety factor of 1.10 according to Soderberg, Goodman-yield, and Gerber-yield criteria (one answer for each criterion).

4.20 An anchor bracket for a hydraulic cylinder is to be designed to support a completely reversed alternating stress of 12 000 psi and a mean stress of 29 040 psi. What hardness would you recommend to provide a safety factor of 1.4 for an as-forged surface and a fatigue strength reduction factor of 1.17? To reduce cost, recommend the minimum hardness that will provide this factor of safety according to the Gerber-yield criterion.

Sections 4.10 and 4.11

4.21 When a stress concentration is present the criteria from Figure 4.10.2 for indefinite life for materials subjected to fluctuating compressive loads are $\sigma_a = \sigma_e/(\zeta K_f)$ and $\sigma_a = (\sigma_y/\zeta + \sigma_m)/K_f$, where K_f is the fatigue strength reduction factor. Use these relations to find the maximum alternating stress that may be applied to a ceramic cutter that must support a steady compressive stress of -300 MPa. The yield stress is -400 MPa, the endurance stress is 270 MPa, $K_f = 2.0$, and the safety factor is 1.2.

4.22 Show that when only a single tensile or compressive stress component acts, say σ_{zz}^*, substitution of equation 4.11.2 into equation 4.11.3 yields an expression that reduces to the Gerber equation 4.9.12.

4.23 Show that when only a single shear stress component, say $\sigma_{r\theta}^*$, acts, substitution of equation 4.11.1 into equation 4.11.4 yields an expression that reduces to the Gerber equation 4.9.12 if we use the results of problem 4.9 and the relation $\sigma_{u_t} = \sqrt{3}\sigma_{u_s}$, where σ_{u_s} represents the ultimate stress in shear and σ_{u_t} the ultimate stress in tension.

4.24 Find the relation between the yield stress in tension σ_y and the yield stress in shear σ_{y_s} according to relations 4.11.3 and 4.11.4 when $\zeta = 1.0$. (*Hint:* Assume only a single steady shear stress $\sigma_{xy} = \sigma_y$ is nonzero.)

4.25 Will a shaft simultaneously subjected to a steady tensile stress of $0.95\sigma_{yt}$ and $0.95\sigma_{ys}$ last indefinitely according to criteria (4.11.5) and (4.11.6)? Assume $\sigma_{us} = \sigma_{ut}/\sqrt{3}$, $\sigma_{ys} = \sigma_{yt}/\sqrt{3}$, and $\sigma_{es} = \sigma_{et}/\sqrt{3}$ and use the stresses given in Table 4.12.1.

4.26 Use equations 4.11.3 through 4.11.6 to find the maximum alternating shear stress $\sigma_{r\theta}$ that may be added to a shaft with a machined surface and hardness BHN 249 when the shaft must support an alternating axial stress of 158 MPa with a safety factor of 1.5. The stress maxima occur at a location that has a fatigue strength of reduction factor of 1.2 in tension and 1.30 in torsion. See Table 4.12.1 for other mechanical properties of the steel.

4.27 Repeat problem 4.26 for the case where the shaft must also support a steady axial stress of 51 MPa. Comment upon the effect of the mean axial stress σ_{zz} upon the maximum alternating stress $\sigma_{r\theta}$.

4.28 Find the safety factor according to criterion (4.11.1) and equation (4.11.2) for the shaft of a circular cutter at the edge of its support bearing where radial, axial, and tangential stresses induced by the load are $\sigma_{zr_a} = 70$ MPa, $\sigma_{zr_m} = 80$ MPa, $\sigma_{z\theta_a} = 200$ MPa, $\sigma_{z\theta_m} = 300$ MPa, $\sigma_{zz_a} = 150$ MPa, and $\sigma_{zz_m} = 0$. Assume that $\sigma_{y_t}/\sigma_{e_t} = \sigma_{y_s}/\sigma_{e_s}$.

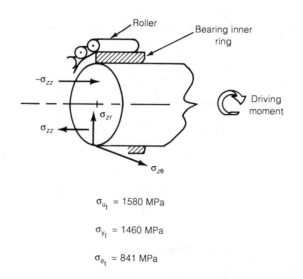

$$\sigma_{u_t} = 1580 \text{ MPa}$$

$$\sigma_{y_t} = 1460 \text{ MPa}$$

$$\sigma_{e_t} = 841 \text{ MPa}$$

4.29 Calculate the safety factor for the contour cutter is problem 4.28 according to criteria (4.11.3) and (4.11.5) using equations 4.11.4 and 4.11.6. Assume that ratios σ_u/σ_e and σ_u/σ_e are equal.

4.30 Recommend a gear shaft diameter in which the gear will induce a fluctuating torsional load that varies from 232 200 Nmm to 680 400 Nmm and will impose a bending moment of 420 700 Nmm on the shaft. The ultimate stress for the steel used is 403.2 MPa, the yield stress is 270 MPa, the endurance stress is 164 MPa, and the desired safety factor is 1.8. Experimental data indicates $K_{f_t} = 1.1$ and $K_{f_s} = 2.1$. $\sigma_{u_t}/\sigma_{e_t} = \sigma_{u_s}/\sigma_{e_s}$

4.31 Find the safety factor according to criteria (4.11.3) and (4.11.5) for the roller pin in the tub drive mechanism in an automatic washer. During each rotation about the z axis of the disk and pin assembly the pin experiences a completely reversed loading with components of 100 lb in the x direction and 310 lb in the y direction. These forces act 0.25 in. above the top of the disk that holds the roller pin. The base of the pin that fits into the plate has a diameter of $\frac{3}{8}$ in. The stress concentration in bending at the base of the pin is 2.5 and in shear is 1.8. Use $q = 0.55$ and assume that this same notch sensitivity factor applies to both tension due to bending and to shear stresses. The limiting stresses for the pin material have been found to be $\sigma_{ut} = 60\,000$ psi, $\sigma_{yt} = 45\,000$ psi, and $\sigma_{et} = 33\,000$ psi. Assume uniform shear stress over the pin cross-section and that

$$\sigma_{u_t}/\sigma_{e_t} = \sigma_{u_s}/\sigma_{e_s}$$

[Optional second part: Show that selection of the pin cross section to satisfy criteria (4.11.3) and (4.11.5) involves numerical solutions to these equations.]

Sections 4.12–4.14

4.32 Find the values of a and b such that $\sigma = b/N^a$ describes the S-N (σ-N) curves for ground, machined, hot-rolled, and as-forged steel having BHN from 187 to 207.

4.33   A machine component was found to have the following lives for the completely reversed stress levels shown below.

σ_{xx} (psi)	N (cycles)
20 000	150 000
28 000	100 000
60 000	7 000

What is its expected life in cycles for the following load schedule for completely reversed loading

σ_{xx} (psi)	Time (min)	rpm
20 000	2.0	20
28 000	1.4	10
60 000	0.7	7

a   according to the Palmgren-Miner relation, and

b   according to the Shanley relation with $k = 1.6$?

4.34   What is the maximum completely reversed load that can be sustained for 1.5 min of a 12-min load sequence if

σ_{xx} (psi)	Time (min)
28 600	10.0
—	1.5
70 000	0.5

The mechanical component has an as-forged surface with BHN 360. Minimum life should be 100 000 cycles, with a slightly greater life (up to 10%) desirable. Use the Shanley criterion with $k = 1.71$.

4.35   To reduce the testing time required to find the life of a part at a completely reversed stress amplitude of 20 000 psi, the laboratory director had directed to run the following two tests:

- Test 1: The machine component failed after 5710 cycles at a completely reversed stress of 70 000 psi.

- Test 2: An identical component was cycled for 2500 cycles at 70 000 psi then cycled at 20 000 psi until it failed after a total of 11 200 cycles.

What is the estimated life at a stress of 20 000 psi according to the Palmgren-Miner criterion?

Sections 4.12–4.15

4.36 A bracket made from 86L20 steel, BHN 187, is subjected to the following duty schedule. The surface is machined and ground and the rotational speed remains constant.

Time (min)	σ_a (psi)	σ_m (psi)
3.0	29 940	0
8.0	22 600	—
3.0	37 390	0

Find the maximum mean stress that can be supported during the 8-min portion of the schedule if the total life is to be 8000 cycles. Use the Palmgren-Miner criterion and the Goodman estimate for σ_{a_e} with a stress concentration factor of $K = 1.40$ and $q = 0.84$.

4.37 Repeat Problem 4.36 using the Gerber estimate for σ_{a_e}.

4.38 Repeat problem 4.37 using the Gerber estimate for σ_{a_e} and using the Shanley relation with exponent $k = 1.30$.

4.39 If the alternating stress in problem 4.19 is increased to 441.741 MPa will the part have indefinite life? If not, what will be its expected life? Use the Gerber-yield criterion to determine whether the life is indefinite and use relation 4.13.1 if its life is finite.

4.40 A machined shaft for a sprocket was surface hardened to BHN 410 and was found to have a residual tensile stress at the surface of 62 110 psi. If the chain tension induces a bending stress of 18 040 psi at the sprocket in an unkeyed shaft and if the stress concentration factor due to the keyway is 5.20, may this shaft be expected to last indefinitely if the notch sensitivity is 0.72? If the shaft will not last indefinitely according to the Gerber-yield condition, find its expected life in cycles if equation 4.13.1, motivated by the Gerber curve, is used to find σ_{a_e}.

4.41 Laboratory testing of an anchor bracket for a hydraulic cylinder indicated a life of approximately 145 000 cycles for completely reversed bending where the maximum stress varied from 40 000 psi in tension to 40 000 psi in compression. After customer modification of the machine the bracket was subjected to a mean stress of 15 000 psi. What life may be expected if the surface is as-forged, the hardness is BHN 370, and $K_f = 1.10$? Use the Gerber-yield and equation 4.13.1 if necessary.

4.42 Calculate the life of a part made from steel having BHN 388 when it is subjected to the following duty schedule.

Time (min)	σ_a (psi)	σ_m (psi)
3.0	70 000	0
7.0	60 100	39 800
4.0	25 300	43 200

The surface is machined and there are no stress concentrations. Compare the Shanley, with $k = 1.90$, and the Palmgren-Miner life estimates. Use equation 4.15.1 to find σ_{a_e}.

4.43 The comptroller has asked that you consider using the above steel in the hot-rolled condition because the price is 70% of that of the machined steel in problem 4.42. Which steel gives the cheaper cost per cycle of life? Use the Palmgren-Miner relation for the comparison.

4.44 The grip on the end of a hydraulic cylinder rod is to be designed to support a completely reversed alternating stress of 15740 psi and a mean stress of 29040 psi. It has an as-forged surface, a stress concentration factor of 1.50, and a notch sensitivity factor of 0.78. What minimum hardness would you recommend to achieve 112000 h of service at 5 cycles/h? Use equation 4.13.1 if necessary.

4.45 Consider the two duty schedules shown below and show that if T_{e1} denotes the life in units of time of the component when loaded according to the schedule in part a and T_{e2} denotes its life when loaded according to the schedule in part b, then

$$\frac{T_{e1}}{T_{e2}} = \frac{t_1 + t_2 + t_3}{t_1 + t_2}$$

so that in general

$$\frac{T_{e1}}{T_{e2}} = \frac{\sum\limits_{i=1}^{m} t_i}{\sum\limits_{i=1}^{n} t_i}$$

where no load acts over $m - n$ intervals in the load schedule corresponding to T_{e1}.

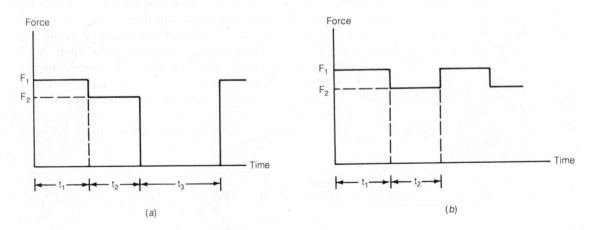

(a)   (b)

4.46 Show that as a consequence of the results in problem 4.45, according to the Palmgren-Miner criterion, the life of a component that is loaded according to the schedule shown in the figure on top of page 233 is equal to that of an identical component when loaded as in part a of the drawing for problem 4.45.

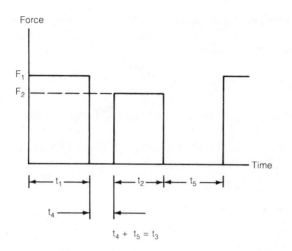

4.47 Show that whenever the Shanley criterion holds and two identical components are loaded as in the figure for problem 4.45, the ratio of the expected lives in units of time may be written as

$$\left(\frac{T_{e1}}{T_{e2}}\right)^k = \frac{t_1 + t_2 + t_3}{t_1 + t_2}$$

4.48 The dimensions and duty schedule for a typical tooth having a machined surface and attached to an adjustable width dado head to be used in furniture manufacture to cut both soft and hard woods are shown in the figure. Actual teeth are triangular in cross section with the cutting surface at an angle to the axis of rotation. The square cross section shown is to simplify the problem.

    a Select the hardness for the tooth to have indefinite life.

    b Since sharpening will eventually shorten the tooth to a length where it will no longer cut a groove of the desired depth, select a hardness for a tooth life as near to 8×10^5 as possible for the Brinell hardness increments given in Table 4.12.1.

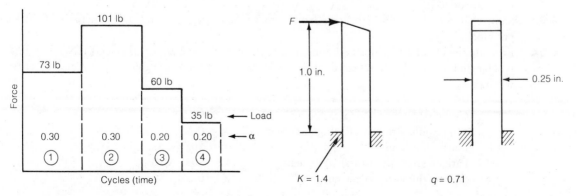

Compare designs based upon Goodman-yield and Gerber-yield criteria, equation 4.13.1, and upon the Palmgren-Miner relations. Give the life in hours for both Goodman and Gerber designs if the dado head rotates at 1725 rpm.

Sections 4.15 and 4.16

4.49 Show that relations similar to those obtained in problems 4.45 and 4.47 also apply to equation 4.16.1.

4.50 Show that equation 4.15.1 reduces to the Shanley equation when the loading schedule consists of a set of loads as pictured in the figure.

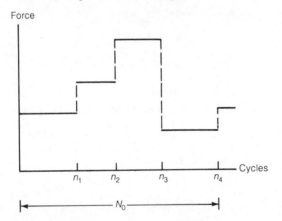

4.51 Using the Shanley relation with $k = 1.2$ and the Gerber criterion to calculate the life in cycles of a hot-rolled link, BHN 245, subjected to a stress load given by

$$\sigma = 45\,000 + 51\,200[\sin(\pi/140)n]^{0.2}$$

for a duty schedule of 140 cycles; i.e., the above loading repeats every 140 cycles.

4.52 Find the life in cycles for the load and link described in problem 4.51 if a mean stress of 39 600 psi is added and if a modification of the link causes a fatigue strength reduction factor of 1.15 to be introduced.

4.53 Show that equation 4.16.10 follows from 4.16.9 when $\alpha_i N_e = n_i$ and when there are β cycles per block.

4.54 Estimate the life of a machine component made from SAE 9262 annealed steel, BHN 260, which is subjected to completely reversed cyclic strain $\Delta\varepsilon = 0.010$. From Tables 1 and 2 of SAE J1099 we find

Cyclic strain hardening exponent	$n' = 0.15$
Monotonic strain hardening exponent	$n = 0.22$
Elastic modulus	$E = 207$ GPa
Fatigue strength (stress) coefficient	$\sigma_f' = 1041$ MPa
Fatigue ductility coefficient	$\varepsilon_f' = 0.16$

Yield stress	$\sigma_y = 455$ MPa
Fatigue strength (stress) exponent	$b = -0.071$
Fatigue ductility exponent	$c = -0.47$
Cyclic strength (stress) coefficient	$K' = 1379$ MPa

where materials are generally strain hardening if $n/n' > 1$ and strain softening if $n/n' < 1$.

[Ans. $N_e = Ne_f = 3998$ cycles]

4.55 Find the expected life of a component made from SAE 9262 steel, whose properties are listed in problem 4.54, when it is subjected to loading blocks in which the maximum strain is 0.020 for 3 cycles and 0.010 for 5 cycles.

[Ans. 924 cycles]

4.56 Show that in situations where there is no plastic strain equation 4.16.4 reduces to the form of given in equation 4.12.1. Be aware that exponent b in equation 4.16.4 is not equal to constant b in equation 4.12.1 and that $2N_f$ in equation 4.16.4 corresponds to N in equation 4.12.1 because N_f represents the number of strain reversals and N represents the number of completely reversed loading cycles. These similar forms emphasize that low-cycle fatigue differs from high-cycle fatigue in that (a) it assumes a plastic region contributes to the overall strain, and (b) that all stresses are above the endurance limit, or endurance stress.

4.57 Demonstrate that the average values for a and b calculated from equations 4.12.2 and 4.12.3 respectively are consistent with the values for the negative of the fatigue strength (stress) exponent b in equation 4.16.2 and for the fatigue strength (stress) coefficient σ'_f in equation 4.16.2 for UNS G10450 steel, BHN 225. That is, the average for a as found from equation 4.12.2 using σ_e for ground and as-forged surfaces agrees well with $-b$ in equation 4.16.2 and the average value for b as found from equation 4.12.3 again using σ_e for ground and as-forged surfaces agrees well the value of σ'_f in equation 4.16.2. Hence, the data for UNS G10450 steel is consistent with the correspondence shown in problem 4.56 between low-cycle and high-cycle fatigue relations. [UNS G10450 is listed as SAE 1045.]

4.58 Find the anticipated life, according to the local strain approach, for a machine component made from SAE 9262 steel, described in problem 4.54, which is subjected to a sequence of identical loading blocks in which the total strain $\Delta\varepsilon$ is 0.010 for 3 cycles and is 0.004 for five cycles in each block. Does the steel yield when the total strain is 0.004? Does this explain the long life calculated?

[Ans. No, Yes, 10 579 cycles]

4.59 Compare the life of the component described in Example 4.16.1 with that of an identical component subjected to a similar loading history except that the smaller hysteresis loop begins from the upper loop when the part is in tension; i.e., the mean stress is thus 250 MPa instead of -250 MPa. Note the marked increase in life when a compressive mean stress acts. This increase has a negligibly small effect upon to total life because of the harmful effects of the larger strains.

[Ans. $N_e = 370$ cycles when $\sigma_m = -250$ MPa]

CHAPTER FIVE

SPRINGS

NOTATION

A	area (l^2)	P	pressure (m/lt^2)
B	shock mount parameter (1)	P_g	gage pressure (m/lt^2)
c	spring index, defined by $c = D/d$ (1)	p	pitch (l)
$C_0, \ldots, C_6$	Belleville spring parameters $(m/l^3t^2, 1, 1, 1, 1, 1, 1)$	q	derating factor (1)
D	mean (average) diameter, helical spring (l)	R	spring rate (spring constant) (m/t^2)
		S_i	initial stress, tension spring (m/lt^2)
d	wire diameter (l)	r	radius, also L/D ratio (l), (1)
E	modulus of elasticity, Young's modulus (m/lt^2)	T	torque (ml^2/t^2)
		t	thickness (l)
F	force (ml/t^2)	u	generic displacement (l)
F_i	initial tension, tension spring (ml/t^2)	u_y	displacement in the y direction (l)
G	shear modulus $G = E/[2(1 + v)]$ (m/lt^2)	u_θ	displacement in the θ direction (l)
		V	volume (l^3)
h	solid height, also height of Belleville spring and height of shock mount elastomer (l)	x,y,z	rectangular coordinates (l)
		z	distance from neutral plane to fiber in tension (l)
ID	inside diameter (l)	δ	deflection (l)
I	moment of area in bending (l^4)	$\varepsilon_{\theta z}$	shear strain, circular cylindrical coordinates (1)
J	moment of area in torsion (l^4)	ε_{xy}	shear strain, rectangular coordinates (1)
L	wire length (l)		
L_a	loop allowance (l)	θ	angle (1)
L_p	length inside the pins, tension spring (l)	v	Poisson's ratio (1)
		σ	generic representation for tensile stress component (m/lt^2)
L_t	coil or helix length along its axis (l)	$\sigma_{\theta z}$	shear stress component (m/lt^2)
l	length (l)	σ_{xx}	tensile stress component due to bending, torsional spring (m/lt^2)
N	total number of turns (l)		
n	number of active turns, also an exponent (1)	σ_{xy}, σ_{zy}	shear stress, rectangular coordinates (m/lt^2)
M	moment (ml^2/t^2)	ϕ	angle (1)
OD	outside diameter (l)		
OD*	outside diameter of helical compression spring when compressed to its solid height (l)		

Spring design depends upon the service environment, the desired life, the loads carried, the desired deflections due to these loads, possible restrictions from the position of adjacent structure or other machine components, and the manufacturing cost of the spring itself. The order of importance of these considerations depends upon the particular machine: whether, for example, it is an aircraft, a production line machine, a home appliance, or a toy.

Properties of spring materials are discussed at the opening of this chapter to aid in delimiting the eligible materials for further consideration. With material selection out of the way, the remainder of the chapter is concerned with relations normally used to complete the design. Our discussion is limited to only a few of the many types of springs manufactured. Therefore, we shall consider torsion bars for a linear force/deflection relation for a limited deflection range; helical torsion, tension, or compression springs for a nearly linear relation over a deflection range dependent upon their dimensions, with a wide range of possibilities; Belleville springs for large forces and small deflections and with the possibility of nonlinear force/deflection relations; elastomeric springs to add damping to the system; and air springs for cases where the force/deflection characteristics need to be changed frequently. Design procedures for most of the other spring types produced in significant quantities may be found in Ref. 1 and 3.

Pursuant to this emphasis on design, recommendations and practices from various sources have influenced the selection of relations and procedures presented. The author thanks, in alphabetical order, Associated Spring, Barnes Group (metal springs), Firestone Rubber Company (air springs), Goodyear Rubber Company (air springs), H.K. Metalcraft (disc springs), Schnorr Corporation (disc springs), The Spring Manufacturers Institute (metal springs), and R.T. Vanderbilt Company (elastomeric materials) for copies of some of their publications and, in several instances, helpful discussions with members of their staffs.

5.1 PROPERTIES OF SPRING MATERIALS

Commonly used spring materials and their properties and uses are displayed in Table 5.1.1 for round wire.[1] The first two columns give the technical designation of each material and its composition, columns three through six provide data explicitly used in design formulas, and the last four columns give the service characteristics of each material as an aid in its selection whenever environmental effects upon the spring must be considered. Since cost is usually a factor (except when performance is of overriding concern, as in military and medical equipment), Figure 5.1.1[2] may be used in conjunction with the data in Table 5.1.1 in material selection. The design formulas based upon spring performance to be described in the following sections will provide the last bit of information, the minimum wire diameter that can be used, in order to complete the material selection.

Spring manufacturers and wire suppliers have also established a standard set of wire diameters which have been found to satisfy most of their customers' requirements

TABLE 5.1.1 PROPERTIES OF COMMON SPRING MATERIALS

Material	Nominal Analysis	Tensile Properties		Torsional Properties		Maximum Temperature		Rockwell Hardness	Method of Manufacture Chief Uses Special Properties
		Minimum Tensile Strength psi × 10³	Modulus of Elasticity E psi × 10⁶	Design Stress[1] %Minimum Tensile	Modulus in Torsion G psi × 10⁶	°F	°C		
Music Wire ASTM A 228	C 0.70–1.00% Mn 0.20–0.60%	230–399	30	45	11.5	250	121	C41–60	Cold drawn high and uniform tensile. High quality springs and wire forms.
Hard Drawn ASTM A 227	C 0.45–0.85% Mn 0.60–1.30%	CLI147–283 CLII171–324	30	40	11.5	250	121	C31–52	Cold drawn. Average stress applications. Lower cost springs and wire forms.
High Tensile Hard Drawn ASTM A 679	C 0.65–1.00% Mn 0.20–1.30%	238–350	30	45	11.5	250	121	C41–60	Cold drawn. Higher quality springs and wire forms.
Oil Tempered ASTM A 229	C 0.55–0.85% Mn 0.60–1.20%	CLI165–293 CLII191–324	30	45	11.5	250	121	C42–55	Cold drawn and heat treated before fabrication. General purpose spring wire.
Carbon Valve ASTM A 230	C 0.60–0.75% Mn 0.60–0.90%	215–240	30	45	11.5	250	121	C45–49	Cold drawn and heat treated before fabrication. Good surface condition and uniform tensile.
Chrome Vanadium ASTM A 231	C 0.48–0.53% Cr 0.80–1.10% V 0.15Min%	190–300	30	45	11.5	425	218.5	C41–55	Cold drawn and heat treated before fabrication. Used for shock loads and moderately elevated temperature.
Chrome Silicon ASTM A 401	C 0.51–0.59% Cr 0.60–0.80% Si 1.20–1.60%	235–300	30	45	11.5	475	246	C48–55	Cold drawn and heat treated before fabrication. Used for shock loads and moderately elevated temperature.

High Carbon Spring Wire

Alloy steel wire

Category	Material / Spec	Composition			Design Stress							Description
Stainless Steel Wire	AISI 302/304 ASTM A 313	Cr 17.0–19.% / Ni 8.0–10.0%			125–325	28	30–40	10.0	550	288	C35–45	Cold drawn general purpose corrosion and heat resistant. Magnetic in spring temper.
Stainless Steel Wire	AISI 316 ASTM A 313	Cr 16.0–18.0% / Ni 10.0–14.0% / Mo 2.0–3.0%			110–245	28	40	10.0	550	288	C35–45	Cold drawn. Heat resistant and better corrosion resistance than 302. Magnetic in spring temper.
Stainless Steel Wire	17-7 PH ASTM A 313 (631)	Cr 16.0–18.0% / Ni 6.5–7.5% / Al 0.75–1.5%			Cond CH 235–335	29.5	45	11.0	650	343	C38–57	Cold drawn and precipitation hardened after fabrication. High strength and general purpose corrosion resistance. Slightly magnetic inspring temper.
Non-Ferrous Alloy Wire	Phosphor Bronze Grade A ASTM B 159	Cu 94.0–96.0% / Sn 4.0–6.0%			105–145	15	40	6.25	200	93.3	B98–104	Cold drawn. Good corrosion resistance and electrical conductivity.
Non-Ferrous Alloy Wire	Beryllium Copper ASTM B 197	Cu 98.9T / Be 2.0T			150–230	18.5	45	7.0	400	204	C35–42	Cold drawn and may be mill hardened before fabrication. Good corrosion resistance and electrical conductivity. High physicals.
Non-Ferrous Alloy Wire	Monel 400 AMS 7233	Ni 66.0T / Cu 31.5% / C/Fe			145–180	26	40	9.5	450	232	C23–32	Cold drawn. Good corrosion resistance at moderately elevated temperature.
Non-Ferrous Alloy Wire	Monel K 500 QQ-N-286	Ni 65.0% / Cu 29.5% / C/Fe/Al/Ti			160–200	26	40	9.5	550	288	C23–35	Excellent corrosion resistance at moderately elevated temperature.
High Temperature Alloy Wire	A 286 Alloy	Ni 26.0T / Cr 15.0T / Fe 53.0T			160–200	29	35	10.4	950	510	C35–42	Cold drawn and precipitation hardened after fabrication. Good corrosion resistance at elevated temperature.
High Temperature Alloy Wire	Inconel 600 QQ-W-390	Ni 76.0T / Cr 15.8% / Fe 7.2%			170–230	31	40	11.0	700	371	C35–45	Cold drawn. Good corrosion resistance at elevated temperature.
High Temperature Alloy Wire	Inconel 718	Ni 52.5% / Cr 18.6% / Fe 18.5%			210–250	29	40	11.2	1100	593	C45–50	Cold drawn and precipitation hardened after fabrication. Good corrosion resistance at elevated temperature.
High Temperature Alloy Wire	Inconel × 750 AMS 5698, 5699	Ni 73.0% / Cr 15.0% / Fe 6.75%			No. IT 155 Min Spg.T190–230	31	40	12.0	750–1100	399–593	C34–39 C42–48	Cold drawn and precipitation hardened after fabrication. Good corrosion resistance at elevated temperature.

1 Design stress is 75% of minimum tensile stress for torsion and flat springs.
Source: Handbook of Spring Design, courtesy Spring Manufacturers Institute, Inc., Wheeling, IL.

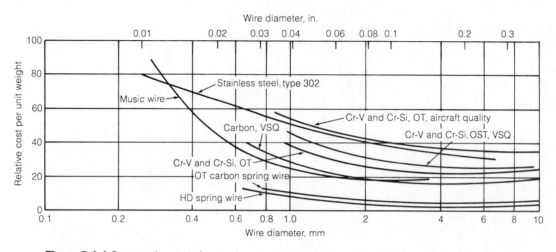

Figure 5.1.1 Comparative cost of several common spring wire materials. *Source*: Metals Handbook, Vol. 1, 9th ed., *American Society for Metals, Metals Park, OH, 1978, p. 305. Reprinted with permission.*

and have built equipment best suited to produce these wires. These wire diameters are listed in Table 5.1.2 for United States manufacturers and in Table 5.1.3 for German manufacturers, as representative of SI standards. Wire grades A, B, C, and II may be better known as materials 1.0500, 1.0600, 1.1200, and 1.211 as defined by the German (DIN) standards.

Note that stress is given in kg/mm^2, a common European practice, rather than in the technically correct N/mm^2, so that conversion is necessary.

Although spring manufacturers will produce any spring the customer is willing to purchase, deviation from these sizes causes a significant increase in spring cost and is strongly discouraged by spring manufacturers.[2,3]

Specification of wire diameter in terms of wire gauge numbers is ambiguous unless the standard is also specified (#1 Brown & Sharp is 0.2893 in., #1 Birmingham is 0.3000 in., and #1 Washburn & Moen is 0.2830 in.) Moreover, these diameters usually differ from those listed in Tables 5.1.2 and 5.1.3.

The term *minimum tensile strength* used in Tables 5.1.1 and 5.1.2 and implied in Table 5.1.3 refers to the minimum of the range of maximum tensile strength for the material specified. Clarification of this somewhat confusing statement may be had from Figure 5.1.2, where the shaded region represents the manufacturing variations for a particular product, such as ASTM A 227 hard-drawn wire.

Elastomeric springs (primarily natural and synthetic rubber), in contrast to metal springs, provide internal damping, which makes them suitable for shock mounts and supports, such as motor mounts, where energy dissipation is desired in addition to shock and vibration isolation.

TABLE 5.1.2   MINIMUM TENSILE STRENGTH OF FERROUS SPRING WIRE MATERIALS (in thousands of pounds per sq. in.)

Wire Size (in.)	Music Wire	Hard Drawn	Oil Temp.	Wire Size (in.)	Music Wire	Hard Drawn	Oil Temp.	Wire Size (in.)	Music Wire	Hard Drawn	Oil Temp.
.008	399	307	315	.046	309	249		.094	274		
.009	393	305	313	.047	309	248	259	.095	274	219	
.010	387	303	311	.048	306	247		.099	274		
.011	382	301	309	.049	306	246		.100	271		
.012	377	299	307	.050	306	245		.101	271		
.013	373	297	305	.051	303	244		.102	270		
.014	369	295	303	.052	303	244		.105	270	216	225
.015	365	293	301	.053	303	243		.106	268		
.016	362	291	300	.054	303	243	253	.109	268		
.017	362	289	298	.055	300	242		.110	267		
.018	356	287	297	.056	300	241		.111	267		
.019	356	285	295	.057	300	240		.112	266		
.020	350	283	293	.058	300	240		.119	266		
.021	350	281		.059	296	239		.120	263	210	220
.022	345	280		.060	296	238		.123	263		
.023	345	278	289	.061	296	237		.124	261		
.024	341	277		.062	296	237	247	.129	261		
.025	341	275	286	.063	293	236		.130	258		
.026	337	274		.064	293	235		.135	258	206	215
.027	337	272		.065	293	235		.139	258		
.028	333	271	283	.066	290			.140	256		
.029	333	267		.067	290	234		.144	256		
.030	330	266		.069	290	233		.145	254		
.031	330	266	280	.070	289			.148	254	203	210
.032	327	265		.071	288			.149	253		
.033	327	264		.072	287	232	241	.150	253		
.034	324	262		.074	287	231		.151	251		
.035	324	216	274	.075	287			.160	251		
.036	321	260		.076	284	230		.161	249		
.037	321	258		.078	284	229		.162	249	200	205
.038	318	257		.079	284			.177	245	195	200
.039	318	256		.080	282	227	235	.192	241	192	195
.040	315	255		.083	282			.207	238	190	190
.041	315	255	266	.084	279			.225	235	186	188
.042	313	254		.085	279	225		.250	230	182	185
.043	313	252		.089	279			.3125		174	183
.044	313	251		.090	276	222		.375		167	180
.045	309	250		.091	276		230	.4375		165	175
				.092	276	220		.500		156	170
				.093	276						

Source: Handbook of Spring Design, courtesy Spring Manufacturers Institute, Inc., Wheeling, IL.

TABLE 5.1.2 (Continued)

STAINLESS STEELS

Wire Size (in.)	Type 302	Type* 17-7 PH	Wire Size (in.)	Type 302	Type* 17-7 PH	Wire Size (in.)	Type 302	Type* 17-7 PH
.008	325	345	.033	276		.060	256	
.009	325		.034	275		.061	255	305
.010	320	345	.035	274		.062	255	297
.011	318	349	.036	273		.063	254	
.012	316		.037	272		.065	254	
.013	314		.038	271		.066	250	297
.014	312		.039	270		.071	250	292
.015	310	340	.040	270		.072	250	
.016	308	335	.041	269	320	.075	250	
.017	306		.042	268	310	.076	245	
.018	304		.043	267		.080	245	292
.019	302		.044	266		.092	240	279
.020	300	335	.045	264		.105	232	274
.021	298	330	.046	263		.120	225	272
.022	296		.047	262		.125		272
.023	294		.048	262		.131		260
.024	292		.049	261		.148	210	256
.025	290	330	.051	261	310	.162	205	256
.026	289	325	.052	260	305	.177	195	
.027	287		.055	260		.192		
.028	286		.056	259		.207	185	
.029	284		.057	258		.225	180	
.030	282	325	.058	258		.250	175	
.031	280	320	.059	257		.375	140	
.032	277							

* After aging

CHROME SILICON/CHROME VANADIUM

Wire Size (in.)	Chrome Silicon	Chrome Vanadium
.020		300
.032	300	290
.041	298	280
.054	292	270
.062	290	265
.080	285	255
.092	280	
.105		245
.120	275	
.135	270	235
.162	265	225
.177	260	
.192	260	220
.218	255	
.250	250	210
.312	245	203
.375	240	200
.437		195
.500		190

COPPER-BASE ALLOYS

Phosphor Bronze (Grade A)

Wire Size Range—in.	
.007-.025	145
.026-.062	135
.063 and over	130

Beryllium Copper (Alloy 25 pretemp)	
.005-.040	180
.041 and over	170

Spring Brass all sizes	120

NICKEL-BASE ALLOYS

Inconel (Spring Temper)

Wire Size Range—in.	
up to .057	185
.057-.114	175
.114-.318	170

Inconel X Spring Temper

	After Aging
190	220

WIRE SIZE TOLERANCE CHARTS

Hard-Drawn/Oil-Tempered

Wire Size Range (in.)	Tolerance ± (in.)
.010-.0199	±.0005
.020-.0347	±.0006
.0348-.051	±.0008
.0511-.075	±.001
.0751-.109	±.0015
.1091-.250	±.002
.2501-.375	±.0025
.3751-.625	±.003

Music Wire

Wire Size Range (in.)	Tolerance ±(in.)
.004-.010	±.0002
.011-.028	±.0003
.029-.063	±.0004
.064-.080	±.0005
.081-.250	±.001

Chrome Silicon

Wire Size Range (in.)	Tolerance ±(in.)
.032-.072	±.001
.0721-.375	±.002

Chrome Vanadium

Wire Size Range (in.)	Tolerance ±(in.)
.020-.0275	±.0008
.0276-.072	±.001
.0721-.375	±.002
.3751-.500	±.003

Type 302 Stainless/ 17-7 PH Stainless

Wire Size Range (in.)	Tolerance ±(in.)
.007-.0079	±.0002
.008-.0119	±.00025
.012-.0239	±.0004
.024-.0329	±.0005
.033-.0439	±.00075
.044 and larger	±.001

TABLE 5.1.3   MINIMUM TENSILE STRENGTH FOR SI WIRE DIMENSIONS

Wire diameter d (nominal) (mm)	Tensile strength of wire grade (kg/mm^2)				Reduction of area for wire grades A, B and C % min.
	A	B	C	II	
0.07	—	—	260–290		
0.08	—	—	259–289		
0.09	—	—	258–288		
0.10	—	—	258–288		
0.11	—	—	257–287		
0.12	—	—	257–287		
0.14	—	—	256–286	275–315	
0.16	—	—	255–285		
0.18	—	—	255–285		
0.20	—	—	254–284		
0.22	—	—	253–283		
0.25	—	—	252–282		
0.28	—	—	251–281		
0.30	175–209	210–250	251–281		
0.32	174–208	209–250	251–281		
0.34	174–200	209–249	250–280	270–310	
0.36	173–207	208–248	249–279		
0.38	173–207	208–247	248–278		—
0.40	173–207	208–247	248–278		
0.43	172–206	207–246	247–277		
0.45	171–205	206–245	246–276	270–300	
0.48	171–205	206–245	246–276		
0.50	170–204	205–244	245–275		
0.53	169–203	204–243	244–274		
0.56	169–203	204–242	243–273		
0.60	168–202	203–241	242–272	260–290	
0.63	167–201	202–240	241–271		
0.65	167–201	202–240	241–271		
0.70	166–200	201–239	240–270		
0.75	165–199	200–237	238–268		
0.80	164–198	199–236	237–262	255–285	
0.85	163–196	197–234	235–260		
0.90	162–195	196–233	234–259		
0.95	161–194	195–231	232–257		
1.00	160–193	194–230	231–256	250–280	
1.05	159–192	193–229	230–255		
1.10	159–191	192–228	229–254		
1.20	157–189	190–225	226–251		
1.25	156–188	189–223	224–249	240–270	
1.30	155–187	188–222	223–248		
1.40	153–185	186–219	220–245		
1.50	152–183	184–216	217–242	230–255	
1.60	150–181	182–214	215–235		
1.70	149–179	180–211	212–232	225–250	
1.80	147–177	178–209	210–230		

(continued)

TABLE 5.1.3   (Continued)

Wire diameter d (nominal) (mm)	Tensile strength of wire grade (kg/mm²)				Reduction of area for wire grades A, B and C % min.
	A	B	C	II	
1.90	146–175	176–206	207–227	215–240	40
2.0	145–174	175–205	206–226		
2.1	144–172	173–202	203–223		
2.25	143–170	171–199	200–220	—	
2.4	141–168	169–196	197–217		
2.5	140–166	167–193	194–214		
2.6	139–164	165–192	193–213		
2.8	137–162	163–188	189–209		
3.0	135–159	160–185	186–206		
3.2	134–157	158–182	183–203	—	
3.4	132–155	156–180	181–201		
3.6	130–153	154–177	178–198		
3.8	129–151	152–174	175–195		40
4.0	128–150	151–172	173–193		
4.25	127–148	149–171	172–192		
4.5	125–146	147–169	170–190		
4.75	123–144	145–165	166–186		
5.0	121–141	142–162	163–183		
5.3	119–139	140–160	161–181		35
5.6	117–137	138–157	158–178		
6.0	115–134	135–154	155–175		
6.3	114–133	134–153	154–174		
6.5	113–131	132–151	152–172		
7.0	111–128	129–147	148–168		
7.5	109–126	127–145	146–166		
8.0	107–123	124–142	143–163	—	
8.5	104–121	122–139	140–160		
9.0	102–119	120–137	138–158		30
9.5	101–116	117–135	136–156		
10.0	100–114	115–134	135–155		
10.5		112–132	133–153		
11.0		110–131	132–152		
12.0		106–127	128–148		
12.5		105–125	126–146		
13.0	—	104–124	125–145		—
14.0		101–121	122–142		
15.0		99–118	119–139		
16.0		98–115	116–136		
17.0		97–112	113–133		

The source company for this information requested anonymity.

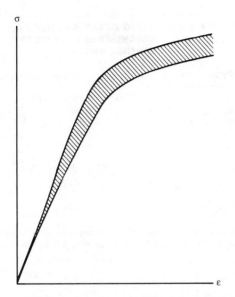

Figure 5.1.2 Possible variation of stress-strain curves for a particular string steel of specified dimensions and metallurgical analysis.

According to the Society of Automotive Engineers (SAE) designation system elastomeric materials are designated by a *type*, based upon resistance to heat aging, and by a *class*, based upon swelling when exposed to various oils.[4] Durometer hardness and tensile strength in MPa are indicated by a three-digit number immediately following the type and class alphabetical code. The first digit is the durometer hardness divided by 10 and the second two digits are the minimum tensile strength in MPa. Thus, 610 represents a rubber having a durometer hardness of 60 ± 5 and a minimum tensile strength of 10 MPa. A typical specification code, or call-out line, with more information than will be used in this text, is[4]

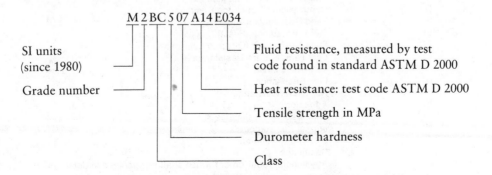

Type, class, and suffix letter meanings are listed in Tables 5.1.4, 5.1.5, and 5.1.6.[4] Representative values of tensile strength for rubber and similar elastomeric materials

TABLE 5.1.4   BASIC REQUIREMENTS FOR ESTABLISHING TYPE BY TEMPERATURE

Type	Test Temperature, °C
A	70
B	100
C	125
D	150
E	175
F	200
G	225
H	250
J	275

Source: SAE J200 MAY87, 1988 SAE Handbook. © 1988 Society of Automotive Engineers, Inc. Reprinted with permission.

TABLE 5.1.5   BASIC REQUIREMENTS FOR ESTABLISHING CLASS BY VOLUME SWELL

Type	Volume Swell, max, %
A	no requirement
B	140
C	120
D	100
E	80
F	60
G	40
H	30
J	20
K	10

Source: SAE J200 MAY87, 1988 SAE Handbook. © 1988 Society of Automotive Engineers, Inc. Reprinted with permission.

TABLE 5.1.6   MEANING OF SUFFIX LETTERS

Suffix Letter	Test Required
A	Heat Resistance
B	Compression Set
C	Ozone or Weather Resistance
D	Compression-Deflection Resistance
EA	Fluid Resistance (Aqueous)
EF	Fluid Resistance (Fuels)
EO	Fluid Resistance (Oils and Lubricants)
F	Low Temperature Resistance
G	Tear Resistance
H	Flex Resistance
J	Abrasion Resistance
K	Adhesion
M	Flammability Resistance
N	Impact Resistance
P	Staining Resistance
R	Resilience
Z	Any special requirement which shall be specified in detail

Source: SAE J200 MAY87, 1988 SAE Handbook. © 1988 Society of Automotive Engineers, Inc. Reprinted with permission.

TABLE 5.1.7   RELATION BETWEEN HARDNESS AND ELASTIC MODULI

Hardness IRHD $\pm$ 2	Elastic Modulus		Shear Modulus	
	psi	MPa	psi	MPa
30	130	0.902	43	0.294
35	168	1.157	53	0.363
40	213	1.471	64	0.441
45	256	1.765	76	0.530
50	310	2.157	90	0.628
55	460	3.187	115	0.794
60	630	4.364	150	1.040
65	830	5.737	195	1.344
70	1040	7.208	245	1.697
75	1340	9.218	317	2.177

are shown in Table 5.1.7 and are indicative of the variation that may be expected for a particular rubber or elastomer with a specified durometer hardness. (*Note:* Readings in International Rubber Hardness Degrees (IRHD) and in Shore hardness—the Shore Durometer A scale—are approximately equal.[5]) This variation, coupled with the wide variation in test results with different *shape factors*[5] clearly indicates that the behavior of rubber and similar elastomers is quite different from that of metals, and much more complicated, as shown in Figure 5.1.3.

5.2   TORSION BARS

Torsion bars have been widely used in vehicles in configurations either similar to that shown in Figure 5.2.1 or with the torsion bar parallel to the longitudinal axis of the vehicle. In either form, one end of the bar is attached to the frame and the other to the lower control arm, or its equivalent.

Angular deflection of one relative to the other may be obtained from the strain–strain relation

$$\varepsilon_{\theta z} = \frac{1}{2}\frac{du_\theta}{dz} = \frac{\sigma_{\theta z}}{2G} = \frac{Tr}{2JG} \tag{5.2.1}$$

which may be integrated with respect to z to get

$$u_\theta = \frac{\sigma_{\theta z}}{G}\,dz = \frac{Trl_0}{JG} \tag{5.2.2}$$

where J is the polar moment of area of the bar cross section. Since $u_\theta = r\theta$ equation 5.2.2 is equivalent to

$$\theta = \frac{Tl_0}{JG} \tag{5.2.3}$$

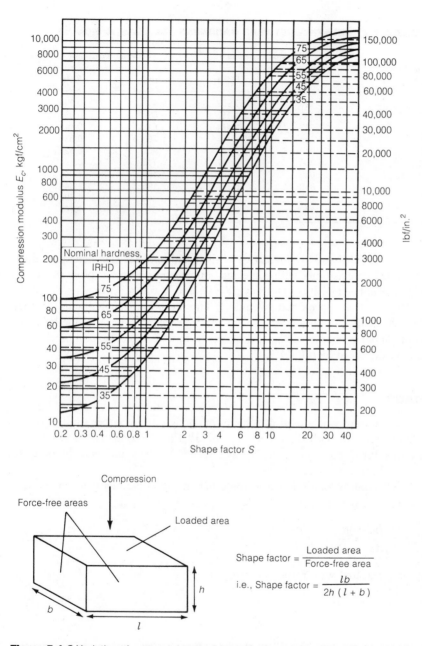

Figure 5.1.3 Variation of compression modulus with shape factor S for natural rubbers of different hardness. *Source: P.W. Allen, P.B. Lindley, and A.R. Payne,* Uses of Rubber in Engineering. *The Natural Rubber Producers Association, London, 1976.*

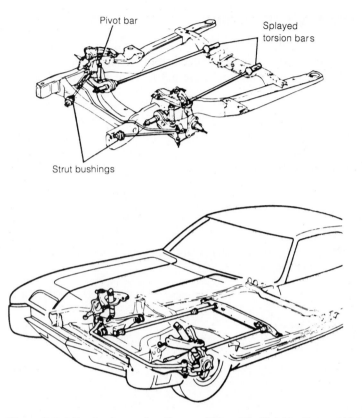

Figure 5.2.1 Torsion bar spring. *Source: Oldsmobile Division, General Motors Corporation.*

which may be used to calculate the angular deflection due to a given torque T. If the end of the torsion bar is connected to a lever of length l_1, the displacement of the free end of the lever may be approximated by $\delta = l_1\theta$ for small angles, so that

$$\delta = \frac{Tl_0l_1}{JG} \tag{5.2.4}$$

where

$$G = \frac{E}{2(1 + v)} \tag{5.2.5}$$

in terms of the elastic modulus E and Poisson's ratio v.

The angular spring rate, or spring constant, may be found from (5.2.3) to be

$$R = \frac{T}{\theta} = \frac{JG}{l} = \frac{\pi r^4 G}{2l} \tag{5.2.6}$$

5.3 HELICAL SPRING RATE AND DEFLECTION

Spring constant, or spring rate, and axial deflection formulas for centrally loaded helical springs made from round wire may be obtained from (5.2.4) by replacing T by $FD/2$, l_0 by πDn, and l_1 by D according to Figure 5.3.1. With these replacements equation 5.2.4 becomes

$$\delta = \frac{8FD^3 n}{d^4 G} \qquad\qquad (5.3.1)$$

and the spring rate, denoted by R, becomes

$$R = \frac{F}{\delta} = \frac{Gd^4}{8D^3 n} \qquad\qquad (5.3.2)$$

It has become industry practice to characterize spring proportions in terms of the dimensionless *spring index c* defined by

$$c = D/d \qquad\qquad (5.3.3)$$

so that the spring rate may be written as

$$R = \frac{Gd}{8c^3 n} \qquad\qquad (5.3.4)$$

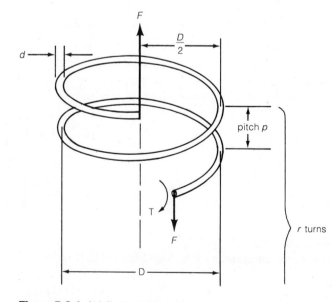

Figure 5.3.1 Axially loaded helical spring.

Analysis of the shear stress in an axially loaded helical spring by Rover,[6] Timoshenko,[7] and Wahl[8] has shown that it may be represented by

$$\sigma_{\theta z_{max}} = \frac{8Fc}{\pi d^2}\left(\frac{4c-1}{4c-4} + \frac{0.615}{c}\right)$$

(5.3.5)

for $c \geq 3$ and subject to the condition that $\tan^{-1}(2p/D) < 10°$. Equation 5.3.5 is used in the *Handbook of Spring Design*[1] for most spring applications, and will, therefore, be used in the following discussion of helical compression and tension springs.

5.4 COMPRESSION SPRINGS

Compression spring terminology and end configurations are shown in Figure 5.4.1. Dimensions of compression springs may be calculated from relations displayed in Table 5.4.1 for the four standard types of ends shown in that figure. Actual spring solid height may differ slightly from the calculated value due to improper coil seating, production variation in wire diameter, and any plating or coating prescribed.

According to the *SMI Handbook*, manufacturing variations in producing the end configurations shown in Figure 5.4.1 may cause the spring rate to be slightly less than calculated during the first 20% of the total deflection. The spring rate (spring constant) is essentially linear and in good agreement with the calculated value over the next 60% of the deflection range and will slightly increase above the calculated value over the last 20% of the range as the coils are progressively pressed into firm contact with one another until the solid height is reached.

Helical compression spring design usually begins by selecting a wire diameter d and a mean coil diameter D and comparing the maximum shear stress for this combination as found from equations 5.3.5 and 5.3.3 with the design shear stress, which is given as a percentage of the minimum tensile stress for standard wire diameters in Table 5.1.2. If we let q denote the ratio of the design shear stress to the minimum tensile stress we can calculate a *comparative* shear stress from a modification of equation 5.3.5; namely

$$\sigma_{z\theta} = \frac{8cF_1}{\pi d^2 q}\left(\frac{4c-1}{4c-4} + \frac{0.615}{c}\right)$$

(5.4.1)

whose magnitude will correspond to the listed minimum tensile stress for the wire diameter selected. This simplifies comparison between calculated and tabulated values. Since either the inside diameter D_i or the outside diameter D_o may be specified, it is convenient to rewrite equation 5.3.3 as

$$c = \begin{cases} (D_i + d)/d \\ (D_o - d)/d \end{cases} = D/d$$

(5.4.2)

Compression Springs Specification Form

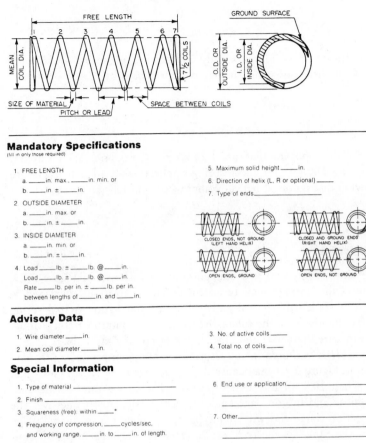

Mandatory Specifications
(fill in only those required)

1. FREE LENGTH
   a. ____ in. max., ____ in. min. or
   b. ____ in. ± ____ in.

2. OUTSIDE DIAMETER
   a. ____ in. max. or
   b. ____ in. ± ____ in.

3. INSIDE DIAMETER
   a. ____ in. min. or
   b. ____ in. ± ____ in.

4. Load ____ lb. ± ____ lb. @ ____ in.
   Load ____ lb. ± ____ lb. @ ____ in.
   Rate ____ lb. per in. ± ____ lb. per in.
   between lengths of ____ in. and ____ in.

5. Maximum solid height ____ in.

6. Direction of helix (L, R or optional) ____

7. Type of ends ____

Advisory Data

1. Wire diameter ____ in.
2. Mean coil diameter ____ in.
3. No. of active coils ____
4. Total no. of coils ____

Special Information

1. Type of material ____
2. Finish ____
3. Squareness (free): within ____ °
4. Frequency of compression, ____ cycles/sec, and working range. ____ in. to ____ in. of length.
5. Operating temp. ____ °F

6. End use or application ____
7. Other ____

Figure 5.4.1 Compression spring specification form recommended by the Spring Manufacturers institute, with standard ends shown. *Source:* Handbook of Spring Design, *courtesy Spring Manufacturers Institute, Inc., Wheeling, IL.*

TABLE 5.4.1 FORMULAS FOR DIMENSIONAL CHARACTERISTICS

Spring Characteristic	Type of Ends			
	Open	**Open & Ground**	**Closed**	**Closed & Ground**
Pitch (p)	$\dfrac{L - d}{n}$	$\dfrac{L}{N}$	$\dfrac{L - 3d}{n}$	$\dfrac{L - 2d}{n}$
Solid Height (H)	$d(N + 1)$	$d \times N$	$d(N + 1)$	$d \times N$
Active Coils (n)	N	$N - 1$	$N - 2$	$N - 2$
Total Coils (N)	$\dfrac{L - d}{p}$	$\dfrac{L}{p}$	$\dfrac{L - 3d}{p} + 2$	$\dfrac{L - 2d}{p} + 2$
Free Length (L)	$(p \times N) + d$	$p \times N$	$(p \times n) + 3d$	$(p \times n) + 2d$

Source: Handbook of Spring Design, courtesy Spring Manufacturers Institute, Inc., Wheeling, IL.

were the upper line holds if the inside diameter is specified and the lower line holds if the outside diameter is specified. If c is less than 3 or greater than 16 the formulas in this section may not give realistic results, so that a different wire diameter should be selected.

Since equations 5.4.1 and 5.4.2 may be repeated for each wire diameter tried, the design process will be much less tedious if at least these two relations are programmed in a computer or a programmable calculator.

Our next step is to calculate spring rate R from

$$R = |F_1 - F_2|/|L_1 - L_2| \qquad\qquad F_1 > F_2 \qquad\qquad (5.4.3)$$

and then calculate the number of active turns n from equation 5.3.4. With n known we then turn to Table 5.4.1 to find the total number of turns N and the spring solid height H. This table is written such that the quantity in the first column is equal to the quantities in the four columns for the end condition at the top of the column. For example, a spring with open and ground ends would have

$$N = n + 1 \qquad\qquad\qquad\qquad (5.4.4)$$

turns because from the table $n = N - 1$.

The pitch and the free length are functions of one another, so that we must know one before we can find the other. If L is known from the statement of the problem we can then find pitch p from the appropriate formula shown in Table 5.4.1. If, in the case of the spring with open and ground ends, the free length is L the pitch is then given by L/N.

The solid height of a compression spring is its height when it is so heavily loaded that all of the turns are pressed against one another. It is often important to find the solid height because the range of deflection of the spring is the free height minus the solid height. Formulas for the solid height of compression springs as a function of the style of their ends are given in Table 5.4.1.

Use of the Goodman-yield condition (the modified Goodman diagram) for estimating fatigue life of the spring is described is Ref. 4.

EXAMPLE 5.4.1*

Design a compression spring with closed ends to support a plunger in a hole 0.203 in. in diameter for a marine application. Minimum length between the end of the plunger and the bottom of the hole is 0.340 in. When the plunger is 0.385 in. from the bottom of the hole the spring force should be 7.20 lb and when it is 0.475 in. from the bottom of the hole the spring force should be zero.

---

* Adapted from the *Handbook of Spring Design*, Spring Manufacturers Institute, Wheeling, IL, 1981.

Because of the marine environment we select stainless steel, type 302 wire as the cheapest of the corrosion resistant wire and use $q = 0.35$, based upon previous laboratory testing of type 302 from this supplier.

Having selected the material, we begin with an OD of 0.195 in. to assure clearance and try wire diameters from $\frac{1}{20}$ of the OD to $\frac{1}{3}$ of the OD, i.e., from 0.008 to 0.065 in., and begin at the mid range to find in which half of the range to search. In this case the search consists of only two trials using the program for solving (5.4.1) and (5.4.2). The results are

| | | σ (psi) | |
Trial Number	d (in.)	Calculated	Listed
1	0.034	285 537	275 000
2	0.035	262 840	274 000

from which it is evident that the minimum wire diameter is 0.35 in. For this wire diameter the mean (average) diameter is given by $D = OD - d$ so $D = 0.195 - 0.035 = 0.160$ in. and $c = D/d = 4.5714$. Turning to (5.4.3),

$$R = (F_{max} - F_{min})/(L_{max} - L_{min}) = 7.2/(0.475 - 0.385) = 80.00 \text{ lb/in.}$$

so that from (5.3.4) we find the active number of turns to be

$$n = \frac{Gd}{8c^3R} = \frac{10 \times 10^6(0.035)}{8(4.5714^3)80} = 5.7244 \text{ turns}$$

From Table 5.4.1 we find that

$$N = n + 2 = 7.7244 \text{ turns}$$

and that the solid height is

$$h = d(N + 1) = 0.3054 \text{ in.}$$

so that the spring will not prevent the plunger from reaching its full stroke. Since the force on the plunger is to vanish when the plunger is 0.475 from the bottom of the hole it is evident that the free length of the spring must be 0.475 in.

According to Table 5.4.1 the pitch is given by

$$p = (L - 3d)/n = [0.475 - 3(0.035)]/5.7244 = 0.0644 \text{ in.}$$

5.5 BUCKLING OF COMPRESSION SPRINGS

Helical compression springs may deflect laterally and collapse under a relatively small load if the load and its reaction do not act along the center line of the spring. This is generally the case for springs that are either open, open and ground, or closed and unground. These springs must, therefore, be prevented from buckling by using a central rod or by placing them in a cylinder, as in the case of compression spring in a hand-operated grease gun.

Springs that are closed and ground are less susceptible to buckling because the applied compressive loads are more uniformly distributed around the first and last turns. These springs will, nevertheless, buckle whenever the axial load is such that the deflection percentage $100\delta/L$ is above the curve in Figure 5.5.1 for the *slenderness ratio* L/D of the spring.

For purposes of computer-aided design this curve may be represented by the following three approximating formulas:

For $1 \leq r \leq 4.6$

$$100\delta/L = 72 - 1.0802\left(r - 0.15 \sin\frac{r-1}{3.6}\pi - 1\right)^2 \tag{5.5.1}$$

For $4.6 \leq r \leq 6.8$

$$100\delta/L = 116.546 - 12.727r \tag{5.5.2}$$

For $6.8 \leq r \leq 8.0$

$$100\delta/L = 17.7499 + 2.7778(8.9 - r)^2 \tag{5.5.3}$$

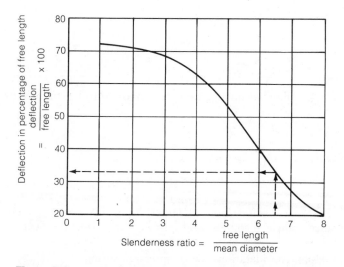

Figure 5.5.1 Buckling region of a compression spring with closed and ground ends. *Source: United States Department of Defense, MIL-STD-29A.*

These formulas produce values lying on the curve to within the accuracy of the grid in Figure 5.5.1 and they agree at transition points to the second decimal place in $100\delta/L$. In these relations $r \equiv L/D$.

Compression spring diameters increase as the spring is compressed because wires originally inclined to the spring axis move toward becoming perpendicular to the axis. The outside diameter OD* of a helical compression spring when compressed solid is often approximated by

$$\text{OD*} = \left[D^2 + \frac{p^2 - d^2}{\pi^2} \right]^{1/2} + d \qquad (5.5.4)$$

Whenever compression springs are enclosed in a sleeve to prevent buckling it is essential that this increased diameter be calculated to allow enough clearance between the spring and the sleeve to prevent any interference between them while the spring is being compressed.

EXAMPLE 5.5.1

Determine the critical buckling length for the spring designed in Example 5.4.1.

From that example the free length was $L = 0.475$ in. and the average diameter was given by $D = 0.160$ in. so that $L/D = 2.9688$. From (5.5.1) we find $100\delta/L = 68.42\%$. Hence, the spring may be expected to buckle after being compressed 0.3249 in. according to Figure 5.5.1. Since the solid height of the spring is 0.3054 in., however, it cannot deflect beyond 35.7%, so that buckling would not be a problem if the spring were operated in a similar fashion outside of the hole.

EXAMPLE 5.5.2

Does the spring designed in Example 5.4.1 allow adequate clearance to allow it to be compressed to its solid height?

Substitution into (5.5.4) yields

$$\text{OD*} = [0.160^2 + (0.0644^2 - 0.035^2)/\pi^2]^{1/2} + 0.035$$
$$= 0.1959 \text{ in.}$$

which leaves a clearance of $(0.203 - 0.1959)/2 = 0.0035$ in. or larger. Since this is of the same order as the published manufacturing tolerance on helical compression springs of this size,[3] the clearance may be too small.

See Refs. 1 and 3 for additional design considerations for compression springs.

5.6 TENSION SPRINGS

Tension spring terminology and common loop types are shown in Figure 5.6.1. Tension springs are easily distinguished from compression springs because their coils are in contact all along the length of the spring. An *initial tension* is introduced in their manufacture to hold the coils together. Consequently, the force–extension diagram for a tension spring is as shown in Figure 5.6.2. Torsional stress values for initial tension as

Extension Springs Specification Form

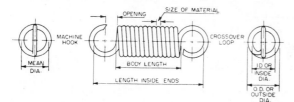

Mandatory Specifications
(fill in only those required)

1. LENGTH INSIDE ENDS
   a. _____ in. max., _____ in. min. or
   b. _____ in. ± _____ in. or
   c. approx. _____ in.

2. OUTSIDE DIAMETER
   a. _____ in. max. or
   b. _____ in. ± _____ in.

3. INSIDE DIAMETER
   a. _____ in. min. or
   b. _____ in. ± _____ in.

4. Load _____ lb. ± _____ lb. @ _____ in.
   Load _____ lb. ± _____ lb. @ _____ in.
   Rate _____ lb. per in. ± _____ lb. per in.
   between _____ in. and _____ in.

5. Maximum extended length (inside ends) without set _____ in.

6. Relative loop position, _____° max. separation of loop planes.

7. Direction of helix (L, R or optional) _____

8. Type of ends_____

Advisory Data

1. Wire diameter _____ in.

2. Mean coil diameter _____ in.

3. No. of active coils _____

4. Body length _____ in.

5. Initial tension _____ lbs.

Special Information

1. Type of material _____

2. Finish _____

3. Frequency of extension, _____ cycles/sec, and working range, _____ in. to _____ in. of length.

4. Operating temp. _____ °F

5. End use or application_____

6. Other_____

Table 1

LOOP TYPE			RECOMMENDED LENGTH[a]	
			Min.	Max.
Machine			½ I.D.[b]	1.1 × I.D.
Crossover			I.D.	I.D.
		Minimum Recommended Index = 7		
Side			I.D.	I.D.
Extended			1.1 × I.D.	As Required
Special: as required by design			As Required	

a Length is distance from last body coil to inside of end
b I.D. is inside diameter of adjacent coil in spring body

Figure 5.6.1 Extension springs specification form and end loop types. *Source*: Handbook of Spring Design, *courtesy Spring Manufacturers Institute, Inc., Wheeling, IL.*

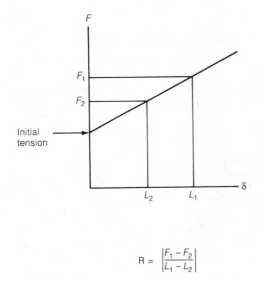

$$R = \left| \frac{F_1 - F_2}{L_1 - L_2} \right|$$

Figure 5.6.2 Force-extension diagram for a tension spring showing the effect of initial tension.

recommended by the Spring Manufacturers Institute are shown in Figure 5.6.3 for the recommended range of the spring index. Initial tension initially holds the coils in contact so that the stretched spring is shorter for a given load.

The first term in equation 5.6.1 describes the middle of the shaded region and the second approximates its width.

$$S_i = \frac{33\,500}{e^{0.105c}} \pm 2 \left(4 - \frac{c - 3}{6.5} \right) \tag{5.6.1}$$

The initial tension is F_i given in terms of S_i according to

$$F_i = \frac{\pi d^3}{8D} S_i \tag{5.6.2}$$

in which F_i is the axial force induced by torsional shear stress placed in the wire by twisting it as the helix is manufactured.

Stress, wire and coil diameter, spring rate, and the number of active turns are again determined by the equations given in Section 5.4. Active and total turns are equal, and the body length is given by

$$b = d(n + 1) \tag{5.6.3}$$

for round wire according to the geometry of Figure 5.6.4 for a two-turn tension spring (two turns for illustration only).

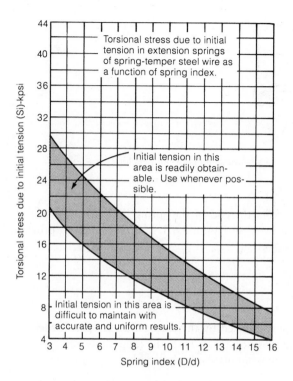

For computer programming purposes, the following is an empirical formula for determining Si as given in the above chart.

$$Si = \frac{33500}{e^{105c}} \quad \text{or} \quad \frac{33500}{10^{0456C}} \quad \text{where } C = \frac{D}{d}$$

Figure 5.6.3 Torsional stress values due to initial tension as recommended by SMI. *Source:* Handbook of Spring Design, *courtesy Spring Manufacturers Institute, Inc., Wheeling, IL.*

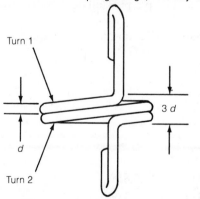

Figure 5.6.4 Body length of a two-turn tension spring.

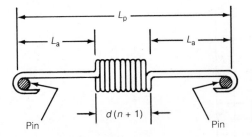

Figure 5.6.5 Loop allowance for a spring initially stretched between pins for $F_0 < F_i$.

Loop allowance is the distance from the end of the spring body to the inside of the end loop, or hook, as illustrated by L_a in Figure 5.6.5. It is calculated to provide a specified tension between the pins, so that

$$L_a = \frac{1}{2}\left(L_p - b - \frac{F_0 - F_i}{R}\right) \tag{5.6.4}$$

where F_0 is the force between the pins induced by the spring for the specified length L_p. Only if the spring is slack at this position ($F_0 \leq F_i$) is the loop allowance equal to one-half of the length outside of the pins minus the body length.

EXAMPLE 5.6.1

Design an extension spring to fit around two pins 0.300 in. in diameter and 8.375 in. apart, outside to outside, and to exert a force of 30 ± 5 lb between pins. When the pins are 9.25 in. apart the spring should exert a force of 45 ± 5 lb. Use the cheapest wire consistent with these requirements and specify the coil diameter, wire size, initial tension, loop allowance, and loop type. Coil outside diameter should be less than 1.00 in. to avoid interference with the frame.

Since hard-drawn wire is the cheapest according to Figure 5.1.3, consider it first, and find for wire 0.109 in. in diameter that

$$D_O = 0.800$$
$$c = 6.339$$
$$d = 0.109$$

Next,

$$\sigma = \frac{8Fc}{\pi d^2 q}\left(\frac{4c-1}{4c-4} + \frac{0.615}{c}\right) = \frac{8(50)6.339}{(0.109)^2 0.40}\left(\frac{4(6.339)-1}{4(6.339)-4} + \frac{0.615}{6.339}\right)$$
$$= 210\,165 \text{ psi} < 214\,400 \text{ psi, obtained by linear interpolation}$$

Thus,

$$n = \frac{Gd^4}{8D^3R} = \frac{11.5 \times 10^6 (0.109)^4}{8(0.691)^3 17.143} = 35.875 \text{ turns}$$

where R was found from

$$R = \left| \frac{F_1 - F_0}{L_1 - L_2} \right| = \frac{50 - 35}{9.25 - 8.375} = 17.143 \text{ lb/in.}$$

Body length is found from

$$b = d(n + 1) = 0.109(36.875) = 4.019 \text{ in.}$$

From 5.6.1 for initial torque find

$$\sigma_{\theta z} = \frac{33\,500}{e^{0.105(6.339)}} = 17\,217.9 \text{ psi}$$

so that 5.6.2 yields

$$F_i = \sigma_{\theta z} \frac{\pi d^2}{8c} = \frac{\pi 17\,217.9 (0.109)^2}{8(6.339)} = 12.673 \text{ lb}$$

Finally, the loop allowance becomes, according to equation 5.6.4,

$$L_a = \frac{1}{2} \left(L_p - b - \frac{F_0 - F_i}{R} \right) = \frac{1}{2} \left(8.375 - 4.019 - \frac{35 - 12.673}{17.143} \right)$$
$$= 1.527 \text{ in.}$$

which implies that an extended loop is required.

Clearly a tension spring designed according to the above considerations will be useless if the end loops cannot support the tensile load. Bending stress in the end loop itself may be approximated by

$$\sigma_{zz_a} = \frac{16F_1}{3\pi d^2} \left[\frac{4D_e}{d} \frac{4c_1^2 - c_1 - 1}{4c_1(c_1 - 1)} + 1 \right] \tag{5.6.5}$$

where D_e represents the mean diameter of the end loop and

$$c_1 = \frac{D_e}{d} \tag{5.6.6}$$

Stress at the upward bend to form the loop as shown in Figure 5.6.6 may be approximated by

$$\sigma_{zz_b} = \frac{32D_e F_1}{3\pi d^3} \left(\frac{4c_2 - 1}{4c_2 - 4} \right) \tag{5.6.7}$$

where

$$c_2 = \frac{2r}{d} \tag{5.6.8}$$

Design stresses in both equations 5.6.5 and 5.6.7 have been taken as 75% of the minimum tensile stress, so that the values computed from these equations may be compared directly with the tabulated minimum tensile stresses for each wire diameter.

EXAMPLE 5.6.2

Design end loops for the tension spring designed in example 5.6.1.

First try end loops with diameters equal to the mean diameter of the spring body. From equation 5.6.5

$$\sigma_{zz_a} = \frac{16(50)}{3\pi(0.109^2)} \left[\frac{4(0.8)}{0.109} \frac{4(6.339)^2 - 6.339 - 1}{4(6.339)5.339} + 1 \right]$$

$$= 244\,802 \text{ psi}$$

Since this stress is larger than the design stress we must select a smaller loop diameter. Try $D_e = 0.60$ in., for an inside diameter of 0.491 in. which is greater than the 0.300 pin diameters. For this value of D_e we find that σ_{zz} at point a is 189 055 psi, which is satisfactory because it is less than the design stress. Next, try $r = 0.25$ in., for which $c_2 = 2(0.25)/0.109$, so

$$\sigma_{zz_b} = \frac{8(0.6)50}{\pi 0.109^3} \frac{4(4.587) - 1}{4(4.587) - 4} = 71\,327 \text{ psi}$$

which is well within the design stress limit.

5.7 SPRINGS IN SERIES AND IN PARALLEL

Two or more springs are said to be in series when their deflections add to give the total deflection of the combination and when all of them experience the same force, as in Figure 5.7.1a. Thus, for n springs in series

$$\delta = \sum_{i=1}^{n} \delta_i \tag{5.7.1}$$

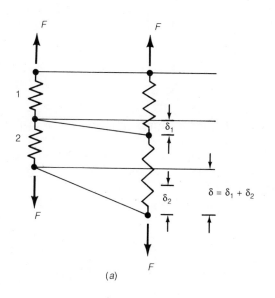

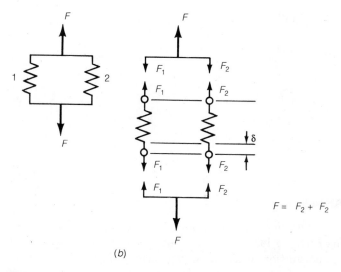

Figure 5.7.1 (a) Springs in series and (b) springs in parallel.

and

$$F_1 = F_2 = \cdots = F_n$$

Two or more springs are said to be in parallel when their forces add to give the total force and when all of them experience the same deflection, as in Figure 5.7.1b.

Thus, for n springs in parallel

$$\delta_1 = \delta_2 = \cdots = \delta_n$$

$$F = \sum_{i=n}^{n} F_n \tag{5.7.2}$$

The equivalent spring constant, or rate, R_e, for n springs in series, where R_i is the spring rate for spring i, is

$$R_e = \frac{F}{\delta} = \frac{F}{\delta_1 + \delta_2 + \cdots + \delta_n} = \frac{1}{\dfrac{1}{R_1} + \dfrac{1}{R_2} + \cdots + \dfrac{1}{R_n}}$$

which may be rewritten as

$$\frac{1}{R_e} = \sum_{i=1}^{n} \frac{1}{R_1} \tag{5.7.3}$$

Similarly, for n springs in parallel

$$R_e = \frac{F}{\delta} = \frac{F_1 + F_2 + \cdots + F_n}{\delta}$$

so that

$$R_e = \sum_{i=1}^{n} R_i \tag{5.7.4}$$

5.8 HELICAL TORSION SPRINGS

As discussed in Section 5.3 the wire in a helical tension or compression spring is stressed in torsion when the spring carries a load. In the case of a helical torsion spring the wire is stressed in bending, not torsion, when the spring carries a load.

Stress and deflection for a helical torsion spring are found from

$$\sigma_{xx} = -\frac{Mz}{I} \tag{5.8.1}$$

and

$$\theta = \frac{M\ell}{EI} \tag{5.8.2}$$

To apply equation 5.8.1 to round wire, replace z by $-d/2$, I by $\pi(d/2)^2/4$, and introduce factor q to obtain

$$\sigma_{xx} = \frac{M_1 \dfrac{d}{2}}{\dfrac{\pi}{4} \dfrac{d^4}{16}} \rightarrow \frac{32M_1}{\pi q d^3} \tag{5.8.3}$$

where M_1 is the maximum moment applied to the spring in bending, as evident from Figure 5.8.1. According to the *SMI Handbook*, $q = 0.75$ is a satisfactory ratio of the recommended design stress to the listed minimum tensile stress for all spring materials in bending, so that after introduction of the stress concentration factor for a coil,[8] equation 5.8.3 for the comparative bending stress may be written as

$$\sigma_{xx} = \frac{42.667M_1}{\pi d^3} \frac{4c^2 - c - 1}{4c(c - 1)} \tag{5.8.4}$$

where σ_{xx} may be compared with listed values of minimum tensile stress to select d and $c = D/d$ as defined for compression and tension springs.

Turning next to equation 5.8.2, adaption to a helical spring is simply a matter of realizing that since we plan to follow the common practice of describing the rotation of the spring ends in terms of turns T rather than in terms of angle, $\theta = 2\pi T$ and that wire length l is given by $l = \pi D N$, where N represents the number of turns in the

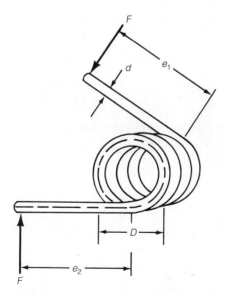

Figure 5.8.1 Helical torsion spring.

helix. Substitution for these values in equation 5.8.2 along with the previous relation for I yields

$$M = \frac{\pi TEd^4}{32ND} = \frac{TEd^4}{10.186\ ND} \tag{5.8.5}$$

Improved agreement with actual moments may be had by writing[1]

$$M_1 - M_2 = \frac{(T_1 - T_2)Ed^4}{10.8\ ND} \tag{5.8.6}$$

in which the M and T in equation 5.8.5 have been replaced by the difference in their maximum and minimum values. Since the spring rate for torsional springs may be written as

$$R_t = \left| \frac{M_1 - M_2}{T_1 - T_2} \right| \tag{5.8.7}$$

if follows from equations 5.8.6 and 5.8.7 that

$$R_t = \frac{Ed^4}{10.8ND} \tag{5.8.8}$$

Once the wire diameter has been selected the required number of turns N may be found from equation 5.8.8, rewritten as

$$N = \frac{Ed^4}{10.8R_tD} \tag{5.8.9}$$

The axial length of the spring, given by

$$L = d(N + 1) \tag{5.8.10}$$

changes as the spring wound or unwound under the torsional load. Thus, L will be largest when N is largest, when the coil is tightly wound.

Flexibility of the end extensions of length e_1 and e_2 shown in Figure 5.8.1 effectively alters to the spring rate according to the relation

$$R_t = \frac{Ed^4}{10.8N_eD} \tag{5.8.11}$$

where

$$N_e = N + \frac{u}{D} \tag{5.8.12}$$

in which

$$u = \frac{e_1 + e_2}{3\pi} \tag{5.8.13}$$

so that whenever end flexure may be significant the number of active turns is given by

$$N = \frac{1}{D}\left(\frac{Ed^4}{10.8R_t} - u\right) \tag{5.8.14}$$

Coil mean diameter D also changes as the torsional spring is loaded or unloaded. If coil diameter is D_1 and the number of turns is N_1 when no load acts, the wire length may be written as $l = \pi D_1 N_1$. Suppose that the ends are loaded as in Figure 5.8.1 so that the number of turns will increase as the load increases. After the load has been applied and one end turned through T turns relative to the other the wire length will be given by $l = \pi D_2(N_1 + T)$. Total wire length is, of course, unchanged, so that

$$D_2 = \frac{N_1}{N_1 + T} D_1 \tag{5.8.15}$$

relates the loaded diameter to the unloaded diameter. If the forces on the ends act to open, or uncoil, the spring, the loaded diameter may be calculated from equation 5.8.11 by replacing T by $-T$ to get

$$D_2 = \frac{N_1}{N_1 - T} D_1 \tag{5.8.16}$$

Equations 5.8.11 and 5.8.12 are equivalent to, but simpler than, solving (5.8.9) for N_1 and diameter D_1 and for D_2 when the number of turns has been changed to N_2 according to $N_2 = N_1 \pm T$, depending upon the direction of the applied load, as shown in Problem 5.21.

Quantities necessary to specify a torsional spring are shown in Figure 5.8.2. Angles in terms of θ are for customer convenience.

EXAMPLE 5.8.1

Design a torsion spring to aid in opening a wooden trap door which weighs 20 lb and measures 20 in. on each side. The door, in a horizontal ceiling frame, should be hinged as in either Figure 5.8.3b or c and it should remain closed from its own weight. A chain attached to the frame prevents the door from rotating more than 100°. It should open with an upward force of 1.0 lb at the handle, which is 18 in. from the hinge line, and it should close easily. Oil-tempered wire is preferred and the spring should fit

Torsion Springs Specification Form

Mandatory Specifications
(fill in only those required)

1. To work over _____ in. dia. shaft.

2. OUTSIDE DIAMETER
   a. _____ in. max. or
   b. _____ in. ± _____ in.

3. INSIDE DIAMETER
   a. _____ in. min. or
   b. _____ in. ± _____ in.

4. Torque _____ in.-lb. ± _____ in-lb. at Θ_1 = _____°.
   Torque _____ in.-lb. ± _____ in.-lb. at Θ_2 = _____°.

5. Length of space available _____ in.

6. Maximum wound position _____ turns or
   _____° from free position.

7. Length of moment arm (R) _____ in.

8. Direction of helix (L, R, or optional). _____

9. Type of ends _____

Advisory Data

1. Wire diameter _____ in.

2. Mean coil diameter _____ in.

3. No. of coils _____

4. Rate _____ in.-lbs. per turn (360°).

5. Θ_F = _____° free angle reference

Special Information

1. Type of material _____

2. Finish _____

3. Frequency of rotation, _____ cycles sec. and
   working range. Θ = _____° to Θ = _____° deflection.

4. Operating temp _____ °F

5. End use or application _____

6. Other _____

Figure 5.8.2 Torsion spring specification form and end types. *Source:* Handbook of Spring Design, *courtesy Spring Manufacturers Institute, Inc., Wheeling, IL.*

around a 0.25-in.-diameter shaft. End extensions should be no shorter than 0.75 in. and no longer than 1.5 in.

Begin by calculating the maximum and minimum moments and the number of turns to open. We find from the stated requirements that

$$M_1 = 20(10) - 18 = 182 \text{ in.-lb} \quad \text{and} \quad T = 100/360 = 0.2778 \text{ turns}$$

When the door is against the stop at $100°$ the weight of the door itself provides a torque of $20(10) \cos 80° = 34.730$ in.-lb to hold the door open. Since this corresponds to a force of about 2 lb

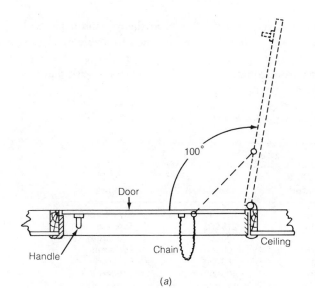

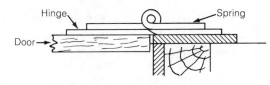

(a)

(b)

(c)

Figure 5.8.3 (a) Trap door, (b) Hinge detail for spring with integer number of turns, and (c) Detail for spring with integer plus three-quarter turns.

(1.737 lb) at the door edge, we set the spring moment to 10 in.-lb at a rotation of 100°, or 0.2778 turns. Our motivation for this choice is to add to the tendency of the door to stay open if it is accidentally bumped. Consequently,

$$M_2 = 10 \text{ in.-lb} \quad \text{and} \quad R = (182 - 10)/(0.2778) = 619.1505 \text{ in.-lb/turn}$$

We arbitrarily select a spring inside diameter of 0.30 in. to clear the rod and begin with $e_1 = e_2 = 1.5$ in. We next turn to Table 5.1.1 to find $E = 30 \times 10^6$ psi. and then to Table 5.1.3 to search for a satisfactory standard round wire size. The search is done with a program that calculates the minimum tensile strength found for each diameter from equation 5.8.4 and compares it with that given in Table 5.1.2 for oil-tempered wire of the same diameter.

Typical trials might be

d (in.)	σ_{xx} Calculated	Tabulated	Remarks
0.100	3 038 256	226 786	(By interpolation)
0.150	976 514	215 000	Too large
0.207	403 267	190 000	Too large
0.375	82 678	180 000	Too small
0.3125	133 513	183 000	Possible
0.250	242 087	185 000	Too large

Clearly $d = 0.3125$ in. must be selected. But this wire diameter is larger than the inside diameter of the helix, and hence difficult to manufacture. Since the *Spring Manufacturers Handbook* suggest a minimum spring index of 4, we turn to the definition of the spring index c to find that

$$c = D/d = \text{ID}/d + 1$$

which yields

$$\text{ID} = d(c - 1) = 0.3125(3) = 0.9375 \text{ in.}$$

as the minimum inside diameter that will give a spring index of 4 with a wire diameter of 0.3125. Upon substituting this value into the stress formula (5.8.4) we find that $d = 0.3125$ in. is still the smallest wire diameter that we can use.

We begin our redesign by assuming an inside diameter of 1.0 in. and end extensions of 1.0 in. (since end extensions of 0.75 in. would not extend beyond the spring's outside diameter) to find from equation 5.8.14 that $N = 3.048$ turns. Although this may be close enough to an integer number of turns to give satisfactory performance, we will investigate the effect of changing the end extensions upon the number of turns. Upon substituting different end extensions into equation 5.8.14 for $e_1 = e_2$ we find

End extension	N	End extension	N
1.50	2.942	1.22	3.001
1.20	3.005		

Hence, a spring for the hinge design in Figure 5.8.3*b* is given by

inside diameter = 1.00 in. mean diameter = 1.3125 in.
wire diameter = 0.3125 in. outside diameter = 1.625 in.
end extensions = 1.220 in. helix length = 0.9375 in.

All but the helix length (which is not critical) are within the tolerances that can normally be met by spring manufacturers.

Although we could also select end extension lengths to have a spring of 2.50 turns, we shall investigate the effect of changing the inside diameter upon the number of active turns. Upon substituting a series of inside diameters into the previous program, which calculates c from equaion 5.4.20 and solves equation 5.8.4 along with equations 5.8.7 through 5.8.16, we find for $e_1 = e_2 = 1.000$ that

ID	N	ID	N
1.000	3.048		
0.930	3.231	0.923	3.2508
0.900	3.316		

Thus, a spring for the hinge design shown in Figure 5.8.2*b* is given by

inside diameter = 0.923 in. outside diameter = 1.548 in.
wire diameter = 0.3125 in. mean diameter = 1.2355 in.
end extensions = 1.000 in. helix length = 1.016 in.

The spring index for this spring is 3.954, which is close enough to 4.00 to be produced commercially.

The largest outside diameters of these springs, found from equations 5.8.15 and 5.8.16, are 1.447 and 1.359 in., so that they may be held parallel to the hinge axis by either placing them over cylinders with outside diameters slightly less than the minimum ID of each spring, or placing them inside of cylinders whose inside diameters are slightly greater than the maximum OD of each spring.

Finally, since the end of each spring extension is loaded by a force F_i which produces the torque exerted by the spring, we can find the largest force acting at the ends of the spring extensions from

$$F_i = T_{max}/e_i$$

Thus, $F_1 = F_2 = 182.0/1.0 = 182$ lb. We, therefore, also recommend that either each spring end press against a small metal plate (to prevent the extensions from pressing into the wood and slightly changing the number of active turns) or that special hinges designed to accept torsion springs and their end extension be used.

5.9 BELLEVILLE (DISC) SPRINGS

These spring elements, which are nothing more than conical washers, are useful where small deflections at large loads are desired. Their physical simplicity belies the complexity of their analysis and their considerable versatility and utility. They are, for example, widely used in multiple disk brakes and clutches to separate the plates when the brake or clutch is to be released, as shown in Figure 5.9.1. They are also used in relief valves, in aircraft arrester cables, and similar equipment.

In addition to providing a wide range of spring constants, disc dimensions may be selected to provide any of the linear or nonlinear load–deflection curves shown in Figure 5.9.2. Moreover, stacking the discs of a particular size and thickness in series, in parallel, or in combination, as pictured in Figure 5.9.3a, enables the engineer to design for a variety of spring constants with just one disc style.[13] Since the variety of spring constants grows exponentially with the number of different discs available, a

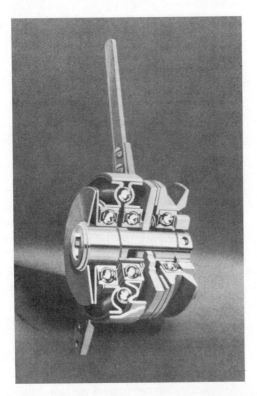

Figure 5.9.1 ROTO-CAM, a mechanical clutch made by Formsprag-Dana in which Belleville springs are used. *Source: Industrial Power Transmission Division, Dana Corporation, Mt. Pleasant, MI.*

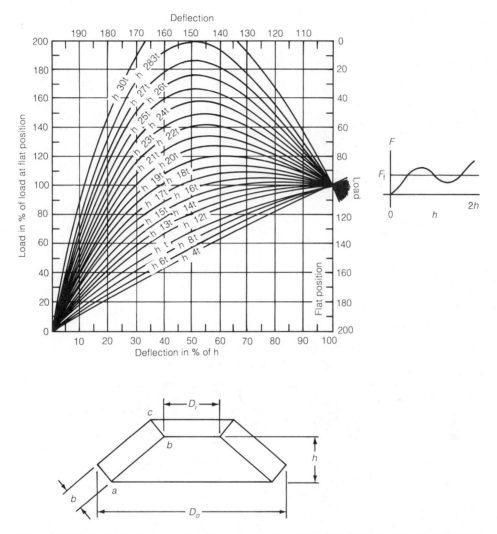

Figure 5.9.2 Load-deflection curves for disc springs and notation for disc springs. *Source:* Handbook of Spring Design, *courtesy Spring Manufacturers Institute, Inc., Wheeling, IL.*

small inventory of discs allows a manufacturer to offer a wide variety of spring characteristics.

The force F required to obtain deflection δ is given by equation 5.9.1[10]

$$F = C_0 \delta \left[(h - \delta) \left(h - \frac{\delta}{2} \right) C_1 t + C_2 t^3 \right] \tag{5.9.1}$$

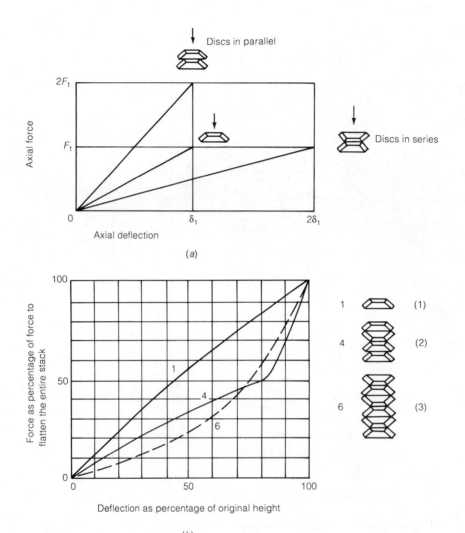

Figure 5.9.3 (a) Three spring rates from single disc, two discs in parallel, and two discs in series. (b) Load–deflection curves for (1) a single disc, (2) two discs in series in combination with two discs in parallel, same disc as in (1), and (3) six different discs in series.

where

$$C_0 = \frac{4E}{D_0^2(1 - v^2)} \qquad C_1 = \pi\left(\frac{c+1}{c-1} - \frac{2}{\ln c}\right)\left(\frac{c}{c-1}\right)^2$$

$$C_2 = \frac{\pi}{6}\left(\frac{c}{c-1}\right)^2 \ln c$$

(5.9.2)

in which v denotes Poisson's ratio and $c = D_0/D_i$. Stresses at points a, b, and c, Figure 5.9.1, are given by

$$\sigma_a = -C_0\delta\left[C_3\left(h - \frac{\delta}{2}\right) + C_4 t\right] \tag{5.9.3}$$

$$\sigma_b = C_0\left[C_4 t - C_3\left(h - \frac{\delta}{2}\right)\right]\delta \tag{5.9.4}$$

$$\sigma_c = C_0\delta\left[C_5\left(h - \frac{\delta}{2}\right) + C_6 t\right] \tag{5.9.5}$$

in which

$$C_3 = \left(\frac{c}{c-1}\right)^2\left(\frac{c-1}{\ln c} - 1\right) \tag{5.9.6}$$

$$C_4 = \frac{c^2}{2(c-1)}$$

$$C_5 = \frac{6}{\pi \ln c}\left(1 - \frac{c-1}{c \ln c}\right) \tag{5.9.7}$$

$$C_6 = \frac{3}{\pi \ln c}\frac{c-1}{c}$$

From equation 5.9.1 the force for the flat position of a disc spring may be found from the condition that $\delta = h$. Thus,

$$F_{\text{flat}} = C_0 C_2 h t^3 \tag{5.9.8}$$

If disc springs are held only at their inner and outer edges they may deflect through twice their height as shown in the small graph to the right of the central graphs in Figure 5.9.2. Due to the symmetry evident from the small graph, the deflection between 100 and 200% may be had by turning the central graphs upside down and reading from the deflection scale which runs from 100 to 200% and from the force scale on the left of the inverted figure.

Nonlinear spring rates may be achieved as shown in Figures 5.9.3b and 5.9.4 for the situation where the axial loads are allowed to increase until all spring elements are flattened against a flat supporting plate.

Calculation of the equivalent spring rate of the stack of disc springs involves solving equation 5.9.1 for the displacement as a function of the applied load for each of the different discs in the stack and then adding them together to get the total displacement. This requires solution of the cubic equation

$$\delta^3 - 3h\delta^3 + 2\left(h^2 + \frac{C_2}{C_1}t^2\right)\delta - \frac{1-v^2}{2E}\frac{D_0^2 F}{C_1 t} = 0 \tag{5.9.9}$$

Details of solution techniques for cubic equations will not be discussed here because they are discussed at length in books on numerical methods and because they are commercially available on diskettes and in math modules.

Load deflection curves in this section have been nondimensionalized by expressing the deflection as a percentage of the free height and the load as a percentage of the load required to flatten the spring so that they may apply to any Belleville spring.

EXAMPLE 5.9.1

Calculate the force–deflection curve for a stack of 3 disc springs in series, for which OD = 1.000 in., $D_i = 0.400$ in., and where

h(in.)	t(in.)
0.010	0.075
0.013	0.050
0.015	0.035

are the heights and thicknesses of the three discs. Also calculate the flattening forces and the stresses at locations a, b, and c for each of the discs. Use $E = 3 \times 10^7$ psi and $v = 0.33$.

From equation 5.9.8 or 5.9.1 we find that the respective flattening forces are

757.13 lb   291.63 lb   115.42 lb

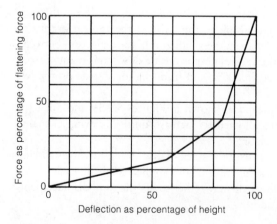

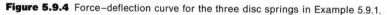

Figure 5.9.4 Force–deflection curve for the three disc springs in Example 5.9.1.

and from equations 5.9.3, 5.9.4, and 5.9.5 the corresponding stresses are

σ_a (psi)	σ_b (psi)	σ_c (psi)
−234 244	198 499	67 999
−222 631	162 222	62 921
−200 906	120 481	55 108

The corresponding force–deflection curve for intermediate forces and deflections is shown in Figure 5.9.4.

5.10   ELASTOMERIC SPRINGS

The term elastomeric spring is incomplete because while elastomeric elements tend to flex as springs, the material is actually more like a spring and a damper, or shock absorber, in combination. This is because of the significant internal damping within the material itself, as mentioned in Section 5.1. Elastomeric elements are used, therefore, primarily where the ability to dissipate energy is desirable, as in shock mounts.

Because of the pronounced sensitivity of stress–strain data for elastomeric materials (rubber and rubberlike materials) to the specimen configuration, to its chemical composition, and to the measurement techniques used, application of the data obtained from one configuration to other configurations may lead to some discrepancy between calculated and actual performance.

Deflection of a shear mount made by bonding plates to opposite sides of a rectangular block of elastomeric material, Figure 5.10.1, is related to the stress by

$$\varepsilon_{zy} = \frac{\sigma_{zy}}{2G} \tag{5.10.1}$$

which, upon substitution for ε_{zy} from (3.9.4) with $u_z = 0$, yields

$$\frac{du_y}{dz} = \frac{\sigma_{zy}}{G} \cong \frac{u_y}{h} \tag{5.10.2}$$

and thus, with $\sigma_{zy} = F/A$ and $u = u_y$, the deflection may be written as

$$u = \frac{Fh}{GA} \tag{5.10.3}$$

where A is the area of the faces supporting the shear load.

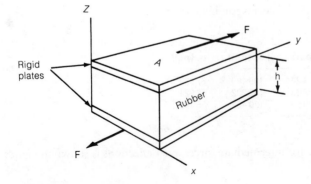

Figure 5.10.1 Rubber shear mount.

The spring rate, or spring constant, is given by

$$R = \frac{F}{u} = \frac{GA}{h}$$

(5.10.4)

in which

$$R = \frac{|F_1 - F_2|}{|u_1 - u_2|}$$

(5.10.5)

Element height h is found from the limiting shear stress according to (5.10.3) to be

$$h = u\,\frac{G}{\sigma_{zy}}$$

(5.10.6)

where u is the maximum deflection required.

Axial deflection of the central shaft or cylinder of a shock mount with the cross section shown in Figure 5.10.2 may be calculated from the stress–strain relation (3.9.4), which is

$$\varepsilon_{rz} = \frac{\sigma_{rz}}{2G}$$

(5.10.7)

Substitution for ε_{rz} from (3.9.6) and for σ_{rz} from

$$\sigma_{rz} = \frac{F}{\pi d_i h}$$

(5.10.8)

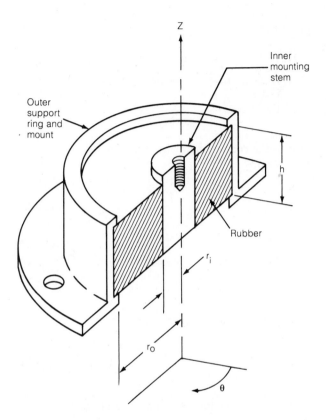

Figure 5.10.2 Cross section of a representative shock mount.

in which d_i is the inner diameter of the rubber cylinder in Figure 5.10.2, yields

$$\frac{du_z}{dr} = \frac{F}{2\pi Grh}$$

for $u_r = 0$. Integrate and let $\delta = u_z$ to get

$$\delta = u_z = \frac{F}{2\pi Gh} \ln \frac{r_o}{r_i} \tag{5.10.9}$$

When the ratio r_o/r_i is greater than 2.4142 the effect of bending may become large enough to influence the force–deflection relations according to

$$F = \frac{2\pi\delta hG}{\ln \dfrac{r_o}{r_i} + 2\pi B} \tag{5.10.10}$$

wherein the bending term B is given by[11]

$$B = 3 \frac{\left[\left(\frac{r_o}{r_i}\right)^2 - 1\right]^2 - 4\left(\frac{r_o}{r_i}\right)^2}{16\left(\frac{r_o}{r_i}\right)^2\left[\left(\frac{r_o}{r_i}\right)^2 - 1\right]}\left(\frac{r_o}{r_i}\ln\frac{r_o}{r_i}\right)^2 \tag{5.10.11}$$

If a torsional moment M is applied to the central shaft of the mount shown in Figure 5.10.2 the pertinent stress–strain relation is

$$\varepsilon_{\theta r} = \frac{\sigma_{\theta r}}{2G} \tag{5.10.12}$$

and the shear stress $\sigma_{r\theta}$ is given by

$$\sigma_{r\theta} = \frac{M}{2\pi r_i^2 h} = \frac{2M}{\pi d_i^2 h} \tag{5.10.13}$$

Substitution for $\varepsilon_{\theta r}$ from (3.9.6) and $\sigma_{\theta r}$ from (5.10.13) yields

$$\frac{du_\theta}{dr} - \frac{u_\theta}{r} = \frac{M}{2\pi r^2 h G} \tag{5.10.14}$$

since there is no displacement in the radial direction. Writing (5.10.14) as

$$r\frac{d}{dr}\left(\frac{u_\theta}{r}\right) = \frac{M}{2\pi G h r^2} = r\frac{d\theta}{dr}$$

and integrating from r_i to r_o, the rotation becomes

$$\theta = \frac{M}{2\pi G h}\int_{r_i}^{r_o}\frac{dx}{x^3} = \frac{M}{4\pi G h}\left(\frac{1}{r_i^2} - \frac{1}{r_o^2}\right) \tag{5.10.15}$$

where θ is the angle of rotation in radians.

Pertinent design equations for combined axial and radial loading are discussed in Ref. 13.

5.11   AIR (PNEUMATIC) SPRINGS

Two types of air springs are commonly available: those using air only and those using a liquid as a transfer and damping agent, as shown in Figure 5.11.1. Analysis of the liquid–pneumatic type is more complicated than that of the air (pneumatic) only type

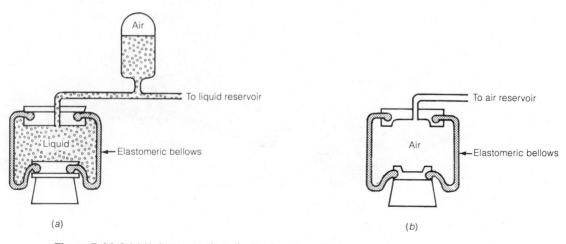

Figure 5.11.1 (a) Hydropneumatic or liquid–air spring, (b) air spring.

even at relatively low-frequency oscillation because of the inertia of the liquid and the energy losses at the orifices and in the liquid lines. Consideration will be limited, therefore, to entirely pneumatic springs and to the air within the spring itself: Neither accumulators nor the negligible effect of air in the supply line will be considered.

The advantage of air springs over metallic springs in vehicular applications is the ability to use an on-board compressor, load sensors, and a microprocessor to adjust the suspension system to the load being carried and to keep the vehicle frame level relative to the road, regardless of the weight distribution within the design load of the vehicle. A typical system for a passenger car is that shown in Figure 5.11.2 for a Lincoln Continental.

Use of air springs may be of greater benefit on buses and truck trailers because of their greater load variation and their tendency to be harder to control when improperly loaded for a nonadjustable suspension is used. A pneumatic suspension system for a Freuhauf trailer is shown in Figure 5.11.3.

Sealing problems in single-chamber air springs may be minimized by using flexible elastomeric sections clamped to the moving portions of the spring, as pictured in Figure 5.11.4. This flexible section is normally reinforced with two bias plies of synthetic cord and may be manufactured in several configurations, as shown in Figure 5.11.4 to obtain different force-displacement characteristics. Construction details and the associated terminology are displayed in Figure 5.11.5.

Because of the elastomeric material used for the lobe they are normally limited to applications where the temperature is between −35 and 135 °F and where they will not come in contact with petroleum based fluids, acids, or hot materials. They are, of course, also limited to locations or equipment where compressed air is available.

Response of an air spring varies to some extent with the rate of loading and unloading. This is because the pressure–volume relation for air depends upon whether

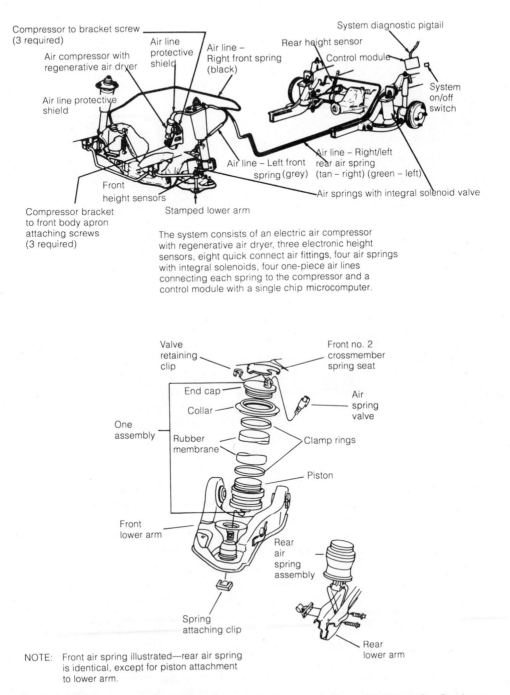

The system consists of an electric air compressor with regenerative air dryer, three electronic height sensors, eight quick connect air fittings, four air springs with integral solenoids, four one-piece air lines connecting each spring to the compressor and a control module with a single chip microcomputer.

NOTE: Front air spring illustrated—rear air spring is identical, except for piston attachment to lower arm.

Figure 5.11.2 Air spring system on the 1984 Lincoln Continental. *Source: Ford Motor Co., Dearborn, MI.*

Figure 5.11.3 Air ride suspension system. *Source: Fruehauf Division, Fruehauf Corporation, Detroit, MI.*

the load is applied slowly enough for the air to remain at ambient temperature as it is compressed (isothermal process), whether it is loaded and unloaded so rapidly and/or insulated so well that the air cannot dissipate the heat generated (adiabatic process), or whether the loading rate and heat transfer rate lie somewhere between these two extreme conditions. In any of these loading conditions, however, the pressure P and the volume V are related as

$$P_1 V_1{}^n = P_0 V_0{}^n = \text{constant} \tag{5.11.1}$$

Exponent n may be taken to be 1.00 for an isothermal process, 1.40 for an adiabatic condition, and between these values for a polytropic (mixed) process. Usually $n = 1.38$ is used for design of air springs for automobiles, buses, and trucks.

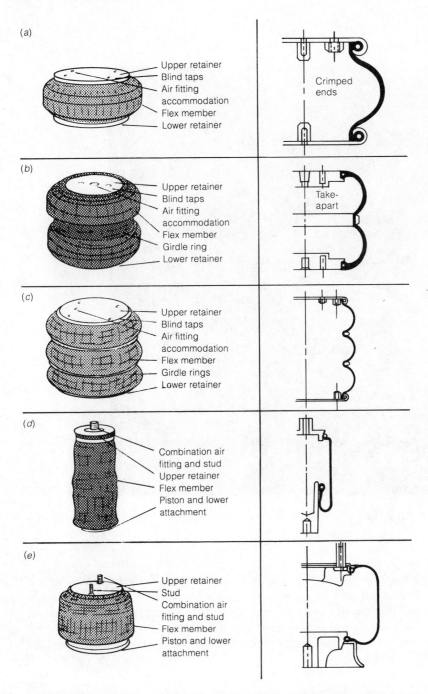

Figure 5.11.4 (*a–c*) Single-, double, and triple-convolute bellows types, (*d*) sleeve-type rolling lobe assembly, and (*e*) conventional rolling lobe assembly. *Source: Industrial Products Division, Goodyear Tire & Rubber Company, Akron, OH.*

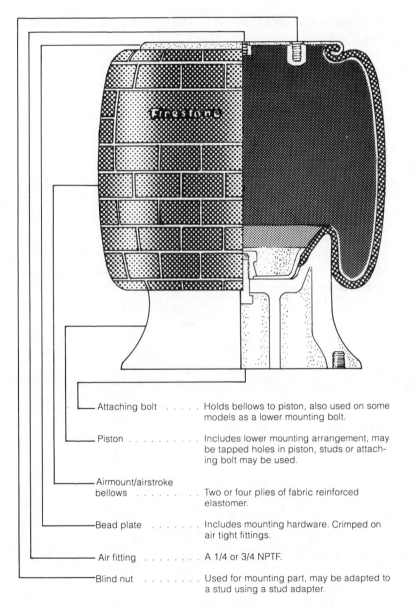

Attaching bolt . . . . . Holds bellows to piston, also used on some
models as a lower mounting bolt.

Piston . . . . Includes lower mounting arrangement, may
be tapped holes in piston, studs or attach-
ing bolt may be used.

Airmount/airstroke
bellows . . . . Two or four plies of fabric reinforced
elastomer.

Bead plate . . . . Includes mounting hardware. Crimped on
air tight fittings.

Air fitting . . . . A 1/4 or 3/4 NPTF.

Blind nut . . . . Used for mounting part, may be adapted to
a stud using a stud adapter.

Figure 5.11.5 Components of a rolling lobe air spring. *Source: Firestone Industrial Products Co.,
Noblesville, IN.*

By definition the spring rate, or spring constant, (which is not constant for a pneumatic spring) is defined by

$$R = \frac{dF}{dz} \tag{5.11.2}$$

where the force in (5.11.2) is given by

$$F = P_g A \tag{5.11.3}$$

where P_g is the gauge pressure. Upon substituting for F from (5.11.3) into (5.11.2) and performing the differentiation indicated, we have

$$R = P_g \frac{\partial A}{\partial z} + A \frac{\partial P_g}{\partial z} \qquad P = P_g + P_a \tag{5.11.4}$$

where P_a denotes the ambient pressure, which changes slowly with changing weather conditions, so that it may be considered constant for vibrations of short duration. Consequently,

$$\frac{\partial P_g}{\partial z} = \frac{\partial P}{\partial z}$$

which may be evaluated by returning to (5.11.1) and differentiating with respect to z to obtain

$$V^n \frac{\partial P}{\partial z} + nPV^{n-1} \frac{\partial V}{\partial z} = 0 \tag{5.11.5}$$

which may be written as

$$-\frac{\partial P}{\partial z} = \frac{nP}{V} \frac{\partial V}{\partial z} \tag{5.11.6}$$

Substitution from (5.11.6) into (5.11.4) enables us to write the spring constant as

$$R = P_g \frac{\partial A}{\partial z} - \frac{nAP}{V} \frac{\partial V}{\partial z} \tag{5.11.7}$$

An alternative form may be found by first solving (5.11.1) for P in terms of some reference pressure P_0 and volume V_0 to get

$$P = P_0 \left(\frac{V_0}{V} \right)^n$$

which may then be differentiated with respect to z to give

$$R = P_g \frac{\partial A}{\partial z} + A P_0 \frac{\partial}{\partial z}\left(\frac{V_0}{V}\right)^n \tag{5.11.8}$$

Typical design curves are shown in Figures 5.11.6 and 5.11.7, which show the static load as a function of the static deflection when the gauge pressure in the spring is held constant. They also show volume as a function of height as an aid in calculating

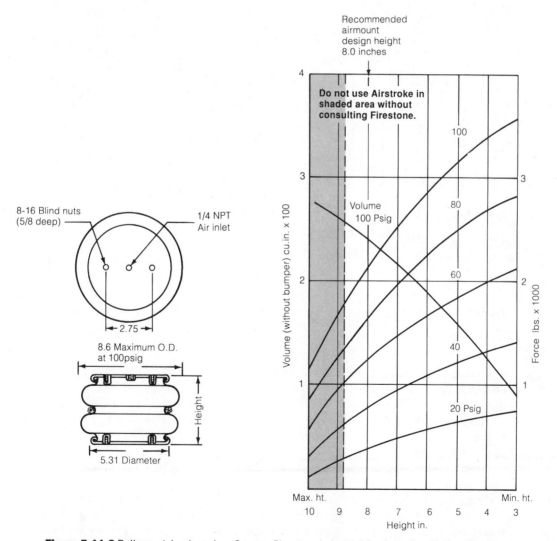

Figure 5.11.6 Bellows-style air spring. *Source: Firestone Industrial Products Co., Noblesville, IN.*

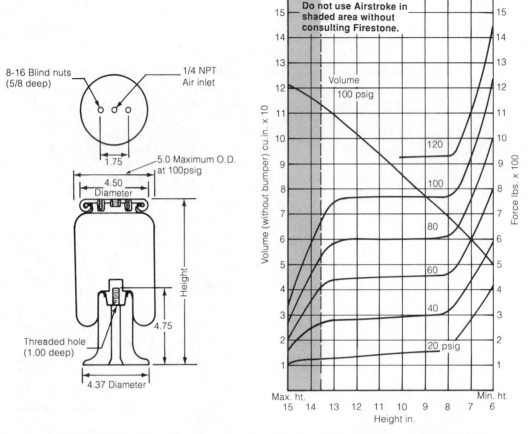

8-16 Blind nuts
(5/8 deep)

1/4 NPT
Air inlet

5.0 Maximum O.D.
at 100psig

1.75

4.50
Diameter

Height

4.75

Threaded hole
(1.00 deep)

4.37 Diameter

Figure 5.11.7 Rolling lobe (also known as reversible sleeve)-style air spring. *Source: Firestone Industrial Products Co., Noblesville, IN.*

the spring constant, or spring rate, according to equation 5.11.7. These static load–deflection curves are all labeled with the constant internal, or gauge, pressure at which they were measured. In Figure 5.11.6 the curves for pressures of 20 psi and larger terminate on the left at one-third of the burst pressure for the heights shown. The curve in Figure 5.11.7 which does not extend from the minimum height to the maximum height is likewise limited by the burst pressure by a factor of three.

The recommended design height shown in Figure 5.11.6 is based upon lateral stability and applies only when the air spring is used for isolation. In spring applications it is assumed that lateral stability is provided by constraints on the load.

The nearly horizontal portion of the design curves in Figure 5.11.7 imply that dA/dz is very small for this type of rolling lobe air spring. From equation 5.11.7 we see that for this spring the spring constant is largely determined by the dV/dz term, which, according to the volume–displacement curve, is also very nearly constant. Hence, this spring behaves much like a helical compression spring in that the spring rate is nearly linear so that the dynamic load and deflection are also linearly related.

EXAMPLE 5.11.1

Recommend the operating pressure for the airspring shown in Figure 5.11.6 to support a load of 1400 lb at a design height of 8.0 in. Also estimate the deflection cause by changing the load by ± 300 lb.

Motivated upon the nearly uniform separation between curves, use linear interpolation to find the pressure of the curve passing through the point whose coordinates are 1400 lb and 8.0 in.; that is, along the $z = 8.0$ in. read 1200 lb at 60 psi gauge pressure and 1670 lb at 80 psi gauge pressure, so that the gauge pressure is estimated to be 68.51 psi at 1400 lb.

The deflection may be estimated from the spring constant given by equation 5.11.7. Area A may be found from the data at hand; namely $A = 1400/68.51 = 20.435$ in.^{2}. Derivative dV/dz may be estimated from Figure 5.11.7 by placing a straight-edge tangent to the volume curve at $z = 8.0$ and reading $dV/dz = 4.4(5) = 22$ in.^{3}/in., approximately, at $V = 293$ in.^{2}.

Finding dA/dx involves more work. We begin by estimating the dA/dz values at 8.00 on the 40-, 60-, and 80-psi curves and then decide whether to use linear or orthogonal interpolation between them to approximate the value on the undrawn curve through 68.51 psi. For ease of reading, record the load on the 40-psi curve at 7.75 in. and at 8.25 in. Divide these values by 40 psi to obtain the effective area at each of these two points as

$$A \text{ at } 8.25 \text{ in.} = 720/40 \qquad A \text{ at } 7.75 \text{ in.} = 800/40$$
$$= 18.000 \text{ in.}^2 \qquad = 20.000 \text{ in.}^2$$

Derivative dA/dz is then approximately

$$dA/dz = (18.000 - 20.000)/0.5 = -4.000 \text{ in.}$$

At 60 psi at the 8.25 and 7.75 points we find

$$A \text{ at } 8.25 \text{ in.} = 1150/60 \qquad A \text{ at } 7.75 = 1260/60$$
$$= 19.167 \text{ in.}^2 \qquad = 21.000 \text{ in.}^2$$

so that

$$dA/dz = (19.167 - 21.000)/0.5 = -3.666 \text{ in.}$$

At 80 psi at the 8.25 and 7.75 points we find

$$A \text{ at } 8.25 \text{ in.} = 1590/80 \qquad A \text{ at } 7.75 \text{ in.} = 1730/80$$
$$= 19.875 \text{ in.}^2 \qquad = 21.625 \text{ in.}^2$$

so that

$$dA/dz = (19.875 - 21.625)/0.5 = -3.500$$

Since the variation of dA/dz is evidently not linear from 40 to 60 to 80 psi, we use orthogonal interpolation to obtain $dA/dz = -3.577$ in. ($P_a = 14.7$ psi). Substitution into (5.11.7) yields

$$R = 68.51(-3.577) - 1.38(20.435)(68.15 + 14.7)(22)/239 = -461.060 \text{ lb/in.}$$

Upon writing (5.11.2) as

$$\Delta z = \frac{\Delta F}{R}$$

we have that $\Delta z = \pm 0.651$ for changes of ± 300 lb in the load. Thus, as the load increases to 1700 lb the spring height decreases from 8.00 to 7.349 in. and, upon reading from the volume curve in Figure 5.11.7, we find that this height corresponds to a volume of 226 in.^{3}.

According to (5.15.1) the absolute pressure in the spring will be given by

$$P_1 = P_0\left(\frac{V_0}{V}\right)^n = 83.21\left(\frac{239}{226}\right)^{1.38} = 89.887 \text{ psi}$$

$$P_g = 89.887 - 14.7 = 75.187 \text{ psig}$$

where a polytropic process has been assumed, for which $n = 1.38$.

If (5.11.8) had been used the second term in the expression for R would have been

$$20.435(83.21)[(253.322/239)1.38 - 1]/0.651 = 218.430 \text{ lb/in.}$$

and the spring rate would have been given as $R = -463.490$. This is satisfactory agreement for the number of significant figures carried in the calculations.

When larger total displacements are involved for a constant amount of air in the spring it may be advisable to calculate the spring constant, or spring rate, in increments to include the effects of the variation of dV/dz and dA/dz with z.

The minus sign included in the value of R is to emphasize that z in Figure 5.11.6 and 5.11.7 is measured in the direction or decreasing F. It did not enter our earlier discussion because z is conventionally measured from the loaded end of the spring in the direction or the applied force in the analysis of helical and Belleville springs. The minus sign is usually omitted in listings of pneumatic spring rates to avoid further explanation.

REFERENCES

1. *Handbook of Spring Design*, Spring Manufacturers Institute, Wheeling, IL, 1981.

2. *Properties and Selections: Irons and Steels. Metal Handbook*, 9th ed., Vol. 1, American Society for Metals, Metals Park, OH, 1978.

3. *Design Handbook, Engineering Guide to Spring Design*, Associated Spring, Barnes Group, Inc., Bristol, CT, 1981.

4. *SAE Handbook*, Vol. 1, Society of Automotive Engineers, Warrendale, PA, 1981.

5. Allen, P.W., Lindley, P.B., and Payne, A.R., *Uses of Rubber in Engineering*, The Natural Rubber Producers Association, London, 1976.

6. Röver, A., Beanspruchung Zylinderrische Schraubenfedern mit Kreisquerschnitt, Zeitschrift des Vereines deutcher Ingenieure.

7. Timoshenko, S., *Strength of Materials*, 3d ed., part 1, van Nostrand, Princeton, NJ, 1956, p. 362.

8. Wahl, A.M., *Mechanical Springs*, McGraw-Hill, New York, p. 235, 1963.

9. Carlson, H., *Spring Designers Handbook*, Dekker, New York, 1978.

10. Almen, J.O., and Laszlo, A., The uniform-section disk spring, *Transactions of the ASME*, 58, 305–313 (1936).

11. Adkins, J.E., and Gent, A.N., Load–deflexion relations of rubber bush mountings, *British Journal of Applied Physics*, 5, 354–358 (1954).

12. Orthwein, W.C., Designing disk springs with a personal computer, *Computers In Mechanical Engineering*, 2 (3), 45–51 (1983).

13. Orthwein, W.C., Shock-mount design for combined loads, *Machine Design*, 55 (5), 284–288 (Mar. 10, 1983).

PROBLEMS

Section 5.2

5.1 A torsion bar 500 mm long has a torsional spring constant of 112 N/m/°. To what length should it be cut to have a torsional spring constant of 220 N/m/°?

5.2 Find an expression for the equivalent spring constant, or spring rate, defined by $R = F/\delta$, of a torsion bar and slotted lever as shown in the drawing. F and δ are measured in the vertical direction as shown. Let $\theta = \theta_0$ when $F = 0$. Show that if θ and θ_0 are less then $5°$

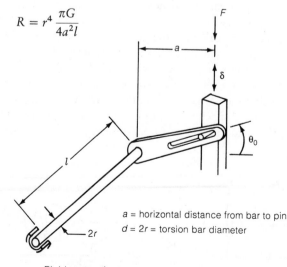

$$R = r^4 \frac{\pi G}{4a^2 l}$$

a = horizontal distance from bar to pin
$d = 2r$ = torsion bar diameter

Rigid connection

5.3   Find the diameter for a solid torsion bar similar to that acting on the lower control arm as shown in the drawing below if each torsion bar should be 1.27 m long and designed to support a load that varies from 225 to 350 kg and acts through a lever arm of 0.30 m. The maximum rotation should be no more than 7.5°. Use $G = 79\,310$ MPa.

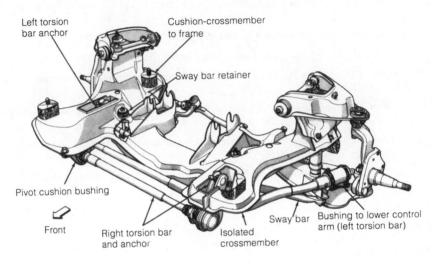

Left torsion bar anchor

Cushion-crossmember to frame

Sway bar retainer

Pivot cushion bushing

Front

Right torsion bar and anchor

Isolated crossmember

Sway bar

Bushing to lower control arm (left torsion bar)

Source: Chrysler Motors

Sections 5.3 and 5.4

5.4   An open, ground, compression spring with a free length of 206 mm and a spring rate of 403 N/m was cut into four equal lengths to obtain four springs with equal spring constants. What is the spring rate of each of the four springs? If each of the three cuts removed 2 mm of the spring (i.e., the kerf is 2 mm), what is the compressed height of each spring when the four together support a 20-kg motor? Each spring is equally loaded. Pitch is 8 mm. (*Hint:* Reduce the uncut spring length by the total of the kerfs, calculate the new spring rate (a small change), and then imagine this spring cut into 4 equal lengths with no loss of material.)

5.5   Design a closed and ground compression spring having an OD of 0.5 in. to provide a spring force of 15 lb against a cone seat in a relief valve when in the closed position and a force of 28 lb when in the full open position, both as shown in the figure. Use music wire (ASTM A228) in the sizes shown in Table 5.1.3 and specify the spring rate, the outside diameter, the wire size, the number of active turns, the total number of turns, the free length, and the pitch.

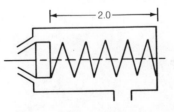

2.0

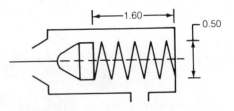

1.60

0.50

5.6   A closed and ground compression spring of oil-tempered wire is to fit around a 0.188-in.-diameter guide rod and to press against a sliding washer having a 0.30-in. outside diameter and a 0.190-in. inside diameter. At a length of 8.0 in. it is to exert a 3.0-lb force and at 5.0 in. it is to exert a 8.0-lb force. Specify the inside diameter, minimum wire size, number of active and total turns, free length, pitch, solid height and outside diameter at the solid height.

5.7   What material should be used if the spring in problem 5.6 is subject to shock loads and temperatures to 460 °F? Can the spring be designed using standard wire sizes of the required material?

5.8   Design a compression spring with an outside diameter of 11.00 in. from round type 302 stainless steel wire to operate a plate dispenser in a cafeteria. When fully loaded the spring is to be compressed 45 in. under a load of 80 lb, and when empty the spring should exert a force of 0.50 lb against the stop. Ends are closed. Use a design stress of 35 % of the minimum tensile stress. Specify the wire diameter, the number of active turns, the total number of turns, the solid height, and the outside diameter at the solid height.

5.9   Design a closed, unground, compression spring for the grease plunger in a hand grease gun for a cylinder 50.0 mm inside diameter which is to provide a load of 9.1 kg when compressed to a length of 45.0 mm and a load of 2.2 kg when extended to 305.0 mm at the end of the grease chamber. Find the wire diameter, the solid height (which must be less than 45 mm), the free length, the active number of turns, the total number of turns, and the pitch. Use Type C wire and let the OD be 48.0 mm for clearance. Use $q = 0.45$ and $G = 8086$ kg/mm^2. (*Note:* Although not technically correct, it is common practice for Europeans to express force in terms of kilograms. This problem, as in Table 5.1.3, is expressed in commonly used European engineering terms.)

5.10   Design a closed and ground compression spring for a palm button on a forming press. Button dimensions are shown below. Force at 3.0 in. should be 5.0 lb and the force at 1.5 in. should be

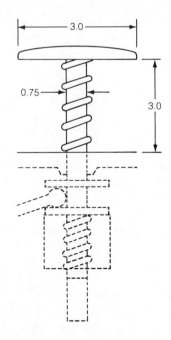

10.0 lb. Use oil-tempered wire and a clearance of 0.02 in. about the rod. Specify the outside diameter, wire diameter, number of total turns, solid height, free length, pitch, and pitch angle.

Sections 5.3–5.6

5.11 Design a tension spring from music wire for an exercise machine which is to exert 25.0 lb at an initial length of 40.0 in. inside the loops and a force of 175.0 lb at a separation of 70.0 in. inside the loops. Outside diameter should be no greater than 2.0 in. Specify the wire diameter, the number of turns, the initial tension, and the loop allowance. Also recommend the largest ID for the end hook.
[*Ans.* $d = 0.207$ in., $N = 91.575$ turns.]

5.12 Design a tension spring for an overhead door to operate as shown in the figure. In position 1, where the door is closed, the user should exert 10.0 lb to initially raise the 160-lb door. At position 2, where the door has been raised 85.0 in., the spring should just become slack and exert zero force. The outside spring diameter should be no greater than 2.00 in. to use the existing spring winding equipment. Specify the number of turns, the free length, and the initial tension. Use hard-drawn spring wire.
[*Ans.* $d = 0.225$ in., $n = 186.66$ turns]

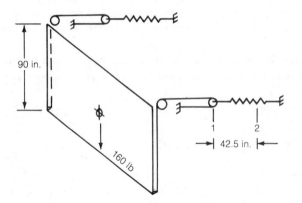

5.13 Find the wire diameter for an instrument tension spring to fit within a tube with an inside diameter of 0.150 in. and to exert a force of 1.50 lb when at a length of 0.779 in. When stretched to 0.893 in. the force should be 3.020 lb. Service temperatures vary from 60 to 500 °F in a humid atmosphere. Also give the coil diameter, the number of turns, the initial tension, and the loop allowance. Allow for a clearance of 0.005 in.
[*Ans.* Use stainless Type 302 wire, $q = 0.30$]

5.14 Design a tension spring having an outside diameter of 1.15 in. to hold a load of 10.0 lb at 8.0 in. inside the loops and a load of 30.0 lb at 12.0 in. inside the loops. Use hard-drawn wire. Give the number of turns, the initial tension, and draw the end loops, with dimensions. Loop ID should be equal to the ID of the body of the spring if stress permit. Use any of the loop styles shown in the text.

Section 5.8

5.15 Design a helical torsion spring from hard-drawn wire (ASTM A 227) to counterbalance a trap door. The door weighs 20 lb and its width is 18 in. A 1-ft-lb moment should be required to open it and the spring should hold it open against a stop 110° from the closed position with a moment of 1.5 ft-lb. The spring must have 1.0 in. end extensions and work over a 0.75-in.-diameter shaft. Specify the wire diameter, the helix length, and the outside diameter of the spring when the door is open and when it is closed.

5.16 Design a helical torsion spring to hold a port cover in the closed position with a torque of 30 in-lb. After rotating 90° to the full open position the torque should be 40 in.-lb. The spring is to have 0.50- and 0.75-in. end extensions and is to fit over a shaft 1.00 in. in diameter. Use chrome vanadium wire and specify the wire diameter, the spring rate in in.-lb per turn, the number of turns, and the length of the helix. Also find the diameter of the spring before it is installed; i.e., when it is free of load.

5.17 Design a helical torsion spring with short hook ends to lift the overhead door described in problem 5.12. Cable drum diameter is 4.5 in. and $\frac{1}{8}$-in.-diameter cable is to be used. Minimum spring ID is 2.00 in. Compare music wire (ASTM A 228) and hard-drawn wire (ASTM A 227). Give wire size, helix length, and the helix diameter when the door is raised and when it is lowered. Use Figure 5.1.1. to find the cheaper spring.

5.18 Design a helical torsional spring with hinge ends and an inside diameter of 155 mm to drive a drum 200 mm in diameter which holds a 3-mm wire rope connected to a shear blade weighing 110 kg. It is to hold the blade 500 mm above the shear table and to provide an upward force of 267 N when the blade is at the table. Find the wire diameter in mm, the number of turns, and the coil length. Also find the coil diameter with the blade in the raised position. Use grade C wire and neglect the weight of the cable. $E = 206.9$ GPa.

5.19 Design a helical torsion spring with hinge ends from round chrome vanadium wire to hold an exhaust cover vent on an industrial oven in the closed position. It is to supply a 12-in.-lb moment when closed and a 100-in.-lb moment when rotated through 120°. The spring is to fit over a 1.50-in.-diameter core. Specify the wire diameter, the number of turns, the coil length, and the outside diameter of the helix when the cover is fully open.

5.20 If a helical torsion spring catalog gives the torsional spring rate of a helical torsion spring 1.0 in. long as 10.0 in.-lb, what is the spring rate of a helical torsion spring of the same wire ($d = 0.07$ in.) and helix diameter but a helix length (length along the axis of the helix) of 1.7 in.?

5.21 Show that if

$$N_1 = \frac{Ed^4}{10.8RD_1}$$

for a helical torsion spring at no load and if it is unwound T turns under load such that

$$N_2 = \frac{Ed^4}{10.8RD_2}$$

that D_1 and D_2 are related as shown in equations 5.8.12. (*Hint:* Set $N_2 = N_1 - T$ and divide.)

Section 5.9

The program to be written in problem 22 will remove the tedium involved in Belleville spring selection, which requires solution of equations 5.9.1 through 5.9.7 for the axial force and stresses associated with a particular deflection. It is, therefore, recommended that problem 22 be completed before attempting problems 23 through 26.

Disc selection for problems 23, 24, and 25 is to be made from those spring washers listed in Table P5.22. Answers to problems 23 through 25 may not be unique; other combinations may also satisfy each problem. Use $E = 30 \times 10^6$ psi, $v = 0.33$.

TABLE P5.22   Representative Stock Belleville Spring Discs, or Washers

Part no.	Hole size	O.D.	I.D.	Thickness	Overall height
BEL-2000-482	3/8″	.750	.380	.060	.125
BEL-2000-483	3/8″	.745	.385	.040	.059
BEL-2000-484	3/8″	.750	.406	.050	.075
BEL-2000-500	3/8″	.781	.378	.037	.065
*BEL-2000-561	3/8″	.875	.406	.062	.112
BEL-2000-562	3/8″	.875	.406	.109	.124
BEL-2000-563	3/8″	.875	.406	.089	.100
BEL-2000-564	3/8″	.875	.375	.020	.100
BEL-2000-565	3/8″	.875	.406	.020	.078
BEL-2001-000	3/8″	1.000	.409	.060	.105
BEL-2001-001	3/8″	1.000	.422	.062	.162
BEL-2001-002	3/8″	1.000	.406	.062	.087
BEL-2001-003	3/8″	1.000	.406	.104	.116
*BEL-2001-040	3/8″	1.062	.405	.032	.045
BEL-2001-041	3/8″	1.060	.390	.080	.119
BEL-2001-160	3/8″	1.250	.409	.074	.100
BEL-2001-162	3/8″	1.250	.406	.060	.235
BEL-2001-163	3/8″	1.250	.400	.040	.187
BEL-2001-320	3/8″	1.500	.382	.104	.125
BEL-2001-480		1.750	.406	.031	.156
*BEL-2200-480	7/16″	.750	.453	.030	.060
BEL-2200-481	7/16″	.752	.440	.020	.035
BEL-2200-520	7/16″	.812	.471	.040	.063
BEL-2200-550	7/16″	.859	.491	.014	.030
BEL-2200-551	7/16″	.859	.484	.010	.030
BEL-2200-552	7/16″	.862	.499	.033	.042
BEL-2200-553	7/16″	.860	.453	.030	.060
BEL-2200-560	7/16″	.875	.475	.093	.118

(continued)

TABLE P5.22 (Continued)

Part no.	Hole size	O.D.	I.D.	Thickness	Overall height
*BEL-2200-561	7/16″	.875	.439	.020	.078
BEL-2200-580	7/16″	.900	.437	.090	.135
BEL-2200-620	7/16″	.968	.463	.018	.081
BEL-2201-000	7/16″	1.000	.437	.062	.105
BEL-2201-001	7/16″	1.000	.455	.035	.102
BEL-2201-002	7/16″	1.000	.458	.093	.162
BEL-2201-003	7/16″	1.000	.450	.062	.140
BEL-2201-120	7/16″	1.187	.500	.062	.152
BEL-2201-130	7/16″	1.203	.500	.020	.082
BEL-2201-160	7/16″	1.250	.453	.088	.168
BEL-2201-320	7/16″	1.500	.471	.062	.102
BEL-2400-440	1/2″	.687	.515	.010	.062
BEL-2400-480	1/2″	.745	.513	.025	.130
BEL-2400-481	1/2″	.745	.515	.025	.052
BEL-2400-530	1/2″	.809	.506	.045	.073
BEL-2400-610	1/2″	.953	.509	.015	.055
BEL-2401-000	1/2″	1.000	.518	.035	.067
*BEL-2401-001	1/2″	1.000	.508	.035	.067
BEL-2401-002	1/2″	1.000	.562	.062	.132
BEL-2401-003	1/2″	1.000	.515	.062	.125
BEL-2401-004	1/2″	1.000	.522	.035	.067
BEL-2401-020	1/2″	1.031	.538	.038	.080
BEL-2401-040	1/2″	1.062	.554	.032	.064
BEL-2401-070	1/2″	1.109	.546	.010	.035
BEL-2401-080	1/2″	1.125	.515	.072	.150
BEL-2401-081	1/2″	1.125	.531	.125	.139
BEL-2401-082	1/2″	1.125	.531	.062	.132

Source: H.K. Metalcraft Corp., Lodi, N.J.

5.22 Write a program to calculate the axial force and the related stresses for a given axial deflection of a disc spring from equations 5.9.1 through 5.9.7. It should first prompt for the elastic modulus and Poisson's ratio and then calculate the axial force and the stresses from a subroutine which prompts for the spring OD, ID, thickness, height, and given deflection. The subroutine may be called repeatedly until satisfactory washer dimensions are selected without monotonously reentering data for the disc material.

5.23 Design a disc spring from stock spring washers to operate the clutch on an electric winch in which the clutch shaft is $\frac{3}{8}$ in. OD. Axial load is to be 140.0 ± 5.0 lb maximum and the spring deflection should be 0.040 ± 0.005 in. at the maximum load. Give the disc part number and the overall free height of the spring, either for a single spring or for a stack, if necessary. Sketch the spring arrangement. Stress limits are 200 000 psi in tension and 250 000 psi in compression. The outside diameter should be no greater than 0.875 in.

[*Ans.* 4 discs, BEL-2000-565, in series]

5.24   Could the disc washer, or washers, selected in problem 22 be used in a larger model in which the load maximum is 435.0 ± 5.0 lb and the deflection is 0.080 ± 0.010 in.? If so draw the disc arrangement and find the overall free height.

[*Ans.* Yes; 8 series of 3 discs in parallel, overall height = 0.944 in.]

5.25   Select a disc spring with an ID of $\frac{7}{16}$ in. from the list which can exert an axial force of 5.60 ± 0.05 lb at a deflection of 0.250h with decreasing spring constant so that at 0.500h the axial force is 9.30 ± 0.50 lb. It should be able to be flattened with a force not to exceed 12.5 ± 0.50 lb with tensile stresses less than 50 000 psi and compressive stresses less than 125 000 psi. If no spring on the list satisfies these requirements, what dimensions may be used? Use a stack if a single disc cannot be used. Also calculate the spring rate, or spring constant, at 0.50h and at 1.00h.

[*Ans.* BEL-2200-550, R = 1480.5 lb/in. at δ = 0.25h, R = 1188.5 lb/in. at δ = 0.5h and R = 799.1 lb/in. at δ = h.]

5.26   Design a Belleville spring with an outside diameter of 23.0 mm and an inside diameter of 11.5 mm to deflect 0.30 ± 0.05 mm when subjected to an axial load of 1500 ± 50 N. Stress should not exceed 1400 MPa in tension or 3000 MPa in compression. Use E = 207 000 MPa and v = 0.30. Hold t constant and vary h and also hold h constant and vary t to demonstrate the effect these parameters on spring performance.

[*Ans.* h = 2.10 mm, t = 0.50 mm.]

Section 5.10

5.27   Select the thickness h of a shear block made from 2BC5018A94E34 rubber such that it will deflect no more then 4.0 mm at the maximum shear stress. Assume that the limiting shear stress is half of the failure stress in tension.

[*Ans.* h = 5.02 mm.]

5.28   The purchasing agent has learned of a supply of shock mounts in which the rubber OD is 1.65 in., the ID is 0.800 in., and the thickness is 1.72 in. What axial load, torsional load, axial deflection, and angle of rotation of the inner shaft can be sustained at the maximum stress if the rubber is 2BC608A94E34? Assume the maximum shear stress is half the maximum tensile stress.

5.29   Find the outer diameter required for a shock mount with an inner diameter of 10.0 mm and a height of 20 mm to deflect 3.0 mm for a load of 35 kg? Use M1BK608A14E34 rubber.

5.30   Find the thickness required for a shock mount having an inside diameter of 8.0 mm and an outside diameter of 17 mm to rotate 20° for a torque of 10.34 Nm. Use M3BK710G18E28 rubber.

5.31   Find the height required for a shock mount to deflect 5.0 mm when subjected to an axial load of 800.0 N if the inner diameter is 8.0 mm and the outer diameter is 14.0 mm. Also find the angular rotation of the inner radius relative to the outer for a moment of 600 N mm is applied to the central rod. Use M3BK710G18E28 rubber.

[*Ans.* h = 8.397 mm, θ = 0.035, rad. = 2.02°.]

Section 5.11

5.32 What pressure should be specified for the air spring in Figure 5.11.7 to support 400 lb at a nominal design height of 9.0 in.? What deflection may be anticipated for a load of 700 lb if the pressure is held constant?

[*Ans. P* = 58.1 psig; *z* = 2.61 in.]

5.33 Derive the formula for the spring rate, or spring constant, for a constant area piston spring (i.e., a friction seal spring) as shown in the figure when the positive direction of motion *z* is in the direction to reduce the volume. Let *h* denote the initial height.

$$\left[Ans.\ R = P\,\frac{nA}{h - z} \right]$$

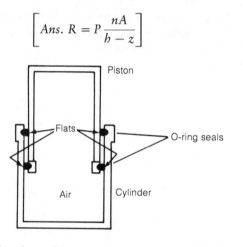

5.34 Show that for a friction seal air spring, as illustrated in the figure for problem 5.33, the piston displacement due to an increase in load from F_0 to F is given by

$$z = h\left[1 - \left(\frac{F_0}{F}\right)^{1/n} \right]$$

where $z = 0$ when $F = F_0$. Let z be positive in the direction to reduce spring volume; i.e., $V = A(h - z)$.

5.35 Using the formula derived in problem 5.34, design a friction seal spring to support a 7000-lb load and to have a spring constant, or spring rate, of 920 lb/in. Use $n = 1.38$ and a pressure limit of 35 ± 5 psig. Give the design height; let $z = 3.0$ in. to find h.

5.36 Find the deflection of the friction seal spring described in problem 5.35 when the load is increased to 8000 lb.

[*Ans. z* = 0.968]

5.37 Show that the spring rate (spring constant) for a friction seal air spring may also be written as $(P = P_0$ and $z = 0)$

$$R = \frac{nAP_0}{h}\left(\frac{1}{1 - \dfrac{z}{h}}\right)^{n+1}$$

5.38   Compare the spring rate for the rolling lobe air spring shown in Figure 5.11.7 for a load of 600 lb at 80 psig and a height of 10 in. with that of the bellow style air spring shown in Figure 5.11.6 for a similar load at 20 psig. The effective support area, given by $A = F/P$, generally changes with the spring height. The force-height curves in Figure 5.11.6 are curved all along their length because the maximum diameter of the bellows always changes with height; the pressure-height curves for the rolling lobe spring shown in Figure 5.11.7 are horizontal at heights from 8 to 12 inches for 80 and 100 psig because the lobe diameter is essentially constant for these pressures and height ranges. Although the volume curve was measured at 100 psig, the volume to height relationship is essentially independent of pressure.

5.39   Would the reversible sleeve-style air spring, shown in Figure 5.11.7, or the convoluted-style air spring, shown in Figure 5.11.6, be better suited for truck and automobile suspensions? Why? Which style would be better suited for vibration isolation where control of the spring rate is of major importance? Why?

CHAPTER SIX

SCREW DESIGN AND THREADED CONNECTIONS

NOTATION

A	Area or tolerance class for external threads (l^2), (l)	N	Number of threads or normal force on the collar (l)
B	Tolerance class (l)	N_e	Number of threads engaged (l)
b	Shear width (l)	n	Revolutions/minute (l/t)
C	Basic dynamic capacity (ml/t^2)	P_1, P_2	Dummy variables $(-)$
D_n	Nut major diameter (l)	p	Pitch (l)
D_s	Screw major diameter (l)	r_c	Collar mean radius (l)
d_c	Collar mean diameter (l)	r_i	Inside collar radius (l)
d_n	Nut minor diameter (l)	r_o	Outside collar radius (l)
d_s	Screw minor diameter (l)	r_s	Minor radius $(r_s = d_s/2)$ (l)
E	Elastic modulus, Young's modulus (m/lt^2)	T	Torque or tolerance (ml^2/t^2), (l)
EI, ES	Internal allowances (l)	TD, Td	Thread variables
e	SI tolerance position for external threads (l)	v	Velocity (l/t)
ei,es	External allowances for SI threads (l)	W	Axial force, axial load (ml/t^2)
F_c	Force of a mechanical component (ml/t^2)	x,y,z	Rectangular Cartesian coordinates (l)
F_n	Force perpendicular to the leading flank (ml/t^2)	α_1, α_2	Leading and following (trailing) flank angles (l)
F_o	Preload force (ml/t^2)	Δ	Pitch tolerance for external threads (l)
f_c	Friction force on the collar (ml/t^2)	δ_b	Bolt extension (l)
f_t	Friction force on the threads (ml/t^2)	δ_c	Component contraction (l)
G,g	Tolerance position for internal and external SI threads (l)	ζ	Safety factor (l)
H,h	Tolerance position for internal and external SI threads (l)	θ_n	Angle between F_n and the xy plane (l)
$\mathbf{i,j,k}$	Unit vectors along x, y, and z directions, respectively (l)	Λ_{10}	Life in terms of inches of travel (l)
		λ	Lead angle (l)
k_b	Bolt spring constant (m/t^2)	μ_c	Collar friction coefficient (l)
k_c	Component spring constant (m/t^2)	μ_t	Thread friction coefficient (l)
L	Lead (l)	v	Poisson's ratio (l)
L_e	Life in revolutions (l)	σ_{rz}	Thread shear stress (m/lt^2)
l	Unconstrained length (l)	σ_{zz}	Axial tensile stress (m/lt^2)
m	Number of starts (l)	ϕ	Friction angle (l)
		ψ	Helix angle (l)
		ω	Angular velocity (l/t)

Although screws may be used as removable fasteners or as devices for moving loads, essentially the same design considerations apply to both, especially if the tightening load is to be determined from applied torque. We shall find that small friction variations can appreciably alter the torque–load relation in the presence of large axial forces, so that other means of measuring bolt tension (deformable load washers, strain gages, bolt inserts, etc.) are recommended if greater accuracy is necessary. The torque–load relation is also important in the design of screw drives in computer-controlled machines which use stepper motors (from automated production units to the read–write heads on disk drives) because the design engineer must calculate the acceleration and deceleration torques and times to provide control pulses that the motors can follow.

As in previous chapters, it is assumed that the reader has access to either a programmable pocket calculator of a personal computer and is sufficiently familiar with them to program formulas and to program iterative procedures that are described by a flowchart.

6.1 TERMINOLOGY

Terminology recommended by the American National Standards Institute (ANSI) for screws is shown in Figure 6.1.1. Angles α_1 and α_2 are the flank angles for the leading and following flanks, respectively. The *major diameter* is the outside diameter of the threaded portion of the screw and defines an imaginary major cylinder which touches the screw only on the thread crests; the *minor diameter* is the diameter of another imaginary cylinder with cuts through the thread roots, both as shown in Figure 6.1.1. The overall usable thread length of the nut is indicated by L_{tn}.

Screw profiles are defined relative to the pitch cylinder and the associated pitch diameter, which differs from the previous quantities in that its location is not clearly visible on the screw itself. According to ANSI standard B1.7-1984, p. 5, the *pitch cylinder* is an imaginary cylinder concentric with the axis of the straight thread on a machine screw (as opposed to the conical thread on a wood screw) of such a diameter (the *pitch diameter*) that the width of the threads and the grooves are equal on this cylinder. The pitch diameter is also known as the simple effective pitch, the effective pitch diameter, or the basic pitch diameter. Although shown as measured on the pitch cylinder in Figure 6.1.1, the pitch p is defined as the distance, parallel to the screw axis, between corresponding points on adjacent threads having uniform spacing. On a dimensionally perfect thread the width of the ridges and grooves on the pitch cylinder would be $p/2$.

The *engagement length* between nut and screw is defined as the axial distance over which two mating threads are designed to contact, denoted by L_e. According to Figure 6.1.1 this distance is measured on the pitch cylinder.

Accepted terminology is ambiguous in the use of the word *threads*. On the one hand, each ridge on the screw profile, as illustrated in Figure 6.1.2, is known as a thread, and the screw is classified in terms of either the number of threads per inch or the millimeters per thread. On the other hand, if the screw is made by cutting a single helical groove, as in Figure 6.1.2a, it is said to be a single-thread screw; if two helical grooves

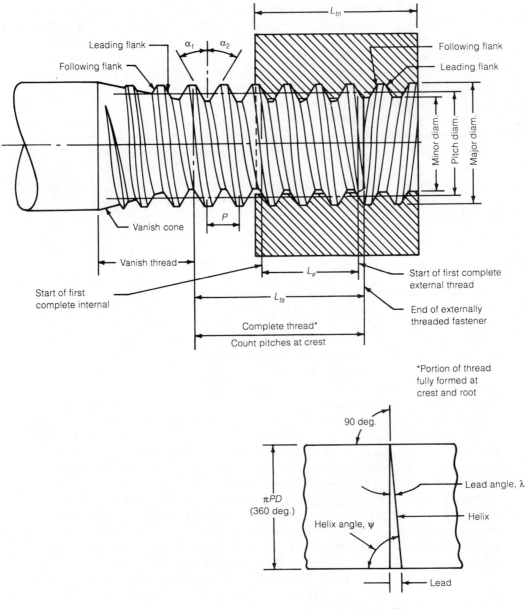

Figure 6.1.1 Screw terminology according to ANSI B1.1-1982. *Source:* Unified Inch Screw Threads (UN and UNR Thread Forms), *ANSI B1.1-1982, American Society of Mechanical Engineers, New York. Reprinted with permission.*

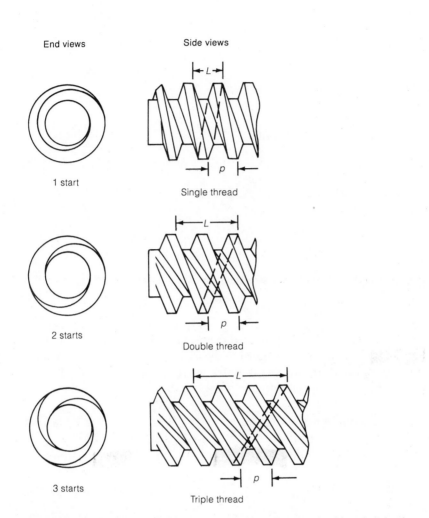

End views

Side views

1 start

Single thread

2 starts

Double thread

3 starts

Triple thread

Figure 6.1.2 Multiple screw thread showing relation between lead L and pitch p for screws with 1, 2, and 3 starts.

are cut, as in Figure 6.1.2b, it is said to be a double-thread screw, and so on. If rotating the screw in the direction of the fingers of the right hand causes it to advance in the direction of the thumb it is said to have a right-hand thread; otherwise it is said to have a left-hand thread.

Suppose that the line of intersection of the leading flank of a single-thread screw is recorded on the imaginary pitch cylinder and that the cylinder is slit along a line parallel to its axis and unrolled. The intersection line would then appear in segments, as in

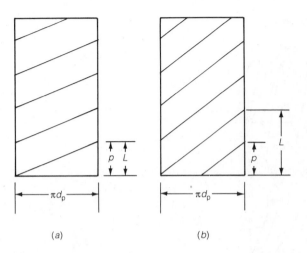

Figure 6.1.3 Developed pitch cylinders for single- and double-start threads.

Figure 6.1.3*a*. If a double-thread screw were used the segments would appear as in Figure 6.1.3*b*. The *lead L* is also measured parallel to the axis of the screw and is equal to the pitch distance multiplied by the number of starts, as written in equation 6.1.1, and, as implied in Figure 6.1.3, it is the distance the screw advances in one revolution.

$$l = mp \qquad m = \text{number of starts} \tag{6.1.1}$$

From Figure 6.1.3 it is evident that the lead angle λ is related to the lead l according to

$$l = \pi d_p \tan \lambda \tag{6.1.2}$$

Some authors, not adhering to the ANSI convention, refer to λ as the helix angle, shown as ψ in Figure 6.1.1, where $\lambda + \psi = \pi/2$.

6.2 STANDARD THREAD PROFILES

Eight standard profiles are shown in Figure 6.2.1. As we shall learn from Section 6.6, the square thread has the advantage of having the smallest bearing stress for a given load and lead angle because the flank angle α is zero. It is, however, the most expensive to manufacture, requiring two or more cuts per thread. (It cannot be cut with a square-ended cutter because of the large stresses induced in the tool itself.) Modified square, Acme, and buttress threads have only slightly greater flank loads and are cheaper to

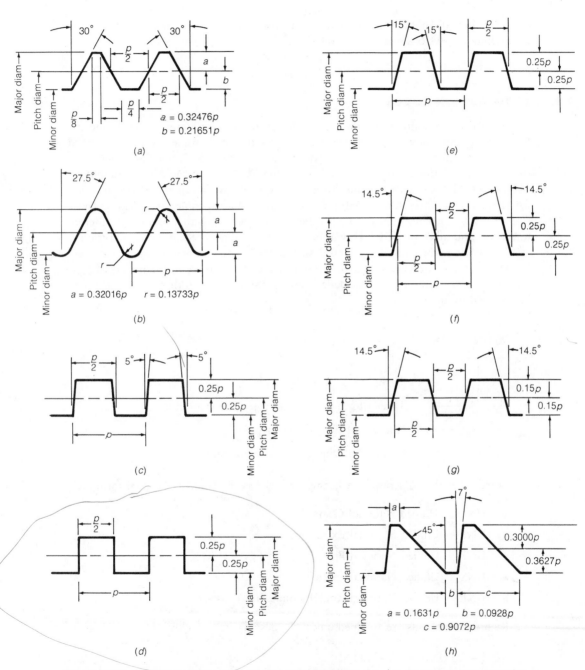

Figure 6.2.1 (a) Unified and M profile; (b) Whitworth; (c) modified square; (d) square; (e) trapezoidal; (f) Acme; (g) stub Acme; (h) buttress thread forms.

produce. Unified (Unified National) and SI (metric) threads are most commonly found on machine screws because of their relatively low manufacturing cost.

6.3   IDENTIFICATION CODES

Screw fasteners commonly use the profile that in the United States, Canada, and the United Kingdom is now known as the Unified or Unified Inch profile, according to ANSI B1.1-1982. In the past this profile was also known as either the Unified National or as the American National profile. This becomes the M Profile in the ANSI terminology (ANSI B1.13M-1983) when SI units are employed, while the International Organization for Standardization (ISO, not IOS!) refers to it as the Basic profile for general purpose screw threads. The importance of these different designations is that they imply different tolerances and allowances, which will be defined and illustrated in the next section.

Screw size in the Unified system is designated by a size number for major diameters less than $\frac{1}{4}$-in., as shown in Table 6.4.2, and by a code sequence for diameters of $\frac{1}{4}$-in. and larger. For example, a screw made from $\frac{1}{4}$-in. stock with 20 threads per inch may be specified by the notation

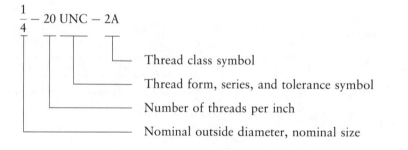

$$\frac{1}{4} - 20\ \text{UNC} - 2A$$

- Thread class symbol
- Thread form, series, and tolerance symbol
- Number of threads per inch
- Nominal outside diameter, nominal size

Some other thread series for what was known as the Unified National form are

- UNC   Unified National Coarse
- UNEF   Unified National Extra Fine
- UNF   Unified National Fine
- UNS   Unified National Special
- UNR   Unified National Round (round root)

Right-hand threads are assumed unless the designation is followed by $-$LH, as in

$$\frac{1}{4} - 20\ \text{UNC} - 2A - \text{LH}$$

SI threads are designated by the letter M preceding the nominal major diameter in millimeters. Pitch in millimeters per thread follows and is succeeded by symbols for tolerance limits. For example,

M 10 × 1.25 − 5h6h

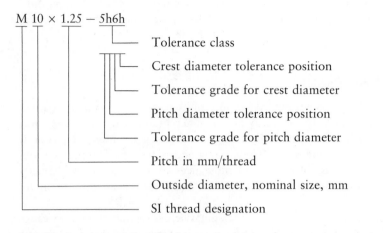

- Tolerance class
- Crest diameter tolerance position
- Tolerance grade for crest diameter
- Pitch diameter tolerance position
- Tolerance grade for pitch diameter
- Pitch in mm/thread
- Outside diameter, nominal size, mm
- SI thread designation

describes an external thread whose major diameter is approximately 10 mm and whose pitch is 1.25 mm per thread.

The Acme thread designation recommended in ANSI B1.5-1977 is

$$1\tfrac{3}{4}\text{-4 Acme-2G}$$

for a single-start Acme thread with a major diameter of $1\tfrac{3}{4}$ in., 4 threads per inch, and class 2 clearance and tolerances, defined in the standard itself. If a two-start thread were used the designation would become

$$1\tfrac{3}{4}\text{-0.25p-0.50L-Acme-2G}$$

for a thread which has a pitch of 0.25 in. and a lead of 0.50 in. Stub Acme threads use a similar designation except that Acme is replaced by STUB ACME (all capital letters).

Buttress thread designation for a single-start thread as recommended by ANSI B1.9-1975 is of the form

2.5-8 PUSH-BUTT-2A-LH-FL

for a screw with a nominal diameter of 2.5 in., 8 threads per inch, with the internal member to push, class 2 allowance and tolerance, external thread, left hand (indicated by −LH), and with a flat root (indicated by −FL). If the internal member were to pull, the PUSH would be deleted, and if an internal thread were to be cut, the 2A would be replace by 2B. The −FL would be omitted if a round root were to be used. If a three-start thread were to be cut instead, the designation would read

2.5-0.2P-0.6L-BUTT-3B (3 START)

for a pitch of 0.2 in. and a lead of 0.6 in.

There are no ANSI standards for square, modified square, or the British Whitworth threads, and consequently no similar thread designations. The 30° trapazoidal thread is included in a German standard, DIN 103, but with no ANSI counterpart.

6.4 TOLERANCES AND ALLOWANCES

Allowances specify the clearance provided for nuts and screws to accomodate coatings and/or foreign material. Tolerances specify limits on acceptable manufacturing errors when allowances are included in producing screws and nuts that are intended to comply with standard profiles and dimensions. These screw and nut profiles and dimensions are given in terms that include tolerances and allowances on the screw major diameter D_s, the screw minor diameter d_s, the nut major diameter D_n, and the nut minor diameter d_n.

Figure 6.4.1a shows a small section of an imaginary cross section of a nut threaded onto a bolt. If the nut and bolt could always be made perfectly, if neither were coated, and if they always remained clean and required no lubricant, both the bolt and nut profiles would contact along the basic profile shown by the wide solid line.

If we are to allow space for a coating or for dirt we may, as in the figure, make the screw slightly smaller than provided by the basic profile. The resulting gap is known as the allowance, as labeled in Figure 6.4.1. It is larger at the root of the screw and nut threads to allow for the radius of the ends of the cutting tools. The difference between the thread profiles after the allowance has been included and the actual profiles is known as the tolerance. It allows for tool wear and measurement errors. Both are measured radially as indicated by the cross-hatched rectangles in Figure 6.4.1b.

Tolerances and allowances for Unified inch screw threads are given in standard ANSI B1.1-1982 for three classes of external threads (screws) which are denoted as 1A, 2A, and 3A and for three classes of internal threads (nuts) which are denoted as 1B, 2B, and 3B. They are compared in Figure 6.4.2 using the representation introduced in Figure 6.4.1b. According to this standard all of the allowance is taken from the screw and tightest fit is obtained from class 3A screws and 3B nuts. Classes 1A and 1B are for screws and nuts that may be used in a dirty environment, such as in road, mining, and farm machinery, and which may be coated. Classes 3A and 3B are for a clean environment and for highly loaded threads where complete contact across the flanks of the screw and nut is important.

Reference tolerence T in Figure 6.4.2 is given in terms of the major screw diameter D_s, thread pitch p and engagement length L_e by

$$T = 0.0015(D_s^{1/3} + L_e^{1/2} + 10p^{2/3}) \tag{6.4.1}$$

SI standards used in the United States are given in the standard ANSI B1.13M-1983 which states that they were selected from international standard ISO 965/1. According

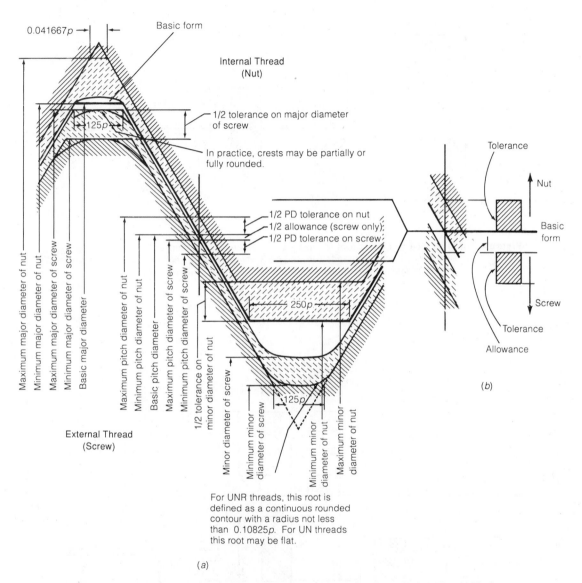

Figure 6.4.1 Tolerances and allowance for Unified screw thread classes 1A, 2A, 1B, and 2B. *Source: Unified Inch Screw Threads (UN and UNR Thread Forms), ANSI B1.1-1982, American Society of Mechanical Engineers, New York. Reprinted with permission.*

to the ANSI standard the tolerances are given in grades 3 through 9 and the allowance is given in terms of 5 positions: two, G and H, for internal threads and three, e, g, and h, for external threads. Relative magnitudes of these allowances, tolerances, and positions are drawn in Figure 6.4.3. From it we find that in the SI system the allowance is taken from both the screw and nut profiles instead of just from the screw.

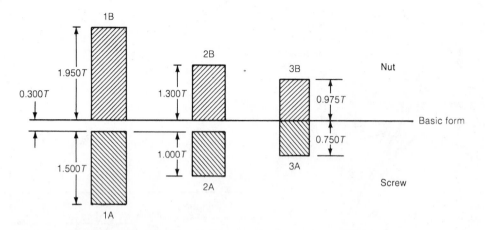

Figure 6.4.2 Allowance and tolerances for classes 1A, 1B, 2A, 2B, 3A, and 3B. *Source: Abstracted from* Unified Screw Threads (UN and UNR Thread Forms), *ANSI B1.1-1982, American Society of Mechanical Engineers, New York. Reprinted with permission.*

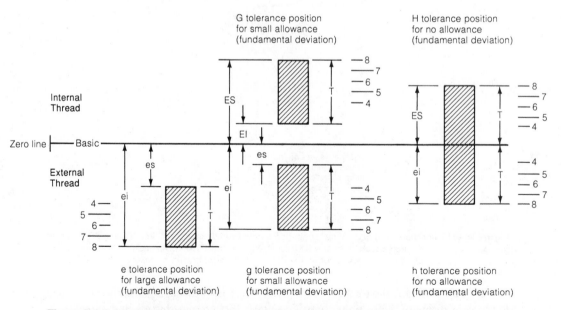

Figure 6.4.3 Allowance and tolerances for M profile threads, tolerance positions G, H, e, g, and h. *Source: Abstracted from* Metric Screw Threads—M Profile, *ANSI B1.13-1983, American Society of Mechanical Engineers, New Yrok. Reprinted with permission.*

TABLE 6.4.1 M Profile (SI thread)

SI allowances and tolerances for the three tolerance positions

Allowance	External Thread	Internal Thread
es_e	$0.05 + 0.011p$	—
es_g	$0.015 + 0.011p$	—
EI_g	—	$0.015 + 0.011p$

Tolerance Grade	External Thread Major Diameter	Internal Thread Minor Diameter
4	0.63 Td	0.63 TD
5	0.80 Td	0.80 TD
6	1.00 Td	1.00 TD
7	1.25 Td	1.25 TD
8	1.60 Td	1.60 TD

For pitches 0.2 to 0.8 mm

$$TD = 0.433p - 0.19p^{1.22}$$

For pitches 1 mm and larger

$$TD = 0.230p^{0.7}$$

For all pitches

$$Td = 0.18p^{2/3} - 0.00315p^{-1/2}$$

Relative magnitudes of these positions and grades are given in Table 6.4.1 for the M profile shown in Figure 6.2.1. Some authors have referred to this profile as the SI profile. Allowances es_e, es_g, and EI_G for the M profile are defined in terms of pitch p as shown in the table. Corresponding tolerances and allowances for all but the square thread profile may be found in the appropriate ANSI standards.

6.5 TORQUE-LOAD RELATIONS

The torque–load relations that follow will be derived for the configurations shown in Figure 6.5.1. Results obtained may be easily adapted to other configurations, such as the horizontal drive screw for a lathe or a compactor, by observing that moving the platform in a direction opposite to the externally applied load, Figure 6.5.1b, is equivalent to raising a load against the force of gravity. Likewise, moving the platform in the same direction as the external force acting on the platform is equivalent to lowering as load, as in Figure 6.5.1b.

In either of the power screw configurations shown in Figure 6.5.1 it is assumed that the platform, or the load, does not rotate with the screw and that it is to be raised or lowered by means of a torque T. Forces on the leading, or load bearing, flank of

TABLE 6.4.2 BASIC DIMENSIONS FOR COARSE THREAD SERIES (UNC/UNRC)

Nominal Size	Basic Major Diameter, D, in.	Threads/ in., n	Basic Pitch Diameter, E, in.	UNR Design Minor Diameter External (Ref.), K_s in.	Basic Minor Diameter Internal, K, in.	Lead Angle at Basic Pitch Diameter, λ Deg.	Min	Section at Minor Diameter at D-2h_b, sq. in.	Tensile Stress Area, sq. in.
1(0.073)	0.0730	64	0.0629	0.0544	0.0561	4	31	0.00218	0.00263
2(0.086)	0.0860	56	0.0744	0.0648	0.0667	4	22	0.00310	0.00370
3(0.099)	0.0990	48	0.0855	0.0741	0.0764	4	26	0.00406	0.00487
4(0.112)	0.1120	40	0.0958	0.0822	0.0849	4	45	0.00496	0.00604
5(0.125)	0.1250	40	0.1088	0.0952	0.0979	4	11	0.00672	0.00796
6(0.138)	0.1380	32	0.1177	0.1008	0.1042	4	50	0.00745	0.00909
8(0.164)	0.1640	32	0.1437	0.1268	0.1302	3	58	0.01196	0.0140
10(0.190)	0.1900	24	0.1629	0.1404	0.1449	4	39	0.01450	0.0175
12(0.216)	0.2160	24	0.1889	0.1664	0.1709	4	1	0.0206	0.0242
$\frac{1}{4}$	0.2500	20	0.2175	0.1905	0.1959	4	11	0.0269	0.0318
$\frac{5}{16}$	0.3125	18	0.2764	0.2464	0.2524	3	40	0.0454	0.0524
$\frac{3}{8}$	0.3750	16	0.3344	0.3005	0.3073	3	24	0.0678	0.0775
$\frac{7}{16}$	0.4375	14	0.3911	0.3525	0.3602	3	20	0.0933	0.1063
$\frac{1}{2}$	0.5000	13	0.4500	0.3334	0.4167	3	7	0.1257	0.1419
$\frac{9}{16}$	0.5625	12	0.5084	0.4633	0.4723	2	59	0.162	0.182
$\frac{5}{8}$	0.6250	11	0.5660	0.5168	0.5266	2	56	0.202	0.226
$\frac{3}{4}$	0.7500	10	0.6850	0.6309	0.6417	2	40	0.302	0.334
$\frac{7}{8}$	0.8750	9	0.8028	0.7427	0.7547	2	31	0.419	0.462
1	1.0000	8	0.9188	0.8512	0.8647	2	29	0.551	0.606
$1\frac{1}{8}$	1.1250	7	1.0322	0.9549	0.9704	2	31	0.693	0.763
$1\frac{1}{4}$	1.2500	7	1.1572	1.0799	1.0954	2	15	0.890	0.969
$1\frac{3}{8}$	1.3750	6	1.2667	1.1766	1.1946	2	24	1.054	1.155
$1\frac{1}{2}$	1.5000	6	1.3917	1.3016	1.3196	2	11	1.294	1.405
$1\frac{3}{4}$	1.7500	5	1.6201	1.5119	1.5335	2	15	1.74	1.90
2	2.0000	$4\frac{1}{2}$	1.8557	1.7353	1.7594	2	11	2.30	2.50
$2\frac{1}{4}$	2.2500	$4\frac{1}{2}$	2.1057	1.9853	2.0094	1	55	3.02	3.25
$2\frac{1}{2}$	2.5000	4	2.3376	2.2023	2.2294	1	57	3.72	4.00
$2\frac{3}{4}$	2.7500	4	2.5876	2.4523	2.4794	1	46	4.62	4.93
3	3.0000	4	2.8376	2.7023	2.7294	1	36	5.62	5.97
$3\frac{1}{4}$	3.2500	4	3.0876	2.9523	2.9794	1	29	6.72	7.10
$3\frac{1}{2}$	3.5000	4	3.3376	3.2023	3.2294	1	22	7.92	8.33
$3\frac{3}{4}$	3.7500	4	3.5876	3.4523	3.4794	1	16	9.21	9.66
4	4.0000	4	3.8376	3.7023	3.7294	1	11	10.61	11.08

Source: Unified Inch Screw Threads (UN and UNR Thread Forms), ANSI B1.1-1982, American Society of Mechanical Engineers, New York. Reprinted with permission.

TABLE 6.4.3 BASIC DIMENSIONS FOR FINE THREAD SERIES (UNF/UNRF)

Nominal Size	Basic Major Diameter, D, in.	Threads/in., n	Basic Pitch Diameter, E, in.	UNR Design Minor Diameter External (Ref.), K_5, in.	Basic Minor Diameter Internal, K, in.	Lead Angle at Basic Pitch Diameter, λ Deg.	Min	Section at Minor Diameter at $D-2h_b$, sq. in.	Tensile Stress Area, sq. in.
0(0.060)	0.0600	80	0.0519	0.0451	0.0465	4	23	0.00151	0.00180
1(0.073)	0.0730	72	0.0640	0.0565	0.0580	3	57	0.00237	0.00278
2(0.086)	0.0860	64	0.0759	0.0674	0.0691	3	45	0.00339	0.00394
3(0.099)	0.0990	56	0.0874	0.0778	0.0797	3	43	0.00451	0.00523
4(0.112)	0.1120	48	0.0985	0.0871	0.0894	3	51	0.00566	0.00661
5(0.125)	0.1250	44	0.1102	0.0979	0.1004	3	45	0.00716	0.00830
6(0.138)	0.1380	40	0.1218	0.1082	0.1109	3	44	0.00874	0.01015
8(0.164)	0.1640	36	0.1460	0.1309	0.1339	3	28	0.01285	0.01474
10(0.190)	0.1900	32	0.1697	0.1528	0.1562	3	21	0.0175	0.0200
12(0.216)	0.2160	28	0.1928	0.1734	0.1773	3	22	0.0226	0.0258
1/4	0.2500	28	0.2268	0.2074	0.2113	2	52	0.0326	0.0364
5/16	0.3125	24	0.2854	0.2629	0.2674	2	40	0.0524	0.0580
3/8	0.3750	24	0.3479	0.3254	0.3299	2	11	0.0809	0.0878
7/16	0.4375	20	0.4050	0.3780	0.3834	2	15	0.1090	0.1187
1/2	0.5000	20	0.4675	0.4405	0.4459	1	57	0.1486	0.1599
9/16	0.5625	18	0.5264	0.4964	0.5024	1	55	0.189	0.203
5/8	0.6250	18	0.5889	0.5589	0.5649	1	43	0.240	0.256
3/4	0.7500	16	0.7094	0.6763	0.6823	1	36	0.351	0.373
7/8	0.8750	14	0.8286	0.7900	0.7977	1	34	0.480	0.509
1	1.0000	12	0.9459	0.9001	0.9098	1	36	0.625	0.663
1 1/8	1.1250	12	1.0709	1.0258	1.0348	1	25	0.812	0.856
1 1/4	1.2500	12	1.1959	1.1508	1.1598	1	16	1.024	1.073
1 3/8	1.3750	12	1.3209	1.2758	1.2848	1	9	1.260	1.315
1 1/2	1.5000	12	1.4459	1.4008	1.4098	1	3	1.521	1.581

Source: Unified Inch Screw Threads (UN and UNR Thread Forms), ANSI B1.1-1982, American Society of Mechanical Engineers, New York. Reprinted with permission.

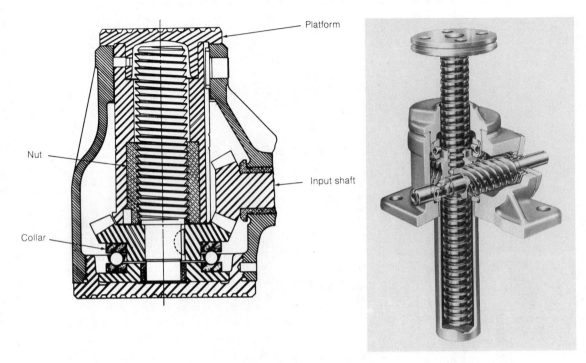

Figure 6.5.1 Power screw geometry; gear driven power screw. *Sources: (left) Aetna Bearing Co., Chicago, IL; (right) Joyce/Dayton Corp., Dayton, OH.*

the screw are shown in Figure 6.5.2. In the following expressions r_o and r_i denote the outer and inner collar radii, as shown in Figure 6.5.1a, and $r_c = (r_o + r_i)/2$

From the force vectors shown in Figure 6.5.2 it follows that the force perpendicular to the thread flank is given by

$$\mathbf{F_n} = F_n(-\hat{i} \cos \theta_n \sin \lambda - \hat{j} \sin \theta_n - \hat{k} \cos \theta_n \cos \lambda) \qquad (6.5.1)$$

in terms of the *xyz*-coordinate system shown and its unit vectors $\hat{i}, \hat{j}, \hat{k}$. Friction force between the thread and the nut is given by

$$\mathbf{f_t} = \mu_t F_n(-\hat{i} \cos \lambda + \hat{k} \sin \lambda) \qquad (6.5.2)$$

Vertical and horizontal reactions on the underside of the collar are given by

$$\mathbf{N} = \hat{k}W \qquad \mathbf{f_c} = -\hat{i}\mu_c W \qquad (6.5.3)$$

Quantities μ_t and μ_c in equations 6.5.2. and 6.5.3 denote the friction coefficients between the screw and the nut and between the collar and its base, in that order. Angle θ_n is between $\mathbf{F_n}$ and the *xy* plane. Its relation to α_1 and λ may be found from the geometry

Z

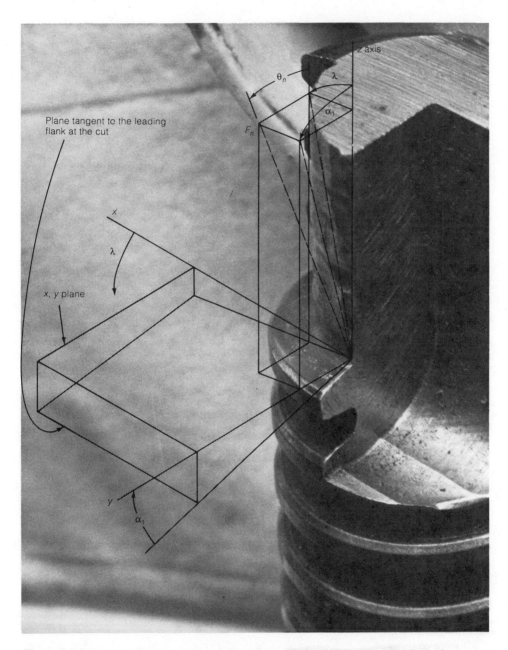

Figure 6.5.2 Forces acting at a representative point on the leading (load bearing) flank of a screw.

in Figure 6.5.2 and the dot products below as

$$\tan \alpha_1 = \frac{F_n \cdot \hat{j}}{F_n \cdot \hat{k}} = \frac{\sin \theta_n}{\cos \theta_n \cos \lambda} \qquad (6.5.4)$$

which may be written as

$$\tan \alpha_1 \cos \lambda = \tan \theta_n \qquad (6.5.5)$$

Torque about the axis of the screw T is given by the equilibrium equation for moments about the z axis, namely,

$$\hat{k} \cdot [r_p \times (F_n + f_t) + (r_c \times f_c) + T] = 0 \qquad (6.5.6)$$

which yields

$$r_p F_n(\cos \theta_n \sin \lambda + \mu_t \cos \lambda) + \mu_c r_c W = T \qquad (6.5.7)$$

Summation of forces in the vertical direction yields

$$\hat{k} \cdot [(F_n + f_t) + N] = 0 \qquad (6.5.8)$$

which may be written as

$$F_n(\cos \theta_n \cos \lambda - \mu_t \sin \lambda) = W \qquad (6.5.9)$$

Substitution from (6.5.9) into (6.5.7) produces the relation

$$W r_p \frac{\cos \theta_n \sin \lambda + \mu_t \cos \lambda}{\cos \theta_n \cos \lambda - \mu_t \sin \lambda} = T - W \mu_c r_c \qquad (6.5.10)$$

Equation 6.5.10 may be simplified by dividing both numerator and denominator by $\cos \theta_n \cos \lambda$ to obtain

$$\frac{\tan \lambda + \dfrac{\mu_t}{\cos \theta_n}}{1 - \dfrac{\mu_t}{\cos \theta_n} \tan \lambda} = \frac{\tan \phi + \tan \lambda}{1 - \tan \phi \tan \lambda} = \tan(\phi + \lambda) \qquad (6.5.11)$$

so that equation 6.5.10 may be written as

$$T = W[r_p \tan(\phi + \lambda) + \mu_c r_c] \qquad (6.5.12)$$

where

$$\tan \phi = \frac{\mu_t}{\cos \theta_n} \qquad (6.5.13)$$

Equation 6.5.12 holds (1) when motion is in a direction opposite to the direction of the force acting on the platform, as in raising a weight, and (2) for constant velocity motion. See problems 6.20 and 6.21 for accelerated motion.

Reversing the direction or rotation to move in the direction of the force acting on the platform, as in lowering a weight, causes the friction forces f_t and f_c to act in directions that oppose the rotation, so that equation 6.5.10 must be replaced by

$$Wr_p \frac{\mu_t \cos \lambda - \cos \theta \sin \lambda}{\cos \theta_n \cos \lambda + \mu_t \sin \lambda} = T - \mu_c r_c W \tag{6.5.14}$$

which likewise simplifies to

$$T = W[r_p \tan(\phi - \lambda) + \mu_c r_c] \tag{6.5.15}$$

for constant velocity motion.

If the lead angle and friction coefficients are such that an axial force on the platform causes it to move and the screw to rotate accordingly, the screw is said to *backdrive* or *overhaul*. The torque required to prevent screw rotation under these conditions may be found from equation 6.5.15 by replacing T by $-T$. Justification for this sign change is that in the derivation of (6.5.15) both f_t and f_c acted to oppose the screw rotation caused by the force on the platform and were assumed to be large enough to prevent rotation. Torque T in the direction to move the load was necessary, therefore, to provide downward motion. If the force on the platform causes screw rotation the friction forces still act to oppose the rotation, but now an applied torque must also act to oppose the motion rather than aid it. Thus,

$$-T = W[r_p \tan(\phi - \lambda) + \mu_c r_c] \tag{6.5.16}$$

applies when the platform moves at a constant velocity when backdriven. Torque T vanishes in both equations 6.5.15 and 6.5.16 when the screw is on the verge of backdriving and μ_t and μ_c take on their static values.

6.6 POWER SCREW DESIGN

Selection of a lead angle to obtain a desired torque/axial load ratio requires simultaneous solution of equations 6.5.5 and 6.5.13, and one of equations 6.5.12, 6.5.14, or 6.5.16. These last three equations pertain to driving the platform against the direction of applied force, driving the platform with the direction of the applied force, or resisting the torque induced by backdriving.

These transcendental equations cannot be solved algebraically but they can be solved by graphical means or by numerical iteration. Explanation of both methods will be simplified if we first rewrite equations 6.5.12 and 6.5.15 in the form

$$P_1 = \tan(\phi \pm \lambda) \tag{6.6.1}$$

where the $(+)$ sign holds for driving against the applied load, the $(-)$ sign holds for driving in the direction of the applied load, and P_1 is given by

$$P_1 = \frac{T}{Wr_p} - \mu_c \frac{r_c}{r_p} \tag{6.6.2}$$

Likewise, equation 6.5.16 may be written in the form

$$P_2 = -\tan(\phi - \lambda) \tag{6.6.3}$$

where

$$P_2 = \frac{T}{Wr_p} + \mu_c \frac{r_c}{r_p} \tag{6.6.4}$$

Variable ϕ in equations 6.6.1 and 6.6.3 is given by

$$\phi = \tan^{-1}\left(\frac{\mu_t}{\cos \theta_n}\right) \tag{6.6.5}$$

where

$$\theta_n = \tan^{-1}(\tan \alpha_1 \cos \lambda) \tag{6.6.6}$$

in terms of the friction coefficient at the threads μ_t, the leading flank angle α_1, and angle θ_n defined in previous sections.

By plotting $f(\lambda) = \tan(\phi + \lambda)$ and $f(\lambda) = \tan(\phi - \lambda)$ against λ as in Figure 6.6.1 we find that the tangent curves intersect the $\lambda = 0$ axis at μ_t for $\alpha_1 = 0$ (square threads) and at $\mu_t/\cos \alpha_1$ for other thread forms. We also see that we may obtain a graphical solution to equation 6.6.1 for a particular value of ϕ by reading down from the intersection point of a horizontal line through $f(\lambda) = P_1$ with the curve $\tan(\phi + \lambda)$ to find the λ value that satisfies the equation. Similarly, a solution to equation 6.6.3 may be had from the intersection of a line through $f(\lambda) = P_1$ and the curve for $\tan(\phi - \lambda)$.

Because the leading $(-)$ sign on the right side of equation 6.6.3 is the only difference between it and the right side of equation 6.6.1 when its argument is $(\phi - \lambda)$, it is evident that equation 6.6.3 holds only when P_2 is negative.

A more accurate solution to equation 6.6.1 may be had by solving it numerically. Take the inverse tangent of both sides to obtain (for $\phi + 2$)

$$\lambda = \tan^{-1} P_1 - \phi(\lambda) \tag{6.6.7}$$

which leads to the iteration formula

$$\lambda_i = \tan^{-1} P_1 - \phi(\lambda_{i-1}) \tag{6.6.8}$$

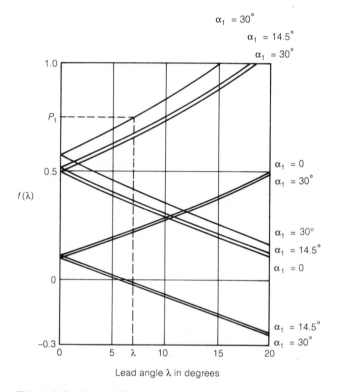

Figure 6.6.1 Plots of $f(\lambda) = \tan(\phi + \lambda)$ (rising to the right) and $f(\lambda) = \tan(\phi - \lambda)$ (falling to the right) as functions of λ.

which may be repeated until

$$\lambda_i - \lambda_{i-1} \leqq \delta \tag{6.6.9}$$

for whatever accuracy δ is required for an acceptable variation in P_1.

Convergence of process may be demonstrated by rewriting equation 6.6.8 in the form $f_1(\lambda) = f_2(\lambda)$ where $f_1(\lambda) = \tan^{-1} P_1 - \lambda$ and $f_2(\lambda) = \phi(\lambda)$ and plotting both functions on the same graph as shown in Figure 6.6.2. In this figure $f_1(\lambda)$ is represented by the $45°$ line down and to the right from P_1 at $\lambda = 0$ and $f_2(\lambda)$ is represented by the curve from $\mu_t/\cos \alpha_1$ at $\gamma = 0$. Thus, equation 6.6.7 is satisfied at point Γ where $f_1(\lambda) - f_2(\lambda) = 0$ and the difference $f_1(\lambda) - f_2(\lambda)$ for $\lambda = \lambda_i$ is represented by length AB as shown.

If we begin iterating at $\lambda = 0$ we find that length ab corresponds to the value of $f_1(\lambda_1)$ so that equation 6.6.8 may be satisfied by associating point b at λ_0 with point c at λ_1. Next, length de represents to the value of $f_1(\lambda_2)$ amd e represents the value of $f_2(\lambda_1)$ so that our equating points e and f will satisfy equation 6.6.8. Upon setting

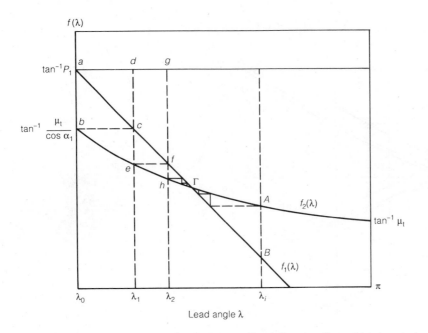

Figure 6.6.2 Plot of $f_1(\lambda) = \tan^{-1} P_1 - \lambda$ and $f_2(\lambda) = \phi(\lambda)$ as functions of lead angle λ.

length gh equal to $f_1(\lambda_3)$ and continuing in this manner we obtain the diminishing stair steps which converge to point Γ. If we were to begin iteration at some λ to the right of Γ, such as $\lambda = \lambda_i$, the iteration will also converge, as illustrated in the figure. Similar demonstrations of convergence can be given for equation 6.6.1 for the $(\phi - \lambda)$ argument and for equation 6.6.3.

A program, written in Microsoft Quick Basic, version 3.10, is shown in Figure 6.6.3. It solves equation 6.6.1 (for either λ or $-\lambda$) or equation 6.6.3 after values 1, 2, or 3 are keyed in answer to the first prompt. The code involved in the iteration is given between lines 40 and 55 for equation 6.6.1 for argument $\phi + \lambda$, between lines 80 and 95 for argument $\phi - \lambda$, and between lines 120 and 135 for equation 6.6.3. It is important to remember that torque T is negative when entering data for expression P_2.

Returning to Figure 6.6.1, we may observe that a solution to equation 6.6.1 exists only for $P_1 \geq \mu_t/\cos \alpha_1$. Hence, we see that to design a screw that can advance a load W with applied torque T we must select a combination of r_p, r_c, μ_t, μ_c, and α_1, such that this condition is satisfied for a value of λ that may be produced with the manufacturing equipment available. We also see that to prevent backdriving the thread and collar friction must be large enough for P_1 to be greater than $\mu_t/\cos \alpha_1$ when $T = 0$.

Since curves for threads with a larger value of α_1 lie above those with a lower value, it is evident that lower torques are required for those screw profiles more closely resembling the square profile. Hence, modified square, Acme, or trapazoidal threads are preferred for bidirectional screw drives, as in lathes and hard-disk memory drives,

```
          'Program LEED to find the lead angle for given load, torque, etc.
          DEFSNG A-M, O-Z
          PI = 3.141592654
          PRINT "Enter 1 if load is to be raised, 2 if it is to be lowered,"
          INPUT "and enter 3 if backdriven.", NFLAG
          INPUT "Enter the load and torque required.", W, T
          INPUT "Enter the pitch diameter and mean collar diameter.", DP, DC
          INPUT "Enter the leading flank angle.", AL: AL = AL*PI/180
          INPUT "Enter the thread and collar friction coefficients.", MT, MC
          INPUT "Enter maximum acceptable error for lambda.", MERR

5     IF NFLAG = 1 THEN
10        GOTO 40
15      ELSEIF NFLAG = 2 THEN
20        GOTO 80
25      ELSE
30        GOTO 120
35    END IF


          'Raise load; drive against applied load
40    LAM = 0: P = 2*T/(DP*W)-MC*DC/DP: IF P < MT/COS(AL) THEN GOTO 190
45    LAMO = LAM: THETA = ATN(TAN(AL)*COS(LAMO)): PHI = ATN(MT/COS(THETA))
50    LAM = ATN(P)-PHI: EP = LAM-LAMO
55    IF ABS(EP) < MERR THEN GOTO 200 ELSE GOTO 35


          'Lower load; drive with applied load
80    LAM = 0: P = 2*T/(DP*W)-MC*DC/DP: IF P > MT/COS(AL) THEN GOTO 190
85    LAMO = LAM: THETA = ATN(TAN(AL)*COS(LAMO)): PHI = ATN(MT/COS(THETA))
90    LAM = PHI-ATN(P): EP = LAM-LAMO
95    IF ABS(EP) < MERR THEN GOTO 200 ELSE GOTO 85


          'Backdrive; load drives screw to provide torque
120   LAM = 0: P = 2*T/(W*DP)+MC*DC/DP
125   LAMO = LAM: THETA = ATN(TAN(AL)*COS(LAMO)): PHI = ATN(MT/COS(THETA))
130   LAM = PHI+ATN(P): EP = LAM-LAMO
135   IF ABS(EP) < MERR THEN GOTO 200 ELSE GOTO 125


190   PRINT "Performance and thread friction incompatibile.": GOTO 210
200   LAM = LAM*180/PI: PRINT "Lead angle =" LAM


210   END
```

Figure 6.6.3 Program listing for LEED which calculates the lead angle for a power screw acting against the applied axial load, with the applied axial load, or backdriving.

while buttress profiles are preferred for one-directional motion, as in pressure vessel closures and breech bolts in heavy artillery. Unified and SI threads are preferred for retaining screws because they are cheaper to manufacture.

The efficiency of a screw is defined as the ratio T'/T where T' is the torque that would be required to raise a weight if there were no friction, namely

$$T' = W r_p \tan \lambda$$

Rather than finding an expression for T'/T, it is easier to calculate T/T', which is

$$\frac{\cos \theta_n + \mu_t \operatorname{ctn} \lambda}{\cos \theta_n - \mu_t \tan \lambda} + \mu_c \frac{r_c}{r_p} \operatorname{ctn} \lambda = \frac{T}{T'} \tag{6.6.10}$$

The relative importance of the thread and collar coefficients of friction upon the efficiency of the screw may be seen from Figure 6.6.4 for Acme threads. Typical friction

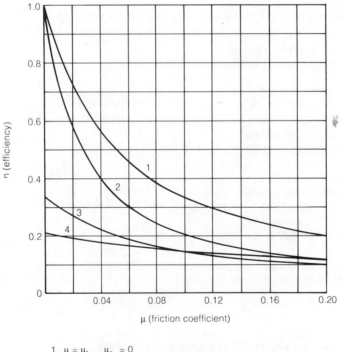

1. $\mu = \mu_t \quad \mu_c = 0$
2. $\mu = \mu_c \quad \mu_t = 0$
3. $\mu = \mu_c \quad \mu_t = 0.10$
4. $\mu = \mu_t \quad \mu_c = 0.10$

$\dfrac{r_c}{r_p} = 2.0$

Figure 6.6.4 Screw efficiency with friction for an Acme profile.

coefficients are[8]

- metal-to-metal contact   0.400 to 0.800
- lubricated metals   0.005 to 0.200
- rolling element bearings   0.0015 to 0.0080

EXAMPLE 6.6.1

Design a crosshead lift screw with Acme threads for a universal testing machine as represented in Figure 6.6.5. The pitch diameter is to be 2.20 in., the collar inside diameter is 2.00 in., and the outside diameter is to be 4.00 in. The thread friction coefficient is to be no greater than 0.02 and the collar friction is to

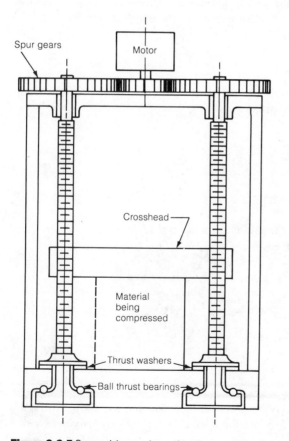

Figure 6.6.5 Screw driven universal testing machine. *Source: Juvinall R.C.,* Fundamentals of Machine Component Design, *Wiley, New York, 1983. Copyright © 1983 John Wiley & Sons, Inc. Reprinted with permission.*

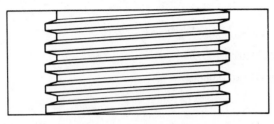

Figure 6.6.6 Sketch of the screw designed in Example 6.6.1 showing the pitch in proportion to the major and minor diameters.

be no greater than 0.05 for the required lubricants to maintain the warranty. Maximum load on each screw is 30 tons and they are to be driven by a 15-hp motor that delivers 7.5 hp to each screw, which turns at 60 rpm. Specify the lead angle, the pitch, and the advance rate of the crosshead. Sketch several threads to show the thread height and width relative to the major diameter.

Solve the formula

$$\text{hp} = \pi T n / 198\,000$$

for torque T to find $T = 7878.17$ in.-lb at 60 rpm. Entering this into the above program along with the mean collar diameter $d_c = 3.00$ in. we find that

$$\lambda = 1.74668°$$

From equation 6.1.2 find

$$L = \pi(2.2) \tan(1.74668°) = 0.2108 \text{ in.}$$

Thus, the major diameter is 2.2632 in. and the minor diameter is 2.1368 in.

The advance rate of the crosshead is given by

$$v = Ln = 0.2108(60) = 12.648 \text{ in./s}$$

A short length of the screw is shown in Figure 6.6.6.

6.7 BALL AND ROLLER SCREWS

The highest efficiency for screws may be obtained with ball screws, shown in Figure 6.7.1, which, because of their very low friction coefficients, closely approximate curve 1 in Figure 6.6.4.

Backlash in a nut and ball screw combination is the maximum amount the nut may rotate without moving axially. It may be eliminated by forcing two nuts together

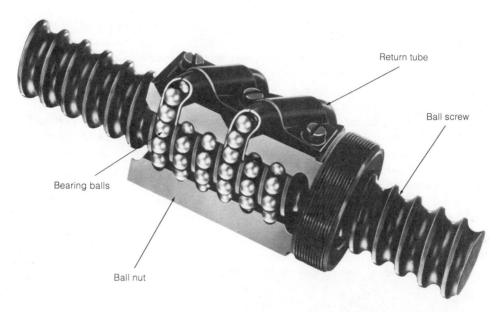

Figure 6.7.1 Ball screw and ball nut. *Source: Thomson-Saginaw Ball Screw Co., Saginaw, MI.*

by a threaded connection between them or by slightly altering the lead angle at each end of a single nut to force the balls at each end toward the middle of the nut. Either mechanism is known as preloading to eliminate backlash.

Their very low friction allows ball nuts to backdrive for very small axial loads that are often less than the weight of the ball nut itself. Thus, braking is often required whenever backdriving is to be prevented.

Ball screws, like ball bearings, have a limited fatigue life, which may be calculated in terms of the number of revolutions L_{10} of the nut as a function of the axial load W from

$$L_{10} = \left(\frac{C}{W}\right)^3 \times 10^6 \qquad (6.7.1)$$

where the basic dynamic capacity C may typically lie within the range from $C = 1500$ N for a small ball screw with a nominal diameter of 8 mm to $C = 136\,300$ N for one with a nominal diameter of 80 mm. Ball screw life is commonly given in terms of length of travel

$$\Lambda_{10} = \pi d_{\mathrm{p}} L_{10} \tan \lambda \qquad (6.7.2)$$

where Λ_{10} denotes the life in inches that may be expected with a probability of 90%. Lead angle λ may typically range from 10° for small screw diameters to 5° for large screw diameters. The large angle for small diameters is due to the limited range of ball diameters.

Roller screws, Figure 6.7.2, have values of thread friction between those of ball screws and conventional (sliding friction) screws. In these screws the nut is internally threaded with the same profile, lead angle, and number of starts as the central screw. The rollers are threaded with the same pitch as the nut and central screw, but with one start and with crowned threads (leading and trailing flanks are slightly convex instead of straight as in Acme screws) so they may roll smoothly between the central screw and the nut. In the planetary roller design shown in Figure 6.7.2*a* gear teeth are cut in the roller ends to mesh with internal gears at each end of the nut to insure that the rollers always roll on the nut without sliding. This prevents the rollers from moving axially out of the nut, as they would if they were allowed to slide along the internal threads in the nut.

These screws generally can support a greater axial load for a given nut length and central screw diameter than can ball screws because of greater contact area per unit length of nut.

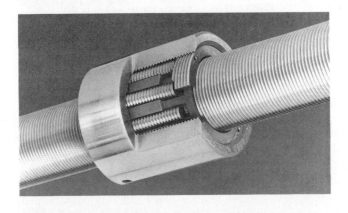

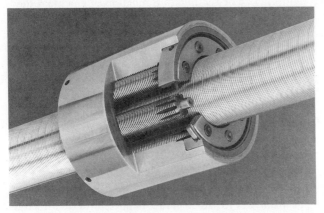

Figure 6.7.2 Recirculating roller screw (*top*) and planetary roller screw design. *Source: SKF Component Systems Co., Allentown, PA.*

The recirculating roller design shown in Figure 6.7.2b differs from the planetary design in that the threads on the rollers are replaced by circumferential grooves with the same profile as the threads on the nut and central screw, but with zero lead angle. Although separated by a cage, they roll freely between the nut and the central screw and advance faster than the nut; after making a complete circuit about the inside of the nut each roller will have advanced one lead relative to the nut. This advance relative to the nut is removed at the end of each circuit by guiding the roller into a slot in the nut where it is moved back to its starting position relative to the nut.

Although it cannot support as great a load for a given nut size nor rotate as fast as the planetary roller design, the recirculating roller nut can be positioned more precisely in the axial direction than can the planetary roller nut because the central screw can be produced with a smaller lead angle.[9] Screw major diameters are available from 8 to 125 mm (0.315 to 4.921 in.) for recirculating roller screws and up to 180 mm (7.087 in.) for planetary roller screws.

Life in revolutions or in length of travel may be calculated from equations 6.7.1 and 6.7.2 using C values from 6200 N(1394 lb) for preloaded planetary roller screws to 1 668 500 N (3 725 433 lb) for non-preloaded planetary screws and from $C = 4800$ N (1079 lb) for preloaded recirculating roller screws to 804 200 N (180 800 lb) for non-preloaded recirculating roller screw and nut combinations.

6.8 BUCKLING AND TENSILE STRESS

Buckling must be considered for power screws in compression, since it is the condition that determines the minor diameter d_s of the screw. In other circumstances this dimension may be determined by the tensile stress in screw fasteners and in power screws used in tension.

Buckling

The classical Euler long-column and the Johnson short-column buckling formulas described in Section 3 of Chapter 3 also apply to screws. To err on the side of caution, the buckling loads are calculated on the basis of a circular cylinder whose radius is the minor radius of the screw.

Condition 3.3.3 for the Johnson formula to hold takes the form

$$\frac{l}{r_s} < \frac{\pi}{k} \left(\frac{E}{2\sigma_y} \right)^{1/2} \tag{6.8.1}$$

where $r_s = d_s/2$ represents the minor radius of the screw profile. When this inequality holds, equation 3.3.4 applies. It may be rewritten in terms of the minor radius of the screw profile as

$$W = \frac{\sigma_y}{\zeta} \pi \left[r_s^2 - \left(\frac{kl}{\pi} \right)^2 \frac{\sigma_y}{E} \right] \tag{6.8.2}$$

When inequality 6.8.1 does not apply, the buckling load is given by Euler formula, equation 3.3.4, which may be rewritten as

$$W = \frac{\pi^3 E r_s^4}{4\zeta k^2 l^2} \tag{6.8.3}$$

In these equations ζ denotes the safety factor and σ_y denotes the yield stress in tension. Since the yield stress in compression is often larger than that in tension, a cautious estimate of the buckling load W is had.

According to Section 3 of Chapter 3, k is determined by the end supports according to

$$
\begin{aligned}
k &= 1/2 && \text{for fixed–fixed} \\
&= 1/\sqrt{2} && \text{fixed–pinned} \\
&= 1 && \text{pinned–pinned} \\
&= 2 && \text{fixed–free}
\end{aligned}
$$

where these representations are associated with the bearing and end conditions shown in Figure 6.8.1. See Chapter 10 on rolling element bearings for further details on bearing mounts that closely approximate fixed or pinned supports.

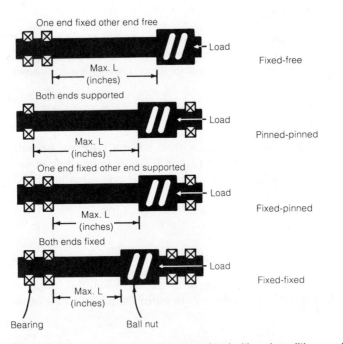

One end fixed other end free

Max. L
(inches)

Load

Fixed-free

Both ends supported

Max. L
(inches)

Load

Pinned-pinned

One end fixed other end supported

Max. L
(inches)

Load

Fixed-pinned

Both ends fixed

Max. L
(inches)

Load

Fixed-fixed

Bearing   Ball nut

Figure 6.8.1 Bearing configurations associated with end conditions, or fixities, as indicated. *Source: Thomson-Saginaw Ball Screw Co., Saginaw, MI.*

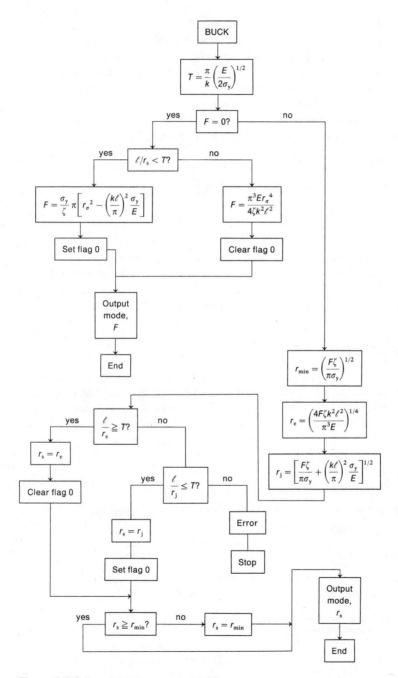

Figure 6.8.2 Flowchart for program BUCK to calculate the buckling load for long or short columns.

Application of criterion 6.8.1 and equations 6.8.2 and 6.8.3 is straightforward when buckling loads are to be calculated. If we wish to find the minor radius of a screw that will support a given axial load, however, we obviously do not know the l/r_s ratio, and so must calculate the radius from both (6.8.2) and (6.8.3) and select the one for which criterion 6.8.1 is satisfied. This is because equation 6.8.3 does not always produce a radius larger than that found from equation 6.8.2.

A flowchart for a program to calculate either the buckling or the required radius to support a given load without buckling is given in Figure 6.8.2.

Tensile Loading

The load limit in tension may be approximated by

$$W = \pi r_s^2 \sigma_y / (\zeta K) \tag{6.8.4}$$

in which K denotes the stress concentration factor for the particular thread profile when the screw is in tension. It should be the largest of the possible stress concentrations where the thread enters the nut, at the fillet at the head of the screw, and in the vicinity of the first thread. Measured values from 2.7 to 6.7 have been reported, based upon the minor diameter; the reference area is given by $A = \pi d_s^2/4.$*

6.9 REQUIRED NUT LENGTH

The required nut length may be estimated by calculating the number of threads required to exert the design force based upon thread failure due to bearing loads or shear loads. Failure due to thread bending is highly improbable for the threads shown in Figure.6.2.1 for standard allowances and tolerances, which, incidentally, usually limit engagement lengths to approximately twice the pitch diameter.

In any event, the required nut length is given by

$$L_e = N_e p \tag{6.9.1}$$

where N_e is the number of threads necessary to sustain the imposed load without exceeding the limits on bearing and shear stresses.

Bearing Stress

To calculate the bearing area we first observe that the area of the leading flank (raising a load is assumed) of one thread is given by

$$A = \frac{\pi}{4} \frac{D_s^2 - d_n^2}{\cos \alpha_1 \cos \lambda} \tag{6.9.2}$$

* M. Hetenyi, The distribution of stress in threaded connections, *Proceedings of the Society for Experimental Stress Analysis,* 1, (1), 147 (1943), A.F.C. Brown, and V.M. Hickson, Study of stresses in screw threads, *Proceedings of the Institution of Mechanical Engineers,* 1B, 605, 608 (1952/53).

in terms of major screw diameter D_s, minor nut diameter d_n and leading flank angle α_1. From equations 6.5.9 and 6.9.2 we have that

$$\sigma = \frac{F_n}{A} = \frac{4W\zeta}{\pi(D_s^2 - d_n^2)} \frac{\cos \alpha_1 \cos \lambda}{\cos \theta_n \cos \lambda - \mu_t \sin \lambda} \tag{6.9.3}$$

If σ_b is the stress at which bearing failure begins, then the number of threads required to avoid bearing failure may be written as

$$N_e = \frac{4W\zeta}{\pi\sigma_b(D_s^2 - d_n^2)} \frac{\cos \alpha_1}{\dfrac{1}{\sqrt{1 + \tan^2 \alpha_1 \cos^2 \lambda}} - \mu_t \tan \lambda} \tag{6.9.4}$$

after trigonometric manipulation.

Shear Stress

Since the exact location of thread shear failure depends upon both the nut and screw materials, we can only estimate shear strength based upon a likely shear surface. If we assume the cylindrical shear surface lies midway between the screw pitch diameter and the nut minor diameter then the width b of the thread cut by that surface may be written as

$$b = \frac{p}{2} + \frac{d_p - d_n}{2} (\tan \alpha_1 + \tan \alpha_2) \tag{6.9.5}$$

where the following flank angle is denoted by α_2. From this we may write that the minimum number of threads required to avoid shear failure is

$$N_e = \frac{4W\zeta}{3\pi d_n \tau_y b} \tag{6.9.6}$$

for a parabolic shear distribution where τ_y is the yield stress in shear. Use of d_p instead of $(d_p - d_n)/2$ affords additional safety.

Bending Stress

Stress at the root of the thread due to bending may be approximated using

$$\sigma = Mc/I \tag{6.9.7}$$

by assuming that the resultant of the load W acts at the pitch radius and by considering the thread to be a wide cantilever beam of width $\pi d_p N_e$ that is loaded at a distance $d_p - d_s/2$ from the supported end. Thus, we have that

$$M = W(d_p - d_s)/2 \qquad (6.9.8)$$

and

$$I = \pi N_e d_p w^3/12 \qquad (6.9.9)$$

where width w of the thread at the minor diameter corresponds to the depth of the cantilever beam at its built-in end. Width w may be expressed in terms of the thread thickness at the minor radius by

$$w = \frac{p}{2} + \frac{d_p - d_s}{2} (\tan \alpha_1 + \tan \alpha_2) \qquad (6.9.10)$$

Substitution for M and I in equation 6.9.7 from equations 6.9.8 and 6.9.9 along with $c = w/2$ yields the following expression for N_e:

$$N_e = \frac{3W\zeta}{\pi w^2 \sigma_y} \left(1 - \frac{d_s}{d_p}\right) \qquad (6.9.11)$$

where w is given by equation 6.9.10. The remaining quantities in equation 6.9.11 have been defined previously.

EXAMPLE 6.9.1

Design a screw drive to raise castings in an automated manufacturing process using the table and guide rails shown in Figure 6.9.1. The maximum weight is 1000 kg and the castings must be elevated 600 mm in two steps of 300 mm each. The first 300 mm step must take no more than 2 s and the second no more than 2.3 s, with a waiting period of 13.55 min between steps. Give the screw minor diameter, the lead, the pitch, the number of threads, nut length, and the kilowatts required for the motor that drives both screws from a common shaft with two sprockets. Use a safety factor of 3.00 and assume that the load is equally divided between the two screws. Use a steel for which $E = 206$ GPa and $\sigma_b = \sigma_y = 415$ MPa.

Our first step is to decide what design problems are involved:

1. Since the screws are in compression, we must first select the minimum minor diameter that will prevent buckling.

2. With the minimum minor diameter known we have two choices for the next step:

   a to select the cheapest thread profile for a screw whose minor diameter equals or exceeds the minimum value, or

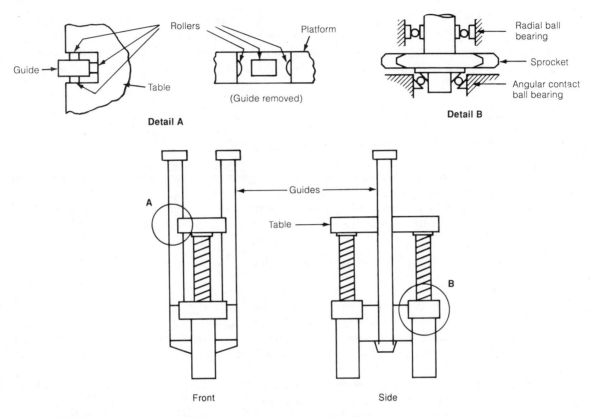

Figure 6.9.1 Power screw lift mechanism and details of the platform guide rollers and the screw drive and bearing mounts.

b to select that screw whose standard thread profile requires the smallest torque, and hence a small motor, to deliver the required force.

This choice is between low first cost, but possibly higher operating cost, as opposed to higher first cost and lower operating costs. We will opt for b, because it involves an additional design step.

3. We must decide whether to use standard screws, ball screws, or roller screws. Since ball and roller screws are subject to backdriving, they are best suited to horizontal drives, where an additional brake is not required. Their vertical applications apply to those cases where power is limited, so that a ball or roller screw is the only choice available. Since precise positioning was not listed as a requirement, we will also consider a ball screw instead of a more precise positioning, but greater friction, roller screw.

4. We must select reasonable friction coefficients if a standard power screw is to be used.

To carry out the first step we turn to a buckling program that solves equations 6.8.2 and 6.8.3 for the diameter, calculates the l/r ratio, and selects the appropriate solution from criterion 6.8.1. According to Figure 6.9.1 and detail A the platform is in contact with the guide rails at only one point on each side. This is equivalent to a pinned fixity at the upper end. Use of an angular contact ball bearing at the base of the screw and a radial ball bearing above the sprocket to keep the chain tension from applying a bending load to the screw is equivalent to a fixed lower condition, so that the fixities are pinned–fixed. Hence $k = 1/\sqrt{2}$, and $\mu_t = 0.008$. The load is given by $W = 1000(9.8067)/2 = 4903.35$ N so that substitution of the values

$$W = 4903.35 \text{ N} \qquad l = 600 \text{ mm} \qquad \zeta = 3.00$$

$$k = 1/\sqrt{2} \qquad E = 206\,000 \text{ MPa} \qquad \sigma_y = 690 \text{ MPa}$$

into the buckling program yields $d_s = 12.792$ mm as the minimum minor diameter to prevent buckling according to the Euler criterion.

To arrive at a minimum torque using standard profiles, turn to Figure 6.6.1 and recall that the curves intersect the $\lambda = 0$ line at μ_t and that this value also equals P_1. Upon solving $\mu_t = P_1$ for T we obtain

$$T_{\min} = Wr_p\left(P_1 + \mu_c \frac{r_c}{r_p}\right) = 4904(8.5)\left(0.1 + 0.008\,\frac{11.5}{8.5}\right) = 4620 \text{ N mm}$$

where we have used $d_p = 17$ mm to allow for a thread depth of about 2 mm below the pitch diameter. We have chosen the thread friction coefficient to be 0.10 for a lubricated surface with some dirt in the lubricant because the screw may be exposed to residual dirt from the foundary where the casting was produced. The mean collar diameter of 23 mm was selected to match an available angular contact ball bearing.

Before moving to the next step, notice that covering the exposed part of the screw with a bellows or a skirt below the platform may reduce the thread friction to 0.05 or less if the threads can be kept clean, which cuts the minimum torque almost in half. We may suggest this as a design modification for a more expensive model.

We now have to resolve conflicting effects of the variables in equation 6.5.15 for torque. To get close to the minimum torque found above we need a small pitch diameter and a small lead angle. But small values for either results in a small pitch, which requires more expensive machining (because of smaller tolerances) and protection from dirt. We shall, therefore, accept a torque increase of perhaps 20 to 25% to get larger threads. We turn to buttress threads and to ANSI B1.9-1973 to search for a satisfactory standard diameter and pitch combination that has a minor diameter greater than 12.8 mm = 0.5093 in. From Table 1 of that standard, reproduced in part in Table 6.9.1 below, and the relations

$$d_s = D_s - 1.3254p$$

derived from Figure 6.2.1, and

$$p = 1/(\text{threads per inch})$$

we select a screw with a major diameter of 0.625 in. and a pitch of 0.0625 in. (16 threads/in.) for which $d_s = 0.5422$ in. is large enough to prevent buckling. A screw with 16 threads/in. was selected

TABLE 6.9.1 PREFERRED DIAMETER–PITCH COMBINATIONS FOR 7°/45° BUTTRESS THREADS

Number of Threads per Inch	Pitch (in.)	Nominal Major Diameters (in.)							
		0.5	0.625	0.75	0.875	1.0	1.25	1.375	1.5
20	0.0500	×	×	×					
16	0.0625	+	+	+	×	×	−		
12	0.0833	×	×	×	+	+	×	×	×
10	0.1000				×	×	+	+	+
8	0.1250						×	×	×
6	0.1667						−	−	−

+, First choice; ×, second choice; −, third choice.
Source: Abstracted from Table 1, ANSI B1.9-1973, reaffirmed 1979.

because it will be easier to clean than one with 20 threads/in. Thus, with $d_p = D_s - 0.6p$ we have $d_p = 0.5875$ in., $r_p = 0.29375$ in. $= 7.4613$ mm, and

$$\lambda = \tan^{-1}[p/(\pi d_p)] = \tan^{-1}[0.0625/(0.5875\pi)] = 1.9395°$$

so that

$$T = Wr_p[\tan(\phi + \lambda) + \mu_c r_c/r_p]$$
$$= 4903.35(7.4613)[\tan(7.6926°) + 0.008(11.5/7.4613)] = 5392.86 \text{ N mm}$$

where

$$\theta_n = \tan^{-1}(\tan \alpha_1 \cos\lambda)\tan^{-1}(\tan 7° \cos 1.9395°) = 6.9960°$$
$$\phi = \tan^{-1}(\mu_t/\cos \theta_n) = \tan^{-1}(0.1/\cos 6.9960) = 5.7531°$$

We must use the larger of the two platform velocities, 150 mm/s, and the pitch 0.0625 in. $= 1.5875$ mm to find that the faster rotational speed of each screw must be

$$n = v/p = (150/1.5875)60 = 5669.29 \text{ rpm}$$

so that the required power is given by

$$kW = \pi Tn/3 \times 10^7 = \pi 5392.86(5669.29)/3 \times 10^7 = 3.20$$

for a total power requirement of 6.40 kW for both screws.

We next consider the possibility of using ball screws. From Table 6.9.2 we find that the smallest ball screw that may be used has a nominal outside diameter of 20 mm and a dynamic load rating of 11 200 N so that the expected life according to equation 6.7.1 is

$$L_{10} = (11\,200/4903.35)^3 10^6 = 11\,913\,514 \text{ rev.}$$

This corresponds to a life in hours of

$$L_{10}(\text{hours}) = 11\,913\,514/[(5669.29)60] = 35.0$$

To obtain a satisfactory life we must select the ball screw with a 63 mm outside diameter to get a basic dynamic load rating of 90 200 N which produces an expected life of

$$L_{10}(\text{hours}) = (90\,200/4903.35)^3 10^6/[(5669.29)60] = 18\,300$$

To calculate the required torque we assume a friction coefficient of the order of 0.004 for both the threads and the collar, an equivalent leading flank angle of 8°, and a pitch diameter that is midway between the major and minor diameters, namely 59.25 mm. From equation 6.5.15 we find that $T = 9030$ N mm and that the torque required to prevent backdriving is 6578 N mm for a collar mean diameter of 65 mm. Reducing this to 40 mm merely reduces the torque required to 8788 N mm.

This torque is larger because the screw and collar diameter increases dominate the effect of the reduced friction coefficients. Thus, we find that although ball screws are excellent for light load they are unsuited for heavy loads.

TABLE 6.9.2 TYPICAL BALL SCREW DYNAMIC AND STATIC LOAD RATINGS WITHOUT PRELOAD

Nominal Diameter (mm)	Lead (mm)	Basic Dynamic Load Rating (N)	Basic Static Load Rating (N)
8	2.5	1 500	2 200
10	3.0	2 300	3 500
12	4.0	3 400	5 400
16	5.0	5 200	8 700
20	5.0	11 200	23 300
25	5.0	17 000	42 100
25	10.0	17 000	42 100
25	2.5	11 100	27 700
32	5.0	19 400	56 200
32	10.0	19 400	56 200
40	5.0	21 200	70 300
40	10.0	55 800	141 700
50	10.0	80 600	240 700
63	10.0	90 200	309 700
80	20.0	136 300	569 700

Source: Abstracted from SKE Transrol High Efficiency Ball and Roller Screws

TABLE 6.9.3 CLASS 2 TOLERANCES AND ALLOWANCES ON MAJOR
AND MINOR DIAMETERS FOR INTERNAL AND EXTERNAL
THREADS FOR 7°/45° BUTTRESS THREADS FOR MAJOR
DIAMETERS FROM 0.5 to 0.7 in.

Threads per Inch	Pitch (in.)	Allowance (in.)	Tolerance (in.)
20	0.0500	0.0037	0.0056
16	0.0625	0.0040	0.0060
12	0.0833	0.0044	0.0067

Source: Abstracted from Table 1, ANSI B1.9-1973, reaffirmed 1979.

The number of threads required to support the load depend upon the tolerances and allowances for the particular thread chosen. Standard values for buttress threads for the major diameter chosen are found in the standard ANSI B1.9-1973 (reaffirmed 1979). Values from those tables which pertain to the screw selected are given in Table 6.9.3. The sum of the tolerance and the allowance will give the least overlap between the threads of the nut and the screw, and, therefore, will be used in our calculation of the required nut length to evaluate the worst condition. From Table 6.9.2 we find that the allowance is 0.0040 in. and the tolerance is 0.0060 in., for a total of 0.0100 in. manufacturing error. Thus,

$$D_s = 0.615 \text{ in.} \qquad D_n = 0.635 \text{ in.}$$

$$d_s = 0.5322 \text{ in.} \qquad d_n = 0.5522 \text{ in.}$$

Substitution into a program to solve equation 6.9.4, 6.9.5, 6.9.6, 6.9.10, and 6.9.11 and compare results yields the shear stress as the most likely failure mode, which requires 0.862 threads, or a nut length of 0.054 in. to avoid failure in the worst case. The shear stress in yield was estimated from $\sigma_{y_s} = \sigma_y/\sqrt{3}$ for use in calculating the minimum nut length to resist shear failure.

EXAMPLE 6.9.2

The chief engineer has commented on the high price of the gear train used to obtain the required rotational speed for the design produced in Example 6.9.1 and suggested that the unit price could be reduced if the gear train were removed and the sprocket driven directly by a 10-kW motor with 5 kW delivered to each screw at 3450 rpm. The addition of electronic speed control would allow speed variation of 1700 to 3500 rpm while maintaining motor torque, thus enabling the unit to be used in other applications. The savings plus the additional uses would justify a nonstandard screw if its major diameter were unchanged. Can a screw be designed to these requirements? If so, specify the lead angle and the pitch.

The torque given by 5 kW to each screw is

$$T = 5(3 \times 10^7)/(\pi 3450) = 13\,840 \text{ N mm}$$

Turn next to program LEED and enter

$$T = 13\,840 \text{ N mm} \qquad W = 4903.35 \text{ N} \qquad \mu_t = 0.1 \qquad \alpha_1 = 7°$$

$$d_c = 23 \text{ mm} \qquad d_p = 14.9226 \text{ mm} \qquad \mu_c = 0.008$$

to find $\lambda = 14.3502°$ and $p = 11.9933$ mm $= 0.4722$ in. Since $D_s = 0.625$ in. and the minimum value for the minor diameter is $d_s = 0.5036$ in. the maximum pitch consistent with these values may be found by solving for p from

$$D_s = d_s + 2(0.3 + 0.3627)p$$

to get $p = (D_s - d_s)/1.3254 = (0.625 - 0.5036)/1.3254 = 0.0916$ in. This corresponds to $1/0.916 = 10.917$ threads/in., which is non-standard. Hence, select 12 threads/in., for which $p = 0.08333$ in./thread or 2.1167 mm/thread.

Upon dividing this pitch into the pitch found from LEED we find $11.9225/2.1167 = 5.6$. Thus, we have sufficient torque to drive a five-start screw that would have a lead angle slightly less than that found from LEED. The advance rate of the five-start screw at the motor's slowest speed would be

$$v = Ln/60 = 5(2.1167)1700/60 = 299.87 \text{ mm/s}$$

This is approximately twice the required advance rate.

We may use a multiple-start screw if the elevation rate must not be exceeded. Since the rotational speed required for a single-start thread is

$$n = v/p = 150(60)/2.1167 = 4252 \text{ rpm}$$

we may use a two-start thread to reduce the rotational speed to 2126 rpm, which is within the proposed speed range.

Turn to the relation $d_p = D_s - 0.6p$, derived from the profile shown in Figure 6.2.1, to find that the pitch diameter for 12 threads/in. is

$$d_p = 0.625 - 0.6(0.08333) = 0.5750 \text{ in.} = 14.605 \text{ mm}$$

Hence, the lead angle for the two start thread is

$$\lambda = \tan^{-1}[2(2.1167)/(14.605\pi)] = 5.2715°$$

Substitution of $d_p = 14.605$ mm and $\lambda = 5.2715°$ into equation (6.5.12) produces a required torque of 7426.01 N mm. The power demand for this combination is given by

$$kW = \frac{\pi T n}{3 \times 10^7} = \frac{\pi 7426.01(2126)}{3 \times 10^7} = 1.653$$

which is well within the proposed 5 kW capability of each screw drive.

Assume SAE Grade 1 specifications, Table 6.11.3, are to be used and that bearing and yield stresses are equal so that $\sigma_b = \sigma_{y_t} = 36\,000$ psi and $\sigma_{y_s} = \sigma_{y_t}/\sqrt{3} = (36\,000)\sqrt{3} = 20\,784$ psi. According to the worst case for class 2 tolerances (ANSI B1.9), $D_s = 0.619$ in., $d_s = 0.510$ in., $D_n = 0.631$ in. and $d_n = 0.520$ in. Substitution of these material properties and screw and nut dimensions into equations 6.9.4, 6.9.6, and 6.9.11 results in a required nut length of 0.6663 in. or greater, corresponding to engagement of at least 8 (7.995) threads, with a safety factor of 3.0, as limited by the shear stress.

Figure 6.9.2 Five-start power screw used in a Chevrolet power steering unit. Three starts are visible.

In this and the previous example it has been assumed that the specifications were maximum performance requirements: that the load included weight plus acceleration forces and that the travel time included acceleration, deceleration, and constant velocity times.

Figure 6.9.2 shows a five lead power screw used in the power steering of a Chevrolet automobile as a common application of multiple start screws.

6.10 CLAMPING FORCE

In the manufacture of assemblies where the proper clamping force is important, as in cylinder heads, pressure vessels, motor mounts, and so on, the clamping force may be found quite accurately by electronically measuring bolt elongation. When the equipment for this is not available the clamping force may be determined less accurately by placing either a Belleville or wave washer whose flattening force is equal to the desired clamping force under the bolt head and then tightening the bolt until the washer is flattened. Loss of the washer during disassembly may be prevented by using a bolt with a wave washer flange, as shown in Figure 6.10.1. Only measurement of bolt length guards against overtightening.

A common alternative to the above procedures is to use a torque wrench and tighten the bolt to a specified torque. This is less accurate than measuring bolt elongation because of the variability of the friction coefficients. Differentiation of equation 6.5.12 with respect to μ_c and μ_t yields

$$\frac{d}{d\mu_c}\left(\frac{T}{W}\right) = r_c \qquad (6.10.1)$$

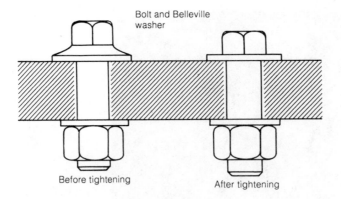

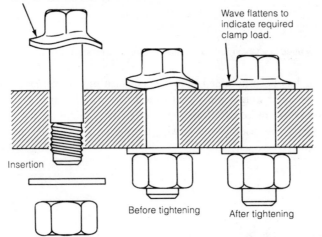

Figure 6.10.1 Bolt tension indicators. *Top*: Belleville washer; *bottom*: bolt with wave washer flange. *Source: Reprinted from* Machine Design, *Nov. 18, 1982. Copyright 1982 by Penton/IPC, Inc., Cleveland, OH.*

and

$$\frac{d}{d\mu_t}\left(\frac{T}{W}\right) = r_p \frac{1}{\cos^2(\phi + \lambda)} \frac{\cos\theta_n}{\mu_t^2 + \cos^2\theta_n} \tag{6.10.2}$$

The first of these equations indicates that a small average collar radius is desirable—as is afforded by a Belleville washer until just before it becomes flat. A plot of the right side of the second equation as a function of μ_t shows that it varies almost linearly with μ_t and changes no more than 24% for μ_t varying from 0 to 1.0, for flank angles no larger than 30°, and lead angles up to and including 5°. This includes the range of standard threaded fasteners. The largest change occurs for a flank angle of 30° and a lead angle of 5°, for which the right side varies from $1.16r_p$ to $1.44r_p$.

Since the right side of (6.10.2) is relatively insensitive to changes in the thread coefficient of friction, and since lubricating the threads may eventually contribute to loosening of bolts and screws if the lubricant seeps out or reduces volume, lubrication is discouraged whenever the clamping force is to be controlled by specifying the tightening torque.

6.11   BOLT PRETENSIONING

Pretensioning of bolts in bolted connections not only assures a tight and perhaps leak-proof connection, but it also increases the fatigue resistance of the bolt. Almen[12] tested $\frac{3}{8}$-in. bolts with a cyclic tensile load varying from 0 to 9220 lb and found the mean life was increased as shown in Table 6.11.1. Data compiled by Heywood from other experimental results, shown in Figure 6.11.1, include the effect of bolt diameter, while Figure 6.11.2 shows that excessive pretension can be detrimental.[11]

Commonly accepted design formulas for estimating the effect of pretensioning of bolts may be derived by considering a cylinder head assembly as illustrated in Figure 6.11.3a in which a Belleville spring has been inserted to indicate the pretension and to provide a relatively low spring rate k_b for the bolt and washer combination. The theory to be developed does not depend upon inclusion of a spring in series with the bolt, however.

Guided by an analysis by Burr,[13] we first apply a pretensioning force F_0 which causes the bolt to extend an amount δ_b and the clamped portions of the cylinder and head to contract an amount δ_c, so

$$F_o = k_b \delta_b = k_c \delta_c \tag{6.11.1}$$

where k_b is the effective spring constant of the bolt and k_c is the effective spring constant of the cylinder and head flanges together, along with any gasket that might be inserted between them. Upon application of a separation force F due to internal cylinder pressure the force F_b on the bolt is given by

$$F_b = F_o + k_b \delta \tag{6.11.2}$$

TABLE 6.11.1

Pretension (lb)	Mean life (cycles)
1420	6 000
5920	36 000
7220	215 000
8420	5 000 000

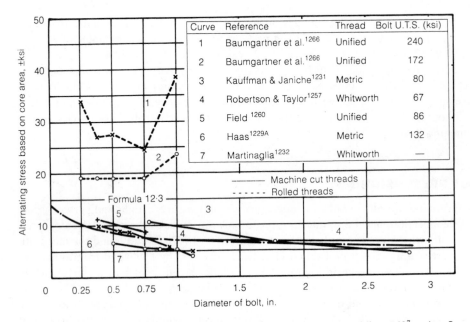

Figure 6.11.1 Effect of bolt diameter on fatigue strength of bolt–nut assemblies at 10^7 cycles. *Source: Heywood*, Designing Against Fatigue of Metals, *Van Nostrand Reinhold, 1962.*

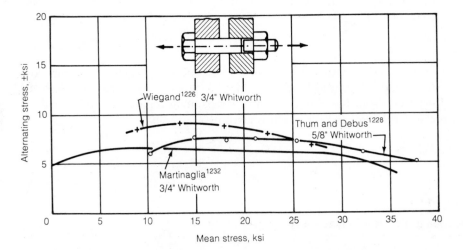

Figure 6.11.2 Effect of mean stress on the fatigue strength of mild steel bolts. Stresses based on core area. *Source: Heywood*, Designing Against Fatigue of Metals, *Van Nostrand Reinhold, 1962.*

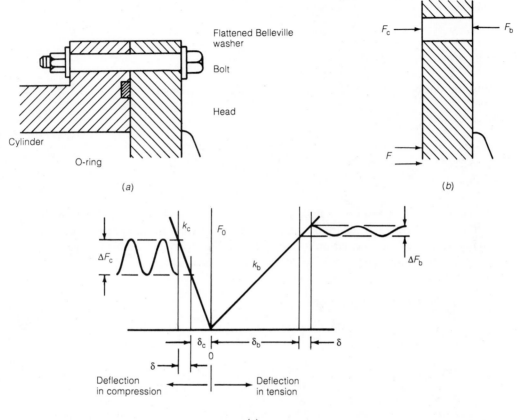

Figure 6.11.3 (a) Representative head bolt; (b) force acting on the head; (c) force-deflection relations for head and bolt.

and the contact force F_c between cylinder and head is given by

$$F_c = F_o - k_c\delta \tag{6.11.3}$$

where

$$F = k_b\delta = k_c\delta$$

Equilibrium of the cylinder head requires that

$$F + F_c = F_b \tag{6.11.4}$$

from which it follows that

$$F + F_o - k_c\delta = F_o + k_b\delta$$

and, hence, that

$$\delta = \frac{F}{k_b + k_c} \tag{6.11.5}$$

Substitution for δ from (6.11.5) into (6.11.2) and (6.11.3) yields

$$F_b = F_o + \frac{k_b}{k_b + k_c} F \tag{6.11.6}$$

$$F_c = F_o - \frac{k_c}{k_b + k_c} F \tag{6.11.7}$$

By selecting the bolt and/or the Belleville or wave washer such that k_b is much smaller than k_c, as illustrated in Figure 6.11.3c, the load fluctuation in the bolt may be made smaller than the load fluctuation in the cylinder. This follows from equations 6.11.6 and 6.11.7 and the definition of the alternating bolt load as

$$\Delta F_b = F_{b_{max}} - F_{b_{min}} = \frac{k_b}{k_b + k_c} \Delta F \tag{6.11.8}$$

Likewise,

$$\Delta F_c = F_{c_{max}} - F_{c_{min}} = -\frac{k_c}{k_b + k_c} \Delta F \tag{6.11.9}$$

Equation 6.11.4 holds only if the external force F is not large enough to separate the clamped surfaces. Once they are separated that equation is invalid and, therefore, the proportioning of the alternating load according to equations 6.11.8 and 6.11.9 is also invalid.

Since the head and cylinder geometry generally give rise to much smaller stress concentrations and notch sensitivities than in the bolt and nut combination, it can better withstand the fluctuating load. Hence, the fatigue life of the connection is lengthened accordingly.

Inasmuch as no approximation for k_c has been generally accepted, we employ a relation given by Shigley and Mitchell[14] and due to C.R. Mischke,

$$k_c = \frac{\pi E d_h}{2 \ln \dfrac{5l + 2.5d_h}{l + 22.5d_h}} \tag{6.11.10}$$

where d_h is the diameter of the bolt head washer and l is the combined thickness of the components held together by the bolt or screw, E is the elastic modulus of the cylinder and head.

EXAMPLE 6.11.1

Recommend a standard Unified screw with a nominal diameter of $1\frac{1}{8}$ in., SAE grade 5, (Table 6.11.3) to provide a clamping force of 8000 lb with a torque of 7310 in.-lb for a hydraulic motor mount made from cast iron. Assume that $\mu_t = 0.5$ and that $\mu_c = 0.8$. Use a standard cap screw head, which for this diameter screw has an outside diameter of 1.688 in. The maximum stress concentration factor for this bolt is 5.00 and a safety factor of 1.5 or greater in tension is required, relative to the proof stress. Assume that $E = 3 \times 10^7$ for both the bolt and the cast iron.

Turn to the lead selection program flowcharted in Section 6.8 and input the values

$$W = 8000 \text{ lb} \qquad\qquad \mu_t = 0.5$$

$$T = 7310 \text{ in.-lb} \qquad\qquad \mu_c = 0.8$$

$$r_c = (1.688 + 1.25)/4 = 0.7345 \text{ in.}$$

and try successive average values for the pitch diameter for the standard screws listed in Table 6.11.2. We find that for $r_p = (1.0844 + 1.0807)/2 = 0.541\,25$ in. the lead angle is $1.073\,69°$ and the pitch is 15.69 threads/in. Thus, select the

$$1\tfrac{1}{8} - 16 - UN - 3A$$

for which $\lambda = 1.0529°$. The cross section at the minor diameter is $\pi(1.0505)^2/4 = 0.8667$ in.^{2} so that the tensile stress is given by

$$\sigma = K_f \frac{F}{A} = 5.0 \frac{8000}{0.8667} = 9230.414(5) = 46\,152 \text{ psi} \tag{6.11.11}$$

Since the proof stress is $\sigma_p = 74\,000$ psi the safety factor becomes

$$\zeta = 74\,000/46\,152 = 1.603 > 1.5$$

which indicates that the screw is sufficiently strong.

EXAMPLE 6.11.2

Estimate the maximum alternating load (due to equipment vibration) that may be sustained by the mounting bolts in example 6.11.1 without eventual failure, based upon the Gerber-yield condition and a safety factor of 1.5. Length l in equation 6.11.10 will be taken as 2.00 in., the sum of the mount thickness and the screw's engagement depth in the tapped mounting hole.

Consistent with the assumption in Chapter 4, we will not apply the stress concentration factor to the mean stress in the Gerber-yield criterion; thus, $\sigma_m = 9230.414$ psi. Since the notch sensitivity factor q has not been given, we assume that it is unity, thus assuring that we err on the side of safety.

TABLE 6.11.2(a) LIMITING DIMENSIONS FOR UNIFIED INCH THREADS FOR NOMINAL 1.000- TO 1.1875 ($1\frac{3}{16}$)-IN. DIAMETER SCREWS/BOLTS

Nominal Size and Threads/in.	Series Designation	Class	Allowance	External							
				Major Diameter			Pitch Diameter			UNR Minor Diam.[3] Max. (Ref.)
				Max.[3]	Min.	Min.[3]	Max.[3]	Min.	Tolerance	
1-32 or 1.000-32	UN	2A	0.0011	0.9989	0.9929	—	0.9786	0.9748	0.0038	0.9617
		3A	0.0000	1.0000	0.9940	—	0.9797	0.9769	0.0028	0.9628
$1\frac{1}{16}$-8 or 1.0625-8	UN	2A	0.0020	1.0605	1.0455	—	0.9793	0.9725	0.0068	0.9117
		3A	0.0000	1.0625	1.0475	—	0.9813	0.9762	0.0051	0.9137
$1\frac{1}{16}$-12 or 1.0625-12	UN	2A	0.0017	1.0608	1.0494	—	1.0067	1.0010	0.0057	0.9616
		3A	0.0000	1.0625	1.0511	—	1.0084	1.0042	0.0042	0.9633
$1\frac{1}{16}$-16 or 1.0625-16	UN	2A	0.0015	1.0610	1.0516	—	1.0204	1.0154	0.0050	0.9865
		3A	0.0000	1.0625	1.0531	—	1.0219	1.0182	0.0037	0.9880
$1\frac{1}{16}$-18 or 1.0625-18	UNEF	2A	0.0014	1.0611	1.0524	—	1.0250	1.0203	0.0047	0.9950
		3A	0.0000	1.0625	1.0538	—	1.0264	1.0228	0.0036	0.9964
$1\frac{1}{16}$-20 or 1.0625-20	UN	2A	0.0014	1.0611	1.0530	—	1.0286	1.0241	0.0045	1.0016
		3A	0.0000	1.0625	1.0544	—	1.0300	1.0266	0.0034	1.0030
$1\frac{1}{16}$-28 or 1.0625-28	UN	2A	0.0012	1.0613	1.0548	—	1.0381	1.0341	0.0040	1.0187
		3A	0.0000	1.0625	1.0560	—	1.0393	1.0363	0.0030	1.0199
$1\frac{1}{8}$-7 or 1.125-7	UNC	1A	0.0022	1.1228	1.0982	—	1.0300	1.0191	0.0109	0.9527
		2A	0.0022	1.1228	1.1064	1.0982	1.0300	1.0228	0.0072	0.9527
		3A	0.0000	1.1250	1.1086	—	1.0322	1.0268	0.0054	0.9549
$1\frac{1}{8}$-8 or 1.125-8	UN	2A	0.0021	1.1229	1.1079	1.1004	1.0417	1.0348	0.0069	0.9741
		3A	0.0000	1.1250	1.1100	—	1.0438	1.0386	0.0052	0.9762
$1\frac{1}{8}$-12 or 1.125-12	UNF	1A	0.0018	1.1232	1.1060	—	1.0691	1.0601	0.0090	1.0240
		2A	0.0018	1.1232	1.1118	—	1.0691	1.0631	0.0060	1.0240
		3A	0.0000	1.1250	1.1136	—	1.0709	1.0664	0.0045	1.0258
$1\frac{1}{8}$-16 or 1.125-16	UN	2A	0.0015	1.1235	1.1141	—	1.0829	1.0779	0.0050	1.0490
		3A	0.0000	1.1250	1.1156	—	1.0844	1.0807	0.0037	1.0505
$1\frac{1}{8}$-18 or 1.125-18	UNEF	2A	0.0014	1.1236	1.1149	—	1.0875	1.0828	0.0047	1.0572
		3A	0.0000	1.1250	1.1163	—	1.0889	1.0853	0.0036	1.0596
$1\frac{1}{8}$-20 or 1.125-20	UN	2A	0.0014	1.1236	1.1155	—	1.0911	1.0866	0.0045	1.0641
		3A	0.0000	1.1250	1.1169	—	1.0925	1.0891	0.0034	1.0655
$1\frac{1}{8}$-28 or 1.125-28	UN	2A	0.0012	1.1238	1.1173	—	1.1006	1.0966	0.0040	1.0812
		3A	0.0000	1.1250	1.1185	—	1.1018	1.0988	0.0030	1.0824
$1\frac{3}{16}$-8 or 1.1875-8	UN	2A	0.0021	1.1854	1.1704	—	1.1042	1.0972	0.0070	1.0366
		3A	0.0000	1.1875	1.1725	—	1.1063	1.1011	0.0052	1.0387
$1\frac{3}{16}$-12 or 1.1875-12	UN	2A	0.0017	1.1858	1.1744	—	1.1317	1.1259	0.0058	1.0866
		3A	0.0000	1.1875	1.1761	—	1.1534	1.1291	0.0043	1.0883

Source: Abstracted from Table 3A, *Unified Inch Screw Threads (UN and UNR Thread Form)*, ANSI B1.1-1982, American Society of Mechanical Engineers, New York. Reprinted with permission.

TABLE 6.11.2(b)   LIMITING DIMENSIONS FOR UNIFIED INCH THREADS FOR NOMINAL 1.000- TO 1.1875 ($1\frac{3}{16}$)-IN. DIAMETER SCREWS/BOLTS

Nominal Size and Threads/in.	Series Designation	Class	Internal						Major Diameter Min.
			Minor Diameter		Pitch Diameter				
			Min.	Max.	Min.	Max.	Tolerance		Min.
1-32 or 1.000-32	UN	2B	0.966	0.974	0.9797	0.9846	0.0049		1.0000
		3B	0.9660	0.9719	0.9797	0.9834	0.0037		1.0000
$1\frac{1}{16}$-8 or 1.0625-8	UN	2B	0.927	0.952	0.9813	0.9902	0.0089		1.0625
		3B	0.9270	0.9422	0.9813	0.9880	0.0067		1.0625
$1\frac{1}{16}$-12 or 1.0625-12	UN	2B	0.972	0.990	1.0084	1.0158	0.0074		1.0625
		3B	0.9720	0.9823	1.0084	1.0139	0.0055		1.0625
$1\frac{1}{16}$-16 or 1.0625-16	UN	2B	0.995	1.009	1.0219	1.0284	0.0065		1.0625
		3B	0.9950	1.0033	1.0219	1.0268	0.0049		1.0625
$1\frac{1}{16}$-18 or 1.0625-18	UNEF	2B	1.002	1.015	1.0264	1.0326	0.0062		1.0625
		3B	1.0020	1.0105	1.0264	1.0310	0.0046		1.0625
$1\frac{1}{16}$-20 or 1.0625-20	UN	2B	1.008	1.020	1.0300	1.0359	0.0059		1.0625
		3B	1.0080	1.0162	1.0300	1.0344	0.0044		1.0625
$1\frac{1}{16}$-28 or 1.0625-28	UN	2B	1.024	1.032	1.0393	1.0445	0.0052		1.0625
		3B	1.0240	1.0301	1.0393	1.0432	0.0039		1.0625
$1\frac{1}{8}$-7 or 1.125-7	UNC	1B	0.970	0.998	1.0322	1.0463	0.0141		1.1250
		2B	0.970	0.998	1.0322	1.0416	0.0094		1.1250
		3B	0.9700	0.9875	1.0322	1.0393	0.0071		1.1250
$1\frac{1}{8}$-8 or 1.125-8	UN	2B	0.990	1.015	1.0438	1.0528	0.0090		1.1250
		3B	0.9900	1.0047	1.0438	1.0505	0.0067		1.1250
$1\frac{1}{8}$-12 or 1.125-12	UNF	1B	1.035	1.053	1.0709	1.0826	0.0117		1.1250
		2B	1.035	1.053	1.0709	1.0787	0.0078		1.1250
		3B	1.0350	1.0448	1.0709	1.0768	0.0059		1.1250
$1\frac{1}{8}$-16 or 1.125-16	UN	2B	1.057	1.071	1.0844	1.0909	0.0065		1.1250
		3B	1.0570	1.0658	1.0844	1.0893	0.0049		1.1250
$1\frac{1}{8}$-18 or 1.125-18	UNEF	2B	1.065	1.078	1.0889	1.0951	0.0062		1.1250
		3B	1.0650	1.0730	1.0889	1.0935	0.0046		1.1250
$1\frac{1}{8}$-20 or 1.125-20	UN	2B	1.071	1.082	1.0925	1.0984	0.0059		1.1250
		3B	1.0710	1.0787	1.0925	1.0969	0.0044		1.1250
$1\frac{1}{8}$-28 or 1.125-28	UN	2B	1.086	1.095	1.1018	1.1070	0.0052		1.1250
		3B	1.0860	1.0926	1.1018	1.1057	0.0039		1.1250
$1\frac{3}{16}$-8 or 1.1875-8	UN	2B	1.052	1.077	1.1063	1.1154	0.0091		1.1875
		3B	1.0520	1.0672	1.1063	1.1131	0.0068		1.1875
$1\frac{3}{16}$-12 or 1.1875-12	UN	2B	1.097	1.115	1.1334	1.1409	0.0075		1.1875
		3B	1.0970	1.1073	1.1334	1.1390	0.0056		1.1875

Source: Abstracted from Table 3A, *Unified Inch Screw Threads (UN and UNR Thread Form)*, ANSI B1.1-1982, American Society of Mechanical Engineers, New York. Reprinted with permission.

Based upon the ultimate strength for grade 5 screws in Table 6.11.3 the endurance limit may be found from Chapter 4 to be approximately 39 000 psi. for a machined surface. From Table 6.11.3 we also find that the yield stress is 81 000 psi, so that from the Gerber-yield condition described in that chapter it follows that, with $\sigma_u = 105\,000$ psi,

$$\sigma_a = \frac{\sigma_e}{\zeta K_f}\left[1 - \left(\zeta \frac{\sigma_m}{\sigma_u}\right)^2\right] = 5110 \text{ psi} \tag{6.11.12}$$

It may be of interest to note that the Goodman-yield and the Soderberg relations would have given smaller values of permissible alternating stress, approximately 4514 and 4311 psi, respectively.

According to elementary strength of materials k_b of the bolt may be estimated to be

$$k_b = \frac{F_b}{\delta_b} = \frac{\sigma A}{\varepsilon l} = \frac{EA}{l}$$

Since the bolt holes are 1.30 in. in diameter, k_c may be estimated from (6.11.10) to be

$$k_c = \frac{\pi E 1.130}{2 \ln \dfrac{10 + 2.825}{2 + 2.825}} = 1.816E$$

Thus, with $k_b = F/\delta = AE/L = 0.8631E/2 = 0.4315E$

$$\frac{k_b}{k_b + k_c} = \frac{0.4315}{0.4315 + 1.816} = 0.1920$$

and, hence, the relation between the alternating load on the bolt and the alternating load on motor mount becomes

$$\Delta F = \frac{1}{0.1920}\,\Delta F_b = 5.208\,\Delta F_b$$

Since the Gerber-yield alternating stress corresponds to a fluctuating bolt load of

$$5110(0.8667) = 4429 \text{ lb}$$

it follows that the motor mount could, from the bolt failure condition, sustain an alternating force of

$$4429(5.2086) = 24\,486 \text{ lb}$$

Note that if the Goodman-yield criterion had been used the estimated limit on the alternating load would have been 3913 lb and if the Soderber criterion had been used it would have been 3736 lb.

TABLE 6.11.3 SAE MECHANICAL REQUIREMENTS AND IDENTIFICATION MARKINGS FOR BOLTS AND SCREWS

Grade Designation	Products	Nominal Size Dia., in.	Full Size Bolts, Screws, Studs, Sems — Proof Load (Stress), psi	Full Size — Tensile Strength (Stress) Min, psi	Full Size — Yield Strength (Stress) Min, psi	Machine Test Specimens — Tensile Strength (Stress) Min, psi	Machine Test — Elongation, % Min	Machine Test — Reduction of Area, % Min	Surface Hardness Rockwell 30N Max	Core Hardness Rockwell Min	Core Hardness Rockwell Max	Grade Identification Marking
1	Bolts, Screws, Studs	$\frac{1}{4}$ thru $1\text{-}\frac{1}{2}$	33 000	60 000	36 000	60 000	18	35	—	B70	B100	None
2	Bolts, Screws, Studs	$\frac{1}{4}$ thru $\frac{3}{4}$	55 000	74 000	57 000	74 000	18	35	—	B80	B100	None
		Over $\frac{3}{4}$ to $1\text{-}\frac{1}{2}$	33 000	60 000	36 000	60 000	18	35	—	B70	B100	
4	Studs	$\frac{1}{4}$ thru $1\text{-}\frac{1}{2}$	65 000	115 000	100 000	115 000	10	35	—	C22	C32	None
5	Bolts, Screws, Studs	$\frac{1}{4}$ thru 1	85 000	120 000	92 000	120 000	14	35	54	C25	C34	(head marking)
		Over 1 to $1\text{-}\frac{1}{2}$	74 000	105 000	81 000	105 000	14	35	50	C19	C30	
5.1	Sems, Bolts, Screws	No. 6 thru $\frac{5}{8}$	85 000	120 000	—	—	—	—	59.5	C25	C40	(head marking)
5.2	Bolts, Screws	No. 6 thru $\frac{1}{2}$	85 000	120 000	92 000	120 000	14	35	56	C26	C36	(head marking)
7	Bolts, Screws	$\frac{1}{4}$ thru $1\text{-}\frac{1}{2}$	105 000	133 000	115 000	133 000	12	35	54	C28	C34	(head marking)
8	Bolts, Screws, Studs	$\frac{1}{4}$ thru $1\text{-}\frac{1}{2}$	120 000	150 000	130 000	150 000	12	35	58.6	C33	C39	(head marking)
8.1	Studs	$\frac{1}{4}$ thru $1\text{-}\frac{1}{2}$	120 000	150 000	130 000	150 000	10	35	—	C32	C38	None
8.2	Bolts, Screws	$\frac{1}{4}$ thru 1	120 000	150 000	130 000	150 000	10	35	58.6	C33	C39	(head marking)

Selected notes: Grade 2 requirements apply to sizes $\frac{1}{4}$ through $\frac{3}{4}$ bolts and screws 6 in. and shorter in length and to all studs. Grade 1 requirements apply to bolts and screws longer than 6 in. Sems denote screw and washer assemblies.

Source: J429 AUG 83, 1988 SAE Handbook, © 1988 Society of Automotive Engineers, Inc. Reprinted with permission.

6.12 HIGH-STRENGTH DESIGNS

Stress distributions in screws and bolts in the vicinity of the nut and the vanish cone are affected by the direction of the forces on the parts held together, by the rigidity of the parts themselves, by the tolerances on the engaging threads, and by the contour of the vanish cone. Screw failure is also affected by the load cycles to which it is subjected, by the temperature and the temperature gradients over its length, the surface finish, the chemical environment at its location, and other such factors. Attention here, however, will be limited to screw, bolt, and nut design configurations, since they are under the direct control of the design engineer.

Bolt and nut design for high-strength applications has been concerned with finding shapes that reduce the stress concentration factors caused by abrupt changes in cross section and in load.

A major stress concentration is often found at the head of a bolt or screw, and another, which is often larger, is located at the beginning of the threaded section in general purpose bolts and screws that are loaded either in tension or in combined tension and torsion. Bolt failures at the second of these locations are shown in Figure 6.12.1.

Examination of the cross section for Metric Heavy Hex Bolts described in the ANSI B18.2.3.6M-1979 standard shows a relatively abrupt change in cross section below the head and at the beginning of the threaded portion as shown in Figure 6.12.2. Both the change in cross section and the associated stress concentration are reminiscent of the stress concentration found in a stepped round bar with a shoulder fillet discussed in Chapter 2. An improved design is shown in the Metric Heavy Hex Structural Bolts described in ANSI B18.2.3.7M-1979, as shown in Figure 6.12.3. In it the fillet radius below the head is increased and the associated stress concentration is thereby reduced.

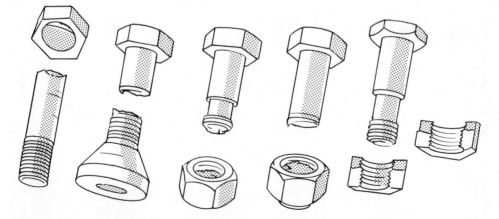

Figure 6.12.1 Bolt failures at the head, near the first thread, and in the vicinity of the nut. *Source: Heywood,* Designing Against Fatigue of Metals, *Van Nostrand Reinhold, 1962.*

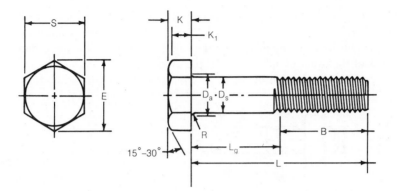

Figure 6.12.2 Metric heavy hex bolt configuration. *Source:* Metric Heavy Hex Bolts, *ANSI B18.2.3.6M, American Society of Mechanical Engineers, New York, 1979. Reprinted with permission.*

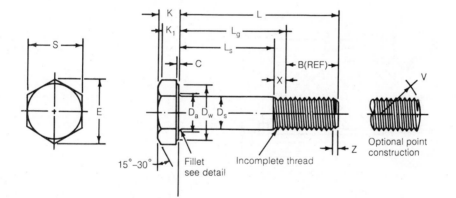

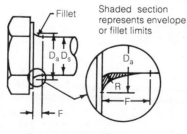

Enlarged detail of fillet

Figure 6.12.3 Metric heavy hex structural bolt with washer-faced hexagon head. *Source:* Metric Heavy Hex Structural Bolts, *ANSI B18.2.3.7M, American Society of Mechanical Engineers, New York, 1979. Reprinted with permission.*

Failure in the vicinity of the first thread is a common type of stress-induced break when bolts and screws produced to either standard. Reduction of the stress concentration at this location may be realized by increasing the length of the vanish cone so that the cone half angle is of the order of 15° or less, as pictured in Figure 6.12.4a, according to Heywood.[11] A greater reduction in the stress concentration factor results from

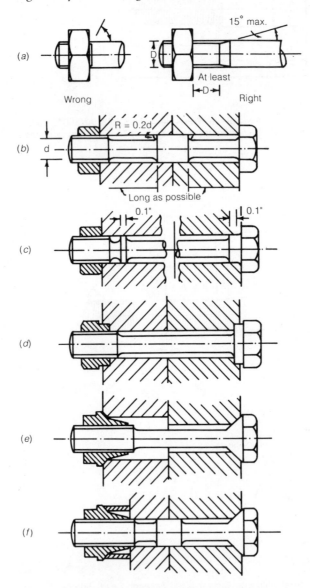

Figure 6.12.4 Shank design for high-strength bolts. *Source: Heywood*, Designing Against Fatigue of Metals, *Van Nostrand Reinhold, 1962.*

reducing the unthreaded shank diameter to the minor diameter of the threaded section, as illustrated in Figures 6.12.4b–f. In other words, a bolt having the larger diameter shank relative to the thread minor diameter as shown in Figure 6.12.2 is weaker in tension than one having any of the reduced cross sections shown in Figures 6.12.4b–f. The larger diameter sections along the shank length in Figures 6.12.4b, c, and f are to hold the bolts concentric with the bolt holes.

Shoulders in Figures 6.12.4b and c are to prevent lateral displacement of the bolt and to eliminate bending of the shank due to lateral wrench forces. Taper below the head is to reduce bending of the bolt shank if the surface against which the head is tightened is not perpendicular to the bolt axis. It also reduces the stress concentration at the bolt head by replacing the fillet with a tapered transition.

Further reduction of the stress concentration in the vicinity of the first thread may be obtained by nut redesign to make load application more gradual. Reduction of nut cross section in the vicinity of the first thread by tapering part of the threaded length, as in Figures 6.12.5a, c, d, and e, allows that end of the nut to elongate slightly under load and thus to transfer part of its load to the threads in the thicker portion of the nut. Comparison of the stress concentration at the first thread of a standard nut, as shown in Figure 6.12.5b, with the reduced stress concentration obtained by adding a

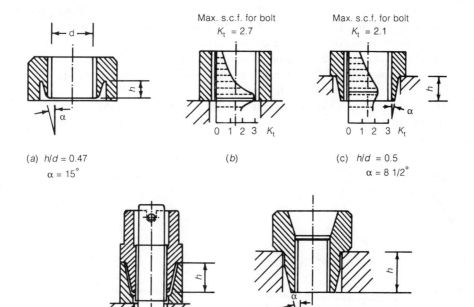

(a) h/d = 0.47
α = 15°

(b)

(c) h/d = 0.5
α = 8 1/2°

(d) h/d = 1.0; α = 11°

(e) h/d = 1.4; α = 15°

Figure 6.12.5 High-strength nut designs. *Source: Heywood,* Designing Against Fatigue of Metals, *Van Nostrand Reinhold, 1962.*

conical section, as shown in Figure 6.12.5c, demonstrates the beneficial effect of the conical portions of these nuts.

Heywood's investigation of thread profiles also indicated that rolled threads produced a stronger bolt than did machine cut threads, and that threads with a 90° included angle $(\alpha_1 + \alpha_2 = \pi/2)$ failed well below the nut face.[11]

6.13 SCREW RETENTION DESIGNS AND FITTINGS

So-called locking screws are screws with either the head or the threads modified to increase the torque required to loosen them. Head modification usually consists of adding a flange to increase μ_c and equipping the flange with ratchet-type teeth, inclined to bite into the mating surface when the screw is initially loosened, thus increasing μ_c in equation 6.5.15 (see Figure 6.13.1). This effective means of bolt retention is limited to applications where the screw need not be removed during routine maintenance because the teeth cut material from the mating surface each time the screw is forcibly removed.

Thread modification, as in Figure 6.13.2, increases both the installation and the removal torque by increasing μ_t, and consequently ϕ, in equation 6.5.15. Although their required loosening torque may not be as great as that of the serrated flange, they have the advantage of not cutting material from the mating face when they are removed. Which of the three means of increasing thread friction shown in Figure 6.13.2 should be used in a particular application depends upon the hardness of the mating threads, the operating temperatures of the machine, and the relative thermal coefficients of the screw and the threaded body into which it fits.

Lock washers, Figure 6.13.3, fit around a screw or bolt between the head and the mating surface and approximate the action of a serrated flange. Locking is achieved

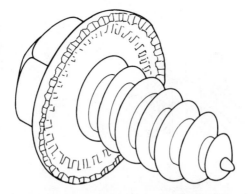

Figure 6.13.1 Bolt head with a serrated flange. *Source: Reprinted from* Machine Design, *Nov. 30, 1980. Copyright 1980 by Penton/IPC, Inc., Cleveland, OH.*

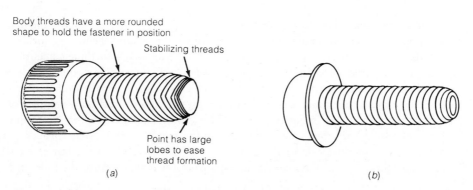

Body threads have a more rounded
shape to hold the fastener in position

Stabilizing threads

Point has large
lobes to ease
thread formation

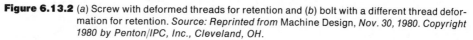

(a)   (b)

Figure 6.13.2 (a) Screw with deformed threads for retention and (b) bolt with a different thread deformation for retention. *Source: Reprinted from* Machine Design, *Nov. 30, 1980. Copyright
1980 by Penton/IPC, Inc., Cleveland, OH.*

Tooth lock washers

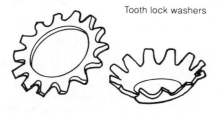

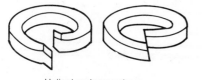

Helical spring washers

Figure 6.13.3 Lock washers. *Source: Reprinted from* Machine Design, *Nov. 30, 1980. Copyright 1980
by Penton/IPC, Inc., Cleveland, OH.*

by having sharp edges inclined to not oppose rotation when the screw is tightened, but to oppose counterrotation by biting into both the bolt head and the mating surface.

Nut designs to increase the removal torque may be divided into three categories: prevailing torque, free spinning, and mixed. Examples of prevailing torque nuts are shown in Figure 6.13.4. They are characterized by thread distortions or by thread inserts which increase the torque required for them to rotate in either direction. Examples of free-spinning nuts are given in Figure 6.13.5. Their common property is that no torque increase is observed until they make contact with the mating surface. The spring nut, Figure 6.13.5, is a modification of the concave face nut discussed in the last section. Not only does this nut transfer some of the load from the first thread to the following

Distorted Shape

Distorted portion of nut thread produces an interference fit. Center dimple allows nut to be assembled with either side up.

Top-crimp nut is easy to start but must be properly oriented before assembly.

Thread profile is distorted, increasing interference. Complex manufacturing process increases cost.

Nut hole, forced into an out-of-round shape after initial forming, produces spring action that maintains an interference fit after assembly.

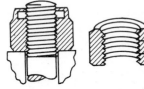

Insert

Plastic or metal section is added to the nut to increase mating-thread friction. In the most commonly used versions, a plastic (usually nylon), is added as a built-in washer, a plug through the side of the nut, or as a sprayed-on patch. Nylon is thought to provide damping action which imparts good vibration resistance. An insert may also be a soft metal such as lead. The added element plastically deforms when the nut is installed.

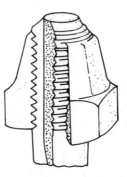

Esna All-Metal flexible slotted beams; applications to 1400°F

Metal-insert locknut has a projecting hardened wire or pin built in to provide a ratchet-like locking action. Reuse is limited by wear of the pin tip.

Figure 6.13.4 Prevailing torque nuts. *Source: Reprinted from* Machine Design, *Nov. 30, 1980. Copyright 1980 by Penton/IPC, Inc., Cleveland, OH.*

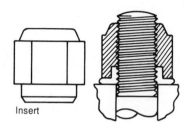

Insert

A plastic or metal washer built into the nut base is permanently deformed and grips the bolt threads when the nut is seated.

Serrated Face

Serrated or grooved face of nut digs into the bearing surface during final tightening.

Spring Head

When fully seated, the concave portion of the nut is forced inward and clamps against the bolt threads.

Captive Washer

Toothed or spring washer attached to the bearing face of the nut increases friction between the bearing surface.

Figure 6.13.5 Free-spinning nuts. *Source: Reprinted from* Machine Design, *Nov. 30, 1980. Copyright 1980 by Penton/IPC, Inc., Cleveland, OH.*

threads, but the flexure of the nut causes the sharp upper corners of the notch to dig into the screw threads, thus increasing the coefficient of friction at the threads.

Mixed types of nuts are displayed in Figure 6.13.6. The split nut, jam nuts, and set screw nuts may be considered as having mixed characteristics because they are free spinning until tightened at their service location. Jam nuts have the disadvantage of consisting of two parts which may require two operations for tightening. Thus, they usually are used where long travel of the nuts along the threaded shank favors free spinning, but where other design considerations, such as frequent disassembly, preclude one-piece free-spinning nuts.

Castle nuts differ from other lock nuts in that they must be used with a bolt which is drilled to receive a cross pin, wire, or cotter pin at approximately the final location

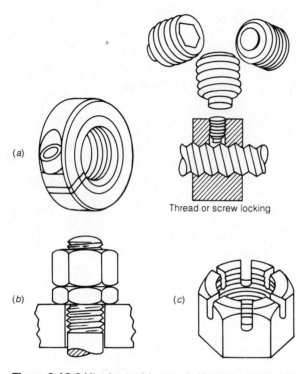

Thread or screw locking

(a)

(b)   (c)

Figure 6.13.6 Mixed nuts: (a) screw held nuts and collars, (b) jam nuts, and (c) castle nut. *Source: Reprinted from* Machine Design, *Nov. 19, 1981. Copyright 1981 by Penton/IPC, Inc., Cleveland, OH.*

of the nut. It is primarily used as a safety nut because it can be tightened in discreet steps only—in 60° increments for the nut pictured.

Anaerobic adhesive thread coatings (materials that become adhesive when deprived of oxygen) have also been used to increase thread friction once the nut and screw are assembled. Threads must be recoated whenever a screw is removed and replaced.

REFERENCES

1. ANSI B1.7-1984, *Nomenclature, Definitions, and Letter Symbols for Screw Threads*, American Society of Mechanical Engineers, New York, 1977.

2. ANSI B1.1-1982, *Unified Inch Screw Threads (UN and UNR Threaded Forms)*, American Society of Mechanical Engineers, New York, 1982.

3. ANSI B1.13-1983, *Metric Screw Threads—M Profile*, American Society of Mechanical Engineers, New York, 1983.

4.   ISO 68-1973(E), *ISO General Purpose Screw Threads—Basic Profile*, International Organization for Standardization, Switzerland, 1973.

5.   ANSI B1.5-1977, *Acme Screw Threads*, American Society of Mechanical Engineers, New York, 1977.

6.   ANSI B1.8-1977, *Stub Acme Screw Threads*, American Society of Mechanical Engineers, New York, 1977.

7.   ANSI B1.9-1973, *Buttress Inch Screw Threads*, American Society of Mechanical Engineers, New York, 1973.

8.   *Product Selection Guide, DSM-110*, Saginaw Steering Gear Div., General Motors Corp. Saginaw, MI, 1982.

9.   *Transrol High Efficiency Ball and Roller Screws*, Transrol, Chambery Cedex, France, undated.

10.   Orthwein, W.C., Designing and selecting screw threads, *Computers in Mechanical Engineering* (5), 49–52 (Mar. 1984).

11.   Heywood, R.B., *Designing against Fatigue of Metals*, Reinhold, New York, 1962.

12.   Almen, J.O., Design of highly stressed studs to improve their fatigue strength, *Product Engineering*, 288–290 (May 1943).

13.   Burr, A.H., *Mechanical Analysis and Design*, Elsevier, New York, 1981.

14.   Shigley, J.E., and Mitchell, L.D., *Mechanical Engineering Design*, McGraw-Hill, New York, 1983.

PROBLEMS

Section 6.1

6.1   Find the pitch and lead for
   a   a single-thread screw with $r_p = 5.22$ mm, $\lambda = 2.184°$, and
   b   a double-thread screw with $r_p = 7.11$ mm, $\lambda = 6.741°$.

6.2   A single Unified screw having 6 threads/in. is to be cut from a bar 1.10 in. in diameter. Find the pitch and the lead angle. What is the pitch diameter for the basic profile, i.e., with zero tolerance and allowance.

6.3   Repeat problem 6.2 for a Unified screw thread having two starts and 6 threads/in. Bar diameter is unchanged.

Sections 6.2 and 6.3

6.4   Find the crest, pitch, and root diameters of a M28 × 2 screw and a $1\frac{1}{2}$-12 UNF screw if the tolerances are set equal to zero. That is, $T = 0$ for H tolerance and for 3A tolerance. (*Hint:* These are the basic profiles shown in Figure 6.2.1*a*.)
   [*Ans.* 28.000 mm, 26.701 mm, 25.835 mm, 1.500 in., 1.4459 in., 1.4098 in.]

6.5   Repeat problem 6.4 for Acme and buttress thread forms.

Section 6.5

6.6 A testing machine of the construction shown applies tensile and compressive loading by means of a hydraulic cylinder acting upon the base plate as indicated in the figure. Since this cylinder has a piston travel of only 8 in. it is necessary to preposition the crosshead by means of the power screws shown. You, as chief engineer, have had complaints from the marketing department that the heavy crosshead often cannot be raised if it is driven full speed against a compressive test specimen in the process of positioning it for a compression test. When this happens either the test specimen must be broken before the crosshead can be moved or a 6-ft pipe wrench must be used to aid the torque motor in raising the crosshead. There is no speed control, the screws are chain driven, and full torque always acts on the power screws. What causes the problem? What relatively inexpensive design change would you order to eliminate the problem? Would the more expensive alternative of changing the lead angle of the power screws eliminate the problem?

6.7 Why is the diameter of the power screw in the pipe flaring tool shown in the figure as large as the largest outside flare diameter in the die plate? (*Hint:* Ease of operation is important.)

Sections 6.6 and 6.7

6.8 Justify the orientation of P_1 relative to the curve for $\phi(\lambda)$ shown in the figure. The figure is to be used to demonstrate convegence of the iteration process for solving

$$P_1 = \tan(\phi - \lambda) \quad \text{and} \quad P_2 = -\tan(\phi - \lambda)$$

Complete the demonstration by showing that the iteration formula for the first of these relations is

$$\lambda = \phi(\lambda) - \tan^{-1} P_1$$

and show that it is equivalent to a rectangular spiral that converges to Γ in the figure.

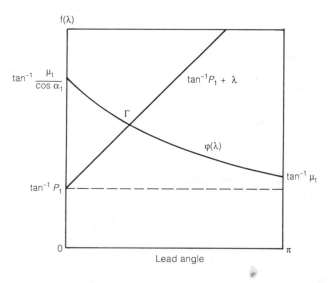

Sections 6.6–6.9

6.9 The frame of a C clamp is designed for a clamping force of 480 lb. Design a screw with a pitch diameter of 0.72 in. and a length of 8.50 in. to provide this clamping force at a torque of 40 in.-lb. The ball socket at the end has an effective mean collar radius of $0.8r_p$. Use Acme threads and expect $\mu_t = \mu_c = 0.10$. Specify the lead angle, the pitch, and the minimum root diameter necessary to prevent buckling to within a safety factor of 3.0. The yield stress is 80 000 psi, $E = 3 \times 10^7$ psi, and the ends are fixed–free.

[*Ans.* $\lambda = 2.717°$]

6.10 Find the minimum length of the threaded sleeve at one end of the C clamp in problem 6.9 for $D_s = 0.780$ in. and $d_s = 0.680$ in. The yield stress in shear for the screw is 41 000 psi and its bearing stress is 80 000 psi. Shear and bearing stress limits for the sleeve are 39 000 and 78 000 psi, respectively. Major diameter D_n of the internal thread in the sleeve is 0.788 in. and its minor diameter d_n is 0.688 in. Safety factor = 8.

[*Ans.* Min length: 0.107 in. for a single thread. Since this will not provide the fixed end condition assumed, increase the length to 1.0 in.]

6.11 Redesign the C clamp in problem 6.9 for the European market, where it will be rated for a load of 220 kg. The preferred pitch diameter is 18.0 mm, the unsupported length is 215 mm, and the torque should be 4.5 Nm. Major diameter D_n of the internal thread in the sleeve is 19.5 mm and the minor diameter d_n is 16.8 mm. The same steel will be used in the European product as in the U.S. product. Specify the lead angle, the pitch, the minimum root diameter to prevent buckling, and the number of threads at the end of the clamp for $D_s = \text{\&}_n$ mm and $d_n = d_s + 0.3$ mm.

[*Ans.* $\lambda = 2.731°$, $p = 2.697$ mm]

6.12 Repeat problem 6.9 for errors of 1.000, 0.1000, 0.0100, 0.0010, 0.0001, 0.000 01, and 0.000 001 and print the value of the counter after each solution. Note the rate of convergence for all five error values. Compare the number of iterations required at an error of 0.000 01 for $T = 480$ in.-lb with that for $T = 300$ in.-lb. The results may depend upon the computer and software used.

6.13 Find the tension in a M8 × 1.25 6h6h screw when tightened to a torque of 250 Nm. Assume the pitch diameter is that which would be calculated from a major diameter of 8.000 mm and calculate the tolerance on the major diameter. Due to lack of lubrication $\mu_t = \mu_c = 0.6$. The mean collar diameter is 10.1 mm.

[*Ans.* $\lambda = 3.168°$, and W $= 42\,917.1$ N]

6.14 Find the initial torque required to loosen the bolt in problem 6.13 for the coefficient of friction given. Also find the torque required if corrosion has increased the collar coefficient of friction to 0.85. Initial tension is 42 917.1 N.

[*Ans.* 214.20 Nm, 278.88 Nm]

6.15 Select a standard single-start Acme power screw to provide a clamping force of 15730 ± 50 lb when tightened with a torque of 1594 in.-lb. for $\mu_t = \mu_c = 0.10$ and a mean collar diameter of 0.8 in. Yield stress in tension for the steel used is 112 000 psi, the bearing stress is 100 000 psi, and the safety factor should be 1.7. (See Table P6.15.)

[*Ans.* $\frac{3}{4}$-6 Acme-5G]

6.16 The drawing shows the ball screw used in the power steering unit of an automobile. A track (hidden in this view) cut into the nut mates with a gear to steer the vehicle. Find the force on the rack applied by the driver through the nut–piston combination if the driver applies a torque of 20 in.-lb to the steering wheel. The pitch diameter of the ball screw is 1.70 in., the lead angle is 15°, $\mu_t = \mu_c = 0.001$, and $r_c = 1.1$ in. Assume the ball angle of contact is equivalent to a 30° flank angle. Hydraulic pressure on the piston augments the force on the rack from the steering wheel.

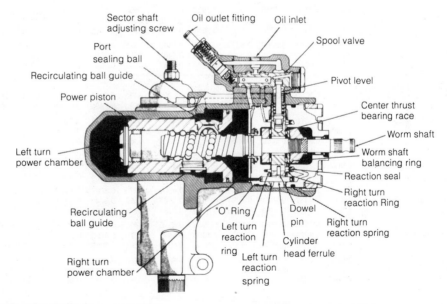

Source: Chrysler Corp.

Sections 6.1–6.9

6.17 Find the minimum minor diameter necessary for a power screw with SAE grade 8 material properties to support a maximum axial load of 14 450 kg at an extended, unconstrained, height

TABLE P6.15 LIMITING DIMENSIONS AND TOLERANCES, GENERAL PURPOSE SINGLE START ACME SCREW THREADS, CLASSES 2G, 3G, 4G, AND 5G

Limiting Diameters and Tolerances		Nominal Diameter (f)											
		$\frac{1}{4}$	$\frac{5}{16}$	$\frac{3}{8}$	$\frac{7}{16}$	$\frac{1}{2}$	$\frac{5}{8}$	$\frac{3}{4}$	$\frac{7}{8}$	1	$1\frac{1}{8}$	$1\frac{1}{4}$	$1\frac{3}{8}$
		Threads per inch*											
		16	14	12	12	10	8	6	6	5	5	5	4
EXTERNAL THREADS													
Classes 2G, 3G, 4G, and 5G, Major Diam	Max (D)	0.2500	0.3125	0.3750	0.4375	0.5000	0.6250	0.7500	0.8750	1.0000	1.1250	1.2500	1.3750
	Min	0.2450	0.3075	0.3700	0.4325	0.4950	0.6188	0.7417	0.8667	0.9900	1.1150	1.2400	1.3625
	Tol	0.0050	0.0050	0.0050	0.0050	0.0050	0.0062	0.0083	0.0083	0.0100	0.0100	0.0100	0.0125
Classes 2G, 3G, 4G, and 5G, Minor Diam	Max	0.1775	0.2311	0.2817	0.3442	0.3800	0.4800	0.5633	0.6883	0.7800	0.9050	1.0300	1.1050
Class 2G, Minor Diameter	Min	0.1618	0.2140	0.2632	0.3253	0.3594	0.4569	0.5372	0.6615	0.7509	0.8753	0.9998	1.0720
Class 3G, Minor Diameter	Min	0.1702	0.2231	0.2730	0.3354	0.3704	0.4692	0.5511	0.6758	0.7664	0.8912	1.0159	1.0896
Class 4G, Minor Diameter	Min	0.1722	0.2254	0.2755	0.3379	0.3731	0.4723	0.5546	0.6794	0.7703	0.8951	1.0199	1.0940
Class 5G, Minor Diameter	Min	0.1733	0.2266	0.2767	0.3391	0.3745	0.4738	0.5563	0.6811	0.7722	0.8971	1.0219	1.0962
Class 2G, Pitch Diameter	Max	0.2148	0.2728	0.3284	0.3909	0.4443	0.5562	0.6598	0.7842	0.8920	1.0165	1.1411	1.2406
	Min	0.2043	0.2614	0.3161	0.3783	0.4306	0.5408	0.6424	0.7663	0.8726	0.9967	1.1210	1.2188
	Tol	0.0105	0.0114	0.0123	0.0126	0.0137	0.0154	0.0174	0.0179	0.0194	0.0198	0.0201	0.0220
Class 3G, Pitch Diameter	Max	0.2158	0.2738	0.3296	0.3921	0.4458	0.5578	0.6615	0.7861	0.8940	1.0186	1.1433	1.2430
	Min	0.2109	0.2685	0.3238	0.3862	0.4394	0.5506	0.6534	0.7778	0.8849	1.0094	1.1339	1.2327
	Tol	0.0049	0.0053	0.0058	0.0059	0.0064	0.0072	0.0081	0.0083	0.0091	0.0092	0.0094	0.0103
Class 4G, Pitch Diameter	Max	0.2168	0.2748	0.3309	0.3934	0.4472	0.5593	0.6632	0.7880	0.8960	1.0208	1.1455	1.2453
	Min	0.2133	0.2710	0.3268	0.3892	0.4426	0.5542	0.6574	0.7820	0.8895	1.0142	1.1388	1.2380
	Tol	0.0035	0.0038	0.0041	0.0042	0.0046	0.0051	0.0058	0.0060	0.0065	0.0066	0.0067	0.0073
Class 5G, Pitch Diameter	Max	0.2188	0.2768	0.3333	0.3958	0.4500	0.5625	0.6667	0.7917	0.9000	1.0250	1.1500	1.2500
	Min	0.2160	0.2738	0.3300	0.3924	0.4463	0.5584	0.6620	0.7869	0.8948	1.0197	1.1446	1.2441
	Tol	0.0028	0.0030	0.0033	0.0034	0.0037	0.0041	0.0047	0.0048	0.0052	0.0053	0.0054	0.0059
INTERNAL THREADS													
Classes 2G, 3G, 4G, and 5G, Major Diam	Min	0.2600	0.3225	0.3850	0.4475	0.5200	0.6450	0.7700	0.8950	1.0200	1.1450	1.2700	1.3950
	Max	0.2700	0.3325	0.3950	0.4575	0.5400	0.6650	0.7900	0.9150	1.0400	1.1650	1.2900	1.4150
Classes 2G, 3G, 4G, and 5G, Minor Diam	Min	0.1875	0.2411	0.2917	0.3542	0.4000	0.5000	0.5833	0.7083	0.8000	0.9250	1.0500	1.1250
	Max	0.1925	0.2461	0.2967	0.3592	0.4050	0.5062	0.5916	0.7166	0.8100	0.9350	1.0600	1.1375
	Tol	0.0050	0.0050	0.0050	0.0050	0.0050	0.0062	0.0083	0.0083	0.0100	0.0100	0.0100	0.0125
Class 2G, Pitch Diameter	Min	0.2188	0.2768	0.3333	0.3958	0.4500	0.5625	0.6667	0.7917	0.9000	1.0250	1.1500	1.2500
	Max	0.2293	0.2882	0.3456	0.4084	0.4637	0.5779	0.6841	0.8096	0.9194	1.0448	1.1701	1.2720
	Tol	0.0105	0.0114	0.0123	0.0126	0.0137	0.0154	0.0174	0.0179	0.0194	0.0198	0.0201	0.0220
Class 3G, Pitch Diameter	Min	0.2188	0.2768	0.3333	0.3958	0.4500	0.5625	0.6667	0.7917	0.9000	1.0250	1.1500	1.2500
	Max	0.2237	0.2821	0.3391	0.4017	0.4564	0.5697	0.6748	0.8000	0.9091	1.0342	1.1594	1.2603
	Tol	0.0049	0.0053	0.0058	0.0059	0.0064	0.0072	0.0081	0.0083	0.0091	0.0092	0.0094	0.0103
Class 4G, Pitch Diameter	Min	0.2188	0.2768	0.3333	0.3958	0.4500	0.5625	0.6667	0.7917	0.9000	1.0250	1.1500	1.2500
	Max	0.2223	0.2806	0.3374	0.4000	0.4546	0.5676	0.6725	0.7977	0.9065	1.0316	1.1567	1.2573
	Tol	0.0035	0.0038	0.0041	0.0042	0.0046	0.0051	0.0058	0.0060	0.0065	0.0066	0.0067	0.0073
Class 5G, Pitch Diameter	Min	0.2188	0.2768	0.3333	0.3958	0.4500	0.5625	0.6667	0.7917	0.9000	1.0250	1.1500	1.2500
	Max	0.2216	0.2798	0.3366	0.3992	0.4537	0.5666	0.6714	0.7965	0.9052	1.0303	1.1554	1.2559
	Tol	0.0028	0.0030	0.0033	0.0034	0.0037	0.0041	0.0047	0.0048	0.0052	0.0053	0.0054	0.0059

* All other dimensions are given in inches. The selection of threads per inch is arbitrary and is intended for the purpose of establishing a standard.

Source: Abstracted from Table 11, *Acme Screw Threads*, ANSI B1.5-1977, American Society of Mechanical Engineers, New York, 1977. Reprinted with permission.

of 1 m if the lower end is fixed and the upper end is pinned. Use a safety factor of 1.8 and $E = 207$ GPa.

[*Ans.* $d_s = 33.581$ mm]

6.18   Repeat problem 6.17 for a maximum load of 4986 kg.

[*Ans.* $d_s = 25.738$]

6.19   Design a single-start Acme thread for the screw described in problem 6.17 for the conditions that the mean collar diameter $d_c = 1.3d_p$ and that a torque of 1900 Nm should raise the load when $\mu_t = 0.4$ and $\mu_c = 0.2$. Since d_p has not been given, the design engineer must select a reasonable pitch as well as a reasonable lead angle ($\lambda < 5°$). (*Hint:* Begin with $p = d_s/10$. Try several other values as well.)

[*Ans.* $\lambda = 2.4796°$, $d_p = 37.0$ mm, $p = 5.034$ mm]

6.20   Find the required nut length for the screw designed in problem 6.19 when $\sigma_y = 897$ MPa, $\sigma_b = 1000$ MPa, and $\tau_y = 517$ MPa. Use a safety factor of 1.2 so that in the event of an overload the nut threads may fail instead of the screw threads. Use $D_s = d_p + 0.70p$ and $d_n = d_p - 0.25p$.

[*Ans.* $L_e = 6.917$ mm, limited by shearing stress]

6.21   Design an Acme thread for the screw minor diameter selected in problem 6.18. Let $\mu_t = 0.2$, $\mu_c = 0.1$, $d_c = 1.3\,d_p$, and $T = 260$ N m. (See the hint following problem 6.19.)

[*Ans.* $p = 3.617$ mm, $d_p = 28$ mm, $\lambda = 2.35488°$]

6.22   Find the required nut length for the screw designed in problem 6.21 using the bearing and shear stresses, safety factor, and tolerances given in problem 6.20.

[*Ans.* $L_e = 6.489$ mm]

6.23   Repeat problem 6.19 using buttress threads. Use the pitch diameter ultimately selected in problem 6.19 and let $d_c = 1.4d_p$.

[*Ans.* $\lambda = 2.0279°$]

6.24   Find the minimum nut length for the screw designed in problem 6.23, a safety factor of 1.2, $\sigma_b = 1000$ MPa, and $\tau_y = 517$ MPa. Use no allowance and a tolerance of 0.15 mm on the screw major diameter and the nut minor diameter.

[*Ans.* $N_e = 4.607$ mm]

6.25   Repeat problem 6.21 when a larger torque of 393 N m is available to raise the load given in problem 6.18. Use $\sigma_b = 1000$ MPa, $\tau_y = 517$ MPa, and a safety factor of 1.2. Also find the required nut length and compare with problem 6.22. Larger threads, as used here, resist mechanical abuse better than small threads, but small threads require shorter nut lengths and less torque. This is the reason for the small threads on the pipe flaring tool in the figure with problem 6.7 and on threaded fittings used in hydraulic systems.

6.26   Find the torque required to lower the load in problem 6.25.

6.27   Design a ball or roller screw that is to have angular contact ball bearing mounts at each end to provide pinned–pinned conditions at each end. The distance between supports is to be 350 mm and the axial force on the ball/roller nut is to be 63 200 N or less. Recommend the screw type (ball or roller), the minor diameter, and the basic dynamic capacity required to achieve a life of 350 000 complete cycles over the full length of the screw. Use a safety factor of 2.0, a yield stress of 690 MPa, and an elastic modulus of 205 GPa. Let the pitch diameter be 6.0 mm larger than the minor diameter and let $p = L = 16$ for ball screws; let the pitch diameter be 3.0 mm larger than the minor diameter and let the pitch be 3.0 mm for roller screws.

[*Ans.* $r_{min} = 10.004$; select $r_s = 10.1$ mm, $r_p = 13.1$ mm; $C = 274\,200$ N for a planetary roller screw]

6.28 Design a ball or roller screw with 400 mm between angular contact ball bearings which is to work against a 8000-N axial force. Material properties and screw proportions are as given in problem 6.27 and the lead $L = 5$ mm.

[*Ans.* $r_{s_{min}} = 6.3354$ mm, $C = 30\,607$ N for a ball screw]

6.29 Find the power in kW required to advance a ball nut at 80 mm/s under a full load of 40 000 N using a lead angle of 5.5940°, and equivalent flank angle of 20°, and screw pitch diameter of 26 mm. The mean collar diameter is 48 mm and the friction coefficients for both collar and screw are 0.0032.

[*Ans.* 3.5 kW]

6.30 Repeat problem 6.29 for a roller screw for which the lead angle is 3.1011° and the friction coefficients for both collar and screw are 0.10. The pitch diameter is 26 mm and the collar mean diameter is 48 mm. The leading flank angle is 30°.

[*Ans.* 21.0 kW]

6.31 Compare the advance of the nut per degree of screw rotation relative to the nut for the ball and roller screws described in problems 6.30 and 6.31.

Sections 6.1–6.10

6.32 Design two power screws to drive a trash compactor which is to have a 2000-lb total compaction force over a distance of 24 in. with a safety factor of 2.00. Ends are fixed and pinned and the screws rotate at 172 rpm. Let $r_c = 1.5r_p$, $E = 3 \times 10^7$ psi, and $\mu_t = \mu_c = 0.10$. Nut major and minor diameters are to be selected according to $D_n = D_s + 0.005$ in. and $d_n = d_s + 0.005$ in. (a) Select the lead angle to allow the compactor to be powered by a $\frac{1}{2}$-hp gear motor: $\frac{1}{4}$-hp per screw. What is the advance rate of the pressure plate? Use Buttress threads. (b) Calculate the number of threads that must engage the screw when the bearing failure stress is 65 400 psi and the yield stress in shear is 20 500 psi. Use $\sigma_y = \sqrt{3}\,\tau_y$.

[*Ans.* $d_p = 0.6$ in., $\lambda = 3.0776°$]

6.33 Design the nut for a double-start (double-thread) power screw with stub Acme threads and having a minor diameter of 30 mm. It is to drive a cutter at 112 mm per minute using a 0.24 kW gear motor having an output rotation of 22 rpm. The gear train in the motor is 95% efficient (i.e., output torque is 95% of the input torque). If $\mu_t = \mu_c = 0.10$ and if $r_c = 20$ mm, find the maximum axial force at the cutter for your design. Use a safety factor of 1.60. Specify the number of threads required and the nut length to support the calculated load when the minor diameter of the nut is given by $d_n = d_s + 0.2$ mm. For the steel used $\sigma_y = 324$ MPa, $\tau_y = 180$ MPa, $\sigma_b = 280$ MPa, and $E = 206.8$ GPa.

6.34 Design a power screw having modified square threads to raise 14 000 lb a distance of 24 in. using a screw and nut of USN G10500 steel whose properties are found in Table 2.12.1 in Chapter 2. The screw is rigidly held at the base and not laterally constrained. Use a safety factor of 2.00, $r_c = 2.0r_p$, $\mu_t = 0.20$, and $\mu_c = 0.01$. Specify the lead angle, the pitch, the thread

depth, and the pitch diameter when the applied torque is to be 3000 in.-lb. Let $d_n = d_s + 0.03$ in.
[*Ans.* $\lambda = 1.3228°$]

6.35 Compare the difference in the required engagement length/1000 lb for a $1\frac{1}{8} - 20$ UN $- 3A$ bolt in a $1\frac{1}{8} - 20$ UN $- 3B$ nut to that of a $1\frac{1}{8} - 18$ UNEF $- 2A$ bolt in a $1\frac{1}{8} - 18$ UNEF $- 2B$ nut. For each $\sigma_b = 107\,000$ psi, $\tau_y = 91\,500$ psi, $\zeta = 1.2$, and $\mu_t = 0.5$. Use average pitch diameters.

6.36 During the repair of a machine a $1\frac{1}{8} - 12$ UNF $- 3A$ bolt was lost and was replaced by a $1\frac{1}{8} - 12$ UNF $- 1A$ bolt in a drilled hole that was tapped to $1\frac{1}{8} - 12$ UNF $- 3B$ standards. Calculate the rating factor for the repaired connection as based upon thread shear strength, where the rating factor is defined as

$$\text{rating factor} = \frac{\text{maximum load after repair}}{\text{maximum load before repair}}$$

Both bolts had the same engagement length and the same bearing and shear stress limits. Assume the loosest fit in each case.

6.37 Find the minimum length of the threaded hole to receive the bolt in problem 6.15 if the limiting bearing stress in the threaded block is $86\,000$ psi and if its yield stress in shear is $50\,000$ psi. The hole is drilled and threaded to class 5G standards.

6.38 Design the ten latch bolts on a pressure vessel access cover which is 600 mm is diameter. The pressure chamber is rated at 0.258 MPa with a safety factor of 2.00. The operator should be able to tighten the latch nuts with a torque of 30 N m or less for $r_c = 1.5r_p$. Specify the lead angle, the pitch, the pitch diameter, and the major and minor diameters for the bolts. Also specify the minimum engagement length of the latch nuts when $d_n = d_s + 0.2p$ and when the minimum value of D_s is given by $D_s = dp + 0.49p$. Use Acme threads and UNS G30400 stainless steel, whose properties are given in Table 2.12.1. Bearing stress is to be no more than 280 MPa, $\mu_t = \mu_c = 0.15$. $K = 4.0$, $\sigma_b = 320$ MPa.

6.39 Show that when acceleration is included the torque–load relation during acceleration or deceleration becomes

$$T = 2\pi I\dot{n} + \dot{n}Pd_p[r_p \tan(\phi + \lambda) + \mu_c r_c] \tan \lambda$$
$$+ W[r_p \tan(\phi + \lambda) + \mu_c r_c] \qquad \dot{n} = d/dt$$

where I denotes the moment of inertia of the rotating element (either the screw or the nut) and any other object rotating with it, m denotes the mass of the translating element (either the screw or the nut) and any other object moving with it. Angular acceleration is denoted by $\dot{n}$, in units of radians/s², and a single start is implied.

6.40 Show that if the screw is backdriving and the nut is accelerated the relation between the torque and the axial load is given by

$$-T = 2\pi I\dot{n} + m\dot{n}Pd_p[r_p \tan(\phi - \lambda) + \mu_c r_c] \tan \lambda + W[r_p \tan(\phi - \lambda) + \mu_c r_c]$$

where m denotes the number of starts and all other notation is as in problem 6.39.

Section 6.11

6.41 An overhead support bracket for a conveyor chain experiences cyclic loading as the product, each weighing approximately 217 kg, passes beneath the bracket. The load on the bracket when no product is under it is 31 kg. What preload, in newtons, would you recommend for the single bracket retainer bolt? Why? Would you recommend that the preload be 10% greater than the minimum load? Why?

6.42 What are the alternating and mean forces in newtons on the bolt in problem 6.41 for a preload of 2500 N if the cross-sectional area of the bolt is 558.11 mm^2 and if its stresses length is 26 mm? Here $E = 206.8$ GPa and the spring constant (spring rate) of the bracket base and beam flange together is $k_c = 12.429 \times 10^6$ N/mm.

   [*Ans.* $F_{b_m} = 2820.06$ N, $F_{b_a} = 240.04$ N]

6.43 After the main bolt on the hook assembly of a construction crane failed after several years of use the manufacturer told the contractor that the single bolt attaching it to the cable should have been preloaded. It was a $1\frac{1}{8} - 16$ UN $- 3$A bolt threaded into a $1\frac{1}{8} - 16$ UN $- 3$B nut with sufficient threads engaged to carry the load. The rated load for the hook assembly is 30 800 lb with a safety factor greater than 1.5 when the support bolt is preloaded with an axial load of 36 000 lb. Was the manufacturer correct in stating that a preload was necessary? What are the safety factors according to the Soderberg relation with and without a preload of 36 000 lb? The endurance limit is 41 000 psi and the yield stress is 80 000. What are they if the Gerber-yield criterion is used if $\sigma_u = 105\,000$ psi? The first thread is 1.25 in. from the bolt head and the head washer diameter is 1.125 in. $K = 3.0$, $q = 0.86$.

6.44 Show that as the preload increases beyond that necessary for the clamped parts to maintain contact at the maximum expected load, the safety factor initially increases, in qualitative agreement with Figure 6.12.2.

CHAPTER SEVEN

FASTENING AND JOINING

NOTATION

A_i — cross-sectional area of rivet/bolt i, area of weld i (l^2)

A_{bX} — bearing area, row X (l^2)

A_{pX} — plate area, row X, any plate (l^2)

A_{sX} — shear area, row X, any plane or planes (l^2)

a_i, b_i — components of r_i (l)

$\mathfrak{a}_i, \mathfrak{b}_i$ — components of r (l)

c_y, c_z — perpendicular distance from a stressed element to bending axes parallel to y and z axes, respectively (l)

E, E_0, E_i, E_c — elastic modulus (Young's modulus) $(m/[lt^2])$

$\mathbf{F}, F_i, F_x, S_{xy}$ — force vectors, force magnitudes, respectively (ml/t^2)

F' — component of $\mathbf{F}$ in the xy plane (ml/t^2)

$\mathfrak{f}_i, \tilde{\mathfrak{f}}_i$ — forces on rivet/bolt i (ml/t^2)

f_s — edge force (ml/t^2)

G — shear modulus $E/[2(1+v)]$ $(m/[lt^2])$

h — nominal fillet weld width (l)

I — moment of area (l^4)

$\mathfrak{I}$ — moment of area bending $(I/[h_0/\sqrt{2}])$

$\hat{\mathbf{i}}, \hat{\mathbf{j}}, \hat{\mathbf{k}}$ — unit vectors in the positive x, y, and z directions (l)

J — polar moment of area (l^4)

$\mathfrak{J}$ — $Jl/A = J/(h_0/\sqrt{2})$, h_0 is reference weld width (l^3)

$k_i, \mathfrak{k}_i$ — equivalent spring constants (m/t^2)

l, l_1, l_p, l_e — length (l)

$\mathbf{M}, \mathbf{M_R}$ — moment vector (ml^2/t^2)

$\mathfrak{m}$ — substitution variable

m, n — integers (1)

$\mathbf{r}_i$ — radius vector to the application point of force i from the center of area (l)

$\mathfrak{r}_i$ — radius vector to center of rivet/bolt i from the center of area (l)

S — in-plane shear force (ml/t^2)

s_1, s_2 — distance (separation) (l)

t_i, t_{o_1}, t_{o_2} — thickness of adherends (l)

u — displacement in the x direction (l)

$\bar{x}, \bar{y}$ — coordinates of the center of area of rivet, bolt, or weld array (l)

x_i, y_i — coordinates of the center of rivet/bolt i (l)

ε^*_{zx} — elastic strain at yield (1)

ε'_{zx} — plastic strain at failure (1)

ζ — safety factor (1)

η — derating factor (1)

θ — angle (1)

v, v_a — Poisson's ratio (1)

σ — direct stress magnitude $(m/[lt^2])$

σ_{u_i} — ultimate stress for inner adherend $(m/[lt^2])$

σ_{u_o} — ultimate stress for outer adherends $(m/[lt^2])$

σ_{xx} — direct stress in the x direction $(m/[lt^2])$

σ_{zx} — shear stress in the x direction $(m/[lt^2])$

τ, τ_D, τ_y — shear stress magnitude $(m/[lt^2])$

ϕ — angle of rotation (1)

Fastening and joining methods are classified either as point methods or as area methods in the following discussion because the analytical techniques are slightly different for the two and because their failure modes are not identical. This chapter will derive and use formulas for calculating the bolt, rivet, weld, and adhesive requirements to support loads in the plane of the plate as well as one case of forces out of the plane of the plate. We shall close by deriving and using formulas for the design of adhesive joints. All of these formulas and methods are important to the design of frames and housings to support loads transmitted from active machine elements such as bearings, brakes, gears, power screws, and motor or engine mounts. They are also used in the manufacture of some machine components such as sheaves, gears, sprockets, and couplings.

7.1   QUALITATIVE COMPARISON OF THE METHODS TO BE CONSIDERED

Before delving into the analytical methods for designing welds, bolted or screwed and riveted connections (point fasteners), and adhesives we shall list some of the primary characteristics of each as a guide to selecting which method to use in a particular application. The selection process does not always point to only one method because there are overlapping applications between welded joints and bolted/riveted joints, between riveted and brazed joints, and between soldered and adhesive joints.

In the following comparison we shall use the word bolt to denote either a bolt or a screw. The distinction between them is that a bolt is threaded into a nut, and a screw is threaded into a hole in something other than a nut. The hole may be threaded or the screw may make the threads as it is inserted.

Bolted Connections

Advantages

1. Thickness of the objects to be joined is limited only by the length of available bolts and the area available for bolt holes.

2. Bolted connections are easily disassembled for machine repair or inspection.

3. Dissimilar materials may be joined.

4. Extremely simple and inexpensive equipment is required (wrench, screw-driver, or power tools).

5. Minimum skills are required of the operator.

6. Bolting is cheaper than welding, brazing, or soldering.

7. Bolting does not introduce residual stresses or warping.

8. Bolting does not change the heat treatment of the objects being joined.

9. Bolts can be used to join laminated, plated, or other treated or composite materials.

Disadvantages

1. Joint is weaker than the object being joined.
2. Introduces stress concentrations at the holes.
3. Bolted joints are not fluid tight unless sealed. Sealant or gasket may fail with age.
4. Bolted joints may have poor electrical conductivity.
5. Bolted joints may loosen and may be vibration sensitive.
6. Corrosion may occur at the edge of the nut or bolt head.
7. Requires overlap or backing plates whenever plates are joined.
8. Bolted joints weaken or loosen under large temperature changes.

Welded Connections

Advantages

1. Welded connections are fluid tight when properly formed.
2. Welded connections have strength equal to that of the materials being joined when properly made.
3. Welding may be used to construct mechanical components that are difficult to form by any other process.
4. Welded components may be lighter than similar components of equal strength made by casting or assembled with bolted connections.
5. Good electrical and thermal conductivity.
6. Weld beads that are ground to a suitable contour may introduce no additional stress concentrations.
7. Strength and rigidity are not affected by temperature changes within the service range of the objects joined.

Disadvantages

1. Welding may change the heat treatment of the metals being joined.
2. Welded joints are either difficult or impossible to disassemble and reassemble.
3. Joined objects may be warped.
4. Welding may introduce residual stresses.
5. Only a very short list of dissimilar metals may be joined (*i.e.*, those whose alloy at the joint is as strong as the weaker of the two metals being joined).
6. Requires
   - Skilled operator
   - Special, relatively expensive equipment
   - Special, expensive equipment to check the quality of the weld

7. Weld defects, such as void or inclusions, may be undetected without the use of special equipment.

Brazing and Soldering

Brazing and soldering differ in the temperatures required for the filler metal used: Solder filler metals melt below 840 °F and brazing filler metals melt above 840 °F. They also differ in that solder is primarily used as a sealer in joints designed such that the sheet metal transfers the load across the joint, as in Figure 7.9.1. Its major applications are oil cooler and radiator seams. Brazed joints use filler metals that melt at temperatures above 840 °F and below the melting temperature of the materials to be joined.

Brazed Connections

Advantages

1. Dissimilar metals may be joined, depending upon the flux used.

2. Joints are liquid tight.

3. Less expensive equipment is required than with welding.

4. Complex components may be built in stages if each stage uses a filler having a lower melting point than that used in the previous stage.

5. Good thermal and electrical conductivity.

Disadvantages

1. Joint strength is usually less than the strength of the metals joined.

2. Service temperatures are limited by the lowest melting temperature of the filler metals used.

3. Metals to be joined must be cleaned before brazing begins.

4. Ventilation is required because some fluxes are toxic.

5. Warping may occur and residual stresses induced if the filler metal melting temperature is too high for the metals being joined.

7.2 FORCES IN THE BOLT/RIVET PLANE

We first consider the analysis of bolted or riveted joints in plates or sheets which are placed in tension. Following that analysis we consider the stresses on bolt/rivet arrays in anchor plates when they are subjected to eccentric loads in the plane of the bolts/rivets. The method of calculation applies equally well to the bolt patterns for equipment mounts and reaction frames.

To begin with the analysis of the potential failure modes for bolted or riveted joints in sheet material, we shall find that the occurrence of these failure patterns, or a combination of them, depends upon the relative strength of the connectors, of the plates,

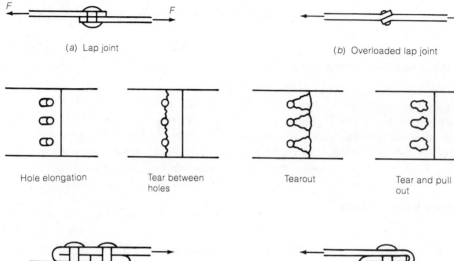

(a) Lap joint

(b) Overloaded lap joint

Hole elongation

Tear between holes

Tearout

Tear and pull out

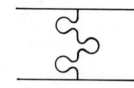

(c)

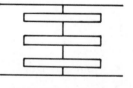

(d)

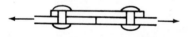

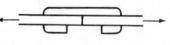

(e) Interlocked butt joint

(f) Stapled butt joint

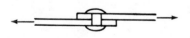

(g)

(h) Butt joint, single lap

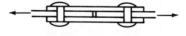

(i) Butt joint, double lap

Figure 7.2.1 Riveted joint failures and joint design.

and upon the hole locations and spacing relative to one another and to the edge of the plates.

Joint reinforcement may be gained by bending the plate, or sheet, back upon itself and joining with two rows of connectors, as in Figure 7.2.1c, or by folding one over the other, as in Figure 7.2.1d. A more expensive but thinner joint may be formed by contouring the mating edges as in Figure 7.2.1e and using a cover plate and connectors to restrain out-of-plane motion at the joint. Strip connectors, Figure 7.2.1f, are usually for light loads that neither cause the strip to bend nor cause the holes in the sheet material to elongate or tear. Greater overlap, Figure 7.2.1g, or addition of one or more plates, as in Figure 7.2.1h and k, may also produce a stronger joint. Reinforcing plates above and below the main plate which are joined by through connectors to eliminate the couple caused by the offset of the load-bearing main plates found in the joints in Figure 7.2.1a, c, and d.

Joints using through connectors (rivets, bolts) are weaker than the main plate even if stress concentrations at the holes are neglected. Load calculations for the possible failure modes identify the mode that is the most likely to fail. The strength of the joint in this weakest mode is taken as the strength of the joint. Comparison of joints may be in terms of *joint efficiency*, defined as the ratio, expressed as a percentage, of the tensile strength of the joint divided by the tensile strength of the plate.

EXAMPLE 7.2.1

Calculate the strength of a butt joint as illustrated in Figure 7.2.1g but with reinforcing plates of different widths. Also find the joint efficiency. Dimensions are given in Figure 7.2.2 in millimeters. Rivets are 6.0 mm in diameter. Failure stresses are to be taken as $\sigma_{zz} = 380$ MPa for tension failure in plates, $\sigma_{zz} = -450$ MPa for compression failure in plates, and $\sigma_{xz} = 300$ MPa for shear failure in connectors.

Failure analysis of riveted joints is based upon a *repeating strip* and *failure modes*. The repeating strip is the narrowest strip parallel to the direction of the applied force which contains a rivet, or point connector, pattern which may be repeated to give the fastener pattern for the entire joint. The width of this strip is known as the *long pitch*. Since the distance between rivets in any one row is known as the *pitch* for that row, it follows that if a joint contains several different pitches the long pitch will be the largest of these.

A failure mode is simply a particular way a joint may fail.

In Figure 7.2.2 both rows in the main plate have the same pitch, so that the pitch and the long pitch are equal. A repeating strip and the pitch are shown near the top of this figure. This joint has 8 failure modes, as itemized below.

1. Shear of connectors. Shear acts on one plane through connectors in row B and on two planes through connectors in row A, for a total shear area of $3\pi r_c^2$, where r_c is the radius of a rivet. Thus, the shear force becomes

$$F_1 = \sigma_{zx}(A_{sA} + A_{sB}) = 300(3)\pi(3)^2 = 25\,466.90 \text{ N}$$

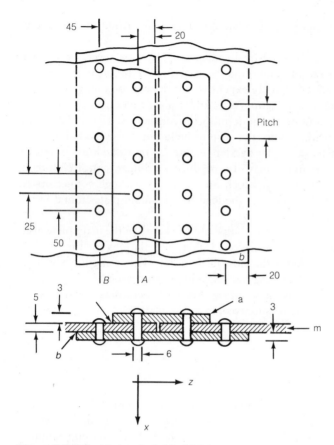

Figure 7.2.2 Double-lap riveted butt joint.

2. Tension failure between holes along row B in the main plate:

$$F_2 = \sigma_{zz}A_{pB} = \sigma_{zz}(p - d)t = 380(50 - 6)5 = 83\,600.00 \text{ N}$$

3. Tension failure in plate b between holes in row B and shear of connectors on row A:

$$F_3 = \sigma_{zz}A_{pB} + \sigma_{zx}A_{sA} = 380(50 - 6)3 + 300(2)\pi(3)^2 = 67\,124.60 \text{ N}$$

4. Shear of rivets along row B between plates b and m and hole elongation (known as bearing fatigue) along row A in plate m:

$$F_4 = \sigma_{zx}A_{sB} + \sigma_{zz}A_{bA} = 300\pi3^2 + 450(5)6 = 21\,982.30 \text{ N}$$

5. Shear of rivets between plates m and a along row A and tension failure in plate b between holes in row A:

$$F_5 = \sigma_{zx}A_{sA} + \sigma_{zz}A_{pA} = 300\pi(9) + 380(3)(50 - 6) = 58\,642.30 \text{ N}$$

6. Tension failure between holes in plate b along row B and hole elongation along row A:

$$F_6 = \sigma_{zz}A_{bB} + \sigma_{zz}A_{pA} = 380(50 - 6)3 + 450(5)6 = 63\,660.00 \text{ N}$$

7. Shear of rivets on row B and tension failure in the main plate between holes in row A. There is no need to calculate this force because it will be greater than that in mode 2.

8. Tearout at each rivet in plate b along row B and at each rivet in plate m along row A. This type of failure, shown at rivets 1, 2, and 3 in Figure 7.2.3, should not occur if the joint is designed according to the ASME boiler code. It is, therefore, seldom mentioned in machine design texts. The formula below has been given as an approximation to the shear tearout force,

$$F_8 = 2\sigma_{zz}(w_A + w_B)t = 2(380)(20 + 20)5 = 152\,000.00 \text{ N}$$

where w_A and w_B represent the distance from rivet centers in rows A and B, respectively, to the near edges of plates m and b. The distances are known as the *margins*.

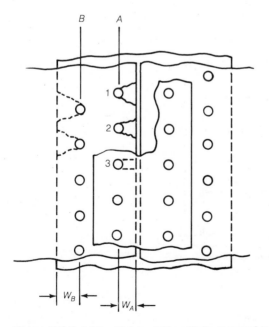

Figure 7.2.3 Joint with top plate partially removed to expose rivets at tearout failure in the main plate (1 and 2).

Based upon a tensile load of 95 000 N per 50-mm strip of plate, the joint efficiency may be calculated from

$$\text{joint efficiency (\%)} = \frac{\text{minimum joint strength}}{\text{main plate strength}} 100$$

$$= \frac{21\,982.30}{95\,000} 100 = 23.14\%$$

In remainder of this section we shall be concerned with forces that act in the plane of an array of bolts or rivets, such as may be used to secure an anchor plate or a machine frame to a flat surface. If the resultant of all of the forces does not pass through the center of area of the array of bolts or rivets its effect may be represented by a force F through the center of area and a moment M about the center of area. In particular, consider the rivet array shown in Figure 7.2.4 with loads F_1 and F_2 applied as shown. The center of area of the rivet array may be found relative to some arbitrarily chosen xy-coordinate system, such as the one shown in Figure 7.2.4, using the relations

$$x = \frac{\sum_{i=1}^{n} A_i x_i}{\sum_{i=1}^{n} A_i} \qquad y = \frac{\sum_{i=1}^{n} A_i y_i}{\sum_{i=1}^{n} A_i} \tag{7.2.1}$$

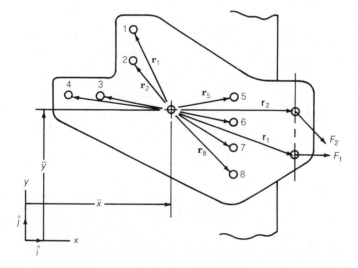

Figure 7.2.4 Eccentric loads on a riveted bracket.

where A_i is the area in shear of rivet i ($i = 1, 2, \ldots, n$) and x_i and y_i are the xy co-ordinates of the center of rivet i. Resultant force vector $\mathbf{F}$ given by $\mathbf{F} = \mathbf{F}_1 + \mathbf{F}_2$ and is assumed to induce a direct force $\mathbf{f}' = \mathbf{F}/n$ on each rivet and the moment $\mathbf{M} = \mathbf{r}_1 \times \mathbf{F}_1 + \mathbf{r}_1 \times \mathbf{F}_2$ is assumed to induce another force $\mathbf{f}_i$ in the plane of the rivets and perpendicular to $\mathbf{r}_i$, as shown in Figure 7.2.5. Since $\mathbf{f}_i$ is perpendicular to the radius vector $\mathbf{r}_i$ it follows that we may write $\mathbf{f}_i = \mathbf{c} \times \mathbf{r}_i$ where $\mathbf{c} = \mathbf{k}c$. This expression for $\mathbf{f}_i$ is based upon the usual assumptions that if the bracket were to rotate slightly about the center of area of the rivet array the shear strain would increase linearly with increasing radial distance from the center of the array. Hence, the force tending to shear the rivets will increase linearly with increasing distance from the array center. The magnitude of $\mathbf{c}$ is the constant of proportionality.

To solve for $\mathbf{c}$ let $\mathbf{r}_i$ be the radius to the rivet number i, where i can take on any value from 1 to n, and let the applied forces be denoted by $\mathbf{F}_i$ for i running from 1 to m. The resultant moment about the center of the array due to all of the applied forces may then be written as

$$\sum_{i=1}^{m} \mathbf{r}_i \times \mathbf{F}_i$$

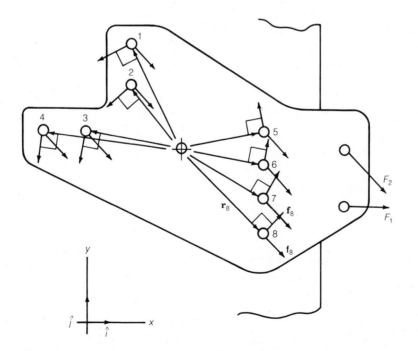

Figure 7.2.5 Display of vectors $\mathbf{r}_i$, $\mathbf{f}_i$, and f_i' at each rivet or bolt location.

and the resultant moment due to all of the reactions to forces $\mathbf{f}_i$ may be written as

$$-\sum_{i=1}^{n} \mathbf{r}_i \times (\mathbf{c} \times \mathbf{r}_i)$$

so that from the equilibrium requirement that the sum of the moments be zero we have that

$$\sum_{i=1}^{m} \mathbf{r}_i \times \mathbf{F}_i = \sum_{i=1}^{n} \mathbf{r}_i \times (\mathbf{c} \times \mathbf{r}_i)$$

Upon recalling the vector identity $\mathbf{A} \times (\mathbf{B} \times \mathbf{C}) = \mathbf{B}(\mathbf{A} \cdot \mathbf{C}) - \mathbf{C}(\mathbf{A} \cdot \mathbf{B})$ and noting that $\mathbf{r}_i \cdot \mathbf{c} = 0$ we may solve for $\mathbf{c}$ to find that

$$\mathbf{c} = \frac{\displaystyle\sum_{i=1}^{m} \mathbf{r}_i \times \mathbf{F}_i}{\displaystyle\sum_{i=1}^{n} \mathbf{r}_i \cdot \mathbf{r}_i}$$

After substituting for $\mathbf{c}$ into $\mathbf{f}_i$ we may write the vector sum of the forces acting on bolt/ rivet i as

$$\mathbf{f}_i = \frac{1}{n} \sum_{i=1}^{m} \mathbf{F}_i + \frac{\displaystyle\sum_{j=1}^{m} (\mathbf{r}_i \times \mathbf{F}_j) \times \mathbf{r}_i}{\displaystyle\sum_{i=1}^{n} \mathbf{r}_i \cdot \mathbf{r}_i} \qquad (7.2.2)$$

where the joint contains n rivets. In equation 7.2.2

$$\mathbf{r}_i = \hat{\mathbf{i}} a_i + \hat{\mathbf{j}} b_i \qquad (7.2.3)$$

where

$$a_i = X_i - \bar{x} \qquad (7.2.4)$$
$$b_i = Y_i - \bar{y} \qquad (7.2.5)$$

and

$$\mathbf{r}_i = \hat{\mathbf{i}} \alpha_i + \hat{\mathbf{j}} \beta_i \qquad (7.2.6)$$

where

$$\alpha_i = x_i - \bar{x} \qquad (7.2.7)$$
$$\beta_i = y_i - \bar{y} \qquad (7.2.8)$$

in terms of the X_i and Y_i coordinates of the points of application of forces $\mathbf{F}_i$ and the coordinates x and y of the center of the rivet array. Coordinates x_i and y_i locate the center of rivet i and $\mathbf{i}$ and $\mathbf{j}$ are unit vectors in the positive x and y directions, respectively.

The shear stress in rivet i having shear area A_i may be calculated from

$$\sigma_{z\theta} = \frac{1}{A_i}\,(\mathbf{f}_i \cdot \mathbf{f}_i)^{1/2} \qquad\qquad (7.2.9)$$

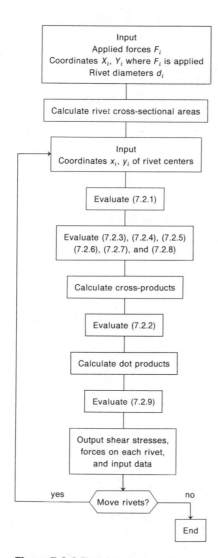

Figure 7.2.6 Flowchart for a program to solve equation 7.2.2 using equations 7.2.1 and 7.2.3 through 7.2.9 inclusive.

The sequence of rules for calculating the shear stress in a riveted connection given by earlier machine design texts is summarized in formulas 7.2.2 and 7.2.9.

An advantage of the explicit formula (7.2.2) is that we can easily see that a rivet joint should be designed to minimize the cross-product in the second term, which represents the moment of the external loads about the center of gravity of the rivets/ bolts. Whenever the bracket supports only a single load in a fixed direction this may be accomplished by arranging the bolts/rivets such that their center of gravity lies along the line of action of the single force so that the moment term vanishes. This is generally not possible when several loads act at different points on the bracket or when the load or loads change direction. In that event the center of gravity should be selected to minimize the moment term for the expected range of loading conditions.

It is easy to write a program to evaluate (7.2.9) using relations 7.2.1 through 7.2.8 and two other subroutines: one for evaluating the cross-product and the other for evaluating the dot product of two vectors, as described by the flowchart in Figure 7.2.6. The query *Move rivet?* is to allow rearrangement of the rivet locations without having to reenter the rivet dimensions and force locations if it is necessary to lower the shear stress on the most heavily loaded rivet.

EXAMPLE 7.2.2

Find the stress on the most heavily loaded rivet for the bracket shown in Figure 7.2.7. Rivet diameters are as shown, with all dimensions in millimeters.

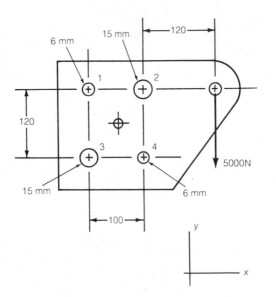

Figure 7.2.7 Riveted bracket.

Since the rivets of different diameter are symmetrically arranged about the center of the rectangle 1234, the center of area will lie at the center of the rectangle. Our calculations may be simplified by placing the origin of coordinates at the center of area so that $\bar{x} = 0$ and $\bar{y} = 0$. Thus,

$$\mathbf{r}_1 = -\hat{i}50 + \hat{j}60 \qquad \mathbf{r}_3 = -\hat{i}50 - \hat{j}60$$
$$\mathbf{r}_2 = \hat{i}50 + \hat{j}60 \qquad \mathbf{r}_4 = \hat{i}50 - \hat{j}60$$

and

$$\mathbf{r}_i = \hat{i}170 + \hat{j}60$$

After writing $\mathbf{F}_1 = -\hat{j}5000$ we find that

$$\sum_{i=1}^{n} r_i^2 = 4(50^2 + 60^2) = 24\,400 \text{ mm}^2$$

Since there is but a single force in this case, the sum in (7.2.2) becomes

$$\sum_{i=1}^{n} \mathbf{r}_i \times \mathbf{F}_i = \begin{vmatrix} \hat{i} & \hat{j} & \hat{k} \\ 170 & 60 & 0 \\ 0 & -5 & 0 \end{vmatrix} = -\hat{k}850 \text{ kN}$$

so that

$$\mathbf{f}_i = -\hat{j}1.250 - \hat{k}850 \times (-\hat{i}50 + \hat{j}60)/24\,400 = \hat{i}2.090 + \hat{j}0.492 \text{ kN}$$

as the force acting on rivet 1. Therefore,

$$\sigma_{z\theta_1} = 2147/(9\pi) = 75.94 \text{ MPa}$$

Similarly,

$$\sigma_{z\theta_2} = 3649/(56.25\pi) = 20.65 \text{ MPa} \qquad \text{for } \mathbf{f}_2 = \hat{i}2.090 - \hat{j}2.992 \text{ MPa}$$
$$\sigma_{z\theta_3} = 2147/(56.25\pi) = 12.15 \text{ MPa} \qquad \text{for } \mathbf{f}_3 = \hat{i}2.090 - \hat{j}0.492$$
$$\sigma_{z\theta_4} = 3649/(9\pi) = 129.06 \text{ MPa} \qquad \text{for } \mathbf{f}_4 = -\hat{i}2.090 + \hat{j}2.992$$

with rivet 4 being the most heavily loaded. Thus, it is evident nothing was accomplished by using larger rivets in positions 2 and 3. The effect of rearranging these rivets is considered in problems 7.12, 7.15, 7.16, and 17.18.

7.3 FORCES NOT IN THE BOLT/RIVET PLANE

Axial stress in bolts or rivets supporting a bracket, as in Figure 7.3.1a, subjected to loads not in the plane of the bolts or rivets may be estimated by assuming that the bracket will rotate as a rigid object. For simplicity we shall assume that the bracket is symmetrically loaded and that the force exerted by the individual rivets or bolts is determined by their individual elastic modulus, their cross-sectional area, and the amount that each rivet/bolt would elongate if the bracket were to rotate as a rigid body. We shall consider only two rows of rivet/bolts in the discussion to follow because little is to be gained in using more than two rows of fasteners.

Based upon the geometry shown in Figure 7.3.1, the equations of force and moment equilibrium may be written as

$$f_1 + f_2 + f_3 + f_4 = F_z + f_s \tag{7.3.1}$$

and

$$(f_1 + f_2)y_1 + (f_3 + f_4)y_3 = F_z y_f + F_y w \tag{7.3.2}$$

respectively.

The force on bolt/rivet i may be represented by

$$f_i = \sigma_{zz_i} A_i = \frac{y_i}{l_i} \phi E A_i = \phi y_i k_i \tag{7.3.3}$$

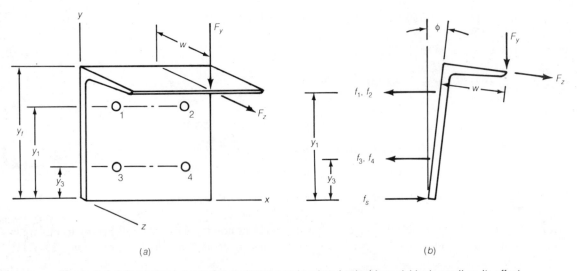

(a)   (b)

Figure 7.3.1 Geometry and out-of-plane forces on a bracket. Angle ϕ is vanishingly small, so its effect upon the geometry is negligible.

here

$$k_i = A_i E_i / l_i \tag{7.3.4}$$

In these expression y_i $(i = 1, 2, 3, 4)$ denotes the perpendicular distance of rivet/bolt i from the axis of rotation of the bracket (at the bottom of the bracket in Figure 7.3.1), E is the elastic modulus of the fasteners, l_i represents their active length (the length over which they may elongate), and ϕ denotes the infinitesimal angle of rotation of the rigid bracket about its edge at $y = 0$. As indicated in the figure, f_i represents the force in rivet/bolt i due to force components F_z and F_y of the resultant force F acting upon the bracket and to reaction f_s along the bottom of the bracket.

If all rivets/bolts are of the same material and have the same dimensions we may write $k_i = k$, so that substitution from equation 7.3.3 into equation 7.3.1 with $y_1 = y_2$ and $y_3 = y_4$ produces

$$\phi k(2y_1 + 2y_2) = F_z + f_s \tag{7.3.5}$$

and substitution into equation 7.3.2 leads to

$$\phi = \frac{F_z y_f + F_y w}{2k(y_1^2 + y_2^2)} \tag{7.3.6}$$

and to

$$f_s = 2k(y_1 + y_2)\phi - F_z \tag{7.3.7}$$

EXAMPLE 7.3.1

Find the reaction at the base of the bracket and the stresses in each of four rivets arranged as in Figure 7.3.2 when subjected to the load shown. All rivets are 4.0 mm in diameter and have an active length of 10 mm.

Upon taking components of the applied load we find $F_y = 100$ N and $F_z = 40$ N. According to the figure $y_1 = y_2 = 32$ mm, $y_3 = y_4 = 16$ mm, and $w = 18$ mm.

Substitution of these values into equations 7.3.4 and 7.3.5 yields

$$k = \frac{4\pi(206.8)10^3}{10} = 259873 \text{ N/mm}$$

and

$$\phi = \frac{40(40) + 100(18)}{2(259873)(1024 + 256)} = 5.1107 \times 10^{-6} \text{ rad}$$

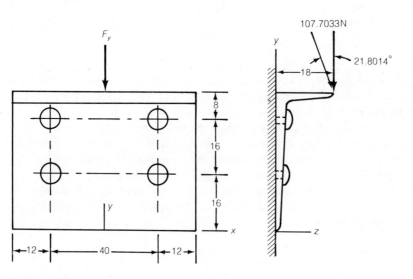

Figure 7.3.2 Bracket for Example 7.3.1. Dimensions are in mm. The origin of the *xyz*-coordinate system is at the center of the lower edge of the bracket.

Substitution into equation 7.3.7 yields

$$f_s = 2(259873)48(5.1107 \times 10^{-6}) - 40 = 127.501 \text{ N}$$

and substitution into equation 7.3.3 gives

$$f_1 = f_2 = y_1 \phi k = 32(5.1107 \times 10^{-6})259873 = 42.500 \text{ N}$$

and

$$f_2 = f_3 = 16(5.1107 \times 10^{-6})259873 = 42.500 \text{ N}$$

The corresponding tensile stresses in rivets 1 and 2 are

$$\sigma_{zz_1} = f_1/A = 42.500/(4\pi) = 3.382 \text{ MPa}$$

and in rivets 3 and 4 are

$$\sigma_{zz_3} = f_3/A = 21.250/(4\pi) = 1.691 \text{ MPa}$$

The shear stress is the same in all of the rivets because force F_y acts through the center of area of the rivets. Hence,

$$\sigma_{zy} = F_y/(4A) = 100/(16\pi) = 1.989 \text{ MPa}$$

7.4 WELDED CONNECTIONS

Welded joints provide the strongest bond of the area processes for joining metals, but they also require the most control in their formation. Many means are used to obtain a welded joint. They include gas (oxyhydrogen and oxyacetylene), electric arc, explosion, diffusion, cold, plasma arc, laser beam, thermit, ultrasonic, inertia, force, friction, electroslag, electron beam, high-frequency, and several less widely used techniques. Of these, only fusion welding using heat and perhaps a filler metal will be described as indicative of the variety of welding techniques available.

Fusion welding refers to a group of processes in which a joint is formed by bringing the contacting surfaces to their molten state, as in gas, electric arc, plasma arc, etc., with or without filler metal.

Filler metals are usually supplied as rods or coils, with the rod material encased in a solid flux sheath which contains one material that melts and another that vaporizes as the rod core is heated to its melting point. The vapor is intended to form an inert gas shield around the region where the molten core is deposited along the joint to prevent oxide formation. Any oxide that forms is expected to float to the surface of the molten flux. Upon cooling the flux forms a hard coating, or slag, which must be removed. Failure to float all oxides to the surface can cause inclusions in the weld that may induce serious stress concentrations.

7.5 ESTIMATION OF JOINT STRENGTH

Strength of a fillet weld, Figure 7.5.1, is usually calculated from the cross section of the weld and the strength of the filler metal used. The maximum shear theory of failure is traditionally assumed and the shear plane is assumed to bisect the angle

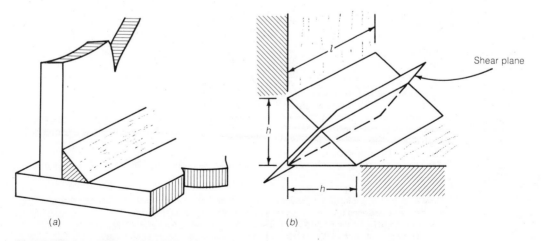

Figure 7.5.1 (a) Fillet weld and (b) shear plane.

TABLE 7.5.1   MECHANICAL PROPERTIES FOR ALL-WELD-METAL ARC WELDING ELECTRODES[a,b] [AWS A5.5-81]

AWS Classification[c]	Tensile strength, min[d]		Yield strength, at 0.2 percent offset[e]		Elongation min. percent
	ksi	MPa	ksi	MPa	
E7010-X					22
E7011-X					22
E7015-X					25
E7016-X	70	480	57[f]	390[f]	25
E7018-X					25
E7020-X					25
E7027-X					25
E8010-X					19
E8011-X					19
E8013-X					16
E8015-X	80	550	67	460	19
E8016-X					19
E8018-X					19
E8016-C3	80	550	68 to 80	470 to 550	24
E8018-C3					
E9010-X					17
E8011-X					17
E9013-X					14
E9015-X	90	620	77	530	17
E9016-X					17
E9018-X					17
E9018-M	90	620	78 to 90	540 to 620	24
E10010-X					16
E10011-X					16
E10013-X					13
E10015-X	100	690	87	600	16
E10016-X					16
E10018-X					16
E10018-M	100	690	88 to 100	610 to 690	20
E11015-X					
E11016-X	110	760	97	670	15
E10018-X					
E11018-M	110	760	98 to 110	680 to 760	20
E12015-X					
E12016-X	120	830	107	740	14
E12018-X					
E12018-M	120[g]	830[g]	108 to 120	745 to 830	18
E12018-M1					

Note: Refer to AWS A5.5-81 for the tables mentioned in the footnotes.

[a]   For the electrode classifications listed in Table 12, the values shown are for specimens which are tested in the as-welded condition. Specimens which are tested for all other electrodes are in the stress-relieved condition (see Table 11). The "M" classifications shall meet the requirements of equivalent strength EXX18-X classifications when tested on AC.

[b]   See Table 3 for sizes to be tested.

[c]   The letter suffix "X" as used in this table stands for all the suffixes (A1, B1, B2, etc.) except the M and C3 suffixes (see Table 2).

[d]   Single values shown are minimums.

[e]   Yield strength may be increased 5000 psi (35 MPa) max for 3/32 in. (2.4 mm) "M" classifications.

[f]   For the as-welded condition, the required yield strength is 60 ksi (415 MPa).

[g]   The minimum UTS for the E12018-M1 classification is nominally 120 ksi (830 MPa). However, the required UTS may be other than 120 ksi (830 MPa) as agreed between supplier and purchaser.

Source: Specification for Low Alloy Steel Covered Arc Welding Electrodes, AWS A5, 5–81, American Welding Society, Miami, FL. Reprinted with permission.

between the planes where the fillet weld joins the parent metal. Normally these planes are 90° apart so the shear plane lies at 45° to the weld interfaces and the area of the shear plane, Figure 7.5.1, also known as the *throat area*, becomes

$$A = \frac{h}{\sqrt{2}} l \tag{7.5.1}$$

for the standard fillet shown in Figure 7.5.1a, where l is the length of the weld, as illustrated. Fillet size is given in terms of dimension h, often termed the *leg* of the weld.

Filler material for welding is classified by the American Welding Society (AWS) according to the welding process used, the chemical composition of the material and the flux, the mechanical strength of the weld, and, in the case of arc welding, the orientation of the weld (flat, vertical, horizontal, overhead), the type of current (alternating or direct), and the polarity of the rod. Those AWS standards adopted by ANSI bear a joint heading, as shown in Ref. 4 and 5. Typical strengths of filler metals are given in Table 7.5.1 taken from Ref. 5. Similar data for low-alloy and aluminum filler material may be found in Ref. 6 and 7 and specifications for metals not listed here may be found in other AWS or ANSI/AWS standards.

7.6 FORCES IN THE WELD PLANE

The butt weld, Figure 7.6.1, with the load in the plane of the plate is the simplest to analyze. In it the weld thickness is equal to or greater than the thickness of the plates being joined. If the finished weld cross section is as shown in Figure 7.6.1a the

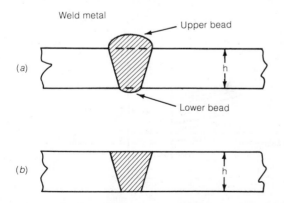

Figure 7.6.1 Butt weld. The ends of the plates are beveled so the entire weld may be made from one side of the plate. Length l is perpendicular to the plane of the page. (a) Weld before bead removed, (b) after bead removed by grinding.

force it can support is given by

$$F = \sigma h l / (\zeta K) \tag{7.6.1}$$

in terms of weld length l, plate thickness h, safety factor ζ, and stress concentration factor K. Since the stress concentration factor is caused by the weld bead (that part of the weld that is thicker than the plate), removal of the bead by grinding strengthens the weld by reducing K to 1.0. The stress concentration factors are usually estimated from Figure 7.6.2 with the caveat that these stress concentration factors may be as

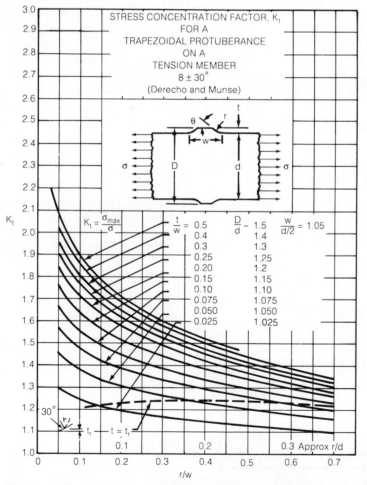

Figure 7.6.2 Stress concentration factors for butt welds. Protrusions were segments of a circle for K below the dashed line. *Source: R.E. Peterson*, Stress Concentration Factors, *Wiley, New York, 1974.* © *1974 by John Wiley & Sons, Inc. Reprinted with permission.*

much as 32% lower. The design engineer may increase the safety factor accordingly.

Next, consider a fillet weld, Figure 7.5.1, to attach a bracket supporting a single load, shown in Figure 7.6.3. The load induces a direct force and a moment about the welds' center of area. Thus, the force diagram in Figure 7.6.3a may be replaced by that in Figure 7.6.3b, where $\mathbf{M} = \mathbf{r} \times \mathbf{F}$.

The center of area may be calculated from the throat area of each weld according to formula 7.6.2 or 7.6.3 wherein x_i and y_i represent the coordinates of the center of weld i

$$\bar{x} = \frac{\sum\limits_{i=1}^{n} A_i x_i}{\sum\limits_{i=1}^{n} A_i} \qquad \bar{y} = \frac{\sum\limits_{i=1}^{n} A_i y_i}{\sum\limits_{i=1}^{n} A_i} \tag{7.6.2}$$

where $A_i = l_i h_i / \sqrt{2}$ is the area of weld i. Upon substitution for A_i in (7.6.2) and using

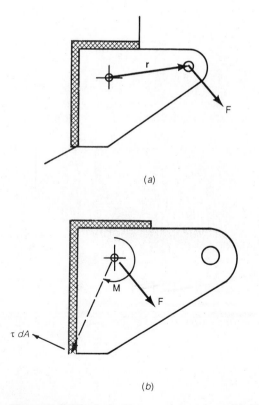

(a)

(b)

Figure 7.6.3 Equivalent force systems on a welded bracket.

$h_i = h_o(h_i/h_o)$ we have that

$$\bar{x} = \frac{\displaystyle\sum_{i=1}^{n} \frac{h_i}{h_o} l_i x_i}{\displaystyle\sum_{i=1}^{n} \frac{h_i}{h_o} l_i} \qquad \bar{y} = \frac{\displaystyle\sum_{i=1}^{n} \frac{h_i}{h_o} l_i y_i}{\displaystyle\sum_{i=1}^{n} \frac{h_i}{h_o} l_i} \tag{7.6.3}$$

in terms of the leg dimension of reference weld 0, which may be any one of welds 1 through n for those cases where the weld array contains welds of different sizes. Following analysis similar to that used for riveted/bolted connections, as shown in problem 7.5 through 7.7, the shear stress may be written as

$$\sigma_{xy} = \left\{ \left[\sum_{i=1}^{m} \left(\frac{\mathbf{F}_i}{A} + \frac{\mathbf{r}_i \times \mathbf{F}_i}{J} \times \mathbf{r} \right) \right] \cdot \left[\sum_{i=1}^{m} \left(\frac{\mathbf{F}_i}{A} + \frac{\mathbf{r}_i \times \mathbf{F}_i}{J} \times \mathbf{r} \right) \right] \right\}^{1/2} \tag{7.6.4}$$

where m forces act on a weld array of n welds. In this expression $\mathbf{r}$ is the radius from the center of area to the point where σ_{xy} is evaluated and

$$A = \frac{b_o}{\sqrt{2}} \sum_{i=1}^{n} \frac{b_i}{b_o} l_i = \frac{1}{\sqrt{2}} \sum_{i=1}^{n} h_i l_i \tag{7.6.5}$$

and

$$J = \frac{b_o}{\sqrt{2}} \sum_{i=1}^{n} \left(\frac{l_i^2}{12} + d_i^2 \right) \left(\frac{h_i}{h_o} \right) l_i = \frac{1}{\sqrt{2}} \sum_{i=1}^{n} \left(\frac{l_i^2}{12} + d_i^2 \right) h_i l_i \tag{7.6.6}$$

where d_i is the distance from the center of weld i to the center of area for all of the welds.

When all of the welds are the same size, as is usually the case, $h_o/h_i = 1$ and (7.6.4) may be written as

$$\sigma_{xy} \frac{b_o}{\sqrt{2}} = \frac{\mathfrak{F}}{l} = \left\{ \left[\sum_{i=1}^{m} \left(\frac{\mathbf{F}_i}{l} + \frac{\mathbf{r}_i \times \mathbf{F}_i}{\mathfrak{J}} \times \mathbf{r} \right) \right] \cdot \left[\sum_{i=1}^{m} \left(\frac{\mathbf{F}_i}{l} + \frac{\mathbf{r}_i \times \mathbf{F}_i}{\mathfrak{J}} \times \mathbf{r} \right) \right] \right\}^{1/2} \tag{7.6.7}$$

where

$$l = \sum_{i=1}^{m} l_i \quad \text{and} \quad J = \frac{b_o}{\sqrt{2}} \mathfrak{J}$$

This form is especially convenient for weld design because the fillet size may be easily determined from

$$b_o = \frac{\sqrt{2}}{\sigma_{xy}} \zeta \left(\frac{\mathfrak{F}}{l} \right) \tag{7.6.8}$$

where $\mathfrak{F}/l$ is the force/unit length of weld and ζ represents the factor of safety.

It is important to notice in equation 7.6.7 that the polar moment of area term $\mathfrak{J}$ has units of length cubed while the numerator has dimensions of force times length squared because this ratio implies that stresses in the weld may be reduced by careful arrangement of the welds. The $\mathfrak{J}$ term may be increased for a given length of weld by using a geometry where the welds are far from their center of gravity.

The numerator, which represents the moment due to external loads about the weld center of gravity, may be reduced by arranging the welds to place their center of gravity as close as possible to the lines of action of the external loads. When a single load acts in a fixed direction it may be possible to eliminate the moment term (cross-product term) by placing the weld center of gravity on the line of action of the force.

Relations 7.6.5, 7.6.6, and 7.6.7 are the relations commonly used in the welding industry, as described in Ref. 8, which also contains design formulas for a wide variety of mechanical and structural applications.

EXAMPLE 7.6.1

Recommend the fillet size and the All-Weld-Metal rod to be used to attach the anchor bracket shown in Figure 7.6.4 to a highway tow truck. It is to be made from $\frac{1}{4}$-in.-thick steel plate and the weld, which is to be X-rayed to guard against voids and inclusions, should be designed with a safety factor of 3. (If

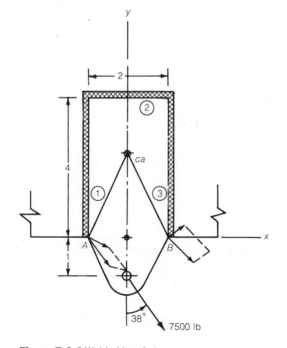

Figure 7.6.4 Welded bracket.

the weld is not to be checked for voids and inclusions it is advisable to use a safety factor of 6 or larger, especially if failure could endanger human life.)

Since the load is given in pounds, specify the weld size in inches. Place the origin of coordinates on the axis of symmetry of the weld array to have $\bar{x} = 0$. Thus

$$\bar{y} = \frac{\sum_{i=1}^{m} l_i y_i}{\sum_{i=1}^{3} l_i} = \frac{4(2) + 4(2) + 2(4)}{4 + 4 + 2} = 2.400 \text{ in.}$$

where, following industry practice, all welds are assumed to have equal throat dimensions. (Arrays of welds having different throat dimensions are significantly more expensive.) From equation 7.6.6 and the definition of $\mathfrak{J}$

$$\mathfrak{J} = \sum_{i=1}^{3} \left(\frac{l_i^2}{12} + d_i^2 \right) l_i = \left[2\frac{16}{12} + 2(0.4^2 + 1) \right] 4 + \left(\frac{4}{12} + 1.6^2 \right) 2 = 25.733 \text{ in.}^3$$

After writing

$$\mathbf{F} = 7500 \, (\hat{\mathbf{i}} \sin 38° - \hat{\mathbf{j}} \cos 38°)$$

we are able to evaluate

$$\mathbf{r} \times \mathbf{F} = \begin{vmatrix} \hat{\mathbf{i}} & \hat{\mathbf{j}} & \hat{\mathbf{k}} \\ 0 & 3.4 & 0 \\ 4617.461 & -5910.081 & 0 \end{vmatrix} = \hat{\mathbf{k}} 15\,699.367 \text{ in.-lb}$$

and then find that the maximum stress is at A, where $x = -1$, $y = -2.4$

$$(\mathbf{r} \times \mathbf{F}) \times \mathbf{r} = \begin{vmatrix} \hat{\mathbf{i}} & \hat{\mathbf{j}} & \hat{\mathbf{k}} \\ 0 & 0 & 15\,699.367 \\ -1 & -2.4 & 0 \end{vmatrix} = \hat{\mathbf{i}} 37\,678.482 - \hat{\mathbf{j}} 15\,699.367 \text{ lb-in.}^2$$

so that

$$\frac{\mathbf{F}}{l} + \frac{(\mathbf{r} \times \mathbf{F}) \times \mathbf{r}}{\mathfrak{J}} = \hat{\mathbf{i}} 461.746 - \hat{\mathbf{j}} 591.008 + \hat{\mathbf{i}} 1464.208 - \hat{\mathbf{j}} 610.087$$

$$= \hat{\mathbf{i}} 1925.954 - \hat{\mathbf{j}} 1201.095 \text{ lb/in.}$$

Substitution into the right side of equation 7.6.8 yields the following expression for the weld size:

$$h_o = \sqrt{2}(3) 2269.769 / \sigma_{xy} = 9629.885 / \sigma_{xy}$$

It is common practice to approximate the yield stress in shear by assuming that it may be obtained from either $\sigma_{xy_y} = \sigma_{xx_y}/2$, $(\tau_y = \sigma_y/2)$ or $\sigma_{xy_y} = \sigma_{xx_y}/\sqrt{3}$, based upon either the Mohr circle or HMH relations, respectively. We shall select the smaller yield stress given by

$$\sigma_{xy_y} = \sigma_{xx_y}/2$$

and calculate the fillet sizes for the following rods selected from Table 7.5.1.

American Welding Society Classification	σ_{xx} (psi)	σ_{xy} (psi)	h_o (in)
E7010-X to E7027-X	57000	28500	0.338
E8010-X to E8018-X	67000	33500	0.287
E9010-X to E9018-X	77000	38500	0.250
E10010-X to E10018-X	87000	43500	0.221

We recommend a rod from the E9010-X to E9018-X range because it provides a fillet having $h_o = 0.250$ in., which corresponds to the plate thickness. A smaller fillet would introduce a larger stress concentration factor because of the discontinuity between the thickness of the fillet and the thickness of the plate and a larger fillet is obviously useless because the plate itself is only 0.250 in. thick. Selection of a particular rod within the selected range depends upon the manufacturer's data for particular rods.

7.7 FORCES NOT IN THE WELD PLANE

To avoid numerical calculations that are beyond the scope of this text, we assume that both the representative bracket and the surface to which it is welded do not deform. For discussion purposes, consider the weld arrays shown in Figure 7.7.1b–d. Force/unit length due to force components in the x-direction is considered to contribute to tensile stress and may be approximated by

$$\frac{h_o}{\sqrt{2}} \sigma_{xx} = \frac{\sum_{i=1}^{n} \mathbf{F}_i \cdot \hat{\mathbf{i}}}{l} \tag{7.7.1}$$

and the force/unit length acting in the plane of the welds may be found from the methods of the previous section. If the resultant forces act through the center of gravity of the weld array, their contribution is

$$\frac{h_o}{\sqrt{2}} \sigma_{yx} = \frac{1}{l} \left[\sum_{i=1}^{n} (\mathbf{F}_i \cdot \hat{\mathbf{k}})^2 + (\mathbf{F}_i \cdot \hat{\mathbf{j}})^2 \right]^{1/2} \tag{7.7.2}$$

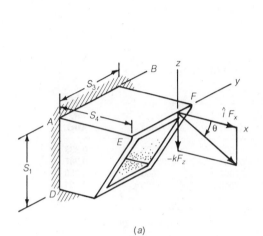

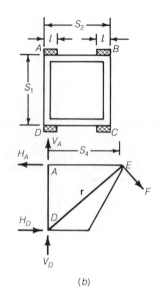

(a)

(b)

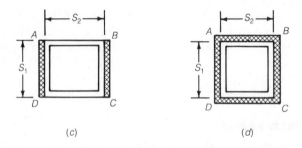

(c)   (d)

Figure 7.7.1 Bracket with weld arrangements.

Tensile stress due to bending may be approximated by

$$\frac{h_o}{\sqrt{2}}\,\sigma_{xx} = \sum_{i=1}^{m}\left(\frac{\mathbf{M}\cdot\hat{\mathbf{k}}c_z}{\mathfrak{I}_z} + \frac{\mathbf{M}\cdot\hat{\mathbf{j}}c_y}{\mathfrak{I}_y}\right) \tag{7.7.3}$$

These stresses may be combined according to the HMH criterion (also known as the distortional-energy theory) to obtain

$$\frac{h_o{}^2}{2}\left(\frac{\sigma_{yp}}{\zeta}\right)^2 = \left[\sum_{i=1}^{n}\frac{\mathbf{F}_i\cdot\hat{\mathbf{i}}}{l} + \sum_{i=1}^{m}\left(\frac{\mathbf{M}_i\cdot\hat{\mathbf{k}}c_z}{\mathfrak{I}_z} + \frac{\mathbf{M}_i\cdot\hat{\mathbf{j}}c_y}{\mathfrak{I}_y}\right)\right]^2$$
$$+ 3\sum_{i=1}^{n}\left[\left(\frac{\mathbf{F}_i\cdot\hat{\mathbf{k}}}{l}\right)^2 + \left(\frac{\mathbf{F}_i\cdot\hat{\mathbf{j}}}{l}\right)^2\right] \tag{7.7.4}$$

where

$$\mathfrak{I}_z = \sum_{i=1}^{k} \mathfrak{I}_{z_i} \qquad \mathfrak{I}_y = \sum_{i=1}^{k} \mathfrak{I}_{y_i} \qquad l = \sum_{i=1}^{k} l_i \tag{7.7.5}$$

in which

$$\mathfrak{I}_{z_i} \frac{h_o}{\sqrt{2}} = \frac{h_o}{\sqrt{2}} \left(\frac{1}{12} l_i^2 + d_i^2 \right) l_i \left(\frac{h_i}{h_o} \right) = I_{z_i} \tag{7.7.6}$$

for these welds parallel to the z axis (i.e., perpendicular to the axis of the bending moment), and

$$\mathfrak{I}_{z_i} \frac{h_o}{\sqrt{2}} = \frac{h_o}{\sqrt{2}} d_i^2 \, l_i \left(\frac{h_i}{h_o} \right) = I_{z_i} \qquad , \tag{7.7.7}$$

for those welds perpendicular to the z axis (i.e., parallel to the axis of the bending moment). Similar definitions hold for welds parallel and perpendicular to the y axis. As implied by the summation limits, these relations hold for n forces, m moments, and k welds, where all of these welds are assumed to be either parallel or perpendicular to the z axis. When equations 7.7.3 and 7.7.4 are applied to the bracket shown in Figures 7.7.1a and b, quantities c_y and c_z are given by $c_y = s_2/2$ and $c_z = s_1/2$ and $l = 4l_1$ where $l_1 = l_2 = l_3 = l_4$ for the array shown. In Figure 7.7.1c $l = 2s_1$ and in Figure 7.7.1d $l = 2s_1 + 2s_2$. Generally $(h_i/h_o) = 1$ because different weld sizes are expensive.

EXAMPLE 7.7.1

Design a bracket to hold the metal end straps for a band brake and to be attached to an anchor plate supplied by the user. The two pins are loaded as shown in Figure 7.7.2 and held in place with a moderate press fit. The base plate is to be bolted to the user's anchor plate using bolts that are SAE Grade 2 or stronger. Use a safety factor of 2.0 relative to the yield stress when calculating plate cross sections and a factor of 4.0 when calculating weld size. The base and cap plates are to be made from UNS G10300 steel and the welding rod will be AWS 7011-X or stronger.

Since the lines of action of the two forces will intersect near the pins, we may eliminate the moment term in equation 7.6.7 by placing the center of gravity of the welds on the line of action of the resultant force through that intersection point. Our first step, therefore, will be to calculate the location of the intersection of the lines of action using the geometry of Figure 7.7.3a and the law of sines

$$\frac{4}{\sin 85°} = \frac{a}{\sin 35°} = \frac{b}{\sin 60°}$$

to obtain

$$a = 2.3031 \text{ in.} \qquad b = 3.4773 \text{ in.}$$

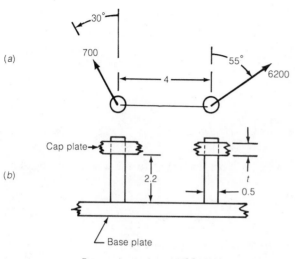

(a)

(b)

Base and cap plates UNSG10300 steel

Figure 7.7.2 Pin height and location requirements for Example 7.7.1. (a) Top view of the pins, (b) side view of the pins. Loads are in pounds, dimensions in inches.

Once we know the direction of the resultant force we may arrange the weld lines parallel and perpendicular to this line to produce a smaller bracket and to simplify our calculation. The resultant force vector $\mathbf{F}_3$, in pounds, is the sum of

$$\mathbf{F}_1 = 700(-\mathbf{i}\cos 60° + \mathbf{j}\sin 60°)$$

$$\mathbf{F}_2 = 6200(\mathbf{i}\cos 35° + \mathbf{j}\sin 35°)$$

so that

$$\mathbf{F}_3 = \mathbf{F}_1 + \mathbf{F}_2 = \mathbf{i}4728.74 + \mathbf{j}4162.39$$
$$= 6299.72(\mathbf{i}\cos 41.3553° + \mathbf{j}\sin 41.3553°)$$

We shall arrange the weld pattern symmetrically about this line.

We shall next turn to the task of selecting the thickness for the upper and lower plates and begin by turning to Appendix A, Figure 3.A.16, to find the relations between the hole diameter, the plate width, and the stress concentration due to a pin in a tightly fitting hole. To keep the bracket fairly narrow, we shall accept a stress concentration factor of 2.0 to have an a/w ratio of 0.4, where a in Figure 3.A.16 is the pin diameter and w is to total width of the plate. Hence, our pin diameter of 0.5 in. requires a plate width of 1.25 in. The corresponding plate thickness may be calculated from the formulas in Figure 3.A.16, where P denotes the load. Thus, after insertion of safety factor ζ, we have

$$\sigma_{\max} = KP\zeta/(ah)$$

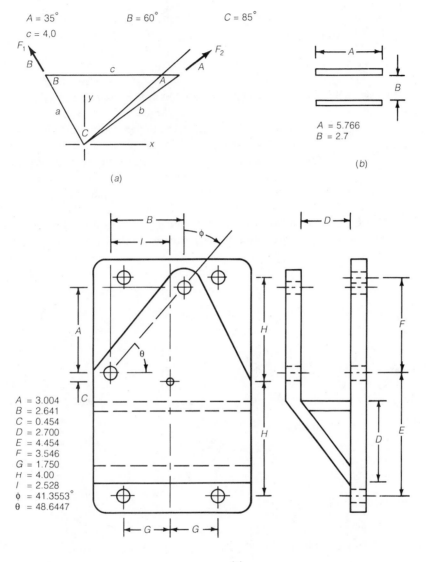

$A = 35°$ $B = 60°$ $C = 85°$
$c = 4,0$

(a)

$A = 5.766$
$B = 2.7$

(b)

$A = 3.004$
$B = 2.641$
$C = 0.454$
$D = 2.700$
$E = 4.454$
$F = 3.546$
$G = 1.750$
$H = 4.00$
$I = 2.528$
$\phi = 41.3553°$
$\theta = 48.6447$

(c)

Figure 7.7.3 (a) Forces and their lines of action, (b) weld distribution on the base plate, (c) brake bracket configuration.

as the formula for σ_{max} in terms of K and plate thickness h. From Table 2.2.3 we find the yield stress of UNS G10300 steel to be 50 000 psi, so that

$$h = KP\zeta/(a\sigma_{max}) = 2(3150)2/[0.5(50\,000)] = 0.504 \simeq 0.500 \text{ in.}$$

where we have reasonably assumed that half of the $6299.72 \simeq 6300$-lb load will be carried by the upper plate and half by the lower plate.

Clearly we are free to choose any weld lengths we wish, with the understanding the longer the length the smaller the cross section. For simplicity of cutting the plates we shall select the dimensions shown in Figure 7.7.3b in order to have the center of the hole $1.25/2 = 0.625$ in. from the edge of the plate, as it would be if it were centered in a plate 1.25 in. wide. We shall transfer tension from the cap plate to the base plate using a vertical plate and a $45°$ plate as shown in the side view in Figure 7.7.3c with welds as shown.

These welds will experience both tension and shear forces, which may be estimated from expressions 7.6.7, 7.7.3, and 7.7.4 after the weld lengths are found from the geometry shown in Figure 7.7.3c. Thus, with $c = 1.35$ in.,

$$L = 2(5.766) = 11.532 \text{ in.}$$

$$I = 5.766(1.35^2)2 = 21.0171 \text{ in.}^3$$

and

$$M = 6299.72(2.2 + 0.25)/2 = 7717.157 \text{ in.-lb}$$

we find that

$$\frac{Mc}{I} = \frac{7717.157(1.35)}{21.0171} = 495.699 \text{ lb/in.}$$

and

$$\frac{F}{l} = \frac{6299.72}{11.532} = 546.282 \text{ lb/in.}$$

Substitution of these values into equation 7.7.4 yields

$$\frac{h_o{}^2}{2}\left(\frac{\sigma_{yt}}{\zeta}\right)^2 = 495.699^2 + 3(546.282^2)$$

so that

$$h_o = \sqrt{2}(4)\sqrt{3}(1068.171)/57\,000 = 0.1836 \text{ in.} \simeq 3/16 \text{ in.}$$

Finally, we need to select the bolt sizes needed to fasten the base plate to the anchor frame on the user's machine. Because the force on the pins acts at 1.6 in. above the bottom surface of the base plate, we must consider both shear and tensile forces on the bolts which fasten the base plate to the user's anchor plate, which we shall assume to be 0.5 in. thick.

We shall use four bolts of the same size to fasten the base plate to the anchor plate. We will first try four $\frac{3}{8}$ – 16 UNC bolts. To estimate the maximum tensile stress we use equation 7.3.4 to find k to be

$$k = 0.0775(3 \times 10^7)/1.0 = 2.325 \times 10^6$$

The bolt's stress area of 0.775 in.^{2} was obtained from Table 6.4.2 and the bolt length is the combined thickness of the base and anchor plates. From Figure 7.7.2b we find that $y_1 = y_2 = 7.86$ and $y_3 = y_4 = 0.7$ so that from equation 7.3.6 we have

$$\phi = \frac{6299.72(1.6)}{2(2.325 \times 10^6)(61.780 + 0.49)} = 3.4810 \times 10^{-5}$$

Substitution into equation 7.3.3 for the largest forces, f_1 and f_2, we find

$$f_1 = f_2 = 3.481 \times 10^{-5}(7.86)2.325 \times 10^6 = 636.14 \text{ lb}$$

so that

$$\sigma_{xx} = 636.143(2.7)/0.0775 = 22\,162.40 \text{ psi}$$

and

$$\sigma_{xy} = 6299.72/[4(0.0775)] = 20\,321.68 \text{ psi}$$

where we have used a stress concentration factor of 2.7 for the screw threads in tension.

We employ the Huber-von Mises-Hencky criterion, equation 4.4.2, for yielding. From it

$$\left(\frac{\sigma_{yt}}{\zeta}\right)^2 = 22\,162.40^2 + 3(20\,321.68^2)$$

or

$$\frac{\sigma_{yt}}{\zeta} = \frac{57\,000}{2} < 41\,594.28$$

Having found that the $\frac{3}{8}$ – 16 UNC is too small, we next try bolts from the UNF series because they have larger tensile stress areas than UNC bolts of the same nominal diameter. We skip the $\frac{7}{16}$ bolt and try a $\frac{1}{2}$ – 20 UNF, for which $A = 0.1599$ in.^{2}. Since k cancels out when ϕ from equation 7.3.6 is substituted into equation 7.3.3 for f_1 of f_2, we can use the values of f_1 and f_2 already found to calculate

$$\sigma_{xx} = 636.143(2.7)/0.1599 = 10\,741.6 \text{ psi}$$

and

$$\sigma_{xy} = 6299.72/[4(0.1599)] = 9849.5 \text{ psi}$$

Substitution of these values into the HMH criterion yields

$$\sigma_{y_t}/\zeta = 50\,000/2 < 20\,159.8 \text{ psi}$$

Thus, our design is complete.

Each step was correct and we did locate the welds and the bolts such that their centers of area were at the intersection of the applied forces. But we could have been more clever. Now that we have finished we note that we could have bolted the base plate to the anchor plate with just one bolt at the center of area of the welds by drilling a bolt hole through the cap and base plates and placing a 2.2 in. length of pipe between the cap and base plate to keep from bending them when we tighten the bolt. But if we do that, why not use a larger pipe and eliminate the vertical and 45° plates that we used to transfer load from the cap plate to the base plate?

To carry out such a redesign we need to find the expression for $\mathfrak{J}$ for a circular weld line. To do this, recall that the polar moment of area J_p of a circular ring is given by

$$J_p = \pi(r_o{}^4 - r_i{}^4)/2 = \pi(r_o{}^2 - r_i{}^2)(r_o{}^2 + r_i{}^2)/2 = Ar^2$$

where

$$r^2 = (r_o{}^2 + r_i{}^2)/2 \simeq r_{av}{}^2$$

When modified to apply to a weld,

$$A = 2\pi r h_o/\sqrt{2} \quad \text{so} \quad J_p = \frac{h_o}{\sqrt{2}}\,2\pi r^3$$

Since $\mathfrak{J} = \sqrt{2}J/h_o$ it follows that

$$\mathfrak{J} = 2\pi r^3$$

and from $I = J/2$ it also follows that

$$\mathfrak{I} = \pi r^3$$

Thus,

$$\frac{h_o}{\sqrt{2}}\,\sigma_{xx} = \frac{Mc}{\mathfrak{J}} \quad \text{and} \quad \frac{h_o}{\sqrt{2}}\,\sigma_{xy} = \frac{F}{2\pi r}$$

Substitution into the HMH criterion, equation 7.7.4, gives

$$\frac{h_o^2}{2}\left(\frac{\sigma_{yt}}{\zeta}\right)^2 = \left(\frac{Mc}{\pi r^3}\right)^2 + \frac{3}{4}\left(\frac{F}{\pi r}\right)^2$$

multiplying both sides by r^6 yields a cubic in r^2; namely,

$$\frac{1}{2}\left(h_o\frac{\sigma_{yt}}{\zeta}\right)^2 r^6 - \frac{3}{4}\left(\frac{F}{\pi}\right)^2 r^4 - \left(\frac{Mc}{\pi}\right)^2 = 0$$

which becomes

$$101\,531\,250h_o^2 r^6 - 3\,015\,810.244 r^4 - 10\,997\,208.60 = 0$$

for $\sigma_y = 57\,000$ psi, $\zeta = 4.0$, $F = 6299.721$ lb, $M = 7717.157$ in.-lb, and $c = 1.35$. From a numerical routine for finding the roots of a cubic we find that for $h_o = 0.25$ in.

$$r = 1.176 \text{ in.}$$

We shall use $r = 1.25$ in. because this is a common steel tube diameter. Buckling and shear strength will be assured if we use an inside diameter of 1.0 in.

The redesigned bracket is shown in Figure 7.7.4. It is not only smaller but it is also easier to manufacture.

Our last task is that of selecting the single bolt that will be used to fasten it to the anchor plate. We may use the moment equilibrium equations of statics to write that

$$6299.72(1.6) = 4.0825F$$

where 4.0825 in. is the distance from the tip of the bracket to the center of area of the circular weld in the direction of the line of action of the resultant force. Hence $F = 2469.0$ lb and the tensile stress on the bolt is given by

$$\sigma_{xx} = 2469.0(2.7)/(\pi r^2)$$

where r is the radius of the tensile area of the bolt and 2.7 is the stress concentration for the screw threads. The shear stress is given by

$$\sigma_{xy} = 6299.72/(\pi r^2)$$

Substitute these values into the HMH criterion (4.4.2) to get

$$r^4(57\,000/4)^2 = (2121.9)^2 + 3(2005.3)^2$$

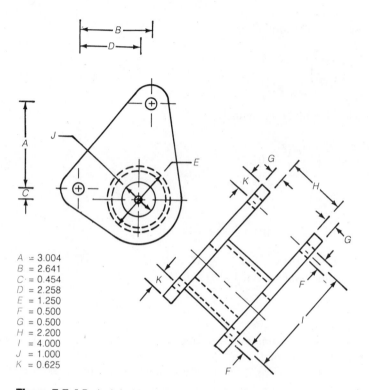

$A \doteq 3.004$
$B = 2.641$
$C \doteq 0.454$
$D = 2.258$
$E = 1.250$
$F = 0.500$
$G = 0.500$
$H = 2.200$
$I = 4.000$
$J = 1.000$
$K = 0.625$

Figure 7.7.4 Redesigned brake bracket. Dimensions are in inches.

which yields a radius $r = 0.5344$ in., which corresponds to a tensile area of 0.8973 in.^{2}. From Table 6.4.2 we can only select a $1\frac{1}{4} - 7$ UNC thread, but the course thread may require a longer nut than one with a smaller pitch. From ANSI B1.1-1982 we find that either $1\frac{3}{16} - 12$ UN or a $1\frac{3}{16} - 16$ UN bolt will support the load. We will select the $1\frac{3}{16} - 12$ UN bolt because on the one hand the pitch is small enough that it may either be threaded into a tapped plate (it will have 6 threads in a $\frac{1}{2}$-in. plate, as opposed to 3.5 threads for a $1\frac{1}{4} - 7$ UNC bolt) or used with a nut, while on the other hand the threads are large enough that they are less likely to be cross-threaded.

7.8 WELDING SYMBOLS

Notation used on shop drawings to specify welds is too extensive for a complete duplication and explanation here. However, the code used is shown in Figure 7.8.1, along with several of the supplementary symbols and a brief explanation of them. See AWS A2.4, *Symbols for Welding and Nondestructive Testing*, published by the American Welding Society, for the complete list.

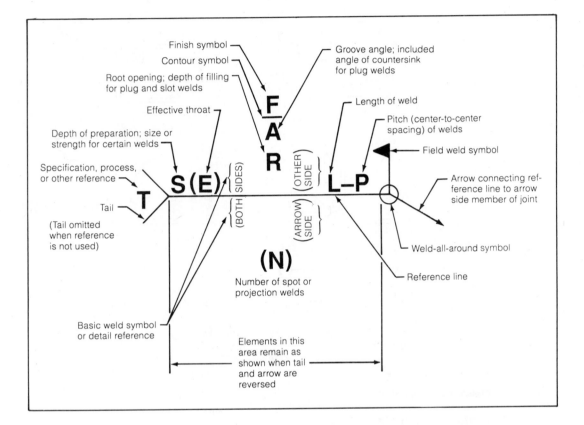

Figure 7.8.1 Weld symbols and their placement. *Source:* Symbols of Welding and Nondestructive Testing, *AWS A2.4–79, American Welding Society, Miami, FL. Reprinted with permission.*

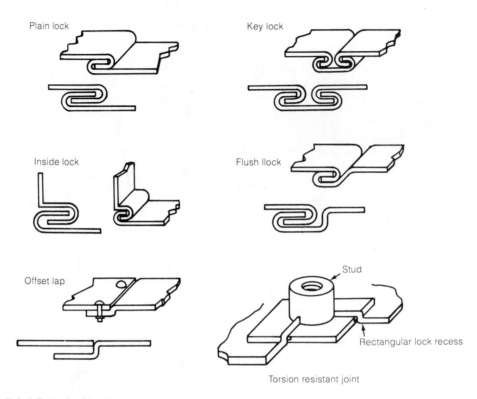

Figure 7.9.1 Typical solder joints.

7.9 SOLDERED CONNECTIONS

Soldered connections between similar or dissimilar metals are usually limited to radiator seams and some housing joints in machine design applications where the primary purpose of the solder is to prevent leakage or to assure electrical contact. These connections employ alloys that are different from the metals to be joined and are classified as having a low melting point, below 840 °F (450 °C).*

These alloys are available in bar, wire, slab, ingot, ribbon, powder, and foil forms and are used to fill the spaces in joints between the materials to be joined. Low melting temperatures produce relatively little expansion in the joined materials and may, therefore, produce little metalurgical change and physical distortion. Low-temperature alloys also permit solder joints to be produced with inexpensive and readily available hand torches.

Low tensile and shear strength of solder alloys require that joints which may experience appreciable tensile or shear loads be designed so that the metal in the components to be joined supports tension or shear while placing the solder in compression. Lock joints, as shown in Figure 7.9.1, satisfy these requirements.

---

*   Listed as 800 °F (426 °C) in some references.

TABLE 7.9.1   WIDELY USED TIN–LEAD SOLDERS AND THEIR APPLICATIONS

Composition, %		Temperature, °F			
Tin	Lead	Solidus	Liquidus	Pasty range	Uses
2	98	518	594	76	Side seams for can manufacturing
5	95	518	594	76	Coating and joining metals
10	90	514	570	56	
15	85	440	550	110	
20	80	361	531	170	Coating and joining metals, or filling dents or seams in automobile bodies
25	75	361	511	150	Machine and torch soldering
30	70	361	491	130	
35	65	361	477	116	General purpose and wiping solder
40	60	361	460	99	Wiping solder for joining lead pipes and cable sheaths. For automobile radiator cores and heating units
45	55	361	441	80	Automobile radiator cores and roofing seams
50	50	361	421	60	General purpose. Most popular of all
60	40	361	374	13	Primarily used in electronic soldering applications where low soldering temperatures are required
63	37	361	361	0	Lowest melting (eutectic) solder for electronic applications

Source: Metals Handbook, Vol. 6, 9th ed., American Society for Metals, Metals Park, OH, 1983, p. 1071. Reprinted with permission.

Before soldering begins, the metal at the joint must be mechanically cleaned and oxides removed by application of a pastelike material called *flux* which melts and flows away as the solid solder alloy is heated to its melting temperature and flows into the joint. Flux prevents formation of oxide films and also wets the surface by lowering the surface tension of the molten solder.

Ultrasonic soldering may be performed without flux because the high-frequency vibration aids penetration of surface films by the solder.

Several typical tin–lead solders are listed in Table 7.9.1 along with their uses, and typical solderable and nonsolderable metals are listed in Table 7.9.2.

7.10   BRAZED CONNECTIONS

Brazing is intermediate between soldering and welding. It is similar to soldering in that a nonferrous filler metal is used, dissimilar metals can be joined, and the melting temperature of the filler metal is below that of the metals to be joined. It differs from

TABLE 7.9.2   METALS AND FLUXES FOR SOLDERING

Metals	Solderability	Rosin Non-Activated	Rosin Midly Activated	Rosin Activated	Organic Water Soluble	Inorganic Water Soluble
Platinum Gold Copper Silver Cadmium Plate Tin (Hot Dipped) Tin Plate Solder Plate	Easy	OK	OK	OK	OK	Not recommended for electrical soldering
Lead Nickel Plate Brass Bronze Rhodium Beryllium Copper	Less Easy	Not Suitable	OK	OK	OK	OK
Galvanized Iron Tin-Nickel Nickel-Iron Mild Steel	Difficult	Not Suitable		OK	OK
Chromium Nickel-Chromium Nickel-Copper Stainless Steel	Very Difficult	Not Suitable		Not Suitable	OK
Aluminum* Aluminum-Bronze*	Most Difficult	Not Suitable		Not Suitable	—
Berylium Titanium	Not Solderable	—		—	—

*   Aluminum and aluminum-bronze require special flux and/or solder.

Source: Reprinted from *Machine Design*, November 15, 1984. Copyright 1984 by Penton Publishing, Inc. Cleveland, OH.

soldering in that the melting temperature of the filler metal is above 840 °F (450 °C), which may cause some of the filler metal to diffuse into the metals to be joined, thus producing surface alloys. It is similar to welding in that the brazing alloys used for filler metal generally have greater tensile and shear strength than solder alloys, so that butt and lap joints may transmit stresses.

Flux is used for wetting and oxide film prevention whenever brazing is done by means of a torch, resistance, inductance, or furnace heating in a uncontrolled atmosphere. No flux may be necessary if the brazing is done in a controlled, nonoxidizing atmosphere.

Dip brazing is a production process that commonly includes two procedures. One may be a bath consisting of a layer of molten filler metal under a layer of liquid flux. The other may be a salt bath which prevents oxide formation and which is heated

TABLE 7.10.1   FILLER METALS FOR TORCH BRAZING LOW-CARBON AND LOW-ALLOY STEELS

AWS Classification	Product Form	Nominal Composition, %										Temperature, °F		
		Ag	Cu	Zn	Cd	Ni	Sn	Fe	Mn	Si	P	Solidus	Liquidus	Brazing
Silver alloys														
BAg-1	Strip, wire, powder	45	15	16	24	...	...	...	...	...	...	1125	1145	1145–1400
BAg-1a	Strip, wire, powder	50	15.5	16.5	18	...	...	...	...	...	...	1160	1175	1175–1400
BAg-2	Strip, wire, powder	35	26	21	18	...	...	...	...	...	...	1125	1295	1295–1400
BAg-2a	Strip, wire, powder	30	27	23	20	...	...	...	...	...	...	1125	1310	1310–1400
BAg-3	Strip, wire, powder	50	15.5	15.5	16	3.0	...	...	...	...	...	1170	1270	1270–1400
BAg-4	Strip, wire, powder	40	30	28	...	2.0	...	...	...	...	...	1220	1435	1435–1650
BAg-5	Strip, wire, powder	45	30	25	...	...	...	...	...	...	...	1225	1370	1370–1550
BAg-6	Strip, wire, powder	50	34	16	...	...	...	...	...	...	...	1250	1425	1425–1600
BAg-7	Strip, wire, powder	56	22	17	...	...	5.0	...	...	...	...	1145	1205	1205–1400
BAg-20	Strip, wire, powder	30	38	32	...	...	...	...	...	...	...	1250	1410	1410–1600
BAg-27	Strip, wire, powder	25	35	26.5	13.5	...	...	...	...	...	...	1125	1375	1300–1400
BAg-28	Strip, wire, powder	40	30	23	...	...	2	...	...	...	...	1200	1310	1310–1500
Copper-zinc alloys														
RBCuZn-A[a]	Strip, rod, wire, powder	...	59	40	...	...	0.6	...	...	...	...	1630	1650	1670–1750
RBCuZn-D[a]	Strip, rod, wired, powder	...	48	41	...	10.0	...	...	...	0.15	0.25	1690	1715	1720–1800
RCuZn-B[b]	Rod	...	58	38	...	0.5	0.95	0.7	0.25	0.08	...	1590	1620	...
RCuZn-C[b]	Rod	...	58	39	...	...	0.95	0.7	0.25	0.08	...	1595	1620	...

[a] Classified for braze welding and brazing. [b] Classified for braze welding.

Source: Metals Handbook, Vol. 6, 9th ed., American Society for Metals, Metals Park, OH, 1983, p. 953. Reprinted with permission.

TABLE 7.10.2 COMMONLY USED FLUXES FOR TORCH BRAZING OF LOW-CARBON AND LOW-ALLOY STEELS

AWS Type	Useful Temperature Range, °F	Principal Constituents	Available Forms	Applicable Filler Metals
3A	1050–1600	Boric acid, borates, fluorides, fluoborates, wetting agent	Paste, liquid, slurry, powder	BAg
3C	1050–1800	Boric acid, borates, boron, fluorides, fluoborates, wetting agent	Paste, slurry, powder	BAg, RBCuZn
3D	1400–2200	Boric acid, borates, borax, fluorides, fluoborates, wetting agent	Paste, slurry, powder	RBCuZn

Source: Metals Handbook, Vol. 6, 9th ed., American Society for Metals, Metals Park, OH, 1983, p. 962. Reprinted with permission.

above the melting temperature of the filler metal. In both cases the parts must be held in a jig while dipped, and in the second case the filler metal must be inserted into the joints before dipping.

Common brazing filler metals are listed in Table 7.10.1 and a guide to the selection of filler metals and fluxes is shown in Table 7.10.2.

Brazing has been widely used in the automotive industry in producing air conditioners, oil coolers, radiators, heaters, and similar equipment. In machine shops high-speed cutting tips have been brazed to carbon steel shanks in making cutting tools without affecting the metallurgy of the tip itself.

7.11 ADHESIVE CONNECTIONS

Joining dissimiliar materials, either metallic or nonmetallic, without the use of holes in either material is one of the main attractions of adhesives because stress concentrations in the load-carrying members are minimized. (A small stress concentration may occur at the joint edge.) A second attraction is that if the parts to be joined are small, lightly loaded, and not subject to excessive moisture, heat, or chemical attack, room temperature adhesives may be used with little or no special equipment.

Disadvantages of adhesives are that they are weakened in varying degrees by moisture and heat, and that some may require meticulous cleaning and surface preparation. A few may release objectionable fumes during cleaning and/or curing, so that venting and treatment of the fumes may be necessary.

Components to be joined, such as the plates in Figure 7.11.1, are termed *adherends* which are joined by an *adhesive* layer between them. Conceptually the adhesive and

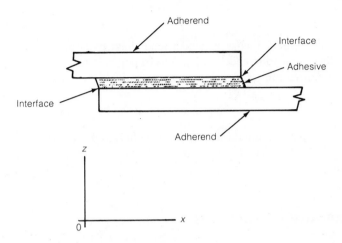

Figure 7.11.1 Adhesive joint components.

the adherends contact along an adhesive–adherend *interface*, also noted in Figure 7.11.1. It is along this interface that the adhesive mechanism is to be found.

Detailed understanding of the adhesion mechanism for all adherend materials and for all adhesives has yet to be achieved. One of the complexities of the adhesion problem is that the interface itself is not well defined on the microscopic level, since the surface of all materials consists of a transition region which is irregular, with hills of height from 1 to 10×10^{-6} in. or greater (0.02 to 0.25 μm; 1 μm $= 1 \times 10^{-6}$ m).[9] The surface may also consist of an oxide of the parent material, a collection of crystal and binder material (grain size in metals may be as large as 120 μm), along with absorbed impurities from the atmosphere.

7.12 PROPOSED ADHESION THEORIES

Acceptability of the following theories may vary from one group of engineers and scientists to another, so that they are listed without comment, except for the notion of polarity, which has largely been dismissed because the term itself has never been clearly defined.[9]

Hook Theory

The hook theory is one in which the adhesive material is believed to penetrate into voids in the adherend surface and locks the adhesive to the adherend. Plating of copper or nickel to plastics (polypropylene, ABS) has been used to demonstrate the hook theory.[10]

Diffusion Theory

This theory is a variation of the hook theory in which long-chain molecules diffuse across the interface from adhesive to adherend.[10] It pertains to polymer adherends and has been demonstrated in bonds to rubber.

Electrostatic Theory

As the name implies, this is related to the polarity or dipole theory, and holds that the adhesive surface and the adherend surface form the plates of a capacitor, with some ill-defined dielectric between.

7.13   REPRESENTATIVE STRENGTH OF SELECTED COMMERCIAL ADHESIVES

Acetylene-terminated polyimide structures, polyphenylquinoxaline resins, and some polyimide resins which have been considered and/or used in aircraft structures are not included in the following table because they are not generally available outside of the aircraft industry. Table 7.13.1 is not intended to be a comprehensive list of industrial adhesives, but rather is an indication of some of the adhesives available to manufacturers.

TABLE 7.13.1   REPRESENTATIVE PROPERTIES OF SEVERAL ONE-PART EPOXY INDUSTRIAL ADHESIVES

Commercial Designation	Overlap Shear Strength (f_D) (psi) at				
	$-67\,°F$	$75\,°F$	$180\,°F$	$250\,°F$	$350\,°F$
Two-Part Urethane Adhesive					
3532 B/A	2500	2000	300	—	—
Two-Part Epoxy Adhesive					
1751	1400	2000	500	200	—
1838 B/A	1500	3000	500	200	—
2158 B/A	1700	2000	400	150	—
2216 B/A Translucent	3000	1200	200	100	—
3533 B/A Gray	3000	2200	190	—	—
3533 B/A Clear	2000	2500	190	—	—
One-Part Epoxy Adhesive					
2214 Regular	3000	4500	4500	1500	200
2214 Nonmetalic	3000	4000	4500	1500	200
2214 Hi-Temp	2800	2800	2800	2500	1200
2214 Hi-Flex	2500	4000	2000	450	100
2214 Hi-Density	3000	4500	4500	1500	200
2290 Adhesive coating	5500	5500	3500	400	100

Source: Adhesives, Coatings and Sealers Division, 3M Corporation, St. Paul, MN.

Many of the widely available industrial adhesives are used where very light loading is expected, such as attaching nameplates, decorative trim, and insulations to a variety of materials. Catalogs may, therefore, not even list the shear strength of these adhesives.

The urethane adhesives in Table 7.13.1 have service temperatures less than or equal to that of two-part epoxy adhesives, and shear strengths near the upper range of that for two-part epoxy adhesives. One-part epoxy adhesives, which require heat curing, have the greatest strengths and highest service temperatures of the materials shown in Table 7.13.1

7.14 JOINT DESIGN*

Adhesive joint design is based upon theoretical calculations and experimental data from specific joints which taken together indicate that stress and strain in the adhesive layer are neither uniform nor linear across the length of the overlap, but are distributed as shown in Figure 7.14.1 for adhesives that are entirely elastic for the applied loads.[10-12]

To calculate the approximate stress distribution along the adhesive in a lap as shown in Figure 7.14.1a imagine reference marks inscribed on the plates at locations numbered from 1 to 5. If the plates are not allowed to bend and if they can only move horizontally—in the x direction—they will shift horizontally under the influence of tensile forces acting beyond point 5 on each plate. Since this force decreases to zero at point 0 the strain increases according to

$$\varepsilon_{xx} = \frac{du}{dx} = \frac{\sigma_{xx}}{E} = \frac{F(x)}{EA} \tag{7.14.1}$$

If $F(x) = Fx^n$ then after integrating (7.14.1) we have

$$u = \frac{F}{EA(n+1)} x^{n+1} \tag{7.14.2}$$

so that u increases nonlinearly with x since $n \neq 0$. When $n = 1$ the strain and deformation of the plates and the adhesive resemble that shown in Figure 7.14.1b.

Let Figure 7.14.1c represent an element of the adhesive and let u represent the displacement in the x direction. From this figure it follows that within the adhesive

$$(1 + \varepsilon_{xx_2})\, dx + t_a \frac{d}{dx}\left(\frac{du}{dz}\right) dx = (1 + \varepsilon_{xx_1})\, dx \tag{7.14.3}$$

* The design methods described here were developed during research sponsored by the Air Force Wright Aeronautical Laboratories, Wright-Patterson Air Force Base, Ohio. The assistance of W. B. Jones of the Air Force Materials Laboratory and of W. L. Shelton of the Air Force Flight Dynamics Laboratory is gratefully acknowledged, as are the enlightening discussions with L. J. Hart-Smith of Douglas Aircraft Co., McDonnell Douglas Corp.

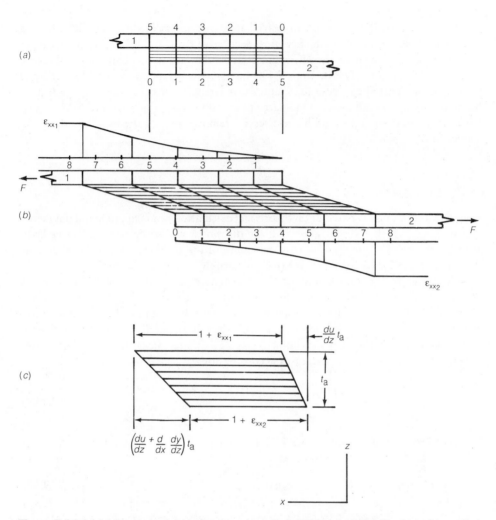

Figure 7.14.1 (a) Adhesive lap joint before force is applied, (b) same lap joint after force F is applied and adherend 2 has been displaced one unit to the right relative to the left end of adherend 1, and (c) displacement of an element of adhesive.

Recall from Chapter 3 that

$$\varepsilon_{xz} = \frac{1}{2}\left(\frac{du}{dz} + \frac{dw}{dx}\right) = \frac{\sigma_{xz}}{2G_a} \qquad (7.14.4)$$

so that (7.14.3) may be written as

$$\varepsilon_{xx_2} - \varepsilon_{xx_1} + \frac{1}{t_a G_a}\frac{d}{dx}\sigma_{xz} = 0 \qquad (7.14.5)$$

inasmuch as $w = 0$ because the plates were not allowed to move in the z direction. In an actual adhesive joint there may be some small motion in z direction; this is one of the reasons this solution is only an approximation to the actual stress distribution.

From the force equilibrium conditions on plates 1 and 2, Figure 7.14.2, it follows that for plates of unit width

$$\frac{dF}{dx} = \frac{d}{dx}(E_1 \varepsilon_{xx_1} t_1) = E_1 t_1 \frac{d}{dx} \varepsilon_{xx_1} = \sigma_{xz} \tag{7.14.6}$$

since F in plate 1 increases with increasing x, and

$$\frac{dF}{dx} = \frac{d}{dx}(E_2 \varepsilon_{xx_2} t_2) = E_2 t_2 \frac{d}{dx} \varepsilon_{xx_2} = -\sigma_{xz} \tag{7.14.7}$$

since F decreases with increasing x in plate 2.

After differentiating (7.14.5) with respect to x and substituting from (7.14.6) and (7.14.7) for the strain derivatives in terms of the shear stess we find that the shear stress is given by

$$\frac{d^2}{dx^2} \sigma_{xz} - k^2 \sigma_{xz} = 0 \tag{7.14.8}$$

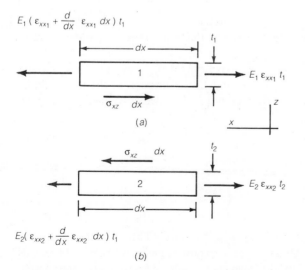

$E_1 (\varepsilon_{xx_1} + \frac{d}{dx} \varepsilon_{xx_1} \, dx) \, t_1$

t_1

dx

1

$E_1 \varepsilon_{xx_1} t_1$

$\sigma_{xz} \quad dx$

(a)

z

x

$\sigma_{xz} \quad dx$

t_2

2

$E_2 \varepsilon_{xx_2} t_2$

dx

$E_2 (\varepsilon_{xx_2} + \frac{d}{dx} \varepsilon_{xx_2} \, dx) \, t_1$

(b)

Figure 7.14.2 Forces acting on plates 1 and 2, where each is of unit width.

where

$$k^2 = \frac{E_a}{2t_a(1 + v_a)} \left(\frac{1}{E_1 t_1} + \frac{1}{E_2 t_2} \right)$$

in terms of the elastic moduli E_a of the adhesive, E_1 of plate 1, and E_2 of plate 2, along with Poisson's ratio v_a for the adhesive, which has a thickness t_a. Plate thicknesses are denoted by t_1 and t_2.

The general solution to (7.14.8) is of the form

$$\sigma_{xz} = A \cosh kx + B \sinh kx$$

Because of the symmetry of displacements and forces in the adhesive lap joint in Figure 7.14.1b the shear stress must be symmetric about the center of the joint. Thus, if we take the origin of coordinates at the center of the joint we have that $B = 0$ because $\sinh kx$ is asymmetric. Hence,

$$\sigma_{xz} = \sigma_{xz\text{min}} \cosh kx \qquad (7.14.9)$$

after A is set equal to the minimum stress, located at $x = 0$.

If the width of the lap joint is one unit the ratio of maximum to minimum stress becomes

$$\frac{\sigma_{xz\text{max}}}{\sigma_{xz\text{min}}} = \cosh \frac{k}{2} \qquad (7.14.10)$$

at the end of the joint where $x = \frac{1}{2}$. By symmetry, this same ratio holds at the other end of the joint; at $x = -\frac{1}{2}$.

By considering a joint of unit width we have obtained a ratio that may be applied to any joint simply by expressing t_1, t_2, and t_a as a fraction of the joint length. Any convenient units may be used for E_1, E_2, and E_a because only their ratios enter into k.

If $E_a = 0.10E_1 = 0.10E_2$, if $v_a = 0.4$, and if $t_a = t_1 = t_2 = 0.10$ the $\sigma_{xz\text{max}}/\sigma_{xz\text{min}}$ ratio is as shown in Figure 7.14.3 for

$$k^2 = \frac{1}{0.2(1.4)} \left(\frac{0.1}{0.1} + \frac{0.1}{0.1} \right) \quad \text{or} \quad k = 2.6726$$

Note that when $E_a = E_1 = E_2$ then $\sigma_{xz\text{max}}/\sigma_{xz\text{min}} = 34.23$. This large stress concentration, in effect, is consistent with the stress concentrations given in Chapter 3 because when these moduli are equal the lap joint becomes a solid with its ends similar to a plate with a change in width and a zero fillet radius at the transition point.

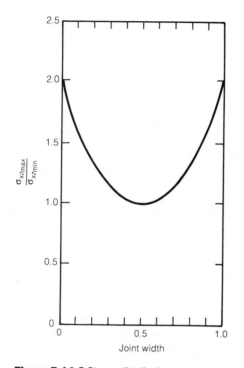

Figure 7.14.3 Stress distribution across a joint of unit width for stress within the elastic range of the adhesive.

Actual adhesive joint stress distributions, Figure 7.14.4b, are qualitatively similar to that shown in Figure 7.14.3 for stresses below the shear yield stress σ_{xz_1} in Figure 7.14.4a. As the shear load increases the shear stress at the ends of the joint moves into the plastic region. At stress level σ_{xz_2}, Figure 7.14.4a, the stress distribution across the joint may appear as in Figure 7.14.4c. The joint is still capable of greater loads, however, because the central elastic region prevents creep. Increased loading further reduces the elastic region, as in Figure 7.12.4d, but does not cause joint failure until the entire joint becomes plastic, Figure 7.14.4e, and the adherends begin to move apart as the adhesive creeps.

Recommended adhesive joint design is based upon the stress distribution shown in Figure 7.14.4c as having the optimum width for the maximum expected load F for the adhesive connection.[14] For this stress distribution the central stress is relatively very small, which in turn implies that the central strain is very small and that while the adherends may have stretched elastically, there has been no shift of the adherend centers relative to one another. Moreover, when the external loads are removed the elasticity of the adherends will restore the joint to its original configuration.

Relatively low yield strength of adhesives, as compared with welds, for instance, requires that adhesives be used only where separation forces are not large if the joint

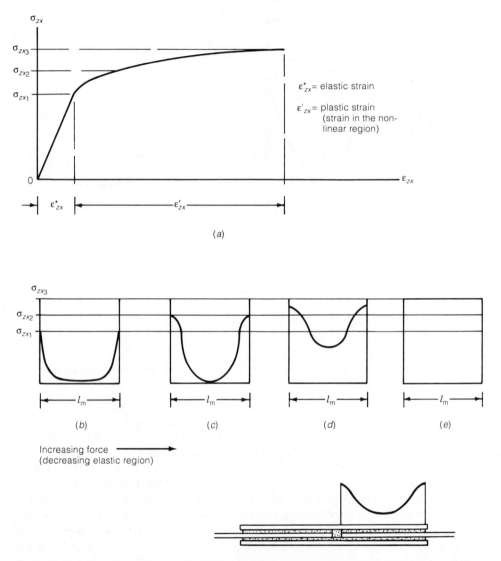

Figure 7.14.4 Stress variation across overlap length *l* with increasing shear force. *Source: Abstracted from L.J. Hart-Smith,* Adhesive-Bonded Joints for Composites—Phenomenological Considerations, Douglas Paper 6707, March 14–16, 1978. *Reprinted by permission of the McDonnell Douglas Corporation.*

is to remain intact. Hence, they are limited at this time to joining sheet and thin plate structures to themselves or to frames and reinforcement structures.

Joint designs for adhesive fastening are, accordingly, of the forms shown in Figure 7.14.5 for sheet metal and plates, and in Figure 7.14.6 for the attachment of stiffeners.

No.	Joint	Comments	No.	Joint	Comments
1.	Single-lap (unsupported) joint	Nonstructural joint having low efficiency (for short overlaps) because of bending of the adherend due to the eccentricity in load path. Thick adherends are associated with failures by peel rather than by shear. For thin adherends, these joints can be given reasonable efficiency by adequate (>80:1) overlaps.	7.	Flush joints	Nonstructural joints suffering from net section loss just outside the joint regions.
2.	Supported signal-lap joint	Practical joint for thin adherends. Needs to be mounted on moment-resistant support to avoid limitations above. Joint load capacity does not increase indefinitely with overlap. Load capacity is limited by single bond surface.	8.	Stepped-lap joint	Used extensively in advanced composite-to-titanium bonded joints. Detail design can be critical. Need to avoid composite net section reduction at end of titanium. End titanium step needs to have lower l/t ratio than other steps.
3.	Supported single-strap joint	Same as for 2. Suitable for flush exterior applications but limited to thin adherends, and need either good moment-resistant supports or very large l/t ratios.	9.	Stiffness-balanced stepped-lap joint	Improvement on no. 8 because both ends of joint are then loaded equally instead of concentrating load transfer through thin end of titanium (or stiffer) adherend in no. 8.
4.	Balanced double-lap joint	Efficient practical joints for thin and moderately thick adherends. Simple joint with tolerant fit requirements. Joint strength limited by adherend thicknesses and independent of overlap beyond very short (uniformly stressed) lengths of bond. Maximum strength limit is set by peel stresses for moderately thick adherends, rather than by adhesive shear stresses. For thin adherends, peel stresses are negligible and shear strength usually exceeds adherend strength.	10.	Double stepped-lap joint	Needed for thick sections beyond the practical capabilities of no. 9.
5.	Balanced double-strap joint	Weaker than no. 4 because only that end of the joint from which the thin (lower Et) adherend extends is loaded to its capacity. Other end has unusable reserve.	11.	Scarf joints	Most efficient of all bonded joints. Necessary for thick adherends, unnecessary for thin. Strength is maximized by balancing adherend stiffnesses at each end of joint. Precise fit requirements for efficient joints can be reduced in some situations by co-cure and bond of composite laminates.
6.	Tapered-lap joint	Efficient practical joints for moderately thick adherends. Overcomes peel-stress limitations of no. 2. Strength limited by adhesive shear strength for thick adherends. Best strengths are obtained when adherends with optimum stiffness imbalance between adherends to compensate for shear strength loss due to taper. Only moderate precision requirements.	12.	Joggled lap joint	Nonstructural joint used (because of aerodynamic smoothness requirements) on exterior skins subjected to normal rather than in-plane loads. See also comments on no. 1.

Figure 7.14.5 Adhesive joints for sheets and plates. *Source: L.J. Hart-Smith,* Advances in the Analysis and Design of Adhesive-Bonded Joints in Composite Aerospace Structures, *Douglas Paper 6224, April 24, 1974. Reprinted by permission of the McDonnell Douglas Corporation.*

423

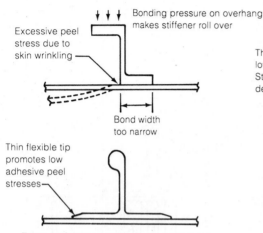

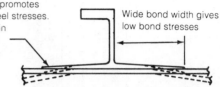

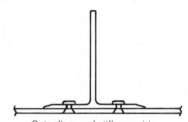

Excessive peel stress due to skin wrinkling

Bonding pressure on overhang makes stiffener roll over

Bond width too narrow

Thin flexible tip promotes low adhesive peel stresses. Stiffener and skin deflect together

Wide bond width gives low bond stresses

Proportions shown are difficult to extrude. Also, stiffener is so thin that a fatigue crack in the skin would not be arrested but would grow in stiffener also

Thin flexible tip promotes low adhesive peel stresses

Taper confined to end has these advantages:

(a) Easier to extrude than continuous taper

(b More rapid buildup in stiffener thickness makes stiffener more effective in stopping skin cracks

(c) Constant thickness over most of width makes bonding of matching details easier

(d) Less handling damage than for long, gradual taper

Outer flange of stiffener wide enough to permit riveted repairs, if necessary

A further consideration for damage tolerance with respect to cracks in skin or stiffeners is that the stiffener bond width must be large enough, in comparison with stiffener area and spacing, to prevent unzipping of the bonds. The adhesive must never be the weak link.

Figure 7.14.6 Considerations for adhesive connection to stiffeners. *Source: L.J. Hart-Smith*, Adhesive Bonding of Aircraft Primary Structures, *SAE Transactions, Sec. 4, Vol. 89, pp. 3718–3732, 1980. Copyright 1980 Society of Automotive Engineers, Inc. Reprinted with permission.*

7.15 ADHESIVE JOINT PROPORTIONS

Joint design consists of calculating the adhesive thickness and the overlap length, or simply the joint length. Satisfactory design formulas may be obtained by assuming that the adhesive is ideally plastic, and, therefore, obeys a stress–strain relation as shown in Figure 7.15.1*a*. Because under the assumption of ideal plasticity there can be no increase in stress beyond the yield stress, the force will give rise to a stress distribution similar to that shown in Figure 7.15.1 according to the reasoning of the previous section. As the joint force increases the elastic region diminishes as indicated in Figure 7.15.1*c* until at the joint's maximum strength the elastic portion disappears and creep failure begins.

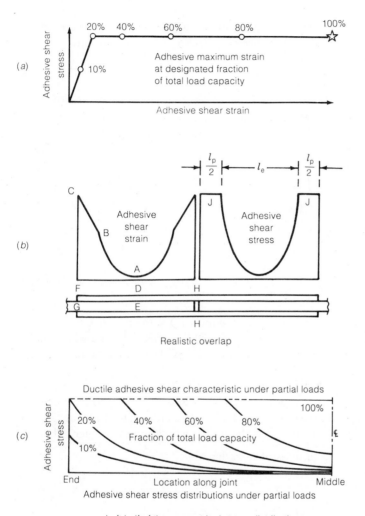

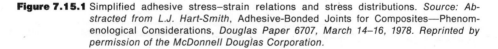

Figure 7.15.1 Simplified adhesive stress–strain relations and stress distributions. *Source: Abstracted from L.J. Hart-Smith*, Adhesive-Bonded Joints for Composites—Phenomenological Considerations, *Douglas Paper 6707, March 14–16, 1978. Reprinted by permission of the McDonnell Douglas Corporation.*

Before presenting formulas for joint design we should emphasize that the terms width and length are used differently in the discussion of an adhesive joint than in the discussion of a welded joint. In the following discussion the length of a joint will refer to the distance its adherends overlap and the width of a joint will refer to the extent of the common boundary of the adherends to be joined; i.e., the width of an adhesive joint corresponds to the length of a welded joint.

Let F denote the maximum tensile (separation) force that the joint can support per unit of joint width, and let it act in the x direction. Hence, F has units of force per unit width and its value for a double lap joint may be found from[13]

$$F^2 = 4\tau_y t_a \left(\frac{\varepsilon_{zx}^*}{2} + \varepsilon_{zx}' \right) \mathbf{m} \qquad (7.15.1)$$

where

$$\mathbf{m} = \min \begin{cases} E_i t_i \left(1 + \dfrac{1}{u} \right) \\ E_o (t_{o_1} + t_{o_2})(1 + u) \end{cases} \qquad (7.15.2)$$

in which u is defined by

$$u = \frac{E_o}{E_i} \left(\frac{t_{o_1}}{t_i} + \frac{t_{o_2}}{t_i} \right) \qquad (7.15.3)$$

for a double lap joint in terms of the elastic modulus E_o of the two outer adherends of thickness t_{o_1} and t_{o_2} and of the elastic modulus E_i of the inner adherend whose thickness is t_i. Use of a single value E_o for the outer adherends clearly implies that they are to be made of identical material.

Maximum joint length for the force given by (7.15.1) is determined by equating the ultimate tensile force per unit width acting on the inner adherend to the force per unit width that will cause the adhesive to yield in shear, as illustrated in Figure 7.15.2. Upon solving the equilibrium equation for the forces acting upon the inner adherend

$$2l\tau_y = t_i \sigma_u = F \qquad \text{(F in units of force/length)}$$

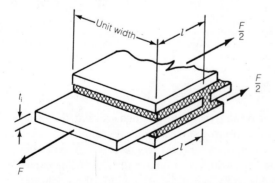

Figure 7.15.2 Forces acting on a double-lap adhesive joint of unit width and length l.

for the joint length l we have

$$l \geq l_p = \frac{t_1 \sigma_u}{2\tau_y} \tag{7.15.4}$$

in which σ_u represents the ultimate strength (ultimate stress) of the inner adherend and τ_y represents the yield stress in shear of the adhesive. It is important to understand that because equation 7.15.4 was based upon the ultimate stress of the inner adherend the main plate may be expected to fail before the joint itself will fail. Consistent with this intended mode of failure, if the joint is loaded to that extent, plate thicknesses are chosen from

$$t_i = \frac{F}{\sigma_{u_i}} \qquad t_{o_1} + t_{o_2} = \frac{F}{\sigma_{u_o}} \tag{7.15.5}$$

These are the minimum permissible values of the thickness of the actual adherends. Although one of the outer adherends may be thicker for a number of reasons (to support screw fasteners, etc.) the strength per unit width of the joint is still given by (7.15.1) and (7.15.4). In equation 7.15.1 ε_{zx}^* denotes the elastic strain of the adhesive in tension at the proportional limit and ε_{zx}' the plastic strain at failure.

The thickness of the adhesive in the joint may be found from (7.15.1) to be

$$t_a = \frac{F^2}{4\tau_y \left(\dfrac{\varepsilon_{zx}^*}{2} + \varepsilon_{zx}' \right) \mathbf{m}} \tag{7.15.6}$$

Full elastic strength of the joint may be realized if the joint length is increased beyond l_p by an amount

$$l_e = 6 \left[t_i t_a (1 + v_a) \frac{E_i}{E_a} \frac{u}{1 + u} \right]^{1/2} \tag{7.15.7}$$

where v_a is Poisson's ratio for the adhesive. Very little added strength results from increasing joint length beyond l_m, given by

$$l_m = l_e + l_p \tag{7.15.8}$$

If all adherends are of the same material and if the outer adherends are both of the same thickness, relation (7.15.7) may be written in terms of adherend thicknesses directly as

$$l_e = 6 \left[t_i t_a (1 + v_a) \frac{E_i}{E_a} \frac{2t_o}{t_i + 2t_o} \right]^{1/2} \tag{7.15.9}$$

Only a double lap joint, Figure 7.15.3, has been considered in the previous discussion because a single lap joint is not recommended for significant tensile loads; the moment induced at the joint by the forces in the adherends may cause the adherends to bend. Adhesive stresses in a single lap joint have been estimated by Goland and Reissner[16] from a stress distribution similar to that shown in Figure 7.14.3 and these stresses have been found to be in fairly good agreement with those calculated using a finite element routine[17].

Examination of Figure 7.15.1b reveals that equations 7.15.7 and 7.15.8 contain an implicit safety factor relative to the joint's ultimate load. Full length l_m is seen to correspond to a load that is 20% of the load that would bring the joint to the threshold of creep, due to the contribution from l_e, the elastic portion. Thus, use of length l_m as the design joint length implies a safety factor of 5 relative to the ultimate load. Inclusion of an explicit, additional, safety factor ζ to increase the magnitude of the design force and, hence, the thickness of the adherends and of the adhesive, would, therefore, provide an effective safety factor of 5ζ.

Whenever the force on the inner adherends acts parallel to the joint (causing the inner adherends to slide in the y direction relative to one another in Figure 7.15.2) the joint is said to experience in-plane shear. The previous relations may be adapted to include this loading condition by replacing F with $\sqrt{3}S$ in equations 7.15.5 and with $\sqrt{2(1+v)}S$ in equations 7.15.1 and 7.15.6.

Adherend thicknesses for joints subjected to combined tension and in-plane shear may be calculated from

$$t_i = \frac{(F^2 + 3S^2)^{1/2}}{\sigma_{u_i}} \qquad t_{o_1} + t_{o_2} = \frac{(F^2 + 3S^2)^{1/2}}{\sigma_{u_o}} \tag{7.15.10}$$

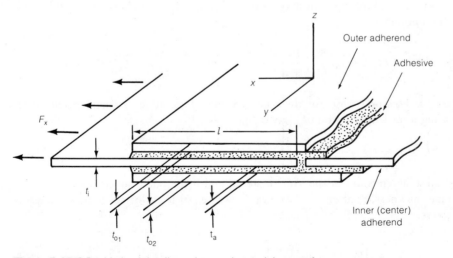

Figure 7.15.3 Double-lap joint dimensions and material properties.

in which σ_{u_i} and σ_{u_o} represent the ultimate stress of the inner and outer adherends, respectively.

Adhesive thickness for combined tension and in-plane shear may be calculated from[14]

$$t_a = \frac{\{F^4 + [2(1+v)S^2]^2\}^{1/2}}{4\tau_y \mathbf{m}\left(\dfrac{\varepsilon^*_{zx}}{2} + \varepsilon'_{zx}\right)} \qquad (7.15.11)$$

in terms of previously defined quantities

Peel stresses may be estimated from

$$\sigma_{zz} = \sigma_y \left[3\frac{E_c t_o}{E_o t_a}(1 - v_a^2) \right]^{1/4} \qquad (7.15.12)$$

in which E_c denotes the effective peel modulus of the adhesive and t_o the thickness of the outer adherend supporting the peel force.

Even though formulas 7.15.1 through 7.15.13 are simplified forms of more detailed expressions, they still involve quantities not presently published by most adhesive manufacturers. Consequently, laboratory determination of quantities τ_y, ε_{zx}, ε'_{zx}, and E_a is necessary if these equations are to be used for the design of adhesive joints. If such laboratory data are not available, approximations to accommodate published data may be made by assuming that

$$\frac{\varepsilon^*_{zx}}{2} + \varepsilon'_{zx} \cong 1 \qquad (7.15.13)$$

and that

$$\tau_y = \eta\tau_D \qquad (7.15.14)$$

where adhesive isotropy is assumed, as has been done implicitly in obtaining all of the equations used thus far. In equation 7.15.14 η is the derating factor (usually from 0.2 to 0.5) and τ_D is the shear stress found from an ASTM D1002 test.

EXAMPLE 7.15.1

Design an adhesive joint to support a tensile load that will not exceed 840 N/mm in an aluminum sheet 5 mm thick for which $E_i = 79$ GPa and $\sigma_u = 559$ MPa. The outer adherends are cut from a different alloy, for which $E_o = 103$ GPa and $\sigma_{u_o} = 621$ MPa. Laboratory tests indicate that for the cured adhesive at the expected service temperature range the minimum properties are that $E = 710$ MPa, $\tau_y = 6.90$ MPa, $\varepsilon_{zx}^* = 0.17$, $\varepsilon_{zx}' = 0.82$, and $v_a = 0.40$.

We begin by calculating the minimum thickness of the adherends according to (7.15.5), from which

$$t_i = 840/559 = 1.503 \text{ mm}$$

$$2t_o = 840/621 = 1.353 \text{ mm, or } 0.676 \text{ mm each adherend}$$

For definition 7.15.3 we have that

$$u = \frac{E_o}{E_i}\left(\frac{2t_o}{t_i}\right) = \frac{103}{79}\left(\frac{1.353}{1.503}\right) = 1.1737$$

so that

$$\mathbf{m} = \min \begin{cases} 79(1.503)\left(1 + \dfrac{1}{1.174}\right) \times 10^3 = 219\,876 \text{ N/mm} \\[2ex] 103(1.353)(1 + 1.174) \times 10^3 = 302\,966 \text{ N/mm} \end{cases}$$

Now turn to (7.15.6) to find that

$$t_a = \frac{840^2}{4(6.90)(0.085 + 0.82)219\,880} = 0.128 \text{ mm}$$

Substitution of the design thickness of the middle adherend, rather than the actual thickness, into equation 7.15.4 for the minimum joint length gives

$$l_p = \frac{1.503(559)}{2(6.9)} = 60.882 \text{ mm}$$

Length l_e is given by

$$l_e = 6\left[1.503(0.128)1.4\,\frac{79\,000}{710}\frac{1.174}{2.174}\right]^{1/2} = 24.137 \text{ mm}$$

so that the total length is 85.39 mm. Because l_e includes an implied safety factor or 5, we may round the joint length to 85 mm, and thus recommend

- adhesive thickness = 0.128 mm
- joint length = 85 mm
- outer adherend thickness = 0.676 mm or greater

No recommendation for the thickness of the middle adherend was made because the minimum 5-mm thickness required for flexural rigidity elsewhere is greater than the minimum plate thickness required for the joint strength.

REFERENCES

1. *Welding Handbook*, Section 3A, 6th ed., S.T. Walter, Ed., American Welding Society, New York, 1970, Chapter 50.

2. *Welding Handbook*, Section 3B, 6th ed., S.T. Walter, Ed., American Welding Society, New York, 1970, Chapter 51, 53.

3. Derecho, A.T., and Munse, W.H., *Stress Concentration at External Notches Subjected to Axial Loadings*, Bulletin 494, Engineering Experiment Station, University of Illinois, 1968.

4. ANSI/AWS A5.2-80, *Specifications for Iron and Steel Gas Welding Rods*, American Welding Society, New York, 1980.

5. ANSI/AWS A5.1-78, *Specifications for Carbon Steel Covered Arc-Welding Electrodes*, American Welding Society, New York, 1978.

6. AWS A5.5-69, *Specifications for Low-Alloy Steel Covered Arc-Welding Electrodes*, American Welding Society, New York, 1969.

7. ANSI/AWS A5.3-80, *Specifications for Aluminum and Aluminum Alloy Arc-Welding Electrodes*, American Welding Society, New York, 1980.

8. Blodgett, O.W., *Design of Welded Structures*, The James F. Lincoln Welding Foundation, Cleveland, OH, 1962, Chapter 7.

9. Bikerman, J.J., *The Science of Adhesive Joints*, 2d. ed., Academic, New York, 1968.

10. Wake, W.C., *Adhesion and the Formulation of Adhesives*, Applied Science, London, 1967.

11. Hart-Smith, L.J., *Adhesive Bonding of Aircraft Primary Structures*, Society of Automotive Engineers, *Transactions*, Section 4, Vol. 89, pp. 3718–3732 (1980).

12. Renton, J.W., and Vinson, J.R., The efficient design of adhesive bonded joints, *Journal of Adhesion* 7, 175–193 (1975).

13. Hart-Smith, L.J., *Further Developments in the Design and Analysis of Adhesive-Bonded Structural Joints*, Douglas Paper 6922, Douglas Aircraft Co., McDonnell Douglas Corp., presented at the Symposium on Joining of Composite Materials, American Society for Testing Materials, Minneapolis, MN, April 1980.

14. Thrall, W.W., *Design Handbook for Adhesive Bonding*, Douglas Aircraft Co., McDonnell Douglas Corp., USAF Technical Report AFFDL-TR-79, 1979.

15. Hart-Smith, L.J., *Adhesive-Bonded Joints for Composites—Phenomenological Considerations*, Douglas Paper 6707, Douglas Aircraft Co., McDonnell Douglas Corp., presented at the Technology Conference Associates Conference on Advanced Composites Technology, El Segundo, CA, March 1978.

16. Goland, M., Reissner, E., The stresses in cemented joints, *Journal of Applied Mechanics* 11, A17–A27 (1944).

17. Anderson, G.P., Bennett, S.J., and DeVries, K.L., *Analysis and Testing of Adhesive Bonds*, Academic, New York, 1977, p. 101.

PROBLEMS

Section 7.1

7.1 Calculate the joint efficiency of the riveted joint shown below. The plates are 15 mm thick and rivets are 10 mm in diameter. All dimensions are in mm. Plate stress limits are $\sigma_{xx} = 400$ MPa in tension, 450 MPa in compression, $\sigma_{xy} = 220$ MPa. Rivet shear stress limit is 275 MPa. Also give the long and short pitches.

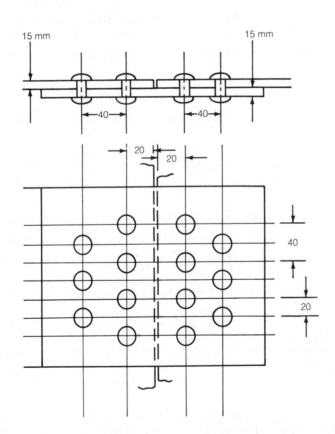

7.2 Calculate the joint efficiency of the joint shown in the figure. Rivet diameters are shown for each row. Plate stress limit is 58.00 ksi in tension, 62.25 ksi in compression, and 31.90 ksi in shear. The stress limit in the rivets is 45.50 ksi.

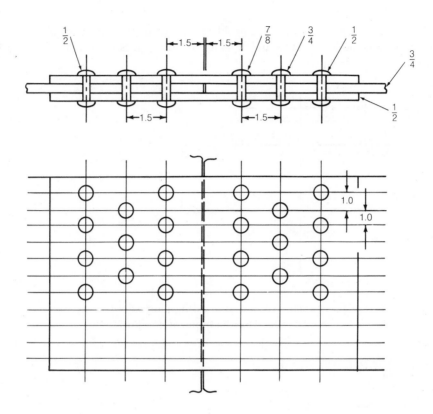

Section 7.2

7.3   Show that formula 7.2.2 yields the obvious (perhaps) result that rivet 1 carries the entire load for the bracket shown in the figure. Rivet diameters are equal.

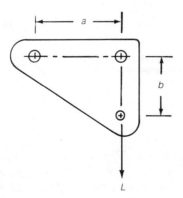

7.4 Show that although adding rivet 2 in problem 7.1 does not reduce the load on rivet 1, modifying the bracket and adding rivets 2 and 3, as in the figure, does reduce the load on rivet 1 to $L/3$. Rivet diameters are equal.

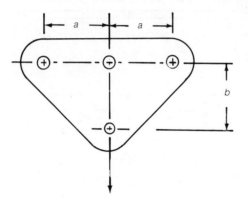

7.5 Forces $\mathbf{f}_i$ (such as $\mathbf{f}_8$) at each rivet are induced by moment M. They may be written in vector form as $\mathbf{f}_i = \mathbf{c} \times \mathbf{r}_i$, where $\mathbf{c} = \hat{\mathbf{k}}c$. Use the moment equilibrium equation

$$\sum_{i=1}^{n} \mathbf{r}_i \times \mathbf{f}_i = \sum_{i=1}^{m} \mathbf{r}_i \times \mathbf{F}_i$$

and Figure 7.2.2 to find

$$\mathbf{f}_i = \frac{\sum\limits_{j=1}^{m} (\mathbf{r}_j \times \mathbf{F}_j) \times \mathbf{r}_i}{\sum\limits_{j=1}^{n} \mathbf{r}_j \cdot \mathbf{r}_j}$$

Hint: Recall the vector identity $\mathbf{A} \times (\mathbf{B} \times \mathbf{C}) = (\mathbf{A} \cdot \mathbf{C})\mathbf{B} - (\mathbf{A} \cdot \mathbf{B})\mathbf{C}$

7.6 Use the results of problem 7.5 to derive equation 7.2.2.

7.7 Show that if a torque $\mathbf{M} = \hat{\mathbf{k}}M$ is also applied to a riveted (or bolted) bracket that equation 7.2.2 becomes

$$\mathbf{f}_i = \frac{1}{n}\sum_{i=1}^{m} \mathbf{F}_i + \frac{\left[\sum\limits_{j=1}^{m} (\mathbf{r}_j \times \mathbf{F}_j) + \mathbf{M}\right] \times \mathbf{r}_i}{\sum\limits_{i=1}^{n} \mathbf{r}_i \cdot \mathbf{r}_i}$$

7.8 As an aid in solving the problems to follow, write a computer program to solve (7.2.1) and (7.2.2) for up to 50 rivets and up 20 forces in the plane of the plate according to the flowchart shown in Figure 7.2.3. This program will eliminate much of the tedium involved in the design or riveted connections. The following problems will, however, involve no more than 2 forces and 5 rivets so that they may be solved in a reasonable time by those who do not have access to a personal computer or programmable calculator.

7.9 Find the shear stress in the most heavily loaded rivet in the hanger bracket shown. Rivet diameters, in millimeters, are as indicated in the figure.

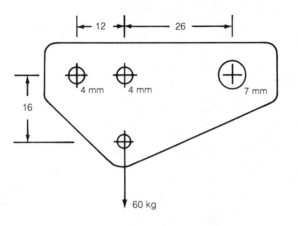

7.10 Find the stress on the most heavily loaded rivet for the bracket shown when the 730-lb force may act 26° on either side of the center line. All rivet diameters are 0.25 in.

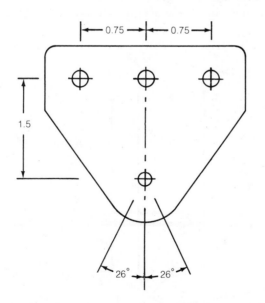

7.11 Find the force and stress for the most heavily loaded rivet for the bracket pictured below. Rivet diameters are 0.125 in.

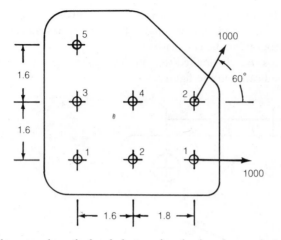

7.12 Find the most heavily loaded rivet for the bracket and rivet array shown in the figure for the following two loadings:

a $F_1 = F_2 = 150$ lb   **b** $F_1 = 0, F_2 = 150$ lb

$\theta_1 = -\theta_2 = 45°$ $\theta_2 = -45°$

where positive angles are measured in the counterclockwise direction and negative angle in the clockwise direction. If the stress on any one rivet should not exceed 4500 psi, would you recommend a warning against use of the bracket with only a single cable? (All rivet diameters = 0.125 in.)

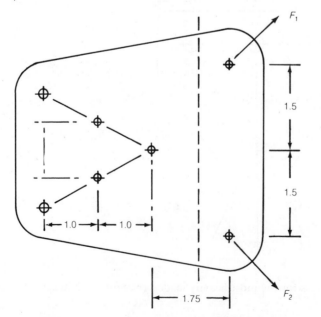

7.13 Is it possible to rearrange the 5 rivets in problem 7.12 such that the maximum rivet load will be no more than 4000 psi when either one or two cables are attached. The bracket itself may be replaced by one whose shape is better suited to the rivet arrangement you suggest. No two rivets should be closer than 0.75 in. and the two holes where forces F_1 and F_2 are applied must be 1.0 in. apart and no closer than 1.0 in. to the nearest rivet. All rivets must lie within a horizontal strip 2.00 in. wide. What principles will guide your design? If no improvement is possible in rivet arrangement in problem 7.12, explain why.

7.14 Find the most heavily loaded rivet in the gusset plate shown. (Dimensions are in inches.) All rivets are $\frac{1}{2}$ in. in diameter. Should all of the rivets shown be used in that application of formulas 7.2.1 and 7.2.2? Rivets 1 through 5 attach the gusset plate to a support frame and rivets 6 through 9 attach the angle iron to the gusset plate.

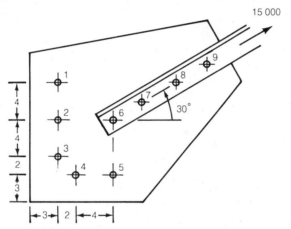

7.15 What rivet placement would you recommend to reduce the stress in the most heavily loaded rivet in problem 7.14 if no two rivets are to be closer than 4 in.? Sketch the redesigned gusset plate and show the calculations used to justify the modification. If no improvement is possible, explain why.

7.16 Is the rivet arrangement shown in the figure safe to within a safety factor of 2.0 if the maximum shear stress for the rivets is 350 MPa when a force of 106 025 N may act at any angle between

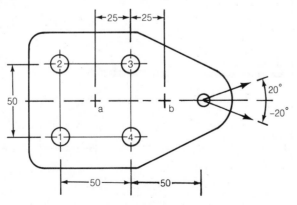

All rivet diameters = 15 mm

$\pm 20°$ relative to the center line shown. If it is not safe, another rivet of the same size as those already used may be placed at either point *a* or point *b*. Which would you recommend? Why?

7.17   The figure shows a bracket which is used to tip a product as it leaves a portion of a semiauto-mated production line. The force at *B* initially acts at 45° and then rotates to the vertical before the product carrier leaves the pin. Find the force on the rivets at the extreme positions of 0° and 45°. The magnitude of the force is approximately 2200 lb throughout its rotation. Rivet diameters are 0.375 in.

[*Ans.* Most heavily stressed rivet: stress at 1 and 4 is 5570 psi at 45°; stress at 2 and 4 is 17 960 psi at 90°].

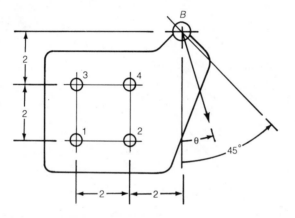

7.18   Two rivet arrangements are possible for the bracket shown in the figure for the vertical load. Rivet diameters are 6.0 mm. Which would you recommend? Why?

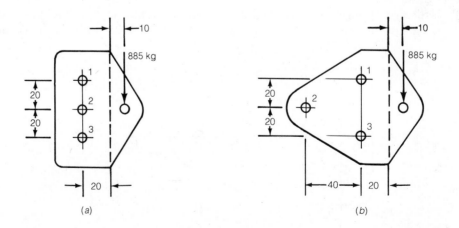

(a)   (b)

Section 7.3

7.19 Show that if the bracket shown in Figure 7.3.1 is loaded such that it tends to rotate about the upper edge the expression for angle ϕ becomes

$$\phi = \frac{F_y w}{k_2(y_f + y_2)^2 + k_1(y_f - y_1)^2}$$

7.20 Generalize equations 7.3.1 and 7.3.2 to the case of n rows of rivets, or bolts, in which case the rotation angle becomes

$$\phi = \frac{F_z y_f + F_y w}{\sum\limits_{i=1}^{n} y_i^2 k_i}$$

where $f_i = y_i k_i \phi$ holds.

7.21 Calculate the stresses in rivets 1 through 4 in the bracket shown in the figure. Rivet spacing is as shown, with each having a diameter of 0.25 in. $E = 3 \times 10^7$ psi and the bracket thickness is 0.25 in., F = 500 lb.

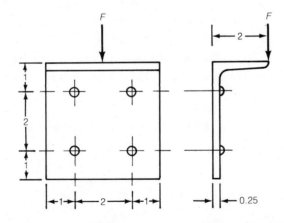

7.22 Remove the rivets 3 and 4 in Example 7.3.1 and compare the stress in each rivet and the reaction f_s at the lower edge of the bracket. The reduced reaction at the lower edge of the bracket is a consequence of the greater distance between the last row of bolts/rivets and the lower edge. It is assumed that deformation of the bracket is negligible.

7.23 It has been suggested that if the rivets in problem 7.21 are replaced by plastic bolts having a diameter of 0.25 in. to better withstand corrosion, the bracket could be lengthened and another bolt placed as illustrated in the figure. The tapered lower edge is to hasten runoff of seawater. Bracket thickness is 0.25 in. Would you agree to this change if the rupture stress for the bolts is 8000 psi in tension, 4200 psi in shear, and if a safety factor of 2.0 is required? $E = 1.7 \times 10^5$ psi. for the bolts.

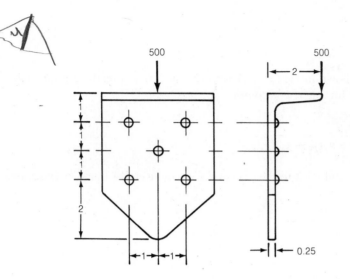

Sections 7.4–7.6

7.24 Write a program to solve equations 7.6.2 for $h_i = h_o$ along with equations 7.6.4, 7.6.5, and 7.6.6 for 10 forces and 20 welds. This program will eliminate much of the tedium involved in weld design.

The following weld problems will involve no more than two forces and five welds to limit the work required of those who do not have access to either a personal computer or a programmable pocket calculator.

7.25 Find the force/length on the weld shown below, where the dimensions are in mm. Recommend the fillet size to provide a safety factor of 2.0 relative to the yield strength of E8013-X welding rod material.

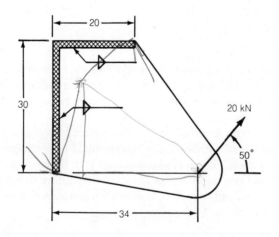

7.26 Hoisting brackets on a mine shuttle car were welded as shown below when the car was delivered. All welds are fillet welds, with the lengths shown. The maintenance foreman has suggested that this was an inspection error and that the 3.0-in.-long welds on each side should be extended to 6.0 in. as on the older cars. Fillet dimensions were unchanged. Do you agree with the maintenance foreman? Compare weld stresses for a 1000-lb load at the recommended load angle.

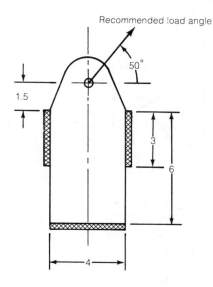

7.27 One week after delivery of the shuttle car described in problem 7.26 a letter arrived from the manufacturer stating that the hoisting brackets were in the process of being redesigned and that the old bracket had mistakenly been installed on the car recently delivered. The new bracket shown below is to be installed free of charge. (Dimensions are in inches. Bracket is $\frac{1}{2}$ in. thick.) (a) Is it an improvement? Base your answer on the stresses induced by the 1000-lb reference load used in problem 7.26. (b) Is use of a welding rod from the E9010-X to E9018-X range satisfactory for if the bracket is $\frac{1}{2}$ in. thick and if the maximum load is 20 000 lb?

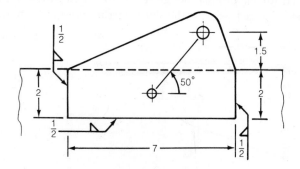

Dimensions are in inches. Bracket is $\frac{1}{2}$ in. thick.

7.28 Recommend the fillet dimensions when using a E8016-X All-Weld-Metal rod for the bracket shown in the figure. All dimensions are in inches and the load directions are constant. Maximum loads are shown. The plate is 0.25 in. thick and a safety factor of 4 is required. The plate may be chamfered at the welds to reduce any stress concentration factors which otherwise would occur if the fillet width were less than the plate thickness.

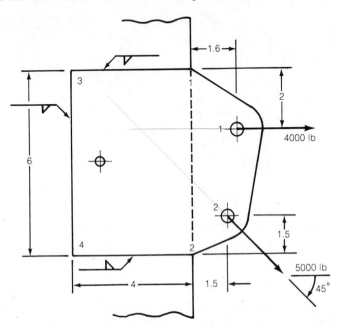

All dimensions in inches.

7.29 A failure occurred in weld A due to an emergency overload which was 100 kg over the rated yield load. The welds were repaired by replacing both 10-mm welds with 14-mm welds even though yielding occurred only in weld A. What is the load limit of the repaired bracket and what are the maximum load for the original design if a E7015-X electrode was used in both? If the rated failure load is unchanged, what is the safety factor with the 14 mm weld?

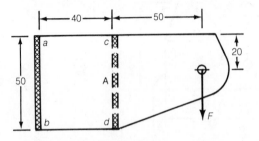

7.30 A customer would like to modify the bracket shown in the figure in problem 7.29 by welding between points *a* and *c* and between *b* and *d* to strengthen the bracket so that a second load of 34 393 N may be applied to it as shown in the figure below. Would you allow this change without voiding the warranty if the weld passes an ultrasonic inspection for freedom from voids and inclusions? An E7015-X electrode is to be used and the safety factor must be greater than 1.45.

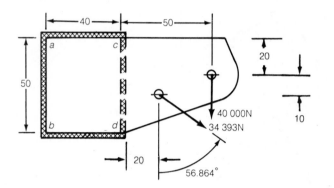

7.31 A bracket for $\frac{1}{4}$-in.-thick plate is to be used to attach a guy wire for a wind generator support tower. If wind loads are expected to induce a 5000-lb force in the steel cable, design a bracket and specify the weld dimensions for a E8016-CS electrode. Fillet welds $\frac{1}{4}$ in. on each side are preferred. Use a safety factor of 2.5.

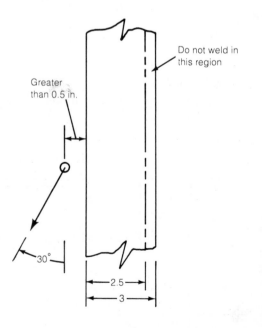

7.32 Compare the force/unit length of the bracket designed for problem 7.29 with the simple, commonly used straight bracket shown in the figure below. The additional weld metal required for the plain straight bracket is generally cheaper than the time and care required to cut a bracket which uses a minimum of weld metal, especially if dimensions are critical.

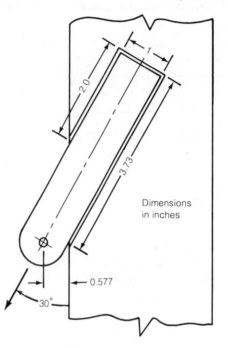

Dimensions
in inches

Sections 7.7–7.8

7.33 Find the force/unit length for the welds on the bracket shown below. Dimensions are in mm and the plates are 10 mm thick. What fillet size would you recommend for a safety factor of 1.5 relative to a yield stress of 482 MPa? $F = 116.5$ kN

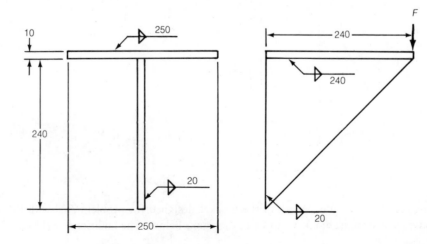

7.34 A rectangular stub box beam is to be welded to a thick steel bulk-head and is to support a vertical load of 1350 lb with a safety factor of 2.00 relative to the yield stress. (Dimensions are in inches.) What fillet width would you specify when a E9015-X rod is to be used? Sketch the beam with weld symbols showing your recommendation.

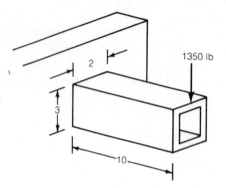

1350 lb

Sections 7.9–7.15

7.35 Recommend an adhesive thickness and an overlap length using the 2214 Regular adhesive shown in Table 7.13.1 for service from −10 to 150 °F. Adherends are annealed UNS 30200 stainless steel, for which $E = 25 \times 10^6$ psi and $v = 0.33$. Use $\varepsilon_{zx}^* = 0.20$, $\varepsilon_{zx}' = 0.50$, $v_a = 0.33$, and $E_a = 95\,000$ psi. Sheet thickness, in inches, is as indicated in the figure below. Sketch the joint and show the dimensions of the adhesive layers and of the upper and lower adherends.

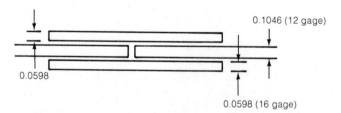

0.1046 (12 gage)

0.0598

0.0598 (16 gage)

7.36 Repeat problem 7.35 using a derating factor of 0.40 for the adhesive.

7.37 Design an adhesive joint using an adhesive similar to the 2214 Regular adhesive, shown Table 7.13.1, in which the main plates are Alclad 2018-T61 aluminum 3.40 mm thick and upper and lower plates are Alclad 2024-T3 aluminum. The upper plate is 1.70 mm thick the lower plate is 1.50 mm thick. For Alclad 2024-T3 nominal mechanical characteristics $E = 73$ GPa and $\sigma_u = 448$ MPa; for Alclad 2018-T61 they are that $E = 74$ GPa and $\sigma_u =$ MPa. Use $E_a = 655$ MPa, $v_a = 0.33$, and $\tau_y = 31$ MPa for the adhesive along with ε_{zx}^* and $\varepsilon_{zx}' = 0.5$. $F_i = \sigma_{ui}t_i$, $v = 0.30$.

7.38 Repeat problem 7.37 when a derating factor of 0.21 is used for the adhesive; i.e., $\tau_D =$ Poisson's ratio is 0.30 for both adherend materials. Also calculate the percent increa joint length compared to that found in problem 7.37.

7.39 Repeat problem 7.37 when the joint is to be subjected to an inplane shear force and use a derating factor of 0.21. Compare the adhesive thickness and joint length with that found in problem 7.38.

7.40 Design an adhesive joint which is to be subjected to a tensile force of 1400 N/mm and an in-plane shear force of 844 N/mm. Use Alclad 2024-T3 for the main plate and Alclad 2018-T61 for the upper and lower plates, which are of equal thickness. Use $v = 0.317$ for both alloys and see problem 7.35 for the remainder of the mechanical properties of these alloys. For the adhesive they are that $E_a = 650$ MPa, $v_a = 0.31$, $\tau_y = 34$ MPa, $\varepsilon_{zx}^* = 0.18$, and $\varepsilon_{zx}' = 0.48$. Give length of the joint, the thickness of the main, upper and lower plates and of the adhesive.

7.41 What is the maximum in-plane shear that may be supported by the joint designed in problem 7.40 if the tension is reduced 40%?

CHAPTER EIGHT

SHAFT DESIGN, CAMS, AND RETAINING RINGS

NOTATION

A	area *(12)*	R_f, R_{fe}, R_{fi}	ring factor, ring factor exterior, interior grooves
$a(i)$	polynomial coefficient	R_b	radius to bottom of groove *(l)*
a,b,c	length *(l)* or labels	R,R	reactions *(ml/t²)*
$\hat{a},\hat{b}$	unit vectors *(l)*	r,t,z	circular cylindrical coordinates *(l,1,l)*
c	wall thickness, thin wall tubes *(l)*	r,r_o,r_i	radii *(l)*
D,d	diameter *(l)*	T	period of rotation *(t)*
d	groove depth *(l)*	T'	duration per cycle of follower motion *(t)*
E	Young's modulus, elastic modulus		
e	length *(l)* or displacement of c.g. or perpendicular distance to line of action of cam follower	T,T_a,T_m	torque *(ml²/t²)*
		t	thickness *(l)*
F	force *(ml/t²)*	u,u_r	displacement *(l)*
G	shear modulus *(ml/t²)*	W	weight *(ml/t²)*
h	stroke of a cam	w	weight/length *(m/t²)* or wall thickness, hollow shaft *(l)*
hp	horsepower *(ml²/t³)*		
I, I_{ij}	moment of inertia *(ml²)* or moment of area *(l⁴)*	x,y,z	rectangular Cartesian coordinates
		Z	edge margin *(l)*
J	polar moment of area *(l⁴)*	α	angle *(1)*
$\hat{i},\hat{j},\hat{k}$	unit vectors in the x, y, and z directions	α_1,α_2	torsion constants *(1)*
i,j	subscripts	β	torsion constant or angle *(1)*
K,K_b,K_t	stress concentration factors	δ	deflection *(l)*
kW	kilowatt *(ml²/t³)*	ζ	safety factor *(1)*
l	length *(l)*	θ	angle *(1)*
$M,M,M,$	moment *(ml²/t²)*	σ_u,τ_u	ultimate stress in tension, shear *(m/lt²)*
n_x,n_y	components of a unit normal vector *(1)*		
		σ_e,τ_e	endurance limit in tension, shear *(m/lt²)*
$\hat{n}$	unit normal vector *(1)*		
P	Power *(ml²/t³)* or a point, or force on cam *(ml/t²)*	σ_{yp},τ_{yp}	yield stress in tension, shear *(m/lt²)*
		σ,σ_{zz}	direct stress *(m/lt²)*
F_g	force on groove wall *(ml/t²)*	$\tau,\sigma_{z\theta}$	stress stress *(m/lt²)*
P	force on cam parallel to follower line of action *(ml/t²)*	Φ	angle of twist *(1)*
		ϕ	angle of twist/length *(1/l)*
q	T'/T or reduction factor (retaining rings)	χ	deflection angle *(1)*
		ψ	angle *(1)*
R_i	inside radius, hollow shaft (retaining rings) *(l)*	ω	angular velocity *(1/t)*
		$(\cdot)$	time derivative

The term *shaft* is normally considered to include *axles* and *spindles*. All three terms are somewhat ill defined. A shaft may be either short or long in terms of its length to diameter (l/d) ratio. The term axle is often associated with vehicles, and in that connection may have an l/d ratio of the order of 60. Short shafts are often termed spindles, although no particular l/d ratio has been agreed upon to distinguish spindles from axles. No particular cross section is implied, although a circular cross section generally is the most practical for manufacture and for the highest stress to weight ratio.

8.1 TORSION OF CIRCULAR SHAFT

Shaft diameter is usually selected on the basis of one of two criteria: stress or deflection. In the first case load-carrying capability is of primary importance and in the second either relative position or deflection is important.

Torsional Stress

Shear stress in a shaft of circular cross section due to torsional loading is given by

$$\sigma_{z\theta} = \frac{Tr}{J} \tag{8.1.1}$$

in which T is the applied torque, r is the radius to the point where $\sigma_{z\theta}$ acts, and J is the polar moment of inertia, defined by

$$J = \frac{\pi}{2}(r_o{}^4 - r_i{}^4) \tag{8.1.2}$$

in terms of outer radius r_o and inner radius r_i. As implied by its subscripts, $\sigma_{z\theta}$ is the shear stress on an element of the cross section as in Figure 8.1.1. It arises due to torque T acting on the shaft and is given by (8.1.1) when the point of application is sufficiently far from either plane A or B in Figure 8.1.1.

Application of torque to a shaft by means of gears, sheaves (pulleys), or cranks, as in internal combustion engines, also adds lateral forces to the shaft so that bending stress must often be considered in shaft design as well. At points sufficiently far from a gear, crank, or sheave (usually three or four diameters is adequate) the stress due to bending may be represented by

$$\sigma_{zz} = \frac{Mr}{I} \tag{8.1.3}$$

where M is the bending moment at the plane on which σ_{zz} acts, r is the radial distance to the point at which σ_{zz} is to be calculated, and I is the transverse moment of inertia,

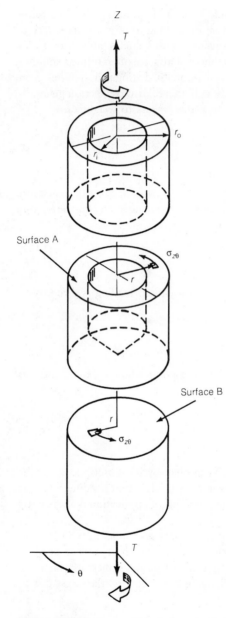

Figure 8.1.1 Torque and stress in a circular shaft.

defined by

$$I = \frac{J}{2} = \frac{\pi}{4} (r_o^4 - r_i^4) \tag{8.1.4}$$

With both bending and torsional stresses acting, the principal stressess must be considered in relation to tensile and shear strength of the shaft. From the equation 3.6.3 find

$$\begin{vmatrix} -\sigma & 0 & 0 \\ 0 & -\sigma & \sigma_{\theta z} \\ 0 & \sigma_{\theta z} & \sigma_{zz} - \sigma \end{vmatrix} = 0$$

or

$$\sigma = \frac{\sigma_{zz}}{2} \pm \left[\left(\frac{\sigma_{zz}}{2} \right)^2 + \sigma_{z\theta}^2 \right]^{1/2} \tag{8.1.5}$$

Substitution for σ_{zz} and $\sigma_{z\theta}$ from (8.1.1) and (8.1.3) into (8.1.5) leads to

$$\sigma = \frac{Mr}{2I} \pm \left[\left(\frac{Mr}{2I} \right)^2 + \left(\frac{Tr}{J} \right)^2 \right]^{1/2} \tag{8.1.6}$$

or

$$\sigma = \frac{Mr}{J} \pm \left[\left(\frac{Mr}{J} \right)^2 + \left(\frac{Tr}{J} \right)^2 \right]^{1/2} \tag{8.1.7}$$

as the largest stress, and to

$$\tau = \frac{1}{2} |\sigma_{max} - \sigma_{min}| = \left[\left(\frac{\sigma_{zz}}{2} \right)^2 + \sigma_{z\theta}^2 \right]^{1/2}$$

$$\tau = \frac{r}{J} \sqrt{M^2 + T^2} \tag{8.1.8}$$

as the maximum shear stress.

Deflection

Torsional deflection may be calculated from the relation

$$\varepsilon_{z\theta} = \frac{\sigma_{z\theta}}{2G} \tag{8.1.9}$$

and the condition that $\partial u_z / \partial \theta = 0$ for a circular shaft (this is not true for a noncircular shaft). Consequently, a rewritten (8.1.9) in the form

$$\frac{\partial u_\theta}{dz} = \frac{Tr}{JG} \tag{8.1.10}$$

and the condition that for a circular shaft

$$u_\theta = r\phi \tag{8.1.11}$$

lead to

$$d\phi = \frac{T}{JG}\, dz \tag{8.1.12}$$

or to

$$\Phi = \int_0^l d\phi = \int_0^l \frac{T}{JG}\, dz \tag{8.1.13}$$

where Φ is expressed in radians and ϕ is expressed in radians/length.

8.2 POWER TRANSMITTED

Power transmitted by a rotating shaft at constant rpm may be derived from the basic definition of power as the rate of doing work. Thus,

$$P = Fv = \frac{T}{r}\, v = T\omega$$

using $T = Fv$, where F is force and peripheral velocity $v = r\omega$ relates the velocity of a point on the shaft to its distance from the axis of rotation and the angular velocity ω of the shaft. Since

$$\omega = 2\pi n$$

relates angular velocity to rpm, the power may be written as

$$P = 2\pi Tn \tag{8.2.1}$$

In Old English units this becomes

$$\text{hp} = \frac{2\pi Tn}{12(33\,000)}$$

$$\text{hp} = \frac{\pi Tn}{198}\, 10^{-3} \cong \frac{Tn}{63}\, 10^{-3} \tag{8.2.2}$$

where hp is the horsepower transmitted by a torque

T in inch-pounds

turning a shaft which rotates at

n revolutions/minute (rpm)

In SI units (8.2.1) becomes

$$kW = \frac{2\pi n T}{60} 10^{-6} = \frac{\pi T n}{30} 10^{-6} \cong \frac{Tn}{9.55} 10^{-6} \tag{8.2.3}$$

where

T is in newton millimeters (N mm)

and

n is in revolutions/minute (rpm)

8.3 SHAFT DEFLECTION IN BENDING

Problems of shaft deflection in machine design are usually those of finding the angular and/or lateral deflection of a shaft of circular cross section in which the diameter changes with position along the shaft, finding these deflections for a shaft in which all forces do not act in the same plane, or a combination of both. Van den Broek's elastic energy method, also known as the unit load method, will be used exclusively for these problems because it is easily programmed.[1]

According to this method the lateral deflection δ, due to bending, is given by

$$\delta(P) = \int_0^l \frac{mM}{EI} dz \tag{8.3.1}$$

and angular rotation in the bending plane is given by

$$\chi(P) = \int_0^l \frac{m*M}{EI} dz \tag{8.3.2}$$

in terms of a coordinate system whose z axis coincides with the longitudinal axis of the shaft. Torsional rotation is, of course, given by equation 8.1.13, already discussed.

In equations 8.3.1 and 8.3.2 $M = M(z)$ is the bending moment due to all forces and moments acting on the shaft (applied loads and reactions) and $m = m(z)$ is the moment due to a unit load acting at point P, the location where the deflection is wanted. If the shaft and its supports form a redundant structure, m may be calculated after enough supports are removed to eliminate redundancy but still maintain stability. Equation 8.3.2 differs only in that m^* is the bending moment due to a unit moment applied at the point where angle χ is to be calculated. A consistent convention must be used for m, m^*, and M in determining the algebraic signs for forces and moments. In this chapter we consider forces positive downward and moments positive if they bend the beam downward, as shown in Figure 8.3.1b.

Before considering more complicated cases, we will demonstrate the use of equation 8.3.1 by calculating the deflection of a simply supported shaft at a distance c to the

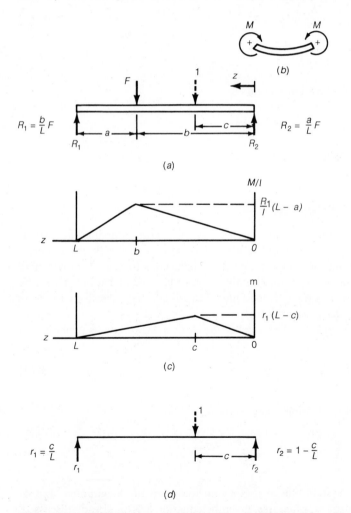

Figure 8.3.1 (a) Simply supported beam with a single load, (b) positive moments, (c) diagrams for $M(z)/I$ and $m(z)$, (d) unit load and its reactions.

left of the right-hand support due to a load at a distance a from the left-hand support, as shown in Figure 8.3.1a. The moment due to reaction $R_1 = Fb/L$ for positions between R_1 and F is given by

$$M(z) = \frac{Fb}{L}(L - z) \qquad \text{for } L \geq z \geq b \tag{8.3.3}$$

and the moment at values of z less than b due to both force F and reaction R_1 is given by

$$M(z) = F\left(1 - \frac{b}{L}\right)z \qquad \text{for } b \geq z \geq 0 \tag{8.3.4}$$

When only the unit load acts, the reaction at the left end of the beam is $r_1 = c/L$. Thus, the moment acting between r_1 and the unit load is given by

$$m(z) = c\left(1 - \frac{z}{L}\right) \qquad \text{for } L \geq z \geq c \tag{8.3.5}$$

while the moment for positions between the unit load and the right-hand support may be written as

$$m(z) = z\left(1 - \frac{c}{L}\right) \qquad \text{for } c \geq z \geq 0 \tag{8.3.6}$$

Selecting the limits on the integral in equation 8.3.1 may be simplified by sketching the moments as illustrated in Figure 8.3.1c. This also serves to check the moment expressions, which were derived algebraically.

Substitution for $M(z)$ and $m(z)$ in equation 8.3.1 from equations 8.3.3 through 8.3.6 leads to

$$\delta = \frac{F}{EI}\left(1 - \frac{b}{L}\right)\left(1 - \frac{c}{L}\right)\int_0^c z^2\, dz + \frac{Fc}{EI}\left(1 - \frac{b}{L}\right)\int_c^b z\left(1 - \frac{z}{L}\right) dz$$
$$+ \frac{Fbc}{EI}\int_b^L \left(1 - \frac{z}{L}\right)^2 dz \tag{8.3.7}$$

After integration, the expression for deflection δ becomes

$$\delta = \frac{Fc^3}{3EIL^3}(L^2 - bL - cL + bc) + \frac{Fc}{EIL^2}(L - b)\left(\frac{Lb^2 - Lc^2}{2} - \frac{b^3 - c^3}{3}\right)$$
$$+ \frac{Fbc}{3EIL^3}(3Lb^2 - 3L^2b + L^3 - b^3)$$

After tedious, but careful, simplification we have

$$\delta = \frac{Fac}{6EIL}(bL + ba - c^2) \tag{8.3.8}$$

To obtain the relation given in panel 7, Figure 3.11.2, in Chapter 3, we must write the sum of the first two terms in parentheses in equation 8.3.8 as

$$b(L + a) = (L - a)(L + a) = L^2 - a^2$$

to get

$$\delta = \frac{Fac}{6EIL}(L^2 - a^2 - c^2) \tag{8.3.9}$$

Graphical integration has not been used here because we plan to write a computer program to solve the integral in equation 8.2.1 with less work than is required for the graphical method. It also has the advantage that it can handle arbitrary, irregular, distributed loads with equal facility.

The computer program consists of a main program which performs trapazoidal integration using an integrand that is calculated by two subroutines, one for moment $M(z)$ and the other for moment $m(z)$.

The main program in FORTRAN is very simple:

```
DIS=0
DO I=1, N
   X=I*DEL
Y=X+DEL
   DIS=DIS+(PMOM(X)+PMOM(Y))*DEL/2
CONTINUE
```

where DIS respresents the displacement, DEL is the distance between calculation points z_i and z_{i+1}, and PMOM is the function which calculates the value of $m(z)M(z)/I(z)$. In Microsoft Quick BASIC (used because of its elimination of the line numbering requirement, is larger vocabulary, and its simplified handling of subroutines compared to primative BASIC) the main program (exclusive of input commands) may appear as

```
DIS= 0 : N=L/DEL : NI=N+2 : I=0
   DO UNTIL I=NI
   X=I*DEL : Y=X+DEL
   DIS=DIS+(FNPMOM(X)+FNPMOM(Y))*DEL/2 : I=I+1
LOOP
```

where L is the distance between the most widely separated forces or reactions on the shaft. To account for shafts that have different diameters along their lengths, we also include a subroutine to calculate any changes in the second moment of area of the a shaft. Thus, the flowchart for a program to calculate shaft deflection may appear as in Figure 8.3.2. Representative subroutines and a program listing will be given in the following example.

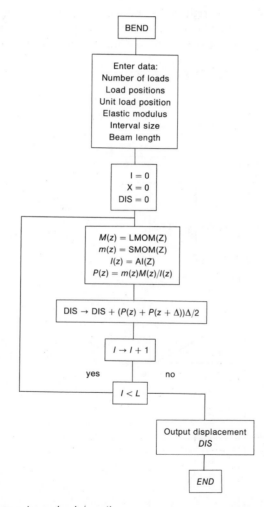

L = shaft length between extreme loads/reactions
Y = product of $m(x)M(x)$
Δ = interval size
DIS = δ = shaft displacement at unit load position

Figure 8.3.2 Flowchart for main program for calculating shaft deflection according to equation 8.3.1.

EXAMPLE 8.3.1

Find the lateral deflection at A of the stepped shaft shown in Figure 8.3.3a. (These changes in diameter are to facilitate press-fitting one gear between stations 2 and 3 and another between stations 3 and 4. The second gear is positioned against the shoulder at station 4 and the first is pressed against it for proper positioning of both. A hub on the first gear may be added if separation of the gear faces is necessary.) Assume the self-aligning bearings at each end are equivalent to a simple support 1-in. inboard from each end. $E = 30 \times 10^6$ psi.

Upon taking moments about the right-hand end we find that $R_1 = Fb/L$ in which b is the distance between point A and station 6, as shown in the Figure 8.3.3a. The second moments of area (area moments of inertia) for shaft cross sections between stations i and j are found from $I_{ij} = \pi r_{ij}^4/4$ to be

$$I_{12} = 0.0491 \text{ in.}^4 \qquad I_{34} = 0.2485 \text{ in.}^4$$

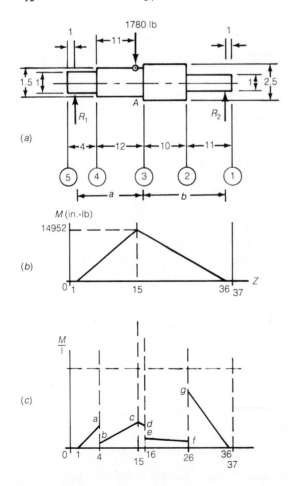

Figure 8.3.3 (a) shaft profile, (b) M diagram, and (c) M/I diagram.

$$I_{23} = 1.9175 \text{ in.}^4 \qquad I_{45} = I_{12} = 0.0491 \text{ in.}^4$$

The bending moment diagram corresponding to the 1780-lb force at A is shown in Figure 8.4.2b, where

$$M(z) = R_1(L - z) = Fb(L - z)/L \qquad \text{between point } A \text{ and station 6}$$

and

$$M(z) = R_1(L - z) - F(b - z) = Fz(1 - b/L) \qquad \text{between } A \text{ and station 1,}$$

where $L = 35$ in. and $a = 21$ in. Since we are concerned with the deflection of the shaft at the load, the bending moment diagram for the unit load is of similar form:

$$m(z) = b(L - z)/L \qquad \text{between station 6 and point A}$$

and

$$m(z) = z(1 - b/L) \qquad \text{between } A \text{ and station 1}$$

The diagram of $M(z)/I(z)$ is shown in Figure 8.3.2c as an aid in checking the magnitude of the contributions of each section of the shaft to the total deflection.

Instead of substituting into the integrals in equation 8.3.1 we will simply write the following three subroutines in Quick BASIC (because FORTRAN programs with adequate checking of coding errors are not available on personal computers at this time) using the DEF FN command which corresponds to a subroutine in primative BASIC. Data input commands for the magnitude of the load, its distance from the left end of the shaft, distances to stations 1 through 6, the distance of the unit load from the left end of the shaft, and the different radii are assumed to be in the main program.

With this data the moment $M(z)$ may be calculated from

```
DEF FNLMOM(Z)
   IF Z> =0 AND Z<LF THEN
      FNLMOM=F*LF*(1-Z/L)
   ELSE
      FNLMOM=F*Z*(1-LF/L)
   END IF
END DEF
```

and moment $m(z)$ for a unit load a distance A from the left end may be calculated from

```
DEF FNSMOM(Z)
   IF Z> =0 AND Z<A THEN
      FNSMOM=A*(1-Z/L)
   ELSE
      FNSMOM=Z*(1-A/L)
   END IF
END DEF
```

The second moment of area I_{ij} for each different cross section may be calculated from

```
DEF FNAI (Z)
    IF Z> =0 AND Z<ST(2) THEN
        FNAI=PI*(R(1)^4)/4
    ELSEIF Z> =ST(2) AND Z<ST(3) THEN
        FNAI=PI*(R(2)^4)/4
    ELSEIF Z> =ST(3) AND Z<ST(5) THEN
        FNAI=PI*(R(3)^4)/4
    ELSE
        FNAI=PI*(R(4)^4)/4
    END IF
END DEF
```

where $z = 0$ at station 1, the location of reaction R_2 through the right-hand bearing. The final function definition required is

```
DEF FNPMOM(Z)=FNLMOM(Z)*FNSMOM(Z)/FNAI(Z)
```

Since the force, its location, the location of the unit load, the radii, and the station locations are all variables in the above program, we may use it to solve the example problem and any other problem involving a shaft which is simply supported with a single load and four different radii. Once the complete main program is written it is obviously easier to rewrite the routines FNLMOM, FNSMOM, and FNAI than it is to calculate the deflection by analytical methods such as the moment area method, the conjugate beam method, or the singularity function method.

Upon substituting

$$F = 1780 \text{ lb} \qquad ST(2) = 10 \text{ in.} \qquad b = 21 \text{ in.}$$

$$ST(3) = 20 \text{ in.} \qquad ST(4) = 32 \text{ in.} \qquad L = 35 \text{ in.}$$

$$R(1) = 0.5 \text{ in.} \qquad R(2) = 1.25 \text{ in.} \qquad R(3) = 0.75 \text{ in.}$$

$$R(4) = 0.5 \text{ in.}$$

into the program we find $DIS = \delta = 0.1756$ in. at A when using $DEL = 0.1$ in.

EXAMPLE 8.3.2

Find the reduction in the deflection of the shaft shown in Figure 8.3.3 when the section between station 2 and 3 is lengthened so that station 2 is now 6 in. to the left of station 1.

Set $ST(2) = 6$ in the program to find $\delta = 0.1258$ in., again using $DEL = 0.1$ in.; i.e., $z_{i+1} - z_i = 0.1$ in.

Deflection out of the vertical plane due to an applied force out of the vertical plane may be found by decomposing the force vector into its components in the vertical and horizontal planes, calculating the deflection in each of these planes, and then taking their vector sum.

EXAMPLE 8.3.3

Find the deflection of the midpoint of the shaft shown in Figure 8.3.3 due to a force of 1800 lb acting at an angle of 25° out of the vertical plane as shown in Figure 8.3.4.

Upon taking components in the vertical and horizontal we find that $F_v = F \cos 25° = 1631.35$ lb and $F_h = F \sin 25° = 760.71$ lb. After placing the unit load at 17.5 in. from the left end of the shaft we find from the above program that $\delta_v = 0.1636$ in. and $\delta_h = 0.0763$ so that

$$\delta = (\delta_v^2 + \delta_h^2)^{1/2} = 0.1805 \text{ in.}$$

The program written for the above problems can easily be extended to the case of two vertical forces acting on the stepped shaft. The listing for the program with this extension is displayed in Figure 8.3.5. It can be applied to a beam with a single force by setting F_1, say, equal to 0 and placing it anywhere to the right of force F_2, say at $z = 1.0$.

Numbered lines 5, 10, 15, and 30 through 150 in the BEND24 listing must be altered if the number of stations is changed. First the dimensions for ST and RA must be changed and then corresponding changes in the ranges of the DO instruction, lines 10 and 15, must be made to prompt for the appropriate stations and radii. Finally, the IF . . . ELSE . . . THEN statement in the DEF FNAI function definition, lines 30 through 150, must include the radii for the modified number of stations.

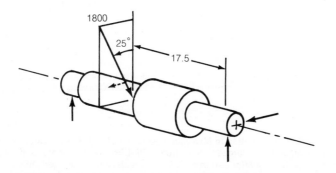

Figure 8.3.4 Shaft loaded out of the vertical plane. Only one load is shown for simplicity. All additional forces are resolved into vertical and horizontal components.

```
      'Program entitles BEND24.BAS for beam deflection due to 2 lateral loads
      'on a simply suported shaft having up to 4 different cross-sections.
      DEFDBL A-H, L-Z
5     DIM F(2), L(2), ST(3), RA(4)
      OPEN "BEMOM.DAT" FOR OUTPUT AS 1 : I=0

      DO UNTIL I = 2
      I=I+1
      PRINT "Enter force " I
      INPUT "and its distance from the right end. ", F(I), L(I)
      LOOP

      I = 1
10    DO UNTIL I=5
15    IF I<4 THEN
         PRINT "Count stations from right to left. "
         PRINT "Enter the distance of station " I
         INPUT "from the right hand end. ", ST(I)
         INPUT "Enter the radius to the right of the above station. ", RA(I)
      ELSE
         INPUT "Enter the radius to the left of the last station. ", RA(I)
      END IF
      I=I+1 : LOOP

      INPUT "Enter shaft length & interval size. ", LS, DEL
      INPUT "Enter the elastic modulus E. ", E
      INPUT "Enter the position of the unit load from the right end. ", U

      DEF FNLMON(X)
        IF X> =0 AND X<L(1) THEN
          FNLMON=F(2)*X*(1−L(2)/LS)+F(1)*X*(1−L(1)/LS)
        ELSEIF X> =L(1) AND X< =L(2) THEN
          FNLMON=F(2)*X*(1−L(2)/LS)+F(1)*L(1)*1−X/LS)
        ELSE
          FNLMON=F(2)*L(2)*(1−X/LS)+F(1)*L(1)*(1−X/LS)
        END IF
      END DEF

      DEF FNSMOM(X)
        IF X> =0 AND X<U THEN
          FNSMOM=X*(1−U/LS)
        ELSE
          FNSMOM=U*(1−X/LS)
        END IF
      END DEF
```

Figure 8.3.5 Program listing for SIMBEND.BAS written in Microsoft BASIC, version 3.06 for bending of a simply supported bean with two forces acting. *Note:* File BEMOM.DAT contains a printout of position z, the values of $M(z)$ and $m(z)$, and the displacement, labeled DIS, for each point as an aid to checking results. X is used in the program to represent position z.

```
30    DEF FNAI(X)
      PI = 3.141592654
        IF X > = 0 AND X < ST(1) THEN
          FNAI = PI*(RA(1)^4)/4
        ELSEIF X > = ST(1) AND X < ST(2) THEN
          FNAI = PI*(RA(2)^4)/4
        ELSEIF X > = ST(2) AND X < ST(3) THEN
          FNAI = PI*(RA(3)^4)/4
        ELSE
          FNAI = PI*(RA(4)^4)/4
        END IF
150   END DEF

      DEF FNPMOM(X) = FNLMOM(X)*FNSMOM(X)/FNAI(X)
      JL = LS/DEL + 1 : J = 0 : DIS = 0 : X = 0
      DO UNTIL J = JL
      J = J + 1 : Y = X + DEL : XA = FNPMOM(X) : XB = FNPMOM(Y)
      DIS = DIS + (XA + XB)/(2*E) : PRINT #1, X; Y; XA; XB
      PRINT #1, DIS : X = Y : LOOP
      PRINT "Displacement is "DIS

      CLOSE #1
      END
```

Figure 8.3.5 (*continued*)

8.4 SHAFT ON THREE SUPPORTS

This problem is statically indeterminant, which means that it cannot be solved directly from the equations of equilibrium because there are more unknowns than equations. It can, however, be solved straightforwardly by means of the superposition of loads and deflections as long as all of the stresses and deflections and their sums are within the elastic range. Problems involving more than one redundant support can be solved by the method of Section 8.3, as demonstrated in Ref. 1.

Our procedure will be to remove one of the redundant supports and its reaction to make the problem determinant and to then solve for the deflection δ at the point P where the support was removed. We will then find the deflection δ_0 at point P by a unit load in the direction of the original reaction at P when the shaft is constrained only by the reactions at the supports that were retained. The original reaction R at P is then given by the product of the unit load and the ratio δ/δ_0 and the actual deflection is given by the superposition principle as the sum of the deflections found with R removed and the deflection due to R.

EXAMPLE 8.4.1

Find the reaction at B on a shaft between self-aligning bearings one meter apart at A and D if the support at B is 1 mm below the line of centers between bearings A and D. Shaft diameter is 20 mm, the modulus of elasticity is 205 GPa, and the transverse load at C is 1017 N.

Following the procedures outlined above, we determine the shaft deflection and the reaction at support B by

a finding the deflection δ at B due to F with the support removed,

b finding the force R_B to move the shaft upward an amount $\delta - \delta_0$ ($\delta_0 = 1$ mm in this example) with force F removed, and

c replacing F and R_B and writing $\delta_0 = \delta - (\delta - \delta_0)$ as the shaft deflection.

According to panel 7, Chapter 3, used in the previous example, with R_B removed

$$\delta = \frac{1017(500)}{6(205)10^3 \frac{\pi}{2} 10^4}\left(1 - \frac{700}{1000}\right)[2(500)10^3 - 300^2 - 500^2] = 5.2111 \text{ mm}$$

With F removed, an upward force of 1.0 N at B will cause a deflection of

$$\delta_1 = \frac{500}{6(205)\frac{\pi}{2} 10^7}(1 - 0.5)[10^6 - 2(500)^2] = 0.00647 \text{ mm/N}$$

so that an upward deflection of 4.2111 mm will require

$$R_B = \frac{1.053}{0.00647} = 650.894 \text{ N}$$

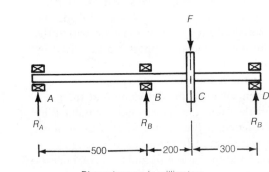

Dimensions are in millimeters

Figure 8.4.1 Shaft on three supports.

EXAMPLE 8.4.2

What will be the reaction at B after the bearing support is repaired to eliminate the initial deflection? With $\delta_0 = 0$ the upward deflection must be equal to δ, so that

$$R_B = \frac{2.053}{0.00647} = 317.31 \text{ N}$$

8.5   STRESS CONCENTRATIONS

Ease of manufacture and repair, in addition to requirements for bearing surfaces to prevent axial motion of shafts, require shaft diameter to change at various locations along the shaft length. Figure 8.5.1 shows the output shaft of a gear reducer in which a spherical roller bearing and thrust shoulder fit on section b to support the left end

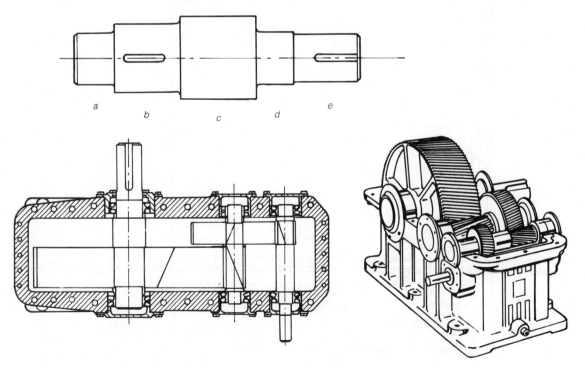

Speed reducer

Figure 8.5.1 Stepped output shaft of a gear train. *Source: The Torrington Company, Torrington, CT.*

of the shaft, to prevent shaft motion to the left side, and to prevent the gear that fits on section *b* from moving to the left. Section *b* is of larger diameter than *a* to make it easier to mount the gear because there is a slight press fit (interference fit) between the gear and the shaft. The slot in the surface of section *b* is the keyway for the output gear. Section *c* is of larger diameter than *b* to prevent motion of the gear to the right and it extends to the housing to provide added rigidity to the shaft. Section *d* is reduced to that of *a* to prevent motion of the shaft to the right and to accept the support bearing on the right side of the shaft. Section *e* is of a smaller diameter than *d* to facilitate mounting the spherical roller bearing on *d* because it too is to be a light press fit. Keyway in section *e* is to accept whatever the customer wishes to attach to the output shaft.

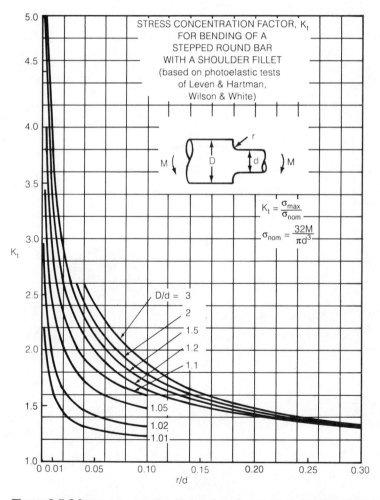

Figure 8.5.2 Stress concentration for bending of a round shaft with a shoulder fillet. *Source: R.E. Peterson,* Stress Concentration Factors, *New York: Wiley, 1974. Copyright © 1974 John Wiley & Sons, Inc. Reprinted by permission.*

Changes in cross section cause a stress concentration in the immediate vicinity of the transition for both bending and torsion. Stress concentration magnitudes are, however, different for torsion than for bending. For this reason symbols K_b and K_t will be used in the remainder of this chapter to distinguish between the one for bending, K_b, and the one for torsion, K_t.

Although there are applications when a shaft may simultaneously experience tension, torsion, and bending (as in a drill string for an offset hole in an oil field) that case will not be considered explicitly because no new principles are involved—just more terms.

Stress concentration factors and their dependence upon shaft parameters shown are given in Figures 8.5.2 through 8.5.6.

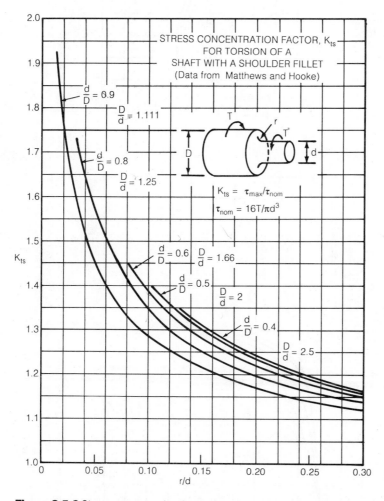

Figure 8.5.3 Stress concentration factor for torsion of a shaft with a shoulder fillet. *Source: R.E. Peterson*, Stress Concentration Factors, *New York: Wiley, 1974. Copyright © 1974 John Wiley & Sons, Inc. Reprinted by permission.*

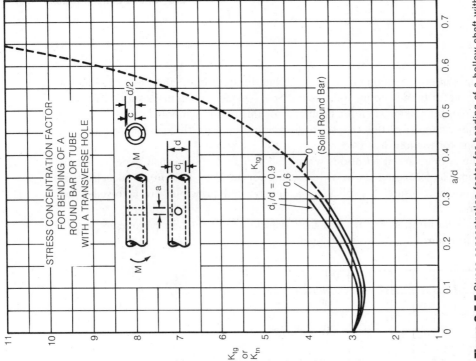

Figure 8.5.4 Stress concentration factor for a shaft with an axial hole and a shoulder fillet. *Note:* K_{tso} obtained from Figure 8.5.3. *Source: R.E. Peterson, Stress Concentration Factors, New York: Wiley, 1974. Copyright © 1974 John Wiley & Sons, Inc. Reprinted by permission.*

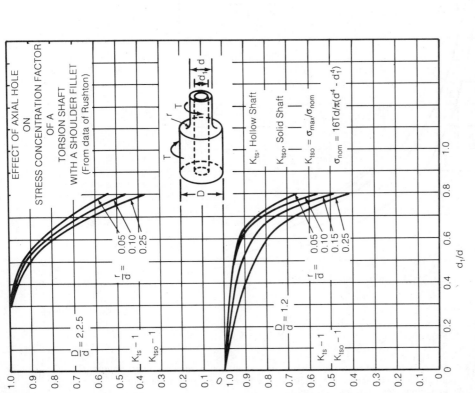

Figure 8.5.5 Stress concentration factor for bending of a hollow shaft with a transverse hole.

$$K_t = K_{tg} = \frac{\sigma_{max}}{32Md_o/[\pi(d_o^4 - d_i^4)]} \qquad d = d_o$$

Source: R.E. Peterson, Stress Concentration Factors, New York: Wiley, 1974. Copyright © 1974 John Wiley & Sons, Inc. Reprinted by permission.

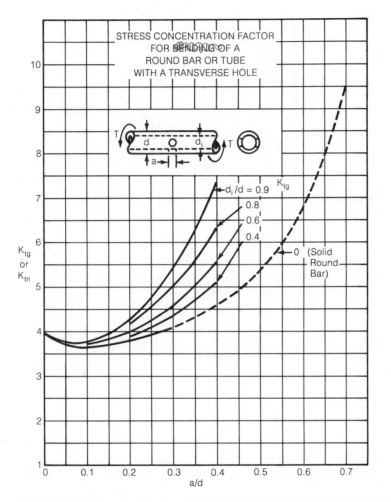

Figure 8.5.6 Stress concentration factor for torsion of a shaft with a transverse hole.

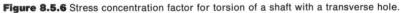

$$K_s = K_{tg} = \frac{\sigma_{max}}{16Td_o/[\pi(d_o^4 - d_i^4)]} \qquad d = d_o$$

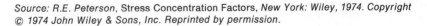

Source: R.E. Peterson, Stress Concentration Factors, *New York: Wiley, 1974. Copyright © 1974 John Wiley & Sons, Inc. Reprinted by permission.*

EXAMPLE 8.5.1

Design a shaft of circular cross section to be 0.90 m long to sustain a torque of 14 433 N m with an angular deflection of less than 0.07° over its length. Shear stress should not exceed 145 MPa. Its profile is to be similar to the shaft shown in Figure 8.5.7, with the restriction that each diametral change should

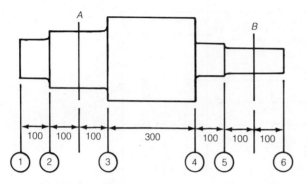

Figure 8.5.7 Location of diametral changes and input and output gears. Dimensions are in millimeters.

be no less than 20 mm and that fillet radii should leave a bearing radius of at least 6 mm at each step to provide adequate bearing surface against the steps. Driving gear is centered at A and the driven gear is centered at B. Neglect bearing friction and use $G = 80$ GPa.

Begin the design by finding diameter B from

$$\sigma_{z\theta} = Tr/J = 2T/(\pi r^3)$$

which yields

$$r^3 = 2T/(\pi\sigma_{z\theta}) = 2(14\,433)10^3/(145\pi)$$

$$r = 39.87 \text{ mm}$$

Thus, use 80 mm as a trial diameter between stations 5 and 6. Minimal diametral change of 20 mm corresponds to a radial step of 10 mm and a fillet radius of 4 mm. If shaft diameter is 100 mm between stations 4 and 5 then

$$D/d = 10/8 = 1.25 \qquad r/d = 4/80 = 0.050$$

and the stress concentration factor at station 5 is 1.565 from Figure 8.5.3. Recalculation of the radius at B yields

$$r = 39.87(1.565)^{1/3} = 46.29 \text{ mm}$$

Set $r = 47.00$ mm and increase the fillet radius to $0.05(2)47.0 = 4.70$ mm to maintain the original r/d ratio. By increasing D to $1.5(2)47 = 141$ mm between stations 4 and 5 we may also maintain the same D/d ratio and hence the same stress concentration factor and a difference in diameters in excess of 20 mm.

Since we have already selected the diameter of the shaft between stations 5 and 6 to satisfy the stress requirement, we need not calculate stress between stations 2 and 4 as long as all diameters exceed

92.58 mm. Thus, let the diameter between stations 3 and 4 be 161 mm and between stations 2 and 3 be 141 mm.

Next, turn to the calculation of the angular twist between locations A and B from equation (8.1.13), which may be written as

$$\Phi = \frac{T}{G} \sum_{i=1}^{4} \frac{l_i}{J_i} = \frac{2(14\,433) \times 10^3}{\pi 80 \times 10^3} \left(\frac{100}{47^4} + \frac{100}{70.5^4} + \frac{300}{80.5^4} + \frac{100}{70.5^4} \right)$$
$$= 4.1041 \times 10^{-3} \text{ rad}$$

This is greater than the limiting twist of $0.07° = 1.2217 \times 10^{-3}$ rad by a ratio of 3.3593. Hence, it is necessary to multiply each radius by $3.3593^{1/4} = 1.3538$ and round upward to reduce the angular deformation to $0.07°$. The resulting diameters of 128 mm between stations 5 and 6 and 192 mm between stations 4 and 5 produces $D/d = 1.5$. Increasing the fillet radius to $r = 7.00$ mm causes little change in the stress concentration, so that no additional stress calculations are necessary.

The design shaft diameters and fillet radii are given in the following table.

Between Stations	Diameters (mm)	Station	Fillet Radius (mm)
2 and 3	192	3	11.00
3 and 4	218	4	11.00
4 and 5	192	5	7.00
5 and 6	128	6	—

The shaft diameter stations 1 and 2 will be selected to accept a stock bearing and the diameter between stations 4 and 5 may be modified slightly to also accept a stock bearing at that location.

In other situations it may also be necessary to design to hold lateral deflections and bending stresses to within specified limits.

8.6 SHAFT FATIGUE

All material elements not on the center line of a drive shaft that experiences lateral loading are subjected to one tension load and one compression load for each rotation of the shaft. This is true for shafts that hold gears, belt pulleys, chain sprockets, grinding wheels, cams, and so on. These shafts will usually be subjected to cyclic torsional stress as the loads change. Moreover, these loads may fluctuate many times per minute, as in the case of shafts driven by internal combustion engines, whose torque pulsates with each power and compression stroke. Torque also fluctuated for shafts driving piston compressors, shakers, hammer mills, punch presses, and similar machinery.

Preliminary design estimates for the allowable fluctuating stresses in power shafts have usually been based upon either a modification of the maximum shear stress, as determined from Mohr's circle, or the HMH criterion involving equivalent static stresses as described in Chapter 4, Section 11. When the maximum shear stress criterion is used and the stress components are replaced by the equivalent static stresses according to the Soderberg criterion, as in Section 4.11, we obtain

$$\frac{\sigma_{y_t}}{\zeta} = \left[\frac{1}{4}\left(\sigma_{zz_m} + K_{f_t}\frac{\sigma_{y_t}}{\sigma_{e_t}}\sigma_{zz_a}\right)^2 + \left(\sigma_{z\theta_m} + K_{f_s}\frac{\sigma_{y_s}}{\sigma_{e_s}}\sigma_{z\theta_a}\right)^2\right]^{1/2} \tag{8.6.1}$$

where σ_{y_t} represents the yield stress in shear and where subscripts t and s denote quantities found from tension or shear data, respectively. Although use of the Soderberg equivalent static stresses insures freedom from yielding, it is unnecessarily cautious and generally results in machine components that are heavier than using more realistic criteria.

These same comments apply to the criterion obtained by substituting the Soderberg equivalent static stresses into the HMH criterion, to obtain

$$\frac{\sigma_{y_t}}{\zeta} = \left[\left(\sigma_{zz_m} + K_{f_t}\frac{\sigma_{y_t}}{\sigma_{e_t}}\sigma_{zz_a}\right)^2 + 3\left(\sigma_{z\theta_m} + K_{f_s}\frac{\sigma_{y_s}}{\sigma_{e_s}}\sigma_{z\theta_a}\right)^2\right]^{1/2} \tag{8.6.2}$$

Generally a lighter but equally safe design may be had by using the Gerber equivalent static stresses in the fatigue life criterion given by equations 4.11.3 through 4.11.6 in Chapter 4, but at the expense of greater computational complexity, since it takes the form shown in equations 8.6.3 and 8.6.4:

$$\frac{\sigma_{u_t}}{\zeta} = \left(\frac{1}{4}\left\{K_{f_t}\frac{\sigma_{u_t}}{\sigma_{e_t}}\sigma_{zz_a} + \left[\left(K_{f_t}\frac{\sigma_{u_t}}{\sigma_{e_t}}\sigma_{zz_a}\right)^2 + 4\sigma_{zz_m}{}^2\right]^{1/2}\right\}^2\right.$$
$$\left. + \frac{3}{4}\left\{K_{f_s}\frac{\sigma_{u_s}}{\sigma_{e_s}}\sigma_{z\theta_a}\right)^2 + \left[\left(K_{f_s}\frac{\sigma_{u_s}}{\sigma_{e_s}}\sigma_{z\theta_\alpha}\right)^2 + 4\sigma_{z\theta_m}{}^2\right]^{1/2}\right\}^2\right)^{1/2} \tag{8.6.3}$$

$$\frac{\sigma_{y_t}}{\zeta} = [(K_{f_t}\sigma_{zz_a} + \sigma_{zz_m})^2 + 3(K_{f_s}\sigma_{z\theta_a} + \sigma_{z\theta_m})^2]^{1/2} \tag{8.6.4}$$

This complexity is immaterial, however, once all three criteria are available in a computer program. Once programmed, the method involving equations 8.6.3 and 8.6.4 may be preferred for component design because of its better agreement with experiment.[3]

Recall from Chapter 4 that these last two equations relate the maximum stresses and the safety factors, so that both must be solved for the particular stress component or safety factor desired. We then accept the smaller of the values so obtained as satisfying the requirement of indefinite life and purely elastic deformation.

These four relations may be more conveniently used in shaft design if they are rewritten in terms of the torques and moments applied to the shaft by replacing the stresses by the applied torques and moments according to

$$\sigma_{zz} = Mc/I \qquad \sigma_{z\theta} = Tr/J$$

where I and J denote the second and polar moments of the cross-sectional area of the shaft at the point where the stresses are to be evaluated. For shafts of circular cross-section $c = r$, $I = \pi r^4/4$, and $J = \pi r^4/2$, so that the above relations become

$$\frac{\sigma_{y_s}}{\zeta} = \frac{16}{\pi d^3} \left[\left(M_m + K_{f_t} \frac{\sigma_{y_t}}{\sigma_{e_t}} M_a \right)^2 + \left(T_m + K_{f_s} \frac{\sigma_{y_s}}{\sigma_{e_s}} T_a \right)^2 \right]^{1/2} \qquad (8.6.5)$$

based upon the maximum shear stress criterion using the Soderberg equivalent static stresses. Use of the HMH criterion and the Soderberg equivalent static stresses leads to

$$\frac{\sigma_{y_t}}{\zeta} = \frac{32}{\pi d^3} \left[\left(M_m + K_{f_t} \frac{\sigma_{y_t}}{\sigma_{e_t}} M_a \right)^2 + \frac{3}{4} \left(T_m + K_{f_s} \frac{\sigma_{y_s}}{\sigma_{e_s}} T_a \right)^2 \right]^{1/2} \qquad (8.6.6)$$

where M_m and M_a denote the mean and alternating applied moments and T_m and T_a denote the mean and alternating torques, respectively. The fatigue strength reduction factor in tension is represented by K_{f_t} and in shear is represented by K_{f_s} in all of the above expressions.

Replacement of the stress components in equations 8.6.3 and 8.6.4 by their associated moments and torques leads to

$$\frac{\sigma_{u_t}}{\zeta} = \frac{16}{\pi d^3} \left(\left\{ K_{f_t} \frac{\sigma_{u_t}}{\sigma_{e_t}} M_a + \left[\left(K_{f_t} \frac{\sigma_{u_t}}{\sigma_{e_t}} M_a \right)^2 + 4 M_m^2 \right]^{1/2} \right\}^2 \right.$$
$$\left. + \frac{3}{4} \left\{ K_{f_s} \frac{\sigma_{u_s}}{\sigma_{e_s}} T_a + \left[\left(K_{f_s} \frac{\sigma_{u_s}}{\sigma_{e_s}} T_a \right)^2 + 4 T_m^2 \right]^{1/2} \right\}^2 \right)^{1/2} \qquad (8.6.7)$$

and to

$$\left(\frac{\sigma_{y_t}}{\zeta} \right) = \frac{32}{\pi d^3} \left[(K_{f_t} M_a + M_m)^2 + \frac{3}{4} (K_{f_s} T_a + T_m)^2 \right]^{1/2} \qquad (8.6.8)$$

EXAMPLE 8.6.1

Find the maximum mean tensile stress that can be superimposed upon a drive shaft that now supports a mean shear stress of 75 MPa, a fluctuating shear stress of 32 MPa, and a completely reversed fluctuating stress of 105 MPa due to bending. A safety factor of 1.5 is required. The notch sensitivity factor is 0.72 and the other mechanical properties of the steel from which the shaft is made are listed below.

$\sigma_{u_t} = 710$ MPa $\sigma_{u_s} = 410$ MPa

$\sigma_{y_t} = 621$ MPa $\sigma_{y_s} = 330$ MPa

$\sigma_{e_t} = 317$ MPa $\sigma_{e_s} = 158.5$ MPa

$K_t = 1.38$ $K_s = 1.73$

The shear properties listed in the second column are frequently missing from tabulated mechanical properties and so must be estimated as either $1/2$ or $1/\sqrt{3}$ times the corresponding value in tension. As indicated in Figure 4.7.4, the $1/\sqrt{3}$ factor appears to be in better agreement with experiment.

We begin by calculating the fatigue strength reduction factor in tension from

$$K_{f_t} = q(K_t - 1) + 1 = 0.72(1.38 - 1) + 1 = 1.274$$

and in shear from

$$K_{f_s} = 0.72(1.73 - 1) + 1 = 1.526$$

Use these results and solve equation 8.6.1 for σ_{zz} to find that

$$\sigma_{zz_m} = 4\left[\left(\frac{\tau_{max}}{\zeta}\right)^2 - \left(\sigma_{z\theta_m} + K_{f_s}\frac{\sigma_{y_s}}{\sigma_{e_s}}\sigma_{z\theta_a}\right)^2\right]^{1/2} - K_{f_t}\frac{\sigma_{y_t}}{\sigma_{e_t}}\sigma_{zz_a}$$

$$\sigma_{zz} = 4\left[\left(\frac{330}{1.5}\right)^2 - \left(75 + 1.526\frac{330}{158.5}32\right)^2\right]^{1/2} - 1.274\frac{621}{317}105$$

$$= 262.4 \text{ MPa} \qquad \text{Max. shear and Soderberg}$$

Next, after solving for σ_{zz} from condition 8.6.2 we have that

$$\sigma_{zz_m} = \left[\left(\frac{\sigma_{y_t}}{\zeta}\right)^2 - 3\left(\sigma_{z\theta_m} + K_{f_s}\frac{\sigma_{y_s}}{\sigma_{e_s}}\sigma_{z\theta_a}\right)^2\right]^{1/2} - K_{f_t}\frac{\sigma_{y_t}}{\sigma_{e_t}}\sigma_{zz_a}$$

$$= \left[\left(\frac{621}{1.5}\right)^2 - 3\left(75 + 1.526\frac{330}{158.5}32\right)^2\right]^{1/2} - 1.274\frac{621}{317}105$$

$$= 16.8 \text{ MPa} \qquad \text{HMH and Soderberg}$$

Last, from equations 8.6.3 and 8.6.4, written as

$$\left(\frac{\sigma_{u_t}}{\zeta}\right)^2 = \sigma_{zz}^{*2} + 3\sigma_{z\theta}^{*2}$$

we have that

$$\sigma_{zz}^* = \left[\left(\frac{\sigma_{u_t}}{\zeta}\right) - 3\sigma_{z\theta}^{*2}\right]^{1/2} \qquad\qquad (8.6.9a)$$

where

$$\sigma_{z\theta}^* = \frac{1}{2}\left(K_{f_s}\sigma_{u_s}\frac{\sigma_{z\theta_a}}{\sigma_{e_s}}\right) + \frac{1}{2}\left[\left(K_{f_s}\sigma_{u_s}\frac{\sigma_{z\theta_a}}{\sigma_{e_s}}\right)^2 + 4\sigma_{z\theta_m}^2\right]^{1/2}$$

$$= \frac{1}{2}\left(1.526(410)\frac{32}{158.5}\right) + \frac{1}{2}\left[\left(1.526(410)\frac{32}{158.5}\right)^2 + 4(75^2)\right]^{1/2}$$

$$= 63.1581 + 81.3771 = 144.5352 \text{ MPa}$$

so that substitution for $\sigma_{z\theta}^*$ into (a) yields

$$\sigma_{zz}^* = \left[\left(\frac{710}{1.5} \right)^2 - 3(161.2088^2) \right]^{1/2} = 401.7128 \text{ MPa}$$

Next,

$$\sigma_{zz_m}^* = \frac{1}{2} \left(K_{f_t} \sigma_{u_t} \frac{\sigma_{zz_a}}{\sigma_{e_t}} \right) + \frac{1}{2} \left[\left(K_{f_t} \sigma_{u_t} \frac{\sigma_{zz_a}}{\sigma_{e_t}} \right)^2 + 4\sigma_{zz_m}^{\ 2} \right]^{1/2} \tag{8.6.9b}$$

so that upon solving for σ_{zz_m} we have that

$$
\begin{aligned}
\sigma_{zz_m} &= \frac{1}{2} \left\{ \left[2\sigma_{zz}^* - K_{f_t} \sigma_{u_t} \frac{\sigma_{zz_a}}{\sigma_{e_t}} \right]^2 - \left(K_{f_t} \sigma_{u_t} \frac{\sigma_{zz_a}}{\sigma_{e_t}} \right)^2 \right\}^{1/2} \\
&= \frac{1}{2} \left\{ \left[2(401.7128) - 1.274(710) \frac{105}{317} \right]^2 - \left[1.274(710) \frac{105}{317} \right]^2 \right\}^{1/2} \\
&= 202.52 \text{ MPa}
\end{aligned}
$$

which satisfies criterion 8.6.3.

Turning next to yield condition 8.6.4,

$$\left(\frac{\sigma_{y_t}}{\zeta} \right)^2 = \sigma_{zz}' + 3\sigma_{z\theta}'$$

we have, upon solving for σ_{zz}', that

$$\sigma_{zz}' = \left[\left(\frac{\sigma_{u_t}}{\zeta} \right)^2 - 3\sigma_{z\theta}'^2 \right]^{1/2} \tag{8.6.9c}$$

where

$$\sigma_{z\theta}' = K_{f_s} \sigma_{z\theta_a} + \sigma_{z\theta_m} = 1.526(32) + 75 = 123.8320 \text{ MPa}$$

and

$$\sigma_{zz}' = K_{f_t} \sigma_{zz_a} + \sigma_{zz_m}$$

so that upon substituting into (c) and solving for σ_{zz_m} we have that

$$\sigma_{zz} = \left[\left(\frac{710}{1.5} \right)^2 - 3(123.832^2) \right]^{1/2} - 1.274(105) = 288.18 \text{ MPa}$$

Since σ_{zz} must take on the smaller of these two values to not exceed the stresses given by equations 8.6.3 and 8.6.4, we must select

$$\sigma_{zz} = 202.5 \text{ MPa}$$

as the maximum mean tensile stress. Its value was rounded down instead of up to avoid violating condition 8.6.3.

The mean stress calculated by this last method was distinctly larger than that from the Soderberg condition. This again demonstrates that the Soderberg criterion is unnecessarily cautious in order to obtain a simple relation.

When using equations 8.6.7 and 8.6.8 to solve for the shaft diameter it is important to remember that it is the larger of the two values for d that must be selected if the shaft is to support the specified torques and moments for an indefinite life and without yielding. This is consistent with the selection of the smaller values of torques, moments, and safety factors because they are all associated with a larger diameter.

8.7 KEY AND KEYWAY DESIGN

Relative rotation between a shaft and a gear, sheave, or flywheel is often prevented by machining a groove in the hub of the gear, for example, and machining a similar groove in the shaft so that when the gear is placed upon the shaft the two grooves may be aligned and a rectangular block of metal fitted into both grooves as shown in Figure 8.7.1. These grooves are known as *keyways* or *keyseats* and the metal block is known as a *key*. Some of the more common styles of keys and keyways are shown in Figure 8.7.2. Open keyways are usually fitted with tapered or gib keys which are forced into the keyways so that friction will keep them in position. Although the enlarged portion of the gib key shown in Figure 8.7.2 is to extend beyond the gear for relatively easy removal, it may be dangerous, unless covered, because of the possibility of catching on clothing when the shaft is rotating.

Figure 8.7.1 Shaft, gear, and key.

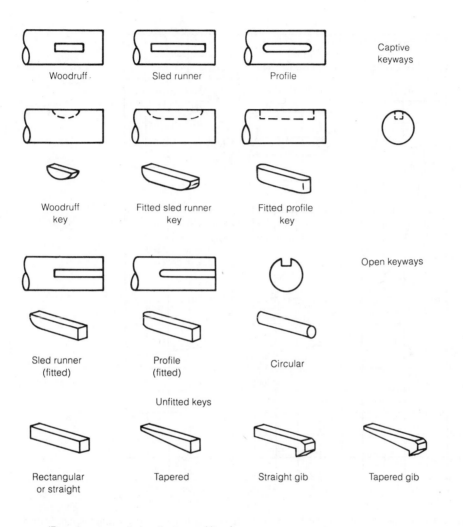

Captive keyways

Woodruff   Sled runner   Profile

Woodruff key   Fitted sled runner key   Fitted profile key

Open keyways

Sled runner (fitted)   Profile (fitted)   Circular

Unfitted keys

Rectangular or straight   Tapered   Straight gib   Tapered gib

(Taper is exaggerated on the tapered keys)

Figure 8.7.2 Keys and keyways.

Because the keyway introduces a change in the geometry of the shaft it causes a stress concentration at the change in cross section, as indicated by the solid curve in Figure 8.7.3. The stress concentration at a keyway is more complicated than that at an external change in shaft diameter, however, because it is three dimensional. Once a key is inserted into the keyway and a torsional load applied to the shaft by means of the key the stress concentration becomes even more complicated because the force on the key introduces another stress concentration that is not on the surface of either the shaft or the keyway but embedded in the shaft near one end of the key. Since this embedded stress concentration was unknown before 1979,[4] the stress concentration factors found in the literature before that date do not represent those associated with

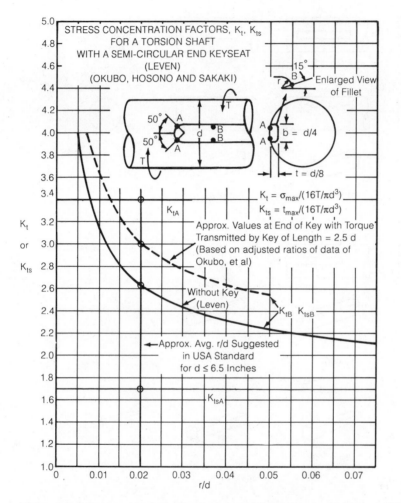

Figure 8.7.3 Stress concentration factors for a profile keyway (empty) in a shaft in torsion. *Source: R.E. Peterson*, Stress Concentration Factors, *New York: Wiley, 1974. Copyright © 1974. John Wiley & Sons, Inc. Reprinted by permission.*

a key and keyway used to transfer torque to or from a shaft. In fact, all of the earlier theoretical and experimental work, with the exception of the work of Okubo, Hosono, and Sakaki,[5] was concerned with the stress concentration at an empty keyway. In other words, the keyway was simply a change in cross section of a shaft placed in torsion by torques applied elsehwere. It was also found[6] that shaft failures at the keyway began as tension failures, so that the stress concentration factors at the keyway listed in Table 8.7.1 are defined by

$$K_t = \frac{\text{tensile stress}}{\text{torsional stress, outer surface, central slice}}$$

TABLE 8.7.1   KEY AND KEYWAY STRESS CONCENTRATION FACTORS

Configuration	K_t
Full length key, profile keyway (round end key)	
At key way end	3.81
Embedded	3.34
Partial length key, profile keyway (round end key)	
At keyway and	2.84
Embedded near end of key	1.94
Partial length key, sled runner keyway (round end key)	
At keyway end	2.79
Embedded near key end	1.68
Full length key, profile keyway (square end key)	
Embedded near key end	5.83*

Note: Changing fillet radii and key and keyway length to width ratios may significantly affect K.
*   To within an uncertainty of 0.3

Redesign of the key and keyway as shown in Figure 8.7.4 was found to reduce both the stress concentration at the end of the key and at the end of the keyway. Rounding off the corner along the inboard end of the key, in the vicinity of point 1 in Figure 8.7.4, has been found to further reduce the stress concentration at the end of the keyway. With these modifications both stress concentrations may be reduced to less than 1.2 for certain fillet radii and key and keyway length to width ratios. It was also found[6]

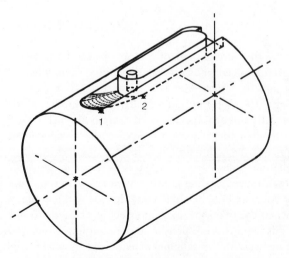

Figure 8.7.4 Detail of redesigned key and keyway for lower stress concentrations. *Source: W.C. Orthwein, A new key and keyway design, Journal of Mechanical Design, vol. 101, pp. 338–341, American Society of Mechanical Engineers, New York (1979).*

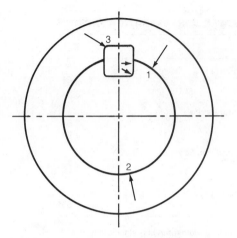

Figure 8.7.5 Forces on a hub containing a key and keyway. *Source: W.C. Orthwein, A new key and keyway design*, Journal of Mechanical Design, *vol. 101, pp. 338–341, American Society of Mechanical Engineers, New York (1979).*

that the forces on the hub of a gear, or any similar attachment, are not smoothly distributed over any part of the hub, but are concentrated at the locations shown in Figure 8.7.5. The force at location 3 may cause hub failure in the vicinity of the keyway if the hub is too thin.

8.8 COUPLINGS

Couplings are a means for connecting the output shaft of a motor to the input shaft, to a clutch, to a geartrain or to a variable speed transmission, depending upon whether the speed ratio between the motor and the driven unit is to have a fixed or variable speed ratio, or to the driven unit itself if the motor output and driven unit inputs speeds are compatible. If the units are carefully aligned and likely to remain aligned a rigid coupling may be used; otherwise a flexible coupling with type of coupling dependent upon the the type of shaft misalignment expected: parallel, in-line, or angular. As illustrated in Figure 8.8.1, parallel misalignment describes the situation where the two shaft center lines are parallel but laterally displaced; in-line misalignment refers to the situation where the two shaft axes are colinear but the shaft ends do not make contact; and angular misalignment occurs when the two shaft axes intersect at an angle to one another. Usually all three types of misalignment are present and a coupling is selected on the basis of its ability to operate with the misalignment expected as well as its ability to reduce vibration from one unit to another.

Some of the more common couplings used in machine design are described in the following paragraphs.

TABLE 8.8.1   SUMMARY AND COUPLING SELECTION GUIDE

Coupling Style	Conditions for Use
Rigid Flange Coupling	When frequent uncoupling is required and angular and parallel misalignment is negligible. Small in-line misalignment can be tolerated. Flanges must be installed in pairs and they must remain paired if moved to another shaft to maintain hole indexing. Index holes should be marked for ease reinstallation. These couplings are usually keyed or splined to each shaft. No vibration or electrical isolation of one shaft from the other is provided.
Rigid Sleeve Coupling	Similar to the rigid flange coupling except easier to remove and install. Requires high-strength bolts and very small tolerance on shaft diameters and on sleeve tolerances.
Rigid Compression Sleeve Couping	Relies upon friction between shaft and coupling, so machining of either shaft is unnecessary. Will not tolerate angular or parallel misalignment but may be used with small in-line misalignment. Not practical for large diameter shafts and high torque and no electrical or vibrational isolation is provided between shafts.
Elastomeric Couplings	Relatively simple mechanically, require little maintenance and may be easily replaced. Angular misalignment up to $2.0°$ or a parallel misalignment up to 7.0 mm may be tolerated, depending upon coupling size. Electrical and partial vibrational isolation is provided. These advantages are accompanied by a reduced torsional capacity as compared to rigid couplings.
Flexible Metal Couplings	Angular, in-line, and parallel misalignment may be accommodated with these couplings at higher torques for a given diameter than with elastomeric units. They provide partial isolation from shock, vibration, and axial loads and can withstand higher temperatures than elastomeric couplings. Used where temperature, torque, or electrical conductivity may rule out elastomeric couplings.
Gear Couplings	Used for large torque and some inline and angular misalignment is present. Parallel misalignment may be accomodated if space permits a pair of gear couplings with a spacer between the couplings.
Spring Couplings	Suitable for low torque and large angular misalignment—up to $60°$. May accept small in-line and parallel misalignment.
Schmidt Couplings	Designed for parallel misalignment with adequate space between shafts for insertion of the coupling. Production models are for low-torque applications associated with fractional hp motors.
Fluid Couplings	Designed to provide vibration isolation and controlled soft starts for a wide range of torques from 1 to 5000 hp (3730 kW) at 1200 rpm. Larger sizes available on order. Usually provided with controlled filling to provide soft start capability. May accept some inline misalignment. Angular and parallel misalignment may require other couplings in series.

Rigid Sleeve Split Couplings

This is simply a split hollow sleeve which is either keyed or clamped to both shafts. Whenever clamping is used high-strength bolts and minimum thread tolerance should be employed because the bolts are placed in tension. Split sleeve construction, as shown in Figure 8.8.2, permits easy installation and removal, and the rigidity of the connection provides power transmission with no power loss, but at the expense of no vibra-

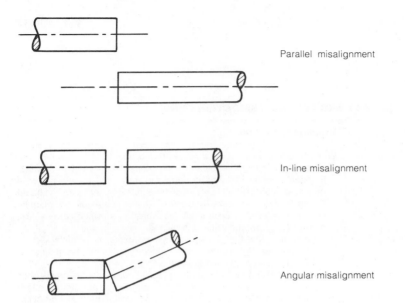

Parallel misalignment

In-line misalignment

Angular misalignment

Figure 8.8.1 Types of shaft misalignment.

Figure 8.8.2 Rigid sleeve coupling. *Source: courtesy of Holo-Krome Co., West Hartford, CT.*

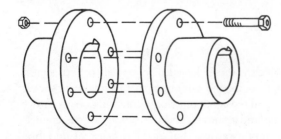

Figure 8.8.3 Rigid flange coupling.

tion isolation, a small tolerance for in-line misalignment, and no tolerance for angular or parallel misalignment.

Rigid Flange Couplings

Hub and flange assemblies on each shaft, Figure 8.8.3, are connected by bolts through the flanges. Minimum tolerance on bolt hole diameter and location are necessary to assure that each bolt carries its share of the torque, which places the bolts in shear. If possible, holes should be drilled with the mating flanges together, and an index mark placed on each flange for proper orientation upon assembly. Alignment tolerances for this coupling are similar to those for the rigid sleeve coupling.

Rigid Compression Sleeve Coupling

These couplings rely upon the wedging action of split conical sections which are clamped between the internally tapered sleeve and the shaft to provide a friction grip upon the two shafts to transmit torque. Since the external sleeve is not split the torsional strength of the sleeve is greater than that of the rigid sleeve split coupling. This advantage is obtained, however, at the cost of a larger external diameter, more expensive machining, and the requirement that each bolt be properly tightened. If a key and keyway are provided, as in Figure 8.8.4, to reduce the circumferential stress in the sleeve the shaft itself is weakened due to the stress concentration at the key and keyway. Tolerance for misalignment of the shafts and vibration isolation between shafts is the same as that of the previous rigid couplings.

Elastomeric Couplings

This is a modification of the flange coupling in that each flange is bolted to an elastomeric section, either a disc or a sleeve, between the two flanges. The flexibility of the elastomer provides tolerance for angular misalignment of the order of 2° and a parallel misalignment of the order of 7 mm. The values may be increased, but usually at the expense of torsional strength. The internal damping of the elastomer also provides some vibrational isolation and the high dielectric constants of most elastomers

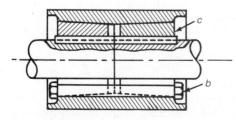

Figure 8.8.4 Compression sleeve coupling, axial bolts, seamless outer sleeve. *b*, bolts; *c*, tapered cylinders.

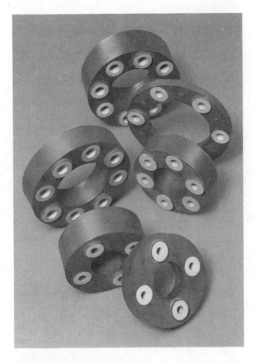

Figure 8.8.5 Annular elastomeric couplings. *Source: Lord Corporation, Erie, PA.*

assures fairly good electrical isolation between shafts. Several styles of these couplings are shown in Figure 8.8.5.

Flexible Metal Couplings

These couplings differ from the elastomeric couplings in that they dissipate very little energy themselves, but may provide greater torsional capacity. The Falk coupling, Figure 8.8.6, consists of a steel spring threaded through axial slots which are contoured at their ends to allow some torsional flexibility. Maximum torque is transmitted when the springs are bent to contact the slot walls along their entire length.

Other versions of a flexible metal coupling are shown in Figure 8.8.7 in which metal disks or diaphrams transmit torque and allow for some angular and in-line misalignment.

Spring Couplings

Greater angular misalignment, perhaps 60° or greater, may be achieved using the helical or springlike couplings illustrated in Figure 8.8.8. Because of the limited capacity of the spring cross sections, these couplings are limited to small torques and may,

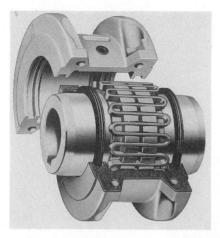

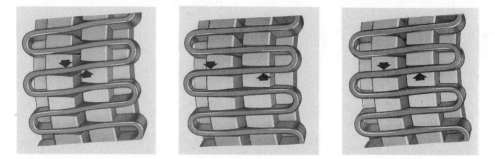

Figure 8.8.6 The Falk coupling. *Source: Reproduced with permission of the Falk Corporation, Milwaukee, WI.*

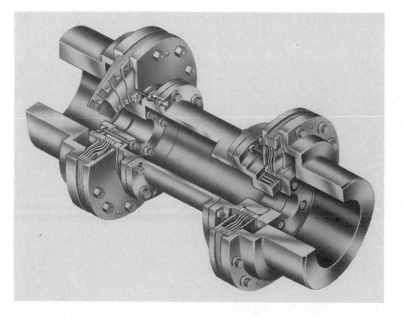

Figure 8.8.7 Flexible metal coupling. *Source: Mechanical Drives Division, Industries, Inc., Erie, PA.*

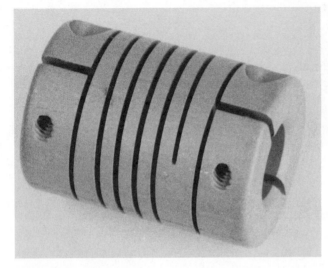

Figure 8.8.8 Helical couplings. *Source: Helical Products Co., Inc., Santa Maria, CA.*

therefore, be attached to the shafts with a set screw or small friction clamp. Some in-line and parallel misalignment may be tolerated as well.

Gear Couplings

Gear couplings, shown in Figure 8.8.9, may provide large torsional capacity along with tolerance for in-line misalignment, dependent upon the length of the central internal

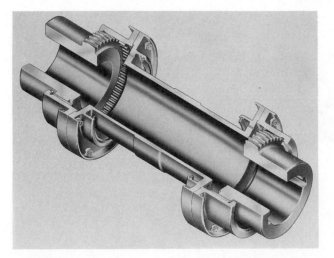

Figure 8.8.9 Gear coupling. *Source: Mechanical Drives Division, Zurn Industries, Inc., Erie, PA.*

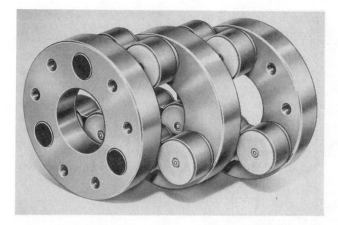

Figure 8.8.10 Schmidt coupling for parallel misalignment. *Source: Zero-Max, Minneapolis, MN.*

gear, and tolerance for angular misalignment of the order of 1.5° when appropriately contoured teeth are used. Part of the price of these advantages is the requirement that they must be well lubricated to prevent wear.

Schmidt Couplings

These mechanical couplings, pictured in Figure 8.8.10, are of interest because they have been designed to accommodate large parallel misalignment between shafts 0.5 in. in diameter or smaller. This capability is obtained at the expense of requiring ample lubrication, since each coupling has 12 journal bearings, and a very small tolerance for in-line misalignment.

Fluid Couplings

Clutch and torsional shock isolation functions are well performed by these couplings, similar to that shown in Figure 8.8.11. They are usually used where the shafts are well aligned and where smooth acceleration is desired. However, these characteristics are purchased at the price of greater energy dissipation during start-up and a very slight increase in energy loss during constant speed operation.

8.9   UNIVERSAL JOINTS

Angular misalignment of more than 3 to 5° prohibits use of the couplings listed in the previous section except for very small torque. Usually universal joints are selected for large torque and intermediate angles of shaft intersection (less than 45°) while flexible shafts are selected for smaller torques and larger angles between shafts. Nevertheless, many intermediate applications exist in which either may be satisfactory.

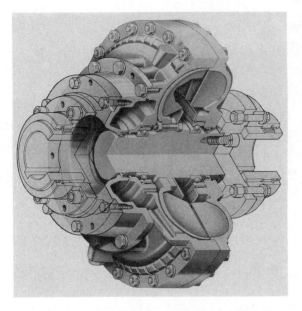

Figure 8.8.11 Fluid coupling. *Source: Reproduced with permission of the Falk Corporation, Milwaukee, WI.*

Universal joints are often classified as either constant or non-constant velocity joints. Unfortunately, the cheapest and most common universal joint, the Cardan joint (actually described by Honnecourt about 300 years earlier), Figure 8.9.1*a*, is not a constant velocity joint. In other words, a constant angular velocity input to one side of the joint does not produce a constant angular velocity output on the other side of the joint whenever the two shafts intersect at an angle greater than zero.

Imagine that a Cardan joint has been oriented with the input shaft and angular velocity vector parallel to the x axis and the output shaft in the xy plane at an angle β with the positive x axis. To reduce clutter, the xyz reference frame and the cross-member of the Cardan joint have been duplicated in Figures 8.9.1*b* and *c*, with the input cross, yoke, and the attached shaft shown in Figure 8.9.1*b* and the output cross, yoke, and shaft displaced as shown in Figure 8.9.1*c*. Since shaft B is along the x axis, it follows that the arm of the cross attached to the input yoke will always be perpendicular to the x axis; hence, a unit vector coincident with the axis of that arm may be given by

$$\hat{\mathbf{b}} = \hat{\mathbf{j}} \sin \theta + \hat{\mathbf{k}} \cos \theta \tag{8.9.1}$$

while a unit vector coincident with that arm of the cross that is perpendicular to arm A, at angle β with the x axis, may be written as

$$\hat{\mathbf{a}} = -\hat{\mathbf{i}} \sin \phi \sin \beta + \hat{\mathbf{j}} \sin \phi \cos \beta + \hat{\mathbf{k}} \cos \phi \tag{8.9.2}$$

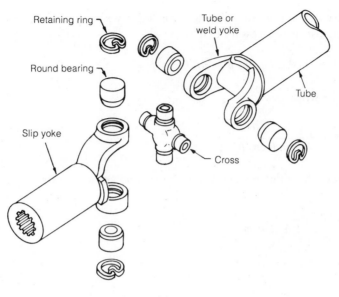

(a)

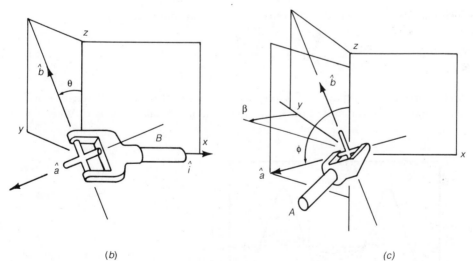

(b)   (c)

Figure 8.9.1 Cardan joint and the coordinate system used in its analysis. *Source:* Universal Joint
and Driveshaft Design Manual, *AE-7.* © *1979 Society of Automotive Engineers, Inc.*
Reprinted with permission.

Because these two unit vectors are each coincident with one of the two perpendicular arms of the cross, it is necessary that

$$\hat{a} \cdot \hat{b} = \sin\theta \sin\phi \cos\beta + \cos\theta \cos\phi = 0 \tag{8.9.3}$$

which may be written as

$$\tan\phi = -\frac{\cos\theta}{\sin\theta \cos\beta} \tag{8.9.4}$$

Substitution of this value for $\tan\phi$ into the trigonometric identity

$$\cos^2\phi = \frac{1}{1 + \tan^2\phi}$$

yields

$$\cos^2\phi = \frac{\tan^2\theta \cos^2\beta}{1 + \tan^2\theta \cos^2\beta}$$

which may be differentiated with respect to time and simplified to get

$$\dot{\phi} = \frac{\cos\beta}{1 - \sin^2\theta \sin^2\beta}\dot{\theta} \tag{8.9.5}$$

which expresses the output angular velocity $\dot{\phi}$ in terms of the angle θ between shafts, the instantaneous position θ of the input shaft, and its angular velocity $\dot{\theta}$. When $\theta = 0$

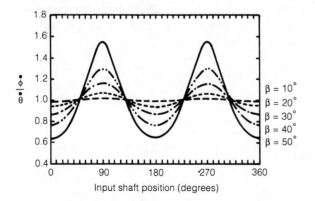

Figure 8.9.2 Angular velocity ratio for a Cardan joint for θ from 0 to 2π for shafts at an angle β as indicated.

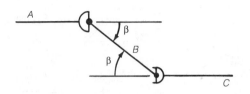

Figure 8.9.3 Cardan universal joints between parallel, offset shafts to achieve constant velocity.

then $\dot{\phi} = \dot{\theta}\cos\beta$ and when $\theta = \pi/2$ then $\dot{\phi} = \dot{\theta}/\cos\beta$. Thus, if $\beta = \pi/4$ the output velocity $\dot{\phi}$ varies from $\dot{\phi} = \dot{\theta}\sqrt{2}$ to $\dot{\phi} = \dot{\theta}/\sqrt{2}$; that is, the maximum angular velocity of the output shaft is twice its minimum angular velocity and this fluctuation occurs twice per rotation of the input shaft, as shown in Figure 8.9.2. Due to this inherent acceleration and deceleration of the output shaft it is standard practice to limit angle B to 3.5° at 5000 rpm and to about 11.3° at 1500 rpm.

If two Cardan joints are used as shown in Figure 8.9.3, however, where shafts A and C are parallel, the rotational speed of shafts A and C will be equal, but that of the intermediate shaft B will oscillate according to equation 8.9.5 and Figure 8.9.2.

Constant velocity joints are available and are used in a number of applications where a double Cardan joint is impractical, such as in front wheel drive automobiles. These include the tripot joint, the Rzeppa, the ball spline Rzeppa, the cross-groove universal joint, the Weiss, and the Devos, or double offset universal joint. One of these, the Rzeppa joint, is shown in Figure 8.9.4.

8.10 FLEXIBLE SHAFTS

Figure 8.10.1 shows several flexible shafts and the nomenclature associated with them.[10] Construction of the rotating *core* begins with a single straight spring wire which is tightly wrapped with bands of parallel round wire as shown in Figure 8.10.1. Wrap directions are alternately clockwise and counterclockwise, with the direction of the outer band determining the core type: clockwise or counterclockwise. This distinction is important in flexible shafts because the applied torque should cause the outer wrap to tighten if the maximum torsional capacity of the shaft is to be obtained. Core ends are terminated in *fittings* which keep the wires from unwinding and which provide a splined, rectangular, keyed, or other stub shaft to fit the couplings at the driving and driven ends.

A nonrotating *casing* usually surrounds the core to protect it from moisture and abrasion, to retain lubricant within the core, and to provide enough flexural stiffness to prevent the core from wrapping around itself when transmitting power along a curved arc. Casings are usually terminated in ferrules which may, for example, attach the casing to the frame of the driving and driven units.

The life of a flexible shaft decreases as the torque increases and as the radius of curvature decreases. A life of 100 000 000 cycles for a core diameter of about 0.125 in.

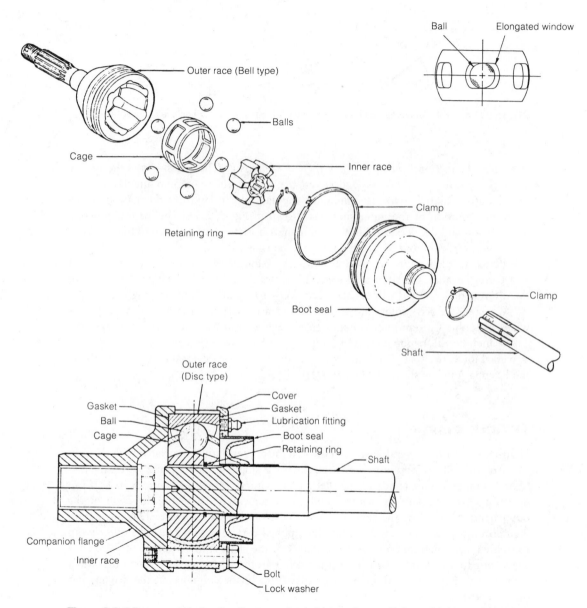

Figure 8.9.4 Rzeppa and ball spline Rzeppa universal joints. *Source:* Universal Joint and Driveshaft Design Manual, *AE-7.* © *1979 Society of Automotive Engineers, Inc. Reprinted with permission.*

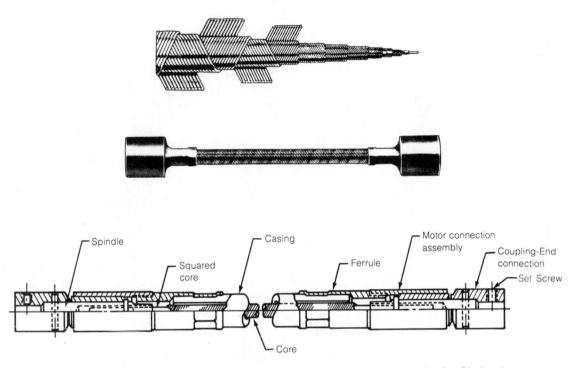

Figure 8.10.1 Flexible shafts and their nomenclature. *Source: Stow Manufacturing Co., Binghamton, NY.*

may be realized, for example, at 4500 rpm if the radius of curvature is no less than 50 in. for a torque of 2.4 lb-in. At a radius of curvature of 3 in. its recommended torque is reduced to 0.5 lb-in. Large diameter cores, about 1.3 in. in diameter, may transmit 1500 lb-in. at a radius of curvature of 25 in. or more.[8]

8.11 CAM DESIGN

Cams provide a simple and reliable means of obtaining periodic translational or angular motion from rotational motion. Perhaps the most common use of cams is in the control of valve opening and closing in four-stroke internal combustion engines. Placement of cams on a single shaft, or use of toothed belts, chains, or gears between camshafts, as shown in Figure 8.11.1 provides an equally simple and reliable means of maintaining the proper phase between the cam controlled elements. These systems are also employed in packing machines, wire-forming machines, textile manufacturing equipment, and numerous other machinery.

Cams have been manufactured in a wide variety of geometries, some of which may be only of academic interest because of the expense of manufacture and very limited

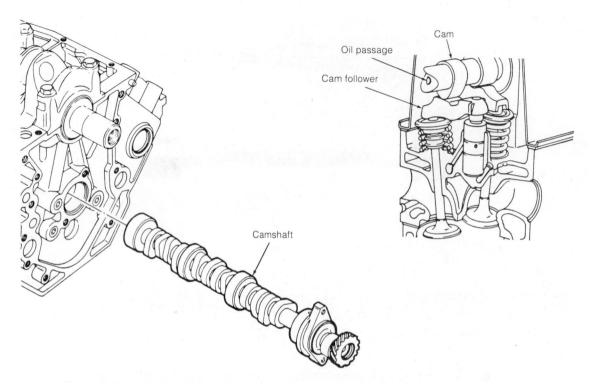

Figure 8.11.1 Cams and flatface followers used in an automobile engine. *Source: Ford Motor Co., Dearborn, MI.*

application. Indicative of the more common types are the conical, globoidal, and spherical cams shown in Figure 8.11.2. The plate cams shown in Figure 8.11.3[12] are generally the most widely used and the easiest to manufacture. Attention will be directed in the following sections to the design of plate cams for use with either a radial or an offset translating roller follower, Figures 8.11.3a and b, respectively. Terminology commonly used in the discussion of cams is illustrated in Figure 8.11.4

The *trace point* is the point of the follower which is used as the reference point in specifying the desired motion of the follower. As shown in the figure, it is conventionally taken as the vertex of the knife cross section in the case of a knife-edge follower, the center of the roller in the case of a roller follower, and is the contact point between cam and follower in the case of a flat-faced follower. The *pitch curve* is the path of the trace point as it appears to rotate about the cam relative to a coordinate system fixed on the cam itself. The contact point between the cam and follower lies on the *cam profile* which defines the physical contour of the cam. The cam profile and the pitch curve are identical only in the case of a cam with a knife-edge follower, although they may occasionally coincide in the case of a cam with a flat-faced follower. The *pitch point* is the point on the pitch curve where the pressure angle is greatest. The *pressure angle* has been defined as the angle between the radius to the trace point and the normal to the pitch curve at the trace point.[10]

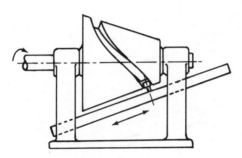

(a) Conical closed-track cam with translating roller. follower

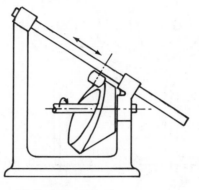

(b) Conical open-track cam with translating roller follower

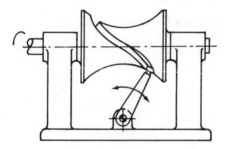

(c) Globoidal closed-track cam with swinging roller follower

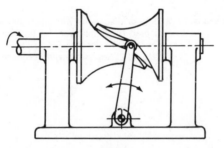

(d) Globoidal closed-track cam with swinging roller follower

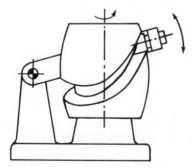

(e) Globoidal closed-track cam with swinging roller follower

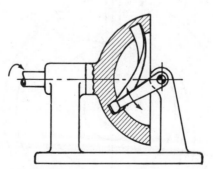

(f) Spherical closed-track cam with swinging roller follower

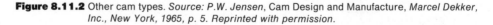

Figure 8.11.2 Other cam types. *Source: P.W. Jensen*, Cam Design and Manufacture, *Marcel Dekker, Inc., New York, 1965, p. 5. Reprinted with permission.*

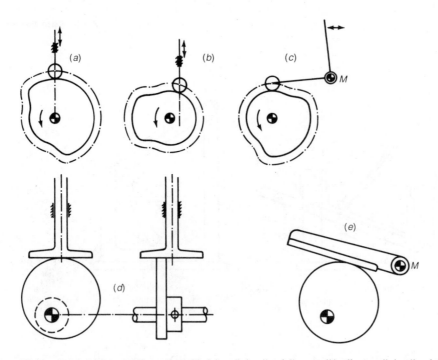

Figure 8.11.3 Plate or disc cams with (a) radial roller follower, (b) offset radial roller follower, (c) pivoted roller follower, (d) flat-faced follower, and (e) pivoted flate faced follower. *Source: P.W. Jensen*, Cam Design and Manufacture, *Marcel Dekker, Inc., New York, 1965, p. 2. Reprinted with permission.*

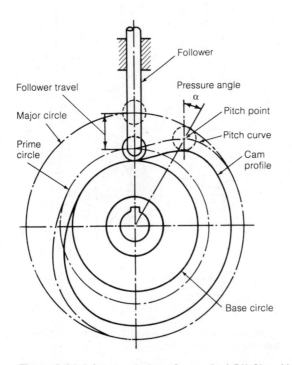

Figure 8.11.4 Cam terminology. *Source: Prof. F.Y. Chen*, Mechanics and Design of Cam Mechanisms, *copyright 1982 Pergamon Press. Reprinted with permission.*

The *pitch circle* is the circle, with center at the center of rotation of the cam, which passes through the pitch point. The *prime circle* is that circle having a radius equal to the smallest radius of the pitch curve and the *major circle* is that circle having a radius equal to the largest radius of the pitch curve. Several definitions of the *base circle* may be found in the literature. The base circle is the smallest circle that is tangent to the cam profile at at least one point, according to Ref. 9. (This also is the definition shown in Figure 1.11 of Ref. 10, although the text on the same page defines the base circle as the smallest circle about the center of cam rotation which is tangent to the pitch curve. For the sake of clarity we shall accept the definition in Ref. 9.) Finally, the last of the definitions given here is that of the *stroke*, or *throw*, of the cam, which is the magnitude of the motion of the follower as it moves from its minimum displacement to its maximum displacement.

Acceleration and, hence, force imparted to the cam and follower mechanism are especially important in cam design because as changes in acceleration become more and more abrupt they tend to approximate an impulse, with a severity in terms of wear and shock related to the magnitude of the impulse. To have a measure of the approach of change in acceleration to an impulse, cam designers consider the rate of change of acceleration, referred to as the *jerk* in cam terminology.

8.12 CONSTRUCTION OF THE CAM PROFILE

Cam design usually begins with the specification of the displacement of the follower rod, Figure 8.11.4, either as a function of time or as a function of the angular position of the cam itself. This information is sufficient if follower accelerations remain small as its velocity and displacement change in direction and magnitude. As follower acceleration and deceleration increase, however, the cam and follower may not always remain in contact unless the dynamic characteristics of the cam, its follower roller, rod, and return spring are all considered in the specification of the pitch curve.

Because follower dynamics are dependent upon its mass, lubrication, and restoring forces, it is not possible to generally specify the dividing line between high- and low-speed operation. In the ensuing paragraphs, however, it is assumed that rotational speeds are low enough that continuous follower and cam contact is assured in the design procedures described.

Suppose that the follower motion is prescribed to be that shown in Figure 8.12.1 in which time T is the period for one revolution of the cam, which is to say that the period is given by

$$T = 60/n \text{ s} \qquad (8.12.1)$$

for rotational speed n in revolutions per minute. If time t is set to zero at the beginning of the follower displacement shown in Figure 8.12.1 then for $0 \le t \le T$ this displacement z of the follower may be represented by

$$z = z(t) \qquad (8.12.2)$$

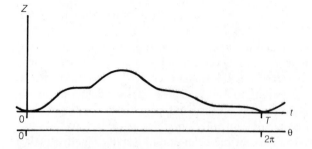

Figure 8.12.1 Representative cam displacement.

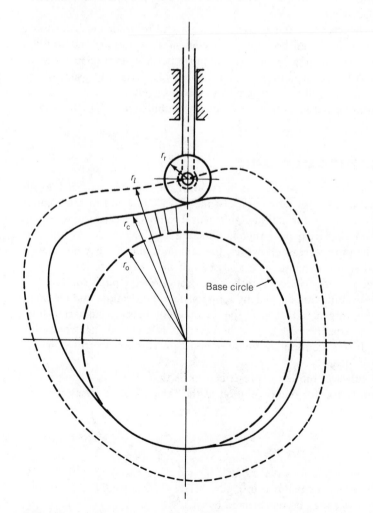

Figure 8.12.2 Radii used in cam analysis.

To construct a cam and follower in which the follower displacement is given by equation 8.12.2 it is only necessary to observe that as time goes from 0 to T the angular position of the follower roller goes from 0 to 2π so that

$$\frac{\theta}{2\pi} = \frac{t}{T}$$

or with $T = 1/f = 60/n$ for T in seconds, f in hertz, and $n = $ rpm,

$$t = \frac{\theta T}{2\pi} = \frac{\theta}{2\pi}\frac{60}{n} = \frac{30}{n\pi}\theta \tag{8.12.3}$$

Thus,

$$z = z\left(\frac{30}{n\pi}\theta\right) = Z(\theta) \tag{8.12.4}$$

gives the same displacement as does equation 8.12.2 except that now it is expressed in terms of θ rather than t. If cam and follower are designed such that the line of action of the follower passes through the center of rotation of the cam, Figure 8.12.2, displacement z may be written as the difference between the magnitudes of the radius to the pitch curve and the radius to the base circle. Thus,

$$r_\ell - r_0 = Z(\theta) \tag{8.12.5}$$

is the desired relation describing the pitch curve as a function of 0.

Further details on the actual determination of the final cam profile from this data may be found in Ref. 11.*

---

* Equations 3, 6, and 20 in Ref. 11 contain misprints. Equation (3) should read

$$r_\ell - r_0 = f\left(\frac{30}{n\pi}\theta\right) = F(\theta)$$

equation 6 should read

$$r_c^2 = r_\ell^2 + r_r^2 + 2r_r r_\ell \sin(\theta - \alpha)$$

and equation 20 should appear as

$$T = -\hat{k}Pr_c\,\frac{\cos(\alpha - \theta - \psi)}{\cos(\alpha - \theta + \zeta)}$$

8.13 OTHER CAM PROFILES

Many cams have been designed in the past by forming a profile composed of segments from several standard geometries. Two commonly used source curves are those in which $z = z(t)$ represents either a parabola or a cycloid over a portion of the cam.

The rise portion of the parabolic cam is defined by the equations

$$
\begin{aligned}
z &= 2h\left(\frac{2t}{T}\right)^2 & 0 \leq t \leq \frac{T}{4} \\
z &= h\left[1 - 2\left(1 - \frac{2t}{T}\right)^2\right] & \frac{T}{4} \leq t \leq \frac{T}{2}
\end{aligned}
\tag{8.13.1}
$$

If

$$
\theta = 2\pi\frac{t}{T} \qquad (\beta = \pi)
$$

then equations (8.13.1) take on the form

$$
\begin{aligned}
z &= 2h\left(\frac{\theta}{\beta}\right)^2 & 0 \leq \theta \leq \frac{\beta}{2} \\
z &= h\left[1 - 2\left(1 - \frac{\theta}{\beta}\right)^2\right] & \frac{\beta}{2} \leq \theta \leq \beta
\end{aligned}
\tag{8.13.2}
$$

Equation 8.13.1 and 8.13.2 describe the curve shown in Figure 8.13.1a. If the curve shown is to occupy only one-fourth of the period, so that it may, for example, be followed by a dwell time of one-fourth of the period (*dwell* is that portion of the profile where the follower remains stationary as the cam rotates), set β equal to $\pi/4$ in equation 8.13.2 to obtain

$$
\begin{aligned}
z &= 2h\left(\frac{8t}{T}\right) & 0 \leq t \leq \frac{T}{16} \quad T = \frac{60}{n} \\
z &= h\left[1 - 2\left(1 - 8\frac{t}{T}\right)^2\right] & \frac{T}{16} \leq t \leq \frac{T}{8}
\end{aligned}
\tag{8.13.3}
$$

The parabolic cam equations, (8.13.1), represent a class of profile curves which may be written as polynomials in θ, represented by expressions of the form.

$$
z = \sum_{i=1}^{n} a_i\theta^i
\tag{8.13.4}
$$

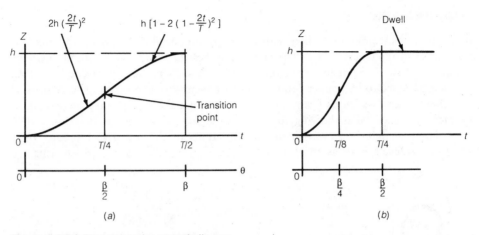

Figure 8.13.1 Rise curves for a parabolic cam.

in which the a_i are constants. These are known as *polydyne* cams, a neologism from polynomial and dynamic.

The second of the two profile types, the cycloid, is defined as the path described by a point on the circumference of a circle which rolls without slipping on a straight line, as shown in Figure 8.13.2.

Although the equation for a cycloid is fairly complicated in rectangular Cartesian coordinates, it simplifies considerably when transformed into polar coordinates, appearing as

$$z = h\left(\frac{\theta}{\beta} - \frac{1}{2\pi} \sin \frac{2\pi\theta}{\beta}\right) \qquad (8.13.5)$$

This is not a polydyne cam because the trigonometric term cannot be represented by a finite number of powers of θ.

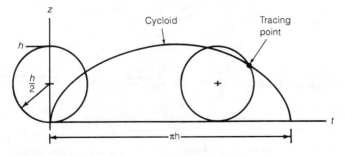

Figure 8.13.2 Formation of a cycloid.

8.14 RETAINING RINGS

Retaining rings, commonly called snap rings, are produced in a variety of configurations, as shown in Figure 8.14.1. The first five of these (*a*–*e* inclusive) have been widely used in the past. The last five are more specialized, such as the high-strength ring (*f*), which is designed to be tamper-proof. The permanent-shoulder series (*g*) may be used on softer materials such as brass and aluminum since these rings may be assembled by radially compressing the ring to collapse the six radial slots as it is inserted in an external retaining groove. Numerous variations of the first five retaining rings are also produced, such as the E-ring (*h*), the klipring (*i*), and the cresent ring (*j*).

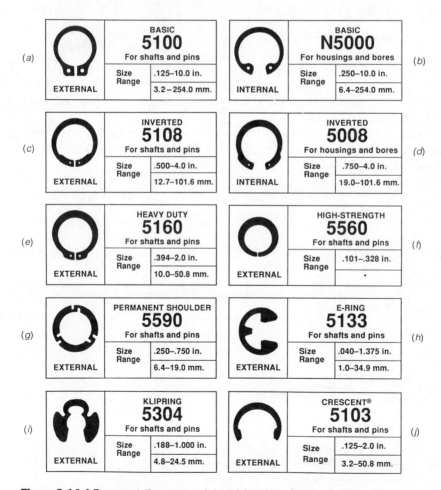

Figure 8.14.1 Representative commercial retaining rings. *Source: © Waldes Truarc, Inc., Long Island City, NY. Reprinted with permission.*

The klipring deserves special mention because it is designed for field use in which it may be inserted with a hammer and removed with a flat blade screwdriver.

Typical applications of retaining rings are illustrated in Figure 8.14.2.

Although carbon-spring steel, SAE 1060 to SAE 1090, is the standard material for retaining ring manufacture, these rings are also made from beryllium copper, stainless steel, AISI 632, and aluminum, Alclad 7070-T6. Ultimate shear strength of the carbon-spring steel rings lies in the range from 100 000 to 150 000 psi, while for beryllium copper, Alloy #25, CDA # 172, it ranges from 95 000 to 110 000 psi.[12]

Only the groove depth and the diameter to the bottom of the groove appear in the design formulas. Groove width is not included because the calculated axial load capability of a ring does not depend upon the ring being clamped between rigid groove walls, although it is dependent upon the walls being perpendicular to the axis of the shaft and without external champfer or large fillets at the bottom. Axial load capability is based upon both the bearing capability of the groove walls and the shear strength of the ring itself. The force F_g that may be supported by the groove walls for an external ring may be estimated from equation 8.14.1 in which R_g is the groove conversion

Figure 8.14.2 Typical applications of retaining rings. *Source: © Waldes Truarc, Inc., Long Island City, NY. Reprinted with permission.*

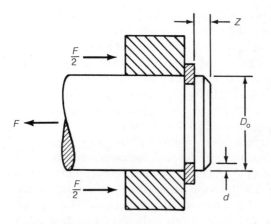

Figure 8.14.3 Axial load restrained by a retaining ring and edge margin Z. Outside diameter is D_o and groove depth is d.

factor read from Table 8.14.1 for the groove, q is a reduction factor given by equation 8.14.2, ζ is the safety factor, D_o is the outside diameter of the shaft, d is the groove depth, and σ_{yp} is the yield stress in tension for the shaft material:

$$F_g = \frac{R_g}{\zeta q}\, \pi D_o \, d\sigma_{yp} \qquad\qquad (8.14.1)$$

TABLE 8.14.1   RETAINING RING CONVERSION FACTORS

Ring Series	Conversion Factor R	
	Ring: R_r	Grove: R_g
N5000, N5001	1.2	1.2
N5002	1.2	1.2^a
5008, 5108	$\frac{2}{3}$	$\frac{1}{2}$
5100, 5101	1	1
5102	1	1^a
5103	$\frac{1}{2}$	$\frac{1}{2}$
5107	$\frac{3}{4}$	$\frac{3}{4}$
5131, 5133	$\frac{1}{3}$	$\frac{1}{3}$
5144	$\frac{1}{4}$	$\frac{1}{4}$
5139	[b]	$\frac{1}{2}$
5160	1.3	2
5304	$\frac{1}{2}$	$\frac{1}{2}$
5560	1.33	2

[a] Use d/2 instead of d.
[b] Use listed data chart values.
© Waldes Truarc, Inc., Long Island City, NY. Reprinted with permission.

where

$$q = 4.2 \left(\frac{Z}{d}\right)^{-1.1} \tag{8.14.2}$$

The recommended ratio $Z/d = 3$ lies within the nearly linear range of the q dependence upon the Z/d ratio; hence, a nearly proportional strength increase may be had by increasing to edge margin Z to $4d$. The increase is enough less than proportional beyond $Z/d = 4$ to suggest that this ratio may be a reasonable limit to the edge margin.

Force F_r that may be supported by the ring itself may be estimated from equation 8.14.3 involving the ring conversion factor R_r, also termed a ring factor, which accounts for ring properties by including effects of thickness, t, width, taper, strength of the metal used, etc. Typical values are listed in Table 8.14.1.

$$F_r = \frac{R_r}{\zeta} \pi D_0 t \sigma_{z\theta} \tag{8.14.3}$$

Similar formulas for external and internal grooves in hollow shafts may be found in design manuals from ring manufacturers, such as Ref. 12, along with information on equipment for both individual and assembly line application of retaining rings.

REFERENCES

1. van den Broek, J.A., *Elastic Energy Theory*, 2d ed., Wiley New York, 1942.

2. Peterson, R.E., *Stress Concentration Factors*, Wiley, New York, 1974.

3. Orthwein, W.C., Estimating fatigue due to cyclic multiaxial stress, *Journal of Vibration, Stress and Reliability in Design* 109, 97–102 (1987).

4. Orthwein, W.C., Keyway stresses when torsional loading is applied by the keys, *Experimental Mechanics* 15, 245–248 (1975).

5. Okubo, H., Hosono, K., and Sakaki, K., The stress concentration in keyways when torque is transmitted through keys, *Experimental Mechanics* 8, 275–280 (1968).

6. Orthwein, W.C., A new key and keyway design, *Journal of Mechanical Design* 101, 338–341 (1979).

7. *Universal Joint and Driveshaft Design Manual*, Advances in Engineering Series, No., 7, Society of Automotive Engineers, Warrendale, PA, 1982.

8. *Flexible Shaft Engineering Handbook*, 9th ed., Stow Mfg. Co., Binghamton, NY, 1982.

9. Rothbart, H.A., *Cams*, Wiley, New York, 1965.

10. Chen, F.Y., *Mechanics and Design of Cam Mechanisms*, Pergamon, New York, 1982.

11. Orthwein, W.C., Numerical design of plate cams, *Computers in Mechanical Engineering* 2 (1), 63–70 (1983).

12. *Waldes Truarc Retaining Rings and Assembly Tools*, Waldes Kohinoor, Inc., Long Island City, NY, 1981.

PROBLEMS

Section 8.1

8.1   Find the principal stresses and principal directions at the outer surface of a shaft in combined tension and torsion if the axial load is 35 600 N, the torsional load is 56 800 N mm, and the shaft diameter is 15.00 mm.

8.2   A hollow mixer shaft having an outside diameter of 20 mm and an inside diameter of 15 mm has operated without failure when driven by torque of 215 N m. Would you approve use of this shaft in a new product where the shaft is to have an internal pressure of 140 MPa in addition to the same torque? Use $\sigma_{y_t} = 510$ MPa and $0.55\sigma_{y_t} = \sigma_{y_s} = 281$ MPa. Yielding must be prevented. The radial stress due to internal pressure is given by $\sigma_{rr} = -p$ and the circumferential stress by $\sigma_{\theta\theta} = p(r_o^2 + r_i^2)/(r_o^2 - r_i^2)$.
Hint: Compare principal stresses.

8.3   As part of the redesign of a motor it is necessary to replace several solid shafts with hollow shafts to obtain a 70% weight reduction without reducing the torsional strength of the shaft. Find the ratio of the hollow shaft's outside diameter r_2 to that of the solid shaft r_0, as well as the ratio of the hollow shaft's outside to inside diameter, r_2/r_1. *Hint*: Let $r_1/r_2 = \lambda$ and solve for r_0/r_2.

8.4   Show that the angular deflection of a tapered shaft as shown in the figure may be calculated from

$$\theta = \frac{2T\ell}{3\pi G(r_1 - r_0)}\left(\frac{1}{r_0^{\,3}} - \frac{1}{r_1^{\,3}}\right)$$

in terms of the quantities shown in the figure.

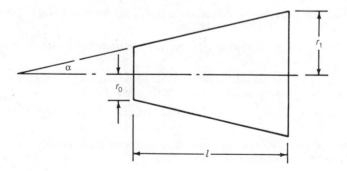

Sections 8.1 and 8.2

8.5   Design a power shaft to drive two sheaves from the gear shown in the figure. The shaft is machined from normalized UNS G10400 steel and the working shear stress should not exceed $0.35\sigma_y$. Angular deflection between sheave should not exceed $1.0°$. The gear delivers 1.5 hp to the shaft and sheave 1 takes 0.5 hp and sheave 2 takes 1.0 hp. The stress concentration at each key and keyway is 3.8. $A = 2.0$ in., $B = 2.5$ in., $G = 11.5 \times 10^6$ psi.

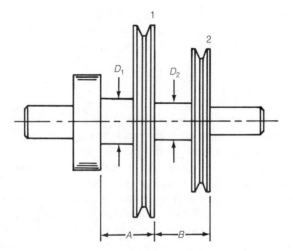

8.6   It is necessary to redesign the shaft in problem 8.5 if each sheave is to be attached to the shaft with a pin through a 0.1 in. hole drilled diametrically through the shaft? If so, what should be the redesigned profile?

8.7   Select the minimum shaft diameter to transmit 30 hp at 980 rpm. It is to be solid and made from annealed UNS G10200 steel using a safety factor of 2.00 according to the maximum shear stress criterion. Assume $2\sigma_{y_s} = \sigma_{y_t}$; i.e., the yield stress in tension is twice the yield stress in shear.

8.8   Would the diameter in problem 8.7 be different if the HMH criterion were used? If so, what would be the new diameter?

8.9   Calculate the maximum shear stress in a shaft whose diameter is 10 mm if it is to transmit 0.5 kW at 1500 rpm. Also find its angular deflection over a length of 22 mm. $G = 79.3$ GPa.

8.10   Design a shaft to transmit a torque of 6250 N m and having a profile proportional to that shown in the figure. It is to be made from normalized UNS G10400 steel with a design safety factor of 1.70 according to the HMH criterion.

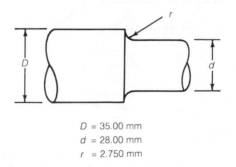

D = 35.00 mm
d = 28.00 mm
r = 2.750 mm

8.11   Select the diameter for a solid shaft to transmit 30 hp at 980 rpm using normalized UNS G10200 steel and a safety factor of 2.00 according to the HMH criterion.

8.12   What outside diameter D would be required at the larger end of a shaft that supports a torque of 88.0 N m if it has a change in diameter similar to that in the figure for problem 8.10 if it is made from normalized UNS G10400 steel and contains a full-length axial hole having an inside diameter $d_i = 0.6d_o$ where $D = 1.20d_o$ and where the fillet radius is $0.11d_o$. Use the HMH criterion with a safety factor of 2.00. Also give the values of d_i, d_o, and the fillet radius.

Section 8.3

8.13 Verify the program result in Example 8.3.1 by any one of the following means for solving equation 8.3.1: direct integration, the moment area method, the conjugate beam method, or the singularity function method. Agreement should be to within 1.1%, with the difference caused by approximation error inherent in the trapazoidal approximation to the integrals.

8.14 Modify the program listing in Figure 8.3.5 to accept 7 different radii. Use the resulting program to find the deflection at the load of the shaft shown in the figure for a load of 31 240 N midway between stations 3 and 4. $E = 207$ GPa. Note that when the program prompts for more stations than there are changes in the shaft radius you may place additional stations between radius changes and specify the same radius on both sides of each artificial station.

   [*Ans.* 0.0546 mm]

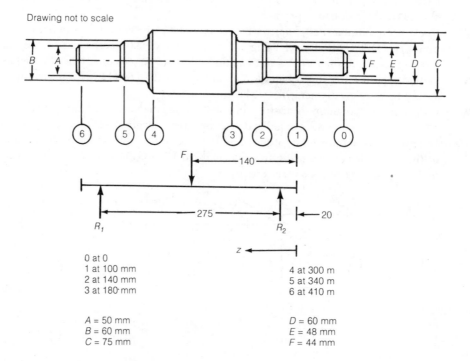

Drawing not to scale

0 at 0
1 at 100 mm
2 at 140 mm
3 at 180 mm

4 at 300 m
5 at 340 m
6 at 410 m

A = 50 mm
B = 60 mm
C = 75 mm

D = 60 mm
E = 48 mm
F = 44 mm

8.15 Calculate the lateral deflection at the midpoint of the largest section (between stations 3 and 4) due to a 56 000 kg load at the midpoint between stations 2 and 3 in the figure for problem 8.14.

8.16 Modify the program listed in Figure 8.3.5 to permit six radius changes anywhere between end reactions R_1 and R_2 and then use the new program to find the deflection at the midpoint between stations 3 and 4 due to a load of 20 111 kN midway between stations 2 and 3 and a load of 17 400 kN midway between stations 4 and 5. $E = 204$ GPa.

[*Ans.* 0.0439 mm]

8.17 Repeat problem 8.16 when the load midway between stations 2 and 3 acts at an angle of 39° with the vertical. Find the magnitude and direction of the displacement.

[*Ans.* 0.0415]

8.18 Rewrite the program listed in Figure 8.3.5 to solve for the deflection at any point along a simply supported shaft with an overhang. The shaft may support a load between reaction R_1 and R_2 and a second load on the overhung length, as shown below.

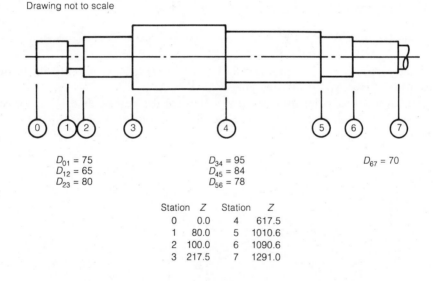

Drawing not to scale

$D_{01} = 75$
$D_{12} = 65$
$D_{23} = 80$

$D_{34} = 95$
$D_{45} = 84$
$D_{56} = 78$

$D_{67} = 70$

Station	Z	Station	Z
0	0.0	4	617.5
1	80.0	5	1010.6
2	100.0	6	1090.6
3	217.5	7	1291.0

8.19 Use the modified program written for problem 8.18 to find the deflection at the load for the shaft shown in the figure for Problem 8.18 when a vertical load of 3880 N acts at a point 40 mm from the left end of the shaft, and when a second load of 4079 N acts at a point 810 mm from the left end of the shaft. The midplanes of the supporting bearings are 420 and 1190 mm from the left end of the shaft. Also find the deflection at the midpoint between the supports with and without the overhung load. $E = 204$ GPa.

[*Ans.* $\delta = 0.3147$ mm at the 3880 N load and $\delta = -0.2574$ mm at the midpoint between the supports to $\delta = 0.0744$ mm.]

8.20 Find the deflection at station 0 in the figure for problem 8.18. Note that the expression for $m(x)$ holds outboard of the cantilevered load if the proper sign change is made in the data by placing the unit load at -40 mm. $E = 205$ GPa.

[*Ans.* $\delta = 0.1990$ mm]

Section 8.4

8.21 Find the reaction at support B in the figure if all supports are in contact with the undeflected shaft before any loads applied. Dimensions are in inches.

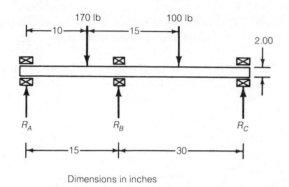

Dimensions in inches

8.22 Find the reactions at C in problem 8.21 if the bearing at C is 0.010 in. below the undeflected shaft before any loads are applied.

8.23 Find the reaction at point B in problem 8.21 when the bearing at B is supported by a beam as shown in the Figure below. The beam is characterized by $EI = 8.4375 \times 10^6$ lb-in.^{2} Dimensions are in inches.

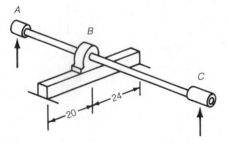

Dimensions in inches

Section 8.6

8.24 Find the largest mean bending stress that can be sustained with a safety factor of 1.7 according to criteria 8.6.3 and 8.6.4 for a rotating shaft whose diameter is 1.625 in. when it is subjected to an alternating bending moment of 2650 in.-lb. The ultimate stress is 71 ksi, the yield stress in shear is 45 ksi, and the endurance limit is 40 ksi.

8.25 Recommend the minimum diameter for a shaft to be made from UNS G10500 normalized steel and having a machined surface it is to be subjected to a bending moment which fluctuates from 0 to 23 500 N m and a completely reversed torque of 12 900 N m. Use Figure 4.7.3 to find σ_e

and Figure 4.7.4 to justify $\sigma_{e_s} = \sigma_{e_t}/\sqrt{3}$. Also $K_t = 1.47$ in tension, $K_s = 1.62$ in torsion, and the notch sensitivity factor is approximately 0.86 for both tension and torsion. Use criteria 8.6.7 and 8.6.8 and a safety factor of 1.2.

8.26   Compare the diameter calculated for problem 8.25 with that obtained from the HMH-Soderberg criterion, equation 8.6.6. Use $\sigma_{y_s} = \sigma_y/\sqrt{3}$.

8.27   Find the safety factor for a 1.0 in. diameter hollow shaft $(d_i/d = 0.6)$ of UNS G10950 steel normalized at 205 °C. The torsional load varies from 1430 to 2349 in.-lb and the bending moment varies from -400 to 1200 in.-lb. It has a ground surface and a 0.25 in. transverse hole. Use a safety factor of 1.5 and a notch sensitivity factor of 0.73. Use the Gerber-yield criterion and assume that $\sigma_{y_s} = \sigma_{y_t}/\sqrt{3}$, $\sigma_{u_s} = \sigma_{u_t}/\sqrt{3}$, and $\sigma_{e_s} = \sigma_{e_t}/2$. Estimated σ_e from Figure 4.7.3.
[*Ans.* $\zeta = 2.85$]

Section 8.8

8.28   What is the torque capacity of a flange coupling with 6 bolt holes 10.1 mm in diameter on a 250 mm bolt circle on a flange 15 mm thick if bolts 10 mm in diameter are used? The limiting stress for the bolts is 279 MPa and for the flange is 252 MPa.

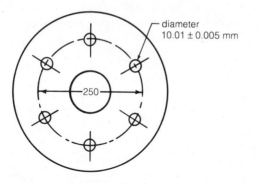

diameter
10.01 ± 0.005 mm

8.29   Specify the radial pressure required between a compression coupling and a 2-in.-diameter shaft if it is to transmit 60 hp at 1000 rpm. Each shaft engages the coupling for a length of 4 in. and the coefficient of friction is expected to be 0.70 or greater.

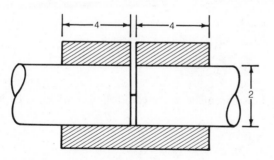

8.30 Sketch the orientation of shaft 2 relative to shaft 1 if their parallel misalignment is 3.5 mm and their angular misalignment is 4°.

Section 8.9

8.31 A single Cardan universal joint is used to deliver 1.0 hp to a small mixing beater turning at 1160 rpm. If $\beta = 9°$, what is the percentage fluctuation in torque at the output shaft for a constant driving speed at a constant input torque?

8.32 If β is increased to 20° and if modification of the connection of the shaft to the Cardan joint introduces a stress concentration $K = 1.8$ and if $q = 0.80$ in torsion, what is the minimum shaft diameter that may be used according to the maximum shear stress criterion for a safety factor of 2.00 in problem 8.31 if the shaft is made from annealed UNS G41400 steel and the surface is machined? Assume $2\sigma_{y_s} = \sigma_{y_t}$, and $\sigma_{e_s} = 0.6\sigma_{e_t}$. Find σ_e from Figure 4.7.3 and let $\tau_{max} = \sigma_{y_s}$.

Section 8.10

8.33 Calculate the forces on the hold-down brackets around the casing of the flexible cable shown in the figure at locations A and B if the cable loss is 0.5%. *Hint*: The cable loss is the power dissipated in friction between the cable and the core as the core tries to turn the casing.

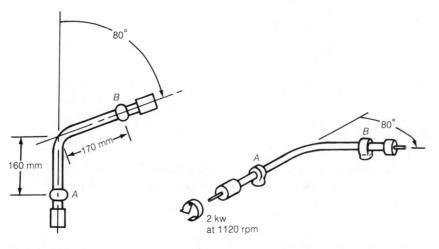

Flexible cable.

Section 8.12

8.34 A cam is to impart a sinusoidal displacement to the cam follower which will cause it to move linearly according to

$$z = a(1 + \sin 4\pi t/T)$$

Show that the radius of the cam profile may be written as

$$r_1 = R + a(1 + \sin 2\theta)$$

where R is the radius of the base circle.

Section 8.14

8.35 Calculate the external groove depth necessary for a series 5100 retaining ring to resist an axial load of 6870 lb when the ring is mounted upon a UNS G10200 steel shaft 1.10 in. in diameter. Use the recommended edge margin and a safety factor of 2.00.

8.36 Find the percent reduction in the load capacity of the groove in problem 8.35 if the edge margin is reduced to $Z = 2d$ and the percent increase if it is increased to $Z = 4d$.

8.37 A repair contractor, finding the stock of 5100 retaining rings depleted, replaced a 5100 retaining ring with a 5103 ring. What effect, if any, will this have on the axial load capacity of the ring and shaft combination described in problem 8.35? The ultimate shear stresses for the 5100 and 5103 rings and 120 000 and 150 000 psi respectively.

CHAPTER NINE

LUBRICATION

NOTATION

a	grid proportion (1)		u	flow velocity (l/t)
b	grid proportion (1)		V	flow velocity (l/t)
C	specific heat (θ/l^3)		W	load on journal bearing (ml/t^2)
c	clearance (l)		x	coordinate or a temporary variable (l)
D	journal diameter (l)		y	coordinate in a direction perpendicular to the fluid flow (l)
e	journal bearing eccentricity (l)			
F	force (ml/t^2)		z	coordinate in a direction perpendicular to the fluid flow, i.e., the axial along journal (l)
f	coefficient of friction or a function (1)			
g	substitution quantity			
H	viscosity at 100 VI (m/lt)		α	angle of load relative to line of centers (1)
h	film thickness (l)			
I	substitution quantity		β	journal bearing angular dimension (1)
J	Joule heat equivalent (ml^2/t^2)		δ	grid dimension (1)
L,l	length or viscosity at 0 VI (l), (m/lt)			substitution quantity
N	revolutions per second $(1/t)$		γ	weight/unit volume in Old English units mass/unit volume in SI units
n	revolution per minute or grid number $(1/t)$, (1)			
P	$W/(DL)$, force per unit of projected bearing area (m/lt^2)		ε	eccentricity ratio (1)
			θ	journal circumferential coordinate (1)
p	pressure (m/lt^2)		μ	viscosity (absolute viscosity, dynamic viscosity (m/lt)
q	quantity of fluid or a solution variable (m)			
			ρ	density (all units)
r	radius (l)		σ_{yx}	explicit shear stress (m/lt^2)
S	Sommerfeld number (1)		τ	generic shear stress (m/lt^2)
SUS,SUV	Seybolt Universal Second (t)		v	kinematic viscosity $(1/lt)$
T	torque or temperature (ml^2/t^2), (θ)		ω	angular velocity in radians per second $(1/t)$
U	flow velocity (l/t)			

Lubrication is used to reduce friction and wear between two surfaces originally in sliding contact by separating them with a lubricating material, either gas, liquid, or solid. Lubricating liquids or solids are chosen from materials that have both a very low shear strength, and, hence, induce a low coefficient of friction, and a sufficient film strength to hold the two wear surfaces apart when the film is thin.

Most of the following discussion will be directed toward fluid (liquid and gas) lubrication because it is the most common lubricant and because it has received more theoretical development than the newer area of solid lubrication.

9.1 LUBRICATION AND WEAR

The effect of the normal load (the load acting perpendicular to the sliding surfaces) on wear is shown in Figure 9.1.1, due to Beerbower.[1] The lubrication regimes in the region from 0 to Y are described in terms of their coefficients of friction by the Stribeck-Hersey curve, Figure 9.1.2,[2] in which μ is the dynamic viscosity, n the rotational speed, and P is the load divided by the projected area of the journal. (This ratio is often written as Zn/p for historical reasons, where $Z = \mu$) Since the ratio is dimensionless, any consistent set of units may be used.

Boundary lubrication, shown in both figures, refers to lubrication by a liquid where the load on the sliding surfaces is large enough so that there is appreciable solid to solid interaction as the surfaces slide on one another.[3] In this regime friction and wear are determined by the metallurgy of the surfaces, their topography, physical and chemical adsorption, any corrosion, catalysis, chemical reaction rates, and any antifriction solids in the lubricant.

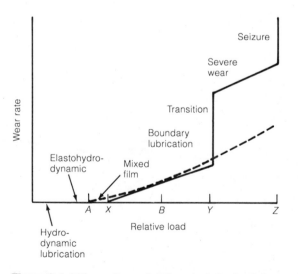

Figure 9.1.1 Wear rate as a function of relative load. *Source:* NASA TN82858, *NASA, Lewis Research Center, Cleveland, OH.*

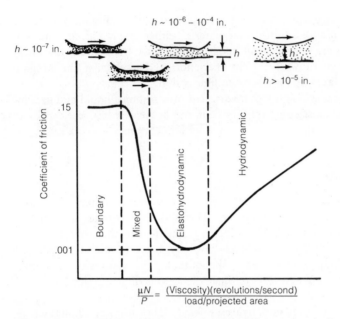

Figure 9.1.2 Coefficient of friction as a function of the speed–velocity–load parameter $\mu N/P$. *Source: NASA TN82858, NASA, Lewis Research Center, Cleveland, OH.*

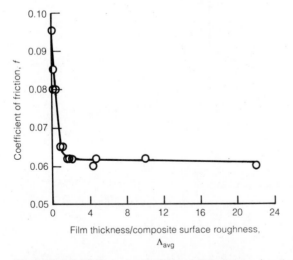

Figure 9.1.3 Friction as a function of the average Λ ratio. (Λ = film thickness/composite surface roughness). *Source: NASA TN82858, NASA, Lewis Research Center, Cleveland, OH.*

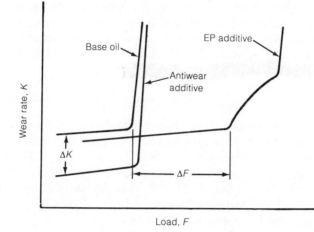

Figure 9.1.4 Wear behavior of boundary–lubrication systems. *Source:* NASA TN82858, *NASA, Lewis Research Center, Cleveland, OH.*

As the film thickness of the lubricant increases, perhaps due to a decreased load or a change in rotational speed n, the solid to solid effects decrease and the lubrication moves into the mixed region. In the mixed region the solid to solid effects decrease and fluid film begins to influence the friction coefficient. As the loads become smaller the $\mu n/P$ ratio moves into the elastohydrodynamic region where solid to solid effects diminish still further and the friction coefficient depends primarily upon the elasticity of the surface and the pressure–viscosity characteristics of the lubricant. At high speeds, low loads, and high viscosities the $\mu n/P$ ratio moves into the hydrodynamic region, as shown in Figure 9.1.2, where the surfaces are completely separated by a lubricating film.

Transition from mixed to elastohydrodynamic to hydrodynamic is clearly shown in Figure 9.1.3, where the coefficient of friction is plotted as a function of the ratio of the film thickness to the composite surface roughness ρ.

Effects of additives are displayed in Figure 9.1.4. Note that although the antiwear additive prevented some wear as the load increased from zero, it did not appreciably increase the load at which serious were was initiated. In contrast to this, the EP (extreme pressure) additive produced a much smaller improvement in wear but doubled the load that could be supported before serious wear began.

9.2 VISCOSITY

Once into the region of hydrodynamic lubrication, the friction force is largely determined by the viscosity of the lubricating fluid according to the relation

$$\sigma_{yx} = \mu \frac{\partial u}{\partial y} \tag{9.2.1}$$

in which σ_{yx} represents the shear stress, μ the fluid velocity parallel to the surface, y the coordinate perpendicular to the lubricated surface, and μ the viscosity, which, as noted above, has also been termed the *absolute* viscosity or the *dynamic* viscosity. Another viscosity, termed the *kinematic* viscosity, relation 9.2.2, may be obtained from (9.2.1) simply by dividing the dynamic viscosity by the density ρ of the fluid.

$$v = \mu/\rho \tag{9.2.2}$$

The term *viscosity* will be used in the remainder of this chapter with the understanding that when used without an adjective it is to denote the absolute or dynamic viscosity.

Almost as if to complicate the easily understood definition of the viscosity, its value has been encumbered with a variety of different units. Commercial units of kinematic viscosity are often Seybolt Universal Seconds (SSU, SUS, or SUV) in the United States, Engler Degrees in Europe and Russia, and Redwood Seconds in Britain. The approximate relation between these three scales is shown in Table 9.2.1.

From (9.2.1) it is evident that the units of viscosity are given by

$$\frac{ml}{t^2}\frac{1}{l^2} = \mu\frac{l}{t}\frac{1}{l} \tag{9.2.3}$$

in terms of mass m, length l, and time t, which may be expressed as

$$\mu \sim \left(\frac{ml}{t^2}\right)\frac{t}{l^2} = \frac{m}{lt} \tag{9.2.4}$$

where both of these relations have been used to select the units of viscosity as

lb-s/in.^{2} = reyn = 6894.757 N-s/m^2 = 6894.757 Pascal-seconds (Pa-s)

TABLE 9.2.1   APPROXIMATE VISCOSITY CONVERSION FACTORS

Centistokes	Seybolt Universal Seconds	Redwood No. 1 Seconds	Engler Degrees
2	33	31	1.1
4	39	36	1.3
6	46	41	1.5
8	52	46	1.7
10	59	52	1.8
15	77	68	2.3
20	98	86	2.9
30	141	125	4.1
50	232	205	6.6
>50	(cS × 4.6)	(cS × 4.1)	(cS × 0.132)

Source: A. Cameron, *The Principles of Lubrication*, New York: Longman Group Ltd, 1966, p. 21.

where

$$\text{Pascal-seconds} = 1000 \text{ centipoise (cP)} \qquad \text{so} \quad \text{cP} = \text{mPa-s}$$

It follows from equations 9.2.2 and 9.2.4 that the units of kinematic viscosity are

$$v \sim \frac{l^3}{m} \frac{m}{lt} = \frac{l^2}{t}$$

and the value of the kinematic viscosity using measured Seybolt Universal Seconds (SUS, SSU, or SUV) is usually approximated by

$$v \text{ (cSt)} = \left(0.22 \text{ SUS} - \frac{180}{\text{SUS}} \right) \tag{9.2.5}$$

where, as indicated, v is in terms of centistokes (cSt = mm^2/s) by virtue to the dimensional constants 0.22 and 180. When multiplied by mass density ρ (mass/unit volume) in units of g/cm^3, equation 9.2.5 yields

$$\mu \text{ (cP)} = \left(0.22 \text{ SUS} - \frac{180}{\text{SUS}} \right) \rho \tag{9.2.6a}$$

and

$$\mu \text{ (reyns)} = 1.4504 \times 10^{-7} \left(0.22 \text{ SUS} - \frac{180}{\text{SUS}} \right) \rho \tag{9.2.6b}$$

where, as indicated, μ is in units of reyns.

The density of lubricating oils may be approximated by

$$\rho = 0.89 - (°C - 15.56)63 \times 10^{-5} \text{ g/cm}^3 \tag{9.2.7a}$$

or

$$\rho = 0.89 - (°F - 60)35 \times 10^{-5} \text{ g/cm}^3 \tag{9.2.7b}$$

as a function of temperature in terms of degrees Celsius or degrees Fahrenheit. In these expressions the density of petroleum oils at 15.56 °C (60 °F) has been approximated by 0.89 g/cm^3 = 890 kg/m^3.

Motor oils are usually sold by the SAE (Society of Automotive Engineers) grades, each of which includes a range of viscosities, as shown in Table 9.2.2. Many European nations use the International Organization for Standardization grades, known as ISO grades (not IOS grades). In the United States the ISO grades are commonly used for

TABLE 9.2.2   SAE VISCOSITY GRADES

Motor Oils

SAE Viscosity Grade	Viscosity Range		Viscosity Range at 100°C	
	Equal to or less than (cP)	Temperature (°C)	Greater than or equal to (cSt)	Less than
0W	3250	−30	3.8	· · ·
5W	3500	−25	3.8	· · ·
10W	3500	−20	4.1	· · ·
15W	3500	−15	5.6	· · ·
20W	4500	−10	5.6	· · ·
25W	6000	−5	9.3	· · ·
20	· · ·	· · ·	5.6	9.3
30	· · ·	· · ·	9.3	12.5
40	· · ·	· · ·	12.5	16.3
50	· · ·	· · ·	16.3	21.9

Axle and Manual Transmission

	Maximum Temperature for Viscosity of 150 Pa-s (°C)	Viscosity Range (cSt at 100 °C)	
		Greater than or equal to	Less than
70W	−55	4.1	· · ·
75W	−40	4.1	· · ·
80W	−26	7.0	· · ·
85W	−12	11.0	· · ·
90	· · ·	13.5	24.0
140	· · ·	24.0	41.0
250	· · ·	41.0	· · ·

Source: Abstracted from SAE J300 JUN87 and SAE J306 MAR85, *1988 SAE Handbook*, Society of Automotive Engineers, Inc., Warrendale, PA.

industrial oils and AGMA (American Gear Manufacturers Association) grades are used for gear oils.

According to SAE standard J300 JUN86, the viscosity of SAE single viscosity motor oils, such as SAE 20, are measured at 100 °C and the viscosity of those with a W suffix are measured at temperatures that depend upon the grade, as shown in Table 9.2.2, and subject to the variations implied. Multigrade motor oils, such as 10W-30, whose viscosity has been modified by Viscosity Index Improvers, must meet the standard at the W temperature.

Axle and manual transmission oil viscosities as specified in SAE standard J306 MAR85 are also listed in Table 9.2.2.

In addition to their viscosity, petroleum base oils are also described by the types of crude oil from which they were refined; paraffinic, naphthenic, asphaltic, or mixed base. Synthetic bases are classed as synthetic hydrocarbons (alkylated aromatics, poly-alphaolefins, polybutenes), organic esters (dibasic acid esters, polyol esters, polyesters), halogenated hydrocarbons, phosphate esters, polyglycols, silicones, etc.

The American Petroleum Institute (API) has still another classification of engine oils based upon additives that protect against sludge, varnish, rust, and high-temperature oil thickening. These classifications are SA, SB, SC, SD, SE, SF, CA, CB, CC, and CD, where the S prefix denotes oils for use in gasoline engines are C prefix designates oils for use in diesel engines. The suffixes indicate levels of service. The SF oil, for example, was designed to meet the 1980 and later warranty requirements of automobile manufacturers. Engine oils that contain so-called friction modifiers that satisfy certain standards of the American Society for Testing and Materials (ASTM) may also be labeled as energy conserving.

Greases are oils that have been thickened by adding a metallic soap or other agent to make them plastic and somewhat adhesive so they will remain on surfaces not easily oiled. Calcium or lime, sodium or soda, lithium, and barium soaps also increase the service temperature from 200 °F for calcium soap to about 500 °F for barium soap. Aluminium soap provides water resistance and lead soaps provide extreme pressure performance, as required in heavy duty gears. Because of their increased viscosity, however, they do not provide the cooling and cleaning functions often achieved with oils in addition to their primary lubrication function.

Because of their plasticity, heavier greases are classed by the National Lubricating Grease Institute (NLGI) according to their penetration rather than according to their viscosity (see Table 9.2.3). Penetration shown in this table refers to the depth of penetration of a cone of prescribed shape and weight dropped from a specified height into the grease after it has been worked by a perforated disk.

TABLE 9.2.3   NATIONAL LUBRICATING GREASE INSTITUTE (NLGI) CONSISTENCY NUMBERS

NLGI Consistency Number	60-Stroke Worked Penetration at 25 °C (77 °F) ASTM D 217	NLGI Consistency Number	60-Stroke Worked Penetration at 25 °C (77 °F) ASTM D 217
NLGI000	445–475	NLGI3	220–250
NLGI00	400–430	NLGI4	175–205
NLGI0	355–385	NLGI5	130–160
NLGI1	310–340	NLGI6	85–115
NLGI2	265–295		

Source: © 1963 National Lubricating Grease Institute.

9.3 VISCOSITY INDEX

According to the American Society for Testing and Materials (ASTM) Practice D2270-86, the viscosity index, abbreviated VI, is "an arbitrary number used to characterize the variation of the kinematic viscosity of a petroleum product with temperature." The viscosity index is not defined for values less than 2.0, it increases with decreased temperature effects, and it is primarily used for temperatures between 40 and 100 °C (104–212 °F).

Its calculation assumes two groups of reference oils, as shown in Figure 9.3.1. One group consists of oils having VI = 100 and viscosities that span the entire range so that an oil can be found in this group that has a kinematic viscosity at 210 °F, which is the same as that of the oil whose VI is to be found. The second group of oils consists of those having VI = 0 and viscosities that also span the range so that this group also contains an oil whose kinematic viscosity v at 210 °F equals that of the oil whose viscosity is to be found.

If $2 \leqq VI \leqq 100$

$$VI = \frac{L - U}{L - H} 100 \qquad (9.3.1)$$

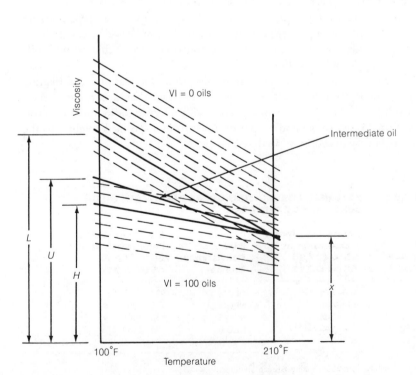

Figure 9.3.1 Plot of viscosity as a function of temperature.

whereas if VI > 100

$$VI = \frac{10^N - 1}{0.00715} + 100 \tag{9.3.2}$$

where

$$Y^N = H/U$$

and where log denotes the logarithm to the base 10. In these equations

$$L = 0.8353 \ Y^2 + 14.67 \ Y - 216$$
$$H = 0.1684 \ Y^2 + 11.85 \ Y - 97 \tag{9.3.3}$$

if $v > 70$ cSt at $100°$ C. If $v \leq 70$ cSt at $100°$ C the values for L and H should be found from Table 1 in ASTM D2270-86. In these empirical relations

- $L = v$ in cSt at $40°$C of a $VI = 0$ reference oil having the same v at $100°$C as the oil whose VI is to be found,
- $H = v$ in cSt at $40°$C of a $VI = 100$ reference oil having the same v at $100°$C as the oil whose VI is to be found,
- $Y = v$ in cSt at $100°$C of the oil whose VI is to be calculated, and
- $U = v$ at $40°$C of the oil whose VI is to be calculated.

Since the VI index used to distinguish between relations (9.3.1) and (9.3.2) is obviously not known before the calculations are made, the usual procedure is to first calculate the VI according to (9.3.1). If the VI from that calculation is equal to or less than 100 no further calculation is necessary. If the VI value from relation (9.3.1) is greater than 100, and, therefore, is not considered indicative of the variation in v with temperature, a more indicative value should be calculated from relation (9.3.3).

9.4 ADDITIVES

These compounds are added to oil and grease to enhance their handling, cleaning ability, and film strength. They commonly are grouped as listed below[5]:

- *Oxidation inhibitors* which impede formation of gums and acids by slowing chemical oxidation through the addition of phosphorus, sulfur, nitrogen, or organic compounds.
- *Defoamers* are used in high-speed bearing and other such mechanisms and owe their behavior to compounds that often are silicon based.
- *Detergents* are added to retain insoluble materials in suspension in the oil used in internal combustion engines and compressors.

- Rust inhibitors often are surface-active compounds that plate a protective layer over ferrous surfaces in contact with the lubricant.

- *EP (extreme pressure)* additives are either fiberous and/or low-friction solids or so-called antiweld compounds often containing phosphorous, lead, sulfur, and the like.

Oils are usually replaced on a schedule based upon estimates of when these additives are depleted; when the rust inhibitors are no longer effective, when the oil is saturated with insoluble materials in suspension, or when the defoamers have reacted with impurities accumulated in the oil from other sources and are not as effective as required. Reprocessing of used oil consists of removing the foreign materials and replacing the additives, or adding others.

9.5 PETROFF'S EQUATION

The torque to drive a lightly loaded journal bearing, shown in Figure 9.5.1, may be estimated from equation 9.2.1 in terms of the quantities shown in the figure. If F denotes the tangential force acting on the surface of the journal (that part of the shaft surrounded by the bearing) of radius r and length l, and if U represents the velocity of the journal

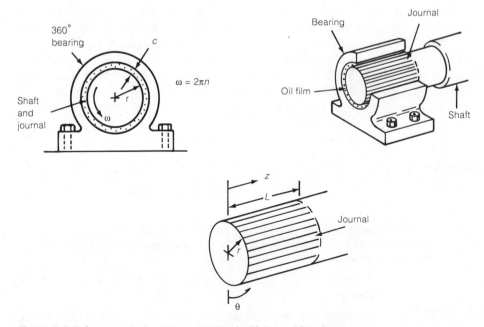

Figure 9.5.1 Geometry and notation associated with journal bearings.

surface relative to the bearing surface, then $U = 2\pi r N$ and

$$\frac{F}{2\pi r l} = \tau = \mu \frac{\partial u}{\partial x} = \mu \frac{U}{c} \tag{9.5.1}$$

from which it follows that

$$\frac{F}{l} = 2\pi \mu \frac{rU}{c} \tag{9.5.2}$$

According to one author this is known as the Petroff equation.[6]

Other authors associate this tangential force with a radial force W on the journal through a coefficient of friction f, written as $f = F/(2rlP)$ where $P = W/(2rl)$ is the radial force per unit of projected journal area. The torque acting on the journal may be written in these terms as

$$T = \tau A r = \frac{4\pi^2}{c} r^3 l \mu N = 2fPlr^2$$

so that

$$f = 2\pi^2 \frac{\mu N}{P} \left(\frac{r}{c}\right) \tag{9.5.3}$$

which is their representation of the Petroff equation.[7,8] In these equations c denotes the clearance between journal and bearing and N denotes the rotational speed in revolutions/second in order to use standard units for viscosity μ.

EXAMPLE 9.5.1

When the dimensionless Sommerfeld number S, defined by*

$$S = \left(\frac{r}{c}\right)^2 \frac{\mu N}{P}$$

is approximately 8 or greater the journal and bearing geometry resembles that assumed in the derivation of the Petroff equation.[9]

Find the rpm necessary to have $S = 10$ and find the power loss for this rotational speed for a journal bearing to support 100 lb on a 2.000-in.-diameter journal which is in a bearing that is 2.500 in. long and has an inside diameter of 2.004 in. Assume $\mu = 6 \times 10^{-5}$ reyns.

---

* 1/S has also been known as the Sommerfeld number.[13]

First calculate

$$\frac{r}{c} = \frac{1.00}{0.002} = 500$$

$$P = \frac{100}{2(2.5)} = 20$$

and solve the Sommerfeld number for N to find that

$$N = \frac{SP}{\mu}\left(\frac{c}{r}\right)^2 = \frac{10(20)}{60}10^6\left(\frac{1}{500}\right)^2 = 13.33 \text{ rps or 800 rpm}$$

From

$$f = 2\pi^2 \frac{N\mu}{P}\left(\frac{r}{c}\right) = 2\pi^2 S\left(\frac{c}{r}\right) = 2\pi^2 \frac{10}{500} = 0.395$$

find that

$$\text{hp} = \frac{Tn}{198}10^{-3} = \frac{\pi(0.395)800}{198000} = 0.0050$$

9.6   FLUID EQUILIBRIUM EQUATIONS

In this section we shall not assume familiarity with the Navier-Stokes equation which describes general Newtonian fluid motion, but will turn instead to somewhat simplified force equilibrium equations for a fluid element and a commonly used simplification of the stress–strain relations as represented by equation 9.2.1 for flow perpendicular to the z direction. These several simplifications have been found to introduce negligible errors in the derived form of the Reynolds equation used in most machine applications.[14] Derivation without all of these simplifying assumptions is necessary only if the analysis is to extend to gas bearings or to bearings of unusual configuration.

If inertia and gravity are neglected, if the flow in the y direction is assumed to be negligible, and if the major shear stress variation is in the y direction, then it follows from Figure 9.6.1 that summation of forces in the x direction leads to

$$\frac{\partial \sigma_{yx}}{\partial y}\,dx\,dy\,dz - \frac{\partial p}{\partial x}\,dx\,dy\,dz = 0 \tag{9.6.1}$$

Substitution for σ_{yx} from (9.2.1) into (9.6.1) enables us to write this as

$$\mu\frac{\partial^2 u}{\partial y^2} - \frac{\partial p}{\partial x} = 0 \tag{9.6.2}$$

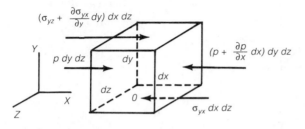

Figure 9.6.1 Forces acting on flow in the x direction.

Similarly, flow in the z direction obeys the equilibrium equation

$$\mu \frac{\partial^2 w}{\partial y^2} - \frac{\partial p}{\partial z} = 0 \tag{9.6.3}$$

From these relations we may deduce that there will be a pressure variation between these surfaces, and hence the possibility of a pressure increase at these surfaces, only if $\partial^2 u / \partial y^2$ is not zero; i.e., if the flow velocity between the two surfaces is neither constant nor varies linearly with y.

9.7   HYDROSTATIC AND SQUEEZE-FILM LUBRICATION

From the equilibrium equations in the previous section it follows that fluid lubrication may be maintained by one or more of the following three means.

The first is known as *hydrostatic lubrication*, which refers to forcing oil between the two surfaces to be lubricated. Pressurized lubrication is necessary whenever the two surfaces involved do not have sufficiently large relative motion to support hydrodynamic lubrication. An example is shown in Figure 9.7.1 where hydrostatic lubrication is used in a thrust bearing. Oil is forced into a pocket at the center of the shaft, where the relative velocity between shaft and bearing is nearly zero, and then flows radially outward under pressure. It is also used during start-up of heavy rotating machinery (steam turbines, steel mill motors) to lift the shaft from the bearing and thereby reduce bearing wear and starting friction until the rotational speed reaches the hydrodynamic lubrication range.

The second is known as *squeeze-film lubrication*, which refers to lubrication in the presence of either lateral vibration or momentary overloads wherein the fluid viscosity and inertia tend to maintain a film between the journal and the bearing.

The third is known as *hydrodynamic lubrication*. It requires rapid relative motion between the surfaces to be lubricated and inclination of one of the surfaces relative to the other so that the relative motion will tend to wedge fluid between the two surfaces and thus hold them apart while providing a low shear force in the direction of motion. We direct our attention to this means of lubrication in the remainder of the chapter.

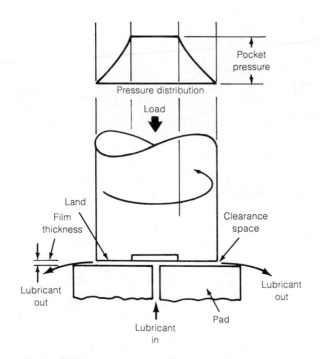

Figure 9.7.1 Hydrostatic lubrication applied to a thrust bearing. *Source: © 1973 Mobil Oil Corporation.*

9.8 HYDRODYNAMIC LUBRICATION

Hydrodynamic lubrication requires that the two surfaces separated by the lubricant move so rapidly relative to one another that one or both may be considered as hydrofoils with lift and drag on one side only. Probably the most common application of hydrodynamic lubrication is found in internal combustion engines, which use journal bearings on both ends of the connecting rods and at the crankshaft supports, as illustrated in Figuure 9.8.1. They are also used to support turbine and generator shafts as well as propeller shafts in ships, often because they are able to support a greater load per length of shaft than rolling element bearings of reasonable size and because they require a bearing radius only slightly larger than the shaft diameter. These advantages are obtained, however, at the expense of requiring a machined and polished surface with a very small roughness height. Use of machined and ground bushings, as pictured in Figure 9.8.2, often eliminates some difficult machining and/or may provide a softer bearing material to reduce journal wear in the event that adequate lubrication may be lost.

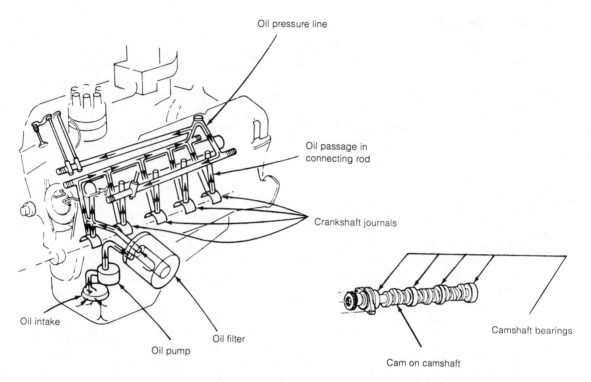

Figure 9.8.1 Journals and oil pressure lines in an V-8 engine. *Source: Ford Motor Co., Dearborn, MI.*

Figure 9.8.2 Bushing to form the bearing surface. *Source: Pioneer Motor Bearing Co., South San Francisco, CA.*

9.9 REYNOLDS' EQUATION

Analysis of the relation between the pressure on the plates and the plate separation, inclination, and fluid velocity for any of the lubrication methods listed above may be accomplished by means of the Reynolds equation. With this information we can integrate the pressure over the area of the bearing surfaces to find the forces acting, which in turn allows us to calculate the load that a fluid bearing can support and the power that will be lost in maintaining this flow within the bearing itself.

If relative motion of the surfaces is in the x direction the boundary conditions on the flow between these surfaces become

$$w = 0 \qquad u = U_1 \qquad v = V_1 \qquad \text{at } y = 0$$
$$w = 0 \qquad u = U_2 \qquad v = V_2 \qquad \text{at } y = h(x) \tag{9.9.1}$$

where the quantities used are illustrated in Figure 9.9.1. We will assume that derivatives $\partial p/\partial x$ and $\partial p/\partial z$ are independent of y and that $\partial p/\partial y = 0$, because these assumptions greatly simplify the analysis at the expense of negligible error in the result. We will use these assumptions when integrating (9.6.2) twice to obtain

$$u = \frac{1}{2\mu} \frac{\partial p}{\partial x} y^2 + f_1(z)y + f_2(z) \tag{9.9.2}$$

and we will use conditions 9.9.1 to evaluate $f_1(z)$ and $f_2(z)$. Thus,

$$U_1 = f_2(z)$$
$$f_1(z) = \frac{U_2 - U_1}{h} - \frac{1}{\mu} \frac{h}{2} \frac{\partial p}{\partial x} \tag{9.9.3}$$

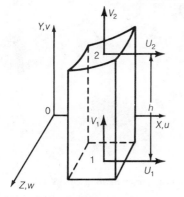

Figure 9.9.1 Assumed velocities of the upper and lower surfaces.

so that velocity $u(x,y)$ may be written as

$$u = \frac{1}{2\mu}\frac{\partial p}{\partial x}(y - h)y + (U_2 - U_1)\frac{y}{h} + U_1 \qquad (9.9.4)$$

Similarly, velocity $w(y,z)$ is given by

$$w = \frac{1}{2\mu}\frac{\partial p}{\partial z}(y - h)y \qquad (9.9.5)$$

Velocity u is associated with the flow through the bearing and velocity w is associated with the velocity of the fluid lost from the bearing through *end leakage*.

To get an equation for the pressure we first turn to the continuity equation for two-dimensional flow of an incompressible fluid, which is

$$\frac{\partial q_x}{\partial x} + \frac{\partial q_z}{\partial z} + (V_2 - V_1) = 0 \qquad (9.9.6)$$

where q_x and q_z represent the quantity of fluid flowing per unit time through a unit area perpendicular to the direction of flow at any instant. This relation may be derived from the infinitesimal element of volume and the flows shown in Figure 9.9.1. If the fluid is incompressible it cannot accumulate in the infinitesimal element of volume. Hence, the sum of the net fluid accumulation in the x direction per unit time, given by the first term in equation 9.9.7, and the net fluid accumulation in the z direction per unit time, given by the second term in equation 9.9.7, must add to zero as shown. Equation 9.9.6 then results from simplification of (9.7.7):

$$\left[\left(q_z + \frac{\partial q_z}{\partial z}dz\right) - q_z\right]dy\,dx + \left[\left(q_x + \frac{\partial q_x}{\partial x}dx\right) - q_x\right]dz\,dy \qquad (9.9.7)$$

Quantities q_x and q_z are defined by

$$q_x = \int_0^h u(x,y)\,dy = \left[\frac{1}{2\mu}\frac{\partial p}{\partial x}\left(\frac{y^3}{3} - \frac{hy^2}{2}\right) + (U_2 - U_1)\frac{y^2}{2h}\right]_0^h$$

$$= -\frac{1}{2\mu}\frac{h^3}{6}\frac{\partial p}{\partial x} + \frac{h}{2}(U_2 - U_1) \qquad (9.9.8)$$

and

$$q_z = \int_0^h w(x,y)\,dy = -\frac{h^3}{12\mu}\frac{\partial p}{\partial z} \qquad (9.9.9)$$

wherein the terms following the integral were calculated by substituting for $u(x,y)$ and $w(z,y)$ from (9.9.4) and (9.9.5) respectively. Substitution from (9.9.8) and (9.9.9) into

(9.9.6) yields the form of the Reynolds equation suitable for most machine design applications:

$$\frac{\partial}{\partial x}\left(\frac{h^3}{\mu}\frac{\partial p}{\partial x}\right) + \frac{\partial}{\partial z}\left(\frac{h^3}{\mu}\frac{\partial p}{\partial z}\right) = 6(U_2 - U_1)\frac{\partial h}{\partial x} + 12(V_2 - V_1) \qquad (9.9.10)$$

where the two terms on the right side of the equation may be simplified by setting $U = U_2 - U_1$ and $V = V_2 - V_1$. The first of these two terms, which represents the velocity of the upper surface relative to the lower, describes hydrodynamic lubrication, and is often the only term displayed on the right side of the Reynolds equation. The second term, which represents the velocity in the y direction of the upper surface relative to the lower, may be omitted in the steady state condition whenever the bearing load is constant. Its inclusion is necessary, however, when the bearing load fluctuates, because it accounts for a type of lubrication that has become known as *squeeze-film* lubrication. The terminology has been attributed to Underwood.[10]

The equation is nonliear because of the dependence of h upon x. Hence, no general analytical solution exists at this time.

Application of the Reynolds equation to journal bearings is clearly much easier if it is written in circular cylindrical coordinates in which the center line of the journal coincides with the z axis of a circular cylindrical coordinate system and the surface of the journal is given by $r = r_o$, where r_o is the radius of the journal. Since circumferential distance around the journal is given by $r_o\theta$, and for other radii by $r\theta$, where θ is measured from a reference direction $\theta = 0$, it follows that the role played by x in the rectangular Cartesian system is now played by $r\theta$ for $0 \leq \theta < 2\pi$. Consequently $\partial/\partial x$ is replaced by $\partial/(r\partial\theta)$, $\partial/\partial z$ remains, and the Reynolds equation now takes on the form

$$\frac{\partial}{\partial \theta}\left(h^3\frac{\partial p}{\partial \theta}\right) + r^2\frac{\partial}{\partial z}\left(h^3\frac{\partial p}{\partial z}\right) = 6\mu r^2\left(\frac{U}{r}\frac{\partial h}{\partial \theta} + 2V\right) \qquad (9.9.11)$$

The physical implications of the coordinate transformation used to obtain (9.9.11) is that we have associated a portion of length $2\pi r_o$ of the plane $x = 0$ with the surface of the journal and have associated the slider with the bearing surface. This coordinate transformation also implied that the slider length was chosen such that its projected length in the x direction was also $2\pi r_o$.

Alternative Derivation: Reduction from Navier-Stokes Relation

Reynolds' equation may be obtained from the Navier-Stokes relation

$$\rho\left[\frac{\partial \mathbf{v}}{\partial t} + (\mathbf{v} \cdot \nabla)\mathbf{v}\right] = \mathbf{F} - \nabla p + \mu\nabla \times (\nabla \times \mathbf{v}) + 2(\nabla \cdot \mu\nabla)\mathbf{v} - \frac{2}{3}\mu\nabla(\nabla \cdot \mathbf{v})$$

$$(9.9.12)$$

which holds for compressible fluids in which viscosity may be dependent upon temperature. The relation between compressibility and the fluid velocity vector $\mathbf{v}$ is given by the continuity equation

$$\frac{\partial \rho}{\partial t} + \nabla \cdot (\rho \mathbf{v}) = 0 \tag{9.9.13}$$

in which ρ denotes the mass per unit volume of the fluid. In these equations

$$\nabla = \hat{\mathbf{i}}\frac{\partial}{\partial x} + \hat{\mathbf{j}}\frac{\partial}{\partial y} + \hat{\mathbf{k}}\frac{\partial}{\partial z}$$

If the viscosity μ is constant, or if changes in viscosity have a negligible effect upon the pressure, we may use the vector identity

$$\nabla \times \nabla \times \mathbf{v} = \nabla(\nabla \cdot \mathbf{v}) - \nabla^2\mathbf{v}$$

to simplify (9.9.12) by combining the last three terms on its right side to obtain

$$\rho\left[\frac{\partial \mathbf{v}}{\partial t} + (\mathbf{v} \cdot \nabla)\mathbf{v}\right] = \mathbf{F} - \nabla p + \frac{\mu}{3}\nabla(\nabla \cdot \mathbf{v}) + \mu\nabla^2\mathbf{v} \tag{9.9.14}$$

If we next add the assumption that inertia and the body force vector $\mathbf{F}$ have negligible effect upon the pressure p then both $\mathbf{F}$ and the left side of (9.9.14) vanish and the Navier-Stokes equation reduces to

$$\nabla p = \frac{\mu}{3}\nabla(\nabla \cdot \mathbf{v}) + \nabla^2\mathbf{v} \tag{9.9.15}$$

Finally, if we assume that the oil is incompressible it follows that ρ is constant. Hence, the time derivative in the continuity equation 9.9.13 vanishes and that equation then yields $\nabla \cdot \mathbf{v} = 0$. Substitution of this condition into (9.9.15) yields

$$\nabla p = \mu\nabla^2\mathbf{v} \tag{9.9.16}$$

If velocity vector $\mathbf{v}$ has components u and w then

$$\mathbf{v} = \hat{\mathbf{i}}u + \hat{\mathbf{k}}w$$

where $\hat{\mathbf{i}}$ and $\hat{\mathbf{k}}$ are unit vectors in the positive x and z directions, respectively. Thus, we may write the components of (9.9.16) as

$$\frac{\partial p}{\partial x} = \mu\frac{\partial^2 u}{\partial y^2} \quad \text{and} \quad \frac{\partial p}{\partial z} = \mu\frac{\partial^2 w}{\partial y^2} \tag{9.9.17}$$

where we have used the conditions that $u = u(y)$, $w = w(y)$, and $p = p(x,z)$ in writing the scalar relations associated with each component. If we integrate these two relations using the boundary conditions (9.9.1) to evaluate the integration constants we obtain

$$u = \frac{1}{2\mu} \frac{\partial p}{\partial x} (y - h)y + U_1 \frac{h - y}{h} + U_2 \frac{y}{h} \tag{9.9.18}$$

and

$$w = \frac{1}{2\mu} \frac{\partial p}{\partial z} (y - h)y \tag{9.9.19}$$

Since these two relations are equal to (9.9.4) and (9.9.5), the steps following (9.9.5) may be repeated to again obtain the Reynolds equation.

9.10 APPROXIMATE SOLUTIONS FOR LONG AND SHORT BEARINGS

Long Bearing Solution

Using a clever substitution Sommerfeld[11] was able to solve (9.7.10) for an infinitely long bearing; one with no ends. If there are no ends there can be no end leakage and hence no z dependence of the pressure. Upon imposition of these conditions equation 9.7.10 becomes

$$\frac{\partial}{\partial x} \left(h^3 \frac{\partial p}{\partial x} \right) = 6\mu U \frac{\partial h}{\partial x} \tag{9.10.1}$$

for which the Sommerfeld solution is (Figure 9.10.1)

$$p = \frac{\mu U r}{c^2} \left[\frac{6\varepsilon(2 + \varepsilon \cos \theta) \sin \theta}{(2 + \varepsilon^2)(1 + \varepsilon \cos \theta)^2} \right] \tag{9.10.2}$$

where

$$\varepsilon = e/c \qquad h = c(1 + \varepsilon \cos \theta) = c + e \cos \theta$$

in which e denotes the *eccentricity* and ε the *eccentricity ratio* as shown in Figure 9.10.1. The angle θ which appears in (9.10.2) is measured from the diameter of the bearing that also passes through the center of the journal, and is zero where h is a maximum, also as illustrated in Figure 9.10.1.

Short Bearing Approximation

This solution, due to Michell[12] and later used by Ocvirck,[13] is predicated upon the assumption that in a very short bearing all of the fluid that impinges upon the in-clined surface is diverted to the end leakage, so that the first term in equation 9.9.11

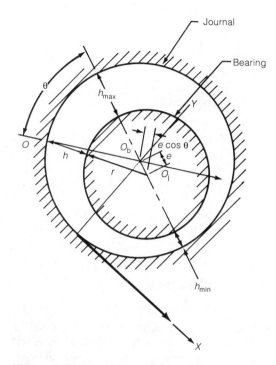

Figure 9.10.1 Local Cartesian coordinates used and other quantities which appear in long and short bearing approximations.

may be neglected. Thus, equation 9.7.11 becomes

$$\frac{\partial}{\partial z}\left(h^3 \frac{\partial p}{\partial z}\right) = 6\mu U \frac{\partial h}{\partial x} \tag{9.10.3}$$

for which the pressure is given by

$$p = \frac{3\mu U}{h^2}\left(z^2 - \frac{l^2}{4}\right)\frac{1}{r}\frac{\partial h}{\partial \theta} \qquad \frac{L}{D} < 4 \tag{9.10.4}$$

where L is the dimension in the axial direction, i.e., perpendicular to the direction of flow.

Load Calculation

The load for either long or short bearings may be calculated from the force components parallel and perpendicular to the line $\theta = 0$. Thus, if z is measured from the transverse plane of symmetry and if P_1 and P_2 represent the components parallel and perpendicular to that line then

$$P_1 = \int_0^{L/2} \int_0^{\pi} (pr\, d\theta\, dz) \cos \theta = \frac{\mu U l^3}{c^2}\frac{\varepsilon^2}{(1 - \varepsilon^2)} \tag{9.10.5}$$

and

$$P_2 = \int_0^{L/2} \int_0^{\pi} (pr\, d\theta\, dz) \sin\theta = \frac{\mu U l^3}{c^2} \frac{\pi\varepsilon}{4(1-\varepsilon^2)^{3/2}} \qquad (9.10.6)$$

Since these force components are mutually perpendicular, their vector sum is

$$P = \frac{\mu U l^3}{c^2} \frac{\varepsilon[\pi^2(1-\varepsilon^2) + 16\varepsilon^2]^{1/2}}{4(1-\varepsilon^2)^2} \qquad (9.10.7)$$

for the case of the short bearing where p is given by equation 9.10.4 and where $\partial h/\partial\theta$ may be calculated from the expression for h given in display (9.10.2). Bearing load for the long bearing approximation may be found from equations 9.10.5 and 9.10.6 by substituting for p from the first of display (9.10.2) and integrating numerically.

9.11 NUMERICAL SOLUTION TO REYNOLDS' EQUATION

Numerical design of a journal bearing to support a given load requires a more precise determination of the pressure distribution and velocity gradients than is provided by the approximate solutions listed above. This is motivation for finding a numerical solution to equation 9.7.11 by either replacing it with a corresponding difference equation or by using the variational method developed by Hayes and described in Refs. 21 and 22. Although it is too long to be described here, the variational method does have the advantage of allowing the pressure to be calculated to any desired accuracy if a sufficient number of terms are evaluated.

Infrequent design of journal bearings may be accomplished using design curves based upon numerical solutions of equation 9.7.11, such as those provided by Raimondi and Boyd, based upon results due to Carter.[15] Although these design curves will be used in the next section because they provide an appreciation of the parameters involved in the Reynolds equation, the procedure is time-consuming and may be less accurate than a direct numerical solution. It should be emphasized, therefore, that in an engineering facility frequently required to design journal bearings it may be preferable to obtain a numerical solution by a method similar to one of those mentioned above.

9.12 DESIGN CURVES

Design curves drawn from pressure distributions calculated from the Reynolds equation and the Reynolds conditions for a common range of bearing proportions were first presented in Refs. 9, 17, and 18 by Raimondi and Boyd. They contain plots of the fluid temperature rise, the minimum fluid film thickness, the effective coefficient of friction, and other values of interest in the design of a journal bearing for length to diameter ratios of 0.25, 0.50, and 1.00 and for 60°, 120°, 180°, and 360° bearings.

Only several of the design charts presented by Raimondi and Boyd for $L/D = 1$ will be considered in this chapter, those shown in Figures 9.12.1 through 9.12.4.

If the viscosity of the fluid at the journal is known, the procedure is to calculate the dimensionless Sommerfeld number and enter each of the design charts at this value to read out the desired quantity.

If the viscosity is not known it must be found by an iterative process using Figures 9.12.1 and 9.12.2 to find a consistent viscosity and temperature combination estimated

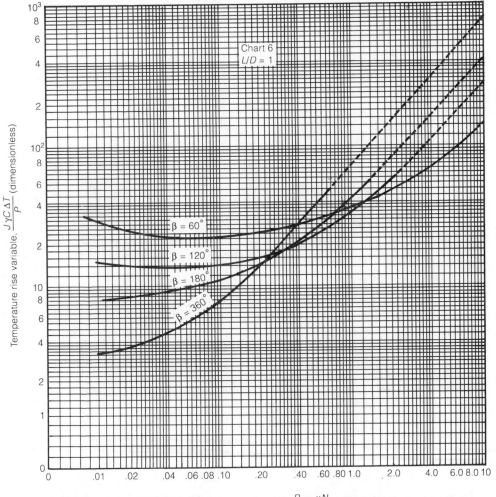

Figure 9.12.1 Temperature as a function of the Sommerfeld number (Chart 6). *Source: Reprinted by permission of the Society of Tribologists and Lubrication Engineers, Park Ridge, IL. All rights reserved.*

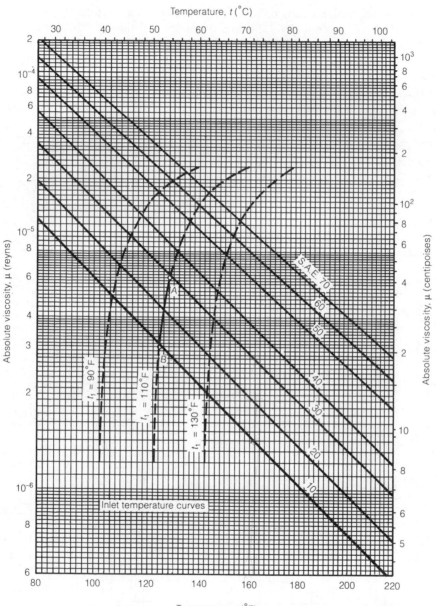

Figure 9.12.2 Viscosity–temperature chart of typical SAE numbered oils. *Source: Reprinted by permission of the Society of Tribologists and Lubrication Engineers, Park Ridge, IL. All rights reserved.*

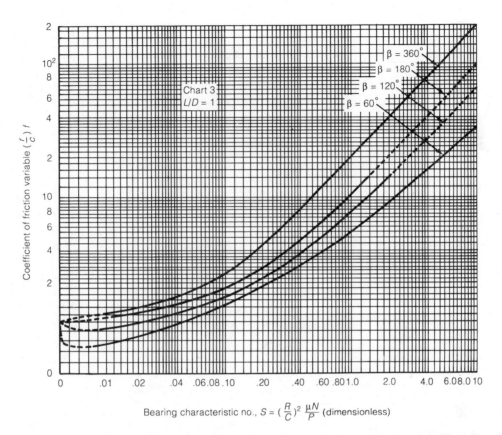

Figure 9.12.3 Coefficient of friction variable as a function of the Sommerfeld number (Chart 3).
Source: Reprinted by permission of the Society of Tribologists and Lubrication Engineers, Park Ridge, IL. All rights reserved.

from the inlet temperature. This process will be demonstrated in the following example and summarized in the design flowchart shown in Figure 9.12.6 at the end of this section.

Recall from (9.5.5) that the Somerfeld number is given by

$$S = \left(\frac{r}{c}\right)^2 \frac{\mu N}{P}$$

in which r denotes the journal radius, c the clearance, μ the absolute viscosity, N the rotational speed in *revolutions per second*, and P the load per unit of projected area $2rL$, where L is the journal length. It is important to realize that since S is dimensionless the units of μ and P must be chosen such that $\mu N/P$ remains dimensionless. Thus,

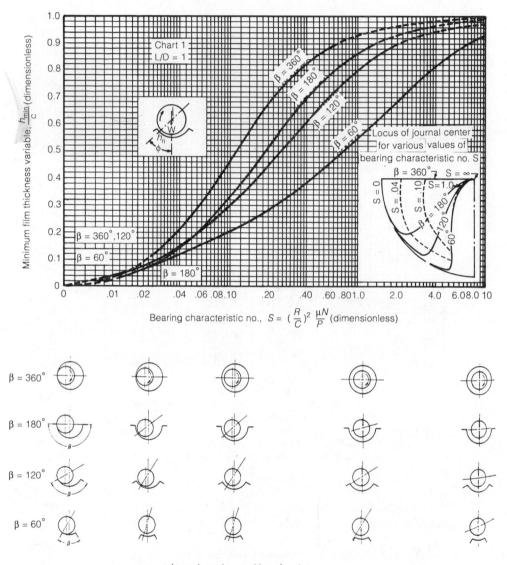

Journal running positions for above
values of characteristic no.

Figure 9.12.4 Minimum film thickness as a function of the Sommerfeld number (Chart 1). *Source: Reprinted by permission of the Society of Tribologists and Lubrication Engineers, Park Ridge, IL. All rights reserved.*

if μ is to remain in Pa-s then from the ratio

$$\frac{\mu N}{P} \sim \frac{\dfrac{N-s}{m^2}\dfrac{1}{s}}{P}$$

it follows that P must be in units of $N/(m^2)$ or Pa. Likewise, the units of

$$\frac{J\gamma C}{P}\Delta T$$

must be selected such that if ΔT is in units of °C or °F the ratio $J\gamma C/P$ is in units of $1/°C$ or $1/°F$, respectively. Upon writing the quantities in this ratio in SI units we have

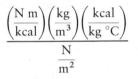

from which we see that the first two quantities in the numerator must have the same units of length, the second two must have the same units of mass, and the first unit in the numerator must have the same unit of force as used in the denominator.

EXAMPLE 9.12.1 (from Ref. 9)

Calculate the minimum film thickness, the frictional power loss, and the temperature rise for a 120° journal bearing for which

$W = 7200$ lb $L = 6$ in.

$N = 60$ rps $c = 0.006$ in.

$r = D/2 = 3$ in.   SAE 20 oil is to be used

Inlet temperature $= 110\,°F$

To begin the semi-iterative process assume an initial oil viscosity of 6×10^{-6} reyns and calculate $P = W/(DL)$ to find

$$P = 7200/(6 \times 6) = 200$$

Insert this value and the assumed viscosity into the expression for the Sommerfeld number, along with the above data to find that

$$S = \left(\frac{r}{c}\right)^2 \frac{\mu N}{P} = \left(\frac{3}{0.006}\right)^2 \frac{6 \times 10^{-6}(60)}{200} = 0.450$$

For this example assume the weight per unit volume of the oil γ is 0.030 lb/in.^{3} (0.830 g/cm^3) and that the specific heat is 0.40 Btu/(lb °F) [Btu/(lb °F) = kcal/(kg °C)]. Use $S = 0.450$ as the entry point in Figure 9.12.1 and read up to the $\beta = 120°$ curve to find that

$$\frac{J\gamma C \,\Delta T}{P} = 20.4 \tag{9.12.1}$$

With the Joule equivalent given by $J = 9336$ in.-lb/Btu (4186 N m/kcal), we may solve this relation for the temperature rise to find

$$\Delta T = \frac{20.4(200)}{9336(0.03)0.4} = 36.4\,°F$$

so that the average temperature of the heated oil may be estimated from

$$T_a = T_i + \frac{\Delta T}{2} \tag{9.12.2}$$

to be

$$T_a = 110 + 36.4/2 = 128.2\,°F$$

This viscosity and temperature combination corresponds to point A in Figure 9.12.2, which is above the line for a SAE 20 oil. Next, choose a lower viscosity, say 3×10^{-6} reyns, recalculate the Sommerfeld number, and repeat the above process to find $T_a = 124.3\,°F$. Thus, point B lies at $T_a = 124.3\,°F$, $\mu = 3 \times 10^{-6}$ reyns in Figure 9.12.2. Although a straight line could be drawn between these two points, we repeat the process several more times until we have sufficient points to draw a curve between A and B as shown in Figure 9.12.2, labeled $t_i = 110\,°F$, the inlet temperature. The inlet temperature has been chosen as the label because T_a changes with μ as we have just demonstrated. (Many more points were calculated to draw the extended curve shown in that figure.) The intersection of the 110 °F line and the line for SAE 20 oil gives the desired viscosity, namely, 4.4×10^{-6} reyns at a temperature of 126.1 °F as either estimated from Figure 9.12.2 or calculated from the above relation and Figure 9.12.1.

Substitution of $\mu = 4.4 \times 10^{-6}$ reyns into the formula for the Sommerfeld number yields $S = 0.330$. Upon entering Figure 9.12.3 at $S = 0.330$ and reading from the $\beta = 120°$ we find $(r/c)f = 3.25$ so that multiplication by c/r yields $f = 0.006\,50$. Likewise, upon entering Figure 9.12.4 at $S = 0.330$ and reading from the $\beta = 120°$ curve we find that $h_{min}/c = 0.561$ so that multiplication by c yields the minimum film thickness $h_{min} = 0.003\,37$ in. This thickness should be compared with the surface finish of the bearing and journal and with the minimum film capacity of the lubricant to be used.

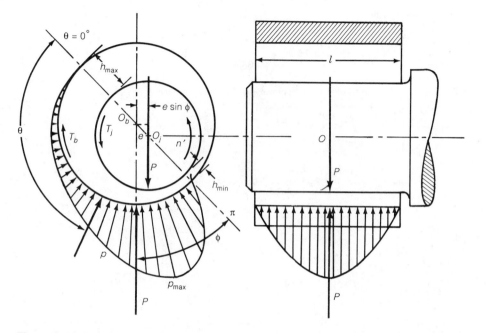

Figure 9.12.5 Radial and axial pressure distributions on the journal of a hydrodynamic journal bearing.

Since the frictional torque is given by Wfr it follows that the power loss due to friction is

$$hp = \frac{\pi W rfn}{198} 10^{-3} = \frac{0.0065(7200)3(3600)10^{-3}}{198} \pi = 8.02$$

Pressure is distributed along the length of the journal and bearing as shown in Figure 9.12.5 Since $\mu N/P$ is itself dimensionless, it is evident from the dimensions of N (1/s)and P (Pa) that μ must be N-s/m^2 when SI units are employed in calculating the Sommerfeld number. Likewise, when SI units are used to calculate the temperature from (9.12.1) the specific heat, denoted by C, must be expressed in kcal/(kg °C). Recall that Btu/(lb °F) = 1.0397 kcal/(kg °C).

The example just discussed was concerned with analyzing a journal bearing already designed and with an oil already selected; it demonstrated the interative procedure required to find the correct value of μ. A more common use of the Raimondi and Boyd curves is in the design of a journal bearing. In this application the rotational speed and the load to be supported are known along with approximate limits on the dimensions and clearance of the journal bearing as determined by machine requirements and

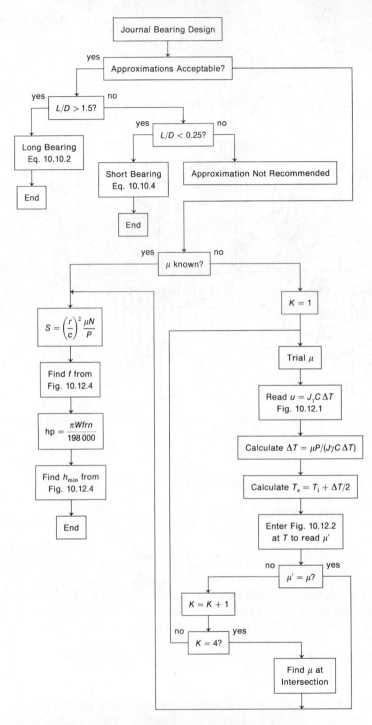

Figure 9.12.6 Flowchart for Raimondi and Boyd procedure.

manufacturing capabilities. We may then select h_{min} and β based upon the expected service environment for the machine, calculate h_{min}/c, enter Chart 1 at this value, read over to the intended β and down to the required Sommerfeld number S. From this value of S we may calculate the required μ since all other quantities in S are known. Since the service ambient temperature range is also known, we may now enter Chart 6 at this Sommerfeld number, calculate the temperature rise, and thus find T_a of the oil. With μ and T_a known, we may enter the absolute viscosity–temperature chart and find the SAE oil required to fulfill these requirements. At this step the viscosity ranges given in Table 9.2.2 are generally needed because the point with coordinates μ and T_a rarely falls upon one of the lines shown. See problems 9.29, 9.30, and 9.31 for applications of this procedure.

A flowchart for the Raimondi and Boyd procedure is given in Figure 9.12.6.

9.13 APPROXIMATION METHOD FOR FULL JOURNAL BEARINGS

Reason and Narang[19] have presented an empirical set of formulas to be used with the values in Table 9.13.1 which closely approximate the results obtained from solution of the Reynolds equation for full journal bearings with the boundary values given by the half Sommerfeld condition. Their relations yield the load-carrying capacity, the angle of the load relative to the line of centers, the effective coefficient of friction, the temperature rise, and the other values of interest produced by the previous method. Input data include the viscosity of the oil, the rotational speed, the bearing dimensions, the clearance, and the eccentricity ratio.

The load capacity is given by

$$W = \mu 6UL \left(\frac{R}{c}\right)^2 (I_c{}^2 + I_s{}^2)^{1/2} \tag{9.13.1}$$

and the angle of the load relative to the line of centers, α, is given by

$$\alpha = \tan^{-1}\left(-\frac{I_s}{I_c}\right) \tag{9.13.2}$$

and the coefficient of friction is given by

$$f = \frac{F}{W} = \frac{\dfrac{\pi}{3\sqrt{1-\varepsilon^2}} + \dfrac{\varepsilon}{2} I_s}{(I_c{}^2 + I_s{}^2)^{1/2}} \left(\frac{c}{r}\right) \tag{9.13.3}$$

all in terms of I_s and I_c listed in Table 9.13.1. Intermediate values are to be found by interpolation, which in this chapter will mean Lagrangian interpolation.

TABLE 9.13.1   TABULATED VALUES FOR I_s AND I_c

Values of Integral $I_s\left(\dfrac{L}{D}, \varepsilon\right)$

ε	$L/D = 2$	$L/D = 1\frac{1}{2}$	$L/D = 1$	$L/D = \frac{3}{4}$	$L/D = \frac{1}{2}$	$L/D = \frac{1}{4}$	$L/D = \infty$
0.1	0.0839	0.0636	0.0380	0.0244	0.0120	0.0032	0.1570
0.2	0.1705	0.1300	0.0783	0.0505	0.0251	0.0067	0.3143
0.3	0.2628	0.2023	0.1236	0.0804	0.0404	0.0109	0.4727
0.4	0.3649	0.2847	0.1776	0.1172	0.0597	0.0164	0.6347
0.5	0.4831	0.3835	0.2462	0.1656	0.0862	0.0241	0.8061
0.6	0.6291	0.5102	0.3396	0.2345	0.1259	0.0363	0.9983
0.7	0.8266	0.6878	0.4793	0.3430	0.1927	0.0582	1.2366
0.8	1.1380	0.9771	0.7220	0.5425	0.3264	0.1071	1.5866
0.9	1.8137	1.6235	1.3002	1.0499	0.7079	0.2761	2.3083
0.95	2.7600	2.5455	2.1632	1.8467	1.3712	0.6429	3.2913
0.99	3.8092	3.5708	3.1293	2.7442	2.1291	1.0719	4.3763

Values of Integral $I_c\left(\dfrac{L}{D}, \varepsilon\right)$

ε	$L/D = 2$	$L/D = 1\frac{1}{2}$	$L/D = 1$	$L/D = \frac{3}{4}$	$L/D = \frac{1}{2}$	$L/D = \frac{1}{4}$	$L/D = \infty$
0.1	−0.0076	−0.0063	−0.0041	−0.0028	−0.0014	−0.0004	−0.0100
0.2	−0.0312	−0.0259	−0.0174	−0.0118	−0.0062	−0.0017	−0.0408
0.3	−0.0733	−0.0615	−0.0419	−0.0289	−0.0153	−0.0043	−0.0946
0.4	−0.1391	−0.1183	−0.0825	−0.0579	−0.0312	−0.0089	−0.1763
0.5	−0.2391	−0.2065	−0.1484	−0.1065	−0.0591	−0.0174	−0.2962
0.6	−0.3949	−0.3474	−0.2590	−0.1917	−0.1105	−0.0338	−0.4766
0.7	−0.6586	−0.5916	−0.4612	−0.3549	−0.2161	−0.0703	−0.7717
0.8	−1.1891	−1.0941	−0.8997	−0.7283	−0.4797	−0.1732	−1.3467
0.9	−2.7932	−2.6461	−2.3269	−2.0172	−1.4990	−0.6644	−3.0339
0.95	−6.0396	−5.8315	−5.3621	−4.8773	−3.9787	−2.1625	−6.3776
0.99	−9.8095	−9.5362	−3.9045	−8.2294	−6.9073	−3.9626	−10.2526

Source: Reprinted by permission of the Society of Tribologists and Lubrication Engineers, Park Ridge, IL. All rights reserved.

Temperature rise is given by

$$\frac{J\gamma C_v \Delta T}{P} = \frac{f}{A_0} \frac{2\pi \dfrac{r}{c}}{1 - \dfrac{1}{2}\left(\dfrac{Q_s}{Q}\right)} \tag{9.13.4}$$

in terms of

$$A_0 = \pi(1 + \varepsilon) - \beta_1 \tag{9.13.5}$$

and

$$A_1 = \pi(1 - \varepsilon) + \beta_2 \tag{9.13.6}$$

where β_1 and β_2 are defined by $(i = 1, 2)$

$$\beta_i = \pi\varepsilon\alpha_i \left\{ 1 - \alpha_i \frac{\left(\dfrac{D}{L}\right)^2}{\left[\left(\dfrac{D}{L}\right)^2 \dfrac{\alpha_i}{2} + \dfrac{1}{4}\right]^{1/2}} \tanh^{-1} \frac{1}{2\left[\left(\dfrac{D}{L}\right)^2 \dfrac{\alpha_i}{2} + \dfrac{1}{4}\right]^{1/2}} \right\} \qquad (9.13.7)$$

in which α_1 and α_2 are in turn defined by

$$\alpha_1 = \frac{(1 + \varepsilon)(2 + \varepsilon)}{(2 + \varepsilon^2)} \qquad (9.13.8)$$

and

$$\alpha_2 = \frac{(1 - \varepsilon)(2 - \varepsilon)}{(2 + \varepsilon^2)} \qquad (9.13.9)$$

Quantity Q_s/Q is defined in term of the above quantities as

$$\frac{Q_s}{Q} = \frac{A_0 - A_1}{A_0} \qquad (9.13.10)$$

EXAMPLE 9.13.1

Calculate the quantities found in Example 9.12.1 but for a 360° instead of a 120° bearing with a specified eccentricity. For convenience the pertinent input data are repeated here:

$r = 3.00$ in.

$c = 0.006$ in., $\qquad\qquad\qquad N = 60$ rps

$e = c - h_{min} = 0.003$ in. (approx.), $U = 1130.973$ in./s

$L = 6$ in., $\qquad\qquad\qquad W = 7200$ lb

For $L/D = 1.00$ and $\varepsilon = 0.003/0.006 = 0.500$ find from Table 9.13.1 that

$I_s = 0.2462$ $\qquad\qquad\qquad I_c = -0.1484$

so that

$(I_s^2 + I_c^2)^{1/2} = 0.2875$

Calculate the required viscosity for a full journal bearing relation 9.13.1. Thus,

$$W = 6\mu 1130.973(6) \left(\frac{3}{0.006}\right)^2 0.2875 = 2.9264 \times 10^9 \mu$$

so that with $W = 7200$ lb the required viscosity is

$$\mu = 2.460 \times 10^{-6} \text{ reyns}$$

According to (9.13.3) the coefficient of friction is

$$f = \frac{F}{W} = \frac{\dfrac{\pi}{3\sqrt{1-0.25}} + 0.25(0.2462)}{0.2875} \frac{0.006}{3} = 0.0088$$

so that multiplication by the load yields a friction force of 63.65 pounds.

$$hp = \pi Tn/198\,000 = 63.36(3600)/198\,000 = 3.62$$

is the estimated power loss due to the effective friction.
    To calculate the temperature rise find that

$$\alpha_1 = \frac{(1+0.5)(2+0.5)}{2+0.25} = 1.667 = 5/3$$

$$\beta_1 = \pi 0.5 \frac{5}{3}\left(1 - \frac{5}{3}\frac{1}{1.041}\tanh^{-1}\frac{1}{2.082}\right) = 0.4242$$

$$A_0 = \pi(1.5) - 0.4242 = 4.2881$$

$$\alpha_2 = 0.5(1.5)/2.25 = 1/3$$

$$\beta_2 = \pi 0.5 \frac{1}{3}\left(1 - \frac{1}{3}\frac{1}{0.6455}\tanh^{-1}\frac{1}{1.2910}\right) = 0.2446$$

$$A_1 = \pi 0.5 + 0.2446 = 1.8154$$

$$\frac{Q_s}{Q} = \frac{4.2881 - 1.8154}{4.2881} = 0.5776$$

$$\frac{J\gamma C_v \Delta T}{P} = \frac{0.0088}{4.2881} \frac{2\pi \dfrac{3}{0.006}}{1 - \frac{1}{2}(0.5776)} = 9.059$$

so that

$$\frac{J\gamma C_v \Delta T}{P} = \frac{9336(0.03)0.40}{200} \Delta T = 0.5602 \Delta T$$

Upon equating the right sides of the last two results the temperature rise is found to be 16.17 °F.

9.14 SOLID LUBRICANTS

The most common solid lubricants appear to be molybdenum disulfide, graphite, PTFE, and nylon. Properties of these and other solid lubricants are listed in Table 9.14.1. The advantages of solid lubricants are good stability at extreme temperatures and in chemically active environments, they require no additional equipment, such as pressure pumps and lines, some are unaffected by nuclear radiation, and some may be used under heavy load conditions. Their disadvantages are that they may be of little aid in cooling and they generally have a higher coefficient of friction than hydrodynamic or rolling element bearings.

Molybdenum disulfide is often used as a paste or an additive in grease. It is limited to temperatures less than about 750 °F in an oxidizing atmosphere, with higher temperatures possible in nonoxidizing atmospheres. Graphite can provide lubrication at higher temperatures and leaves no abrasive residue if oxidized.

Nylon and similar rigid lubricants may be molded or machined to provide low-friction surfaces for irregularly shaped parts or to be mechanically retained where low friction is desired. Its bearing properties may be improved by filling with molybdenum disulfide or graphite.

PTFE (polytetrafluoroethylene) lubricating properties are best at loads less than 5000 psi. Chatter is almost eliminated by PTFE because its static and dynamic friction coefficients are nearly equal. It appears to provide less friction than either molybdenum disulfide or graphite as an additive to greases to be used on some soft metals, such as zinc and similar die-casting metals.

Because little is known about the behavior of solid lubricants, they are usually compared on the basis of their R value, which is defined by the empirical formula[20]

$$R = KTPV \tag{9.14.1}$$

where R is the amount of material removed from the contacting surfaces per unit of width, i.e., the dimension of the surface perpendicular to the direction of motion. Its units are determined by the units of coefficient K. In definition 9.14.1 P denotes the load and V denotes the relative velocity of the sliding surfaces. Proportionality constant K

TABLE 9.14.1   FRICTION DATA FOR COMPOUNDS CONSIDERED AS SOLID LUBRICANTS

Material	Load g	Coefficient of Friction		
		80 °F	500 °F	1000 °F
Ag_2Se	326	0.30–0.35	—	0.8–0.9
LiF	''	0.3–0.4	0.9	0.65–0.75
Si_3N_4	''	0.3–0.9	—	0.9
SrS	''	0.7–0.9	—	0.7–0.9
Tl_2S	''	0.25–0.5	—	—(melted)
ZnSe	''	0.4–0.6	—	0.55–0.70
ZrN	''	0.2–0.3	—	0.55–0.75
AgI	128	1.0	1.0	1.0
Ag_2S	''	0.36	—	—
Ag_2Te	''	0.44	—	—
$AlPO_4$	''	1.33	1.31	0.31–0.37
$AlPO_4$	''	1.0	1.0	0.80
$AlPO_4$	326	0.60	—	0.51
Bi_2S_3	128	0.23–0.60	0.21–0.77	0.18–0.38
Bi_2S_3	326	—	0.24–0.39	0.21–0.27
Bi_2S_3	''	0.56–0.58	0.49–0.62	0.20–0.32
$Bi_2S_3 + Bi_2O_3$	128	0.38	0.20	0.20
$Bi_2S_3 + Bi_2O_3$	''	—	0.42	0.31–0.37
CdSe	128	0.23–0.33	0.58	0.27–0.38
CdS	''	0.58–1.0	0.84	0.55
CdTe	''	0.46	0.40	0.33–0.60
CoS	''	0.50–1.15	0.72	0.58
Cu_2S	128	1.0	1.2	1.2
GeO_2	''	0.48–0.58	—	—
InSe	''	0.46–0.60	0.41–0.60	0.60 (at 750 °F)
$MoSe_2$	''	0.20–0.33	0.31–0.40	—
NiS	''	0.29–1.0	—	—
PbSe	''	0.40–0.67	0.25	0.25
PbS	''	0.08	0.47	0.21
PbS	''	0.30–0.68	—	0.29
PbS(I)	326	0.27–0.39	0.48	0.20
PbS	''	0.47	0.27–0.47	0.15–0.19
PbS-MoS_2	''	0.16–0.38	0.13	0.37
PbS-graphite	''	0.20	0.29	0.21
Sb_2O_5	128	0.21–0.96	—	—
Sb_2S_3	''	0.38	0.21–0.49	0.49
Sb_2S_5	''	0.31–0.80	0.35–1.0	—
Sb_2S_5	''	0.50	1.0 (at 300 °F)	—
$TiTe_2$	128	0.36–0.58	0.95	0.77 (at 670 °F)
$TiTe_2$	''	0.48–0.58	0.86	0.38–0.48 (at 800 °F)
ZnTe	''	0.60–0.68	0.32–0.40	0.58
ZrCl	''	0.45	0.48–0.96	0.27
Graphite no. 2	128	0.15	0.18–0.22	0.50–0.60
Graphite no. 1	''	0.15	0.10	0.13–0.23
Graphite	326	0.14–0.30	0.06–0.12	0.20–0.27
MoS_2 + graphite (7% by weight)	128	0.20–0.25	0.11	0.22–0.56
PbS(II)	—	0.55–0.70	0.3–0.7	—
MoS_2 + graphite (31% C by weight)	128	0.16–0.21	0.12–0.14	0.13–0.16
MoS_2 + graphite (73% C by weight)	''	0.15–0.21	0.08–0.11	0.11–0.17
MoS_2	326	0.34	0.10	—

Test Conditions: Mark III Pellet Machine
Load: 326 g (avg approx 50 psi)
Speed: 600 ft/min
Track: polished, Rex AAA steel.
Tests were conducted at Midwest Research Institute (MRI).

Source: M.E. Campbell, *Solid Lubricants: A Survey, NASA SP-5059*, Washington, D.C.: NASA, 1972.

TABLE 9.14.2 COMPARATIVE PERFORMANCE DATA FOR SELF-LUBRICATING PLASTIC BEARING MATERIALS

Properties	Modified polymers					Unmodified polymers				
	Nylon	Acetal	Fluoro-carbons	Poly-imides	Phenolics	Nylon, graphite filled	Acetal TFE, fiber filled	Fluoro-carbon, wide range of fillers used	Polyimide, graphite filled	Phenolic, TFE filled
Maximum load, projected area, zero speed, psi	4900	5200	1000	>10,000	>4000	>1000	1800	2000	>10,000	>4000
Speed, max ft/min[1]	200–400	500	100	>1000	1000	200–400	800	>1000	>1000	>1000
PV for continuous service, 0.005 in. wear in 1000 hr, psi × ft/min	1000	1000	200	300	100	1000	2500	2500 [2]50,000	3000	5000
"Limiting PV" at 100 ft/min, psi × ft/min	4000	3000	1800	>100,000	5000	4000	5500	30,000	>100,000	>40,000
Coefficient friction	0.20–0.40	0.15–0.30	0.04–0.13	0.1–0.3	0.90–1.1	0.1–0.25	0.05–0.15	0.04–0.25	0.1–0.3	0.05–0.45
Wear Factor, $K \times 10^{-10}$ in.^{3}min/ft-lb-hr	50	50	2500	150	250 2000	50	20	1–20	15	10
Elastic modulus, bending, psi × 10^6	0.3	0.4	.08	0.45	5	0.4	0.4	0.4	0.63	5
Critical temp. at bearing surface, °F	400	300	500	>600	300–400	400	300	500	600	300–400
Thermal conductivity Btu/in./hr/sq ft/°F	1.7	1.6	2.2	2.2					2.2	
Thermal expansion in./in./°F × 10^6	55	45	80	30	25	35–50	45	3.0–5.6	30	25
Resistance to humidity	fair	good	excellent	good	good	fair	good	excellent	good	good
Resistance to chemicals	good	good	excellent	good	good	good	good	excellent	good	good
Density, g/cc	1.2	1.43	2.15–2.20	1.42	1.4	1.2	1.54	2.15–2.25	1.49	1.4
Cost index for base material	1.4	1	5	15		1.5	6	5	15	

[1] Continuous operation under 5 lb load.
[2] Exceeds limiting PV.

Source: M.E. Campbell, *Solid Lubricants: A Survey, NASA SP-5059,* Washington, D.C.: NASA, 1972.

and time T are often set equal to 1.00 for comparison of commercially available solid lubricants. These values were assumed in Table 9.14.2 for the comparison of common solid lubricants. Other values may apply to special tests and unusual applications.

REFERENCES

1. Beerbower, A., *Boundary Lubrication, Scientific and Technical Application Forcast*, Dept. of the Army, DAHC-19-69-C-0033 (1972).

2. Stribeck, R., Characteristics of plain and roller bearings, *Zeit. V.D.I.* 46 (1902)

3. Jones, W.R., Jr., *Boundary Lubrication—Revisited*, NASA, TM-82858 (March 9, 1982)

4. Dean, E.W., and Davis, G.H.B., Viscosity of oils with temperature, *Chemical and Metallurgical Engineering* 36, 618–619 (1929); also *Industrial and Engineering Chemistry* 32, 102–107 (1940).

5. Mechanical drives issue, *Machine Design* 55 (June 30, 1983).

6. Spotts, M.F., *Design of Machine Elements*, 5th ed., Prentice-Hall, Englewood Cliffs, NJ, 1978.

7. Juvinall, R.C., *Fundamentals of Machine Component Design*, Wiley, New York, 1983.

8. Shigley, J.E., and Mitchell, L.D., *Mechanical Engineering Design*, 4th ed., McGraw-Hill, New York, 1983.

9. Raimondi, A.A., and Boyd, J., A solution for the finite journal bearing and its application to analysis and design, I, *ASLE Transactions* 1, 159–174 (1958).

10. Underwood, A.F., Rotating load bearings, *Automotive and Aviation Industries* 92, 26–28, 60, 64 (1945).

11. Cameron, A., The *Principles of Lubrication*, Wiley, New York, 1966, p. 286.

12. Michell, A.G.M., Progress in fluid film lubrication, *Transactions of the ASME* 51, 153–163 (1929).

13. Ocvirk, F.W., and DuBois, G.B., *Analytical Derivation and Short Bearing Approximation for Full Journal Bearings*, NASA Report 1157 (1953).

14. Trumpler, R.T., *Design of Film Bearings*, Macmillan, New York, 1966.

15. Carter, D.S., An electrical method for determining journal-bearing characteristics, *Journal of Applied Mechanics* 19, 114–118 (1952).

16. Kunz, K.S., *Numerical Analysis*, McGraw-Hill, New York, 1957.

17. Raimondi, A.A., and Boyd, J., A solution for the finite journal bearing and its application to analysis and design, II, *ASLE Transactions* 1, 175–193 (1958).

18. Raimondi, A.A., and Boyd, J., A solution for the finite journal bearing and its applications to analysis and design, III, *ASLE Transactions* 1, 194–209 (1958).

19. Reason, B.R., and Narang, I.P., Rapid design and performance evaluation of steady-state journal bearings—a technique ameanable to programmable hand calculators, *ASLE Transactions* 25, 429–444 (1982).

20. Campbell, M.E., *Solid Lubricants, A Survey*, NASA SP-5059(01) Washington DC (1972).

21.   Hays, D.F., A variational approach to lubrication problems and the solution of the finite journal bearing, *Journal of Basic Engineering, ASME Transactions* 89, 13–22 (1959).

22.   Li, D.R., Allaire, P.E., and Barrett, L.E., Analytical dynamics of partial journal bearings with applications, *ASLE Transactions* 22, 99–112 (1980).

PROBLEMS

Section 9.2

9.1   Find the conversion factor, or multiplier, to be used in converting from reyns to millipascal-seconds, abbreviated mPa-s. Also find the multiplier to be used in converting from mPa-s to reyns.

[*Ans.* reyns × (6.89476×10^6) = mPa-s, mPa-s × $1.450\,38 \times 10^{-7}$ = reyns]

9.2   (a) Verify relation 9.2.6a from relation 9.2.5 using cSt = mm^2/s. Mass density ρ is to remain in units of g/cm^3. (b) Find the kinematic viscosity in centistokes (cSt) of an oil whose viscosity was measured as 143 SUS at 65.5 °C. Also find its dynamic viscosity in reyns and centipoise.

9.3   Calculate the viscosity in centipoise at 100 °C of an oil whose kinematic viscosity was measured as 62 SUS at that temperature. Also find the dimensions of the dimensional constants 0.22 and 180 in equation 9.2.6a. The mass density of the oil is 0.89 g/cm^3 at 15.6 °C.

9.4   A proprietary synthetic oil which as a mass density of 0.78 g/cm^3 at 15.6 °C, so that

$$\rho = 0.78 - (°C - 15.6)51 \times 10^{-5} \text{ g/cm}^3$$

Find its viscosity in reyns at 60 and 90°C if the corresponding SUS values are 37 and 32 s, respectively.

Section 9.3

9.5   Find the viscosity index of an oil that has a viscosity of 70 SUS at 210 °F and 489.0 SUS at 100 °F.

9.6   What should be the viscosity in mPa-s of an oil at 100 °F for which VI = 82 if its viscosity at 210 °F is 24.8 mPa-s?

Section 9.4

9.7   What class of additives should be added to mineral oil (petroleum oil with no additives) for use in the oil sump for a reduction gear if accumulated water is to be drained from the sump on a weekly schedule. Gear speeds may induce foam.

Section 9.5

9.8   It has been proposed that the Seybolt viscosity measurement system be replaced by a test based upon a journal bearing that operates at a Sommerfeld number of 10 or greater. Indicate which quantities should be measured and recommend available instrumentation to obtain these measurements.

9.9   A squirrel cage blower is to be designed for a curing chamber and is to be chain driven at a rotational speed of about 2200 rpm. If the chain tension produces an equivalent radial load of 8 lb, design journal bearings at locations *A* and *B* in the schematic below according to the Petroff equation for a journal having diameter and length 0.825 in. Use the Sommerfeld number of 10 for validity of the Petroff relation (9.5.3) and specify the SAE oil to be used at 100 °F (see Figure 9.12.2), the radius of the bearing (which may be between 0.4145 and 0.4185 in.), the clearance, and the power loss in the bearings. Does this problem have a unique answer? Why?

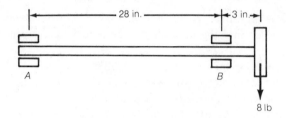

9.10   Find the change in the power loss and the change in recommended oil viscosity if the diameter of the journal in problem 9.9 is doubled and if the clearance is to be 0.005 in. Journal length to remain unchanged.

Section 9.12

9.11   If *P* is to remain in N/mm^2, what units must be chosen for μ in the calculation of the Sommerfeld number? Does this restrict the choice of units for *r* and *c*?

9.12   What must be the units of *P* in the Sommerfeld number if μ is to remain in *cP*?

9.13   If *P* is to remain in $lbs/in.^2$ what units must be selected for μ in the calculation of the Sommerfeld number *S*?

9.14   If γ is 0.830 g/cm^3, find its value in kg/m^3 and in kg/mm^3. If $J = 4186$ N m/kcal in SI units, what is its value in N mm/kcal?

9.15   Find the minimum film thickness and the power loss due to friction for a 120° partial bearing to support a load of 4000 lb rotating at 3500 rpm, The journal diameter is 2.50 in., the length is 2.50 in., and the clearance is 0.004 in. Also find the temperature rise for an inlet temperature of 120 °F when SAE 30 oil is used.

9.16   Repeat problem 9.15 for a partial bearing of 180°.

9.17   Repeat problem 9.15 for a full journal bearing.

9.18   After completing problems 9.15, 9.16, and 9.17, which bearing would you recommend if frictional losses are to be minimized? Which would you recommend if oil changes are to be infrequent and if the machine operates in a dirty environment? Why?

9.19   Find the minimum film thickness and the power loss due to friction for a complete journal bearing to support a load of 3740 N using a journal which is 50 mm in diameter, 50 mm long, rotates at 4000 rpm, and has a clearance of 0.120 mm. SAE oil is to be used and the inlet temperature is to be 50 °C. Assume the density of the oil is 0.830 g/cm^3 and that $C = 0.40$ kcal/(kg °C). Note that the temperature in degrees Celsius is shown at the top of Figure 9.12.2 and that the viscosity in centipoise is shown on the right side.

9.20   Repeat problem 10.19 for a 180° bearing.

9.21   Use the figure below (from Ref. 9) to find the maximum pressure on the journal in problem 9.19 and compare with the long bearing approximation if the ambient pressure is 0.10 MPa and if $\theta = 90°$.

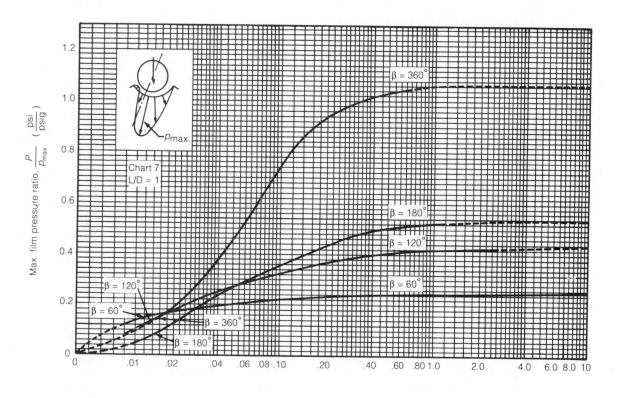

$$p'_{max} = \frac{P}{\left(\dfrac{P}{p_{max}}\right)} + p_{ambient}$$

p'_{max} = absolute maximum oil pressure

$P = W/(DL)$

$p_{ambient}$ = ambient pressure in the chamber containing the journal bearing

9.22   Find the film thickness, power loss, and the temperature rise of a 180° bearing with a clearance of 0.050 mm in which the journal is 24 mm in diameter, 24 mm long, and rotates at 3900 rpm.

The inlet oil temperature is to be 40 °C and SAE 30 oil is be used for a load of 3150 N. $C = 0.40$ kcal/(kg °C) and $\gamma = 830$ kg/m^3.

9.23 Repeat problem 9.22 for SAE 40 oil. What should be the maximum diameter of solid impurities that can be passed by the filter in the oil circulation system which provides cleaning and cooling? Use the criterion that the particle diameter should be one-tenth or less of the minimum clearance.

9.24 What is the change in the power loss and the minimum clearance in problem 9.23 when a 25% overload is encountered?

9.25 The journals of a used machine are 35 mm long, 35 mm in diameter, and appear to have been scored by overloads in dirty oil. If the effect of surface roughness is to increase the effective oil viscosity by 10%, decide whether to replace the bearings, based upon the criterion that a 5% increase in power loss will justify the expense. Clearance is 0.020 mm, the rotational speed is 3200 rpm, the radial load is 850 N. SAE 10 oil is to be used and the inlet temperature will be 60 °C. Bearings are 360°.

9.26 Recalculate the effect of worn bearings in problem 9.25 to decide if bearing replacement is necessary if SAE 30 oil is to be used. Compare the power loss with new bearings when SAE 10 and SAE 30 oil is used.

9.27 Justify the statement in Section 9.5 that the journal bearing geometry resembles that assumed in the derivation of the Petroff equation if $S > 8$, where S denotes the Sommerfeld number. If $S \geq 10$ may the power loss be calculated from the Petroff equation with an error less than 5%? *Hint*: Refer to two of the curves in Section 9.12.

9.28 A complete journal bearing 100 mm long and 100 mm in diameter supports a shaft for a total load of 500 N on the bearing. May its power loss be calculated from the Petroff equation if the clearance is 0.10 mm, if the journal rotates at 50 rps, and if the viscosity of the bearing is 40 cP at a temperature of 72.22 °C? Demonstrate by calculating the power loss by both the Petroff equation and the charts in Section 9.12. Also find the oil inlet temperature if SAE 50 oil is used.

9.29 Recommend an SAE grade of oil for use in a 360° journal bearing that is 60 mm long, has a 60 mm diameter, and a clearance of 0.060 mm. It is to operate at 5250 rpm and support a radial load of 8560 N when the inlet temperature is 30 °C. The minimum oil film thickness h_{min} must be no smaller than 0.040 mm because of possible contaminant sizes. Assume an oil density of 830 kg/m^3.
[*Ans*. SAE 10]

9.30 Recommend an SAE grade oil to be used in a 180° journal bearing that is 2.10 in. long, 2.10 in. in diameter, and has a clearance of 0.010 in. It is to operate at 6000 rpm when supporting a radial load of 1210 lb at an inlet temperature of 80 °F with a minimum film thickness h_{min} of 0.004 in.
[*Ans*. SAE 30 oil]

9.31 Would you change your recommendation for problem 9.30 if an oil filter is used to reduce the size of suspended particles such that a clearance of 0.002 in. could be achieved. Would use of a different oil have any advantages?
[*Ans*. SAE 20 oil. Yes, a reduction in power loss.]

9.32 Calculate the Sommerfeld number for the viscosity calculated in Example 9.13.1 and use it to find the minimum clearance h_{min} and the friction coefficient f from equations 9.12.3 and 9.12.4.

Compare with the values found in Example 9.13.1. What inlet temperature is required for the viscosity found in Example 9.13.1 to correspond to SAE 20 oil?

9.33   Select a self-lubricating bearing material for bearings on a ram piston guide and on a retraction lock, both in an explosive environment. The continuous load on the piston guide is 20 psi and on the retraction lock is 120 psi. Piston guide thickness may be 0.10 in. and it should have a minimum life of 10 000 h. The retraction block bearing may have a continuous life of only 500 h and its thickness may be 0.05 in. Average speed of the piston is 5 in./s and of the lock is 12 in./s. Modified polymer bearings are preferred.

CHAPTER TEN

ROLLING ELEMENT BEARINGS

NOTATION

A	axial load (ml/t^2)		j	subscript
a_{ij}	life adjustment factors (1)		L_{ab}	bearing life in cycles for which no more than ab% have failed
C	basic dynamic load rating (basic load rating) (ml/t^2)		N,N'	number (1)
C_o	basic static load rating (ml/t^2)		n	subscript (1) or rpm $(1/t)$
D	rolling element diameter or outside diameter (mlt^{-2})		P	equivalent radial or thrust force (ml/t^2)
			R,r	radius (l)
d	bore diameter (l)		t	time (t)
d_m	pitch diameter of the rolling element compliment (diameter of a circle passing through the centers of all rolling elements)		U	dummy variable
			X,Y	coefficients (1)
			Z	number of rolling elements per row (1)
e	reference number (l)		α	contact angle (1)
F,F^*	force (ml/t^2)		κ	exponent (1) or kinematic viscosity ratio (1)
F_a	axial load (ml/t^2)			
F_r	radial load (ml/t^2)		v,v_1	kinematic viscosity $(1/lt)$
f_c,f_o	dimensionless coefficients (1)		Φ	angle ratio (1)
I	induced axial force (ml/t^2)		ϕ	angle of rotation (1)
i	number of rows of rolling elements, subscript (1)		ψ	angle (1)
			Ω,ω	angular velocities $(1/t)$

As the name implies, rolling element bearings are those in which sliding motion is largely replaced by rolling motion. Sliding is not entirely eliminated because of contact between the rolling elements and their retainers, known as *cages*, and because the rolling elements may not actually roll everywhere in the contact area between them and the surfaces on which they roll, known as *rings*. Energy is also lost in rolling resistance due to local deformation of both the rolling elements and the races.

These bearings have also have been known as antifriction bearings and the organization that establishes manufacturing and performance standards for them in the United States is known as the Anti-Friction Bearing Manufacturers Association (AFBMA).

Most of this chapter is devoted to the selection of rolling element bearings because most of the engineers outside of the bearing manufacturing industry seldom, if ever, design custom bearings. This is because it is a highly specialized industry which requires specialized processes and expensive equipment if good quality bearings are to be produced.

This chapter differs from that of other machine design texts used at this time in the United States in that it includes both ball and roller bearings and demonstrates the importance of accounting for the axial forces induced by radial loads on angular contact ball and roller bearings. It also presents formulas used in industry for calculating bearing life in the presence of realistic time dependent forces rather than the highly idealized step-function loading often assumed.

Because of the time restrictions often placed upon an undergraduate mechanical engineering program, it may be impossible to cover all of the material in this chapter. The breadth of material included is, therefore, intended to permit the professor to select those portions which fit the program emphasis at the particular university where it is used while permitting the student easy access to additional material.

10.1 BEARING FRICTION

Although no accurate methods exist for predicting a bearing's friction coefficient under all service conditions, the following values, based upon radial (or axial) load and the tangential force at the bore diameter, are indicative of performance under normal loads and ideal lubrication.[1,2] All bearings are without seals.

Bearing Type	Friction Coefficient
Self-aligning radial ball bearing	0.0010–0.0020
Cylindrical roller thrust bearing	0.0011–0.0080
Ball thrust bearing	0.0013–0.0015
Single row radial ball bearing	0.0015–0.0020
Spherical roller bearing	0.0018–0.0020
Tapered roller bearing	0.0018–0.0020

Steel on steel dry sliding friction coefficients are of the order of 0.40 to 0.80 and with grease of the order of 0.005 to 0.20. These are handbook values, and are included for comparison.

10.2 BEARING TYPES

Rolling element bearings are classified as either roller or ball bearings, with roller bearings subdivided into radial and thrust bearings, depending upon whether the load is primarily perpendicular or parallel to the bearing axis. Ball bearings are subdivided into radial, thrust, and angular contact, with definitions of radial and thrust ball bearings identical to that used for roller bearings. Angular contact ball bearings have their inner and outer rings ground in such a manner as to allow them to support a combination of radial and thrust loads without damage to the bearing.

Cylindrical Roller Bearings

By convention a cylindrical roller bearing is defined to be one constructed as in Figure 10.2.1. Some manufacturers refer to the double row bearing as a journal roller bearing. The surface of the *outer* and *inner rings* which contact the rollers are termed *races* or *raceways*. The structure that separates the rollers is termed the *separator*, *cage*, or *retainer*. Although it keeps rollers from contacting, its main purpose is to prevent the rollers from moving away from the load-bearing portion of each ring. Some needle bearings do not have a retainer because they are tightly packed.

Cylindrical and Conical Thrust Bearings

Cylindrical roller thrust bearings, Figure 10.2.2, usually consist of three separate units: two thrust washers, or rings, and the separator and rollers. One-piece cylindrical rollers

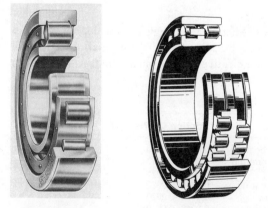

Figure 10.2.1 Cylindrical roller bearings. *Source: SKF Industries, Inc., Bearings Group, King of Prussia, PA.*

Figure 10.2.2 Cylindrical roller thrust bearing. *Source:* Courtesy of The Torrington Co., Torrington, CT.

are the least efficient of the thrust roller bearing styles because the rollers actually roll along only one circumferential ring at best, and skid everywhere else. In particular, suppose there is no slip between a roller of radius r and its raceway at some radius R_o, Figure 10.2.3. Hence, it must be that

$$\omega r = \Omega R_o \qquad\qquad\qquad (10.2.1)$$

where ω is the rotational speed of the roller about its axis and Ω is the rotational speed of the separator about the bearing axis. If (10.2.1) continues to hold true, then at some other radius $R_o + \Delta R_o$

$$\omega r \neq \Omega(R_o + \Delta R_o)$$

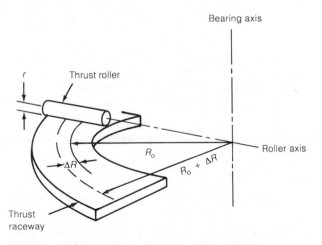

Figure 10.2.3 Cylindrical roller bearing kinematics.

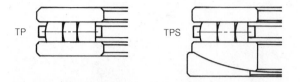

Figure 10.2.4 Segmented cylinder roller thrust bearings. *Source: Courtesy of The Torrington Company, Torrington, CT.*

Thus, there is sliding friction between the roller and the raceway at any other radius $R_o + \Delta R_o$. It is this skidding that largely accounts for the coefficient of friction of a cylindrical roller thrust bearing being greater than the others.

Skidding may be reduced by dividing the roller into segments, as in Figure 10.2.4, or by replacing the cylindrical roller with a conical roller, Figure 10.2.5, so that for every r and R in their intervals of definition

$$r = \frac{\Omega}{\omega} R$$

Spherical thrust roller bearings, Figure 10.2.6, are manufactured with the outer raceway and the rollers ground as a portion of a sphere to accomodate shaft misalignment. Because of this geometry they can withstand higher radial loads than conical, or tapered, roller thrust bearings. Since spherical rollers slide over most of their length, these advantages are obtained at the price of increased friction.

Tapered Roller Bearings

Inclination of the roller axis provides a roller bearing that can sustain both axial and radial loads at moderate speeds. To assure true rolling the rollers are tapered, as shown in Figure 10.2.7. These bearings have a slightly different terminology in that the inner ring is called the *cone* and the outer ring is called the *cup*.

Double-row tapered roller bearings (tapered radial roller bearings) are available in two styles as represented in Figure 10.2.8 by the TDI and TDO bearings. Convergence of the contact lines in Figure 10.2.8 for the TDI style provides a less rigid bearing than in the TDO style where the contact lines diverge. Two single-row tapered

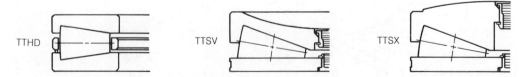

Figure 10.2.5 Tapered roller thrust bearings. *Source: Courtesy of The Torrington Company Torrington, CT.*

Figure 10.2.6 Spherical roller thrust bearing. *Source: SKF Industries, Bearings Group, King of Prussia, PA.*

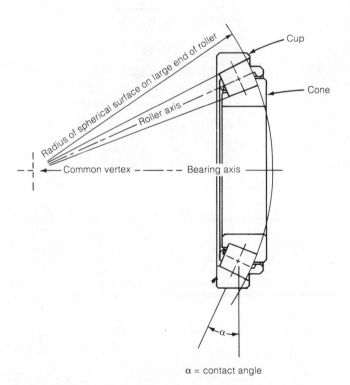

Figure 10.2.7 Tapered roller bearings and their terminology. *Source: Courtesy of The Torrington Company, Torrington, CT.*

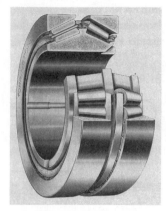

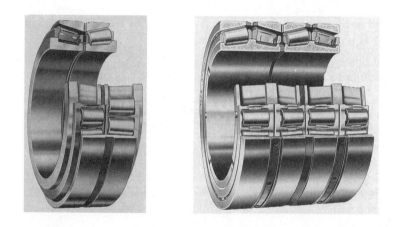

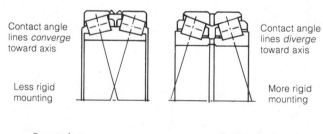

Contact angle
lines *converge*
toward axis

Less rigid
mounting

Contact angle
lines *diverge*
toward axis

More rigid
mounting

Face-to-face   Back-to-back

Figure 10.2.8 Double-row tapered roller bearings. *Source: Courtesy of The Torrington Company, Torrington, CT.*

roller bearings placed on a shaft so their contact lines converge are said to be mounted *face to face*; if their lines diverge they are said to be mounted *back to back*. If their contact lines are parallel they are said to be mounted in tandem. Tapered roller bearings are seldom mounted in tandem.

Spherical Radial Roller Bearings (Self-aligning)

As in the case of spherical thrust bearings, the outer raceway is ground as a portion of a sphere, as shown in Figure 10.2.9a, to accomodate angular misalignment between

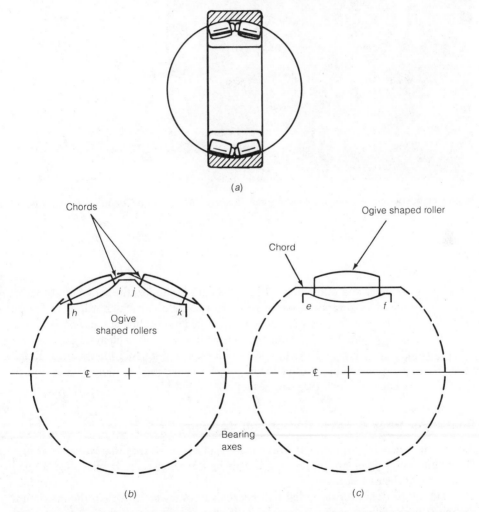

Figure 10.2.9 (a) Spherical surface for the outer race, (b) inner race and ogive (pronounced ohjive) shaped double rollers, and (c) ogive shaped single roller and its inner race. *Source: (a) Courtesy of The Torrington Company, Torrington, CT (modified).*

Figure 10.2.10 Spherical radial roller bearings. *Source: SKF Industries, Bearings Group, King of Prussia, PA.*

the axes of the inner and outer rings. The rollers are ground to the shape of an ogive, which is produced by rotating a sector of a great circle about its chord, as in Figure 10.2.9b and c. The inner raceway surfaces are obtained by rotating curves hijk in Figure 10.2.9b and ef in Figure 10.2.9c about the indicated axes of the spheres.

Both single-row, Figure 10.2.10, and double-row, Figure 10.2.11, spherical radial roller bearings are available and each can support some axial loading in addition to the radial loading for which they are designed.

Conrad, or Deep Groove, Radial Ball Bearings

Ball bearings generally operate at lower loads and higher speeds than roller bearings. Depending upon inner and outer ring design ball bearings may be classified as ball radial or ball thrust bearings.

Conrad, or deep groove, radial ball bearings have fewer balls than the maximum capacity, or slot filling, type because the balls are placed between the inner and outer rings when the inner ring is pushed to one side, Figure 10.2.12a, so that no filling slot is cut in either ring. Once no more balls can be slipped in they are evenly spaced by

Figure 10.2.11 Single row spherical radial roller bearing. *Source: McGill Manufacturing Co., Valpariso, IN.*

a two-piece separator, Figure 10.2.12b. Deep grooves enable this radial bearing to resist greater axial loads than the filling slot type.

Maximum Capacity (Filling Slot) Ball Radial Bearings

These bearings have a slot in one ring through which balls are placed between the inner and outer rings. An increased number of balls (compare Figures 10.2.12 and 10.2.13) gives it a larger radial load capability than the deep groove type. Because of the filling slot this bearing can withstand axial loads in one direction only.

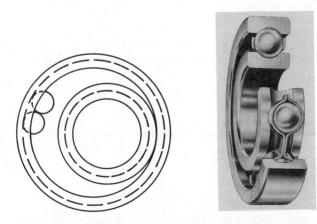

Figure 10.2.12 Deep groove (Conrad) radial ball bearing and filling method. *Source: SKF Industries, Bearings Group, King of Prussia, PA.*

Figure 10.2.13 Maximum capacity (filling slot) ball radial bearing. Compare with Figure 10.2.12. Note that balls are closer together than in the deep groove, or Conrad, radial ball bearing. *Source: SKF Industries, Bearings Group, King of Prussia, PA.*

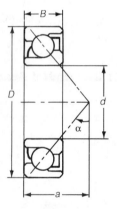

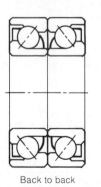

Back to back

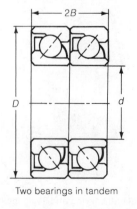

Two bearings in tandem

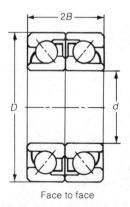

Face to face

Figure 10.2.14 Angular contact ball bearings. *Source: SKF Industries, Bearings Group, King of Prussia, PA.*

Cylindrical bore   Tapered bore   Adapter mountings

Figure 10.2.15 Self-aligning ball radial bearings. *Source: SKF Industries, Bearings Group, King of Prussia, PA.*

Angular Contact Ball Radial Bearings

Angular contact ball bearings have inner and outer rings that are specifically designed to support moderate thrust loads in one direction combined with moderate radial loads, as in Figure 10.2.14.

They are often mounted in tandem for greater load capacity and back to back or face to face to obtain bidirectional axial load capability and greater or lesser axial rigidity, respectively, Figure 10.2.14.

Self-aligning Ball Radial Bearings

These are double-row ball bearings with circumferential raceways on the inner ring and a spherically ground raceway on the outer ring. Figure 10.2.15 shows them to be maximum capacity bearings which may be filled without a filling slot since the inner ring may be rotated about a diameter for filling. They are for moderate radial loads and high speeds where shaft misalignment or bending is expected.

Ball Thrust Bearings

Ball thrust bearings are produced with thrust washers and raceways designed for pure thrust, Figure 10.2.16a or for combined thrust and low radial force, Figure 10.2.16b.

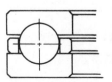

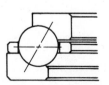

Figure 10.2.16 Ball thrust and angular contact ball thrust bearings. *Source: Courtesy of The Torrington Co., Torrington, CT.*

TABLE 10.2.1   RELATIVE OPERATING CHARACTERISTICS OF ANTIFRICTION BEARINGS

	Bearing Type	Torrington Type Code	Radial Capacity	Thrust Capacity	Limiting Speed	Resistance to Elastic Deformation Radial	Axial
Ball Radial	Conrad	BC, BIC	Moderate	Moderate in both directions	High	Moderate	Low
	Maximum Capacity	BH, BIH	Moderate+	Moderate in one direction	High	Moderate+	Low+
	Angular Contact	BA, BIA	Moderate	Moderate+ in one direction	High−	Moderate	Moderate
Roller Radial	Cylindrical Roller	RN, RIN, RU, RIU	High	None	Moderate+	High	None
		RF, RIF, RJ, RIJ	High	Light in one direction	Moderate+	High	Not Recommended
		RP, RIP, RT, RIT	High	Light in both directions	Moderate+	High	Not Recommended
	Journal	MTE, JTE, MTED, JTED, MTD, MTDD, JTD, JTDD	High+	None	Low	High	None
	Spherical Roller	SD	High	Moderate in both directions	Moderate	High−	Moderate
Tapered Roller	Single Row	TS, TSF	High−	Moderate+ in one direction	Moderate	High−	Moderate
	Single Row Steep Angle	TSS	Moderate+	High in one direction	Moderate−	Moderate	High
	Double Row	TDI, TDIK	High	Moderate+ in both directions	Moderate	High	Moderate
		TDIS, TDIE	Moderate+	High in both directions	Moderate−	Moderate	High
		TDO, TDOD, TNA, TNAU, TNAD	High	Moderate+ in both directions	Moderate	High	Moderate
	4 Row	TQO, TQOK	High+	High in both directions	Moderate−	High+	High
Thrust	Angular Contact Ball Thrust	TVL	Low+	High− one direction	Moderate	Low	High−
	Ball Thrust	TVB	None	High one direction	Moderate−	None	High
	Roller Thrust	TP	None	High+ one direction	Low	None	High+
	Self Aligning Roller Thrust	TPS	None	High+ one direction	Low	None	High+
	Tapered Roller Thrust	TTHD, TTSV, TTSX	Locational Only	High+ one direction	Low	None	High+

Source: Courtesy of The Torrington Co., Torrington, CT.

Comparison of radial thrust, and speed capacity for Torrington bearings, Table 10.2.1, is characteristic of bearings from other manufacturers of quality bearings.

Needle Bearings

These bearings are characterized by long cylindrical rollers having length/diameter ratios of the order of 4 or greater, as in Figure 10.2.17, and are designed for applications requiring small radial clearance.

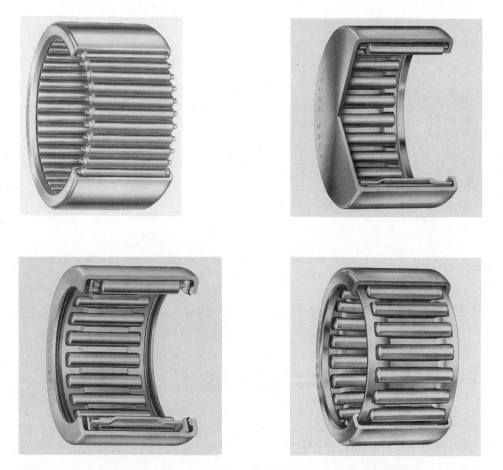

Figure 10.2.17 Drawn cup needle roller bearings. *Source: Courtesy of The Torrington Co., Torrington, CT.*

10.3 BEARING LIFE

Bearing life calculations in this and all following sections is based upon the formulas and tables found in the American National Standard ANSI/AFBMA Std. 9-1978 (reaffirmed in 1987) and Std. 11-1978 (reaffirmed in 1987) that are in essential agreement with the international standard ISO 281/1-1977(E).

After becoming familiar with the methods recommended in these standards, as described in the following sections of this chapter, the reader should, before selecting a bearing for actual service, consult the latest version of these standards themselves for revisions that include new developments and for particular limitations and restrictions on the use of these formulas; the presence of proper lubrication, the assumption that bearing speeds are low enough for centrifugal and gyroscopic effects to be negligible, and so on.

Both of the ANSI/AFBMA standards open their introduction with statements on the standard's purpose, the life criteria, and the static load criterion which read as follows in the case of ball bearings.

> **1.1 Purpose of standard.** Ball bearing performance is a function of many variables. These include the bearing design, the characteristics of the material from which the bearing are made, the way in which they are manufacturered, as well as many variables associated with their application. The only sure way to establish the satisfactory operation of a bearing selected for a specific application is by actual performance in the application. As this is often impractical, another basis is required to estimate the suitability of a particular bearing for a given application. This is the purpose of this standard. The major portion of this standard is concerned with calculation of bearing fatigue life. The bearing's ability to withstand static loading is also considered.
>
> **1.2 Life Criterion.** Even if ball bearings are properly mounted, adequately lubricated, protected from foreign matter, and are not subjected to extreme operating conditions, they can ultimately fatigue. Under ideal conditions, the repeated stresses developed in the contact areas between the balls and the raceways eventually can result in fatigue of the material which manifests itself as spalling of the loading carrying surfaces. In most applications the fatigue life is the maximum useful life of a bearing. This fatigue is the criterion of life used as the basis for the first part of this standard. [ANSI/AFBMA Std. 9-1978 and Std. 11-1978]
>
> **1.3 Static Load.** A static load is a load acting on a non-rotating bearing. Permanent deformations appear in balls and raceways under static load of moderate magnitude and increase gradually with increasing load. The permissible static load is, therefore, dependent upon the permissible magnitude of permanent deformation. It has been found that for ball bearings suitably manufacturered from hardened alloy steel, deformations occurring under maximum contact stress of 4000 megapascals (580,000 pounds per square inch) acting at the center of contact do not greatly impair smoothness or friction. Depending on requirements for smoothness of operation, friction, or sound level, higher or lower static load limits may be tolerated."

Identical comments apply to roller bearings.

Many test by bearing manufacturers have all produced life curves which on the average agree with that reproduced in Figure 10.3.1. ANSI, AFBMA, and ISO have all

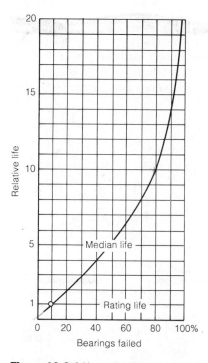

Figure 10.3.1 Normalized bearing life relative to percentage failed. *Source: SKF Industries, Bearings Group, King of Prussia, PA.*

agreed that *the life of a particular group of bearings is defined as the number of millions of revolutions at a given constant speed that 90% of the bearings will complete or exceed before the first evidence of fatigue develops in either rings or rolling elements.*

According to this definition the number of millions of revolutions at which 10% of the bearings have displayed some evidence of fatigue is taken as one unit of the ordinate in Figure 10.3.1. From this graph and this scale it is evident that after five times as many revolutions 50% of the bearings will show signs of metal fatigue and that after 14 times as many revolutions 90% of the bearings will evidence fatigue.

In bearing industry terminology the number of revolutions necessary to cause failure of 10% of the bearings is known as the L_{10} life and so on. Figure 10.3.1 shows that the L_{50} life is 5 times the L_{10} life.

By agreement the L_{10} life is termed the *rating life*, and, of course, the L_{50} life is termed the *median*, or *average*, life. Data from the life tests referenced in the beginning of this section correspond quite well to the relations[3,4]

$$L_{10} = \left(\frac{C}{P}\right)^3 10^6 \text{ revolutions for ball bearings} \tag{10.3.1}$$

$$L_{10} = \left(\frac{C}{P}\right)^{10/3} 10^6 \text{ revolutions for roller bearings} \tag{10.3.2}$$

In these relations P is the equivalent radial load and C is the basic dynamic load rating, commonly called the basic load rating. It is calculated from the formulas and tables listed below. These tables and formulas are not often used directly in bearing selection. They may, however, be incorporated into some computer-aided design programs as an internal aid in selecting the next group of bearings to be displayed in the design menu.

The dynamic load rating C which appears in formulas 10.3.1 and 10.3.2 is given by relations of the form

$$C = \gamma f_c (i \cos \alpha)^{0.7} Z^{2/3} D \beta \qquad (10.3.3)$$

TABLE 10.3.1 VALUES OF f_c FOR EQUATION 10.3.3

$\dfrac{D \cos \alpha}{d_m}$	Single Row Radial Contact; Single & Double Row Angular Contract, Groove Type[1]		Double Row Radial Contact Groove Type		Self-Aligning	
	Metric[2]	Inch[3]	Metric[2]	Inch[3]	Metric[2]	Inch[3]
0.05	46.7	3550	44.2	3360	17.3	1310
0.06	49.1	3730	46.5	3530	18.6	1420
0.07	51.1	3880	48.4	3680	19.9	1510
0.08	52.8	4020	50.0	3810	21.1	1600
0.09	54.3	4130	51.4	3900	22.3	1690
0.10	55.5	4220	52.6	4000	23.4	1770
0.12	57.5	4370	54.5	4140	25.6	1940
0.14	58.8	4470	55.7	4230	27.7	2100
0.16	59.6	4530	56.5	4290	29.7	2260
0.18	59.9	4550	56.8	4310	31.7	2410
0.20	59.9	4550	56.8	4310	33.5	2550
0.22	59.6	4530	56.5	4290	35.2	2680
0.24	59.0	4480	55.9	4250	36.8	2790
0.26	58.2	4420	55.1	4190	38.2	2910
0.28	57.1	4340	54.1	4110	39.4	3000
0.30	56.0	4250	53.0	4030	40.3	3060
0.32	54.6	4160	51.8	3950	40.9	3110
0.34	53.2	4050	50.4	3840	41.2	3130
0.36	51.7	3930	48.9	3730	41.3	3140
0.38	50.0	3800	47.4	3610	41.0	3110
0.40	48.4	3670	45.8	3480	40.4	3070

[1] (a) When calculating the basic load rating for a unit consisting of two similar, single row, radial contact ball bearings, in a duplex mounting, the pair is considered as one, double row, radial contact ball bearing.
 (b) When calculating the basic load rating for a unit consisting of two, similar, single row, angular contact ball bearings in a duplex mounting, "Face-to-Face" or "Back-to-Back", the pair is considered as one, double row, angular contact ball bearing.
[2] Use to obtain C in newtons when D is given in mm.
[3] Use to obtain C in pounds when D is given in inches.

Source: Copyright 1978 by the Anti-Friction Bearing Manufacturers Association, Inc.

Equation 10.3.3, with the coefficients and exponents shown, holds for radial and angular contact ball bearings, where

$$\beta = 1.8 \qquad \text{for } 0 < D \le 1 \text{ in. (25.4 mm)}$$
$$ = 1.4 \qquad \text{for } D > 1 \text{ in.}$$

$$\gamma = 1.0 \qquad \text{for } 0 < D \le 1 \text{ in.}$$
$$ = 3.647 \qquad \text{if SI units are used and if } D > 1 \text{ in.}$$

and where f_c is found from Table 10.3.1 in which d_m is the diameter of a circle passing through the centers of the balls in any one raceway. Quantity i is the number of rows of balls in the bearing, α is the nominal contact angle in degrees (see Figure 10.2.14), Z is the number of balls per row, and D denotes the ball diameter.

The static load rating C_0 for radial and angular contact ball bearings is given in terms of the above quantities by

$$C_o = f_o i Z D^2 \cos \alpha \tag{10.3.4}$$

where values of f_o are given in Table 10.3.2.

Similar formulas and tables are given for the dynamic and static load ratings for all other common types of rolling element bearings listed in Ref. 3 and 4.

Table 10.3.3 lists typical lives usually selected for a number of common machines.

According to Ref. 9, ball bearings are considered to be lightly loaded for loads equal to or less than $0.07C$ and to be heavily loaded for loads in excess of $0.15C$. Roller bearings are said to be lightly loaded for loads equal to or less than $0.08C$ and to be heavily loaded for loads in excess of $0.18C$, where C is the basic dynamic load rating of the bearing involved.

10.4   LOAD CAPACITY OF BALL BEARINGS, EQUIVALENT LOADS

Ball bearings may be grouped into three types for design purposes: radial, angular contact, and thrust. Radial ball bearings are designed to support radial loads with little or no axial loads; angular contact ball bearings are designed to support combined radial and axial loads using races ground such that the line of contact between balls and races is as shown in Figure 10.4.1; and thrust ball bearings are designed to support axial loads with little or no radial loads.

The contact angle α for angular contact ball bearings is measured between the line of action of the bearing and the plane of the bearing as shown in Figure 10.2.14. If $0 \le \alpha < 45°$ the bearing is said to be an angular contact radial ball bearing and if $45° < \alpha \le 90°$ it is said to be an angular contact thrust ball bearing.

Selection of a ball bearing to support a specified maximum dynamic radial load P throughout its design life is a matter of substituting P into equation 10.3.1, solving for the basic dynamic load rating C and then searching the manufacturers' catalogs

TABLE 10.3.2   VALUES OF F_o FOR EQUATIONS 10.3.4 AND 10.3.6

$\dfrac{D \cos \alpha}{d_m}$	Radial and angular Contact Groove Type		Radial Self-Aligning		Thrust	
	Metric	Inch	Metric	Inch	Metric	Inch
0.00	12.7	1850	1.3	187	51.9	7730
0.01	13.0	1880	1.3	191	52.6	7620
0.02	13.2	1920	1.3	195	51.7	7500
0.03	13.5	1960	1.4	198	50.9	7380
0.04	13.7	1990	1.4	202	50.2	7280
0.05	14.0	2030	1.4	206	49.6	7190
0.06	14.3	2070	1.5	210	48.9	7090
0.07	14.5	2100	1.5	214	48.3	7000
0.08	14.7	2140	1.5	218	47.6	6900
0.09	14.5	2110	1.5	222	46.9	6800
0.10	14.3	2080	1.6	226	46.4	6730
0.11	14.1	2050	1.6	231	45.9	6660
0.12	13.9	2020	1.6	235	45.5	6590
0.13	13.6	1980	1.7	239	44.7	6480
0.14	13.4	1950	1.7	243	44.0	6380
0.15	13.2	1920	1.7	247	43.3	6280
0.16	13.0	1890	1.7	252	42.6	6180
0.17	12.7	1850	1.8	256	41.9	6070
0.18	12.5	1820	1.8	261	41.2	5970
0.19	12.3	1790	1.8	265	40.4	5860
0.20	12.1	1760	1.9	269	39.7	5760
0.21	11.9	1730	1.9	274	39.0	5650
0.22	11.6	1690	1.9	278	38.3	5550
0.23	11.4	1660	2.0	283	37.5	5440
0.24	11.2	1630	2.0	288	37.0	5360
0.25	11.0	1600	2.0	293	36.4	5280
0.26	10.8	1570	2.1	297	35.8	5190
0.27	10.6	1540	2.1	302	35.0	5080
0.28	10.4	1510	2.1	307	34.4	4980
0.29	10.3	1490	2.1	311	33.7	4890
0.30	10.1	1460	2.2	316	33.2	4810
0.31	9.9	1440	2.2	321	32.7	4740
0.32	9.7	1410	2.3	326	32.0	4640
0.33	9.5	1380	2.3	331	31.2	4530
0.34	9.3	1350	2.3	336	30.5	4420
0.35	9.1	1320	2.4	341	30.0	4350
0.36	8.9	1290	2.4	346	29.5	4270
0.37	8.7	1260	2.4	351	28.8	4170
0.38	8.5	1240	2.5	356	28.0	4060
0.39	8.3	1210	2.5	361	27.2	3950
0.40	8.1	1180	2.5	367	26.8	3880
0.41	8.0	1160	2.6	372	26.2	3800
0.42	7.8	1130	2.6	377	25.7	3720
0.43	7.6	1100	2.6	383	25.1	3640
0.44	7.4	1080	2.7	388	24.6	3560
0.45	7.2	1050	2.7	393	24.0	3480
0.46	7.1	1030	2.8	399	23.5	3400
0.47	6.9	1000	2.8	404	22.9	3320
0.48	6.7	977	2.8	410	22.4	3240
0.49	6.6	952	2.9	415	21.8	3160
0.50	6.4	927	2.9	421	21.2	3080

Source: Copyright 1978 by The Anti-Friction Bearing Manufacturers Association, Inc.

TABLE 10.3.3 GUIDE TO VALUES OF RECOMMENDED BASE RATING LIFE

Class of Machine	L_{10} Hours of Service		
Domestic machines, agricultural machines, instruments, technical apparatus for medical use	300	to	3 000
Machines used for short periods or intermittently: Electric hand tools, lifting tackle in workshops, construction machines	3 000	to	8 000
Machines to work with high operational reliability during short periods or intermittently: Lifts, cranes for packaged goods or slings of drums, bales etc.	8 000	to	12 000
Machines for use 8 hours per day and not always fully utilized: Gear drives for general purposes, electric motors for industrial use, rotary crushers	10 000	to	25 000
Machines for use 8 hours per day and fully utilized: Machine tools, woodworking machines, machines for the engineering industry, cranes for bulk materials, ventilator fans, conveyor belts, printing equipment, separators and centrifuges	20 000	to	30 000
Machines for continuous use 24 hours per day: Rolling mill gear units, medium sized electrical machinery, compressors, mine hoists, pumps, textile machinery	40 000	to	50 000
Water works machinery, rotary furnaces, cable stranding machines, propulsion machinery for ocean-going vessels	60 000	to	100 000
Pulp and papermaking industry, large electric machinery, power station plant, mine pumps and mine ventilator fans, tunnel shaft bearings for ocean-going vessels			$\approx 100\,000$

Source: SKF Industries, Inc., Bearings Group, King of Prussia, PA.

for a bearing having a C value equal to or larger than the calculated value and also having the desired bore and outside dimensions. Since static loading must usually be considered as well, the bearing selected must also have a basic static load rating equal to or greater than the maximum static radial load which the bearing must support at any time during its life.

Whenever the ball bearings must support combined radial and axial loads another step is involved, that of replacing the combined load with an equivalent radial load P, defined by

$$P = XF_r + YF_a \tag{10.4.1}$$

in terms of radial load F_r, axial load F_a, and constants X and Y which are obtained from Table 10.4.1 based upon the F_a/F_r ratio and either the F_a/C_0 ratio or the $F_a/(iZD^3)$ ratio. This last ratio involving the number i of rows of balls in the bearing, the number Z of balls per row, and the ball diameter D is less convenient to use when working with catalog data, and will, therefore, not be considered further.

TABLE 10.4.1 DYNAMIC VALUES OF X AND Y FOR RADIAL BALL BEARINGS FOR USE IN EQUATION 10.4.1

Bearing Type				Single Row Bearings $\frac{F_a}{F_r} > e$		Double Row Bearings $\frac{F_a}{F_r} \leq e$		Double Row Bearings $\frac{F_a}{F_r} > e$		e
				X	Y	X	Y	X	Y	

Radial Contact Groove Ball Bearings

$\dfrac{F_a}{iZD^2}$

$\dfrac{F_a}{C_o}$	Units Newtons, mm	Units lbs. in	Single Row Y	Double Row ≤ X	Double Row ≤ Y	Double Row > X	Double Row > Y	e
0.014	0.172	25	2.30				2.30	0.19
0.028	0.345	50	1.99				1.99	0.22
0.056	0.689	100	1.71				1.71	0.26
0.084	1.03	150	1.56 (X = 0.56)	1	0	0.56	1.55	0.28
0.11	1.38	200	1.45				1.45	0.30
0.17	2.07	300	1.31				1.31	0.34
0.28	3.45	500	1.15				1.15	0.38
0.42	5.17	750	1.04				1.04	0.42
0.56	6.89	1000	1.00				1.00	0.44

Angular Contact Groove Ball Bearings with Contact Angle: 5°

$\dfrac{F_a}{ZD^2}$

For single row ($\frac{F_a}{F_r} > e$): For this type use the X, Y and e values applicable to single row radial contact bearings.

$\dfrac{F_a}{C_o}$	Units Newtons, mm	Units lbs. in	Double Row ≤ X	Double Row ≤ Y	Double Row > X	Double Row > Y	e
0.014	0.172	25		2.78		3.74	0.23
0.028	0.345	50		2.40		3.23	0.26
0.056	0.689	100		2.07		2.78	0.30
0.085	1.03	150	1	1.87	0.78	2.52	0.34
0.11	1.38	200		1.75		2.36	0.36
0.17	2.07	300		1.58		2.13	0.40
0.28	3.45	500		1.39		1.87	0.45
0.42	5.17	750		1.26		1.69	0.50
0.56	6.89	1000		1.21		1.63	0.52

10°

$\dfrac{F_a}{C_o}$	Units Newtons, mm	Units lbs. in	Single Row X	Single Row Y	Double Row ≤ X	Double Row ≤ Y	Double Row > X	Double Row > Y	e
0.014	0.172	25		1.88		2.18		3.06	0.29
0.029	0.345	50		1.71		1.98		2.78	0.32
0.057	0.689	100		1.52		1.76		2.47	0.36
0.086	1.03	150		1.41		1.63		2.20	0.38
0.11	1.38	200	0.46	1.34	1	1.55	0.75	2.18	0.40
0.17	2.07	300		1.23		1.42		2.00	0.44
0.29	3.45	500		1.10		1.27		1.79	0.49
0.43	5.17	750		1.01		1.17		1.64	0.54
0.57	6.89	1000		1.00		1.16		1.63	0.54

15°

$\dfrac{F_a}{C_o}$	Units Newtons, mm	Units lbs. in	Single Row X	Single Row Y	Double Row ≤ X	Double Row ≤ Y	Double Row > X	Double Row > Y	e
0.015	0.172	25		1.47		1.65		2.39	0.38
0.029	0.345	50		1.40		1.57		2.28	0.40
0.058	0.689	100		1.30		1.46		2.11	0.43
0.087	1.03	150		1.23		1.38		2.00	0.46
0.12	1.38	200	0.44	1.19	1	1.34	0.72	1.93	0.47
0.17	2.07	300		1.12		1.26		1.82	0.50
0.29	3.45	500		1.02		1.14		1.66	0.55
0.44	5.17	750		1.00		1.12		1.63	0.56
0.58	6.89	1000		1.00		1.12		1.63	0.56

Contact Angle	Single Row X	Single Row Y	Double Row ≤ X	Double Row ≤ Y	Double Row > X	Double Row > Y	e
20°	0.43	1.00	1	1.09	0.70	1.63	0.57
25°	0.41	0.87	1	0.92	0.67	1.41	0.68
30°	0.39	0.76	1	0.78	0.63	1.24	0.80
35°	0.37	0.66	1	0.66	0.60	1.07	0.95
40°	0.35	0.57	1	0.55	0.57	0.98	1.14
Self-aligning Ball Bearings	0.40	0.4 cotα	1	0.42 cotα	0.65	0.65 cotα	1.5 tanα

1. Two similar, single row, angular contact ball bearings mounted "Face-to-Face" or "Back-to-Back" are considered as one, double row, angular contact bearing.

2. Values of X, Y and e for a load or contact angle other than shown in Table 2 are obtained by linear interpolation.

3. Values of X, Y and e shown in Table 2 do not apply to filling slot bearings for applications in which ball-raceway contact areas project substantially into the filling slot under load.

4. For single row bearings when $F_a/F_r \leq e$, use X = 1, Y = 0.

Source: Copyright 1979 by The Anti-Friction Bearing Manufacturers Association, Inc.

A bearing should not be used without consulting the manufacturer if its F/C_0 ratio falls outside of the range listed in Table 10.4.1.

Two similar single-row bearings mounted either face to face or back to back (see Figure 10.2.14) are to be considered as a single double-row bearing when selecting X and Y from Table 10.4.2. When $N(N = 2,3, \ldots)$ identical single-row bearings are mounted back to back, face to face, or in tandem (see Figure 10.2.14) the basic dynamic load rating C^* of the combination according to definition 10.3.3, is given by

$$C^* = CN^{0.7} \tag{10.4.2}$$

where C denotes the basic dynamic load rating of any one of the individual bearings.

Static equivalent loads for radial and angular contact ball bearings are given by

$$P = \max \begin{cases} X_o F_r + Y_o F_a \\ F_r \end{cases} \tag{10.4.3}$$

where X_o and Y_o are found from Table 10.4.2.

The basic static load rating C_o^* of any combination of N identical single row ball bearings is given by

$$C_o^* = NC_o \tag{10.4.4}$$

TABLE 10.4.2   STATIC VALUES OF X_o AND Y_o FOR RADIAL BALL BEARINGS FOR USE IN EQUATION 10.4.2

Bearing Type		Single Row[a] Bearings		Double Row Bearings	
		X_o	Y_o	X_o	Y_o
Radial Contact Groove Ball Bearing[a,b]		0.6	0.5	0.6	0.5
Angular Contact Groove Ball Bearings	$\alpha = 15°$	0.5	0.47	1	0.94
	$\alpha = 20°$	0.5	0.42	1	0.84
	$\alpha = 25°$	0.5	0.38	1	0.76
	$\alpha = 30°$	0.5	0.33	1	0.66
	$\alpha = 35°$	0.5	0.29	1	0.58
	$\alpha = 40°$	0.5	0.26	1	0.52
Self-aligning Ball Bearings		0.5	0.22 cot α	1	0.44 cot α

[a] Permissible maximum value of F_a/C_o depends on the bearing design (groove depth and internal clearance.
[b] P_o is always $\geq F_r$.
[c] Values of Y_o for intermediate contact angles are obtained by linear interpolation.

Source: Copyright 1978 by The Anti-Friction Bearing Manufacturers Association, Inc.

Equivalent axial thrust for thrust ball bearings under combined radial and thrust loading is given by formula 10.4.1 except that in thrust applications P represents the equivalent thrust and X and Y are found from Table 10.4.3.

Static equivalent axial thrust for thrust ball bearings is calculated from

$$
\begin{aligned}
P &= 2.3F_r \tan \alpha + F_a \qquad \alpha < 90° \\
&= F_r \qquad\qquad\qquad\quad \alpha = 90°
\end{aligned}
\tag{10.4.5}
$$

The formula for $\alpha < 90°$ in display (10.4.5) is valid for all load directions for double-direction bearings, but holds only for $F_a/F_r \leq 0.44 \cot \alpha$ for single-direction bearings. According to ANSI/AFBM Std 9-1978 (in effect in 1987) this relation is satisfactory, but less cautious, by an unspecified amount, for F_a/F_r ratios up to $0.67 \cot \alpha$.

Angular contact radial ball bearings differ from radial ball bearings in that they induce an axial load whenever they are subjected to a radial load. This is a consequence of the normal to the contacting surfaces at the points of contact between the balls and races making an angle α with the plane of the bearing. The relation between the radial load F_r and the induced axial force I is given by

$$
I = \frac{F_r}{2Y}
\tag{10.4.6}
$$

where the value of Y is given in Table 10.4.1. It follows from the criteria for selecting X and Y for F_a/F_r relative to e that condition 10.4.6 does not affect the selection procedure used for single angular contact radial ball bearings to support an applied radial load with no externally applied axial load. The effect of this induced axial force upon

TABLE 10.4.3 VALUES OF X AND Y FOR EQUIVALENT THRUST LOADS USED WITH EQUATION 10.4.1

Bearing Type	Single Direction Bearings $\frac{F_a}{F_r} > e$		Double Direction Bearings $\frac{F_a}{F_r} \leq e$		$\frac{F_a}{F_r} > e$		e
	X	Y	X	Y	X	Y	
Thrust Ball[a] Bearings with Contact Angle							
$\alpha = 45°$	0.66	1	1.18	0.59	0.66	1	1.25
$\alpha = 60°$	0.92	1	1.90	0.54	0.92	1	2.17
$\alpha = 75°$	1.66	1	3.89	0.52	1.66	1	4.67

[a] For $\alpha = 90°$, $F_r = 0$ and $Y = 1$

Source: Copyright 1979 by The Anti-Friction Bearing Manufacturers Association, Inc.

other bearings on the shaft upon which it is mounted will be discussed further in Section 10.8 and in several problems at the end of the chapter.

Representative data from manufacturers' engineering catalogs (not bearing suppliers' catalogs) for radial, angular contact radial, self-aligning radial, thrust, and angular contact thrust bearings are displayed in Tables 10.4.4 through 10.4.9. Since catalog data are frequently up-dated to reflect design changes and increased performance information, these tables may not represent the latest values. Hence, they are to be used only to demonstrate bearing selection procedures with representative bearing parameters.

EXAMPLE 10.4.1

Select a radial ball bearing to fit a portion of a shaft where the design diameter may be from 15 to 25 mm, according to the bearing selected. It must withstand a radial load of 85.6 kg and should have a L_{10} life of 2000 h at 5500 rpm.

Since there is no axial load, $P = F_r$ and for L_{10} in hours and

$$C = (L_{10})^{1/3}P = [2000(60)5500/10^6]^{1/3}85.6(9.80665) = 7309 \text{ N}$$
$$= 1643 \text{ lb}$$

Perusal of Table 10.4.4 suggests a bearing from the 6203 series, with the particular choice dependent upon cleanliness and lubrication. The shaft diameter should be 17 mm.

EXAMPLE 10.4.2

After the machine using the shaft and bearing selected in Example 10.4.1 was in production for 20 months, it was suggested that the same shaft and the same bearing, be used in another product operating at the same speed. The shaft would experience the same radial load, but would also support an axial load of 310 N. What L_{10} life may be expected for this loading condition? Would replacement with a different bearing provide an L_{10} life of 2000 h? If so, what bearing would you recommend?

First, calculate $F_a/C_o = 310/4448.3 = 0.070$ using $C_o = 1000 \text{ lb} = 4448.3 \text{ N}$. Thus, enter Table 10.4.1 in the Radial Contact Groove Ball Bearings level (the ANSI terminology for radial contact ball bearings) with $F_r = 85.6(9.80665) = 839.45 \text{ N}$ to find after linear interpolation that

$$F_a/F_r = 310.0/839.45 = 0.369 > e = 0.27$$

Next, enter the Single Row Bearings columns, read the X value and linearly interpolate for Y to find that

$$X = 0.56 \qquad Y = 1.63$$

TABLE 10.4.4 CATALOG DATA FOR RADIAL BALL BEARINGS

Bearing Number	Z	2Z	RS	2RS	Nominal d (mm)	Nominal d (in.)	Bearing Dim. D (mm)	Bearing Dim. D (in.)	Balls No.	Balls Diam. (in.)	Static Load Rating C₀ (lb)	Basic Load Rating C (lb)	Approx. Speed Limit (rpm)	Basic Brg. No.
6200	6200 Z	6200 2Z	6200 RS	6200 2RS	10	.3937	30	1.1811	7	$\frac{3}{16}$	440	805	25000	6200
6201	6201 Z	6201 2Z	6201 RS	6201 2RS	12	.4724	32	1.2598	7	$\frac{15}{64}$	685	1180	23000	6201
6202	6202 Z	6202 2Z	6202 RS	6202 2RS	15	.5906	35	1.3780	8	$\frac{15}{64}$	790	1320	20000	6202
6203	6203 Z	6203 2Z	6203 RS	6203 2RS	17	.6693	40	1.5748	8	$\frac{17}{64}$	1000	1650	18000	6203
6204	6204 Z	6204 2Z	6204 RS	6204 2RS	20	.7874	47	1.8504	8	$\frac{5}{16}$	1390	2210	15000	6204
6205	6205 Z	6205 2Z	6205 RS	6205 2RS	25	.9843	52	2.0472	9	$\frac{5}{16}$	1560	2420	13000	6205
6206	6206 Z	6206 2Z	6206 RS	6206 2RS	30	1.1811	62	2.4409	9	$\frac{3}{8}$	2250	3360	11000	6206
6207	6207 Z	6207 2Z	6207 RS	6207 2RS	35	1.3780	72	2.8346	9	$\frac{7}{16}$	3070	4440	9400	6207
6208	6208 Z	6208 2Z	6208 RS	6208 2RS	40	1.5748	80	3.1496	9	$\frac{15}{32}$	3520	5040	8400	6208
6209	6209 Z	6209 2Z	6209 RS	6209 2RS	45	1.7717	85	3.3465	9	$\frac{1}{2}$	4010	5660	7700	6209
6210	6210 Z	6210 2Z	6210 RS	6210 2RS	50	1.9685	90	3.5433	10	$\frac{1}{2}$	4450	6070	7100	6210
6211	6211 Z	6211 2Z	6211 RS	6211 2RS	55	2.1654	100	3.9370	10	$\frac{9}{16}$	5630	7500	6500	6211
6212	6212 Z	6212 2Z	6212 RS	6212 2RS	60	2.3622	110	4.3307	10	$\frac{5}{8}$	6950	9070	5900	6212
6213	6213 Z	6213 2Z	6213 RS	6213 2RS	65	2.5591	120	4.7244	10	$\frac{21}{32}$	7670	9900	5400	6213
6214	6214 Z	6214 2Z	6214 RS	6214 2RS	70	2.7559	125	4.9213	10	$\frac{11}{16}$	8410	10800	5100	6214
6215	6215 Z	6215 2Z	6215 RS	6215 2RS	75	2.9528	130	5.1181	11	$\frac{11}{16}$	9250	11400	4800	6215
6216	6216 Z	6216 2Z	6216 RS	6216 2RS	80	3.1496	140	5.5118	10	$\frac{3}{4}$	10000	12600	4500	6216
6217	6217 Z	6217 2Z	6217 RS	6217 2RS	85	3.3465	150	5.9055	11	$\frac{25}{32}$	12000	14400	4200	6217
6218	6218 Z	6218 2Z			90	3.5433	160	6.2992	10	$\frac{7}{8}$	13600	16600	3900	6218
6219	6219 Z	6219 2Z			95	3.7402	170	6.6929	10	$\frac{15}{16}$	15600	18800	3700	6219
6220	6220 Z	6220 2Z	6220 RS	6220 2RS	100	3.9370	180	7.0866	10	1	17800	21100	3500	6220
6221	6221 Z	6221 2Z			105	4.1339	190	7.4803	10	$1\frac{1}{16}$	20100	23000	3300	6221
6222	6222 Z	6222 2Z			110	4.3307	200	7.8740	10	$1\frac{1}{8}$	22500	24900	3100	6222
6224					120	4.7244	215	8.4646	9	$1\frac{3}{16}$	22600	25100	2900	6224
6226					130	5.1181	230	9.0551	9	$1\frac{1}{4}$	25000	26900	2600	6226
6228					140	5.5118	250	9.8425	10	$1\frac{1}{4}$	27800	28800	2400	6228
6230					150	5.9055	270	10.6299	11	$1\frac{1}{4}$	30600	30400	2200	6230
6232					160	6.2992	290	11.4173	12	$1\frac{1}{4}$	33400	32000	2000	6232
6234					170	6.6929	310	12.2047	12	$1\frac{3}{8}$	40400	36700	1900	6234
6236					180	7.0866	320	12.5984	11	$1\frac{1}{2}$	44100	37500	1800	6236
6238					190	7.4803	340	13.3858	11	$1\frac{5}{8}$	51700	44100	1700	6238

TABLE 10.4.5 CATALOG DATA FOR ANGULAR CONTACT BALL BEARINGS

Bearing Number	d (mm)	d (in.)	D (mm)	D (in.)	Balls No.	Balls Diam. in.	When $F_a/F_r > e$ Use X_1	Y_1	When $F_a/F_r > e$ Use X_2	Y_2	e	Static Load Rating C lb	Basic Load Rating C lb	Approx. Speed Limit rpm	Basic Brg. No.
7304 B	20	.7874	52	2.0472	10	$\frac{3}{8}$	1	0	0.35	0.57	1.14	1920	3000	14000	7304 B
7305 B	25	.9843	62	2.4409	11	$\frac{7}{16}$	1	0	0.35	0.57	1.14	2870	4220	12000	7305 B
7306 B	30	1.1811	72	2.8346	12	$\frac{31}{64}$	1	0	0.35	0.57	1.14	3840	5370	10000	7306 B
7307 B	35	1.3780	80	3.1496	12	$\frac{17}{32}$	1	0	0.35	0.57	1.14	4620	6340	8700	7307 B
7308 B	40	1.5748	90	3.5433	12	$\frac{19}{32}$	1	0	0.35	0.57	1.14	5770	7740	7700	7308 B
7309 B	45	1.7717	100	3.9370	12	$\frac{11}{16}$	1	0	0.35	0.57	1.14	7730	10100	6900	7309 B
7310 B	50	1.9685	110	4.3307	12	$\frac{3}{4}$	1	0	0.35	0.57	1.14	9200	11800	6300	7310 B
7311 B	55	2.1654	120	4.7244	12	$\frac{13}{16}$	1	0	0.35	0.57	1.14	10800	13600	5700	7311 B
7312 B	60	2.3622	130	5.1181	12	$\frac{7}{8}$	1	0	0.35	0.57	1.14	12500	15600	5200	7312 B
7313 B	65	2.5591	140	5.5118	12	$\frac{15}{16}$	1	0	0.35	0.57	1.14	14400	17600	4800	7313 B
7314 B	70	2.7559	150	5.9055	12	1	1	0	0.35	0.57	1.14	16400	19800	4500	7314 B
7315 B	75	2.9528	160	6.2992	12	$1\frac{1}{16}$	1	0	0.35	0.57	1.14	18500	21500	4200	7315 B
7316 B	80	3.1496	170	6.6929	12	$1\frac{1}{8}$	1	0	0.35	0.57	1.14	20700	23300	3900	7316 B
7317 B	85	3.3465	180	7.0866	12	$1\frac{3}{16}$	1	0	0.35	0.57	1.14	23100	25200	3700	7317 B
7318 B	90	3.5433	190	7.4803	12	$1\frac{1}{4}$	1	0	0.35	0.57	1.14	25600	27000	3500	7318 B
7319 B	95	3.7402	200	7.8740	12	$1\frac{5}{16}$	1	0	0.35	0.57	1.14	28200	28900	3300	7319B
7320	100	3.9370	215	8.4646	11	$1\frac{1}{2}$	1	0	0.36	0.62	1.03	35200	33900	3100	7320
7321	105	4.1339	225	8.8583	11	$1\frac{9}{16}$	1	0	0.36	0.62	1.03	38200	35900	2900	7321
7322	110	4.3307	240	9.4488	11	$1\frac{11}{16}$	1	0	0.36	0.62	1.03	44500	40000	2700	7322
7324	120	4.7244	260	10.2362	12	$1\frac{3}{4}$	1	0	0.36	0.62	1.03	52200	44600	2500	7324
7326	130	5.1181	280	11.0236	12	$1\frac{7}{8}$	1	0	0.36	0.62	1.03	60000	49100	2300	7326
7328	140	5.5118	300	11.8110	12	2	1	0	0.36	0.62	1.03	68200	53800	2100	7328
7330	150	5.9055	320	12.5984	12	$2\frac{1}{8}$	1	0	0.36	0.62	1.03	77000	58600	1900	7330
7334	170	6.6929	360	14.1732	13	$2\frac{5}{16}$	1	0	0.36	0.62	1.03	98800	69500	1700	7334

Nominal contact angle for 73 B is 40°.
Nominal contact angle for 73 is 37°.

Source: SKF Industries, Inc., Bearings Group, King of Prussia, PA.

TABLE 10.4.6 CATALOG DATA FOR ANGULAR CONTACT BALL BEARINGS

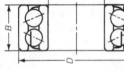

Bearing Number mm	d mm	d in.	D mm	D in.	Balls No. Per Row	Balls Diam. in.	When $\frac{F_a}{F_r} \le e$ Use X₁	Y₁	When $\frac{F_a}{F_r} > e$ Use X₂	Y₂	e	Static Load Rating C₀ lb	Basic Load Rating C lb	Approx. Speed Limit rpm	Basic Brg. No.
5200	10	.3937	30	1.1811	7	3/16	1	0.95	0.67	1.45	0.65	800	1240	18000	5200
5201	12	.4724	32	1.2598	7	15/64	1	0.95	0.67	1.45	0.65	1250	1820	16000	5201
5202	15	.5906	35	1.3780	8	15/64	1	0.95	0.67	1.45	0.65	1430	2030	14000	5202
5203	17	.6693	40	1.5748	8	17/64	1	0.95	0.67	1.45	0.65	1840	2540	12000	5203
5204	20	.7874	47	1.8504	8	5/16	1	0.95	0.67	1.45	0.65	2540	3410	10000	5204
5205	25	.9843	52	2.0472	9	5/16	1	0.95	0.67	1.45	0.65	2860	3700	9100	5205
5206	30	1.1811	62	2.4409	9	3/8	1	0.95	0.67	1.45	0.65	4120	5140	7600	5206
5207	35	1.3780	72	2.8346	9	7/16	1	0.95	0.67	1.45	0.65	5600	6780	6600	5207
5208	40	1.5748	80	3.1496	9	15/32	1	0.95	0.67	1.45	0.65	6430	7680	5800	5208
5209	45	1.7717	85	3.3465	9	1/2	1	0.95	0.67	1.45	0.65	7320	8620	5400	5209
5210	50	1.9685	90	3.5433	10	1/2	1	0.95	0.67	1.45	0.65	8130	9220	5000	5210
5211	55	2.1654	100	3.9370	10	9/16	1	0.95	0.67	1.45	0.65	10300	11400	4500	5211
5212	60	2.3622	110	4.3307	10	5/8	1	0.95	0.67	1.45	0.65	12700	13800	4100	5212
5213	65	2.5591	120	4.7244	10	21/32	1	0.95	0.67	1.45	0.65	14000	15000	3800	5213
5214	70	2.7559	125	4.9213	10	11/16	1	0.95	0.67	1.45	0.65	15400	16300	3600	5214
5215	75	2.9528	130	5.1181	11	11/16	1	0.95	0.67	1.45	0.65	16900	17300	3400	5215
5216	80	3.1496	140	5.5118	10	3/4	1	0.95	0.67	1.45	0.65	18300	19100	3200	5216
5217	85	3.3465	150	5.9055	11	3/4	1	0.85	0.65	1.32	0.74	19500	19700	2900	5217
5218	90	3.5433	160	6.2992	11	13/16	1	0.80	0.64	1.27	0.77	22100	22600	2700	5218
5219	95	3.7402	170	6.6929	10	15/16	1	0.95	0.67	1.45	0.65	28600	28600	2600	5219
5220	100	3.9370	180	7.0866	10	1	1	0.95	0.67	1.45	0.65	32500	32100	2400	5220
5221	105	4.1339	190	7.4803	10	31/32	1	0.95	0.67	1.45	0.65	30500	30100	2300	5221
5222	110	4.3307	200	7.8740	10	1	1	0.80	0.64	1.27	0.77	31100	30800	2200	5222

Source: SKF Industries, Inc., Bearings Group, King of Prussia, PA.

TABLE 10.4.7 CATALOG DATA FOR SELF-ALIGNING BALL BEARINGS

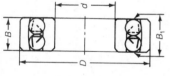

Bearing Number		d mm	d in.	D mm	D in.	Balls No. Per Row	Balls Diam. in.	When $F_a \le e$ Use F_r X_1	Y_1	When $F_a > e$ Use F_r X_2	Y_2	e	Static Load Rating C_0 lb	Basic Load Rating C lb	Approx. Speed Limit rpm	Basic Brg. No.
a1301		12	.4724	37	1.4567	9	1/4	1	1.81	0.65	2.81	0.35	530	1630	20000	1301
1302		15	.5906	42	1.6535	10	1/4	1	1.89	0.65	2.92	0.33	590	1650	18000	1302
1303		17	.6693	47	1.8504	11	9/32	1	1.91	0.65	2.96	0.33	820	2170	16000	1303
1304		20	.7874	52	2.0472	12	9/32	1	2.16	0.65	3.34	0.29	900	2150	14000	1304
1305		25	.9843	62	2.4409	12	11/32	1	2.26	0.65	3.50	0.28	1350	3110	12000	1305
1306	1306 K	30	1.1811	72	2.8346	13	3/8	1	2.47	0.65	3.82	0.26	1740	3700	10000	1306
	1307 K	35	1.3780	80	3.1496	14	13/32	1	2.57	0.65	3.97	0.25	2210	4350	8700	1307
	1308 K	40	1.5748	90	3.5433	15	7/16	1	2.61	0.65	4.04	0.24	2740	5110	7700	1308
	1309 K	45	1.7717	100	3.9370	15	1/2	1	2.53	0.65	3.92	0.25	3580	6600	7000	1309
	1310 K	50	1.9685	110	4.3307	13	9/16	1	2.68	0.65	4.14	0.25	3930	7510	6300	1310
	1311 K	55	2.1654	120	4.7244	15	19/32	1	2.70	0.65	4.17	0.24	5060	8800	5700	1311
1312		60	2.3622	130	5.1181	16	5/8	1	2.79	0.65	4.32	0.23	5980	9860	5200	1312
1313 K		65	2.5591	140	5.5118	16	21/32	1	2.78	0.65	4.31	0.23	6600	10700	4800	1313
1314		70	2.7559	150	5.9055	16	23/32	1	2.81	0.65	4.35	0.23	7910	12900	4500	1314
1315 K		75	2.9528	160	6.2992	16	3/4	1	2.83	0.65	4.38	0.22	8620	13700	4200	1315
1316 K		80	3.1496	170	6.6929	15	13/16	1	2.92	0.65	4.52	0.22	9490	15300	3900	1316
1317 K		85	3.3465	180	7.0866	16	27/32	1	2.91	0.65	4.50	0.22	10900	16900	3700	1317
1318 K		90	3.5433	190	7.4803	15	15/16	1	2.81	0.65	4.35	0.22	12600	20200	3500	1318
1319		95	3.7402	200	7.8740	15	1	1	2.77	0.65	4.29	0.23	14400	22800	3300	1319
1320 K		100	3.9370	215	8.4646	15	1 1/16	1	2.68	0.65	4.15	0.24	16200	24800	3100	1320
1321 K		105	4.1339	225	8.8583	14	1 1/16	1	2.76	0.65	4.27	0.23	17000	25700	2900	1321
1322 K		110	4.3307	240	9.4488	17	1 1/8	1	2.83	0.65	4.38	0.22	20600	28200	2700	1322

Source: SKF Industries, Inc., Bearings Group, King of Prussia, PA.

TABLE 10.4.8 CATALOG DATA FOR BALL THRUST BEARINGS

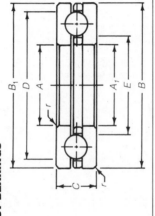

Bore A	o.d. B	Height C	Basic Dynamic Thrust Capacity (BDTC) lb	Basic Static Thrust Capacity (BSTC) lb	Bearing Number	Max Fillet r	Small Bore Washer o.d B₁	Shaft Shoulder Diameter D	Large Bore Washer i.d. A₁	Housing Shoulder Diameter E
5.0000	7.2500	1.6250	37,900	131,000	50 TVB 190	$\frac{3}{32}$	7.2500	$6\frac{23}{32}$	5.0000	$5\frac{17}{32}$
5.0000	7.3125	2.0000	37,900	131,000	50 TVB 190 AA175	$\frac{3}{32}$	7.3125	$6\frac{23}{32}$	5.0000	$5\frac{3}{32}$
5.0000	7.3125	2.0000	37,900	131,000	50 TVB 190 BB283	$\frac{3}{32}$	7.2500	$6\frac{23}{32}$	5.0625	$5\frac{17}{32}$
5.1181	7.4803	1.9685	45,800	149,800	51 TVB 240 AA167	$\frac{3}{32}$	7.4803	$6\frac{31}{32}$	5.1181	$5\frac{5}{32}$
5.1250	7.3750	1.6250	39,100	128,700	51 TVB 241	$\frac{3}{32}$	7.3750	$6\frac{23}{32}$	5.1250	$5\frac{21}{32}$
5.2500	7.5000	1.6250	40,700	135,500	52 TVB 242	$\frac{3}{32}$	7.5000	$6\frac{21}{32}$	5.2500	$5\frac{25}{32}$
5.2500	8.0000	2.0000	50,000	170,000	52 TVB 253	$\frac{3}{32}$	8.0000	$7\frac{5}{16}$	5.2500	$5\frac{15}{32}$
5.3750	7.6250	1.6250	40,000	135,000	53 TVB 243	$\frac{3}{32}$	7.6250	$7\frac{3}{8}$	5.3750	$5\frac{19}{32}$
5.5000	8.0625	2.1875	55,600	186,000	55 TVB 244	$\frac{3}{32}$	8.0625	$7\frac{17}{32}$	5.5000	$6\frac{1}{32}$
5.5000	8.0625	2.1870	55,600	186,000	55 TVB 244 AA285	$\frac{3}{32}$	8.0000	$6\frac{23}{32}$	5.5620	$5\frac{1}{32}$
5.5000	8.2500	1.8750	53,100	178,600	55 TVB 245	$\frac{3}{32}$	8.2500	$7\frac{9}{16}$	5.5000	$6\frac{3}{16}$
5.5118	7.8740	2.0472	38,200	137,300	55 TVB 246 AA130	$\frac{3}{32}$	7.8740	$7\frac{3}{8}$	5.5118	$6\frac{1}{16}$
5.5118	7.8740	2.0472	38,200	137,300	55 TVB 246 AB130	$\frac{3}{32}$	7.8740	$7\frac{3}{8}$	5.6300	$6\frac{1}{32}$
5.6250	8.3750	1.8750	52,500	173,200	56 TVB 247	$\frac{3}{32}$	8.3750	$7\frac{1}{8}$	5.6250	$6\frac{5}{16}$
5.7500	8.5000	1.8750	54,600	182,300	57 TVB 248	$\frac{3}{32}$	8.5000	$7\frac{13}{16}$	5.7500	$6\frac{7}{16}$

5.8750	8.6250	1.8750	182,300	54,600	3/32	58 TVB 249	8 3/32	8.6250	6 13/32	5.8750
5.9055	7.4803	1.2205	89,000	22,300	3/32	59 TVB 250 AA167	7 3/32	7.4803	6 9/32	5.9055
5.9055	7.4803	1.2205	89,000	22,300	3/32	59 TVB 250 AB286	7 3/32	7.4803	6 3/32	6.0240
5.9061	7.4803	1.2205	89,000	22,300	3/32	59 TVB 250 BO167	7 3/32	7.4803	6 3/32	5.9370
6.0000	8.6875	2.3750	182,300	54,600	3/32	60 TVB 251	8 1/8	8.6875	6 9/16	6.0000
6.0000	8.6875	2.3750	182,300	54,600	3/32	60 TVB 251 AA285	8 1/8	8.6250	6 9/16	6.0625
6.0000	8.7500	1.8750	188,000	53,400	3/32	60 TVB 252	8 1/8	8.7500	6 11/16	6.0000
6.1250	8.8750	1.8750	191,500	55,400	3/32	61 TVB 290	8 11/32	8.8750	6 21/32	6.1250
6.2500	9.0000	1.8750	239,000	71,400	3/32	62 TVB 291	8 15/32	9.0000	6 25/32	6.2500
6.3750	9.1250	1.8750	196,000	55,200	3/32	63 TVB 292	8 7/16	9.1250	7 1/16	6.3750
6.5000	9.5000	2.2500	237,800	72,500	1/8	65 TVB 293	8 27/32	9.5000	7 5/32	6.5000
6.6250	9.6250	2.2500	251,000	73,900	1/8	66 TVB 295	8 29/32	9.6250	7 11/32	6.6250
6.7500	9.7500	2.2500	251,000	73,900	1/8	67 TVB 296	9 3/32	9.7500	7 3/16	6.7500
6.8750	9.8750	2.2500	251,000	73,900	1/8	68 TVB 297	9 3/32	9.8750	7 19/32	6.8750
7.0000	10.0000	2.2500	262,800	77,400	1/8	70 TVB 298	9 9/32	10.0000	7 23/32	7.0000
7.1250	10.1250	2.2500	244,200	70,200	1/8	71 TVB 340	9 15/32	10.1250	7 25/32	7.1250
7.2500	10.2500	2.2500	255,300	72,100	1/8	72 TVB 341	9 19/32	10.2500	7 29/32	7.2500
7.3750	10.3750	2.2500	267,000	73,900	1/8	73 TVB 342	9 21/32	10.3750	8 3/32	7.3750
7.5000	10.5000	2.2500	255,300	72,100	1/8	75 TVB 343	9 13/32	10.5000	8 9/32	7.5000
7.5000	10.8750	2.7500	313,000	93,000	1/8	75 TVB 344	10 3/16	10.8750	8 3/16	7.5000
7.5000	10.8750	2.7500	313,000	93,000	1/8	75 TVB 344 AA285	10 3/32	10.7500	8 3/16	7.6250
7.7500	10.7500	2.2500	266,400	74,200	1/8	77 TVB 345	10 13/32	10.7500	8 13/32	7.7500
8.0000	11.0000	2.2500	308,000	89,000	1/8	80 TVB 346	10 11/32	11.0000	8 21/32	8.0000
8.0000	11.6250	3.0000	382,500	113,500	1/4	80 TVB 347	10 3/32	11.6250	8 3/32	8.0000
8.2500	11.2500	2.2500	277,500	76,200	1/8	82 TVB 390	10 17/32	11.2500	8 3/32	8.2500
8.5000	11.5000	2.2500	288,600	78,500	1/8	85 TVB 391	10 27/32	11.5000	9 5/32	8.5000
8.7500	11.7500	2.2500	288,600	78,500	1/8	87 TVB 392	11 3/32	11.7500	9 13/32	8.7500
9.0000	12.0000	2.2500	366,000	100,000	1/8	90 TVB 393	11 13/32	12.0000	9 3/32	9.0000
9.2500	12.2500	2.2500	310,800	82,200	1/8	92 TVB 430	11 19/32	12.2500	9 29/32	9.2500
9.5000	12.5000	2.2500	310,800	82,200	1/8	95 TVB 431	11 27/32	12.5000	10 5/32	9.5000

r is the maximum radius of the fillet on the shaft and in the housing that the bearing will clear.

A step must be provided on the shaft to clear the lower bearing washer.

Courtesy The Torrington Co., South Bend, IN.

TABLE 10.4.9 CATALOG DATA FOR ANGULAR CONTACT THRUST BALL BEARINGS ($\alpha = 45°$)

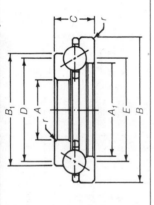

Bore A	o.d. B	Height C	Basic Dynamic Thrust Capacity (BDTC) lb	Basic Static Thrust Capacity (BSTC) lb	T_c Column No.	Bearing Number	Max Fillet r	Small Bore Washer o.d. B_1	Shaft Shoulder Diameter D	Large Bore Washer i.d. A_1	Housing Shoulder Diameter E
9.000	11.625	1.500	41,000	162,000	4	90 TVL 710 AA642	$\frac{1}{8}$	10.937	$10\frac{5}{16}$	9.687	$10\frac{5}{16}$
11.500	14.500	2.500	71,800	282,000	4	115 TVL 650 AA648	$\frac{1}{8}$	11.750	13	14.250	13
11.500	15.000	2.000	64,800	259,000	3	115 TVL 450 AA643	$\frac{1}{8}$	14.500	$13\frac{1}{4}$	12.000	$13\frac{1}{4}$
12.000	16.000	2.250	108,000	403,000	4	120 TVL 700	$\frac{1}{8}$	14.500	14	13.500	14
13.875	16.935	1.875	87,000	365,000	4	138 TVL 655	$\frac{1}{16}$	16.250	$15\frac{1}{2}$	14.750	$15\frac{1}{2}$
14.625	18.750	2.250	126,000	510,000	4	146 TVL 660 AA272	$\frac{3}{16}$	17.562	$16\frac{11}{16}$	15.812	$16\frac{11}{16}$
15.000	20.500	3.312	182,000	686,000	4	150 TVL 701	$\frac{3}{16}$	19.000	$17\frac{3}{16}$	16.500	$17\frac{3}{16}$
16.812	22.250	2.750	156,800	636,000	4	167 TVL 702 A0130	$\frac{1}{16}$	20.312	$19\frac{1}{4}$	18.250	$19\frac{1}{2}$
17.000	25.000	3.500	233,000	903,000	3	170 TVL 500	$\frac{5}{16}$	22.250	$19\frac{1}{2}$	19.250	21
17.254	22.750	2.755	124,000	481,000	3	172 TVL 460 AA644	$\frac{1}{4}$	20.250	20	19.750	20
17.750	22.000	2.312	140,000	578,000	3	177 TVL 610 A0313	$\frac{1}{8}$	20.675	$19\frac{7}{8}$	19.000	$19\frac{7}{8}$
18.000	23.000	3.000	182,000	730,000	4	180 TVL 703	$\frac{1}{4}$	21.625	$20\frac{1}{2}$	19.000	$20\frac{1}{2}$
18.000	24.625	3.625	288,000	1,076,000	3	180 TVL 605 A0313	$\frac{1}{8}$	21.625	$21\frac{5}{16}$	20.000	$21\frac{5}{16}$
19.250	29.250	5.000	452,000	1,685,000	3	192 TVL 607	$\frac{1}{4}$	25.000	$24\frac{1}{4}$	23.500	$24\frac{1}{16}$
19.250	29.250	5.000	297,000	1,099,000	3	192 TVL 300	$\frac{1}{4}$	25.000	$24\frac{1}{4}$	23.500	$24\frac{1}{4}$

19.500	23.000	2.250	123,000	539,000	3	195 TVL 470 AA643	1/8	22.500	21 1/4	20.000	21 1/4
20.000	27.750	4.625	381,000	1,465,000	5	200 TVL 850	1/4	24.750	23 7/8	22.250	23 7/8
20.000	27.750	4.625	409,000	1,550,000	4	200 TVL 704	1/4	24.750	23 7/8	22.250	23 7/8
20.125	24.750	2.625	176,000	746,000	3	201 TVL 615 AA410	1/8	23.250	22 7/16	21.625	22 7/16
20.247	27.750	4.500	229,000	867,000	3	202 TVL 301 AA344	1/4	24.500	24	22.500	24
20.250	27.750	4.500	350,000	1,328,000	3	202 TVL 620	1/4	24.500	24	22.500	24
22.750	30.500	4.625	378,000	1,489,000	3	227 TVL 621	1/4	27.750	26 5/8	24.500	26 5/8
22.750	30.500	4.625	247,000	971,000	3	227 TVL 302 AA341	1/4	27.750	26 5/8	24.500	26 5/8
23.375	31.125	4.625	388,000	1,540,000	3	233 TVL 622	1/4	28.375	27 1/4	25.625	27 1/4
23.375	31.125	4.625	252,000	1,005,000	3	233 TVL 303 AA341	1/4	28.375	27 1/4	25.625	27 1/4
23.875	33.375	5.250	493,000	1,915,000	3	238 TVL 623	1/4	29.125	28 5/8	27.125	28 5/8
23.875	33.375	5.250	322,000	1,248,000	3	238 TVL 304 AA341	1/4	29.125	28 5/8	27.125	28 5/8
24.000	33.500	5.250	558,000	2,162,000	4	240 TVL 715 AA645	1/8	29.250	28 13/16	27.250	28 13/16
24.000	32.750	4.625	354,000	1,410,000	3	245 TVL 501	1/4	29.250	28 5/8	26.750	28 5/8
24.500	32.750	4.625	396,000	1,594,000	3	245 TVL 612 AA360	1/4	29.250	28 5/8	26.750	28 5/8
25.250	31.250	3.500	270,000	1,130,000	3	252 TVL 505 0A368	1/4	29.375	28 1/4	27.875	28 1/4
26.000	35.250	5.250	531,000	2,142,000	3	260 TVL 625 AA366	1/4	31.125	30 5/8	28.625	30 5/8
26.287	36.000	5.000	446,000	1,780,000	3	262 TVL 502 AA130	1/4	31.750	31 1/8	30.250	31 1/8
28.500	38.500	6.625	697,000	2,820,000	4	285 TVL 705	1/4	34.500	33 1/2	32.500	33 1/2
30.240	36.250	3.500	300,000	1,314,000	3	302 TVL 510 AA422	1/4	34.375	33 3/4	32.875	33 3/4
30.250	39.625	5.500	566,000	2,375,000	3	302 TVL 624	1/4	35.500	34 15/16	33.000	34 15/16
30.260	39.625	5.500	372,000	1,548,000	3	302 TVL 305 AA342	1/4	35.500	34 5/8	33.000	34 5/8
30.938	37.500	3.750	280,000	1,076,000	4	309 TVL 707 AA646	1/4	34.750	34 3/4	33.750	34 3/4
31.000	39.625	5.000	476,000	1,970,000	3	310 TVL 503	1/4	35.750	35 5/16	33.500	35 5/16
31.000	40.375	5.500	380,000	1,594,000	3	310 TVL 306 AA343	1/4	36.125	25 11/16	35.187	35 11/16
31.000	40.375	5.500	582,000	2,448,000	3	310 TVL 625 AA411	1/4	36.125	35 11/16	35.187	35 11/16
31.750	40.375	5.000	413,000	1,742,000	3	317 TVL 307 AA470	1/4	36.750	36	34.375	36
42.250	51.000	4.500	519,000	2,331,000	4	422 TVL 680 AA647	1/4	47.000	46 1/2	44.750	46 1/2

r is the maximum radius of the fillet on the shaft and in the housing that the bearing will clear.

T_e column is used in determining Thrust Equivalent Load. For explanation see page 26.

Source: Courtesy of The Torrington Co., Torrington, CT.

With this, find that

$$P = 0.56(839.45) + 1.635(310.0) = 976.942 \text{ N}$$

so that for the 6203 series bearing with $C = 1650$ lb we find that

$$L_{10} = \left(\frac{C}{P}\right)^3 \frac{10^6}{60n} = \left(\frac{1650 \times 4.4483}{976.942}\right)^3 \frac{16\,666.67}{5500} = 1285 \text{ h}$$

To achieve a 2000-h life a basic load rating of

$$\frac{2000}{1285} = \left(\frac{C}{1650}\right)^3$$

or

$$C = \left(\frac{2000}{1285}\right)^{1/3} 1650 = 1912.17 \rightarrow 1913 \text{ lb}$$

is needed. (Result is rounded up to assure desired life.) We could recommend a bearing from the 6204 series if increasing the shaft diameter to 20 mm at the bearing location is acceptable. If it is not, we may turn to equation 10.4.2 to find that the basic dynamic load rating for two 6203 series bearings in tandem is given by

$$C^* = CN^{0.7} = 1650(2^{0.7}) = 2680 \text{ lb}$$

Hence, two of the original bearings mounted in tandem will support the load if sufficient axial space is available.

10.5 BALL BEARING SELECTION PROGRAM

Ball bearing selection according to the methods outlined by the Anti-Friction Bearing Manufacturers Association (AFBMA) may be programmed on either a personal computer or a programmable pocket calculator. If a personal computer is used the tables shown in ANSI/AFBMA Std. 9-1972 (Redesignation of ANSI B3.15-1972) may be stored as files and interpolation for X and Y between tabulated values for a particular F_a/C_o ratio may be included in the program. File recall, interpolation, and output subprograms are not explicitly shown in the flowchart in Figure 10.5.1 for such a program in order to emphasize the bearing selection process. It is assumed that the programmer will add these according to the operating requirements of the computer at hand.

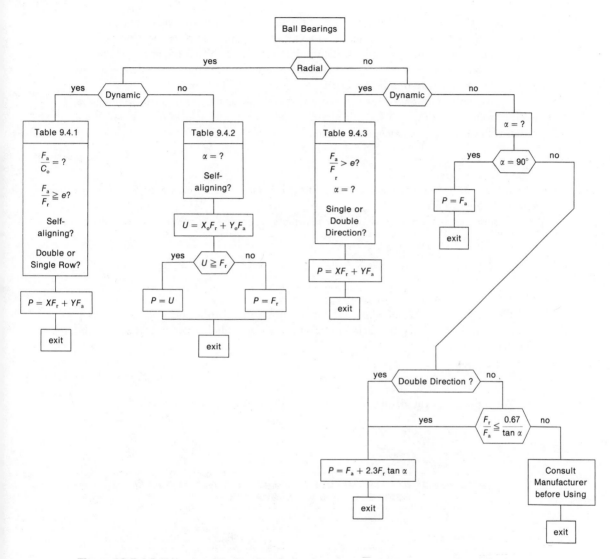

Figure 10.5.1 Ball bearing selection flowchart.

10.6   LOAD CAPACITY OF ROLLER BEARINGS, EQUIVALENT LOADS

Selection of roller bearings is simpler than selection of ball bearings because the associated tables have fewer entries, and, therefore, fewer decisions. One important simplification is the use of double-row tapered roller bearings to resist both radial and axial load without inducing an associated radial load because the induced axial forces are internal to the bearings inner and outer rings.

Coefficients X and Y listed in the following tables apply to single- and double-row self-aligning and tapered roller bearings. Radial cylindrical roller bearings, Figure 10.2.1, and thrust cylindrical roller bearings, Figures 10.2.2 and 10.2.4, are not included in these tables because $X = 1$, $F_a = 0$ is required for radial cylindrical roller bearings and $F_r = 0$, $Y = 1$ is required for thrust cylindrical roller bearings.

Equivalent radial load P for tapered radial roller bearings, also known as angular contact roller bearings, and for self-aligning roller bearings, also known as spherical radial roller bearings, is calculated from

$$P = XF_r + YF_a \tag{10.6.1}$$

where F_r again denotes the radial load and F_a the axial load. Coefficients X and Y are listed in Table 10.6.1.

Under static loading of radial roller bearings the equivalent radial load is given by

$$P = \max \begin{cases} X_o F_r + Y_o F_a \\ F_r \end{cases} \tag{10.6.2}$$

where X and Y are obtained from Table 10.6.2.

The basic dynamic load rating C^* for N similar radial roller bearings mounted in tandem (also face to face or back to back for tapered radial roller bearings) may be calculated from

$$C^* = CN^{7/9} \tag{10.6.3}$$

as a consequence of the definition of C for roller bearings, which is that

$$C = f_c (i\ell_{\text{eff}} \cos \alpha)^{7/9} Z^{3/4} D^{29/27}$$

TABLE 10.6.1   DYNAMIC VALUES OF X AND Y FOR RADIAL ROLLER BEARINGS FOR USE IN EQUATION 10.6.2

Bearing Type	$\dfrac{F_a}{F_r} \leq e$		$\dfrac{F_a}{F_r} > e$		e
Self-aligning and Tapered Roller Bearings $\alpha \neq 0^{\circ a}$	X	Y	X	Y	
	Single Row Bearings				
	1	0	0.4	0.4 cotα	1.5 tanα
	Double Row Bearings				
	1	0.45 cotα	0.67	0.67 cotα	1.5 tanα

[a] For $\alpha = 0^{\circ}$; $F_a = 0$; and $X = 1$.

Source: Copyright 1978 by The Anti-Friction Bearing Manufacturers Association, Inc.

TABLE 10.6.2 STATIC VALUES OF X_o AND Y_o FOR RADIAL ROLLER BEARINGS FOR USE IN EQUATION 10.6.3

Bearing Type	Single Row Bearings[a]		Double Row Bearings	
	X_o	Y_o	X_o	Y_o
Self-aligning and Tapered Roller Bearings $\alpha \neq 0°$	0.5	0.22 cotα	1	0.44 cotα

[a] P_o is always $\geq F_r$

The static equivalent radial load for radial roller bearings with $\alpha = 0°$, and subjected to radial load only is

$$P_{or} = F_r$$

Note: The ability of radial roller bearings with $\alpha = 0°$ to support axial loads varies considerably with bearing design and execution. The bearing user should therefore consult the bearing manufacturer for recommendations regarding the evaluation of equivalent load in cases where bearings with $\alpha = 0°$ are subjected to axial load.

Source: Copyright 1978 by The Anti-Friction Bearing Manufacturers Association, Inc.

where i denotes the number of rows, ℓ_{eff} the roller effective length, Z the number of rollers per row, and D the roller diameter for straight roller, the mean (average) diameter for tapered rollers, and the major diameter for spherical rollers. A table of values for coefficient f_c is given in ANSI/AFBMA Std 11-1978, reaffirmed, and effective in 1988. Two tapered radial roller bearings mounted face to face or back to back are to be considered as a single double-row tapered radial roller bearing in selecting the X and Y values from Table 10.6.2.

The basic static load rating C_o^* for N identical radial roller bearings mounted in tandem may be calculated from

$$C_o^* = C_o N \tag{10.6.4}$$

Equation 10.6.1 also applies to thrust roller bearings except that X and Y are obtained from Table 10.6.3 and P represents the equivalent thrust load.

Equivalent static thrust load P for thrust roller bearings under static loads is given by

$$P = 2.3 F_r \tan \alpha + F_a \quad \text{for } \alpha \neq 90°$$
$$P = F_a \quad \text{for } \alpha = 90° \tag{10.6.5}$$

The accuracy of these relations decreases for single-direction bearings if $F_r \tan \alpha > 0.44 F_a$.

Tapered radial roller bearings induce an axial load in response to a radial load because the force between races and rollers is at an angle to the plane of the bearing.

TABLE 10.6.3 DYNAMIC VALUES OF X AND Y FOR THRUST ROLLER BEARINGS FOR USE IN EQUATION 10.6.2

Bearing Type	Single Direction Bearings $\dfrac{F_a}{F_r} > e$		Double Direction Bearings $\dfrac{F_a}{F_r} \leq e$		$\dfrac{F_a}{F_r} > e$		e
	X	Y	X	Y	X	Y	
Self-aligning and Tapered Thurst Roller	$\tan \alpha$	1	$1.5 \tan \alpha$	0.67	$\tan \alpha$	1	$1.5 \tan \alpha$

For $\alpha = 90°$; $F_r = 0$; and $Y = 1$.

Source: Copyright 1978 by The Anti-Friction Bearing Manufacturers Association, Inc.

This induced axial load I is related to the radial load according to

$$I = \frac{F_r}{2Y} \tag{10.6.6}$$

where $Y = 0.4/\tan \alpha$. Its effect upon other bearings on the same shaft and upon its own L_{10} life will be discussed in Section 10.8 and Problem 10.33.

Typical data from manufacturers' engineering catalogs are reproduced in Tables 10.6.4 through 10.6.8. As noted earlier, these values are frequently up-dated, so values in these tables are used for demonstration only.

Although applications of ball and roller bearings overlap for given bore and outside diameters, roller bearings generally will support larger loads, but usually at lower speeds. Radial roller bearings have one capability not found in ball bearings: they may be designed to permit small axial motion of one ring relative to the other, as may be seen by close examination of the rings in Figure 10.2.1. This is an important feature which enables them to be used on shafts which, due to heating or loading, may change length to an extent that would damage radial ball bearings.

EXAMPLE 10.6.1

Select a radial roller bearing for a 4-in.-diameter shaft to support a steady radial load of 9200 lb at 550 rpm for an L_{10} life of 2 years for continuous service at 8 h/day, 5 days/week, 50 weeks/year.

Write (10.3.1) in terms of rpm, hours, and years as

$$L_{10} = \left(\frac{C}{P}\right)^{10/3} 10^6 \text{ cycles} \rightarrow \left(\frac{C}{P}\right)^{10/3} \frac{10^6}{60n} = \left(\frac{C}{P}\right)^{10/3} \frac{16\,667.67}{n} \text{ h} \tag{10.6.7}$$

TABLE 10.6.4 CATALOG DATA FOR RADIAL ROLLER BEARINGS

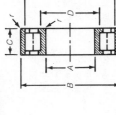

Bore A		o.d. B		Width C		Basic Dynamic Capacity (BDC) lb	Basic Static Capacity (BSC) lb	Bearing Number						Max Fillet r inch	Shoulder Diam	
								RN, RIN	RF, RIF	RT, RIT	RU, RIU	RJ, RIJ	RP, RIP		Shaft D inch	Housing E inch
mm	inch	mm	inch	mm	inch	lb	lb							inch		
100	3.9370	180	7.0866	34	1.3386	44,600	47,300	100 RN 02	100 RF 02	100 RT 02	100 RU 02	100 RJ 02	100 RP 02	.08	$4\frac{1}{2}$	$6\frac{1}{4}$
100	3.9370	180	7.0866	60.3	2.3750	77,000	91,700	100 RN 32	100 RF 32	100 RT 32	100 RU 32	100 RJ 32	100 RP 32	.08	$4\frac{1}{2}$	$6\frac{7}{32}$
100	3.9370	215	8.4646	47	1.8504	77,600	80,500	100 RN 03	100 RF 03	100 RT 03	100 RU 03	100 RJ 03	100 RP 03	.10	$4\frac{25}{32}$	$7\frac{5}{32}$
100	3.9370	215	8.4646	82.6	3.2500	136,000	158,000	100 RN 33	100 RF 33	100 RT 33	100 RU 33	100 RJ 33	100 RP 33	.10	$4\frac{13}{16}$	$7\frac{1}{8}$
100	3.9370	250	9.8425	58	2.2835	99,500	100,000	100 RN 04	100 RF 04	100 RT 04	100 RU 04	100 RJ 04	100 RP 04	.12	5	$8\frac{3}{4}$
100	3.9370	265	10.4431	60	2.3622	115,000	114,000	100 RN 05	100 RF 05	100 RT 05	100 RU 05	100 RJ 05	100 RP 05	.12	$5\frac{3}{32}$	$9\frac{3}{32}$
	4.0000		5.6250		.8750	20,500	27,000	40 RIN 130	40 RIF 130	40 RIT 130	40 RIU 130	40 RIJ 130	40 RIP 130	.10	$4\frac{5}{16}$	$5\frac{5}{16}$
	4.0000		7.2500		1.2500	40,400	43,500	40 RIN 131	40 RIF 131	40 RIT 131	40 RIU 131	40 RIJ 131	40 RIP 131	.12	$4\frac{5}{8}$	$6\frac{5}{8}$
	4.0000		7.2500		1.9375	72,500	83,300	40 RIN 132	40 RIF 132	40 RIT 132	40 RIU 132	40 RIJ 132	40 RIP 132	.12	$4\frac{19}{32}$	$6\frac{21}{32}$
	4.0000		8.5000		1.7500	64,700	65,000	40 RIN 133	40 RIF 133	40 RIT 133	40 RIU 133	40 RIJ 133	40 RIP 133	.16	$4\frac{7}{16}$	$7\frac{1}{4}$
	4.0000		8.5000		2.7500	112,000	123,000	40 RIN 134	40 RIF 134	40 RIT 134	40 RIU 134	40 RIJ 134	40 RIP 134	.16	$4\frac{13}{16}$	$7\frac{11}{16}$
	4.1250		6.0000		.8750	24,900	33,400	41 RIN 190	41 RIF 190	41 RIT 190	41 RIU 190	41 RIJ 190	41 RIP 190	.10	$4\frac{19}{32}$	$5\frac{21}{32}$
105	4.1339	190	7.4803	36	1.4173	46,000	49,000	105 RN 02	105 RF 02	105 RT 02	105 RU 02	105 RJ 02	105 RP 02	.08	$4\frac{3}{4}$	$6\frac{7}{8}$
105	4.1339	190	7.4803	65.1	2.5625	96,300	119,000	105 RN 32	105 RF 32	105 RT 32	105 RU 32	105 RJ 32	105 RP 32	.08	$4\frac{3}{4}$	$6\frac{7}{8}$
105	4.1339	225	8.8583	49	1.9291	81,300	84,300	105 RN 03	105 RF 03	105 RT 03	105 RU 03	105 RJ 03	105 RP 03	.10	5	8
105	4.1339	225	8.8583	87.3	3.4375	147,000	174,000	105 RN 33	105 RF 33	105 RT 33	105 RU 33	105 RJ 33	105 RP 33	.10	5	8

r is the maximum radius of the fillet on the shaft and in the housing that the bearing will clear.

Courtesy The Torrington Co., South Bend, IN.

Source: Courtesy of The Torrington Co., Torrington, CT.

TABLE 10.6.4 CATALOG DATA FOR RADIAL ROLLER BEARINGS (Continued)

Bore A (mm)	Bore A (inch)	o.d. B (mm)	o.d. B (inch)	Width C (mm)	Width C (inch)	Basic Dynamic Capacity (BDC) lb	Basic Static Capacity (BSC) lb	RN, RIN	RF, RIF	RT, RIT	RU, RIU	RJ, RIJ	RP, RIP	Max Fillet r inch	Shaft D inch	Housing E inch
	4.2500		6.0000		.8750	24,900	33,400	42 RIN 190	42 RIF 190	42 RIT 190	42 RIU 190	42 RIJ 190	42 RIP 190	.10	$4\frac{19}{32}$	$5\frac{21}{32}$
	4.2500		7.5000		1.2500	40,500	43,700	42 RIN 191	42 RIF 191	42 RIT 191	42 RIU 191	42 RIJ 191	42 RIP 191	.12	$4\frac{27}{32}$	$6\frac{29}{32}$
	4.2500		7.5000		1.9375	69,300	80,200	42 RIN 192	42 RIF 192	42 RIT 192	42 RIU 192	42 RIJ 192	42 RIP 192	.12	$4\frac{3}{32}$	$6\frac{3}{32}$
	4.2500		8.7500		1.7500	69,600	71,000	42 RIN 193	42 RIF 193	42 RIT 193	42 RIU 193	42 RIJ 193	42 RIP 193	.12	$5\frac{3}{32}$	$7\frac{31}{32}$
	4.2500		8.7500		2.7500	118,000	135,000	42 RIF 194	42 RIF 194	42 RIT 194	42 RIU 194	42 RIJ 194	42 RIP 194	.16	5	8
110	4.3307	200	7.8740	38	1.4961	55,100	60,500	110 RN 02	110 RF 02	110 RT 02	110 RU 02	110 RJ 02	110 RP 02	.08	$4\frac{15}{16}$	$7\frac{9}{32}$
110	4.3307	200	7.8740	69.8	2.7500	110,000	136,000	110 RN 32	110 RF 32	110 RT 32	110 RU 32	110 RJ 32	110 RP 32	.08	$4\frac{1}{16}$	$7\frac{3}{32}$
110	4.3307	240	9.4488	50	1.9685	85,800	86,800	110 RN 03	110 RF 03	110 RT 03	110 RU 03	110 RJ 03	110 RP 03	.10	$4\frac{5}{16}$	$8\frac{15}{32}$
110	4.3307	240	9.4488	92.1	3.6250	172,000	197,000	110 RN 33	110 RF 33	110 RT 33	110 RU 33	110 RJ 33	110 RP 33	.10	$4\frac{3}{16}$	$8\frac{3}{32}$
	4.3750		6.2500		.8750	24,500	33,200	43 RIN 195	43 RIF 195	43 RIT 195	43 RIU 195	43 RIJ 195	43 RIP 195	.10	$4\frac{3}{16}$	$5\frac{15}{16}$
	4.5000		6.2500		.8750	24,500	33,200	45 RIN 195	45 RIF 195	45 RIT 195	45 RIU 195	45 RIJ 195	45 RIP 195	.10	$4\frac{1}{2}$	$5\frac{15}{16}$
	4.5000		8.0000		1.3125	44,400	48,500	45 RIN 196	45 RIF 196	45 RIT 196	45 RIU 196	45 RIJ 196	45 RIP 196	.12	$5\frac{1}{16}$	$7\frac{3}{8}$
	4.5000		8.0000		2.0625	79,000	93,500	45 RIN 197	45 RIF 197	45 RIT 197	45 RIU 197	45 RIJ 197	45 RIP 197	.12	$5\frac{1}{8}$	$7\frac{3}{8}$
	4.5000		9.3750		2.0000	88,000	93,000	45 RIN 198	45 RIF 198	45 RIT 198	45 RIU 198	45 RIJ 198	45 RIP 198	.20	$5\frac{3}{8}$	$8\frac{1}{8}$
	4.5000		9.3750		3.1250	146,000	171,000	45 RIF 199	45 RIF 199	45 RIT 199	45 RIU 199	45 RIJ 199	45 RIP 199	.20	$5\frac{3}{32}$	$8\frac{1}{32}$
	4.6250		6.5000		.8750	23,700	32,800	46 RIN 201	46 RIF 201	46 RIT 201	46 RIU 201	46 RIJ 201	46 RIP 201	.10	$5\frac{1}{16}$	$6\frac{3}{16}$
120	4.7244	180	7.0866	46	1.8110	53,400	71,500	120 RN 30	120 RF 30	120 RT 30	120 RU 30	120 RJ 30	120 RP 30	.08	$5\frac{1}{8}$	$6\frac{11}{16}$
120	4.7244	215	8.4646	40	1.5748	58,000	62,600	120 RN 02	120 RF 02	120 RT 02	120 RU 02	120 RJ 02	120 RP 02	.08	$5\frac{3}{8}$	$7\frac{1}{16}$
120	4.7244	215	8.4646	76.2	3.0000	125,000	157,000	120 RN 92	120 RF 92	120 RT 92	120 RU 92	120 RJ 92	120 RP 92	.08	$5\frac{3}{8}$	$7\frac{13}{16}$
120	4.7244	260	10.2362	55	2.1654	100,000	104,000	120 RN 03	120 RF 03	120 RT 03	120 RU 03	120 RJ 03	120 RP 03	.10	$5\frac{23}{32}$	$9\frac{1}{4}$
120	4.7244	260	10.2362	104.8	4.1250	202,000	234,000	120 RN 93	120 RF 93	120 RT 93	120 RU 93	120 RJ 93	120 RP 93	.10	$5\frac{1}{4}$	$9\frac{3}{32}$

TABLE 10.6.5 CATAOLOG DATA FOR SPHERICAL ROLLER BEARINGS

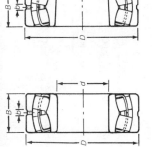

Cylindrical bore   Tapered bore   Adapter mountings

Bore d		Outside Dia. D		Width B		Groove Width b		Basic Load Rating C		Static Load Rating C_0		Speed Limit (2) rpm		Bearing Mass		Bearing[3] Designation		Contact Angle degrees
mm in.		mm in.		mm in.		mm in.		N lb.		N lb.		Grease	Oil	kg lb.		Cylindrical Bore	Tapered Bore	
20	0.7874	52	2.0402	15	0.5906	—	—	26 500	6 000	18 000	4 000	8500	11 000	0.16	0.35	21304 CC	—	11.68
25	0.9843	52	2.0472	18	0.7087	—	—	31 000	6 950	21 600	4 800	8500	11 000	0.18	0.40	22205 CC	—	13.13
		62	2.4409	17	0.6693	—	—	36 000	8 150	24 000	5 400	6700	8 500	0.25	0.55	21305 CC	—	9.09
30	1.1811	62	2.4409	20	0.7874	—	—	42 500	9 500	30 000	6 700	7500	9 500	0.28	0.62	22206 CC	—	12.41
		72	2.8346	19	0.7480	—	—	48 000	10 800	35 500	8 000	6000	7 500	0.38	0.84	21306 CC	21306 CCK	8.72
35	1.3780	72	2.8346	23	0.9055	5.56	0.219	55 000	12 200	40 500	9 150	6300	8 000	0.43	0.95	22207 CC/W33	22207 CCK/W33	11.68
		80	3.1496	21	0.8268	—	—	57 000	12 700	41 500	9 300	5300	6 700	0.51	1.10	21307 CC	—	8.34
40	1.5748	80	3.1496	23	0.9055	5.56	0.219	64 000	14 300	47 500	10 600	6000	7 500	0.52	1.15	22208 CC/W33	22208 CCK/W33	10.57
		90	3.5433	23	0.9055	—	—	72 000	16 300	55 000	12 200	4500	5 600	0.71	1.55	21308 CC	21308 CCK	8.34
		90	3.5433	33	1.2992	5.56	0.219	100 000	24 000	73 500	16 600	4500	5 600	1.00	2.20	22308 CC/W33	22308 CCK/W33	13.86
45	1.7717	85	3.3465	23	0.9055	5.56	0.219	67 000	15 000	51 000	11 400	5300	6 700	0.56	1.25	22209 CC/W33	22209 CCK/W33	9.83

Source: SKF Industries, Inc., Bearings Group, King of Prussia, PA.

TABLE 10.6.6 CATALOG DATA FOR TAPERED ROLLER BEARINGS

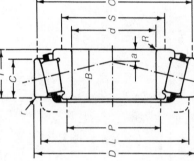

Nominal External Dimensions inches & millimetres			Part Numbers		Axial Load Factor Y	Static Load Rating Co pounds & newtons	Basic Load Rating C pounds & newtons	Approx Speed Limit rpm	Mass pounds & kilograms		Series Number
Bore d	Outside Diam D	Width T	Cone	Cup					Cone	Cup	
0.6875 17.463	1.5700 39.878	0.5450 13.843	LM11749	LM11710	2.10	3050 13 400	4250 19 000	14 000	0.12 0.05	0.06 0.03	LM11700
0.7500 19.050	1.7810 45.237	0.6100 15.494	LM11949	LM11910	2.00	3900 17 600	5500 24 500	13 000	0.17 0.08	0.10 0.05	LM11900
	1.9380 49.225	0.7100 18.034	09067	09195	2.20	5200 23 200	7200 32 000	12 000	0.23 0.10	0.15 0.07	09000
	1.9380 49.225	0.7813 19.845	09078	09195	2.20	5200 23 200	7200 32 000	12 000	0.25 0.11	0.15 0.07	09000
0.8120 20.625	1.9380 49.225	0.7813 19.845	09081	09195	2.20	5200 23 200	7200 32 000	12 000	0.23 0.10	0.15 0.07	09000

d (in / mm)	D (in / mm)	Cone	B (in / mm)	Cup	r	(in / mm)	(in / mm)		(in / mm)	(in / mm)	
0.8120 / 20.625	1.9380 / 49.225	09081	0.7813 / 19.845	09195	2.20	5 200 / 23 200	7 200 / 32 000	12 000	0.23 / 0.10	0.15 / 0.07	09000
	1.9380 / 49.225	09081	0.9063 / 23.020	09196	2.20	5 200 / 23 200	7 200 / 32 000	12 000	0.23 / 0.10	0.19 / 0.09	09000
	1.9380 / 49.225	09081	0.9063 / 23.020	09194	2.20	5 200 / 23 200	7 200 / 32 000	12 000	0.23 / 0.10	0.18 / 0.08	09000
0.8437 / 21.430	1.9687 / 50.005	M12649	0.6900 / 17.526	M12610	2.10	5 400 / 24 000	7 350 / 32 500	11 000	0.24 / 0.11	0.13 / 0.06	M12600
0.8661 / 22.000	1.7810 / 45.237	LM12749	0.6100 / 15.494	LM12710	1.90	4 250 / 19 000	5 500 / 24 500	12 000	0.16 / 0.07	0.08 / 0.04	LM12700
	1.8100 / 45.974	LM12749	0.6100 / 15.494	LM12711	1.90	4 250 / 19 000	5 500 / 24 500	12 000	0.16 / 0.07	0.09 / 0.04	LM12700
0.8750 / 22.225	1.9687 / 50.005	07087	0.5313 / 13.495	07196	1.50	3 400 / 15 300	4 750 / 21 200	11 000	0.22 / 0.10	0.07 / 0.03	07000
	2.0470 / 51.994	07087	0.5910 / 15.011	07204	1.50	3 400 / 15 300	4 750 / 21 200	11 000	0.22 / 0.10	0.13 / 0.06	07000
0.9375 / 23.813	1.9800 / 50.292	L44640	0.5600 / 14.224	L44610	1.60	4 000 / 18 000	5 200 / 23 200	10 000	0.16 / 0.07	0.09 / 0.04	L44600
0.9835 / 24.981	1.9687 / 50.005	07098	0.5313 / 13.495	07196	1.50	3 400 / 15 300	4 750 / 21 200	11 000	0.18 / 0.08	0.07 / 0.03	07000
	2.0470 / 51.994	07098	0.5910 / 15.011	07204	1.50	3 400 / 15 300	4 750 / 21 200	11 000	0.18 / 0.08	0.13 / 0.06	07000
0.9842 / 25.000	1.9687 / 50.005	07097	0.5313 / 13.495	07196	1.50	3 400 / 15 300	4 750 / 21 200	11 000	0.20 / 0.09	0.07 / 0.03	07000
	2.0470 / 51.994	07097	0.5910 / 15.011	07204	1.50	3 400 / 15 300	4 750 / 21 200	11 000	0.20 / 0.09	0.13 / 0.06	07000
1.0000 / 25.400	1.9687 / 50.005	07100	0.5313 / 13.495	07196	1.50	3 400 / 15 300	4 750 / 21 200	11 000	0.18 / 0.08	0.07 / 0.03	07000
	1.9800 / 50.292	L44643	0.5600 / 14.224	L44610	1.60	4 000 / 18 000	5 200 / 23 200	10 000	0.19 / 0.09	0.09 / 0.04	L44600
	2.0470 / 51.994	07100	0.5910 / 15.011	07204	1.50	3 400 / 15 300	4 750 / 21 200	11 000	0.18 / 0.08	0.13 / 0.06	07000

Source: SKF Industries, Inc., Bearings Group, King of Prussia, PA.

TABLE 10.6.7 CATALOG DATA FOR CYLINDRICAL ROLLER THRUST BEARINGS

Bore A	o.d. B	Height C	Basic Dynamic Thrust Capacity (BDTC) lb	Basic Static Thrust Capacity (BSTC) lb	Bearing Number	Washer Thickness C_i	Max Fillet r	Small Bore Washer o.d. B_1	Shaft Shoulder Diameter D	Large Bore Washer i.d. A_1	Housing Shoulder Diameter E	Reference Number
2.000	6.000	1.375	74,500	183,000	20 TP 103	$\frac{3}{8}$	.062	$5\frac{15}{16}$	$5\frac{9}{16}$	$2\frac{1}{16}$	$2\frac{7}{16}$	T-727
2.000	7.000	1.375	89,500	227,000	20 TP 104	$\frac{3}{8}$	.062	$6\frac{15}{16}$	$6\frac{7}{16}$	$2\frac{1}{16}$	$2\frac{9}{16}$	T-728
2.000	8.000	1.375	114,500	304,000	20 TP 105	$\frac{3}{8}$	.062	$7\frac{15}{16}$	$7\frac{3}{16}$	$2\frac{1}{16}$	$2\frac{5}{16}$	T-729
3.000	6.000	1.375	76,500	168,000	30 TP 106	$\frac{3}{8}$	.062	$5\frac{15}{16}$	$5\frac{5}{8}$	$3\frac{1}{16}$	$3\frac{3}{8}$	T-730
3.000	7.000	1.375	93,200	234,000	30 TP 107	$\frac{3}{8}$	.062	$6\frac{15}{16}$	$6\frac{9}{16}$	$3\frac{1}{16}$	$3\frac{7}{16}$	T-731
3.000	8.000	1.375	117,000	331,000	30 TP 108	$\frac{3}{8}$	.062	$7\frac{15}{16}$	$7\frac{7}{16}$	$3\frac{1}{16}$	$3\frac{9}{16}$	T-732
3.000	9.000	1.375	143,000	405,000	30 TP 109	$\frac{3}{8}$	.062	$8\frac{15}{16}$	$8\frac{3}{8}$	$3\frac{1}{16}$	$3\frac{5}{16}$	T-733
3.500	5.218	1.000	40,450	85,600	35 TP 113	$\frac{9}{32}$	.062	$5\frac{5}{32}$	$4\frac{7}{8}$	$3\frac{1}{16}$	$3\frac{27}{32}$	T-626

						r						
4.000	7.000	1.750	94,000	231,000	40 TP 114	$\frac{1}{2}$	$6\frac{15}{16}$	$6\frac{5}{8}$	.062	$4\frac{1}{16}$	$4\frac{3}{8}$	T-734
4.000	8.000	1.750	132,500	308,000	40 TP 115	$\frac{1}{2}$	$7\frac{15}{16}$	$7\frac{1}{8}$	.062	$4\frac{1}{16}$	$4\frac{1}{2}$	T-735
4.000	9.000	1.750	152,000	397,000	40 TP 116	$\frac{1}{2}$	$8\frac{15}{16}$	$8\frac{7}{16}$	.062	$4\frac{1}{16}$	$4\frac{9}{16}$	T-736
4.000	10.000	1.750	201,500	498,000	40 TP 117	$\frac{1}{2}$	$9\frac{15}{16}$	$9\frac{3}{8}$	.062	$4\frac{1}{16}$	$4\frac{5}{8}$	T-737
5.000	8.000	1.750	136,000	288,000	50 TP 119	1	$7\frac{15}{16}$	$7\frac{1}{2}$	.062	$5\frac{1}{16}$	$5\frac{1}{2}$	T-738
5.000	9.000	1.750	161,000	385,000	50 TP 120	1	$8\frac{15}{16}$	$8\frac{1}{2}$	.062	$5\frac{1}{16}$	$5\frac{1}{2}$	T-739
5.000	10.000	2.000	189,000	491,000	50 TP 121	$\frac{9}{16}$	$9\frac{15}{16}$	$9\frac{7}{16}$	.125	$5\frac{1}{16}$	$5\frac{9}{16}$	T-740
5.000	11.000	2.000	224,000	620,000	50 TP 122	$\frac{9}{16}$	$10\frac{15}{16}$	$10\frac{5}{16}$	.125	$5\frac{1}{16}$	$5\frac{11}{16}$	T-741
5.000	12.000	2.000	262,000	739,000	50 TP 123	$\frac{9}{16}$	$11\frac{15}{16}$	$11\frac{3}{8}$	.125	$5\frac{1}{16}$	$5\frac{3}{4}$	T-742
6.000	9.000	2.000	135,000	317,000	60 TP 124	$\frac{9}{16}$	$8\frac{15}{16}$	$8\frac{9}{16}$	.125	$6\frac{1}{16}$	$6\frac{7}{16}$	T-743
6.000	10.000	2.000	190,000	449,000	60 TP 125	$\frac{9}{16}$	$9\frac{15}{16}$	$9\frac{1}{2}$	.125	$6\frac{1}{16}$	$6\frac{1}{2}$	T-744
6.000	11.000	2.000	225,500	607,500	60 TP 126	$\frac{9}{16}$	$10\frac{15}{16}$	$10\frac{7}{16}$	.125	$6\frac{1}{16}$	$6\frac{9}{16}$	T-745
6.000	12.000	2.000	250,000	725,000	60 TP 127	$\frac{9}{16}$	$11\frac{15}{16}$	$11\frac{5}{16}$	.125	$6\frac{1}{16}$	$6\frac{11}{16}$	T-746
7.000	10.000	2.000	149,000	365,000	70 TP 129	$\frac{9}{16}$	$9\frac{29}{32}$	$9\frac{9}{16}$	.125	$7\frac{3}{32}$	$7\frac{7}{16}$	T-747
7.000	11.000	2.000	209,000	540,000	70 TP 130	$\frac{9}{16}$	$10\frac{29}{32}$	$10\frac{1}{2}$	.125	$7\frac{3}{32}$	$7\frac{1}{2}$	T-748
7.000	12.000	2.000	242,000	695,000	70 TP 131	$\frac{9}{16}$	$11\frac{29}{32}$	$11\frac{1}{8}$	.125	$7\frac{3}{32}$	$7\frac{5}{8}$	T-749
7.000	14.000	3.000	394,000	1,010,000	70 TP 132	$\frac{13}{16}$	$13\frac{29}{32}$	$13\frac{3}{16}$	.250	$7\frac{3}{32}$	$7\frac{13}{16}$	T-750
8.000	12.000	3.000	258,000	599,000	80 TP 134	$\frac{13}{16}$	$11\frac{29}{32}$	$11\frac{1}{16}$	.250	$8\frac{3}{32}$	$8\frac{1}{8}$	T-751
8.000	14.000	3.000	389,000	950,000	80 TP 135	$\frac{13}{16}$	$13\frac{29}{32}$	$13\frac{1}{2}$	.250	$8\frac{3}{32}$	$8\frac{3}{4}$	T-752
8.000	16.000	3.000	520,000	1,290,000	80 TP 136	$\frac{13}{16}$	$15\frac{29}{32}$	$15\frac{1}{16}$	.250	$8\frac{3}{32}$	$8\frac{15}{16}$	T-753

r is the maximum radius of the fillet on the shaft and in the housing that the bearing will clear.

Shaft seat should extend below the cage to provide full contact in the cage bore for effective piloting.

Source: Courtesy of The Torrington Co., Torrington, CT.

TABLE 10.6.8 CATALOG DATA FOR TAPERED ROLLER THRUST BEARINGS

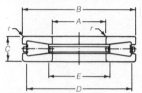

Bore A	o.d. B	Height C	Basic Dynamic Thrust Capacity (BDTC) lb	Basic Static Thrust Capacity (BSTC) lb	Bearing Number	Min Radius r	Shaft Shoulder Diameter D	Housing Shoulder Diameter E	Reference Number
3.000	6.375	1.312	89,000	225,000	30 TTHD 013	.125	$5\frac{15}{16}$	$3\frac{7}{16}$	T-311
3.000	6.375	1.312	114,000	310,000	30 TTHD 013 00278	.125	$5\frac{15}{16}$	$3\frac{7}{16}$	T-311F
4.000	8.500	1.812	173,500	430,000	40 TTHD 015	.125	$7\frac{15}{16}$	$4\frac{9}{16}$	T-411
4.000	8.500	1.812	210,000	563,000	40 TTHD 015 00278	.125	$7\frac{15}{16}$	$4\frac{9}{16}$	T-411F
4.400	8.800	2.200	184,000	450,000	43 TTHD 019 AA226	.125	$8\frac{5}{32}$	$5\frac{1}{16}$	T-441
4.500	9.875	2.125	252,300	634,000	45 TTHD 020	.156	$9\frac{1}{4}$	$5\frac{1}{8}$	T-451
5.000	9.875	2.187	208,000	504,000	50 TTHD 024	.187	$9\frac{7}{32}$	$5\frac{21}{32}$	T-520
5.000	10.500	2.312	275,000	685,000	50 TTHD 025	.187	$9\frac{13}{16}$	$5\frac{11}{16}$	T-511
5.000	10.500	2.312	320,000	845,000	50 TTHD 025 00278	.187	$9\frac{13}{16}$	$5\frac{11}{16}$	T-511F
5.062	10.437	2.500	275,000	685,000	50 TTHD 025 AA277	.250	$9\frac{13}{16}$	$5\frac{11}{16}$	
6.000	12.500	2.750	392,000	960,000	60 TTHD 026	.250	$11\frac{11}{16}$	$6\frac{13}{16}$	T-611
6.000	12.500	2.750	459,000	1,185,000	60 TTHD 026 00278	.250	$11\frac{11}{16}$	$6\frac{13}{16}$	T-611F
6.500	12.250	3.500	321,000	735,000	65 TTHD 029	.250	$11\frac{3}{16}$	$7\frac{9}{16}$	T-651
6.625	12.000	2.750	304,000	745,000	66 TTHD 030	.250	$11\frac{3}{16}$	$7\frac{7}{16}$	T-661
6.875	14.125	3.250	495,000	1,190,000	68 TTHD 031	.250	$13\frac{5}{32}$	$7\frac{27}{32}$	T-691
7.000	14.500	3.250	537,000	1,300,000	70 TTHD 032	.312	$13\frac{15}{32}$	$8\frac{1}{32}$	T-711
7.000	14.500	3.250	631,000	1,610,000	70 TTHD 032 00278	.312	$13\frac{15}{32}$	$8\frac{1}{32}$	T-711F
8.000	16.500	3.625	689,000	1,665,000	80 TTHD 034	.375	$15\frac{13}{32}$	$9\frac{3}{32}$	T-811
8.000	16.500	3.625	810,000	2,060,000	80 TTHD 034 00278	.375	$15\frac{13}{32}$	$9\frac{3}{32}$	T-811F
9.000	19.000	4.125	970,000	2,380,000	90 TTHD 039	.437	$17\frac{3}{4}$	$10\frac{1}{4}$	T-911
9.000	19.000	4.125	1,090,000	2,780,000	90 TTHD 039 00278	.437	$17\frac{3}{4}$	$10\frac{1}{4}$	T-911F
9.250	19.000	4.125	970,000	2,380,000	92 TTHD 039	.437	$17\frac{3}{4}$	$10\frac{1}{4}$	T-911A
9.250	21.500	5.000	1,350,000	3,150,000	92 TTHD 043	.625	20	$10\frac{3}{4}$	T-921
11.000	23.750	5.375	1,570,000	3,700,000	110 TTHD 047	.437	$22\frac{1}{8}$	$12\frac{5}{8}$	T-1120
11.000	23.750	5.375	1,770,000	4,350,000	110 TTHD 047 00278	.437	$22\frac{1}{8}$	$12\frac{5}{8}$	T-1120F
16.000	28.000	5.750	1,690,000	4,130,000	160 TTHD 064	.375	$26\frac{1}{4}$	$17\frac{3}{4}$	T-16021
16.000	33.000	7.000	3,040,000	7,100,000	160 TTHD 065	.500	$30\frac{29}{32}$	$18\frac{3}{32}$	T-16050
16.000	33.000	7.000	3,420,000	8,300,000	160 TTHD 065 00278	.500	$30\frac{29}{32}$	$18\frac{3}{32}$	T-16050F

Suffix 278 in the bearing number indicates a full complement of rollers without cage. These full complement bearings are used only in slow speed applications.

r is minimum radius on the washers.

The fillet radius for the shaft or housing shoulder should be a minimum of $\frac{1}{64}''$ less than the nominal radius listed for the washers.

A step must be provided on the shaft to clear the lower bearing washer.

Source: The Torrington Co., Torrington, CT.

or for 2000 h/year

$$L_{10} = \left(\frac{C}{P}\right)^{10/3} \frac{16\,667.67}{2\,000\,n}\,\mathrm{h} = \left(\frac{C}{P}\right)^{10/3} \frac{8.333}{n}\ \text{shift-years} \qquad (10.6.8)$$

where n denotes rpm. Thus,

$$P\left(\frac{nL_{10}}{8.33}\right)^{0.3} = 9200\left(\frac{550 \times 2}{8.333}\right)^{0.3} = 39\,808\ \text{lb} = C$$

Turn to Table 10.6.4 to find the closest bearing listed is the 40 RXX 131 series, for which $C = 40\,400$ lb, so that for 40 RIP 131, say,

$$L_{10} = \left(\frac{40\,400}{9200}\right)^{10/3} \frac{8.333}{550} = 2.10\ \text{shift-years}$$

Note from a close examination of the cross section shown above the column of bearing numbers that neither the inner nor outer rings will slip from the RIP series bearing when the bearing is removed. In contrast, the outer race will slip from the bearings in the RIN series.

EXAMPLE 10.6.2

Rollers at the loading section of a belt conveyor system, Figure 10.6.1, are subjected to an axial load of 2270 N and a radial load of 5120 N. Shaft flexure may be 0.10°, so self-aligning roller bearings are to be used. Select a self-aligning roller bearing to provide a minimum L_{50} (average) life of 25 years at 147 rpm and 2000 h/year. Shaft diameter is to be between 20 and 30 mm. Recommend the maximum static load that may be permitted.

Assume $\alpha = 10°$ for the initial bearing selection. Since only double row spherical roller bearing data are shown in this chapter, use Table 10.6.1 to find

$$X = 0.67 \qquad Y = 0.67/\tan 10° = 3.7998$$

Next

$$F_a/F_r = 2270/5120 = 0.443 > 1.5 \tan 10° = 0.264$$

so from (10.6.1)

$$P = 12\,056\ \text{N}$$

Recall that $L_{50} = 5L_{10}$ before substituting into (10.6.7) and solving for C to find

$$C = [147(5)/8.333]^{0.3}12\,056 = 46\,222\ \text{N}$$

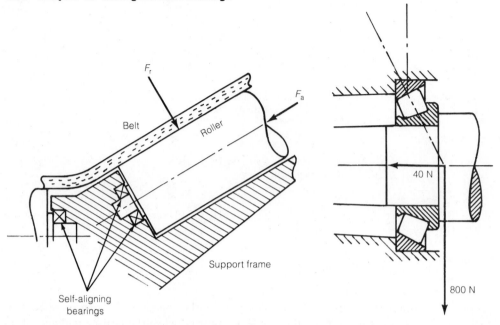

Figure 10.6.1 Roller detail for belt conveyor, Example 10.6.2.   **Figure 10.6.2** Single-row tapered roller bearing.

Guided by this value of C, first try bearing 21306CC from Table 10.6.5, for which $C = 48\,000$ N and $\alpha = 8.72°$. Since $F_a/F_r > e$ for this value of α,

$$P = 0.67(5120) + [0.67 \cot 8.72°]2270 = 13\,346 \text{ N}$$

Substitution into (10.6.8) yields an L_{10} life of 4.04 years. Hence, try bearing 21308CC for which $C = 55\,000$ N and $\alpha = 8.34°$. This value of α yields an equivalent radial load of $P = 13\,805$ N which in turn yields

$$L_{10} = \left(\frac{55\,000}{13\,805}\right)^{10/3} \frac{8.333}{147} = 5.68 \text{ years}$$

which is satisfactory, since $L_{50} = 5L_{10} = 5(5.68) = 28.40$ years.

According to Table 10.6.2 and equations 10.6.2 we find that

$$P = 5120 + 2270(0.44) \cot 8.34° = 11\,933 \text{ N}$$

is the recommended maximum static load, corresponding to the larger value of P as determined by the two equations displayed in (10.6.2).

EXAMPLE 10.6.3

Select a single-row tapered roller bearing with a 0.75 in. bore to support a radial load of 855 lb and an axial load of 87 lb as illustrated in Figure 10.6.2. The shaft turns at 2000 rpm and the bearing L_{10} life should be at least 5 years at 2000 hours/year.

According to Table 10.6.1 $Y = 0.4/\tan \alpha$ for a single-row tapered roller bearing for $F_a/F_r > e$. Hence, the Y values listed in Table 10.6.6 imply that for these bearings $10.3° \leq \alpha \leq 14.4°$ and 0.4, because Table 10.6.1 gives $e = 1.5 \tan \alpha$. For the loading at hand $F_a/F_r = 87/855 = 0.102$, so that $X = 1$ and $Y = 0$. Hence, $P = F_r$ so that from (10.6.8) we find that

$$C = \left(\frac{2000(5)}{8.333}\right)^{0.3} 8.55 = 7173 \text{ lb}$$

is the minimum required basic dynamic capacity. Turn to Table 10.6.6 to find that bearing cone and cup combination 09081/09195 have $C = 7200$ lb that will provide that L_{10} life of

$$C = \left(\frac{7.200}{8.333}\right)^{10/3} \frac{8.333}{2000} = 5.06 \text{ years}$$

The induced axial load was not considered because it was greater than the applied axial load and was equal to the reactions by the shaft and frame that held the cone and cup together before the axial load was applied. The added axial load simply redistributes the reactions between the shaft and the housing. The axial load plays no part in determining the equivalent radial load until it become greater than the induced load and, therefore, alters the axial load applied to the bearing.

10.7 ROLLER BEARING SELECTION ROUTINE

Roller bearing selection according to the methods outlined by the Anti-Friction Bearing Manufacturers Association (AFBMA) may be programmed on either a personal computer or a programmable pocket calculator. If a personal computer is used the tables shown in ANSI/AFBMA Std. 11-1978 (Revision and Redesignation of ANSI B3.16-1972) may be stored as files. Programming for roller bearing selection is simpler than that for ball bearings because no interpolation is required. As stated in the corresponding selection section for ball bearings, file recall and output subprograms are not explicitly shown in the flowchart in Figure 10.7.1 in order to emphasize the selection process. Again, it is assumed that the programmer will add these according to the operating requirements of the particular calculator or computer being used.

Reference to Section 10.8 found in the flowchart is concerned with what has been termed *opposition mounting* of a pair of angular contact tapered roller bearings and the associated auxiliary formulas to be used before calculating bearing life. Use of a computer routine is especially helpful in selecting bearings for such a mounting because of the repeated trials that may be necessary.

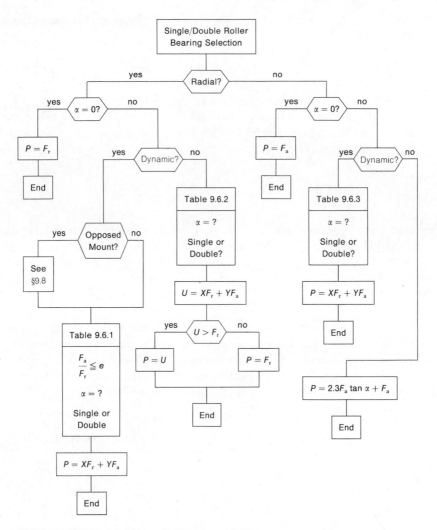

Figure 10.7.1 Flowchart for roller bearing selection.

10.8 OPPOSED MOUNTS, PRELOADING

Opposed Mounts

Both angular contact ball and tapered roller bearings may be used in opposed mounts, as pictured in Figure 10.8.1, to maintain the axial position of a shaft when it is acted upon by axial forces. They may be mounted either back to back, as shown, or face to face.

We begin our analysis of the forces on each bearing by adopting the convention that bearing 1 will be the bearing that resists the axial load, as labeled in Figure 10.8.1. Its radial force and induced axial load will be designated by F_{r_1} and I_1 and the radial force and induced axial load on bearing 2 by F_{r_2} and I_2, respectively. We first suppose

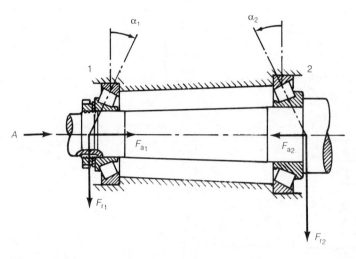

Figure 10.8.1 Angular contact roller bearings in opposed mounting (back-to-back arrangement).

that

$$I_1 > I_2 \tag{10.8.1}$$

where I_1 and I_2 are given by equation 10.4.6 for angular contact ball bearings and by equation 10.6.5 for tapered roller bearings.

If I_1 acts on one part of the shaft before the application of an axial load then bearing 2 must resist axial motion of the shaft by exerting an axial load I_1 as well.

If the applied axial load A is equal to or smaller than $I_1 - I_2$ it will relieve some or all of the excess load on bearing 2 while leaving the load on bearing 1 unchanged. Thus, we may write the axial load on each bearing as

$$\left.\begin{array}{l} F_{a_1} = I_1 \\ F_{a_2} = I_1 - A \end{array}\right\} \quad \text{for } A \leq I_1 - I_2 \tag{10.8.2}$$

which may be verified by drawing a free-body diagram as in Figure 10.8.2a. When A becomes larger than $I_1 - I_2$ it increases the load on bearing 1 to $I_2 + A > I_1$ without further reducing to load on bearing 2, so that, as may be verified from the force diagram in Figure 10.8.2b,

$$\left.\begin{array}{l} F_{a_1} = I_2 + A \\ F_{a_2} = I_2 \end{array}\right\} \quad \text{for } A > I_1 - I_a \tag{10.8.3}$$

The tolerances and strength of the housing and shaft must be such that the axial load on bearing 2 does not fall below the induced load because a lower force would allow the inner and outer races to move axially and possibly damage the bearing as a result.

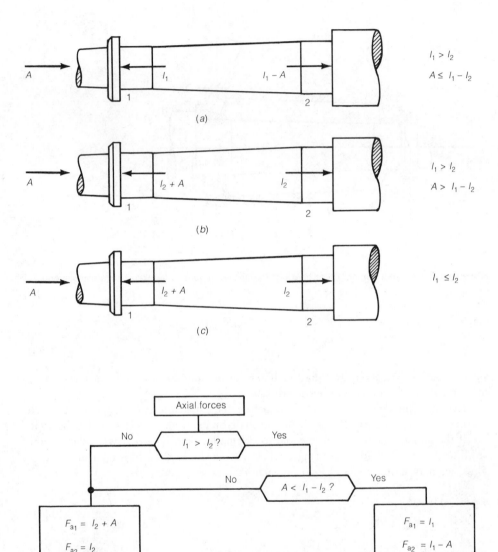

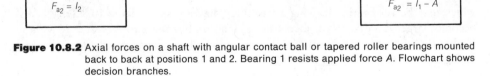

Figure 10.8.2 Axial forces on a shaft with angular contact ball or tapered roller bearings mounted back to back at positions 1 and 2. Bearing 1 resists applied force A. Flowchart shows decision branches.

Whenever $I_1 \leq I_2$ the axial loads on the bearings for all A values are, for properly designed shaft and housing, given by

$$\left. \begin{array}{l} F_{a_1} = I_2 + A \\ F_{a_2} = I_2 \end{array} \right\} \quad \text{for } I_1 \leqq I_2 \qquad (10.8.4)$$

as may be verified from Figure 10.8.2c.

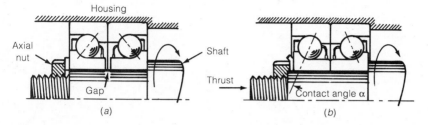

Figure 10.8.3 Angular contact ball bearings mounted back-to-back (*a*) before preloading, and (*b*) after preloading by tightening the nut on the shaft itself.

Preloading

Although angular contact ball bearings are often mounted either face to face or back to back to prevent axial motion of the shaft due to axial loads, as in Figure 10.8.3, some axial motion of the shaft will occur due to compression of the bearing combination itself. For ball bearings the axial force vs. axial displacement is nonlinear, as represented by the curve in Figure 10.8.4.

Since the slope of the curve in Figure 10.8.4 increases with increasing load, it follows that a given load would produce a smaller bearing deflection, and, hence, a smaller axial deflection, if we could start at point *A* instead of at the origin. We can very

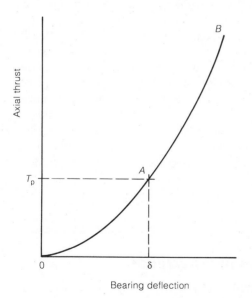

Figure 10.8.4 Thrust–deflection curve for a typical angular contact ball bearing. *Source: T. A. Harris, Rolling Bearing Analysis, 2nd ed., New York: Wiley-Interscience, 1984, p. 285. Copyright 1984 John Wiley and Sons, Inc. Reprinted with permission.*

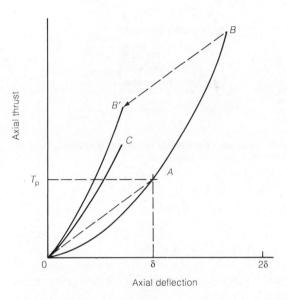

Figure 10.8.5 Thrust–deflection curve OB for an angular contact ball bearing, the associated thrust–deflection curve OB' relative to the applied load for a preload T_p, and the thrust–deflection curve OC for an opposed pair of angular contact ball bearings.

nearly do this by tightening the nut in Figure 10.8.3 to put both bearings in compression due to preload T_p, which compresses each bearing by an amount δ.

At this point we consider that the deflection shown in Figure 10.8.4 was due to deformation of the balls and races but not to deformation of the rings. We further assume that the shaft, housing, and rings are essentially rigid. Thus, axial loads applied to the shaft will only compress the left-hand bearing. Hence, the force displacement relation will follow section AB of the curve in Figure 10.8.4, which now moves to the origin of a plot of axial thrust versus the axial displacement of the left side of the shaft relative to the housing as shown by curve $0B'$ in Figure 10.8.5. If we relax our assumption that the rings, housing, and shaft are rigid, the thrust–displacement curve may follow curve $0C$ rather than $0B'$. The thrust displacement curve $0AB$ without preload has been included for comparison.

Tapered roller bearings are usually not preloaded because their axial force–axial displacement relation is essentially linear.

10.9 SINGLE- AND DOUBLE-DIRECTION THRUST BEARINGS

Thrust bearings are said to be single or double direction depending upon whether the bearing can resist axial forces in one or two directions when only one ring or thrust washer is restrained. Tapered roller thrust bearings in Figure 10.2.5 are all single-direction bearings, as illustrated in Figure 10.9.1*a*. Double-direction thrust bearings

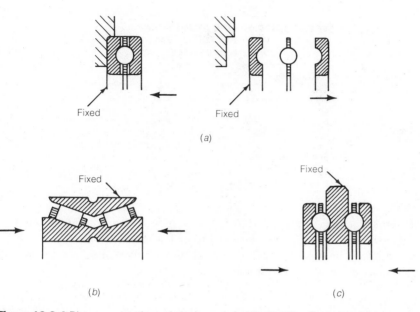

Figure 10.9.1 Ring cross sections of single- and double-direction thrust bearings.

with $\alpha = 90°$ are not common. One may be designed as in Figure 10.9.1c, having three thrust washers and two rows of balls (or rollers).

Although not so labeled, a double-row tapered roller bearing can serve as a double-direction thrust bearing, especially if α is large, as illustrated in Figure 10.9.1b.

Angular contact ball bearings may be considered as radial bearings for $\alpha \leqq 45°$ and as thrust bearings for $\alpha > 45°$. One angular contact thrust ball bearing by itself is a single direction thrust bearing; two angular contact thrust ball bearings mounted either face to face or back to back together made a double-direction ball thrust bearing.

10.10   LIFE ADJUSTMENT FACTORS

A life adjustment factor a_{1_j} is defined as the ratio between L_j life and L_{10} life. Thus

$$a_{1_j} = \frac{L_j}{L_{10}} \tag{10.10.1}$$

We have already mentioned that $a_{1_{50}} = 5$ for both ball and roller bearings. Ball and roller bearings also have equal life adjustment factors for L_1 through L_5, which are listed in Table 10.10.1 as given in Ref. 3 and 4.

The double subscripts have been used to refer to specific entries in Table 10.10.1. The industry practice is to refer to all of these a_{1_j} simply as a_1 and to denote to life adjustment factors for other effects by successive subscripts: a_2 for bearing material

TABLE 10.10.1	LIFE ADJUSTMENT FACTOR a_{1_j} For RELIABILITY	
Reliability (%)	L_j	a_{1_j} a_1
90	L_{10}	1.00
95	L_5	0.62
96	L_4	0.53
97	L_3	0.44
98	L_2	0.33
99	L_1	0.21

(sometimes termed bearing quality) and a_3 for lubrication. Factor a_2 may be equal to or greater than 1.0 and factor a_3 may be equal to or less than 1.0.

These factors are multiplicative, so that if all three of those mentioned were applicable the expected life would be estimated from

$$L'_j = a_{1_j} a_2 a_3 L_{10} \tag{10.10.2}$$

Since life adjustment factors a_2 and a_3 are provided by the manufacturer we shall use only a_1 in the remainder of this chapter and shall refer to it as a_{1_j}.

10.11 UNSTEADY LOADING

Bearings frequently operate under conditions of fluctuating loads in machines such as engines, motors, reciprocating pumps, vehicles, and many home appliances. Moreover, both radial and axial loads often repeat after a certain number of cycles, represented by N_o. The sequence and values of the loads during these N_o cycles will be referred to as the *load schedule*. Upon combining axial and radial loads to form the load vector **F** shown in Figure 10.11.1 and combining X and Y into bearing vector **B** the equivalent radial load for a fluctuating **F** vector may be calculated from[2]

$$P^\kappa = \frac{1}{N_o} \int_0^{N_o} (\mathbf{F} \cdot \mathbf{B})^\kappa \, dN \tag{10.11.1}$$

where

$$\mathbf{F} = F(\hat{\mathbf{i}}_r |\cos \psi| + \hat{\mathbf{i}}_z |\sin \psi|) \tag{10.11.2}$$

and

$$\mathbf{B} = \hat{\mathbf{i}}_r X + \hat{\mathbf{i}}_z Y \tag{10.11.3}$$

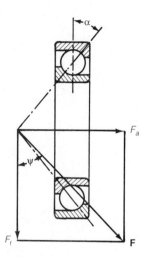

Figure 10.11.1 Force **F** whose components are F_r and F_a and its angle ψ with F_r.

where $\hat{\mathbf{i}}_r$ and $\hat{\mathbf{i}}_z$ represent unit vectors in the positive radial and axial directions, respectively, and $\kappa = 3$ for ball and $\frac{10}{3}$ for roller bearings. Absolute values of the sine and cosine are used in **F** because all radial forces have a similar effect on bearing life and all axial forces have a similar effect on bearing life. Since there is no directional effect, the radial and axial components of **F** should not change sign as the force acting on the bearing changes direction and magnitude during the load schedule.

If ψ is constant equation 10.11.1 may be written as

$$P = (X|\cos \psi| + Y|\sin \psi|)(F^*)^{1/\kappa} \tag{10.11.4}$$

where

$$F^* = \frac{1}{N_o} \int_0^{N_o} F^\kappa \, dN \tag{10.11.5}$$

Whenever F may be idealized as in Figure 10.11.2a, the integral in equation 10.11.5 may be approximated by

$$\sum_{i=1}^{m} F_i^\kappa N_i$$

so that substitution from (10.11.5) into (10.11.4) yields

$$P^\kappa = (X|\cos \psi| + Y|\sin \psi|) \frac{1}{N_o} \sum_{i=1}^{m} F_i^\kappa N_i \tag{10.11.6}$$

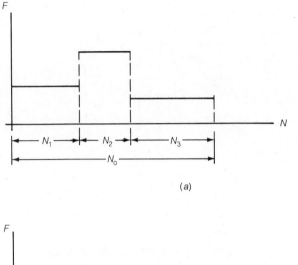

(a)

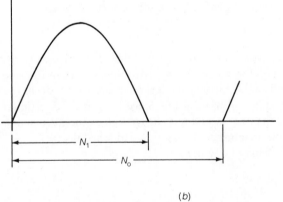

(b)

Figure 10.11.2 Idealized loads: (a) step-function profile, (b) sine profile.

EXAMPLE 10.11.1

Find the equivalent radial load for the load schedule shown in Figure 10.11.2b for a radial ball and for a radial roller bearing. The force is radial and follows and varies as a half sine wave.

We will assume the load is radial because a radial roller (cylindrical roller) bearing can support no axial load. Thus, set $\psi = 0$ and from the figure write $F = \mathcal{F} \sin(N\pi/N_1)$ so that the radial and axial components of the force are given by

$$F_r = F \cos \psi = \mathcal{F} \sin \frac{N\pi}{N_1}$$

$$F_a = F \sin \psi = 0$$

Substitution into (10.11.1) for F yields

$$P^\kappa = \frac{\mathscr{F}^\kappa}{N_o} \int_0^{N_1} \left(\sin \frac{N\pi}{N_1}\right)^\kappa dN = \frac{\mathscr{F}^\kappa N_1}{\pi N_o} \int_0^\pi \sin^\kappa x\, dx$$

after making the substitution $x = \pi N/N_1$. Analytical evaluation for the case when $\kappa = 3$ for ball bearings produces.

$$P^3 = \mathscr{F}^3 \frac{4}{3\pi} \frac{N_1}{N_o}$$

Upon taking the cube root we find that

$$P = 0.752 \left(\frac{N_1}{N_o}\right)^{1/3} \mathscr{F} \tag{10.11.7}$$

Numerical methods must be used when $\kappa = \frac{10}{3}$ to obtain

$$P = 0.620 \left(\frac{N_1}{N_o}\right)^{0.30} \mathscr{F} \tag{10.11.8}$$

The value given in (10.11.8) was obtained by trapazoidal integration. In many engineering applications both F and ψ are functions of N, as in the problems at the end of this chapter.

When the shaft on which a rolling element bearing is mounted does not make a complete revolution, but rotates from $+\phi$ to $-\phi$, where ϕ is less than π (or 180°), as does the robot arm of an automatic spot welding machine, the equivalent radial load may be satisfactorily approximated by multiplying equation 10.11.1 by ϕ, where

$$\phi = 2\phi/\pi \qquad \phi \text{ in radians}$$

or $\hspace{10cm} (10.11.9)$

$$\phi = \phi/90 \qquad \phi \text{ in degrees}$$

and where ϕ denotes the angular amplitude of the oscillation. The resulting expression is a valid approximation only if angle ϕ also obeys the conditions that[2]

$$\phi > 90/Z \qquad \phi \text{ in degrees}$$

or $\hspace{10cm} (10.11.10)$

$$\phi > \pi/(2Z) \qquad \phi \text{ in radians}$$

where Z represents the number of rolling elements in each row of the bearing. Thus, for oscillatory motion in which the amplitude of the rotation satisfies (10.11.10) we have that

$$P^{\kappa} = \frac{\phi}{N_o} \int_0^{N_o} (XF_r + YF_a)^{\kappa} \, dN \qquad (10.11.11)$$

Whenever the condition on ϕ given by inequality 10.11.10 is not satisfied the surfaces of the races and rolling elements will deteriorate due to a type of wear known as false brinelling unless the surfaces are lubricated with either oil or a light grease.

10.12 BOUNDARY DIMENSIONS

Rolling element bearings are relatively low priced in relation to the high-precision grinding, polishing, and metallurgical control needed for their manufacture largely because of national and international agreement on a Basic Plan for the Boundary Dimension of Metric Bearings. This standard prescribes a series of standard radial dimensions for the outside diameter and the bore (the inside diameter) and a series of width dimensions for each radial dimension. Common members of the radial series for the outside diameter are designated as 8, 9, 0, 1, 2, 3, 4, with 8 the smallest, and common members of the width series are designated 0, 1, 2, 3, 4, 5, 6, with 0 the smallest. Bore, outside diameter, and width of a bearing may then be defined by three digits, such that a 203 bearing has a series 2 bore, a series 0 outside diameter, and a series 3 width. Unfortunately there is no simple relation between the series designation and the actual dimensions: A table is required to relate the two. Because of this, and the practice of most bearing manufacturers to list bearing according to their own numbering system, boundary dimension designation of bearings has not been used in this text.

10.13 NEEDLE BEARINGS

Needle bearings are long roller bearings which are generally supplied without the inner ring and with or without a cage to minimize the radial dimensions of the housing which encloses the bearing. The load bearing capability of needle bearings that have no inner ring will be reduced and depends upon the hardness of the shaft upon which they roll. Consequently, the basic (dynamic) load factor may be supplemented with a table of *hardness factors*, where the hardness factor is defined as the ratio of the reduced basic load rating to the basic load rating. Similarly, the static load rating may be supplemented with a table of *working load limits*, where the working load limit is defined as the ratio of the reduced static load rating to the static load rating. Tables of hardness factor and working load limits as a function of the shaft surface hardness are given in Ref. 7.

10.14 LUBRICATION OF ROLLING ELEMENT BEARINGS

The primary functions of the bearing lubricant are to reduce friction between the retainer, or cage, and other parts of the bearing and between the race and the rolling elements when true rolling does not take place, as may occur in cylindrical roller thrust bearings. Secondary functions are to protect the highly finished surfaces from corrosion, to aid in cooling, and where heavier lubricants are used, to obstruct entry of foreign matter. Oil is the preferred lubricant because it can be filtered, heated, cooled, circulated, and applied as a bath, an oil jet, or a mist.

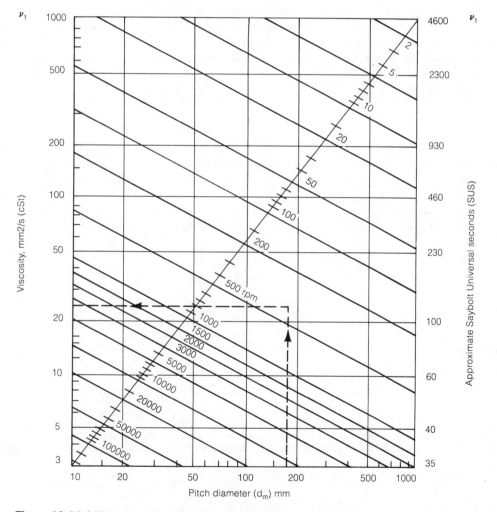

Figure 10.14.1 Kinematic viscosity ν_1 as a function of mean bearing diameter d_m and shaft rotational speed. Use only mm²/s scale. The SUS scale is approximate because of temperature variation of oil density ρ. *Source: SKF Industries, Bearings Group, King of Prussia, PA.*

The Annular Bearing Engineers' Committee has written that[2]

> It has been found that the friction torque in a bearing is lowest with a very small quantity of oil, just sufficient to form a thin film over the contacting surfaces, and that the friction will increase with greater quantity and with higher viscosity of the oil. With more oil than just enough to make a film, the friction torque will also increase with the speed.

One bearing manufacturer, SKF,[3] has conducted research that indicates that factors a_2 and a_3 in equation 10.10.2 may be combined into a single factor a_{23}, at least for their bearings. Factor a_{23} may be estimated from the following figures.

The minimum required lubricant viscosity may be found from Figure 10.14.1, the kinematic viscosity as a function of temperature may be found from Figure 10.14.2 for the oil to be used, and the combined life adjustment factor a_{23} may be read from Figure 10.14.3. Figures 10.14.1 and 10.14.2 include a consideration of film thickness and surface finish and, therefore, may vary from one manufacturer to another.

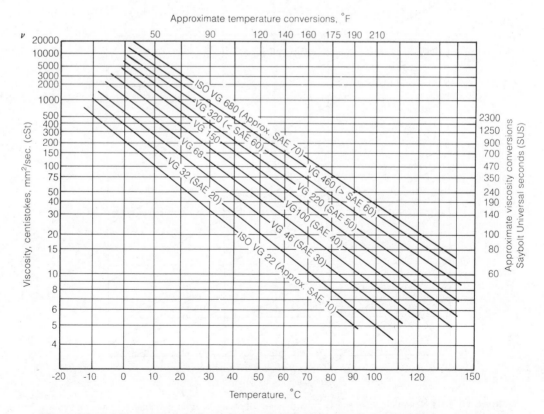

Figure 10.14.2 Kinematic viscosity v of lubricating oils with viscosity index of 95 as a function of temperature. Use mm²/s scale only. *Source: SKF Industries, Bearings Group, King of Prussia, PA.*

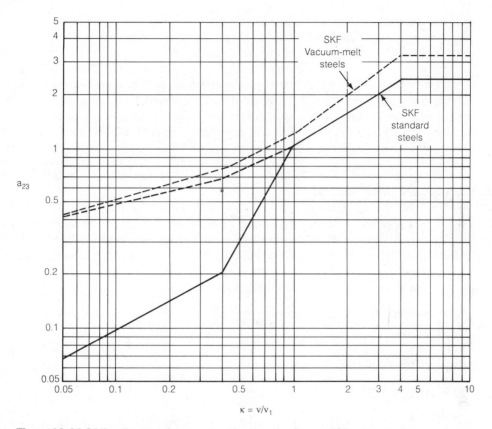

Figure 10.14.3 Life adjustment factor a_{23} as a function of κ. *Source: SKF Industries, Bearings Group, King of Prussia, PA.*

The procedure for relating a_{23} to the lubricant selected is to calculate the mean bearing d_m from its defining relation

$$d_m = (D + d)/2$$

where D and d denote the outside and bore diameters respectively. Next, enter Figure 10.14.1 at this value, read up to the shaft speed in rpm, and then read the kinematic viscosity v_1 in mm²/s from the left-hand scale. The third step is to enter Figure 10.14.2 at the operating temperature of the bearing, read up to the particular oil to be used (given in ISO grades and their approximte SAE equivalents), and then read the kinematic viscosity v at that temperature from the left-hand scale. Calculate the ratio

$$\kappa = v/v_1$$

and enter Figure 10.14.3 at this value, read up to the appropriate curve, and then read the appropriate a_{23} value from the left-hand scale. The upper dashed curve is for special steels, as labeled, and the solid curve is for standard steels.

Oils when an EP additive are recommended for v/v_1 ratios < 1, and must be used for v/v_1 ratios < 0.4. The lower viscosity ratio decreases the life of the bearing by an amount that depends upon the effectiveness of the EP additive. Thus, factor a_{23} would be read from the lower dashed curve AB for an effective EP additive and somewhere between the dashed curve AB and the solid curve below it for EP additives that are less effective to entirely ineffective.

EXAMPLE 10.14.1 (from Ref. 8)

Find the L_{10} life of a spherical roller bearing having a basic load rating of 466 000 N when supporting an equivalent static radial load of 45 000 N when lubricated with an SAE 40 (VG 100) oil with EP additives and operating at 440 rpm and 62 °C. The bore diameter is 120 mm and the outside diameter is 215 mm.

According to equation 10.3.2 the L_{10} life is given by

$$L_{10} = \left(\frac{466\,000}{45\,000}\right)^{10/3} \times 10^6 = 2420.535 \times 10^6 \text{ cycles}$$

After calculating the mean bearing diameter to be

$$(120 + 215)/2 = 167.5 \text{ mm}$$

enter Figure 10.14.1 (tics on the diagonal line mark intermediate rpm) at 167.5 mm, read up to the 440 rpm line (supplied with a straight edge), and then read to the right to find $v_1 = 23$ mm²/s. Enter Figure 10.14.2 at 62 °C to read a kinematic viscosity of 35 mm²/s for SAE 40 oil. Calculate

$$\kappa = v/v_1 = \tfrac{35}{23} = 1.522$$

and enter Figure 10.14.3 at $\kappa = 1.5$, read up to the

$$a_{23} = 1.2$$

with the aid Figure 1.6.1 in Chapter 1. According to equation 10.10.2 the L_{10} life becomes

$$L_{10} = 1.2(2420.535) \times 10^6 = 2904.642 \times 10^6 \text{ cycles}$$

for a 20% increase in rated life due to good lubrication

We did not use $\kappa = 1.522$ because the additional accuracy is not only useless for the scale involved in Figure 10.14.3 but also implies more precision than the method provides because of the viscosity variation permitted for SAE 40 oil.

10.15 MOUNTING DESIGN

Mount design refers to the design of the shaft that holds the inner race of the bearing, the design of the housing that holds the outer race, and the selection of the bearing itself. Because of the wide variety of applications, bearing types, maintenance requirements to be considered, and similar factors, it is not possible to give a short, general rule for the design of all bearing mounts. With this reservation in mind, however, we can observe that many bearing mounts may be classified as either a fixed mount or a float mount.

A fixed mount is one that locates the shaft relative to the housing by resisting any axial thrust expected, while a float mount is one that allows small axial motion between shaft and housing. These are often used in fixed–float, fixed–fixed, and float–float combinations. The fixed–float combination should be used for long shafts, for cases where thermal expansion of the housing and shaft are different, or for separate housings on a common shaft. Float–float combinations are usually used when some component on the shaft determines the axial position of the shaft, so that more than one fixed bearing could cause additional stresses in the shaft and housing. Figure 10.15.1 shows an application where a float–fixed mounting on one shaft in a herringbone gear train is used to allow for thermal expansion while fixing the position of the shaft relative

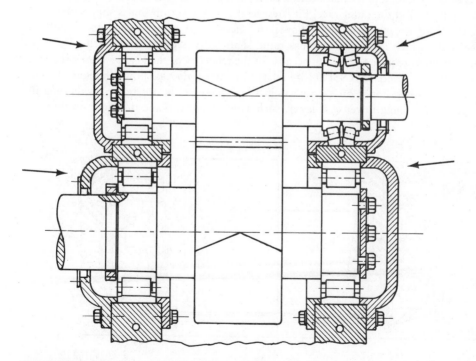

Figure 10.15.1 Representative fixed–float (upper) and float–float(below) bearing mounts.
Source: Courtesy of The Torrington Co., Torrington, CT.

to one side of the housing. Proper meshing of the gears is assured by using a float–float mounting on the second shaft so that it may move slightly in either axial direction.

The two-row tapered roller bearing at the right of the upper shaft fixes the position of that shaft because the outer race is held in place in the housing by restraining shoulders on either side, the inner race is held in place on the shaft by shoulders on either side, and the races cannot move relative to one another because of the oppositely inclined tapered rollers. In contrast to this, the cylindrical roller bearing on the left of the shaft allows some axial motion because close examination of the figure shows that the outer race has no shoulder to restrain the axial motion of the rollers, which are held within the inner race by shoulders on either side. Likewise, similar, but larger, cylindrical roller bearings used on each end of the lower shaft allow for small axial motion in either direction.

Figure 10.15.2 shows a fixed–fixed mounting using two angular contact ball bearings of different size and capacity in a relatively high-speed application which experiences a large thrust load in one direction, a small thrust load in the other direction, and some radial load from the bevel gears (right and top). In this configuration thermal expansion is not a problem and the fixed–fixed mounting assures accurate mating of the gears.

Manufacturer's specifications should be followed for the clearance of interference fits to be used for a particular bearing between the shaft and the inner race and between the housing and the outer race. Fit is especially important on shafts supporting heavy radial loads because a loose fit may allow the inner ring to slowly rotate relative to the shaft, which in turn may cause rapid shaft wear, scoring, and cracks. Generally heavier loads require tighter fits because of the tendency of the load to loosen the inner bearing ring. Standard ANSI fits are related to one another as shown in Figure 10.15.3, where a leading upper case letter, as in K6, represents the tolerance on the housing bore and a leading lower case letter, such as h6, represents the tolerance on the shaft

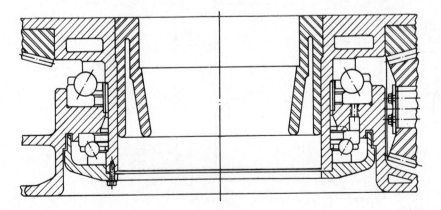

Figure 10.15.2 Representative fixed–fixed mounting using angular contact ball bearings. *Source: Courtesy of the Torrington Co., Torrington, CT.*

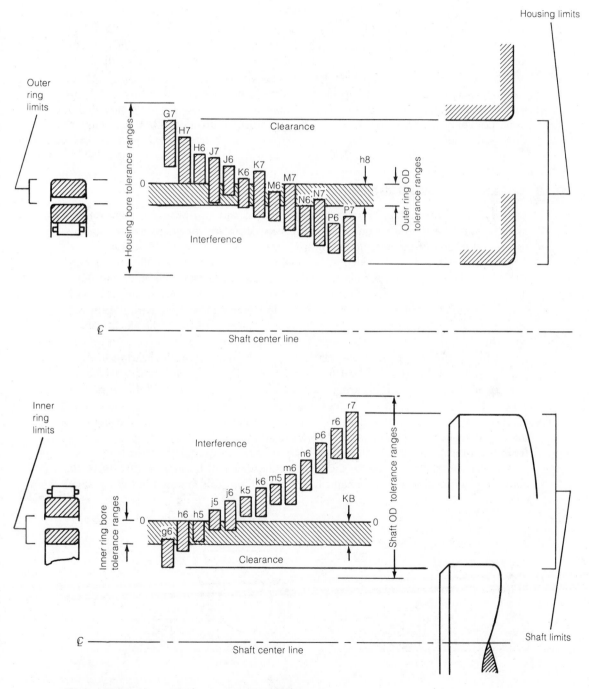

Figure 10.15.3 Interrelation of standard interference fits.

diameter. The numerical values associated with each tolerance is given in tables in Ref. 9, which are too extensive to be reproduced here. As indicated in Figure 10.15.3, a g6 fit is one in which the inner ring is easily displaced on the shaft, but a j5 fit is a light interference fit. Likewise a J7 fit is one in which the outer ring may be easily displaced axially, but a K7 fit is one in which it normally cannot.

Although angular contact ball bearings are often mounted either face to face or back to back to prevent axial motion of the shaft under axial load, as in Figure 10.8.4, some axial motion of the shaft will occur under thrust loading due to bearing compression by the load.

10.16 SEALS AND ENCLOSURES

Seals and enclosures retain lubricants in the bearing and also protect it from foreign matter—water, grit, and so on. The choice between a seal or an enclosure depends upon the location of the bearing mounting.

Typical bearing seals provided by one manufacturer, McGill,[9] are shown in Figure 10.16.1. These seals consist of plastic discs that fit in grooves in the inner and outer bearing rings and are equipped with a lip as shown in Figure 10.16.1a to provide both labyrinth (longer distance) and lip wipe, or friction, sealing. Harsher environments may require an additional metal shield which, as in Figure 10.16.1b. If the shield rotates with the shaft the shield also serves as a slinger because centrifugal force on foreign matter on the outside of the shield will cause it to be flung from the bearing.

The advantage of shielded bearings is that they may be exposed directly to the environment. If both sides of a bearing are sealed, the factory-installed lubricant may be expected to last for the life of the bearing if the seals are not mechanically damaged and if the bearings are not overheated to either damage the seals or cause the lubricant to expand enough to open the seal. The disadvantage of shielded bearings is that seal friction is often greater than the rolling friction of the bearing without seals.

Use of either a half housing, Figure 10.16.2a, or a full housing, Figure 10.16.2b, allows replenishment of the lubricant. Bearings with no seals are used in full housings

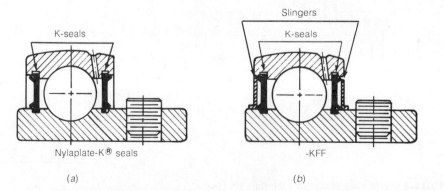

(a)   (b)

Figure 10.16.1 Representative bearing seals. *Source: McGill Manufacturing Co., Inc., Valparaiso, IN.*

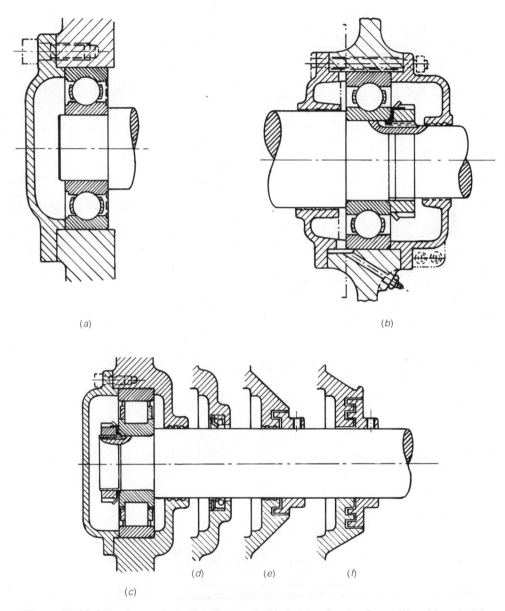

(a)   *(b)*

(d)   *(e)*   *(f)*

(c)

Figure 10.16.2 Housing configurations. *Source: SKF Industries, Bearing Group, King of Prussia, PA.*

to insure unrestricted access of the bearing to the lubricant in the housing. Labyrinth or wipe seal are placed between the enclosure and the shaft as shown in Figures 10.16.2*c* through *f*. Oil lubrication may be provided by a bath as shown in 10.16.3*a*, where the oil pool should not be above the middle of the rolling elements at their lowest position, or by an oil circulation system as in Figure 10.16.3*b*. Medium and large-size spherical

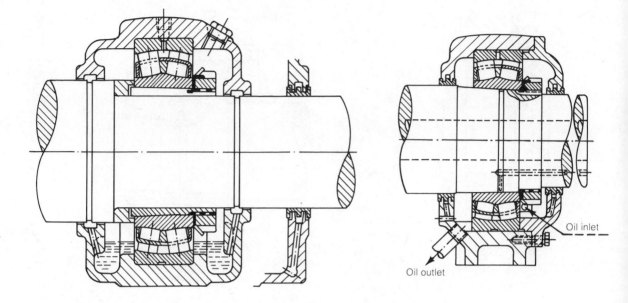

(a)

(b)

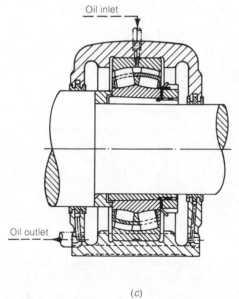

(c)

Figure 10.16.3 Enclosures and their associated lubrication systems. *Source: SKF Industries, Bearings Group, King of Prussia, PA.*

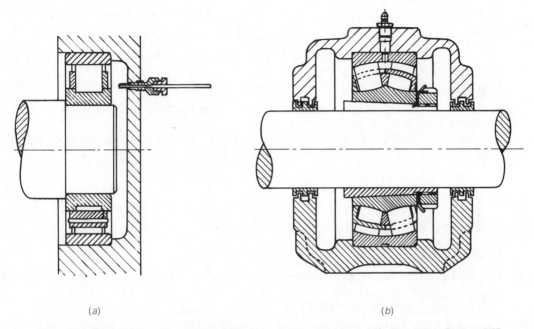

Figure 10.16.4 (a) Spray lubrication and (b) grease lubrication, heavy duty pillow block. *Source: SKF Industries, Bearings Group, King of Prussia, PA.*

roller bearings are usually designed for this type of lubrication and are fitted with oil grooves and holes in the outer ring, as in Figure 10.16.3c.

Ball and roller bearings operating at high speed are often lubricated with a mist from a spray nozzle as in Figure 10.16.4a. Grease fittings are provided on low-speed bearings and on bearings at remote locations whose service is severe enough that sealed, permanently lubricated bearings cannot be used. Pillow block bearing, shown in Figure 10.16.4b, exemplify bearing of this type. The housing usually contains the grease fitting and heavy duty models usually contain a spherical bearing to allow for shaft misalignment. The outer ring of the bearing does not rotate relative to the housing so the grease hole in the outer ring will remain aligned with the grease fitting in the housing. With this arrangement new grease entering the bearing forces old grease out through the labyrinth seal and thus purges any dirt or water that may have begun to penetrate the seal.

REFERENCES

1. *Torrington Bearings, Catalog 1269*, The Torrington Co., South Bend, IN, 1969.

2. *SKF Engineering Data*, SKF Industries, Philadelphia, PA, 1973.

3. *Load Ratings and Fatigue Life for Roller Bearings*, ANSI/AFBMA Std. 11-1978 (in effect in 1987).

4. *Load Rating and Fatigue Life for Ball Bearings*, ANSI/AFBMA Std. 9-1978 (in effect in 1987).

5. *Bearing Technical Journal*, FMC Corporation, Bearings Division, Indianapolis, IN, 1973.

6. *Rolling Bearings—Dynamic Load Ratings and Rating Life*, ISO Std. 281/1-1977, International Organization for Standardization.

7. *Torrington Bearings*, Catalog 576, The Torrington Co., South Bend, IN, 1976.

8. *Spherical Roller Bearings*, Catalog 310-11, SKF Industries, Bearing Group, King of Prussia, PA, June 1983.

9. *Shaft and Housing Fits for Metric Radial Ball and Roller Bearings (Except Tapered Roller Bearings) Conforming to Basic Boundary Plans*, ANSI/AFBMA B3.17-1973.

PROBLEMS

Section 10.3

10.1 Figure 10.3.1 is the basis for the relation that $L_{50} = 5L_{10}$. Use this figure to find the relation between L_{80} and L_{10} and between L_{90} and L_{10}.
[*Ans.* $L_{80} = 10L_{10}$, $L_{90} = 14L_{10}$]

10.2 Are ball or roller bearing lives more affected by a 20% increase in their radial loads? Compare ball and roller bearings with the same basic dynamic load ratings and radial loads.

10.3 What value of the dynamic load rating is required for a radial ball bearing to support a radial load of 770 N for 2.15×10^9 revolutions? Repeat for a radial roller bearing.
[*Ans.* 9939 N, 7696 N]

10.4 Find the value of C required for a radial roller bearing to support a radial load of 1050 N for 5 years when used on a machine in continuous service at 1725 rpm, 8 h/day, 5 days/week, 50 weeks/year.
[*Ans.* C = 8427 N]

10.5 A manufaturing company has collected detailed information on ball diameters and the number of balls per row for all radial ball bearings supplied by vendors who do business with them. This is to be cataloged in their design computer as an aid in selecting bearing size and manufacturer. What criteria should be programmed to decide whether to seek larger or smaller values of i, α, Z, or D to obtain a desired basic load rating?

10.6 Select a radial, single-row ball bearing having a bore between 65 and 90 mm to support a steady radial load of 1525 lb at 1725 rpm for an L_{10} life of two years at 50 weeks/year, 5 days/week, 8 h/day.
[*Ans.* 6215, Table 10.4.4]

Sections 10.3 and 10.4

10.7 Find the life of the bearing in problem 10.6 if the machine is modified by the user for an application in which an axial load of 925 lb is combined with the radial load.
[*Ans.* 7 mo., 21 days]

10.8 The sales department would like to offer an extended service model of the machine in problem 10.6 which is warranted for 6 years without a bearing change. (L_{10} life of 6 years) Can this be done without changing the shaft or housing diameters?

[*Ans.* Yes, bearing 5215, Table 10.4.5]

10.9 Since the warranty in problem 10.8 is based upon L_{10} life, the company must be prepared to repair 10% of the machines. The financial vice-president has object to this cost and has asked for a warranty period during which only 5% of the machines would need repair. Give your answer in integer years.

[*Ans.* 4 years]

10.10 What radial static load limit should be listed in the operator's manual for the machine in problem 10.8, based upon a safety factor of 4.0? Assume the bearing supports the entire load.

10.11 Would you approve using a single-row angular contact ball bearing with a contact angle of 15° to support a steady combined axial load of 320 lb and a radial load of 670 lb if the bearing's basic dynamic load rating is 4410 lb and its basic static load rating is 3750 lb if it is to have an average life (L_{50} of 6.5 years when used in a two-shift operation. Shaft speed is 870 rpm. Assume a 5 day week and a 50 week year.

[*Ans.* No. Its L_{50} life is 6.3 years.]

10.12 Select a self-aligning, double-row ball bearing for an L_{50} life of 15 000 h or slightly longer for a bench grinder and buffer operating at 1500 rpm. The double-ended rotor has a grinder on one end and a buffer on the other. Radial loads may reach 720 N. Although axial loading is to void the warranty, design for an axial load of 215 N and assume these forces act simultaneously for bearing selection. Bore diameter should not exceed 20 mm and the outside diameter should not exceed 55 mm.

[*Ans.* 1301 bearing, Table 10.4.7]

10.13 Customer relations has been asked if the unit described in problem 10.12 may be used in an application where the bearing will be subjected to a static radial load of 700 N and a static axial load of 810 N. Would this application void the warranty provided with the unit? (It will not void the warranty if the load does not damage the bearing.)

[*Ans.* It may be used.]

10.14 Select a ball thrust bearing with a bore ranging from 5.00 to 5.50 in. for the drive shaft of a coastal fishing boat to withstand a thrust of 7350 lb with an L_{10} life of 150 million cycles.

[*Ans.* 51 TVB 241, Table 10.4.8]

10.15 Modification of the boat originally designed for coastal service as in Problem 10.14 requires that the ball thrust bearing be replaced with an angular contact thrust ball bearing. What bearing would you suggest and what should be the new bore diameter?

[*Ans.* bore = 9.0 in., 90 TVL 710 AA642 bearing]

10.16 Estimate the reduction in life of the bearing in problem 10.15 due to the boat having been leased to an operator who repeatedly grounded the boat and/or fouled the propeller in seaweed and mangrove beds if the effect is equivalent to a radial load of 4000 lb on the thrust bearing.

[*Ans.* 60.2%]

10.17 Select a tapered radial roller bearing to support a radial load of 2010 N and an axial load of 1600 N for an L_{10} life of 360×10^6 revolutions or somewhat longer. Excessive life is undesireable because of added cost.

10.18 What is the L_2 life of the bearing selected in problem 10.17?

10.19 Select a 15° angular contact ball bearing to support an axial load of 490 lb, a radial load of 940 lb, and to provide an L_{10} life of 3100 h at 1050 rpm. Bore diameter may be from 40 to 65 mm and the outside diameter may be from 60 to 95 mm. See table below.

TABLE P10.19   ANGULAR CONTACT BALL BEARING BASIC DYNAMIC AND STATIC LOAD RATINGS

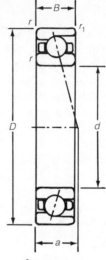

15° Contact angle

Bearing No.	d		D		B		a		Basic Load Ratings lbs	
	mm	in.	mm	in.	mm	in.	mm	in.	Dynamic C	Static C_o
7005C	25	.9843	47	1.8504	12	.4724	11	.43	2020	1470
06C	30	1.1811	55	2.1654	13	.5118	12.5	.49	2620	2010
07C	35	1.3780	62	2.4409	14	.5512	13.5	.53	3300	2660
7008C	40	1.5748	68	2.6772	15	.5906	14.5	.57	3550	3050
09C	45	1.7717	75	2.9528	16	.6299	16	.63	4200	3700
10C	50	1.9685	80	3.1496	16	.6299	16.5	.65	4500	4150
7011C	55	2.1654	90	3.5433	18	.7087	18.5	.73	5900	5450
12C	60	2.3622	95	3.7402	18	.7087	19.5	.77	6050	5800
13C	65	2.5591	100	3.9370	18	.7087	20	.79	6400	6450

Source: NTN Bearing Corp. of America, Des Plaines, IL.

10.20 Find the life of two bearings identical to that selected in problem 10.19 when mounted in tandem to support the load specified in that problem.

10.21 Show that if no additional axial load is applied to a single-row angular contact radial ball bearing its equivalent radial load is given by $P = F_r$ when the induced load is considered.

10.22 Find the L_{10} life of a double-row angular contact ball bearing having a contact angle of 12° when used to support a radial load of 295 lb and an axial load of 115 lb. Its static load rating is 1225 lb and its dynamic load rating is 1916 lb. *Hint:* Use linear interpolation to find X and Y for 10° and for 15° angular contact bearings for the F_a/C_o ratio for this bearing and then linearly interpolate for the 12° contact angle.

10.23 Compare the L_{10} life of the angular contact bearing described in problem 10.22 with one having a contact angle of 15° when subjected to the same loads and having the same basic static and dynamic load ratings.

Sections 10.3–10.6

10.24 Select a single-row radial roller bearing having a 100-mm bore to provide an L_{10} life of 3 years for 8 h/day, 5 days/week, 50 weeks/year at 135 rpm when supporting an axial load of 51 460 lb either alone or in conjunction with one or more identical bearings mounted in tandem.
[*Ans.* 100RN04, Table 10.6.4]

10.25 Select a roller thrust bearing for an L_{10} life of 200 000 h at 1800 rpm and an axial load of 4665 lb with a bore of 3.00 in.
[*Ans.* 30TP107, Table 10.6.7]

10.26 Find the maximum dynamic and static radial loads that may be supported by the bearing selected in problem 10.25.
[*Ans.* 0, 0]

10.27 It has been proposed that a self-aligning, double-row, spherical radial roller bearing be used to support a combined radial load of 47 100 N and an axial load of 9020 N. For the bearing suggested the basic dynamic load rating is 180 000 N and the catalog values of Y are 4.37 if F_a/F_r is greater than e and 2.94 if F_a/F_r is less than or equal to e, where $e = 0.23$. Find the approximate (because input data have been rounded) contact angle and the average life of the bearing.
[*Ans.* $\alpha = 8.71°$, $L_{50} = 98\,457\,600$ revolutions]

10.28 Would a static load of 80 000 N acting at 34° to the plane of the bearing damage it if its basic static load rating is 140 000 N?
[*Ans.* yes]

10.29 Find the L_{10} life of a single-direction, self-aligning roller thrust bearing, Figure 10.2.6, for which $\alpha = 70°$ and $C = 6440$ N when it is to support an axial load of 930 N and a radial load of 220 N.
[*Ans.* 119.259 million revolutions]

10.30 An automated, remote control, drilling rig is to be installed in the antarctic. During its operation it will place an axial load of 1 830 000 N and a simultaneous radial load of 841 000 N on a single-direction spherical roller thrust bearing. Would you approve the use of a bearing having a basic dynamic load rating of 4 650 000 N if the L_1 life is to be at least 1700 h with the bearing rotating at 10 rpm? The catalog entries for e, X, and Y are 1.818, 1.20, and 1.00 for $F_a/F_r > e$. Also calculate the contact angle for the bearing.
[*Ans.* $L_1 = 1812$ hrs, $\alpha = 50.475°$]

10.31 What is the maximum static load that the bearing in problem 10.30 can withstand at a load angle of 85° if $C_o = 3\,394\,000$ N?

10.32 Due to competitive pressure it has been decided that bearing 50TP121, Table 10.6.7, should be replaced by a tapered roller thrust bearing and that the L_{10} life should be increased by a factor of about 3.4 while holding the bore diameter to 5.00 in. and restricting the outside diameter to under 10.90 in., if possible. Which bearing in Table 10.6.8 should be selected to come closest to this goal? The static load capacity should not decrease.

10.33 Show that if no additional axial load is applied to a single-row tapered (angular contact) radial roller bearing that its equivalent radial load is given by $P = F_r$. *Hint:* Recall equation 10.6.1 and use the relations in Table 10.6.1.

10.34 Part *a* the figure below shows the head of a proposed rock drill in which the drill is held in chuck *D*. The drill extends 4 ft and may support an axial load of 11 300 lb and a radial load of 200 lb, as in part *b*. Which bearing would you select at *A*, *B*, and *C*? Bore diameter should be between 40 and 60 mm inclusive. Justify your choice.

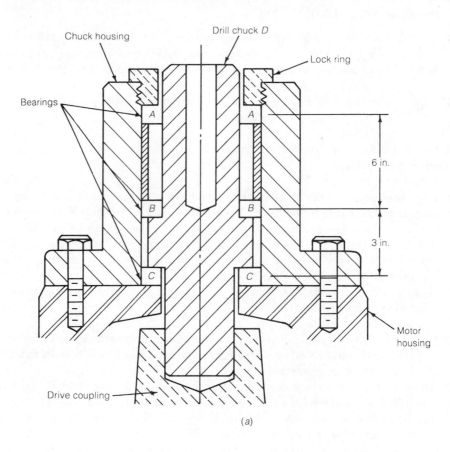

(a)

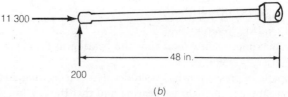

(b)

Choice	A	B	C
1	Ball thrust	Ball radial*	Ball radial*
2	Spherical roller	Tapered roller	Tapered roller
3	Radial roller	Tapered roller	Spherical thrust

* Support collar moved to A.

10.35 Is there another acceptable combination of the bearing types shown in problem 10.34? If so, what is it?

10.36 Select bearings from the following table for choice 3, problem 10.34, and calculate the L_{10} life of each. The tapered roller bearings at B is mounted facing the thrust bearing to hold the chuck

TABLE P10.36   DIMENSIONS AND LOAD RATINGS FOR AVAILABLE BEARINGS[a]

Bearing Type	Bearing Number	Bore (mm)	Outside Diameter (mm)	Basic Static Rating (lb)	Basic Dynamic Rating (lb)	α (°)
Ball thrust	51108	40	60	11 700	4 650	·
	51109	45	65	12 800	4 800	·
	51110	50	70	14 000	4 950	·
	51111	55	78	17 300	6 000	·
	51112	60	85	20 900	7 150	·
Radial ball	63/32	40	90	5 050	7 050	·
	6309	45	100	6 800	9 150	·
	6310	50	110	8 100	10 700	·
	6311	55	120	9 500	12 400	·
	6312	60	130	11 000	14 200	·
Spherical roller	22211B	55	100	15 600	18 300	10.5
	22212B	60	110	20 800	22 400	10.25
Radial roller or Cylindrical roller	208	40	80	6 200	8 950	·
	209	45	85	6 700	9 400	·
	210	50	90	7 750	10 300	·
	211	55	100	9 450	12 400	·
	212	60	110	11 400	14 700	·
Tapered roller or Angular contact roller	4T-30208	40	80	9 300	12 100	12.1
	30209U	45	85	10 800	13 500	14.2
	30210U	50	90	12 800	15 400	14.2
	30211U	55	100	15 300	18 600	14.2
	30212U	60	110	17 300	20 900	14.2
Spherical thrust	29421	60	130	90 500	49 500	51.0

[a] Abstracted from NTN Ball & Roller Bearings, cat. A1000-IV. Mixed Old English and SI units commonly appear in catalogs at this time.

in place. The chuck holder diameter at each bearing location may be altered to accomodate the bearing chosen. Design for an L_{10} life of 15 000 h at 120 rpm.

Section 10.8

10.37 It has been proposed that cones 09081 and cups 09196 (designated as bearing 09081/09196) be selected for a pair of angular contact tapered roller bearings to support a radial load of 3080 N on bearing 1 and of 3310 N on bearing 2. Axial load on the shaft is 1100 N. Find the induced axial load for each bearing and calculate the expected life of bearings 1 and 2 when mounted back to back as in Figure 10.8.1.

10.38 Repeat problem 10.37 with the bearings arranged face to face as shown in the figure. The radial load on bearing 1 is 3310 N and on bearing 2 is 3080 N. (Shaft loads are unchanged; bearings are interchanged.)

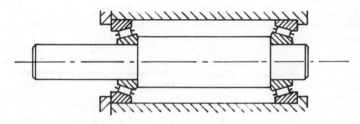

Opposed angular contact tapered roller bearings mounted face-to-face.

10.39 Find the L_{10} life of tapered radial roller bearings 09078/09195 and 070787/07204 when mounted face to face and loaded as shown in the figure.

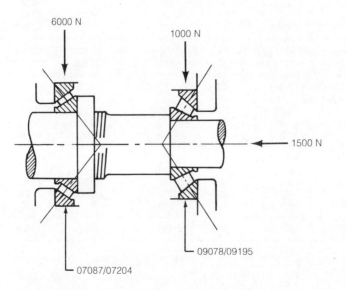

10.40 Which of the following two bearings should be selected to replace bearing 1 in problem 10.37 in order to have similar L_{10} lives for both bearings? This demonstrates the combined influence of α and C.

Number	Y	C
M88048/M88010	1.1	53 900 N
14123A/14283	1.6	44 600 N

10.41 Show that the equivalent radial load P for single- and double-row radial roller bearings varies continuously as the axial force F_a increases from $(F_a/F_r) \le 1.5 \tan \alpha$ to $(F_a/F_r) > 1.5 \tan \alpha$. *Hint:* Consider two cases: where $(F_a/F_r) = 1.5 \tan \alpha + \varepsilon$ and where $(F_a/F_r) = 1.5 \tan \alpha + \varepsilon$ and write $P = F_r(X + F_a Y/F_r)$.

Section 10.11

10.42 The buffing wheel in problem 10.12 was replaced by a truncated grinding cone which induced an axial loading of 400 N and a radial load of 340 N for 50 h. If the grinder had accumulated 1000 h before this violation of the warranty, how many hours of L_{50} life remain if it is to be used with the grinding cone? Base your calculations upon the bearing selected in problem 10.12.

10.43 Show that equation 10.11.4 is equivalent to the Palmgren-Miner relation. Recall that $N_i'/N_o = \alpha_i$ and that $L_{10} \times 10^6$ is the number of cycles to failure. Set $\psi = 0$ for simplicity.

10.44 Show that when n different radial loads act, and when each different radial load is constant during the time it acts, the equivalent radial load as given by equation 10.11.6 reduces to

$$P^\kappa = \frac{1}{N_o} \sum_{i=1}^{n} F_r^\kappa N_i$$

Use this result to select a radial roller bearing with 100-mm bore for an L_{10} life of 2.25 shift years when subjected to the loading shown in the figure. The shaft rotates at 575 rpm.

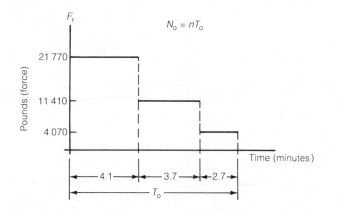

10.45 Show that for a radial ball bearing subjected to the radial load that varies as shown in the figure the equivalent radial load is given by

$$P^3 = \frac{N_1}{4N_o} (F_0 + F_1)(F_0{}^2 + F_1{}^2)$$

and that for a radial roller bearing it is given by

$$P^{10/3} = \frac{3N_1}{13N_o} \frac{F_1{}^{13/3} - F_0{}^{13/3}}{F_1 - F_0}$$

where F_0 and F_1 refer to radial forces. Would this relation hold for a ball thrust bearing where F_0 and F_1 refer to axial forces?

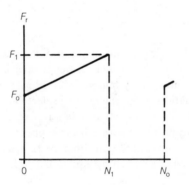

10.46 Show that for a radial ball bearing subjected to a radial load that varies as in the figure the equivalent radial load is given by

$$P = 2.504 \left(\frac{A}{4\pi} \frac{N_1}{N_o} \right)^{1/3}$$

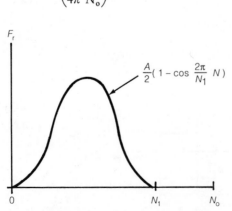

and that for a radial roller bearing subjected to the load shown in the figure the equivalent radial load is given by

$$P = 2.413 \left(\frac{A}{4\pi} \frac{N_1}{N_o} \right)^{0.3}$$

10.47 Show that the equivalent radial loads for radial ball and roller bearings subjected to a fluctuating radial load as illustrated in the figure may be found from

$$P = 1.988 \left(\frac{A}{4\pi} \frac{N_1}{N_o} \right)^{1/3} \quad \text{and} \quad P = 1.960 \left(\frac{A}{4\pi} \frac{N_1}{N_o} \right)^{0.3}$$

respectively.

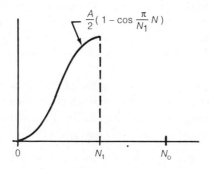

$$\frac{A}{2} \left(1 - \cos \frac{\pi}{N_1} N \right)$$

10.48 Find the minimum required basic dynamic capacity for a single-row, angular contact radial ball bearing with a contact angle of $20°$ to support the radial and axial loads for the duty schedule shown in the figure if it is to have a 30-year L_{10} life when in continuous operation at 300 rpm. Assume 365 day years (neglect the additional day during leap years).

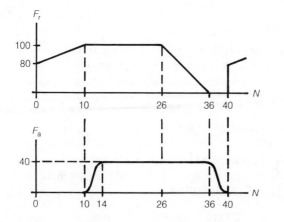

10.49 Select a single-direction, angular contact ball thrust bearing to support a steady axial load of 18 800 lb and a radial load which is zero from 0 to 21 revolutions of a 44 revolution duty cycle and is 10 800 lb from 21 revolutions to the remainder of the duty schedule. Its L_{10} life is to be 100 million revolutions.

[*Ans.* 120 TVL 700]

10.50 Ball thrust bearing 150 TYL 701, Table 10.4.9, has been proposed as a thrust bearing for the end of a horizontal boring machine. If the load acts at $\psi = 67°$ and if $F = 29\,330 \sin^2(2\pi N/700)$ lb, find the estimated life in hours when operated at 870 rpm. The contact angle is 45°.

10.51 Find the equivalent radial load for a single-row tapered radial roller bearing with a contact angle of 16° when it is subjected to a load given by $F = 500(100 - N)$ N acting at an angle $\psi = 2\pi N/100$. (The first N in the expression for F denotes the number of revolutions into the duty schedule; the second is the abbreviation for newtons. This is an example of conflicting conventional symbols.)

10.52 After repeated service calls resulting in replacement of double-row spherical roller radial (self-aligning) bearings the chief engineer has suggested that the L_{10} life may have been miscalculated and the overload switches improperly set. Specify the maximum load at 870 rpm for bearing 21308CC, Table 10.6.5, for an L_{10} life of 2200 h when the load angle is given by $\psi = 0.83 + 0.53 \cos(2\pi N/1740)$ radians, where N is the number of revolutions in a duty schedule in which $N_o = 1740$ revolutions.

10.53 Select a 100-mm-bore radial roller bearing to provide an L_{10} life of 5 years when operated at 690 rpm one shift/day, 5 days/week, 50 weeks/year on a ball mill which causes a radial load that may be represented by a half-sine wave having a peak value of 25 200 lb, as shown in the figure. The bearing outside diameter should be no greater than 250 mm.

[*Ans.* 100 RN 03]

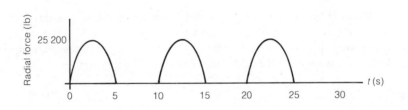

10.54 Select a double-row angular contact ball bearing for the swing arm of a workpiece transfer arm on a numerically controlled machine in which the bearing turns from $-20°$ to $+20°$ as given by $\phi = 20° \cos(360N)°$, where ϕ represents the angular position of the shaft. The force on the bearing is radial only, as given by

$$F_r = 3130 \cos\{35[1 - \cos(720N)°]\}° \text{ lb}$$

where N refers to the number of cycles. The shaft diameter may be from 20 to 60 mm and the L_{10} life should be in 5 years in continuous operation at 30 cycles/min. Which relation implies the number of cycles in the duty schedule? Why?

[*Ans.* ϕ, bearing 5207]

10.55 Find the minimum required basic dynamic capacity for a double-row angular contact ball bearing to have an L_{10} life of 60,000 h at 60 cycles/min where the shaft angle ϕ is given by $\phi = 30° \cos(360N)°$ and the force acting on the bearing is given by

$$F_r = 101 \cos\{35[1 - \cos(20N)°]\}° \text{ lb}$$

where N represents the number of cycles. Which relation implies the number of cycles in the duty schedule? Why?

[*Ans. F_r, C = 342 lb*]

Section 10.14

10.56 Recommend an ISO viscosity grade (VG) lubricating oil without an EP additive for bearing 22206CC, Table 10.6.5, when operating at 5000 rpm at 70°C. What is the resulting life adjustment factor a_{23}?

CLUTCHES AND BRAKES

NOTATION

A	area (l^2) and substitution variable	R,r	radius (l)
a	lever arm (l)	r_c	radius to the center of gravity (l)
B	substitution variable	S	slip for fluid couplings and torque converters
b	lever arm (l)	T	torque
C	constant	t	time
c	lever arm (l)	W	weight
d	diameter (l)	w	width
F	force (ml/t^2)	$\dot{x}$	dx/dt
I	mass moment of inertia (ml^2)	$\ddot{x}$	d^2x/dt^2
k	equivalent spring constant (m/t^2)	α	angle (l) or angular accelerating (l/t^2)
KE	kinetic energy (ml^2/t^2)	Γ	ratio of radii (l)
l	length (l)	Δ,δ	displacement (l)
M_a	moment required to activate a brake or clutch (ml^2/t^2)	$\Delta\omega$	increment of angular velocity (l/t)
		ε	error
M_f	moment due to friction ($ml^2 t^2$)	θ	angle (l)
M_p	moment due to pressure (ml^2/t^2)	λ	friction factor
m	mass (m)	μ	friction coefficient
N	number (l)	ρ	density (m/l^3)
n	revolutions/min; rpm (l/t)	Φ	angular momentum (ml/t)
$\mathfrak{n}$	kinetic energy	ϕ	angle (l)
q	angular speed ratio (l)	ω	angular velocity in radian/second (l/t)
P	force (ml/t^2)	Φ_i	impeller angular momentum (ml/t)
p	pressure (m/lt^2)	Φ_t	turbine angular momentum (ml/t)

Design of devices that can serve only as friction brakes are discussed in the first five sections, which include a section on computer design of a double-shoe drum brake. Design of friction-dependent devices that can serve both as clutches and as brakes are discussed in the next three sections. After studying how to design a clutch or brake to deliver a specified torque we turn our attention to procedures for deciding what size brake or clutch is needed for a particular application. The next three sections deal with one-way clutches, brakes and clutches not based upon friction, and torque converters. The chapter then closes with a comparison of the brakes and clutches discussed.

Brakes, clutches, and torque converters are grouped together because all three are often used together in power trains, because brakes and clutches operate on the same principles, and because torque converters, brakes, and one-way clutches form a symbiotic unit in a very common application, the automatic transmission, shown in Figure 11.0.1

In the following discussions of friction brakes and clutches we shall for simplicity, assume the coefficient of friction is constant. This assumption is generally true if the temperature of the brake lining varies only over a small range. Because of the effects of moisture, dirt, chemicals, and installation differences, it is difficult to duplicate service conditions in the laboratory. Consequently, standard tests of lining material, such as SAE J661a by the Society of Automotive Engineers, provide a standard means for both users and manufacturers to specify lining material characteristics. They are not for design purposes. Design data are often found by manufacturers based upon a combination of laboratory and field tests and user experience.

Representative data from brake manufacturers are shown in Figures 11.0.2 and 11.0.3 as an aid in selecting a material for its coefficient of friction and wear as a function of temperature. These properties depend both upon the primary material used and upon additives. The curves themselves were obtained from square samples 1×1 in. at approximately 100 psi and 1200 ft/min at controlled temperatures. Wear was calculated from weight loss rather than from direct thickness measurements because of swelling during the course of the tests.

As these graphs show, the choice of a lining is generally not clear-cut. From Figure 11.0.2 we find that a relatively high coefficient of friction may be associated with a relatively high wear, especially at higher temperatures, while improved wear performance is obtained at the expense of a lower friction coefficient which becomes even lower above about 450 °F.

Lining materials are often of asbestos-based materials where they present no health hazard. Other materials include wood, cast iron, polymers, and molded materials, such as phenolic resin, which may contain asbestos fibers, brass and /or zinc chips to aid heat conduction and stabilize friction, and other proprietary additives to improve friction and strength.

Heat conduction is important because as the lining temperature increases the friction usually decreases as shown in Figures 11.0.2 and 11.0.3, at which time the brakes are said to *fade*. As the brake cools the friction coefficient increases and the brake is said to *recover*. Similar comments apply to friction clutches.

Permissible lining pressure obviously depends upon the material used: It is of the order of 1.03 MPa (150 psi) for dry nonasbestos linings, of the order of 1.38 (200 psi)

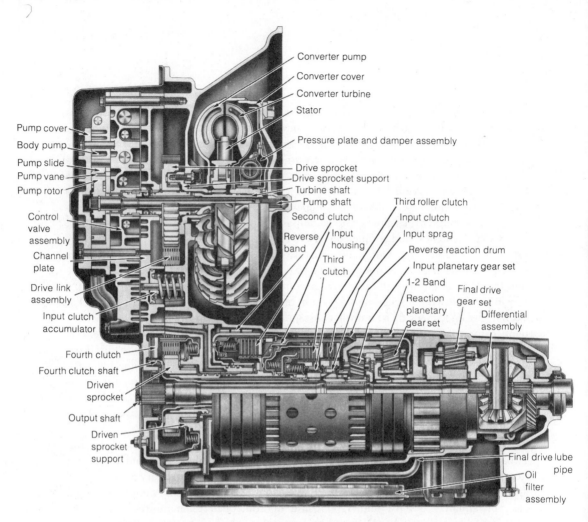

Figure 11.0.1 Cutaway view of an automatic transmission (transaxle). *Source: Hydramatic Division. General Motors Corp., Ypsilanti, MI.*

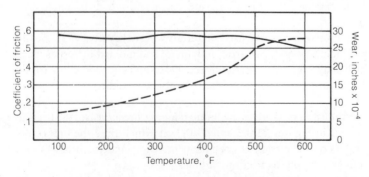

Figure 11.0.2 Representative friction and wear characteristics for a rigid molded lining material. *Source: Friction Material Company, Huntington, IN.*

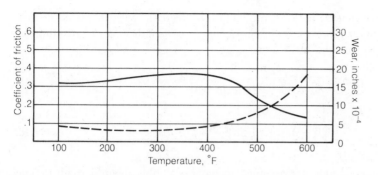

Figure 11.0.3 Representative friction and war characteristics for another rigid molded lining material. *Source: Friction Material Company, Hungtington, IN.*

for dry asbestos-filled material, and may rise to between 20 and 30 MPa (approximately 3000 to 4500 psi) for sintered metal pads for disc brakes.

11.1 BAND BRAKES

Band brakes, Figure 11.1.1, are simpler and less expensive than most other braking devices. Shoe brakes, to be discussed in the next section, are their nearest rivals for manufacturing simplicity and low cost.

Consistent with the direction of rotation ω shown in Figure 11.1.1, it follows that because of the friction force on an element of the brake lining due to pressure p over

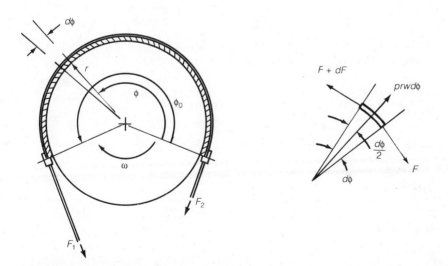

Figure 11.1.1 Band brake geometry and forces.

width w that the tangential force in the band will decrease with increased distance from F_1 by an amount $\mu pwr\,d\phi$ for each increment of length $r\,d\phi$. The equilibrium equations for an element are thus

$$(F + dF) \cos \frac{d\phi}{2} - F \cos \frac{d\phi}{2} - \mu prw\,d\phi = 0 \qquad (11.1.1)$$

$$[(F + dF) + F] \sin \frac{d\phi}{2} = prw\,d\phi \qquad (11.1.2)$$

where band and brake lining are assumed to have negligible flexural rigidity. Since $d\phi$ is very small, these relations become

$$dF = \mu prw\,d\phi \qquad (11.1.3)$$

and

$$F = prw \qquad (11.1.4)$$

after observing that $dF \sin d\phi/2 \to dF/(d\phi/2)$ is negligible compared to

$$2F \frac{d\phi}{2} = prw\,d\phi$$

in (11.1.2). Substitution for prw from (11.1.4) into (11.1.3) yields

$$dF/F = \mu\,d\phi$$

where μ is the coefficient of friction between the brake lining and the drum. Using the end conditions that $F = F_2$ when $\phi = 0$, write

$$\int_{F_2}^{F_1} \frac{dF}{F} = \mu \int_0^\phi d\phi$$

which leads to

$$\frac{F}{F_2} = e^{\mu\phi} \qquad (11.1.5)$$

at position ϕ, and to

$$\frac{F_1}{F_2} = e^{\mu\phi_0} \qquad (11.1.6)$$

at $\phi = \phi_0$, at which point $F = F_1$. Equation 11.1.4 may be used to find the maximum required band pressure p_{max} from

$$p_{max} = \frac{F_1}{rw} \tag{11.1.7}$$

Torque T exerted by the brake is related to the band force by

$$T = (F_1 - F_2)r \tag{11.1.8}$$

which may be rewritten as

$$T_{max} = (1 - e^{-\mu\phi_0})p_{max}r^2w \tag{11.1.9}$$

to give the maximum restraining torque as limited by the compressive strength of the lining, assuming that the band and links can withstand force $F_1 = F_{max}$.

A manually operated band brake may have a control level as pictured in Figure 11.1.2. When the direction of rotation is in the direction shown the moment equilibrium equation about the pivot point becomes

$$P(a + c) + F_2b = F_1a \tag{11.1.10}$$

Substitution from (11.1.6) for F_2 in terms of F_1 yields an expression for P in terms of the brake parameters and F_1:

$$P = \frac{a - be^{-\mu\phi_0}}{a + c}F_1 \qquad a > b \tag{11.1.11}$$

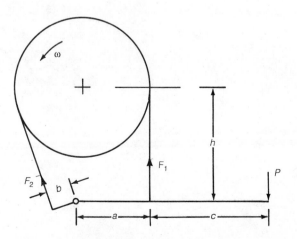

Figure 11.1.2 Geometry of a manually operated band brake.

Braking torque obtained by the application of force P may also be written in terms of F_1 as

$$T = (1 - e^{-\mu\phi_0})F_1 r \qquad (11.1.12)$$

Elimination of F_1 between (11.1.11) and (11.1.12) yields

$$P = \frac{a - be^{-\mu\phi_0}}{r(1 - e^{-\mu\phi_0})} \frac{T}{a + c} \qquad (11.1.13)$$

According to (11.1.13) P will become zero if $a = be^{-\mu\phi_0}$ and will become negative if $a < be^{-\mu\phi_0}$. In other words, no force will be required to apply the brake for the direction of rotation illustrated if $a = be^{-\mu\phi_0}$ and a force in the opposite direction from that shown will be required to prevent the brake from acting if $a < be^{-\mu\phi_0}$. Thus, the brake will be *self-locking* whenever

$$a \leqq be^{-\mu\phi_0} \qquad (11.1.14)$$

A self-locking brake is also known as a *backstop*. The simplest form of backstop is that shown in Figure 11.1.3, in which lengths a and b are chosen to satisfy (11.1.14) to resist the rotation pictured.

A more compact configuration is that shown in Figure 11.1.4, which also affords a greater wrap angle, thus increasing the braking torque available for the given drum diameter and band width. Both levers are on the triangular plate which is pivoted at the support bracket. See problem 11.1 for a schematic of this lever arrangement.

Figure 11.1.5 shows a typical application for a backstop, which is usually located at the foot end of the bucket elevator. It prevents the loaded buckets from reversing

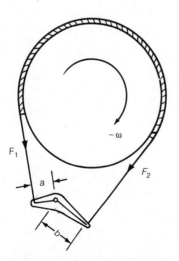

Figure 11.1.3 Single lever backstop (self-locking band brake).

Figure 11.1.4 Alternative backstop linkage. *Source: American Chain Association, Rockville, MD.*

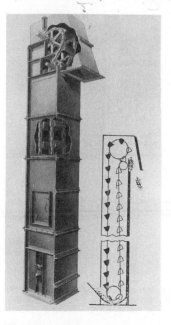

Figure 11.1.5 Bucket lift, or vertical elevator. Backstop and motor may be placed at head or foot end of the elevator. *Source: American Chain Association, Rockville, MD.*

the direction of the elevator and dumping their contents at its foot when the power is turned off at the end of a work day.

Simplicity is the attractive feature of the band brakes used in automatic transmissions. Typical brake band construction is that shown in Figure 11.1.6a and the actuation mechanism is that shown in Figure 11.1.6b, which employs a reaction rod and a hydraulic cylinder. The metal band that supports the lining material is also a one-turn spring which disengages the brake when hydraulic pressure is relieved.

Pressure variation over the length of the lining may be found to be

$$\frac{p_{max}}{p_{min}} = e^{\mu \phi_0} \tag{11.1.15}$$

from equations 11.1.4 and 11.1.6. This implies that lining wear will be greater at one end than at the other. Uneven lining wear is the price paid for band brake simplicity.

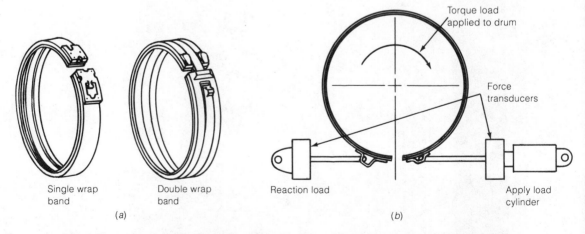

Single wrap band   Double wrap band   Reaction load   Apply load cylinder

(a)   (b)

Figure 11.1.6 (a) Typical brake band construction for an automatic transmission, and (b) representative schematic of the band brake actuation system used. *Source: Friction Material Company, Huntington, IN.*

11.2 PIVOTED EXTERNAL DRUM BRAKES

Figure 11.2.1 shows a representative external shoe brake. Analyses of lining pressure, torque exerted on the drum, and reactive moments on the brake differ from those of the previous section because of the rigidity of the brake shoe to which the lining is fastened.

Because of the elasticity of the lining itself, however, the pressure at any point will be proportional to its compressive displacement. It is evident from the geometry illustrated in Figure 11.2.1 that the motion in the radial direction of a point at location ϕ on the shoe due to a small rotation $\delta\alpha$ of the shoe about the pivot may be written as $R \sin \phi \, \delta\alpha$. Hence, lining pressure p may be expressed as

$$p = kR \sin \phi \, \delta\alpha \qquad (11.2.1)$$

Since $\sin \phi$ is the only variable in (11.2.1) it follows that the maximum pressure may be written as

$$p_{max} = kR(\sin \phi)_{max} \, \delta\alpha$$

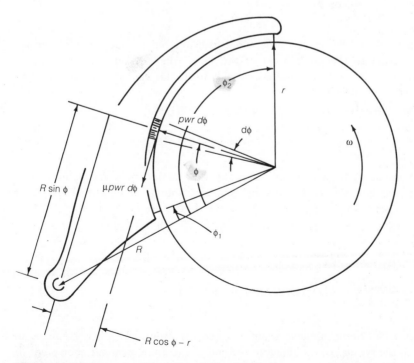

Figure 11.2.1 Geometry of an external shoe drum brake.

Proportionality constant k and rotation $\delta\alpha$ may be eliminated by dividing p by p_{max} to obtain

$$\frac{p}{p_{max}} = \frac{\sin\phi}{(\sin\phi)_{max}}$$

which may be rewritten as

$$p = \frac{p_{max}}{(\sin\phi)_{max}} \sin\phi \tag{11.2.2}$$

Torque exerted on a drum of radius r by the tangential force $\mu prw \, d\phi$ acting from heel to toe, i.e., from ϕ_1 to ϕ_2, is given by the integral

$$T = \mu r^2 w \frac{p_{max}}{(\sin\phi)_{max}} \int_{\phi_1}^{\phi_2} \sin\phi \, d\phi$$

which integrates to

$$T = \frac{\mu p_{max} r^2 w}{(\sin\phi)_{max}} (\cos\phi_1 - \cos\phi_2) \tag{11.2.3}$$

in which angles ϕ_1 and ϕ_2 are both measured from the radius that extends from the center of the drum to the pivot point of the shoe. Angle ϕ_1 extends to the near end, or heel, of the brake lining and ϕ_2 extends to the far end, or toe, of the brake lining, as shown in Figure 11.2.1.

The angle subtended by the brake lining may be denoted by ϕ_0, where

$$\phi_0 = \phi_2 - \phi_1 \tag{11.2.4}$$

Reactive moment M_f acting on the brake shoe about its pivot point due to friction may also be calculated from the geometry shown in Figure 11.2.1. After writing the incremental contribution from the lining element at ϕ as $dM_f = \mu pwr \, d\phi(R\cos\phi - r)$ we may integrate over the entire lining to obtain

$$M_f = \int_{\phi_1}^{\phi_2} \mu pwr \, d\phi(R\cos\phi - r)$$

$$= \frac{\mu p_{max} wr}{(\sin\phi)_{max}} \int_{\phi_1}^{\phi_2} (R\cos\phi\sin\phi - r\sin\phi) \, d\phi \tag{11.2.5}$$

which defines a *clockwise* moment about A when the drum rotation is from toe to heel, i.e., from the far end of the lining to the end of the lining near the shoe pivot. If the direction of drum rotation is reversed a positive M_f indicates a *counterclockwise* moment acting on the shoe about its pivot. Evaluation of (11.2.5) yields

$$M_f = \frac{\mu p_{max} w r}{4(\sin \phi)_{max}} [R(\cos 2\phi_1 - \cos 2\phi_2) - 4r(\cos \phi_1 - \cos \phi_2)] \qquad (11.2.6)$$

Radial force $prw\,d\phi$ due to pressure on the lining also contributes to a moment M_p about pivot A. This moment may be expressed in terms of the geometry shown in Figure 11.2.1 as

$$M_p = \int_{\phi_1}^{\phi_2} (pwr\,d\phi)R \sin \phi = \frac{p_{max} w r R}{(\sin \phi)_{max}} \int_{\phi_1}^{\phi_2} \sin^2 \phi\,d\phi$$

after substitution for p from (11.2.2). Thus,

$$M_p = \frac{p_{max} w r R}{4(\sin \phi)_{max}} (2\phi_0 - \sin 2\phi_2 + \sin 2\phi_1) \qquad (11.2.7)$$

where ϕ_0 was defined by (11.2.4). Positive M_p acts in the *counterclockwise* direction and is, of course, independent of the direction of rotation of the brake drum.

An external shoe brake as described will, therefore, not be self-locking if the applied moment M_a must be positive; i.e., if

$$M_a = M_p \pm M_f > 0 \qquad (11.2.8)$$

where the (+) sign holds for rotation away from the pivot and the (−) sign holds for rotation toward the pivot.

Short Shoe Approximation

If angle ϕ_0 is small enough that $\sin 2\phi_0 \simeq 2\phi_0$ and $\cos 2\phi_0 \simeq 1$ then $p_{max} \simeq p$, $(\sin \phi)_{max} = \sin \phi_2$ for $\phi_2 \leq \pi/2$, $(\sin \phi)_{max} = \phi_1$ for $\phi_2 > \phi_1 \geq \pi/2$ and

$$\cos \phi_1 = \cos(\phi_2 - \phi_o) \simeq \cos \phi_2 + \phi_0 \sin \phi_2$$

Whenever ϕ_0 satisfies these conditions equation 11.2.3 approaches

$$T = \mu p r^2 w \phi_0 \qquad (11.2.9)$$
$$= \mu r F$$

where

$$F = prw\phi_0 \qquad (11.2.10)$$

These are known as the *short shoe* equations in which F is the force exerted on the short shoe and T is the braking torque it can provide. Likewise, from (11.2.6) we find that M_f becomes

$$M_f = \mu F \left[R \left(\cos \phi_1 - \frac{\phi_o}{2} \sin \phi_1 \right) - r \right] \qquad (11.2.11)$$

and M_p becomes

$$M_p = FR \left[\sin \phi_1 + \frac{\phi_o}{2} \cos \phi_1 \right] \qquad (11.2.12)$$

Substitution into (11.2.8) indicates that a short shoe brake will not be self-locking if

$$\sin \phi_1 + \frac{\phi_o}{2} \cos \phi_1 \pm \mu \left(\cos \phi_1 - \frac{\phi_o}{2} \sin \phi_1 - \frac{r}{R} \right) > 0$$

Symmetrically Pivoted External Shoe (or Drum) Brakes

The friction moment about the pivot of an external shoe will vanish if the shoe is symmetrical about the pivot, as shown in Figure 11.2.2. Since the brake is engaged by

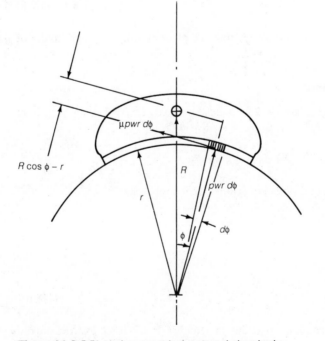

Figure 11.2.2 Pivoted, symmetrical, external shoe brake.

a radial force acting on the pivot, the pressure distribution may be written as

$$p = k \, \delta x \cos \phi$$

and the maximum pressure occurs at $\phi = 0$. Thus, we have that

$$p_{\max} = kx$$

so that upon dividing p by $p_{\max}$ and rewriting the result,

$$p = p_{\max} \cos \phi \tag{11.2.13}$$

The incremental moment about the pivot may, therefore, be written as $dM_f = \mu(P_{\max} \cos \phi)(R \cos \phi - r) wr \, d\phi$, so that upon integration over the length of the shoe, and taking advantage of symmetry, we have

$$M_f = 2\mu p_{\max} \, rw \int_0^{\phi_0/2} (R \cos^2 \phi - r \cos \phi) \, d\phi \tag{11.2.14}$$

which may be integrated directly. By requiring that $M_f = 0$ we have

$$M_f = 2\mu p_{\max} wt \left[R\left(\frac{\phi_0}{4} + \frac{1}{4} \sin \phi_0 \right) - r \sin \frac{\phi_0}{2} \right] = 0 \tag{11.2.15}$$

which may be satisfied by

$$\frac{R}{r} = \frac{4 \sin(\phi_0/2)}{\phi_0 + \sin \phi_0} \tag{11.2.16}$$

which gives the necessary proportion between R and r for the shoe to have zero moment due to friction. Clearly $M_p = 0$ due to symmetry. Because of (11.2.13) the braking torque provided may be written as

$$T = 2\mu p_{\max} wr^2 \int_0^{\phi_0/2} \cos \phi \, d\phi = 2\mu p_{\max} wr^2 \sin(\phi_0/2) \tag{11.2.17}$$

Finally, the radial force required at the pivot to activate the brake is

$$F = 2wr p_{\max} \int_0^{\phi_0/2} \cos^2 \phi \, d\phi = p_{\max} wr[\phi_0 + \sin(\phi_0/2)] \tag{11.2.18}$$

11.3 PIVOTED INTERNAL DRUM BRAKES

An internal shoe or automotive type brake provides torque T and moments M_f and M_p also given by (11.2.3), (11.2.6), and (11.2.7). The geometry used to obtain these relations for internal shoe brakes is shown in Figure 11.3.1. Rotation of the drum from

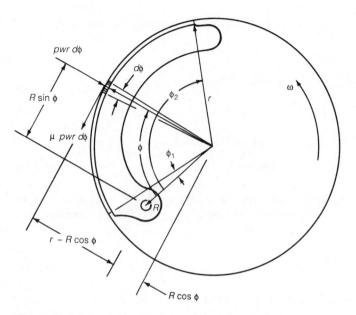

Figure 11.3.1 Geometry of an internal shoe drum brake.

the far end of the shoes to the end near the pivot (from toe to heel) corresponds to positive M_f given by

$$M_f = \int_{\phi_1}^{\phi_2} (\mu p w r \, d\phi)(r - R \cos \phi) \, d\phi \qquad (11.3.1)$$

which is the *negative* of the integral in (11.2.5). Hence a negative value of M_f calculated from (11.2.5) corresponds to a counterclockwise rotation of an internal shoe when the rotation is toward the pivot.

The moment due to lining pressure M_p may be calculated from (11.2.7). A positive value of M_p again implies a moment that tends to move the shoe away from the drum except that in the case of internal shoes this motion results from a clockwise moment applied to the shoe. This is because pressure p acting on the shoe is positive toward the center of the drum when acting on an internal shoe.

Moment relations about the pivot for external and internal shoes and for drum rotation toward or from the pivot relative to the far end (or toe) of the shoe are displayed in Table 11.3.1.

Self-locking or self-activating is generally to be avoided except in the case of a backstop, because the braking torque is out of the operator's control as it jumps to its maximum value. Conditions for self-locking of external and internal drum brakes vary with the angular dimensions of the shoes and the friction coefficient as shown in Figure 11.3.2.

Self-locking may be avoided by selecting curves in Figure 11.3.2 that lie between $M_f/(\mu M_p) > -1$ and $M_f/(\mu M_p) < 1$. These limits follow from equations 11.3.1 and

TABLE 11.3.1 MOMENT RELATIONS FOR INTERNAL AND EXTERNAL DRUM BRAKES[a]

| Rotation[a] | Moment | External Shoe | | Internal Shoe | |
		Implied Braking Action	Implied Shoe Rotation	Implied Braking Action	Implied Shoe Rotation
$p \rightarrow$	$M_p > 0$	open	ccw	open	cw
$p \leftarrow$	$M_p > 0$	open	ccw	open	cw
$p \rightarrow$	$M_f > 0$	open	ccw	close	ccw
$p \leftarrow$	$M_f > 0$	close	cw	open	cw
		Applied Moment Relations		**Applied Moment Relations**	
$p \rightarrow$	—	$M_p + M_f = M_a$		$M_p - M_f = M_a$	
$p \leftarrow$	—	$M_p - M_f = M_a$		$M_p + M_f = M_a$	

[a] $p \rightarrow$, rotation away from the pivot; $p \leftarrow$, rotation toward the pivot; cw, clockwise rotation; ccw, counterclockwise rotation.

Table 11.3.1. Their derivation may be simplified if we first consider the requirements for self-locking.

Begin by considering an external leading shoe for which the rotation is *toward* the pivot. In this case M_a will be zero or negative and the brake will be self-locking whenever $M_f > 0$ and $M_f > M_p$. Upon dividing both sides by μM_p we may write this condition in the form

$$M_f/(\mu M_p) \geqq 1/\mu \tag{11.3.2}$$

The advantage of this form is that although μ appears in the expression of the left side of the inequality, the ratio itself is not a function of μ; it is a function of only the drum and shoe geometry. Thus, the relation between the geometry and the friction coefficient is easily displayed in Figure 11.3.2 in dimensionless form.

Next consider the trailing shoe (rotation *away from* the pivot) to find that the brake will self-lock if $M_f < 0$ and if $M_f \leqq -M_p$. Again divide both sides of μM_p to obtain

$$M_f/(\mu M_p) \leqq -1/\mu \tag{11.3.3}$$

Thus, either (11.3.2) or (11.3.3) must be satisfied if a dual external shoe drum brake is to be self-locking.

Turning to internal brakes we find that for the leading shoe to be self-locking it is necessary that $M_f < 0$ and $M_f < -M_p$ and for the trailing shoe to be self-locking it is necessary that $M_f > 0$ and that $M_f \geqq M_p$. Thus (11.3.2) and (11.3.3) again apply.

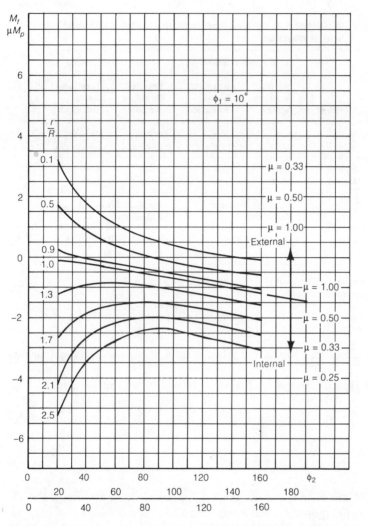

Figure 11.3.2 Curves of the self-locking ratio $M_f/(\mu M_p)$ for $\phi_1 = 10°$ as a function of ϕ_2 for selected values of r/R and coefficient of friction μ.

Since equations 11.3.2 and 11.3.3 must be satisfied for a dual shoe drum brake to be self-locking, it follows that the brake will *not* be self-locking if

$$-1/\mu \le M_f/(\mu M_p) \le 1/\mu \qquad (11.3.4)$$

These regions are indicated by their bounding $\pm 1/\mu$ values in Figure 11.3.2 but they are labeled with their corresponding values of μ.

Force on the Anchor Pin

Once the angular extent of the lining has been selected to meet the design requirements, it is necessary to size the anchor pin to withstand the force exerted on it by the shoe. It is easiest to calculate this by calculating its components parallel and perpendicular to the radius vector from the center of the drum to the pivot point of the shoe for either external or internal brakes.

From Figure 11.2.1 for an external brake shoe we find that the component of the force is the radial direction, taken positive outward, may be written as

$$F_r = \int_{\phi_1}^{\phi_2} prw \cos \phi \, d\phi + \int_{\phi_1}^{\phi_2} \mu prw \sin \phi \, d\phi \qquad (11.3.5)$$

Integration after substitution for p yields

$$F_r = \frac{p_{max} rw}{4(\sin \phi)_{max}} [\pm (\cos 2\phi_1 - \cos 2\phi_2) + 2\mu(\phi_2 - \phi_1)$$
$$+ \mu(\sin 2\phi_1 - \sin 2\phi_2)] \qquad (11.3.6)$$

where comparison of Figures 11.2.1 and 11.3.1 indicates that the $(+)$ sign holds for external shoes and the $(-)$ sign holds for internal shoes.

Likewise, from Figures 11.2.1 and 11.3.1 we find that

$$F_\theta = -\int_{\phi_1}^{\phi_2} prw \sin \phi \, d\phi + \int_{\phi_1}^{\phi_2} \mu prw \cos \phi \, d\phi \qquad (11.3.7)$$

and upon again substituting for p and integrating we obtain

$$F_\theta = \frac{p_{max} rw}{4(\sin \phi)_{max}} \{\pm [2(\phi_2 - \phi_1) + \sin 2\phi_1 - \sin 2\phi_2]$$
$$+ \mu(\cos 2\phi_1 - \cos 2\phi_2)\} \qquad (11.3.8)$$

where in this expression the $(+)$ sign holds for an internal shoe and the $(-)$ sign holds for an external shoe. Force F_θ is positive in the angular direction from the toe to the heel of the shoe.

Since the anchor force F_a is the vector sum of its components, we have that

$$F_a = (F_r{}^2 + F_\theta{}^2)^{1/2} \qquad (11.3.9)$$

11.4 DESIGN OF DOUBLE-SHOE DRUM BRAKES

Design of a double-shoe, twin anchor drum brake, similar to that shown in Figure 11.4.1, to deliver a specified torque for a given activating moment to each shoe as supplied by a single hydraulic cylinder, or cam, requires repeated trials because each

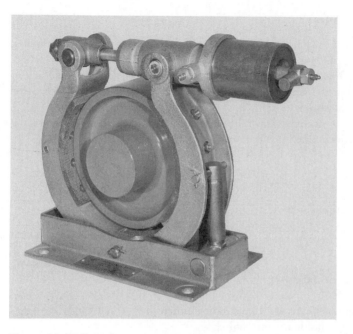

Figure 11.4.1 Twin shoe external drum brake. *Source: Wagner Industrial Brake Products, McGraw-Edison Co., St. Louis, MO.*

shoe does not have the same maximum pressure. The pressure difference arises because one shoe trails while the other leads.

It is necessary, therefore, to choose a common value for the angle ϕ_1 between the radius from the center of the drum to the center of the shoe pivot and the radius to the trailing (heel) end of the shoe, to select a common value of ϕ_2 for both shoes, to calculate the maximum pressure for each, to use this pressure to calculate the torque contributed by each, and to add these torques together to get the braking capability of the two shoes together. This torque is compared to the specified torque, an appropriately smaller or larger value of ϕ_2 is selected for the next trial, and the process is repeated until the difference between the calculated and specified torques is satisfactorily small.

The tedium in this process may be eliminated by a computer or programmable calculator program that employs the bisection technique described in Chapter 1 to arrive at a value of ϕ_2 that makes the difference in the design and calculated torques less than a suitable error ε.

The flowchart in Figure 11.4.2 is based upon writing M_p and M_f as

$$M_{p_a} = b_a A \qquad M_{f_a} = b_a B \tag{11.4.1}$$

for the shoe with the larger pressure and as

$$M_{p_b} = b_b A \qquad M_{f_b} = b_b B \tag{11.4.2}$$

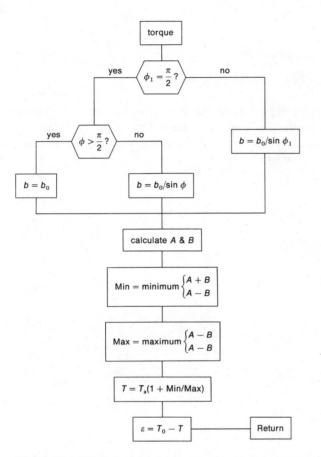

Figure 11.4.2 Flowchart for subprogram to calculate torque when used with main program BIS.

Note: $b_0 = \dfrac{rw}{4}$

for the shoe with the smaller pressure. In these terms, the torque on the shoe with the larger pressure may be written as

$$T_a = 4\mu b_a r(\cos \phi_1 - \cos \phi_2) \qquad (11.4.3)$$

where

$$b_a = p_a b \qquad b_b = p_b b \qquad b = \frac{\dfrac{rw}{4}}{(\sin \phi)_{max}} \qquad (11.4.4)$$

and

$$A = R(2\phi_2 - 2\phi_1 - \sin 2\phi_2 + \sin 2\phi_1)$$
$$B = \mu[R(\cos 2\phi_1 - \cos 2\phi_2) - 4r(\cos \phi_1 - \cos \phi_2)]$$

It follows from Table 11.3.1 that regardless of the direction of rotation for either internal or external shoe brakes, the maximum pressure will occur on that shoe for which values of either $(A + B)$ or $(A - B)$ are such that for a specified activating moment the maximum pressure will be given by

$$M_a = b_a \text{ Min} \begin{cases} A + B \\ A - B \end{cases} \qquad (11.4.5)$$

and the minimum pressure will be given by

$$M_a = b_b \text{ Max} \begin{cases} A + B \\ A - B \end{cases} \qquad (11.4.6)$$

After calculating T_a from (11.4.3) T_b may be calculated from

$$T_b = \frac{p_b}{p_a} T_a \qquad (11.4.7)$$

and the total braking torque from

$$T = T_a \left(1 + \frac{p_b}{P_a}\right) \qquad (11.4.8)$$

Use of these relations in a subroutine that calculates the difference between the total brake torque T and the prescribed torque T_0 as

$$\varepsilon = T_0 - T$$

for a bisection routine that enables the design engineer to easily find angle ϕ_2 to within a prescribed error ε. A listing for a brake design program is given in Ref. 12.

Many automotive brakes are of the double-shoe, single anchor Bendix, or servo-brake, type in which the leading shoe pivots against the trailing shoe which pivots against the anchor pin. Thus, the force from the leading shoe adds to the braking moment for the trailing shoe, known as servo action, and further increases the braking torque obtained from the activating moment. Design of these brakes, which is entirely computer based, and too lengthy to be discussed here, is described in Ref. 13.

EXAMPLE 11.4.1

Design an external drum brake with a leading and trailing shoe to provide 3000 in.-lb of torque on an 8.0-in.-diameter drum. Lining width should be 1.5 ± 0.3 in. on a shoe having $\phi_1 = 20°$. Maximum lining pressure is 140 psi, the radius of the shoe pivot is 5.5 in., and $\mu = 0.3$.

Substitute

$$R = 5.5 \text{ in.} \qquad w = 1.5 \text{ in.} \qquad T = 3000 \text{ in.-lb}$$

$$r = 4.0 \text{ in.} \qquad \mu = 0.3 \qquad p_{max} = 140 \text{ psi}$$

$$\phi_1 = 20° \qquad \varepsilon = 0.001$$

into a program defined by Figures 11.4.2 and 1.8.1 to find $\phi_2 = 161.183°$. Reducing e to 0.01 does not affect the third place in ϕ_2, but reducing it to 0.10 produces $\phi_2 = 161.221°$. Recommend $\phi_2 = 162°$ to err on the side of safety.

11.5 CENTRIFUGAL CLUTCHES

Centrifugal clutches are used in a wide variety of applications to either disconnect the driver (electric motor, internal combustion engine, etc.) when an overload occurs, as applied to grinders, mixers, and power shears, or to assure that the motor is up to its operating speed before the load is applied, as in chain saw applications. They also allow greater motor acceleration during starting, which in the case of a NEMA B electric motor, for example, results in longer motor life by limiting its operation in the high-current, low-torque part of its power curve.

Both load engagement and overload protection are combined in the centrifugal clutch used in most chain saws, as shown in Figure 11.5.1, where the clutch is designed to allow the chain to remain stationary while the motor idles, to engage the chain as the motor is speeded up, and to slip if the chain is jammed. This provides safety for the user and prevents damage to the motor.

The lining pressure on a shoe may be written as

$$p = k\Delta \cos \phi \tag{11.5.1}$$

where Δ is the deflection of the lining and k is its effective spring constant. Maximum lining compression, and hence maximum pressure, occur at $\phi = 0$, where $p_{max} = k\Delta$. Thus, (11.5.1) may be written as

$$p = p_{max} \cos \phi \tag{11.5.2}$$

Figure 11.5.1 Centrifugal clutch in a chain saw.

From the geometry in Figure 11.5.2 we find that incremental torque may be written as

$$dT = \mu(p_{max} \cos \phi)wr^2 \, d\phi \tag{11.5.3}$$

for a drum of radius r. Integration over the lining, which subtends angle ϕ_0 at the center of the drum, leads to

$$T = \mu p_{max} wr^2 \int_{-\phi_0/2}^{\phi_0/2} \cos \phi \, d\phi$$

$$= 2\mu p_{max} wr^2 \sin \frac{\phi_0}{2} \tag{11.5.4}$$

Equilibrium in the direction of the centerline of the shoe requires that

$$mr_c\omega^2 - k_s\delta = p_{max}rw \int_{-\phi_0/2}^{\phi_0/2} \cos^2 \phi \, d\phi \tag{11.5.5}$$

where r_c is the radius to the center of gravity of the shoe, k_s is the effective spring constant of the retraction springs attached to the shoe, and δ is the clearance between drum and lining when the clutch is fully retracted. Upon solving (11.5.5) we find the maximum pressure may be written as (using $\omega = \pi n/30$, n = rpm)

$$p_{max} = 2\frac{\left(\dfrac{\pi n}{30}\right)^2 mr_c - k_s\delta}{rw(\phi_0 + \sin \phi_0)} \tag{11.5.6}$$

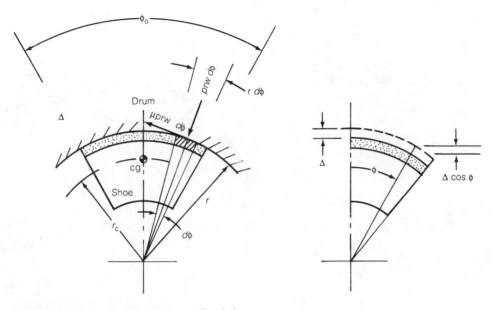

Figure 11.5.2 Forces acting on a centrifugal shoe.

Substitution into (11.5.4) gives an expression for the torque capability of a centrifugal clutch with N shoes each of mass m:

$$T = 4\mu r N \left[\left(\frac{\pi n}{30} \right)^2 mr_c - k_s \delta \right] \frac{\sin \dfrac{\phi_0}{2}}{\phi_0 + \sin \phi_0} \tag{11.5.7}$$

11.6 DISK CLUTCHES AND BRAKES

Disk (or disc) clutches and brakes are often produced with several disks in parallel, as illustrated in Figure 11.6.1, to obtain a large torque capacity in a relatively small volume. Even larger torques may be transmitted with wet (oil bath) disk brakes and clutches that are connected to a central oil circulation and cooling system. Single-plate clutches are used in most automotive applications because of the relatively light torques required. Likewise, the lighter braking loads required of the average automobile enable the designer to use caliper disk brakes with brake pads that extend over only a small portion of the disk, as in Figure 11.6.2. The large disk to pad area provides air cooling and thus reduces brake *fade*, in which the friction coefficient may decrease as the lining temperature increases. Air cooling of the disk may be improved by using a ventilated disk, shown in Figure 11.6.3, wherein the solid disk is replaced by two disks separated by radial spacers in an attempt to increase the cooling surface without increasing the radius of the disk.

Figure 11.6.1 Multiple-disk face contact clutch. *Source: Twin Disc, Inc., Racine, WI.*

Figure 11.6.2 Caliper disk brake. *Source: Twiflex Corp., Horseheads, NY.*

Figure 11.6.3 Ventilated disk brake. *Source: Auto Specialties Mfg. Co., Ausco Brake Div., Benton Harbor, MI.*

Braking torque for all disk brakes may be calculated from

$$T = \mu \int_{r_i}^{r_o} dr \int_0^\theta pr \, d\theta \tag{11.6.1}$$

Because of the larger contact area in face contact disk clutches and brakes, the pressure distribution over the disk face may have a greater influence on the torque capability of the clutch or brake than in the case of a caliper brake. Two distinct pressure distributions are commonly assumed: the *uniform wear* distribution and the *uniform pressure* distribution.

Uniform Wear

From the moment of contact until both faces of the disk or plate rotate at the same speed the relative velocity will vary linearly with the radius. If the thickness of the lining material removed, δ, is dependent upon both the relative velocity and the pressure, then

$$\delta = Kpr \tag{11.6.2}$$

where K is a constant of proportionality. If δ is assumed to be constant over the lining face, then the maximum pressure p_{max} is found at the inner radius r_i and

$$\delta = Kp_{max}r_i = Kpr$$

or

$$p = p_{max} \frac{r_i}{r} \qquad (11.6.3)$$

This condition is assumed to hold after the lining has been worn enough for its shape to conform to the pressure distribution implied.

Substitution from (11.6.3) into the expression for the related axial force F_a, where

$$F_a = \int_{r_i}^{r_o} 2\pi p r \, dr$$

yields

$$F_a = 2\pi p_{max} r_i (r_o - r_i) \qquad (11.6.4)$$

Transmitted torque is given by

$$T = \int_{r_i}^{r_o} 2\pi \mu p r^2 \, dr = 2\pi \mu p_{max} r_i \int_{r_i}^{r_o} r \, dr$$

$$T = \pi \mu p_{max} r_i (r_o^2 - r_i^2) \qquad (11.6.5)$$

Since $T = 0$ for $r_i = 0$ and for $r_i = r_o$ it is of interest to calculate dT/dr_i to find from

$$\frac{dT}{dr_i} = 2\pi \mu p_{max}(r_o^2 - 3r_i^2) = 0$$

that to obtain the maximum torque for a given outer diameter the inner radius should be

$$r_i = \frac{1}{\sqrt{3}} r_o = 0.5774 r_o. \qquad (11.6.6)$$

If uniform wear is assumed for caliper disk brakes (11.6.1) yields

$$T = \mu p_{max} r_i \theta \int_{r_i}^{r_o} r \, dr = \mu p_{max} \frac{\theta}{2} r_i (r_o^2 - r_i^2) \qquad (11.6.7)$$

at each brake pad. The activating force on each pad may be found from

$$F_a = p_{max} \theta r_i (r_o - r_i) \qquad (11.6.8)$$

for pads that are segments of an annular ring. Since caliper disk brakes have similar pads on each side of the disk, the force given by (11.6.8) acts on each side of the disk to yield a total torgue T which is twice that given by (11.6.7), namely

$$T = \mu p_{max}\theta r_i(r_o^2 - r_i^2)$$ (11.6.9)

Uniform Pressure

If the clutch or brake plate supporting the lining is flexible and properly spring loaded the pressure may be assumed to be uniform, so that

$$F_a = \pi p(r_o^2 - r_i^2)$$ (11.6.10)

and

$$T = 2\pi\mu p \int_{r_i}^{r_o} r^2\, dr = \frac{2}{3}\pi\mu p(r_o^3 - r_i^3)$$ (11.6.11)

These relations become

$$F_a = p\frac{\theta}{2}(r_o^2 - r_i^2)$$

and (11.6.12)

$$T = \frac{\mu}{3}p\theta(r_o^3 - r_i^3)$$

for caliper brake linings that are segments of an annular ring.

11.7   CONE CLUTCHES AND BRAKES

Cone clutches and brakes provide large lining pressure and correspondingly large torque with smaller axial forces than required for plate clutches and brakes. If the cone angle is α and if uniform wear is assumed, so that the lining pressure is given by (11.6.3), which is,

$$p = p_{max}r_i/r$$

then from the equilibrium of forces in the axial direction, Figure 11.7.1, find

$$F_a = \mu P \cos\frac{\alpha}{2} + P\sin\frac{\alpha}{2}$$ (11.7.1)

where

$$P = \int_A p\, da$$ (11.7.2)

in which

$$da = 2\pi r\,du \tag{11.7.3}$$

and

$$r = r_i + u\sin\frac{\alpha}{2}$$

so that

$$dr = \left(\sin\frac{\alpha}{2}\right)du \tag{11.7.4}$$

Substitution for du from (11.7.4) into (11.7.3) followed by substitution from (11.7.3) into (11.7.2) and then into (11.7.1) and integration from inner radius r_i to outer radius r_o yields

$$F_a = 2\pi\left(\mu\cos\frac{\alpha}{2} + \sin\frac{\alpha}{2}\right)r_i p_{max}\int_{r_i}^{r_o}\frac{dr}{\sin\dfrac{\alpha}{2}}$$

$$= 2\pi\left(1 + \frac{\mu}{\tan\dfrac{\alpha}{2}}\right)r_i p_{max}(r_i - r_o) \tag{11.7.5}$$

as the actuation force required to realize the maximum torque capability of the clutch.

Before calculating the torque that may be obtained, it is of interest to calculate the disengagement force, again with reference to Figure 11.7.1. Thus,

$$F_d = \left(\frac{\mu}{\tan\dfrac{\alpha}{2}} - 1\right)2\pi r_i p_{max}(r_o - r_i) \tag{11.7.6}$$

so that the extraction force vanishes if

$$\mu \leq \tan\frac{\alpha}{2} \tag{11.7.7}$$

Transmitted torque capability may be found from

$$T = \int_A \mu r p\,da = \frac{2\mu\pi r_i p_{max}}{\sin\dfrac{\alpha}{2}}\int_{r_i}^{r_o} r\,dr$$

$$= \mu\pi r_i p_{max}\frac{r_o^2 - r_i^2}{\sin\dfrac{\alpha}{2}} \tag{11.7.8}$$

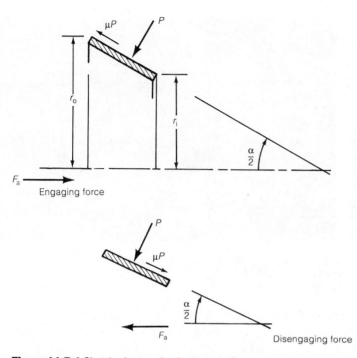

Figure 11.7.1 Sketch of cone clutch cross section.

after again substituting for p and da from equations 11.6.3, 11.7.3, and 11.7.4. The ratio of the torque transmitted by a cone clutch to that of a plate clutch is thus

$$\frac{T_{\text{cone}}}{T_{\text{plate}}} = \frac{1}{\sin\dfrac{\alpha}{2}} \tag{11.7.9}$$

The ratio of the braking torque to the axial force

$$\frac{T}{F_a} = \frac{\mu}{2}\,\frac{r_o + r_i}{\left(1 + \dfrac{\mu}{\tan\dfrac{\alpha}{2}}\right)\sin\dfrac{\alpha}{2}} \tag{11.7.10}$$

implies that a cone brake would be less sensitive to changes in the coefficient of friction than a disk brake because the denominator tends to reduce the effect of a change in the numerator when the friction coefficient changes, due to heat or moisture.

Prototype cone brakes have been tried on a front-wheel-drive subcompact automobile, as shown in Figure 11.7.2. Their advantage of fewer parts than conventional

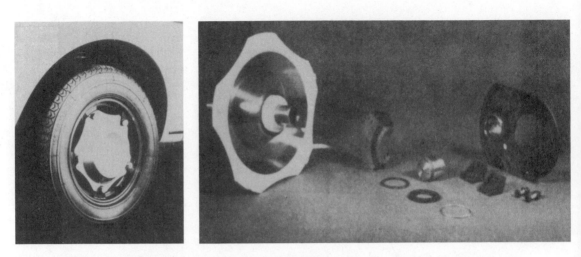

Figure 11.7.2 Cone brake on a front wheel drive subcompact. *Source: M.E. Johnson*, Testing the Cone Brake Design, *SAE Technical Paper 790465,* © *1979 Society of Automotive Engineers, Inc. Reprinted with permission.*

drum or disc brakes[6] appears to be not worth the disadvantage of requiring outboard wheel bearings. The bright steel wheel in the left photo in Figure 11.7.2 forms the outer cone (actually a folded cone) which must be restrained by the outboard wheel bearings against the outward force of the inner folded cone segment, shown in the right photo, when the brake is engaged. The only uses of cone brakes at the present seem to be in washing machines and automotive twin traction differentials, where metal to metal contact is acceptable, so that no linings are necessary.

The expense of producing, shipping, and storing conical lining material is only one of two objections to lined cone brakes. The other is that greater work per unit volume may be had from multiple disk brakes or from rim brakes.[14]

11.8 REQUIRED TORQUE AND HEAT DISSIPATED

The design of a brake or clutch system or the selection of one that is commercially available depends not only upon the torque capabilities of the unit but also upon its ability to dissipate heat. The latter consideration is important because the coefficient of friction of the lining or facing generally decreases with increasing temperature according to curves similar to that shown in Figure 11.8.1. These curves are representative of brake lining response when tested according to the SAE J661a standard in which the lining is subjected to two controlled braking and cooling cycles in which the friction coefficient is measured during braking as the brake is cooled (recovery) and heated (fade). The first fade run is generally not shown because the lining and drum may not be in contact during the entire fade run.

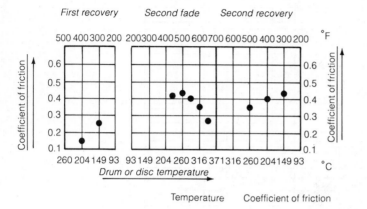

10-point average for hot friction coefficient

Figure 11.8.1 Brake fade curves. *Source: M.N. Josi,* Disc Brake Linings, *SAE Technical Paper 800782,* © *1980 Society of Automotive Engineers, Inc. Reprinted with permission.*

Brake life also decreases as the temperature increases, as illustrated in Figure 11.8.2 for representative disk brake materials.

Heat generated by the clutch in bringing machinery up to speed or by a brake in slowing or stopping it may be conveniently estimated from energy considerations; namely, the change in energy of the system is equal to the work done by the brake or the clutch. Although the heat equivalent of the work done may be found from the joule equivalent that 1 kilocalorie = 4186.8 N m or that 1 Btu = 778.169 ft·lb*, brakes are usually rated in terms of mechanical units of horsepower of kilowatts that can be dissipated. Pertinent calculations for either units are the torque required and the acceleration and deceleration times. Required torque may be found from

$$W_0 = \int_{t_0}^{t_1} T\omega \, dt = \Delta KE + \Delta PE + W_1 \tag{11.8.1}$$

in which W_0 is the work done by the clutch ($W_0 > 0$) or brake ($W_0 < 0$), ΔKE is the change in kinetic energy, ΔPE is the change in potential energy during braking, and W_1 includes the work done on the system by motors or by other brakes and clutches during the time from t_0 to t_1. W_1 and ΔPE will be taken to be zero in the following discussion.

If constant acceleration or deceleration is assumed it follows that during braking, for example,

$$\omega = \omega_0 - \alpha t \tag{11.8.2}$$

---

* Chemistry and Physics Handbook, 66th ed., CRC Press, Inc., Boca Raton, FL, 1986.

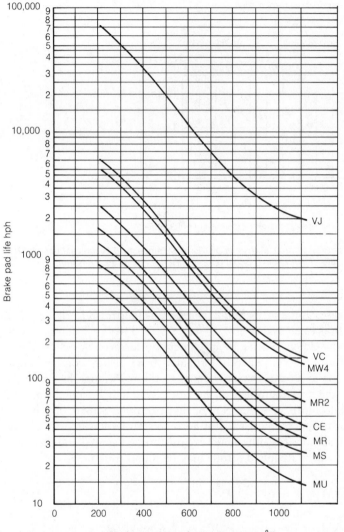

Figure 11.8.2 Brake life as a function of temperature. Code on right side of the graph refers to various disk brake models (different sizes and shapes of brake pads). *Source: Twiflex Corp., Horseheads, NY.*

in terms of rotational velocity ω_0 at time $t = 0$, deceleration α, and time t. Thus, (11.8.1) becomes

$$T \int_0^t (\omega_0 - \alpha s)\, ds = \tfrac{1}{2} I (\omega_0{}^2 - \omega^2). \tag{11.8.3}$$

When stopped $\omega = 0$ so that (11.8.3) becomes

$$T\left(\omega_0 - \frac{\alpha t}{2}\right)t = \tfrac{1}{2}I\omega_0{}^2 \tag{11.8.4}$$

when T and α are constant. Substitution of α from (11.8.2) into (11.8.4) yields

$$T = \frac{I\omega_0}{t} \tag{11.8.5}$$

as the torque necessary to stop the rotation in time t of an object having moment of inertia I from an initial angular velocity ω_0.

Inertia I is often expressed in terms of its *radius of gyration R*, where

$$I = mR^2 = \frac{W}{g}R^2$$

and where W is the weight of the rotating part. Angular velocity ω_0 is often written in terms of the revolutions per minute n as

$$\omega_0 = 2\pi n/60$$

so that (11.8.5) appears as

$$T = \frac{WR^2 2\pi n}{60gt} = \frac{WR^2 n}{307.487t} \cong \frac{WR^2 n}{307t}$$

in the formula section of brake and clutch manufacturers' catalogs.[1,2] In this formula T is in foot-pounds, W in pounds, R in feet, and time t in seconds.

Geared Rotation

Application of (11.8.1) to several rotating components, as in braking a motor turning at n_1 rpm and geared to a generator that turns at n_2 rpm, simply requires an additional kinetic energy term, so that

$$\tfrac{1}{2}T\omega_1 t = \tfrac{1}{2}I_1\omega_1{}^2 + \tfrac{1}{2}I_2\omega_2{}^2 = \tfrac{1}{2}(I_1 + \mathfrak{n}^2 I_2)\omega_1{}^2$$

which may be written as

$$T = \frac{(I_1 + \mathfrak{n}^2 I_2)\omega_1}{t}$$

$$\cong \frac{W_1 R_1{}^2 + W_2(\mathfrak{n}R_2)^2}{307t} n_1 \tag{11.8.6}$$

in which $\mathfrak{n} = n_2/n_1$, where the brake, or clutch, acts on the shaft rotating at n_1 rpm.

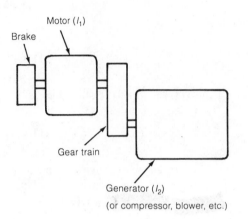

Brake

Motor (I_1)

Gear train

Generator (I_2)

(or compressor, blower, etc.)

Figure 11.8.3 Brake for motor and generator turning at different angular velocities.

Vehicle Braking, Clutch-Started Conveyors

Kinetic energy due to linear motion must be included in these applications. Thus, (11.8.1) becomes

$$\frac{1}{2} T\omega_1 t = \frac{1}{2}(I_1 + \sum_{i=2}^{q} I_i n_i{}^2)\omega_1{}^2 + \tfrac{1}{2}mv^2 \tag{11.8.7}$$

According to the geometry in Figure 11.8.4, $v = r\omega_1$ so that (11.8.1) takes the form

$$T = \frac{1}{t}[(I_1 + \sum_{i=2}^{q} I_i n_i{}^2) + \sum_{j=1}^{s} m_j r_j{}^2 n_j{}^2]\,\omega_1{}^2 \tag{11.8.8}$$

for q rotating components and s translating components. Torque T is minimized if the brake is on the fastest shaft to minimize n_i for all i.

Whenever the brake or clutch is to effect a speed change, (11.8.5) obviously becomes

$$T = \frac{I(\omega_2{}^2 - \omega_1{}^2)}{\omega_1 t} \tag{11.8.9}$$

with similar modification in equations 11.8.6 through 11.8.8.

Because the time to dissipate heat is important when repetitive braking is necessary, a nominal heat dissipation capability is often given, in foot-pounds/minute (or N m/min), as in Figure 11.8.5, even though heat dissipation is strongly dependent upon the thermal gradients at the brake. Curves shown in Figure 11.8.5 are typical of the heat transfer rate to air at 70 °F when the brake or clutch surface temperature is about 95 °F in the steady state condition.

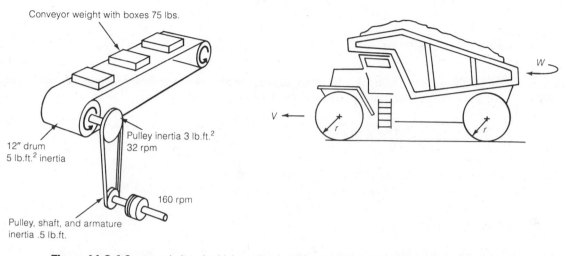

Figure 11.8.4 Conveyor belt and vehicle applications. *Source: Warner Electric/DANA, South Beloit, IL.*

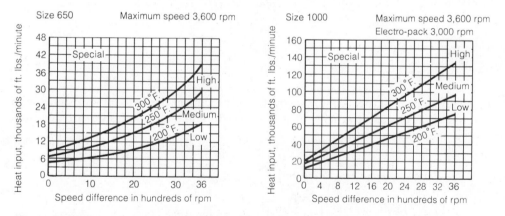

Figure 11.8.5 Typical heat dissipation curves. *Source: Warner Electric/DANA, South Beloit, IL.*

EXAMPLE 11.8.1

Estimate the time to stop, the required torque per brake, and the average heat that must be dissipated per second by the brakes in stopping a truck initially moving at 70 mph if the friction coefficient between highway and tires is 0.80 and if there is no skidding. Inertia of motor and wheels is 70 slug-feet^2 when converted to wheel rpm. Wheel outside diameter is 50 in. Assume two brakes on each of 5 axles. Truck gross weight is 32 000 lbs.

From elementary dynamics the initial value problem may be written as

$$\mu m g = -m \frac{\partial^2 x}{\partial x^2} = -m \ddot{x}$$

where m denotes the mass of the truck and where the brakes are applied at time $t = 0$, to that $\dot{x}(0) = v_0$ and $x(0) = 0$. Since the deceleration is constant, the stopping time is

$$t = \frac{v_0}{\mu g} = \frac{70(5280)}{0.8(32.2)3600} = 3.99 \text{ s}$$

in terms of the friction coefficient μ and the acceleration of gravity g.

The required torque may be estimated from equation 11.8.7, which reduces to

$$\tfrac{1}{2} T \omega_0 t = \tfrac{1}{2} I \omega_0^2 + \tfrac{1}{2} m v_0^2 \qquad\qquad (11.8.10)$$

If the truck wheels are of radius r, then $\omega_0 r = v_0$ so that substitution for t and ω_0 in equation 11.8.10 yields a total torque of

$$T = \mu r g \left(\frac{I}{r^2} + m \right) = \frac{25}{12} 0.8(32.2) \left(\frac{70(144)}{25^2} + \frac{32\,000}{32.2} \right) = 54199 \text{ ft-lb}$$

so that the required torque per wheel is

$$T = 5419.9 \text{ ft-lb}$$

The energy to be dissipated as heat is equal to the energy in the form of work done on the brake. Thus, the total heat dissipated in units of ft-lb for all 10 brakes together is given by equation 11.8.7, which in this circumstance becomes

$$W = \frac{1}{2} I \omega_0^2 + \frac{1}{2} m v_0^2 = \frac{1}{2} \left(\frac{I}{r^2} + m \right) v_0^2$$

$$W = \frac{1}{2} \left[70 \left(\frac{12}{25} \right)^2 + \frac{32\,000}{32.2} \right] \left[70 \frac{5280}{3600} \right]^2 = 5\,322\,486 \text{ ft-lb}$$

or $532\,248.6$ ft-lb per brake. The thermal equivalent is 683.98 Btu in 3.99 s.

11.9 ONE-WAY (OVERRUNNING) CLUTCHES

These are clutches that will transmit torque when $n_1 > n_2$ but will transmit little or no torque when $n_1 < n_2$ where n_1 and n_2 represent the rpm of shafts 1 and 2, respectively. Eight typical industrial applications for one way clutches are depicted in Figure 11.9.1.

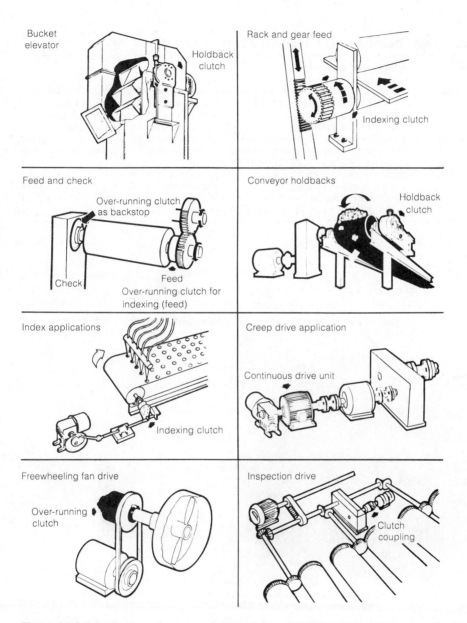

Figure 11.9.1 Applications of one-way clutches. *Source: DANA Corporation-Formsprag, Warren, MI.*

The most commonly commercially available one-way clutches are listed in Table 11.9.1. Spring one-way clutches are usually employed to engage and disengage depending upon relative directions of rotation. Roller and sprag one-way clutches are usually found in applications where engagement and disengagement depend upon relative speeds. They

TABLE 11.9.1 COMMERCIALLY AVAILABLE ONE-WAY CLUTCHES

Type	Principal of Operation
Spring clutch	Coil spring 4 in Figure 11.9.2 is wrapped loosely around input hub 2 and output hub 5. When hub 2 rotates in the clockwise direction the spring wraps more tightly about both hubs and causes hub 5 to rotate with hub 2. When hub 2 rotates in the counterclockwise direction the spring loosens and hub 5 does not rotate.
Roller clutch	Rollers are wedged between cams on shafts 1 and 2 for rotation in one direction, thus forcing the cams to rotate together. No wedging occurs when driving shaft reverses direction. See Figure 11.9.3.
Sprag clutch	Similar to the roller clutch except that the rollers are replaced by sprags which respond more rapidly to a changes in relative rotation. See Figure 11.9.4, where no torque is transmitted if the outer race rotates faster in the clockwise direction than the inner race. Widely used in automatic transmissions.

are, for example, used in starting steam turbines where an electric motor accelerates the turbine up to speed through a sprag clutch. When the blades begin to drive the turbine the clutch will automatically disengage as the turbine accelerates beyond the speed of the motor. Sprag clutches perform a similar function in an automatic transmission.

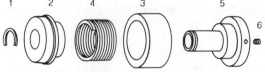

Figure 11.9.2 Spring clutch and an exploded view of its construction. *Source: Warner Electric/DANA, South Beloit, IL.*

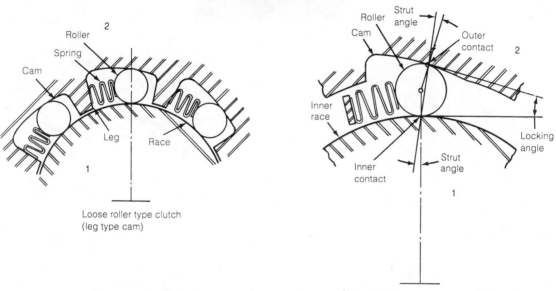

Loose roller type clutch
(leg type cam)

Outer cam type roller one-way clutch

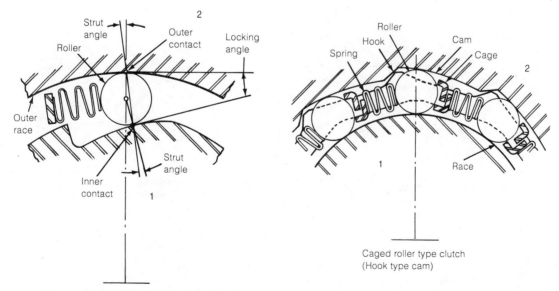

Inner cam type roller one-way clutch

Caged roller type clutch
(Hook type cam)

Figure 11.9.3 Four common types of roller clutches. *Sources:* Illustrations—*SAE J1087*, 1988 SAE Handbook. © *1988 Society of Automotive Engineers. Reprinted with permission.* Photo—*The Torrington Co., Torrington, CT.*

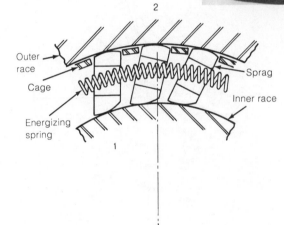

Typical single-cage one-way clutch

Typical full complement sprag one-way clutch

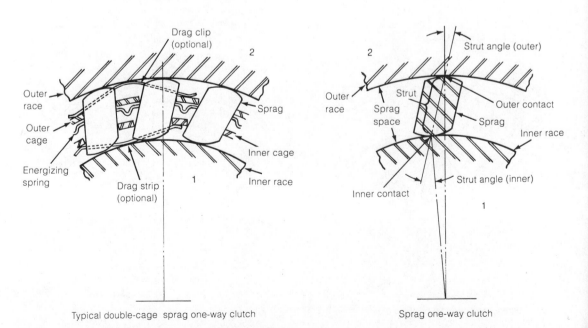

Typical double-cage sprag one-way clutch

Sprag one-way clutch

Figure 11.9.4 Four varieties of sprag clutches and sprags. *Sources:* Illustrations—*SAE J1087*, 1988 SAE Handbook. © *1988 Society of Automotive Engineers. Reprinted with permission.* Photo—*DANA Corp., Toledo, OH.*

11.10 MAGNETIC PARTICLE, HYSTERESIS, AND EDDY CURRENT CLUTCHES AND BRAKES

All three of these clutches/brakes (clutches when both input and output shafts turn; brakes when the output shaft is replaced by a rigid connection) have the advantage of transmiting torque without physical contact between input and output shafts. They have the advantage of no lining wear and they may be computer controlled from remote locations by regulating their coil currents. They all have the disadvantage that in the larger sizes they require a source of electrical power and they have a relatively large weight to torque ratio. Very small hysteresis and eddy current clutches/brakes may rely upon permanent magnets and, therefore, require no external electrical power.

Magnetic particle clutches/brakes, similar to that shown in Figure 11.10.1a, transmit torque by means of magnetic attraction between the rotor and drive cylinder, shown in Figure 11.10.1b. The magnetic lines of force are concentrated along radial arms of magnetic particles that form between the drive cylinder and the rotor when the electromagnetic coil is energized to create a magnetic field. (Magnetic particles reduce the magnetic resistance between the rotor and the drive cylinder.) This clutch provides a constant torque over its entire speed range regardless of the speed difference between the input and output shaft speeds (this difference is known as the slip speed), as displayed in Figure 11.10.4. It is, therefore, well suited for dynamometers (devices for measuring motor and brake torque at various speeds), soft starts, and tension control in tape drives and tape manufacture. Commercial magnetic particle clutch capacities range from about 1 ft-1b to 200 ft-1b and their weights range from about 4 to 230 1b.

Hysteresis clutches/brakes, shown in Figure 11.10.2, also transmit torque by means of magnetic attraction, but without magnetic particles between a rotor and a drive cylinder. Magnetic hysteresis in the material of the rotor causes an angular displacement of magnetic poles such that a torque is developed as it rotates relative to the cup, or drive cylinder. This clutch/brake displays a constant torque for a range of slip speeds below a critical value, as shown in Figure 11.10.4. This critical value is dependent upon the heat dissipation capability of the clutch/brake because the cup material loses its magnetism when it is heated beyond the Curie temperature. These clutches or brakes are smaller than magnetic particle clutches/brakes. They have torques ranging from 6 in-oz to 26 ft-1b and weights from about 3 oz to 125 1b.

Eddy current clutches/brakes, pictured in Figure 11.10.3, have a construction similar to that of a hysteresis clutch/brake except that different materials are used for the rotor so that the magnetic attraction is formed by eddy currents rather than by hysteresis. The important characteristic of this clutch/brake is that its torque capability increases as the slip speed increases, as shown in Figure 11.10.4. It provides a torque to bring the speed of the output shaft to that of the input shaft when the two are different, but transmits no torque when the two are running at the same speed. few Torques for these units range from several in.-oz to 55 000 ft-1b. Their weights range from a few ounces to 31 000 1b.

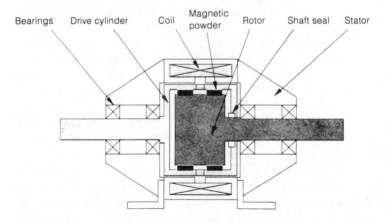

| Bearings | Drive cylinder | Coil | Magnetic powder | Rotor | Shaft seal | Stator |

Figure 11.10.1 Appearance and construction of a magnetic clutch or brake. *Source: Magnecor Corp., St. Louis, MO.*

Torque–slip-speed curves for magnetic particle, hysteresis, and eddy current clutches/brakes are shown in Figure 11.10.4. They show why magnetic particle and hysteresis clutches are well suited for acceleration control during starting, for tension control in filament and tape manufacture, and similar processes. They are also valuable whenever either consant or controlled braking torque is required. The eddy current curves show why these clutches are used in extruders, pumps, and other applications where a par-

Figure 11.10.2 Representative configurations of commercial hysteresis clutch/brakes with torque capacities from (a) 0.1 to 60 oz in. and (b) 60 to 3200 oz in. *Source: Magtrol, Inc. Buffalo, NY.*

ticular speed difference is wanted between the driving and driven unit. They automatically provides a torque to accelerate or retard the driven unit whenever the speed deviates from the design speed. As brakes they provide the largest braking torque at high speeds and automatically reduce the braking torque as the speed slows.

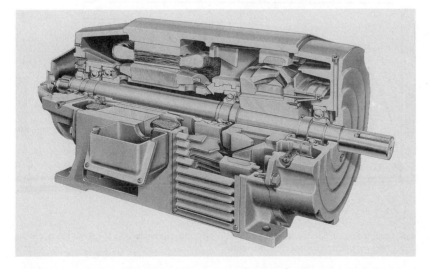

Figure 11.10.3 Eddy current motor and clutch combination in models available with rated power from 15 to 300 hp. *Source: Eaton Corp., Industrial Drives Operations, Kenosha, WI.*

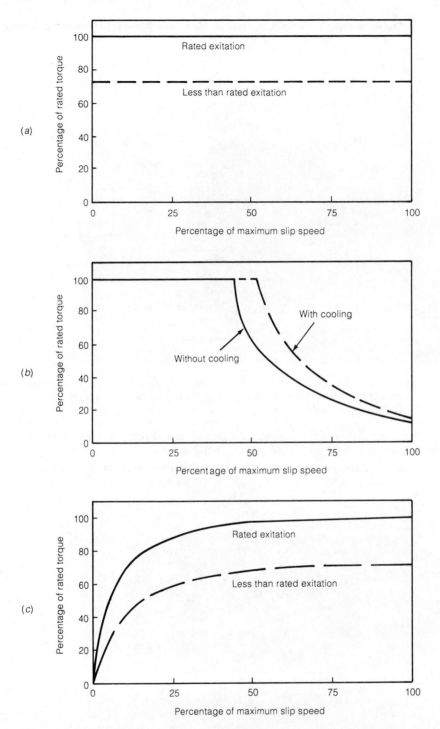

Figure 11.10.4 Torque-speed curves for (a) magnetic particle clutch/brake, (b) hysteresis clutch/brake, (c) eddy current clutch/brake.

11.11 TORQUE CONVERTERS

A torque converter is a hydrodynamic device that can serve both as a clutch and as a variable speed device in that it can be used to bring one shaft up to the speed of another and can transmit torque at low slip to drive the output shaft. It can also cause the input shaft to operate at a torque and angular velocity different from the output shaft. The most common application of a torque converter is in automobile transmissions, where it serves as a starting aid, as a speed and torque changing device, and as a retarder. A typical three-element torque converter is shown in Figure 11.11.1. The three elements referred to are the impeller, the turbine, and the reactor, or stator, all of which are in the fluid chamber, as shown in the operational schematic in Figure 11.11.2.

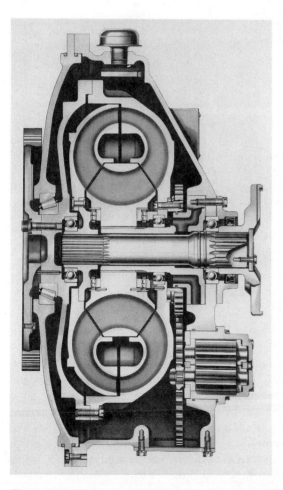

Figure 11.11.1 Cross section of a typical three-element converter. *Source: Borg-Warner Automotive Inc., Off-Highway Systems, Rockford, IL.*

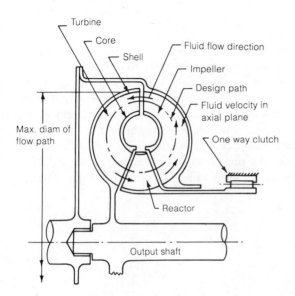

Figure 11.11.2 Schematic of a three-element torque converter. *Source: Advances in Engineering Series, No. 5, © 1973 Society of Automotive Engineers, Inc. Reprinted with permission.*

As is apparent from this figure, the torque converter is a fluid coupling with its blades modified to make space for the reactor.

In the case of a fluid coupling the angular momentum Φ_i of the impeller is equal to the angular momentum Φ_t of the turbine if the small viscous losses in the coupling are neglected. Thus

$$\Phi_i = \Phi_t \tag{11.11.1}$$

Addition of the reactor blades to redirect the flow from the impeller to the turbine to give it a greater tangential component adds another force to the fluid and changes equation 11.11.1 to

$$\Phi_i + \Phi_r = \Phi_t \tag{11.11.2}$$

where Φ_t has been increased by an amount Φ_r, the angular momentum contributed by the reactor blades. Normally these blades are set to give the greatest torque increase at relatively low impeller speeds. As the impeller speed increases the direction of flow from the impeller to the reactor changes until a speed is reached where the flow strikes the rear of the reactor blades, which produces a negative effect on the torque. To eliminate this negative effect the reactor shaft is held in place by a one way clutch, shown in Figure 11.11.2, which prevents it from turning when the fluid strikes the front of the blades but which allows the shaft to turn slightly before the flow has changed enough to strike the rear of the blades. At this point the torque converter begins to act as a fluid coupling.

At least one automobile manufacturer has used variable pitch reactor blades to extend the acceleration range of the automatic transmission.

An approximate expression for the transmitted torque T given by Hunsaker and Rightmire[9] is

$$T = Cn^2d^5 \tag{11.11.3}$$

where n represents the input rpm and

$$C = \frac{\pi^2}{2} \rho \frac{A}{d^2} \left[\frac{s}{\lambda} (2 - s)(1 - \Gamma^2) \right]^{1/2} [1 - (1 - s)\Gamma^2] \tag{11.11.4}$$

in which $\Gamma = r_1/r_2$, $d = 2r_2$, $s = (\omega_i - \omega_t)/\omega_i$, ρ is the fluid density, and λ is a dimensionless factor similar to a pipe-friction factor. Here r_1 and r_2 represent the respective center radii of the flow field at the impeller inlet and outlet as shown in Figure 11.11.2, and ω_i and ω_t denote the angular velocities of the impeller and turbine, in that order.

It is important to note that (1) the torque increases as the fifth power of the reactor diameter, as implied by the d^5 term in (11.11.3), (2) if the slip factors s vanishes the torque vanishes because $C = 0$ in that event, and (3) the torque is proportional to the square of the input angular velocity. From the first of these observations it follows that a large torque may be transferred by a comparatively small torque converter (i.e., doubling the diameter of the unit increases its torque capacity 32 fold), making it suitable for vehicular use. From the second we find that some loss of power in the unit itself is necessary if torque is to be transmitted, and from the last we find that little torque is transmitted at low speed and that doubling the input rpm quadruples the torque transfer.

Consult Ref. 7 for a discussion by the inventor of the torque converter, Ref. 8 and 9 for further explanation of the theory, and Ref. 11 for design procedures.

11.12 COMPARISON OF CLUTCHES AND BRAKES

This section is devoted to a brief comparison of the various types of clutches and clutch/brakes discussed in the preceding sections plus several for which little analysis is required, such as the toothed clutch and the detent or overload clutch (see Table 11.12.1 and Figures 11.12.1–11.12.5). Design formulas for these clutches are, therefore, left to problems 11.30 through 11.35.

Perhaps the only additional comment needed is that the terminology for electric clutches and brakes is ambiguous in that it does not distinguish between friction clutches, which are electrically activated, and clutches that depend upon electric or magnetic phenomena for torque transfer, namely, magnetic particle, eddy current, and hysteresis clutches and brakes.

TABLE 11.12.1 COMPARISON OF CLUTCHES AND BRAKES

Clutch Designation	Description	Application
Serrated tooth or postive lock-up clutch	Small teeth with angular faces. Once engaged there is no slip until the load reaches the disengagement torque, which is determined by the angle of the teeth and the axial force applied. See Figure 11.12.1.	Positive connection where angular position between shafts on either side of the clutch may be controlled to within one degree or less. May be engaged only with shafts stationary or with very small speed differences. Some shock occurs when moving shaft are engaged.
Square jaw of positive lock-up clutch	Usually 2, 3, or 4 teeth. No slip unless teeth are broken. See Figure 11.12.2a.	Used where positive lock-up is required. The relative angle between shafts is determined by the number of teeth. Must be engaged when the shafts are either stationary or when the speed difference is too small for the shock upon engagement to cause damage.
Spiral law clutch	Usually 2, 3, or 4 teeth. Will transmit high torque in one direction only. See Figure 11.12.2b.	Used for positive lock-up in one direction only. May be engaged at slightly higher speeds than the square jaw clutch.
Detent overload clutch	Relative rotation between shaft prevented by detents in mating flanges until torque exceeds the design value. Operation is independent of the direction of the applied torque or of the rotational speed. See Figure 11.12.3.	Used where overload release is to be independent of speed and direction of rotation. Use in packing and bottle filling machines, in textile machines, and in conveyors.
Dry disk clutch/brake	Perhaps the most common type of clutch/brake, it consists of two or more disks alternately connected to the input and output shafts. Torque is dependent upon friction between plates and upon the axial load. Largest at zero slip; when input shafts rotate at the same speed. May be manually, hydraulically, pneumatically, or electrically activated. See Figure 11.6.1.	These are used to engage shafts initially turning at speeds which may differ 4000 rpm or somewhat more, depend upon the lining material and the cooling capacity. Used in light and medium duty vehicles, presses, ball mills, and other industrial equipment. Easily disengaged.
Rim clutch brake	High torque friction clutch with segmented lining which is hydraulically or pneumatically forced against a concentric drum. See Figure 11.12.4 and problem 11.34.	Used with high torque motors and heavy industral equipment, and in propeller drives for tugboats, towboats, and large ships. Relatively easy to maintain.
Cone clutch/brake	Axially loaded concentric friction cones with the torque capacity dependent upon axial force and friction coefficient. Construction limited to one set of cones. See Figure 11.7.2.	Used where small axial force is to provide somewhat larger torque than available from a single plate clutch of similar diameter. Cone clutch/brake with small angle may require disengagement force. Not readily available commercially.

TABLE 11.12.1 COMPARISON OF CLUTCHES AND BRAKES (continued)

Clutch Designation	Description	Application
Centrifugal clutch/brake	Rim-type clutch using angularly constrained radially free masses to move outward under centrifugal force. Torque proportional to square of rotational speed. May be used as brake to prevent overspeeding. See Figure 11.5.1.	Used for automatic clutching during startup, for gradual acceleration of high inertia loads, and for automatic release at low speeds. Also used for overload protection.
Wet disk clutch/brake	Clutch plates and lining are immersed in oil for protection against dirt and water and to provide greater cooling with circulation through an oil cooler. Special lining material provides friction in an oil bath. Internal construction is similar to that of dry multiple disk clutches and brakes.	Used in dirty and wet environments and for large equipment with heavy braking loads, such as earth moving equipment and mining trucks and equipment.
Controlled slip clutch/brake	Plate clutch/brake with hard, smooth, pressure plates. Lubricant may be used with a grooved lining. Uses in an oil bath. See Figure 11.12.5.	Used as a speed control device: speed regulated by axial load on disks. Used in large fans, conveyors, pumps, etc. May require an oil cooler.
Eddy-current clutch/brake	These units have no wear surfaces. Energy is dissipated by heating of the plates or cylinders. Torque capacity increases with speed difference between input and output shafts. Larger sizes require external electrical power. See Figure 11.10.3.	Used in material handling, conveyors, tape manufacturing, machine tools, ball mills, mixers. Small units used in tape recorders and computers. Used with computer control of manufacing processes. May be used for slip control.
Hysteresis clutch/brake	Torque controlled by coil current and independent of speed differential for values less than critical. Larger sizes require external electrical power. Also classes as controlled slip clutches/brakes. See Figure 11.10.2.	Used in tape drives, and for tension control in tape, foil, wire, synthetic fiber, and paper manufacture. Also used with coil winding machines.
Magnetic particle clutch/brake	Torque proportional to coil current and independent of speed difference between input and output shafts throughout its entire speed range. All sizes require external electrical power. See Figure 11.10.1.	Used in extrusion machines for plastics, bar soap, etc. Also for conveyors and for soft starts and stops. Widely used in tension control. Suitable for remote computer control.
One-way clutch (overrunning clutch)	Either roller and ramp or sprag type. Wedging action allows relative rotation in one direction but not the other. See Figures 11.9.3 and 11.9.4.	Used in automotive automatic transmissions as well as tractors and heavy equipment. Also for bucket elevators, rack and gear feeds, and turbine starting.

(continued)

TABLE 11.12.1   COMPARISON OF CLUTCHES AND BRAKES (continued)

Clutch Designation	Description	Application
Wrap spring clutch	Friction between helical spring and input and output shafts transmits torque in one direction but not the other. Needs no lubrication. May be used with electrical control to engage and release wrap spring. See Figure 11.9.2.	Used for rotary indexing, jogging, and step positioning in farm machinery, copying and sorting equipment. Used with electrical control to give start and stop motion from a steadily turning shaft.
Band brakes	Large torque may be had from small activation force, but with braking for only one direction of rotation. Lining wear is uneven. May be used where self-locking is desired. See Figure 11.1.4.	Used for truck and automotive emergency brakes, for safety stop on chain saws, and for backstops (anti-reversing brakes) on elevators and hoists.
Pivoted shoe drum brakes	Large torque may be had from a small activation force. Requires two shoes to obtain equal braking in for both directions of rotation. May be self-locking. More subject to brake fade than caliper disk brakes when similar cooling is available. See Figure 11.4.1.	Widely used for both industrial and vehicular braking for light to medium loads. Simple construction, relatively easy to maintain.
Caliper disk brakes	Requires relative larger activating force than drum brakes of the same diameter and normal width proportion. Easily air cooled and largely self-cleaning. Will not self-lock. Easily maintained See Figure 11.6.2.	Primarily used on vehicles and light industrial equipment where air cooling and ease of maintenance are important.

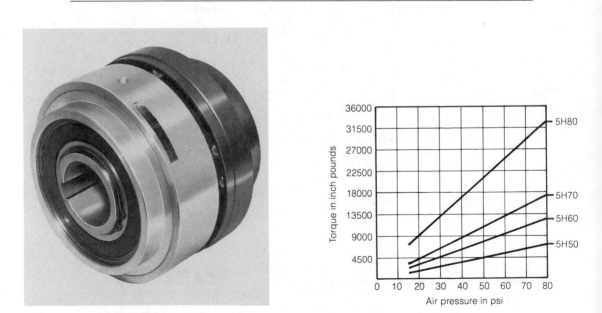

Figure 11.12.1 Tooth clutch—air actuated—and representative torque–air pressure relation. *Source: Horton Manufacturing Co., Minneapolis, MN.*

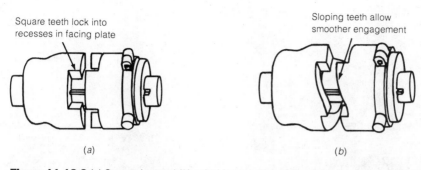

Square teeth lock into
recesses in facing plate

Sloping teeth allow
smoother engagement

(a)   (b)

Figure 11.12.2 (a) Square jaw and (b) spiral jaw positive lock-up clutches. *Source:* Machine Design, *June 28, 1984. Copyright 1984 Penton Publishing, Inc., Cleveland, OH.*

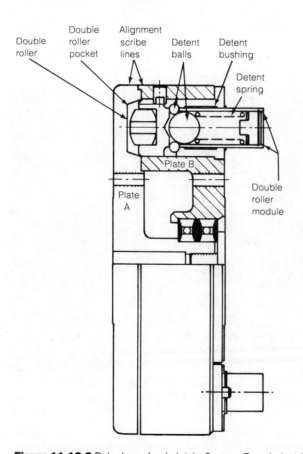

Double
roller

Double
roller
pocket

Alignment
scribe
lines

Detent
balls

Detent
bushing

Detent
spring

Plate B

Plate
A

Double
roller
module

Figure 11.12.3 Detent overload clutch. *Source: Zurn Industries, Inc., Erie, PA.*

Figure 11.12.4 (a) Air-activated drum clutch, (b) detail of the external shoes, and (c) an application of a drum clutch to a grinding mill in a Canadian refractories plant. *Source: Eaton Corp., Airflex Division, Cleveland, OH.*

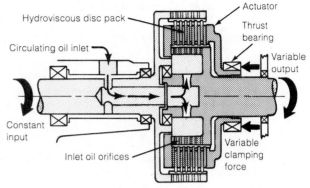

Figure 11.12.5 (a) One variety of controlled slip clutch, (b) a schematic of its operation, and (c) its use at a pipeline pumping station. *Source: Philadelphia Gear Corp., King of Prussia, PA.*

REFERENCES

1. Catalog P-137-16, 16th ed., Warner Electric Brake and Clutch Co., South Beloit, IL, 1981.

2. *Twinflex Industrial and Marine Disc Brakes*, Bulletin TC8/81, Twinflex Corp., Horseheads, NY, undated.

3. Wiebusch, C.F., The spring clutch, *Journal of Applied Mechanics*, 6 (3), A103–A108 (1939).

4. Wahl, A.M., Discussion of the spring clutch, *Journal of Applied Mechanics*, 7 (2), A89–A91 (1940).

5. Kaplan, J., and Marshall, D., Spring clutches, *Machine Design*, 28, (8), 107–111 (1956).

6. Johnson, M.E., *Testing the Cone Brake Design*, SAE Technical Paper 790465, Society of Automotive Engineers, Warrendale, PA, 1979.

7. Föttinger, H., Die Hydrodynamische Arbeitsubertragen, *Jahrbuch der Schiffbautechnischen Gesellschaft*, 31, 171–214 (1930).

8. Eksergian, R., The fluid torque converter and coupling, *Journal of the Franklin Institute*, 235, 441–478 (1943).

9. Hunsaker, J.C., and Rightmire, B.G., *Engineering Applications of Fluid Mechanics*, Mc-Graw-Hill, New York, 1947, Chapter 17.

10. Jandasek, V.J., *Design of Single-Stage, Three-Element Torque Converter*, Design Practices—Passenger Car Automatic Transmissions, AE-5, Society of Automotive Engineers, Warrendale, PA, 1973, pp. 201–226.

11. Upton, E.W., *Application of Hydrodynamic Drive Units to Passenger Car Automatic Transmissions*, Design Practices—Passenger Car Automatic Transmissions, AE-5, Society of Automotive Engineers, Warrendale, PA, 1973.

12. Orthwein, W.C., Designing double-shoe drum brakes with a programmable calculator, *Computers in Mechanical Engineering*, 1, 51–56 (1983).

13. Orthwein, W.C., *Estimating Torque and Lining Pressure for Bendix-Type Drum Brakes*, SAE Technical Paper 841234, Society of Automotive Engineers, Warrendale, PA, 1984.

14. Orthwein, W.C., *Clutches and Brakes, Design and Selection*, Dekker, New York, 1986.

PROBLEMS

Section 11.1

11.1 Show that equation 11.1.14 holds for a backstop to resist the rotation indicated if levers *a* and *b* are arranged as in Figure 11.1.4 and shown schematically in the figure below.

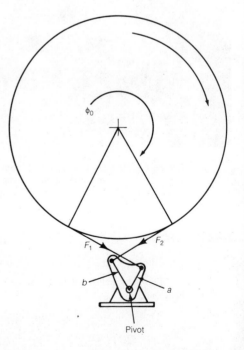

11.2   Find the force P required to hold a torque of 1800 N mm using a band brake as shown in the figure. Dimensions are in mm, $\mu = 0.41$, and $\phi = 200°$.

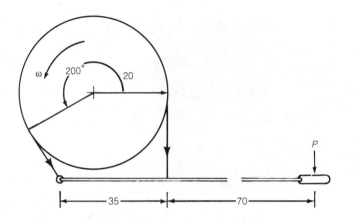

11.3   Design a band brake to resist a torque of 17 800 N m when the maximum lining pressure must not exceed 2.2 MPa, the friction coefficient is 0.33, and the drum width must not exceed 710 mm. If the drum radius is less than 710 mm let the width be equal to or less than the drum radius for cooling and uniform wear across the width. The angle of wrap must not exceed 300° because of the linkage intended. Give the drum width, drum diameter, and forces F_1 and F_2.

11.4   Calculate the required force P on the brake as shown in Figure 11.1.2 if the brake is to provide a torque of 3800 in.-lb for a wrap angle of 250°, a drum radius of 8.0 in., and a friction coefficient is 0.40. Length a is 4.00 in., length b is 0.90 in., and length c is 5.20 in.
  [*Ans.* 240.38 lb.]

11.5   Derive an expression for P as a function of T for the band brake and lever arrangement shown in Figure 11.1.2 when the drum rotates in the opposite direction and $b = 4.50$ in.

11.6   Calculate $dT/d\phi_0$ from equation 11.1.12 to show that changes in the friction coefficient between 0.4 and 0.2 have less effect on the brake torque for $\phi_0 = 3.0$ radians than for $\phi_0 = 2.0$ radians. This is an added benefit of the design shown in Figure 11.1.6 in addition to force equilibrium relative to the shoe.

11.7   Design a self-locking band brake with a lever as in the figure with problem 11.1 such that it may be prevented from locking by an upward force on the lever. It should exert a torque of 5000 ft-lb on a drum 28 in. in diameter with a lining pressure no greater than 210 psi when the drum rotates in the self-locking direction. The wrap angle should be no larger than 310°. The control handle is to be welded to the plate shown in the figure. Use $a = 1.5$ in. and $\mu = 0.34$. Answer depends upon choices for a and b.

11.8   A chain saw manufacturer has used a lever in front of the upper handle, as shown in Figure 11.5.1, to activate a band brake which acts on the flywheel. If the drum diameter is 4.0 in. and if the torque is 300 in.-lb, design a similar band brake to stop the chain saw in the event of a kickback (where the saw rises, the lever strikes the user's arm, and the brake is applied). Give the dimensions of the lever, the width and thickness of the band, force F_1, and p_{max}. Let the wrap angle be 270°, the yield stress of the available spring steel used for the band be 170 000 psi, and the metal-to-metal friction coefficient be 0.25. A typical band brake attached to the inside of the housing with M on it is shown in the figure below. Assume a 25-lb force between the lever and the user's arm.

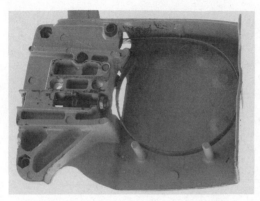

11.9 Design a wet band brake (operating in an oil bath) as shown in Figure 11.1.6 to operate on a drum 8 in. in diameter and to exert a torque of 2500 in.-lb. Although p_{max} can exceed 400 psi for wet brake lining material, use $p_{max} = 300$ psi and use $\mu = 0.14$ and $\phi_0 = 350°$. Find F_1, F_2, w, and the minimum thickness of the brake band based upon a safety factor of 2.0 relative to the tensile stress of 160 000 psi for ASTM medium carbon steel strip.

[*Ans.* $t_{min} = 0.015$ in.]

11.10 Find the lining width w and sketch the pivot position, with dimensions, relative to the lining ends for a single-shoe external drum brake to supply a torque of 800 N m when acting upon a drum having a 350 mm diameter if the expected friction coefficient is 0.35 and if the lining pressure is to be no more than 1.13 MPa. An earlier design, which had a lining that was too narrow, had $\phi_1 = 25°$ and $\phi_2 = 110°$. (Single-shoe brakes are of academic interest only.)

[*Ans.* $w = 52.91$ mm, $R_{max} = 620.5$ mm. Recommend $w = 53$ mm, $R = 615$ mm.]

11.11 What force must be applied by a hydraulic cylinder to close the brake and provide the required torque in problem 11.10 if it acts at a distance $l = 2(R - r)/3$ from the pivot?

11.12 Show that the torque calculated from equation 11.2.3 for a non-symmetrically pivoted external shoe and that calculated from (11.2.17) for a symmetrically pivoted external shoe are equal when the lining in both shoes is symmetrical about $\phi = \pi/2$ and when ϕ_1 and ϕ_2 are the same for both shoes.

11.13 By evaluating the derivatives of torque with respect to ϕ_1 and ϕ_2 at the heel and toe of a shoe, respectively, show that relatively little torque may be gained from a drum brake by either extending ϕ_1 below 20° or extending ϕ_2 beyond 160°.

Sections 11.3 and 11.4

11.14 Design a pivoted internal drum brake with two shoes to supply a torque of 600 in.-lb and calculate the minimum anchor diameter for a safety factor of 1.5 relative to a yield stress in shear of 35 000 psi. Shoe width should be no greater than one-third of the drum diameter for diameters less than 6.0 in. For larger diameters the shoe width should not be greater than one-fifth of the drum diameter. Use $\mu = 0.20$, $p_{max} = 250$ psi, and $\phi_1 = 15°$. Small drum diameters are preferred and $1.4 \leq r/R \leq 1.7$. [A computer/calculator program is recommended.]

[*Ans.* $R = 1.6$ in., $\phi_2 = 139.057°$, $d = 5.0$ in., $w = .75$ in. is one of many possible designs.]

11.15 Design a pivoted external drum brake with two shoes for problem 11.14 for $1.4 \leq R/r \leq 1.7$. It is to be used to control the speed of a drum on which the shoes act, so it should not be self-locking. Compare the anchor pin force with that found in problem 11.14.

[*Ans.* $R = 3.5$ in., $\phi_2 = 116.770°$ is one of many possible designs.]

11.16 Design a pivoted internal drum brake with two shoes to supply a torque of 10 340 N m. Because of other mechanisms in the drum, the shoes should not extend beyond 140° and the shoe pivots must be no less than 0.18 of the radius from the inside surface of the drum. The preferred inside diameter is 500 mm and the preferred lining width is 100 mm. If these restrictions are in conflict, ignore the inside diameter preference. The expected friction coefficient is 0.37 and $p_{max} = 2.0$ MPa. Use $\phi_1 = 20°$.

[*Ans.* $r = 250$ mm, $R = 200$ mm, $\phi_2 = 137.020°$, $F = 76\,866$ N at the anchor.]

11.17 Design an external drum brake to satisfy the torque in problem 11.16 if the minimum outside diameter of the drum is 500 mm and the drum width should be no more than 100 mm. The friction coefficient is unchanged, $\phi_1 = 30°$, and the shoe pivots should be 40 mm from the drum surface.

[*Ans.* $r = 250$ mm, $R = 290$ mm, $w = 90$ mm, $\phi_1 = 30°$, $\phi_2 = 145.50°$ $F_a = 69\,970$ N.]

Section 11.5

11.18 Find the torque that may be transmitted by a centrifugal clutch having 9 shoes as shown in the figure when it rotates at 3650 rpm. Use $r_c = 140$ mm, $r = 180$ mm, $\mu = 0.55$, and $k_s = 840.0$ N/mm.

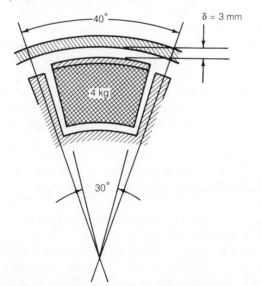

11.19 Find the speed in rpm at which the shoes in problem 11.18 first make contact with the drum.

Section 11.6

11.20 Show that the torque capability of a multiple plate clutch or brake with N friction surfaces may be expressed as

$$T = \mu\pi N p_{max} r_i (r_o^2 - r_i^2)$$

and that if N' is the number of plates mounted on the central shaft for a clutch or brake construction similar to that shown in Figure 11.6.1, that

$$N = 2N'$$

11.21 Show that when equation 11.6.6 holds the maximum torque capacity for a clutch/brake with N friction surfaces is given by

$$T = \frac{2\pi N}{3\sqrt{3}} \mu p_{max} r_o^3$$

and that the activation force becomes

$$F = 2\pi p_{max} \left(1 - \frac{1}{\sqrt{3}}\right) \frac{r_o^2}{\sqrt{3}}$$

where uniform wear is assumed.

11.22 Design an annular plate clutch for a rated torque of 117 910 in.-lb with a maximum plate diameter of 11.0 in. using a lining material with a friction coefficient of 0.44. Assume uniform wear and a maximum lining pressure of 360 psi. Use results from problem 11.21 and indicate the number of plates on the central shaft and the activation force required.
  [*Ans.* 2 plates (4 contact surfaces), $F_a = 16\,697$ lb.]

11.23 Re-rate the clutch/brake described in problem 11.22 if the lining coefficient may drop to 0.20 due to careless handling and maintenance and if the effective maximum lining pressure may drop to 175 psi due to line obstructions in the hydraulic control system.
  [*Ans.* 28 165 in.-lb.]

11.24 Design an annular plate clutch rated at 9600 N m using a friction coefficient of 0.40, a lining pressure of 1.21 MPa, and a maximum plate diameter is 280 mm. Give the number of friction surfaces and the number of plates on the central shaft along with the required activation force. Assume uniform wear.
  [*Ans.* 3 plates on the central shaft.]

11.25 Redesign the clutch in problem 11.24 if the pressure is assumed uniform over the lining. Give the number of plates on the central shaft and the required activation force. Central shaft radius is 30 mm.

11.26 Find the diameter of a single-side, single-plate clutch (one contact surface) to provide the torque required in problem 11.24. Assume uniform wear.

11.27 Your company has decided to enter the caliper disk brake market using a 280-mm disk. Using the same lining material as in problem 11.24, estimate the braking torque that can be obtained from a caliper with two pads for which $\phi_0 = 20°$ and $r_i = 115$ mm. Also find the required activating force.

Section 11.7

11.28 Compare the torque capacity of a cone clutch with $\alpha = 70°$, $r_o = 5.50$ in., and $r_i = 3.12$ in. with that of plate clutches having 1 and 2 plates on the central shaft (2 and 4 contact surfaces repectfully) for a friction coefficient of 0.44.

11.29 Your company plans to propose use of a cone clutch to replace a coupling between motor and gearbox in a portable sawmill for use in Alaska. Recommend α, r_o, r_i, and μ if the torque is to be 10 500 N m and if the extraction (disengagement) force is to be between 250 and 300 N. Design for minimum r_o by using the result of problem 11.21. p_{max} is 1.3 MPa for the lining material to be used. *Note:* $\alpha = 32°$ is easiest to produce with the company's machines. Can it be used? $\mu = 0.4$.

Section 11.8

11.30 Estimate the torque required by each of six brakes to stop a loaded mine shuttle car weighing 37.5 tons and moving at 6.2 mph if the stopping distance must be no more than 15 ft. Assume constant deceleration. Also estimate the heat that must be dissipated. Neglect rotary inertia of the motor, the wheels, and other rotating components. Wheel diameter is 44.0 in.

11.31 Estimate the torque required to stop a conveyor in 1.5 s if it moves at 131 mm/s and usually carries a load of 107 700 kg, not including the weight of the chain itself, which is 2400 kg. The sprocket at each end weighs 44.2 kg and each has a radius of gyration of 50 mm. The gear train is as shown in the figure where the weights and radii of gyration of the gears are listed below. Each of the 78 rollers which support the chain along its length weigh 2.3 kg, has an outside diameter of 30 mm and has a radius of gyration of 11 mm.

Gear number	rpm	Radius of Gyration (mm)	Mass (kg)
1	575.00	65	2.1
2	122.42	306	46.3
3	122.42	54	1.5
4	12.51	537	144.8

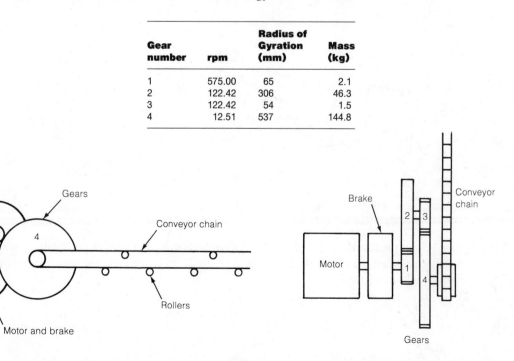

11.32 Repeat problem 11.31 but with the brake attached to gear 4. The increased torque requirement is the reason that the brake is placed on the fastest shaft.

11.33 Show that the axial load for a tooth clutch with triangular teeth required to hold the jaws in contact is given by

$$F_a = \frac{2T}{r_o + r_i} \tan(\theta + \alpha)$$

where α is defined in terms of the coefficient of friction between teeth by $\mu = \tan \alpha$, T denotes the applied torque, and the remaining quantities are defined in the figure.

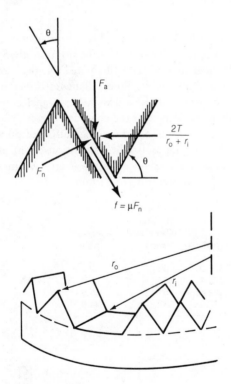

Tooth geometry for triangular teeth

11.34 Based upon the results of problem 11.33, show that if δ represents the amount an axial spring is deflected when the clutch described in problem 11.33 is engaged, the spring constant (spring rate) should be given by

$$k \geq \frac{T}{r_i} \frac{\tan \theta}{\delta}$$

where T is the maximum design torque and θ and r_i are as shown in the figure with problem 11.33. Justify setting $\alpha = 0$ in the expression for F_a obtained in problem 11.33. Is there an implied safety factor in this formula? Explain.

11.35 Show that for a tooth clutch with rectangular teeth as in Figure 11.12.2 the tooth proportions should be given by

$$\frac{b}{\dfrac{\pi}{2N}(r_o + r_i)} = \frac{\tau_y}{\sigma_y}$$

where h is determined from the relation

$$h = \frac{2T}{N\sigma_y}\frac{\zeta}{r_o^2 - r_i^2}$$

where T denotes the transmitted torque, ζ the safety factor, N the number of teeth, σ_y the yield stress in tension (assumed equal to or greater than the yield in compression), and τ_y the yield stress in shear. The remaining quantities are pictured in the figure below.

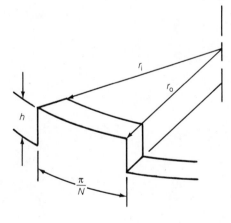

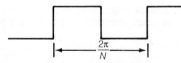

Tooth geometry for rectangular teeth.

11.36 Show that the torque capability of a rim clutch/brake is given by the relation

$$T = 2\pi\mu pwr^2$$

where r denotes the drum radius, w the lining width in the direction of the drum axis, μ the friction coefficient, and p the pressure between the lining and the drum. Pressure p is applied hydraulically or pneumatically. Also show that the ratio of the torque capacity of a rim clutch to that of a multiple disk clutch is given by

$$\frac{T_{\text{rim}}}{T_{\text{disk}}} = \frac{3\sqrt{3}}{N}\left(\frac{w}{r_o}\right)\left(\frac{r_1}{r_o}\right)^2$$

where r_1 represents the diameter of the rim clutch and r_o the outside diameter of the lining in a plate clutch having N contact surfaces in which the inner lining diameter r_i is given by $r_i = r_o/\sqrt{3}$. This shows the advantage of rim clutches in installations where space permits w/r_o to be greater than 1 and $(r_1/r_o)^2$ to be greater than 10. Rim clutches are used in equipment where large diameters are acceptable, as in hammer, ball mill, and ship and tugboat drives.

CHAPTER TWELVE

BELT AND CHAIN DRIVES

NOTATION

A	area (l^2)		k,k'	coefficient, integer (1) or exponent
b	belt constant (1)			$(\mu/\sin \alpha/2)$ (1)
C	center distance (l)		L	belt or chain length (l)
c	perpendicular distance from the neutral surface (l)		M	bending moment (ml^2/t^2) or constant (m/l)
c_1,c_2,c_3	coefficients		m	integer (1) or mass/unit length (m/l)
D	outside diameter, large sheave (l)		N,N_i	number of teeth or number of cycles (1)
D_p	pitch diameter, large sheave (l)		N_1,N_2	number of teeth on small and large sprocket, respectively (1)
d	outside diameter, small sheave (l)		n	revolutions/minute (rpm) $(1/t)$
d_p	pitch diameter, small sheave (l)		$\tilde{n}$	$n/1000$ $(1/t)$
F_1,F_2	tight and slack side belt tensions (ml/t^2)		p	pitch (l)
F_s	static belt tension (ml/t^2)		R,r	radius (l)
hp	horsepower (ml^2/t^3)		T	torque (ml^2/t^2)
I	second moment of area (area moment of inertia) (l^4)		v	velocity (l/t)
$\hat{i},\hat{j},\hat{k}$	unit vectors in x, y, and z directions, respectively (1)		α	groove included angle (1)
			θ	angle of contact between belt and sheave (1)
K_0,K_m	dimensional constants for belts (m/lt^2)		μ	friction coefficient (1)
K_1,K_4	dimensional constants for belts (ml/t^2), $(ml^{0.7+k}/t^{2-k})$		ρ	radius of curvature (l)
K_2,K_{sr}	dimensional constants for belts (ml^2/t^3)		σ	stress (m/lt^2)
K_3	dimensional constant for belts (m/t^2)		ϕ	$\sin^{-1}[D-d)/2C] = \sin^{-1}[D_p - d_p)/2C]$ (1)
K_r	dimensional constant for chains $(ml^{1.2}/t^3)$		ψ	angle (1)
K_s	dimensional constant for chains (ml^x/t^3)		ω	angular velocity $(1/t)$

Belt and chain selection is a matter of choosing a standard belt or chain size that will transmit the design power for an expected life of 18 000 h or longer for belts and 15 000 h or longer for chains. Empirical formulas for belt and chain power ratings per belt or per strand of chain imply these respective fatigue lives. Accordingly, the purpose of this chapter is to present these formulas and to demonstrate their use in designing belt and chain drives.

American National Standards Institute (ANSI) standards for chains employ Old English units and employ both Old English and SI for most belts. For brevity, however, neither examples nor problems in this chapter will involve SI units because that would double the number of tables required to explain the same method.

12.1 COMPARISON OF BELT, CHAIN, AND GEAR DRIVES[1]

None of these three mechanical drives is best suited for all machine systems. The following comparison is to enumerate the advantages of each and thus describe the rather ill-defined and partially overlapping regions in which each of the three may be the first choice of the designer. Once an engineering selection has been made on the basis of strength and life, economic factors such as original cost, maintenance direct costs, and the cost of production lost during down time are usually considered in arriving at the final selection.

Advantages of belts

1. Electrical insulation is provided because there is no metal-to-metal contact between driver and driven units.
2. There is less noise than with a chain drive.
3. Flat belts can be used for extremely long center distances where chain weight would be excessive.
4. Flat belts can be used at extremely high-speeds where chain inertia must be considered as influencing chain fit at the sprocket and chain tension.
5. No lubrication is required.
6. Shaft center distance variation and shaft alignment is much less critical than for gear drives or chain drives.

Chain drive advantages

1. Shaft center distance variation may be more easily accomodated than with gear drives.
2. Chains are easier to install and replace than belts because the center distance between driven and driver units need not be reduced for installation (splice and link belts overcome this objection at the cost of lower power ratings).
3. Chains require no tension on the slack side, so that bearing loads are reduced.
4. Chains do not slip or creep as do belt drives (except for toothed belts).

5. Chain drives are more compact because sprocket diameters are smaller and chains are narrower than sheaves and belts for the same power transfer.

6. Chains do not develop static charges.

7. Chains do not deteriorate with age, heat, or oil and grease.

8. Chains can operate at higher temperatures than belts.

Advantages of gear drives

1. Gear drives are more compact than chain or belt drives because center distances are minimum.

2. Gears have a greater speed capability than either belts or chains.

3. Gears have a greater range of speed ratios than chains.

4. Gears can better transfer high power at high speed than can either belts or chains.

5. Metal gears do not deteriorate with age, heat, or oil and grease.

6. Metal gears do not develop static electric charges.

12.2 SERVICE FACTORS AND DESIGN POWER

Transmitted power, as represented by the rated power of the driving unit, is often increased for design purposes by a multiplicative factor called a *service factor* prior to the actual selection of the belts or chains to be used in the drive. The increased power is known as the *design power*. Its purpose is to account for momentary overloads (i.e., starting torque) and load fluctuations from the driven unit (punch press, piston pumps, hammer mills, etc.) as well as from the driver, such as an internal combustion engine. Service factors are provided by belt and chain manufacturers and/or their associations based upon accumulated experience in applications represented by that equipment listed in their design guides. In this chapter we use both sources: service factors from a manufacturer (Gates Rubber Co.) for belts and from an association (American Chain Association) for chains.

Although service factors used for belts and chains may range from 1.0 to 1.8, it should be emphasized that differences of 0.1 represent a precision that is more apparent than real. Distinctions of 0.1 in the service factors in Figure 12.5.1 will be obeyed, however, because boundaries must be set somewhere and because the divisions shown have generally proven to be satisfactory.

12.3 V-BELT DESCRIPTION

Five different organizations have established three different standards for V-belt design. Three of them, the Rubber Manufacturers Association, the Mechanical Power Transmission Association, and the Rubber Association of Canada, have cooperated to form the RMA-MPTA-RAC standard. The other two standards are the SAE standard by the Society of Automotive Engineers and the ASAE standard by the American Society of Agricultural Engineers. The RMA-MPTA-RAC has established different belt specifications for

1. *Classical Multiple V-Belts*. These obsolete A, B, C, D, and E cross-sections, shown in Figure 12.3.1a, are not recommended for new designs.

2. *Double V-Belts* (see Figure 12.3.1b). Use where the belt must flex in two directions.

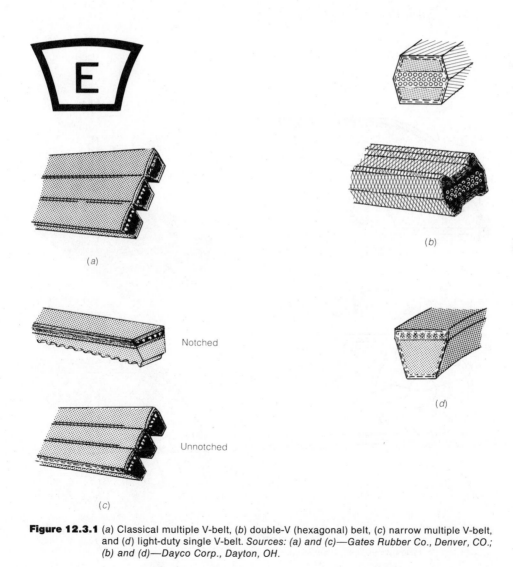

(a)

(b)

Notched

Unnotched

(c)

(d)

Figure 12.3.1 (a) Classical multiple V-belt, (b) double-V (hexagonal) belt, (c) narrow multiple V-belt, and (d) light-duty single V-belt. *Sources: (a) and (c)—Gates Rubber Co., Denver, CO.; (b) and (d)—Dayco Corp., Dayton, OH.*

3. *Narrow Multiple V-Belts.* These 3V, 5V, and 8V cross sections have replaced the five obsolete A, B, C, D, and E cross sections because they cover the same power range with only three belt sizes.

4. *Single V-Belts.* Light duty belts (Figure 12.3.1*d*).

5. *Synchronous Belts.* Toothed belts, teeth on one or two sides to reduce slip; used for timing etc. (Figure 12.3.2*a*).

6. *Variable Speed Belts.* Wide belts to be used with variable speed sheaves; shown in Figure 12.3.2*b*).

7. *V-Ribbed Belts.* For high-speed drives.

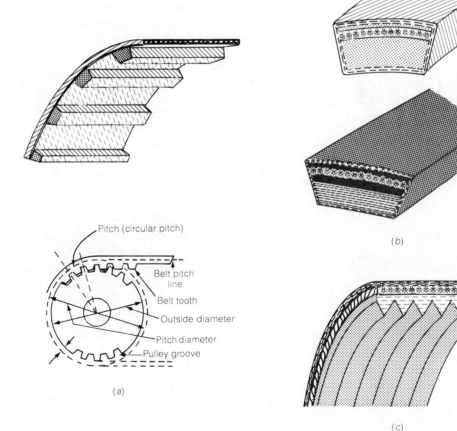

(b)

Pitch (circular pitch)

Belt pitch line

Belt tooth

Outside diameter

Pitch diameter

Pulley groove

(a)

(c)

Figure 12.3.2 (a) Synchronous belt (cog belt), (b) variable speed belts, and (c) V-ribbed belt. *Sources: Dayco Corp., Dayton, OH.; pulley illustration courtesy The Rubber Manufacturers Assn., Inc., Washington, DC.*

The SAE standards do not specify particular belt cross sections, but rather specify sheave dimensions and require only that the V-belts be functional in sheaves manufactured according to SAE specifications, that standard lengths up to and including 80 in. be in $\frac{1}{2}$-in. increments, and that lengths over 80 in. be in 1-in. increments without fractions.

Just as each of these three associations have different standards, they also differ on the use of the words sheave and pulley. Although none of these standards contain a glossary, it appears that RMA-MPTA-RAC and ASAE use sheave in connection with V-belts and pulley in connection with flat belts; in other words, a sheave has grooves and a pulley does not. We shall adopt this terminology for the remainder of this chapter.

We will be concerned with only one of the seven standards listed above, that for Narrow Multiple V-Belts. Their cross sections are designated as 3V, 5V, and 8V for specifications in inches and pounds and as 9N, 15N, and 25N for specifications in SI units. Similar design procedures hold for the remaining six belt standards.

Since the RMA-MPTA-RAC design method we shall describe implies an 18 000-h life for the belts involved, but provides no formulas for particular life estimation, we shall append the life predictions given in Refs. 3 and 5 as a means of estimating particular belt life.

Drive design formulas for agricultural and automotive systems involving more than two sheaves may be found in the excellent collection of formulas and tables listed in Refs. 4 and 5 published by the Dayco Corporation.

12.4 VARIABLE PITCH SHEAVES

Variable pitch sheaves are used with the wide variable speed V-belts shown in Figure 12.3.2b. Their design may be divided into two types: those for driveN sheaves and those for driveR sheaves. (Capitalization of the last character is an industry practice for emphasis.) Some mechanism must be available to maintain belt tension, such as changing the center distance or using idlers, as the sheave pitch diameter changes. Idlers are discouraged by belt manufacturers because they shorten belt life.

Representative driveN sheaves, which are tension sensitive, are shown in Figures 12.4.1 and 12.4.2d. As the tension increases the sheave's spring loaded side flange moves outward, which causes the pitch diameter to decrease. This reduces the speed ratio between the driveN and driveR when the driveN sheave is the larger of the two, as is generally the case, thus enabling the driveR to deliver power at a nearly constant speed. The Belleville spring version is preferred over the helical spring because it may apply less side force to the belt, and thus lengthen belt life.

Since decreasing the driveN speed as the belt tension increases tends to hold the torque constant, these sheaves are useful when the driveR delivers maximum torque over a relatively small speed range, as in the case of an internal combustion engine.

Speed-sensitive driveR sheaves are shown in Figure 12.4.2(a), (b), and (c). The weights attached to the Belleville spring rotate outward as the speed increases, thus widening the groove and reducing the pitch diameter, as shown in Figure 12.4.2b.

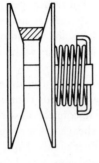
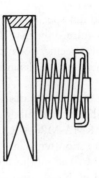

(a)

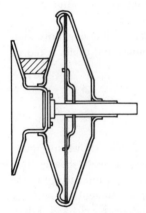
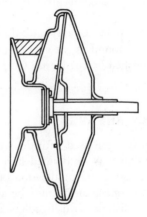

(b)

Figure 12.4.1 Variable pitch sheaves for drive shafts: (a) helical spring type and (b) Belleville spring type. *Source: Dayco Corp., Dayton, OH.*

Consequently, sheaves of this type may be used to drive nearly constant speed fluctuating torque machines with a driveR whose speed changes with the torque demand; i.e., a piston compressor with pressure relief (little torque is required when the tank is at rated pressure) when driven by an internal combustion engine.

The sheave shown in Figure 12.4.2c behaves just the opposite: Centrifugal force on the weights causes them to move outward and narrow the groove with increased

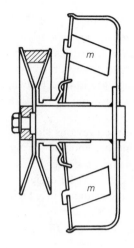

(a)

(b)

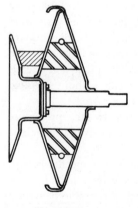

(c)

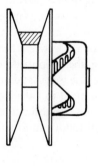

(d)

Figure 12.4.2 Variable pitch sheaves for drive shaft. (a) Belleville spring with weights *m*, (b) which move outward with increasing speed, (c) internal centrifugal weights, and (d) torque–tension type with cam activation; cam and spring at right side. *Source: Dayco Corp., Dayton, OH.*

sheave speed, which forces the belt to a larger pitch diameter. A spring, such as the garter spring shown, retracts the weights as the speed reduces. When used with a driveN machine which tends to maintain constant belt tension, such as a centrifugal pump, the increased torque demand on the driveR causes it to reduce speed. Hence, the combination tends toward speed stability. This and the previous design are limited to belt tensions that are not large enough to prevent the weights from narrowing the groove width.

The torque tensioning sheave in Figure 12.4.2d differs from the previous sheaves in that the pitch radius depends upon the torque, the cam contour, and the spring rather than upon tension or speed. It acts to force the sheave flanges together as the torque increases, thus increasing the pitch diameter. Spring and cam may be designed to function together to give the proper tension for a range of torque changes, in which case the sheave may contribute to longer belt life.

12.5 V-BELT SELECTION

Belt selection is a matter of finding a standard belt that will closely match the required speed ratio between input and output shafts, the required power transfer, and the required center distance, which is the distance between center lines of the input and output shafts. In this statement of the problem we have assumed only two sheaves are involved. For cases with more than two sheaves, see Refs. 3, 4, and 8.

As stated earlier, we shall demonstrate the design procedure for the 3V, 5V, 8V cross sections as being representative of the other six series. Although the basic design procedure to be followed is described in ANSI IP-22/1983, dated March 17, 1983, we shall use the version that appears in Ref. 2 because it has augmented the procedure with tables limiting the small sheave diameter based upon bearing loads for standard NEMA (National Electrical Manufacturers Association) motor sizes and has provided tables relating sheave outside diameter to the *pitch diameter*. The pitch diameter is important because it is that diameter at which the belt appears to ride without slipping; it is, therefore, used to calculate the speed ratio between shafts.

Because of standardization, belt selection has become a step by step procedure which we shall follow and then summarize at the end of this section with a flowchart of the process.

At the outset of the belt selection procedure it is assumed that a specific motor, or engine, has been selected and that a specific speed ratio and a specific center distance between input and output shafts have been chosen, along with acceptable deviation limits for the last two.

The first step is to select a service factor from Table 12.5.1, which lists machines that represent a variety of starting, running, and shock characteristics. If none of these machine characteristics are equivalent to the driveR and driveN equipment being designed, Figure 12.5.1 may be used as a guide.

Once the service factor has been selected the next step is to multiply it by the rated power of the driveR to obtain the design power.

TABLE 12.5.1   SERVICE FACTORS

DriveN Machine				DriveR		
The machines listed below are representative samples only. Select the group listed below whose load characteristics most closely approximate those of the machine being considered.	AC Motors: Normal Torque, Squirrel Cage, Synchronous, Split Phase. DC Motors: Shunt Wound Engines: Mutliple Cylinder Internal Combustion.*			AC Motors: High Torque, High Slip, Repulsion-Induction, Single Phase, Series Wound, Slip Ring. DC Motors: Series Wound, Compound Wound. Engines: Single Cylinder internal Combustion.* Line shafts Clutches		
	Intermittent Service	Normal Service	Continuous Service	Intermittent Service	Normal Service	Continuous Service
	3-5 Hours Daily or Seasonal	8-10 Hours Daily	16-24 Hours Daily	3-5 Hours Daily or Seasonal	8-10 Hours Daily	16-24 Hours Daily
Agitators for Liquids Blowers and Exhausters Centrifugal Pumps & Compressors Fans up to 10 Horsepower Light Duty Conveyors	1.0	1.1	1.2	1.1	1.2	1.3
Belt Conveyors For Sand, Grain, Etc. Dough Mixers Fans-Over 10 Horsepower Generators Line Shafts Laundry Machinery Machine Tools Punches-Presses-Shears Printing Machinery Positive Displacement Rotary Pumps Revolving and Vibrating Screens	1.1	1.2	1.3	1.2	1.3	1.4
Brick Machinery Bucket Elevators Exciters Piston Compressors Conveyors (Drag-Pan-Screw) Hammer Mills Paper Mill Beaters Piston Pumps Positive Displacement Blowers Pulverizers Saw Mill and Woodworking Machinery Textile Machinery	1.2	1.3	1.4	1.4	1.5	1.6
Crushers (Gyratory-Jaw-Roll) Mills (Ball-Rod-Tube) Hoists Rubber Calenders-Extruders-Mills	1.3	1.4	1.5	1.5	1.6	1.8

* Apply indicated service factor to continuous engine rating. Deduct 0.2 (with a minimum service factor of 1.0) when applying to maximum intermittant rating.
  The use of a service factor of 2.0 is recommended for equipment subject to choking.
  For Grain Milling and Elevator Equipment, see Mill Mutual Bulletin No. VB-601-62.
  For Oil Field Machinery, see API specification for Oil Field V-Belting, API Standard 1B.

  Source: Gates Rubber Co., Denver, CO.

Step three is selection of the belt cross section from Figure 12.5.1 by finding the intersection of a line representing the rpm of the faster shaft, given along the ordinate, and a line representing the design power, given in horsepower along the abscissa. The label in the region of intersection indicates the cross section to be used. The 3VX and 5VX belts, incidentally, are 3V and 5V belts with shallow notches, as shown in Figure 12.3.1c, which allow the belts to better fit small sheaves. They are not timing belts. It is recommended that if the intersection point falls near a boundary that two selections be made, one for each. For example, if $n = 690$ and DP = 10, one design should use a 3VX belt and the other a 5VX belt.

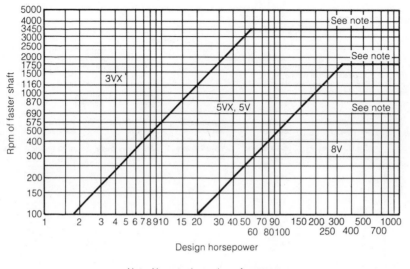

Figure 12.5.1 Cross-section selection chart. *Source: Gates Rubber Co., Denver, CO.*

The fourth step is to attempt to select standard sheaves for this belt cross section from Table 12.5.2 to provide the desired speed ratio, which is calculated after finding the pitch diameter with the aid of Table 12.5.3. Use of standard sheaves is recommended because they are much less expensive than custom sheaves made to ANSI standards.

Table 12.5.4 lists the minimum sheave diameters for standard electric motors, based upon bearing loads, which increase as the sheave diameter decreases for a given power output. If the motor sheave selected is smaller than that listed in Table 12.5.4 another set of sheaves should be selected. Failure to do so will shorten the life of the motor bearings. This table is entered at the rated power of the motor, not the design horsepower, because bearing load is related to the power rating. This is the only step after step two (calculating the design horsepower) where the design power is not used.

If other driveRs are used, the bearing load may be calculated from the belt tension, which, as shown in problem 12.17, may be found from

$$\frac{F_1}{F_2} = e^{k\theta} \tag{12.5.1}$$

where θ is the wrap angle around the small sheave, as shown in Figure 12.5.2. Forces F_1 and F_2 are related to the tension T according to

$$T = (F_1 - F_2)r_p \tag{12.5.2}$$

TABLE 12.5.2 STANDARD (STOCK) SHEAVE OUTSIDE DIAMETERS FOR 3VX, 5VX, 5V, AND 8V BELT SERIES

	3V/3VX	5V/5VX	8V
	—	*4.40	—
	—	*4.65	—
	—	*4.90	—
	—	*5.20	—
	—	*5.50	—
	—	*5.90	—
	—	*6.30	—
	—	*6.70	—
	*2.20	7.10	—
	*2.35	7.50	—
Recommended	*2.50	8.00	—
Range of	2.65	8.50	—
Small Sheave	2.80	9.00	—
Diameters	3.00	9.25	12.5
	3.15	9.75	13.2
	3.35	10.3	14.0
	3.65	10.9	15.0
	4.12	11.3	16.0
	4.50	11.8	17.0
	4.75	12.5	18.0
	5.00	13.2	19.0
	5.30	14.0	20.0
	5.60	15.0	21.2
	6.00	16.0	22.4
	6.50	18.7	24.8
	6.90	21.2	30.0
	8.00	23.6	35.5
	10.6	28.0	40.0
	14.0	31.5	44.5
	19.0	37.5	53.0
	25.0	50.0	71.0
	33.5	67.0	95.0

Note. Outside diameters in inches
*   Use with notched V-belts only.
Source: Gates Rubber Co., Denver, CO.

TABLE 12.5.3 AMOUNT TO SUBTRACT FROM THE OUTSIDE DIAMETER TO FIND THE PITCH DIAMETER OF A GROOVED SHEAVE

V-Belt Cross Section	3V (in.)	5V (in.)	8V (in.)
Standard	0.05	0.10	0.20
Deep Groove	0.268	0.420	0.772

TABLE 12.5.4 RECOMMENDED MINIMUM SHEAVE OUTSIDE DIAMETERS FOR GENERAL PURPOSE ELECTRIC MOTORS

Motor Horsepower	Motor RPM (60 cycle and 50 cycle Electric Motors)						Motor Horsepower
	575 485[a]	690 575[a]	870 725[a]	1160 950[a]	1750 1425[a]	3450 2850[a]	
$\frac{1}{2}$	—	—	2.2	—	—	—	$\frac{1}{2}$
$\frac{3}{4}$	—	—	2.4	2.2	—	—	$\frac{3}{4}$
1	3.0	2.5	2.4	2.4	2.2	—	1
$1\frac{1}{2}$	3.0	3.0	2.4	2.4	2.4	2.2	$1\frac{1}{2}$
2	3.8	3.0	3.0	2.4	2.4	2.4	2
3	4.5	3.8	3.0	3.0	2.4	2.4	3
5	4.5	4.5	3.8	3.0	3.0	2.4	5
$7\frac{1}{2}$	5.2	4.5	4.4	3.8	3.0	3.0	$7\frac{1}{2}$
10	6.0	5.2	4.4	4.4	3.8	3.0	10
15	6.8	6.0	5.2	4.4	4.4	3.8	15
20	8.2	6.8	6.0	5.2	4.4	4.4	20
25	9.0	8.2	6.8	6.0	4.4	4.4	25
30	10	9.0	6.8	6.8	5.2	—	30
40	10	10	8.2	6.8	6.0	—	40
50	11	10	8.4	8.2	6.8	—	50
60	12	11	10	8.0	7.4	—	60
75	14	13	9.5	10	8.6	—	75
100	18	15	12	10	8.6	—	100
125	20	18	15	12	10.5[b]	—	125
150	22	20	18	13	10.5	—	150
200	22	22	22	—	13.2	—	200
250	22	22	—	—	—	—	250
300	27	27	—	—	—	—	300

[a] These RPM are for 50 cycle electric motors.
[b] 9.5 for Frame Number 444T.

Data in the white area of the table are from NEMA Standard MG-1-14.42, June, 1972. Data in the light gray area are from MG-1-14.43. January, 1968. Data in the dark gray area are a composite of electric motor manufacturers data. They are generally conservative, and specific motors and bearings may permit the use of a smaller motor sheave. Consult the motor manufacturer.
Source: Gates Rubber Co., Denver, CO

where r_p is the radius of the small sheave. From equation 12.5.2 we find that increasing the size of the smaller sheave will increase belt life by reducing belt tensions F_1 and F_2 for a given torque; and from equation 12.5.1 we see that we must have a slack side tension if we are to have a tight side tension. The design guidelines derived from these observations are that we should select the largest sheaves and the longest center distance permitted by space constraints to minimize F_1 and F_2 and to maximize θ.

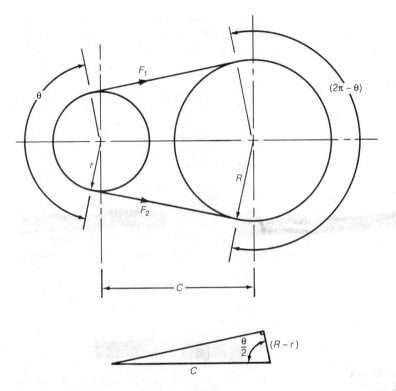

Figure 12.5.2 Sheave and belt geometry.

Tension F_1 may be found from the power requirement. After torque T is found from the required power, relation 12.5.2 becomes, after substituting for F_2 from equation 12.5.1,

$$T = \frac{\text{hp}}{\pi n} \, 198\,000 = F_1(1 - e^{-k\theta})r_\text{p} \qquad (12.5.3)$$

which leads to

$$F_1 = \frac{396\,000}{\pi n d_\text{p}} \, \frac{\text{hp}}{1 - e^{-k\theta}} \qquad (12.5.4)$$

Tension F_2 may be calculated from equation 12.5.1 now that F_1 is known. In these equations $d_\text{p} = 2r_\text{p}$ denotes the pitch diameter of the smaller sheave and $k = 0.5123$ is normally used for V-belts in sheaves designed according to ANSI/RMA standards for these belts. (RMA is the Rubber Manufacturers Association acronym).

In practice, forces F_1 and F_2 are increased by the centrifugal force acting at each sheave, which, as shown in problem 12.18, adds the term mv^2/r_p to both. Thus, the actual tight and slack side tensions are represented by

$$F_1 = \frac{396\,000}{\pi n d_p} \frac{\text{hp}}{1 - e^{-k\theta}} + \frac{mv^2}{r_p} \tag{12.5.5}$$

and

$$F_2 = \frac{396\,000}{\pi n d_p} \frac{\text{hp}}{1 - e^{-k\theta}} e^{-k\theta} + \frac{mv^2}{r_p} \tag{12.5.6}$$

in which m represents the mass/unit length of the belt and v denotes the belt velocity. Addition of the centrifugal force term is to counteract the effect of the centrifugal force in decreasing the effective tension difference at the small sheave and thereby reducing the power transmitted.

Belt tension is usually measured with the belts stationary. In most machinery the belt tensions equalize as the machine is stopped, so that the measured tension is the average of F_1 and F_2, denoted here by F_s.

Industrial practice, as represented by design formulas presented by the Gates Rubber Company, is to replace the horsepower by the design horsepower, Dhp, and to replace the second term in equation 12.5.6 by $Mv^2/10^6$, where M is given in Table 12.5.5 and velocity v in expressed in feet/minute. Upon introducing these changes we have that the static tension F_s may be closely approximated by

$$F_s = \frac{F_1 + F_2}{2} = 198\,000 \frac{1 + e^{-k\theta}}{1 - e^{-k\theta}} \frac{\text{Dhp}}{\pi n d_p} + \frac{Mv^2}{10^6} \tag{12.5.7}$$

where the centrifugal force term is retained to provide enough tension to maintain proper contact force between the belt and the sheaves at the operating speed.

TABLE 12.5.5 M FACTOR FOR SINGLE BELTS AND POWER BANDS

Belt Cross Section	M per Belt	Belt Cross Section Power Bands	M per Belt
3VX	0.29	3VX	0.39
5VX	0.78	5VX	0.98
5V	1.0	5V	1.2
8V	2.6	8V	3.0

Source: Abstracted from Gates Heavy-Duty V-Belt Drive Design Manual, Gates Rubber Co., Denver, Co.

Radial load on the shaft bearings adjacent to the sheaves on the driveR and driveN machines may be approximated by $2F_s$. This approximation contains an inherent safety factor because it replaces the vector sum of F_1 and F_2 by their algebraic sum.

Static tension is usually measured by deflecting the belt with a known force. A spring scale is commonly used to measure a force applied at the midpoint between two sheaves and the deflection at that point is measured from a straight edge tangent to the belt at each sheave. Tension may then be found from these measurements using tables or formulas supplied by the belt manufacturer.

Belt stiffness has been neglected in the discussion thus far. It effectively adds to the radial pressure between the belt and sheave near the center of the contact length, so that a smaller tension is required to maintain this pressure. For simplicity, the effect of subtracting the several-term stiffness relation may be approximated by mutliplying (12.5.7) by 0.9, which is said to produce no more than a 10% error for all standard belts.[2] Thus, relation 12.5.7 is replaced by

$$F_s = 178\,200 \frac{1 + e^{-k\theta}}{1 - e^{-k\theta}} \frac{Dhp}{\pi n d_p} + \frac{Mv^2}{10^6} \tag{12.5.8}$$

for the calculation of the required static tension for a specified horsepower.

Step five is to calculate the belt length for the desired center distance and the sheaves that have been selected. Returning to the geometry displayed in Figure 12.5.2 we easily see that

$$L = \frac{\theta}{2} d + (2\pi - \theta) \frac{D}{2} + 2[C^2 - (R - r)^2]^{1/2} \tag{12.5.9}$$

where

$$\theta = 2 \cos^{-1} \left(\frac{D - d}{2C} \right) \tag{12.5.9a}$$

Use of series expansions for the square root and for the arccosine, as in problem 12.1, enables us to approximate equation 12.5.9 by

$$L = 2C + \frac{\pi}{2} (D + d) + \frac{(D - d)^2}{4C} \tag{12.5.10}$$

where C denotes the distance between shaft centers and D and d denote the outside diameters of the large and small sheaves, respectively. Belt length for this cross section is measured along the ridge on the belt that contacts the outside of the sheaves.

Formula 12.5.10 is exactly correct when $D = d$ and errs by 1.33% for $d = 2$ in., $D = 40$ in., and $C = 21$ in. Since this is the largest error that would be encountered in practically all belt drives because of the small contact angle at the small sheave, use of (12.5.10) is well justified.

TABLE 12.5.6   SERIES 3VX, 5VX, 5V, AND 8V BELTS AVAILABLE

3VX			5VX & 5V			8V		
RMA Nomenclature Lengths (mm)**	Outside Circumference (in.)	Super HC Molded Notch V-Belt Number	RMA Nomenclature Lengths (mm)**	Outside Circumference (in.)	Super HC Molded Notch V-Belt Number	RMA Nomenclature Lengths (mm)**	Outside Circumference (in.)	Super HC V-Belt Number
630	25	*3VX250	1270	50	5VX500	2540	100	8V1000
670	26.5	*3VX265	1345	53	5VX530	2690	106	8V1060
710	28	3VX280	1420	56	5VX560	2840	112	8V1120
760	30	3VX300	1525	60	5VX600	3000	118	8V1180
800	31.5	3VX315	1600	63	5VX630	3180	125	8V1250
850	33.5	3VX335	1700	67	5VX670	3350	132	8V1320
900	35.5	3VX355	1800	71	5VX710	3550	140	8V1400
950	37.5	3VX375	1900	75	5VX750	3810	150	8V1500
1015	40	3VX400	2030	80	5VX800	4060	160	8V1600
1080	42.5	3VX425	2160	85	5VX850	4320	170	8V1700
1145	45	3VX450	2290	90	5VX900	4570	180	8V1800
1205	47.5	3VX475	2410	95	5VX950	4830	190	8V1900
1270	50	3VX500	2540	100	5VX1000	5080	200	8V2000
1345	53	3VX530	2690	106	5VX1060	5380	212	8V2120
1420	56	3VX560	2840	112	5VX1120	5690	224	8V2240
1525	60	3VX600	3000	118	5VX1180	6000	236	8V2360
1600	63	3VX630	3180	125	5VX1250	6350	250	8V2500
1700	67	3VX670	3350	132	5VX1320	6730	265	8V2650
1800	71	3VX710	3550	140	5VX1400	7100	280	8V2800
1900	75	3VX750	3810	150	5VX1500	7620	300	8V3000
2030	80	3VX800	4060	160	5VX1600	8000	315	8V3150
2160	85	3VX850	4320	170	5VX1700	8500	335	8V3350
2290	90	3VX900	4570	180	5VX1800	9000	355	8V3550
2410	95	3VX950	4830	190	5VX1900	9500	375	8V3750
2540	100	3VX1000	5080	200	5VX2000	10160	400	8V4000
2690	106	3VX1060	5380	212	5V2120	10800	425	8V4250
2840	112	3VX1120	5690	224	5V2240	11430	450	8V4500
3000	118	3VX1180	6000	236	5V2360	12060	475	8V4750
3180	125	3VX1250	6350	250	5V2500	12700	500	8V5000
3350	132	3VX1320	6730	265	5V2650	14220	560	8V5600
3550	140	3VX1400	7100	280	5V2800	15240	600	*8V6000
			7620	300	5V3000	5V3000		
			8000	315	5V3150			
			8500	335	5V3350			
			9000	355	5V3550			

*   For PowerBand Stock availability, check with your local Gates Field Engineer
** The metric length designation corresponds to the joint industry standard IP-22 established by RMA/MPTA/RAC. The conventional V-belt numbers listed correspond to these established metric lengths. The metric conversion will also encompass a V-belt number change which will be: 3VX to 9NX, 5VX to 15NX, 5V to 15N and 8V to 25N. An example of this change is 5V2500 to 15N6350.

Source: Gates Rubber Co., Denver, CO.

The sixth step is to select a belt from Table 12.5.6 whose length is closest to length L calculated from equation 12.5.10 and to calculate the modified center distance corresponding to this length from

$$C = \frac{1}{4}\left(L - \pi\frac{D+d}{2}\right)\left\{1 + \left[1 - \frac{2(D-d)^2}{\left(L - \pi\dfrac{D+d}{2}\right)^2}\right]^{1/2}\right\} \qquad (12.5.11)$$

If this center distance is satisfactory we shall move to the last step, that of finding the number of belts required. Otherwise, we must repeat the process until we find an acceptable belt length and center distance. If no standard belts provide an acceptable center distance we must consider other drive systems, such as chains or gears. It is well to remember at this point that the center distance may change several inches before the belt is replaced in order to maintain the correct tension as the belt stretches. The extent of this change is indicated by Table 12.5.7 which gives the required installation

TABLE 12.5.7 MINIMUM CENTER DISTANCE ALLOWANCES FOR BELT INSTALLATION AND TAKE-UP

| V-Belt Number | Minimum Center Distance Allowance For Installation (inches) | | | | | | Minimum Center Distance Allowance For Initial Tensioning and Subsequent Take-up (inches) |
| | 3V/3VX | | 5V/5VX | | 8V | | All Cross Sections |
	Super HC® V-Belt	Super HC Power Band® Belt*	Super HC V-Belt	Super HC Power Band Belt*	Super HC V-Belt	Super HC Power Band Belt*	All Types
Up To and Incl. 475	0.5	1.2					1.0
Over 475 To and Incl. 710	0.8	1.4	1.0	2.1			1.2
Over 710 To and Incl. 1060	0.8	1.4	1.0	2.1	1.5	3.4	1.5
Over 1060 To and Incl. 1250	0.8	1.4	1.0	2.1	1.5	3.4	1.8
Over 1250 To and Incl. 1700	0.8	1.4	1.0	2.1	1.5	3.4	2.2
Over 1700 To and Incl. 2000			1.0	2.1	1.8	3.6	2.5
Over 2000 To and Incl. 2360			1.2	2.4	1.8	3.6	3.0
Over 2360 To and Incl. 2650			1.2	2.4	1.8	3.6	3.2
Over 2650 To and Incl. 3000			1.2	2.4	1.8	3.6	3.5
Over 3000 To and Incl. 3550			1.2	2.4	2.0	4.0	4.0
Over 3550 To and Incl. 3750					2.0	4.0	4.5
Over 3750 To and Incl. 5000					2.0	4.0	5.5
5600					2.0	4.0	6.0

Source: Gates Rubber Co., Denver, CO.

* Power Band is the Gates term for joined V-belts that consist of three or more V-belts fastened together (as shown in Figure 12.31c) to prevent belt roll-over.

and take-up distances for standard belt lengths. Use of a spring-loaded idler to maintain tension eliminates the task of periodically adjusting the center distance at the expense of shorter belt life.

The seventh step, that of calculating the power per belt to ascertain the number of belts to use, may be accomplished with the aid of formula 12.5.12:

$$\frac{\text{hp}}{\text{belt}} = \{\tilde{n}d_p\, [K_1 - \frac{K_2}{d_p} - K_3(\tilde{n}d_p)^2 - K_4 \log(\tilde{n}d_p)] + K_{sr}\tilde{n}\}\, Gb \tag{12.5.12}$$

where $\tilde{n} = n/1000$ in which n denotes the rpm of the faster shaft and d_p denotes the pitch diameter of that sheave. Constants K_1 through K_4, K_{sr}, and b are found from Table 12.5.8. G is obtained from the defining equation

$$G = 1.25(1 - e^{-k\theta}) \tag{12.5.13}$$

where, as before, $k = 0.5123$.

The eighth and last step is that of dividing the power per belt into the design power and rounding up to the nearest integer to find the number of belts to be used.

This procedure may be summarized in a flowchart shown in Figure 12.5.3, which may be useful in writing a program to perform the belt selection process automatically for standard sheaves. By including provisions for a subprogram at the custom design location it may be extended to custom designing that can be categorized enough to be programmed. Examples of V-belt selection programs may be found in Refs. 6 and 7.

EXAMPLE 12.5.1

Recommend a belt drive for driving a centrifugal pump with a 10-hp squirrel cage motor operating continuously at 1750 rpm. Pump speed should be 1635 ± 5 rpm and the mounting frame should accomodate a center distance from 34 to 40 in., with 38 in. preferred to allow access for maintenance.

Since an electric motor is used, begin with Table 12.5.4 to find that the minimum sheave diameter is 3.8 in. From Table 12.5.1 the service factor is 1.2 so the design horsepower is 12.0. Entering the chart in Figure 12.5.1 at 12 hp and reading up to 1750 rpm places the intersection well within the 3VX region. If it had been near the 3VX, 5VX boundary (within 3 mm, say) we would execute two designs, one with the 3VX belt and one with the 5VX belt, to find which is cheaper.

The smallest acceptable sheave diameter from Tables 12.5.2 and 12.5.3 is 4.12 in. with a pitch diameter of 4.07 in. Since the speed ratio is from

$$\tfrac{1750}{1640} = 1.067 \quad \text{to} \quad \tfrac{1750}{1630} = 1.074$$

the larger sheave diameter should be between 4.40 and 4.42 in. if we use the smallest standard sheave. Unfortunately, these diameters are nonstandard. Before ordering custom sheaves we shall consider a

TABLE 12.5.8 CONSTANTS FOR FORMULA 12.3.10 FOR HORSEPOWER PER BELT

Cross Section	K_1	K_2	K_3	K_4
3VX	1.1691	1.5295	1.5229×10^{-4}	0.15960
5VX	3.3038	7.7810	3.6432×10^{-4}	0.43343
8V	8.6628	49.323	1.5804×10^{-3}	1.1669

	b					
	3VX		**5VX & 5V**		**8V**	
	V-Belt Number	Correction Factor b	V-Belt Number	Correction Factor	V-Belt Number	Correction Factor b
	3VX250	0.83	5VX500	0.85	8V1000	0.87
	3VX265	0.84	5VX530	0.86	8V1060	0.88
	3VX280	0.85	5VX560	0.87	8V1120	0.88
	3VX300	0.86	5VX600	0.88	8V1180	0.89
	3VX315	0.87	5VX630	0.89	8V1250	0.90
	3VX335	0.88	5VX670	0.90	8V1320	0.91
	3VX355	0.90	5VX710	0.91	8V1400	0.92
	3VX375	0.91	5VX750	0.92	8V1500	0.93
	3VX400	0.92	5VX800	0.93	8V1600	0.94
	3VX425	0.93	5VX850	0.94	8V1700	0.94
	3VX450	0.94	5VX900	0.95	8V1800	0.95
	3VX475	0.95	5VX950	0.96	8V1900	0.96
	3VX500	0.96	5VX1000	0.96	8V2000	0.97
	3VX530	0.97	5VX1060	0.97	8V2120	0.98
	3VX560	0.98	5VX1120	0.98	8V2240	0.98
	3VX600	0.99	5VX1180	0.99	8V2360	0.99
	3VX630	1.00	5VX1250	1.00	8V2500	1.00
	3VX670	1.01	5VX1320	1.01	8V2650	1.01
	3VX710	1.02	5VX1400	1.02	8V2800	1.02
	3VX750	1.03	5VX1500	1.03	8V3000	1.03
	3VX800	1.04	5VX1600	1.04	8V3150	1.03
	3VX850	1.06	5VX1700	1.05	8V3350	1.04
	3VX900	1.07	5VX1800	1.06	8V3550	1.05
	3VX950	1.08	5VX1900	1.07	8V3750	1.06
	3VX1000	1.09	5VX2000	1.08	8V4000	1.07
	3VX1060	1.10	5V2120	1.09	8V4250	1.08
	3VX1120	1.11	5V2240	1.09	8V4500	1.09
	3VX1180	1.12	5V2360	1.10	8V4750	1.09
	3VX1250	1.13	5V2500	1.11	8V5000	1.10
	3VX1320	1.14	5V2650	1.12	8V5600	1.12
	3VX1400	1.15	5V2800	1.13		
			5V3000	1.14		
			5V3150	1.15		
			5V3350	1.16		
			5V3550	1.17		

K_{SR} Values

Speed Ratio Range	Cross Section	
	3VX	5VX
1.00–1.01	0.0000	0.0000
1.02–1.03	0.0157	0.0801
1.04–1.06	0.0315	0.1600
1.07–1.09	0.0471	0.2398
1.10–1.13	0.0629	0.3201
1.14–1.18	0.0786	0.4001
1.19–1.25	0.0944	0.4804
1.26–1.35	0.1101	0.5603
1.36–1.57	0.1259	0.6405
1.58 & Over	0.1416	0.7202

K_{SR} Values

Speed Ratio Range	Cross Section	
	5V	8V
1.00–1.01	0.0000	0.0000
1.02–1.05	0.0963	0.4690
1.06–1.11	0.2623	1.2780
1.12–1.18	0.4572	2.2276
1.19–1.26	0.6223	3.0321
1.27–1.38	0.7542	3.6747
1.39–1.57	0.8833	4.3038
1.58–1.94	0.9941	4.8438
1.95–3.38	1.0830	5.2767
3.39 & Over	1.1471	5.5892

b = belt length correction factor
Source: Gates Rubber Co., Denver, CO.

larger diameter small sheave if space permits. We shall examine successively larger diameters using a simple program that calculates the pitch diameter corresponding to each outside diameter and then finds the ratio of the pitch diameters. We find that for sheaves with outside diameters

$$d = 5.60 \text{ in.} \quad \text{and} \quad D = 6.00 \text{ in.}$$

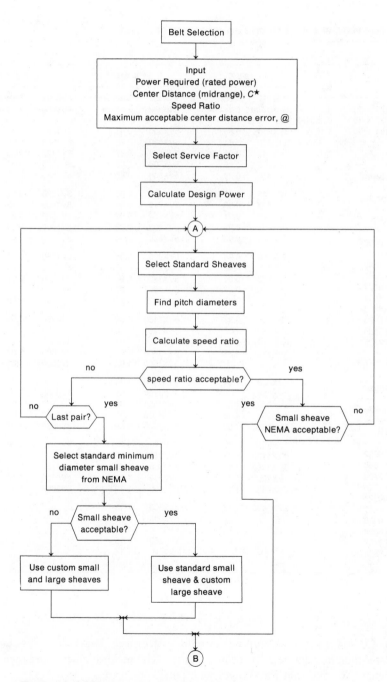

Figure 12.5.3 Flowchart for belt drive design.

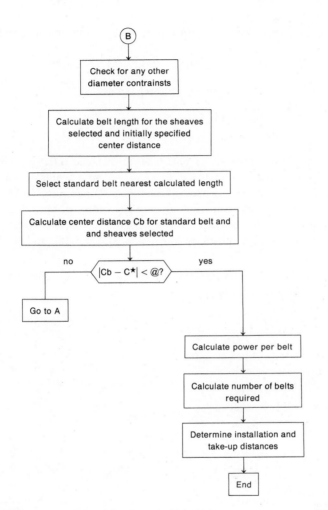

Figure 12.5.3 (continued) Flowchart for belt drive design

the ratio of 1.071 is within the acceptable range. Find belt speed from the small sheave outside diameter using

$$V = r\omega = \pi n d = \pi 5.6(1750) = 30\,787.61 \text{ in./min} = 2565.63 \text{ ft/min}$$

Since this is less than the 6500-ft/min limit for these sheaves, continue with this sheave combination to find initial belt length from (12.5.10) to be

$$L = 2C + \frac{\pi}{2}(D + d) + \frac{(D - d)^2}{4C} = 2(38) + \frac{\pi}{2}(11.60) + \frac{(0.4)^2}{4(38)}$$

$$= 94.22 \text{ in.}$$

which is closest to the length of belt 3VX950. Substitution of $L = 95$ in. into (12.5.11) yields

$$C = \frac{1}{4}\left(95 - \frac{\pi}{2}11.60\right)\left\{1 + \left[1 - 2\left(\frac{0.40}{95 - \frac{\pi}{2}11.60}\right)^2\right]^{1/2}\right\}$$

$$= \tfrac{1}{4}(76.779)(1 - 0.005^2)^{1/2} = 38.39 \text{ in.}$$

Installation distance is 0.8 in. from Table 12.5.5 and take-up distance is 1.5 in. This means that the motor or the pump must be mounted to allow them to be moved to a center distance of $38.4 - 0.8 = 37.6$ in. for belt installlation and then moved back to 38.4 in. for operation. As the belt stretches the center distance must be increased to maintain tension until the center distance becomes $38.4 + 1.5 = 39.9$ in. After sufficient operation at this distance to lose tension the belt is considered worn out.

To find how many belts are needed, substitute into (12.5.9a), (12.5.12), and (12.5.13) to find

$$\begin{aligned}
\text{hp/belt} = \{&5.55(1.750)[1.1691 - 1.5295/5.55 - 1.5229 \times 10^{-4} \\
&\times (1.750 \times 5.55)^2 - 0.15960 \log(1.750 \times 5.55)] \\
&+ 0.0471(1.750)1\}0.999(1.08) \\
= \;&7.64
\end{aligned}$$

Division into the design horsepower yields

$$12/7.64 = 1.57 \text{ belts} \rightarrow 2 \text{ belts needed}$$

12.6 BELT STRESSES

The recommendation that the largest sheaves compatible with space restrictions be used in a belt drive design is motivated by the observation that bending stresses arise as the belt is bent around a sheave, and that these stresses are greater for smaller sheaves. If these stresses and deflections obey the Bernoulli-Euler beam formulas

$$\sigma = \frac{Mc}{I} \quad \text{and} \quad \frac{1}{\rho} = \frac{M}{EI} \tag{12.6.1}$$

then tensile stress σ at a distance c from the neutral surface is given by

$$\sigma = \frac{Ec}{\rho} \tag{12.6.2}$$

where E denotes the equivalent elastic modulus of the belt and ρ denotes its radius of curvature, which may be approximated by the pitch radius of the sheave. Clearly, smaller sheaves induce larger stresses.

The relation between belt stresses and belt life is represented by the curve in Figure 12.6.1, where belt stress between sheaves may be approximated by

$$\sigma = F_1/A \qquad \text{tight side}$$
$$\sigma = F_2/A \qquad \text{slack side}$$

(12.6.3)

in terms of the belt cross-sectional area A and the respective tensions. The abrupt stress increases at points A, B, C, and D are due to the stress changing from that given by (12.6.3) to that given by the sum of (12.6.1) and (12.6.3) as the belt is bent around the sheave. Stress decrease between points D and A and between C and B are due to the decrease in tension because of the friction between the belt and sheave, as given by

$$\frac{F_1}{F_2} = e^{k\theta}$$

(12.6.4)

if the full torque capacity of the smaller sheave is exerted, and by

$$\frac{F_1}{F_2} = e^{k'(2\pi - \theta)}$$

(12.6.5)

on the large sheave, where $0 \le k' \le k$ for a two-sheave system. This inequality holds whenever one sheave is larger than the other; it becomes an equality when both sheaves have the same pitch diameter. The working values of k are less than k and k' respectively if the small sheave does not transmit its maximum torque.

The unscaled σ versus life curve on the right side of Figure 12.6.1 indicates that as the stresses increase a greater portion of the remaining life is consumed with each revolution of the belt. Adding an idler for automatic take-up is, therefore, a convenience that is purchased at the price of reduced belt life.

12.7 BELT LIFE ESTIMATION

No formulas or other methods for estimating belt life are included in the RMA-MPTA-RAC standards, although it has been implied elsewhere that the power per belt constants have been selected for a 18 000-h life under ideal conditions. We will, therefore, turn to the formula presented by Oliver, Johnson, and Breig[3] which has been given by the Dayco Rubber Products Company in its current design guide. Although Dayco provides experimental constants for only their belts, designed according to SAE standards that allow innovation, their formulation provides quantitative information on belt life for one manufacturer and qualitative behaviour for all belts.

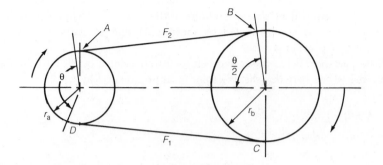

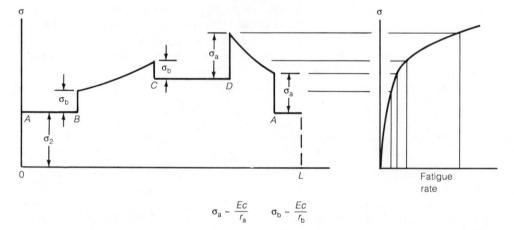

$$\sigma_a \sim \frac{Ec}{r_a} \qquad \sigma_b \sim \frac{Ec}{r_b}$$

Figure 12.6.1 Belt stresses as a function of position along the belt and their effect upon belt life. *Source: Abstracted from* Gates Agricultural V-Belt Drive Design Manual, *1976, Gates Rubber Co., Denver, CO.*

It is good engineering practice to contact the belt manufacturer for the recommended design formulas for the belts to be used—Dayco, RMA, or a modification, as used by the Gates Rubber Company.

Oliver, Johnson, and Breig found that belt life could be satisfactorily predicted from the Palmgren-Miner equation and the stress cycles induced at each sheave. After analyzing a number of laboratory tests they found that the life in cycles of each cross section due to tension j at one sheave, say sheave i, could be represented by

$$N_{ij} = K_1(K_0 - \sigma_{0ij})^2(K_m - \sigma_{mij})^2 L^{1.75} v^k \tag{12.7.1}$$

in which $k = 0$ or -1, depending upon the belt cross section, and

$$\sigma_{mij} = (F_{1e} + F_{be} + 2F_{ce} + F_{2e})_{ij}/(2A_c)$$
$$\sigma_{0ij} = (F_{1e} + F_{be} - F_{2e})_{ij}/(2A_c) \tag{12.7.2}$$

where

$$F_{1e} = \text{tight side edge cord tension} \simeq F_1/n_c$$

$$F_{2e} = \text{slack side edge cord tension} = F_2/n_c$$

$$F_{be} = \text{edge cord tension due to bending, Table 12.7.1}$$

$$F_{ce} = \text{edge cord tension due to centrifugal force, Table 12.7.1}$$

$$A_c = \text{nominal area, each cord, Table 12.7.1}$$

If there are m sheaves in the belt drive, and if belt tension F_j is constant for a period of time, then each time a point on the belt passes over sheave i during the time that F_j is constant it loses $1/N_{ij}$th of its life. If there are m sheaves in the belt loop, then each time the belt makes one revolution all sheaves have contributed and the life lost is given by

$$\frac{1}{N_j} = \sum_{i=1}^{m} \frac{1}{N_{ij}} \qquad (12.7.3)$$

Since N_j denotes the life of the belt when supporting tension F_j, it follows from the Palmgren-Miner relation that when the belt supports a duty schedule of n different

TABLE 12.7.1 V-BELT LIFE PREDICTION CONSTANTS

Belt	K_1	K_0 (lb/In.^{2})	K_m (lb/In.^{2})	$F_{be} = K_2/D$ (D-In.) K_2	$F_{ce} = K_3V^2$ (V-ft/min) K_3	A (In.^{2})	n_c
.380 LAM &	$\times 10^{-10}$				$\times 10^{-6}$	$\times 10^{-3}$	
PLAIN HD	.408	18750	34000	71	.058	.99	7.6
COG	.560	17750	36000	65	.061	.99	7.4
CNA	.423	17900	34000	71	.069	.99	6.8
.470 LAM & (.380 GROOVE)							
PLAIN HD	.648	18300	34000	71	.067	.99	8.4
COG	.354	18150	36000	65	.060	.99	8.6
.440 LAM	.205	18400	34000	71	.067	.99	8.4
.500 LAM &							
PLAIN HD	7.85	13700	21500	115	.085	1.73	8.0
COG	1.12	13550	32000	110	.083	1.73	8.2
.600 PLAIN HD	6.74	13150	21500	125	.104	1.73	10.5
COG	1.18	13250	32000	115	.093	1.73	10.5
.660 PLAIN HD	5.96	12700	21500	125	.102	1.73	11.8
COG	1.17	12800	32000	115	.091	1.73	11.9

Source: Dayco Corp., Dayton, OH.

loads, where each is constant over the time it acts, the life of the belt may be found from

$$\frac{1}{N_t} = \sum_{j=1}^{n} \frac{\beta_j}{N_j} \qquad (12.7.4)$$

where β_j is the proportion of the duty schedule during which tension F_j acts. Substitution for $1/N_j$ from equation 12.7.3 into equation 12.7.4 yields

$$\frac{1}{N_t} = \sum_{j=1}^{n} \sum_{i=1}^{m} \frac{\beta_j}{N_{ij}} \qquad (12.7.5)$$

To convert from life in cycles to life in hours, denoted by H_t, we simply observe that $H_t/N_t = L/v$, where L is the length of the belt and v is its linear velocity. Upon inserting the conversion factors to feet and hours,

$$\frac{H_t}{N_t} = \frac{L}{720v} \qquad (12.7.6)$$

where $v = \pi d_p n/12$ is in feet/minute, pitch diameter d_p is in inches, and n is in revolutions/minute.

According to Ref. 4, the effect on belt life of changes in belt length, belt tension, sheave diameter, and velocity may be expressed as

$$\frac{H_{t_1}}{H_{t_2}} = \left(\frac{d_1}{d_2}\right)^{c_1} \left(\frac{F_{1_2}}{F_{1_1}}\right)^{c_2} \left(\frac{L_1}{L_2}\right)^{c_3} \left(\frac{v_2}{v_1}\right) \qquad (12.7.7)$$

where exponents c_1, c_2, c_3 are given by

c_1	c_2	c_3	
5.00	4.00	2.75	V-belts
3.00	2.00	1.00	Synchronous and multiribbed belts

Values of the coefficients in equations 12.7.1 and 12.7.3 for belts designed according to SAE standards are listed in Table 12.7.1 for the cross sections shown in Figure 12.1.1. Only the 0.380 and 0.600 Plain HD cross sections resemble RMA-MPTA-RAC cross sections; i.e., 3V and 5V, respectively.

Figure 12.8.1 Other chain types. *Source: American Chain Association*, Chains for Power Transmission and Material Handling, *Marcel Dekker, Inc., New York, 1982, pp. 4, 6, 16, 28.*

12.8 CHAIN DESCRIPTION, ROLLER CHAINS

The variety of chains is far too extensive to be considered in just one chapter: the ANSI standard alone listed 11 different types in 1988. In addition, there are many other varieties used in foreign countries and still more for which no standards have been established. Five of the types listed by ANSI are pictured in Figure 12.8.1. Only the roller chains shown in Figures 12.8.2 and 12.8.3 will be considered in the remainder of the chapter because they are the type most commonly used for power transfer. Chains of

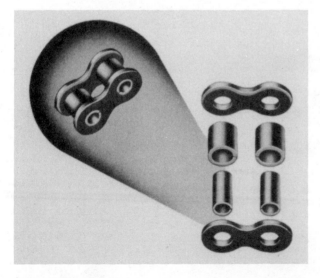

Figure 12.8.2 Roller link. *Source: American Chain Association*, chains for Power Transmission and Material Handling, *Marcel Dekker, Inc., New York, 1982, p. 50.*

Figure 12.8.3 Pin links. *Source: American Chain Association,* chains for Power Transmission and Material Handling, *Marcel Dekker, Inc., New York, 1982, p. 50.*

this type contain two kinds of links: *roller links* (Figure 12.8.3) and *pin links* (Figure 12.8.4). Roller links consist of two rollers which roll on two bushings that are press-fitted into the link plates. Pin links consist of two pins that are press-fitted into tow link plates. Roller links and pin links alternate as shown in Figure 12.8.2, so that roller chains usually consist of an even number of links, with one pin link replaced by a *connecting link*, Figure 12.8.4, for easy installation or removal. This connecting link

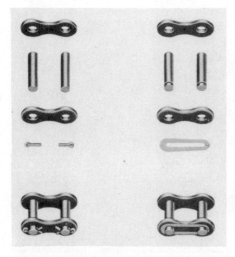

Figure 12.8.4 Connecting links. *Source: American Chain Association,* chains for Power Transmission and Material Handling, *Marcel Dekker, Inc., New York, 1982, p. 51.*

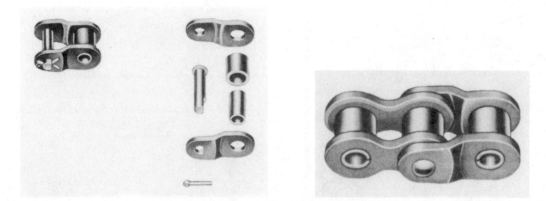

Figure 12.8.5 Offset link and offset section. *Source: American Chain Association,* chains for Power Transmission and Material Handling, *Marcel Dekker, Inc., New York, 1982, p. 51.*

differs from the pin links in that two pins are press-fitted into only one link, with a slip or light press fit in the other link, which is secured by cotter pins, a spring clip, or snap rings.

In those unusual cases where an odd number of links are needed an *offset link* or an *offset section*, Figure 12.8.5, is used as a connecting link. An offset section does not provide connection capability.

Roller chains drive and are driven by *sprockets*, which are toothed wheels machined to fit the chain rollers. Four types of sprockets are considered standard. Type A has no hub, type B has a hub on one side, type C has a hub on both sides, and type D has a detachable hub. These types and a split sprocket are shown in Figure 12.8.6. Figure 12.8.7 shows the assembled chain components and the proper meshing of chain and sprocket.

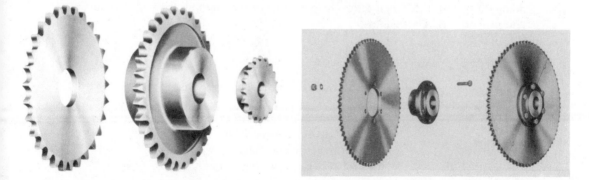

Figure 12.8.6 Four types of standard sprockets: A, B, C, and D. *Source: American Chain Association,* chains for Power Transmission and Material Handling, *Marcel Dekker, Inc., New York, 1982, p. 80.*

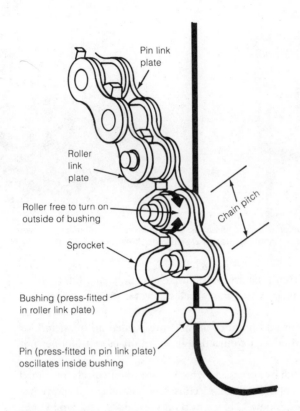

Pin link plate

Roller link plate

Roller free to turn on outside of bushing

Sprocket

Bushing (press-fitted in roller link plate)

Pin (press-fitted in pin link plate) oscillates inside bushing

Chain pitch

Figure 12.8.7 Chain and sprocket in mesh. *Source: American Chain Association,* chains for Power Transmission and Material Handling, *Marcel Dekker, Inc., New York, 1982, p. 11.*

12.9   ROLLER CHAIN DESIGNATION

Standard two- or three-digit roller chain numbers are used to identify roller chain sizes and proportions. The code is

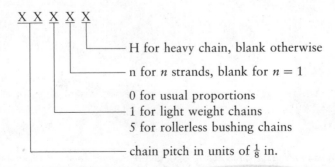

X X X X

H for heavy chain, blank otherwise

n for *n* strands, blank for $n = 1$

0 for usual proportions
1 for light weight chains
5 for rollerless bushing chains

chain pitch in units of $\frac{1}{8}$ in.

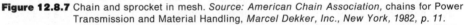

where the chain *pitch* is the distance between centers of adjacent pins, as illustrated in Figure 12.8.7 and in Table 12.9.1.

TABLE 12.9.1   GENERAL CHAIN DIMENSIONS FOR ROLLER CHAIN (in inches)

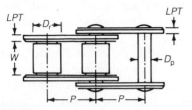

| ANSI Standard Chain No. | | Pitch P | Max Roller Diameter Dr | Width W | Pin Diameter Dp | Link Plate Thickness LPT | | Measuring Load. LB |
Std.	Heavy					Std.	Heavy	
25*	—	$\frac{1}{4}$	0.130*	$\frac{1}{8}$	0.0905	0.030	—	18
35*	—	$\frac{3}{8}$	0.200*	$\frac{3}{16}$	0.141	0.050	—	18
41†	—	$\frac{1}{2}$	0.306	$\frac{1}{4}$	0.141	0.050	—	18
40	—	$\frac{1}{2}$	$\frac{5}{16}$	$\frac{5}{16}$	0.156	0.060	—	31
50	—	$\frac{5}{8}$	0.400	$\frac{3}{8}$	0.200	0.080	—	49
60	60H	$\frac{3}{4}$	$\frac{15}{32}$	$\frac{1}{2}$	0.234	0.094	.125	70
80	80H	1	$\frac{5}{8}$	$\frac{5}{8}$	0.312	0.125	.156	125
100	100H	$1\frac{1}{4}$	$\frac{3}{4}$	$\frac{3}{4}$	0.375	0.156	.187	195
120	120H	$1\frac{1}{2}$	$\frac{7}{8}$	1	0.437	0.187	.219	281
140	140H	$1\frac{3}{4}$	1	1	0.500	0.219	.250	383
160	160H	2	$1\frac{1}{8}$	$1\frac{1}{4}$	0.562	0.250	.281	500
180	180H	$2\frac{1}{4}$	$1\frac{13}{32}$	$1\frac{13}{32}$	0.687	0.281	.312	633
200	200H	$2\frac{1}{2}$	$1\frac{9}{16}$	$1\frac{1}{2}$	0.781	0.312	.375	781
240	240H	3	$1\frac{7}{8}$	$1\frac{7}{8}$	0.937	0.375	.500	1125

Bushing Diameter. Chain is rollerless.
* Without rollers.
† Light Machinery Series.
Source: American Chain Association Chains for Power Transmission and Material Handling, Marcel Dekker, Inc., New York, 1982, p. 31.

For example, a number 25 chain is a rollerless bushing chain with a pitch of $\frac{1}{8}$ in. Suffixes may be added to indicate the number of strands; thus, a 41-2 designation denotes a light weight chain with a pitch of $\frac{1}{2}$ in. and having two strands. The letter H following the chain number designation signifies a heavy chain with link plates $\frac{1}{32}$ in. thicker than the corresponding basic chain. Hence, a 50-2H designation calls for a chain having a $\frac{5}{8}$ pitch, two strands, with link plates $\frac{1}{32}$ in. thicker than those on a standard 50 chain.

Obviously 41 H is a contradictory designation.

Light, standard, and heavy chain proportions are listed in Table 12.9.1.

12.10 CHAIN SELECTION FORMULAS

Chain selection for drives such as those shown in Figure 12.10.1 depend upon the rated power of the driver, the specified speed ratio the center distance and its allowable variation, the shaft diameters, and the service conditions.

The formulas involved are those for the chain length, the center distance, and the power that may be transmitted per strand of chain. Center distance and chain length formulas differ from those for belts because the chain links form part of a polygon as they contact the sprockets. The vertices of the polygon coincide with the centers of the chain's pins and rollers. One segment of such a polygon of length p, the chain pitch, is shown in Figure 12.10.2a. As shown, it subtends an angle of $2\pi/N$ at the center of the sprocket. The circle through the pin centers is known as the pitch circle. Its diameter d' in units of inches or millimeters is given in terms of pitches by

$$d' = \frac{p}{\sin \dfrac{\pi}{N}}$$

Because it will ultimately prove more convenient, it is common practice to define the

Figure 12.10.1 Typical roller chain applications. *Source: American Chain Association,* chains for Power Transmission and Material Handling, *Marcel Dekker, Inc., New York, 1982, pp. 130, 141.*

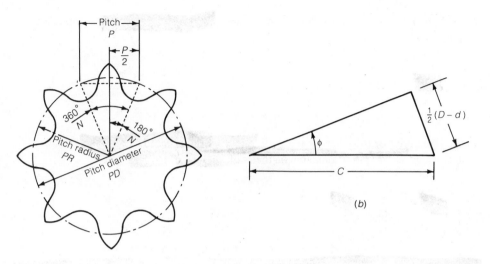

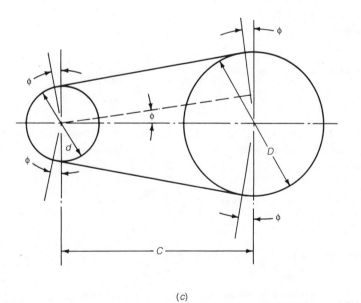

Figure 12.10.2 Sprocket and chain geometry. *Source: American Chain Association*, chains for Power Transmission and Material Handling, *Marcel Dekker, Inc., New York, 1982, p. 99.*

pitch diameter in terms of pitches. Thus, if we let d denote the pitch diameter in units of pitch we have that

$$d \equiv \frac{d'}{p} = \frac{1}{\sin \dfrac{\pi}{N}} \qquad (12.10.1)$$

In the remainder of this chapter we shall let d represent the pitch diameter of the small sprocket and D the pitch diameter of the large sprocket, so that

$$d = \frac{1}{\sin \dfrac{\pi}{N_1}} \qquad D = \frac{1}{\sin \dfrac{\pi}{N_2}} \qquad (12.10.2)$$

in *units of pitch*, where N_1 and N_2 denote the number of teeth on the small and large sprocket, respectively.

Deriving a formula for the center distance according to the geometry in Figure 12.10.2 is easier if we make two observations. First, from Figure 12.10.2 *b* we see that with the aid of (12.10.2)

$$\phi = \sin^{-1} \frac{D - d}{2C} = \sin^{-1} \frac{1}{2C} \left(\frac{1}{\sin \dfrac{\pi}{N_2}} - \frac{1}{\sin \dfrac{\pi}{N_1}} \right) \qquad (12.10.3)$$

Second, since there are N_1 and N_2 teeth in the small and large sprockets, the number of pitches and fractions thereof, N_ϕ, along each of the contact arcs on each sprocket may be written as

$$\frac{N_{\phi_1}}{N_1} = \frac{\pi - 2\phi}{2\pi} \quad \text{and} \quad \frac{N_{\phi_2}}{N_2} = \frac{\pi + 2\phi}{2\pi}$$

so that upon solving for N_{ϕ_1} and N_{ϕ_2} we have

$$N_{\phi_1} = N_1 \left(\frac{1}{2} - \frac{\phi}{\pi} \right) \quad \text{and} \quad N_{\phi_2} = N_2 \left(\frac{1}{2} + \frac{\phi}{\pi} \right) \qquad (12.10.4)$$

as the number of pitches in the arc of contact on the small and large sprockets, respectively. Consequently, the chain length L may be written in terms of pitches as

$$L = 2C \cos \phi + N_{\phi_1} + N_{\phi_2}$$

With these formulas at hand we can immediately write that

$$L_0 = 2C \cos \phi + \frac{N_2 + N_1}{2} + (N_2 - N_1) \frac{\phi}{\pi} \quad \text{rad} \tag{12.10.5}$$

where ϕ may be calculated from (12.10.3) which also involves C. Substitution of the desired approximate center distance C_0 into (12.10.5) yields an initial estimate L_0 of the chain length. Since L_0 will generally not be an even integer, the next step is to select the actual chain length L as an even number of pitches. Whether L is longer or shorter than L_0 is usually not critical.

We must next find the actual center distance corresponding to L. This cannot be done by solving (12.10.5) for C because ϕ also depends upon C, as shown in (12.10.3). Instead, we must find C from

$$\frac{\pi}{N_2 - N_1} \left(L - 2C \cos \phi - \frac{N_2 + N_1}{2} \right) - \phi = 0 \quad \text{rad} \tag{12.10.6}$$

in which ϕ must be found from (12.10.3). After substituting for ϕ from (12.10.3) it is painfully clear that an algebraic solution is impossible except for the trivial case where $N_1 = N_2$. Hence, a numerical solution is in order. Fortunately (12.10.6) may be solved for C quite rapidly by means of a bisection routine.

Whenever either the driveR or driveN units may be moved slightly after the chain is installed the center distance may be approximated for initial positioning from

$$C = \frac{L - N_1 \dfrac{90 - \phi}{180} - N_2 \dfrac{90 + \phi}{180}}{2 \cos \phi} \quad \text{deg} \tag{12.10.6a}$$

where ϕ is calculated from equation 12.10.3 using the initially specified center distance. As demonstrated in problem 12.37, the final center distances calculated from equations 12.10.6 and 12.10.6a often agree to within 0.001 pitches or less when the final chain length is that even number of pitches closest to the length found from equation 12.10.5.

The power that may be transmitted per strand as limited by fatigue of the link plates and by roller and bushing impact life for a chain life of approximately 15 000 h may be calculated from

$$\text{hp/strand} = \text{minimum of} \begin{cases} \text{hp}_s \\ \text{hp}_r \end{cases} \tag{12.10.7}$$

where

$$\text{hp}_s = K_s N_1^{1.08} n^{0.9} p^{(3.00 - 0.07p)} \tag{12.10.8}$$

is based upon link plate fatigue and where

$$hp_r = K_r p^{0.8}\left(\frac{100N_1}{n}\right)^{1.5} \tag{12.10.9}$$

is based upon roller and bushing impact life. Dimensional constants K_s and K_r are defined by

$$K_s = 0.004 \quad \text{for all but number 41 chains}$$
$$\quad = 0.0022 \quad \text{for number 41 chain}$$

and by

$$K_r = 24 \quad \text{for number 25 and 35 chains}$$
$$\quad = 3.4 \quad \text{for number 41 chains}$$
$$\quad = 17 \quad \text{for chains 40 through 240 inclusive except for chain 41}$$

Localized cold welding (galling) on the bearing surface between a roller pin and its bushing results in surface roughening as the weld is broken and may appear at speeds n (rpm) of the small sprocket greater than

$$\left(\frac{n}{1000}\right)^{1.59\log p + 1.873} = \frac{82.5}{7.95^p (1.0278)^{N_1} 1.323^{F/1000}} \tag{12.10.10}$$

where n (rpm) denotes the speed of the small sprocket in equations 12.10.9 and 12.10.8 as well. In equation 12.10.10 F denotes the force/strand and may be found from $F = 396\,000\,\mathrm{Dhp}/(\pi dn\,\mathrm{Smf})$ in which d represents the pitch diameter of the small sprocket and Smf is the strand multiplication factor to be used in the next section and listed in Table 12.11.2. It accounts for unequal loading of strands of chain due to slight bending of cantilevered input and output shafts.

Since the chain links form part of a polygon about the center of each sprocket, it follows that the radius at which the chain acts on each sprocket may vary from $d_p/2$ to $(d_p \cos \phi)/2$ for the small sprocket and from $d_p/2$ to $(d_p \cos \phi)/2$ for the large sprocket. These radii may occur in any combination, depending upon the individual chain length and the sprocket sizes. Hence, it follows that if the driveR sprocket operates at constant speed the driveN sprocket will operate at a fluctuating speed. This is known as polygonal action. Its effect decreases as the number of teeth on both sprockets increases and it is minimum for given sprocket sizes whenever $C \cos \phi$ is an integer number of pitches.[12] It vanishes entirely whenever $C \cos \phi$ is an integer and both driver and driveN sprockets have the same number of teeth, i.e., if they are both the same size.

Polygonal action must be considered whenever sprocket position is important or whenever the load has a large moment of inertia. The large moment of inertia is more important when it is driven by either an internal combustion engine or a piston type

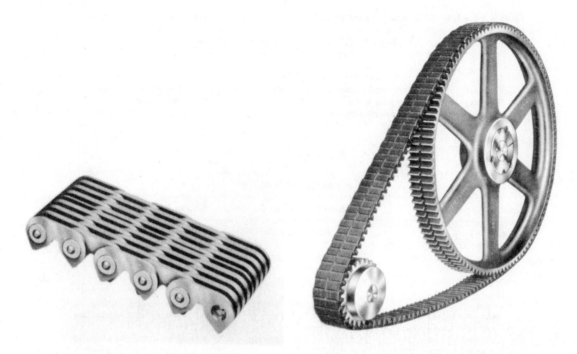

Figure 12.10.3 One of several varieties of silent chain. *Source: American Chain Association,* chains for Power Transmission and Material Handling, *Marcel Dekker, Inc., New York, 1982, pp. 55, 189.*

hydraulic motor than when it is driven by an electric motor. This is because the inertia of the load resists acceleration when the speed increases and a piston driveR resists deceleration by the load when the speed decreases. At certain chain lengths and chain weights this may induce harmful transverse chain vibration, especially on the slack side.

The effects of polygonal action for a given power per strand and speed ratio may be minimized by changing to a *silent chain*, as shown in Figure 12.10.3, because the pitch tends to be smaller. Notice that the chain also tends to be quieter because the links are designed to be in contact with both sides of the sprocket teeth.

12.11 CHAIN SELECTION PROCEDURES

The first step in chain selection is that of determining the classification of the load according to its shock characteristics as guided by List 2 of Table 12.11.1 and then determining the service factor from List 1 of Table 12.11.1 which is dependent upon the characteristics of the input power.

TABLE 12.11.1   SERVICE FACTORS AND LOADS FOR ROLLER CHAIN DESIGN

Service Factors for Roller Chain Drives

Type of Driven Load	Type of Input Power		
	Internal Combustion Engine with Hydraulic Drive	Electric Motor or Turbine	Internal Combustion Engine with Mechanical Drive
Smooth	1.0	1.0	1.2
Moderate Shock	1.2	1.3	1.4
Heavy Shock	1.4	1.5	1.7

Load Classifications

Smooth Load	Moderate Shock Load	Heavy Shock Load
Agitators—Pure liquid Conveyors—Uniformly loaded or fed (apron, assembly, belt, flight, oven, screw) Fans—Centrifugal and light, small diameter Line shafts—Light Machines—All types with uniform non-reversing loads Screens—Rotary (uniformly fed), traveling water intake Sewage disposal equipment—Inside service (uniformly fed)	Clay working machinery—Pug mills Conveyors—Heavy duty and NOT uniformly loaded (apron, assembly, belt, bucket, flight, oven, screw) Cranes and hoists—Medium duty, skip hoists (travel motion and trolley motion) Dredges—Cable, reel and conveyor Food industry—Beet slicers, dough mixers, meat grinders Grinders Laundry industry—Washers, tumblers Line shafts—Heavy service Machine tools—Main and auxiliary drives Machine—All types with moderate shock and nonreversing loads Screens—Rotary (stone or gravel) Textile industry—Calenders, dyeing machinery, mangles, nappers, soapers, spinners, tenter frames	Clay working machinery—Brick press, briquetting machinery Conveyors—Reciprocating and shaker Cranes and Hoists—Heavy duty, including logging, lumbering, and rotary drilling rigs Dredges—Cutter head drives, jig drives, screen drives Hammer mills Machine tools—Punch press, shears, plate planers Machines—All types with severe impact shock loads or speed variations, and reversing service Metal mills—Draw bench, forming machines, slitters, small rolling mill drives, wire drawing or flattening Mills—(rotary type) ball, cement kilns, rod mills, tumbling mills Paper industry—Mixers, calendars, rubber mill sheeters Textile industry—Carding machinery

---

Source: American Chain Association, *Chains for Power Transmission and Material Handling*, Marcel Dekker, Inc., New York, 1982, pp. 128–29.

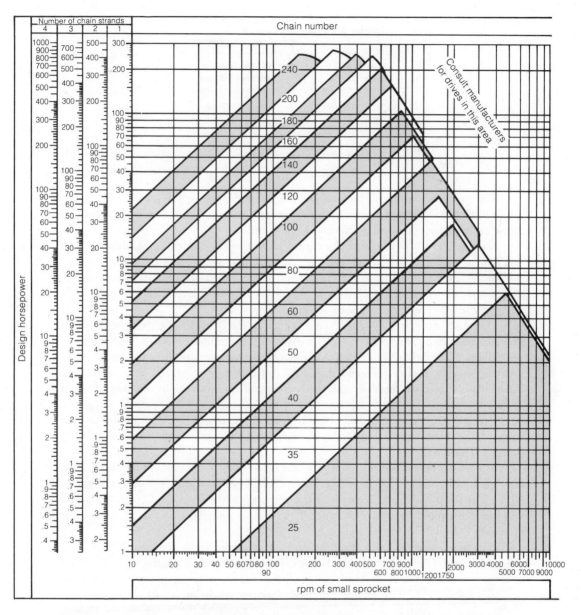

Figure 12.11.1 Roller chain selection chart. *Source: American Chain Association,* chains for Power Transmission and Material Handling, *Marcel Dekker, Inc., New York, 1982, p. 136.*

TABLE 12.11.2 MULTIPLE STRAND FACTORS

Number of Strands	Multiple Strand Factor
2	1.7
3	2.5
4	3.3

Source: American Chain Association

Step two is to multiply the power rating of the input power unit by the service factor to obtain the design power. In this chapter we will use design horsepower because all of the formulas involved are in Old English units.

The third step is that of entering Figure 12.11.1 at the rpm of the smaller sprocket and reading up to the design horsepower and reading the chain number at the intersection of the small sprocket rpm with the design horsepower. The boundary lines on this chart were, of course, calculated from equations 12.10.8, 12.10.9, and 12.10.10. The left vertical axis has four scales to show horsepower when 1, 2, 3, or 4 strands are used. Because of flexing of cantilevered sprocket shafts under the chain tension the power transmitted does not increase linearly with the number of chains. The empirical multipliers given in Table 12.11.2 account for power per strand reduction with increasing strands. It is recommended that no more than four strands be used because of the loads placed on the shaft and the corresponding reduction in the load rating of additional strands.

Multiple strands are generally used to increase the power that may be transmitted without increasing the pitch in order to minimize speed and torque variations due to polygonal action.

Step four is that of turning to Table 12.11.3 to find the minimum number of teeth on a standard sprocket for the chain size (number) selected that will fit the shaft on which it is to be mounted. These dimensions were standardized after considering hub and key dimensions necessary to support the torque and the clearance necessary for the chain links.

Steps five is that of calculating the power per strand from criterion 12.10.8 for a range of numbers of teeth on the small sprocket in order to select a suitable combination of large sprocket diameter and number of strands for the required power transmission.

Step six is calculation of the required chain length from (12.10.5), rounding to the nearest even number of pitches (to avoid using weaker offset links, Figure 12.8.5), finding the final center distance from either (12.10.6) or (12.10.6a), selecting the lubrication type, and finding the galling limit.

Figure 12.11.2 shows a representative flowchart suitable for automating the chain selection process. See Ref. 9 for a program that may be used with a programmable pocket calculator.

TABLE 12.11.3 MAXIMUM BORE AND HUB DIAMETERS (with Standard Keyways)

No. of Teeth	$\frac{3}{8}''$ Pitch		$\frac{1}{2}''$		$\frac{5}{8}''$		$\frac{3}{4}''$		$1''$	
	Max. Bore	Maximum Hub Dia.	Max. Bore	Maximum Hub Dia.	Max. Bore	Maximum Hub Dia.	Max. Bore	Maximum Hub Dia.	Max. Bore	Maximum Hub Dia.
11	$\frac{19}{32}$	$\frac{55}{64}$	$\frac{25}{32}$	$1\frac{11}{64}$	$\frac{31}{32}$	$1\frac{15}{32}$	$1\frac{1}{4}$	$1\frac{49}{64}$	$1\frac{5}{8}$	$2\frac{3}{8}$
12	$\frac{5}{8}$	$\frac{63}{64}$	$\frac{7}{8}$	$1\frac{21}{64}$	$1\frac{5}{32}$	$1\frac{43}{64}$	$1\frac{9}{32}$	$2\frac{1}{64}$	$1\frac{25}{32}$	$2\frac{45}{64}$
13	$\frac{3}{4}$	$1\frac{7}{64}$	1	$1\frac{1}{2}$	$1\frac{9}{32}$	$1\frac{7}{8}$	$1\frac{1}{2}$	$2\frac{1}{4}$	2	$3\frac{1}{64}$
14	$\frac{27}{32}$	$1\frac{15}{64}$	$1\frac{5}{32}$	$1\frac{21}{32}$	$1\frac{5}{16}$	$2\frac{5}{64}$	$1\frac{3}{4}$	$2\frac{1}{2}$	$2\frac{9}{32}$	$3\frac{11}{32}$
15	$\frac{7}{8}$	$1\frac{23}{64}$	$1\frac{1}{4}$	$1\frac{13}{16}$	$1\frac{17}{32}$	$2\frac{9}{32}$	$1\frac{25}{32}$	$2\frac{3}{4}$	$2\frac{13}{32}$	$3\frac{43}{64}$
16	$\frac{31}{32}$	$1\frac{15}{32}$	$1\frac{9}{32}$	$1\frac{63}{64}$	$1\frac{11}{16}$	$2\frac{31}{64}$	$1\frac{31}{32}$	$2\frac{63}{64}$	$2\frac{23}{32}$	$3\frac{63}{64}$
17	$1\frac{3}{32}$	$1\frac{19}{32}$	$1\frac{3}{8}$	$2\frac{9}{64}$	$1\frac{25}{32}$	$2\frac{11}{16}$	$2\frac{7}{32}$	$3\frac{7}{32}$	$2\frac{13}{16}$	$4\frac{5}{16}$
18	$1\frac{7}{32}$	$1\frac{23}{32}$	$1\frac{17}{32}$	$2\frac{19}{64}$	$1\frac{7}{8}$	$2\frac{57}{64}$	$2\frac{9}{32}$	$3\frac{15}{32}$	$3\frac{1}{8}$	$4\frac{41}{64}$
19	$1\frac{1}{4}$	$1\frac{27}{32}$	$1\frac{11}{16}$	$2\frac{29}{64}$	$2\frac{1}{16}$	$3\frac{5}{64}$	$2\frac{7}{16}$	$3\frac{45}{64}$	$3\frac{5}{16}$	$4\frac{61}{64}$
20	$1\frac{9}{32}$	$1\frac{61}{64}$	$1\frac{25}{32}$	$2\frac{5}{8}$	$2\frac{1}{4}$	$3\frac{9}{32}$	$2\frac{11}{16}$	$3\frac{61}{64}$	$3\frac{1}{2}$	$5\frac{9}{32}$
21	$1\frac{5}{16}$	$2\frac{5}{64}$	$1\frac{25}{32}$	$2\frac{25}{32}$	$2\frac{9}{32}$	$3\frac{31}{64}$	$2\frac{13}{16}$	$4\frac{3}{16}$	$3\frac{3}{4}$	$5\frac{19}{32}$
22	$1\frac{7}{16}$	$2\frac{13}{64}$	$1\frac{15}{16}$	$2\frac{15}{16}$	$2\frac{7}{16}$	$3\frac{11}{16}$	$2\frac{15}{16}$	$4\frac{7}{16}$	$3\frac{7}{8}$	$5\frac{59}{64}$
23	$1\frac{9}{16}$	$2\frac{5}{16}$	$2\frac{3}{32}$	$3\frac{3}{32}$	$2\frac{5}{8}$	$3\frac{57}{64}$	$3\frac{1}{8}$	$4\frac{43}{64}$	$4\frac{3}{16}$	$6\frac{15}{64}$
24	$1\frac{11}{16}$	$2\frac{7}{16}$	$2\frac{1}{4}$	$3\frac{17}{64}$	$2\frac{13}{16}$	$4\frac{5}{64}$	$3\frac{1}{4}$	$4\frac{29}{32}$	$4\frac{9}{16}$	$6\frac{9}{16}$
25	$1\frac{3}{4}$	$2\frac{9}{16}$	$2\frac{9}{32}$	$3\frac{27}{64}$	$2\frac{27}{32}$	$4\frac{9}{32}$	$3\frac{3}{8}$	$5\frac{5}{32}$	$4\frac{11}{16}$	$6\frac{7}{8}$

No. of Teeth	$1\frac{1}{4}''$		$1\frac{1}{2}''$		$1\frac{3}{4}''$		$2''$		$2\frac{1}{2}''$	
	Max. Bore	Maximum Hub Dia.	Max. Bore	Maximum Hub Dia.	Max. Bore	Maximum Hub Dia.	Max. Bore	Maximum Hub Dia.	Max. Bore	Maximum Hub Dia.
11	$1\frac{31}{32}$	$2\frac{31}{32}$	$2\frac{5}{16}$	$3\frac{37}{64}$	$2\frac{13}{16}$	$4\frac{11}{64}$	$3\frac{9}{32}$	$4\frac{25}{64}$	$3\frac{15}{16}$	$5\frac{63}{64}$
12	$2\frac{9}{32}$	$3\frac{3}{8}$	$2\frac{3}{4}$	$4\frac{1}{16}$	$3\frac{1}{4}$	$4\frac{3}{4}$	$3\frac{5}{8}$	$5\frac{27}{64}$	$4\frac{23}{32}$	$6\frac{51}{64}$
13	$2\frac{17}{32}$	$3\frac{25}{32}$	$3\frac{1}{16}$	$4\frac{35}{64}$	$3\frac{9}{16}$	$5\frac{5}{16}$	$4\frac{1}{16}$	$6\frac{5}{64}$	$5\frac{3}{32}$	$7\frac{39}{64}$
14	$2\frac{11}{16}$	$4\frac{3}{16}$	$3\frac{5}{16}$	$5\frac{1}{32}$	$3\frac{7}{8}$	$5\frac{7}{8}$	$4\frac{11}{16}$	$6\frac{23}{64}$	$5\frac{23}{32}$	$8\frac{27}{64}$
15	$3\frac{3}{32}$	$4\frac{19}{32}$	$3\frac{3}{4}$	$5\frac{33}{64}$	$4\frac{7}{16}$	$6\frac{29}{64}$	$4\frac{7}{8}$	$7\frac{3}{64}$	$6\frac{1}{4}$	$9\frac{7}{32}$
16	$3\frac{9}{32}$	5	4	6	$4\frac{11}{16}$	$7\frac{1}{64}$	$5\frac{1}{2}$	$8\frac{1}{64}$	7	$10\frac{1}{32}$
17	$3\frac{21}{32}$	$5\frac{13}{32}$	$4\frac{15}{32}$	$6\frac{31}{64}$	$5\frac{1}{16}$	$7\frac{37}{64}$	$5\frac{11}{16}$	$8\frac{21}{32}$	$7\frac{7}{16}$	$10\frac{27}{32}$
18	$3\frac{25}{32}$	$5\frac{51}{64}$	$4\frac{21}{32}$	$6\frac{31}{32}$	$5\frac{5}{8}$	$8\frac{9}{32}$	$6\frac{1}{4}$	$9\frac{5}{64}$	$8\frac{1}{4}$	$11\frac{41}{64}$
19	$4\frac{3}{16}$	$6\frac{13}{64}$	$4\frac{15}{16}$	$7\frac{29}{64}$	$5\frac{11}{16}$	$8\frac{45}{64}$	$6\frac{7}{8}$	$9\frac{61}{64}$	9	$12\frac{7}{16}$
20	$4\frac{19}{32}$	$6\frac{30}{64}$	$5\frac{7}{16}$	$7\frac{15}{16}$	$6\frac{1}{4}$	$9\frac{17}{64}$	7	$10\frac{19}{32}$	$9\frac{3}{4}$	$13\frac{1}{4}$
21	$4\frac{11}{16}$	7	$5\frac{11}{16}$	$8\frac{27}{64}$	$6\frac{13}{16}$	$9\frac{53}{64}$	$7\frac{3}{4}$	$11\frac{15}{64}$	10	$14\frac{3}{64}$
22	$4\frac{7}{8}$	$7\frac{13}{32}$	$5\frac{7}{8}$	$8\frac{57}{64}$	$7\frac{1}{4}$	$10\frac{25}{64}$	$8\frac{3}{8}$	$11\frac{7}{8}$	$10\frac{7}{8}$	$14\frac{27}{32}$
23	$5\frac{5}{16}$	$7\frac{13}{16}$	$6\frac{3}{8}$	$9\frac{3}{8}$	$7\frac{7}{16}$	$10\frac{15}{16}$	9	$12\frac{33}{64}$	$11\frac{5}{8}$	$15\frac{21}{32}$
24	$5\frac{11}{16}$	$8\frac{13}{64}$	$6\frac{13}{16}$	$9\frac{55}{64}$	8	$11\frac{1}{2}$	$9\frac{5}{8}$	$13\frac{5}{32}$	13	$16\frac{29}{64}$
25	$5\frac{23}{32}$	$8\frac{39}{64}$	$7\frac{1}{4}$	$10\frac{11}{32}$	$8\frac{9}{16}$	$12\frac{1}{16}$	$10\frac{1}{4}$	$13\frac{51}{64}$	$13\frac{1}{2}$	$17\frac{1}{4}$

Note: Maximum bore and hub diameters for a specified number of teeth are equivalent to the minimum number of teeth for a selected bore and associated hub diameter. All dimensions are in inches.

Source: American Chain Association, *chains for Power Transmission and Material Handling*, Marcel Dekker, Inc., New York, 1982, p. 118.

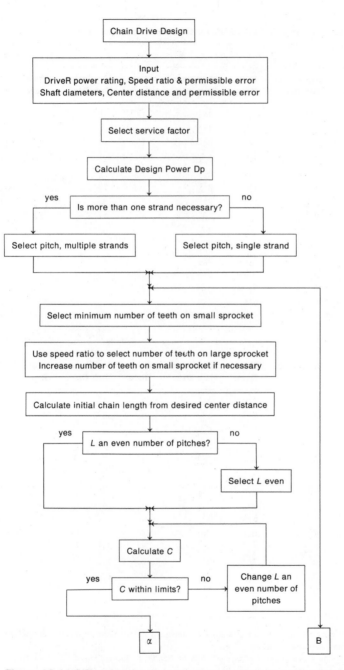

Figure 12.11.2 Flowchart for chain selection. No explicit provision has been made in this flowchart to decrease the pitch in order to reduce polygonal action. It is implicit in the selection of the number of strands. If reduction of polygonal action is important, steps may be inserted after the calculation of C for an even number of links to check $C \cos \phi$ to find if it is also an even number and iterate sprocket diameters and center distance if it is not.

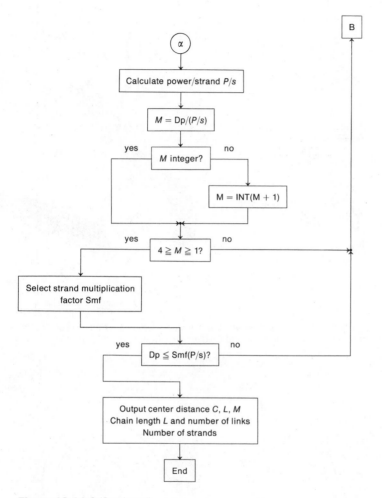

Figure 12.11.2 (Continued)

12.12 CHAIN LUBRICATION

Type A lubrication includes manual lubrication with a brush or an oil can before each 8 h of operation, or drip lubrication as pictured in Figure 12.12.1. It is generally used for low rpm applications. It is important to emphasize that the brush or oil can should be held outside of the chain loop to prevent either the hand or the brush from being quickly pulled between chain and sprocket.

Type B lubrication consists of either having the lowest portion of the chain pitch line immersed in an oil sump in the chain housing, as in Figure 12.12.2, or, if the chain is not immersed in oil, having a disk on the lower shaft that flings oil onto a collector plate from which it drips either onto the chain directly, as in the case of a single strand,

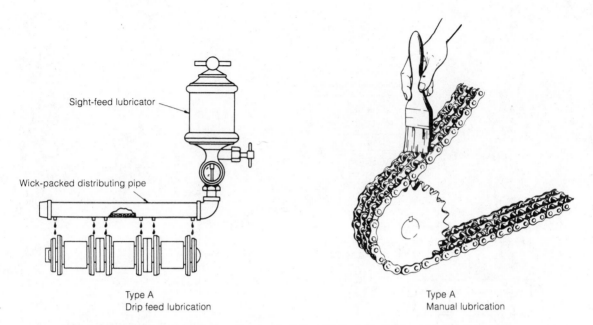

Sight-feed lubricator

Wick-packed distributing pipe

Type A
Drip feed lubrication

Type A
Manual lubrication

Figure 12.12.1 Type A lubrication. *Source: American Chain Association*, chains for Power Transmission and Material Handling, *Marcel Dekker, Inc., New York, 1982, p. 324.*

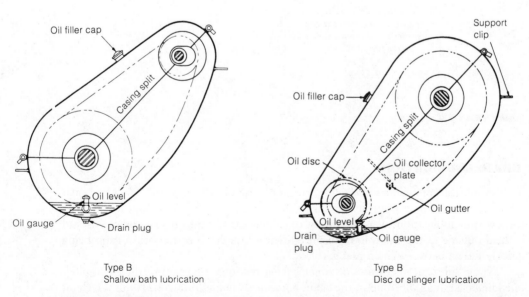

Oil filler cap

Casing split

Oil level

Oil gauge

Drain plug

Type B
Shallow bath lubrication

Support clip

Oil filler cap

Casing split

Oil disc

Oil collector plate

Oil gutter

Oil level

Drain plug

Oil gauge

Type B
Disc or slinger lubrication

Figure 12.12.2 Type B lubrication. *Source: American Chain Association*, chains for Power Transmission and Material Handling, *Marcel Dekker, Inc., New York, 1982, p. 325.*

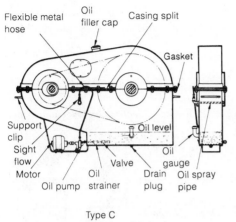

Type C
Oil stream lubrication

Figure 12.12.3 Type C lubrication. *Source: American Chain Association*, chains for Power Transmission and Material Handling, *Marcel Dekker, Inc., New York, 1982, p. 326.*

or onto an oil gutter which causes it to drip onto several strands of chain. This type of lubrication is often found on chain drives operating at intermediate speed and power.

Type C lubrication involves an oil stream or a spray from a pump that draws oil from a sump in the chain housing, Figure 12.12.3. This is the most expensive of the three common systems described, but it also provides the most thorough lubrication. It is frequently used for high speed and/or high-power drives. Cooling fins or oil coolers may be needed in some instances to keep the oil within its recommended temperature range.

Chain size and chain speed are used to select the type of lubricant according to Table 12.12.1. Since these recommendations have changed frequently in the past and

TABLE 12.12.1 LUBRICATION LIMITATIONS BASED UPON MAXIMUM CHAIN SPEED (in feet/minute)

Lubrication System	Chain Numbers											
	35	40	41	50	60	80	100	120	140	160	200	240
Type A	370	300	300	250	220	170	150	130	115	100	85	75
Type B	2800	2300	2300	2000	1800	2500	1300	1200	1100	1000	900	800
Type C	Up to manufacturers' maximum recommended chain speed for all sizes											

Type A: Manual or drip lubricaton. Recommended for a 60 chain for speeds up to and including 220 ft/min
Type B: Bath or disk lubrication. Recommended for a 60 chain for speeds greater than 220 ft/min and up to and including 1800 ft/min
Type C: Oil stream lubrication. Recommended for a 60 chain for speeds greater than 1800 ft/min but less than a maximum speed recommended by the manufacturer
Source: Emerson Power Transmission, Ithaca, NY.

may continue to do so in the future as improved materials and lubricants become available, the engineer should check with the chain manufacturer to obtain the latest guidelines.

REFERENCES

1. *Design Manual Roller and Silent Chain Drives, Roller and Silent Chain Selection*, American Chain Association, Washington, DC, 1974.

2. *Heavy Duty V-Belt Drive Design Manual, 14995-A*, The Gates Rubber Co., Denver, CO, 1982.

3. Oliver, L.R., Johnson, C.O., and Breig, W.E., V-belt life prediction and power rating, *Journal of Engineering for Industry*, 98, 340–347 (1976).

4. Waugh, D.L., Fisher, D.G., and Gayer, M.D., *Dayco Automotive Belt Drive Design Handbook*, Dayco Corp., Springfield, MO, 1980.

5. Oliver, L.R., Johnson, C.O., and Breig, W.F., *Agricultural V-Belt Drive Design*, Dayco Corp., Dayton, OH, 1977.

6. Kirk, J.A., and Tahmasebi, F., Selecting V-belts by the book, *Computers in Mechanical Engineering*, 2(6), 47–57 (1984).

7. Orthwein, W.C., V-belt selection using a programmable pocket calculator, *Computers in Engineering* 1, 91–97 (1983).

8. Orthwein, W.C., Calculating belt length for more than two sheaves, *Computers in Engineering* 2, 490–494 (1984).

9. Orthwein, W.C., Chain drive design, *Computers in Mechanical Engineering*, 1 (2) 55–61 (1982).

10. Bouillon, G., and Tordion, G.V., On polygonal action in roller chain drives, *Transactions of the ASME, Series B, Journal of Engineering for Industry*, 87 (2), 243–250 (May 1965)

PROBLEMS

Section 12.5

12.1 Express belt length in terms of sheave diameters, distance between centers, and angle θ, shown in Figure 12.5.2. Obtain approximation 12.5.10 by using series expansions for the square root and arc cosine terms.

12.2 Change the static belt tension required to transmit 5 hp at 1160 rpm using a motor sheave 3.35 in. in diameter. The contact angle is 150° and a 3VX cross section is to be used.

12.3 Calculate the decrease in belt tension for problem 12.2 if the small sheave diameter is increased to 5.60 in. with $\theta = 150°$. It is recommended that belt drives be designed using the largest sheaves that are compatible with available space and other design constraints. (The largest sheave pairs will not always be used in the following examples.)

12.4 Design a belt drive for a light duty conveyor for 8-h/day service. Input speed to the reduction gears driving the conveyor should be 880 ± 5 rpm. It is to be driven by a 5-hp synchronous ac motor that operates at 1725 rpm. The center distance should be no greater than 18 in., with

10 in. preferred. Specify the belt number, sheave diameters, center distance, the number of belts, and the installation and take-up allowances.

[*Ans.* Two 3VX400 belts required.]

12.5 Specify sheave diameters, belt number, and the number of belts needed to drive a dough mixer whose input shaft turns at 400 rpm and is in continuous service, driven by a high-torque ac motor rated at 10 hp at 870 rpm. Center distance is 20 ± 3 in. and the speed ratio should be within 10% of that requested.

12.6 Conversion to a specialty food requires that the mixer in problem 12.5 be converted to 250 ± 10 rpm. Can this be done simply by changing one sheave? Can it be done if both sheaves are replaced? Can the same belts be used if the center distance is modified by 2.0 in.?

12.7 Select belts and sheaves to drive a fan at 300 rpm with a 12-hp two-cylinder internal combustion engine whose output shaft rotates at 700 rpm. The center distance should be 35 ± 5 in. Fan speed may be increased to 360 rpm if necessary to use standard sheaves. Expected service is 9 h/day. Also specify the installation and take-up distances.

[*Ans.* One 5 V × 1250 belt.]

12.8 Calculate the expected static load on the sheaves for problem 12.7.

[*Ans.* $F_s = 371$ 1b.]

12.9 Select belts and sheaves for an air compressor on an industrial sprayer in which the input speed to the reduction gearing is 140 ± 10 rpm, the center distance is 26 ± 3 in., and power is supplied by a 1-hp single-phase ac motor at 1750 rpm. Design for intermittant service. Specify position A and length B of the motor mount slots shown in the figure to allow for installation and take-up. Mounting holes on the motor base are 2.75 in. on either side of the shaft center line. Also give belt number and number of belts.

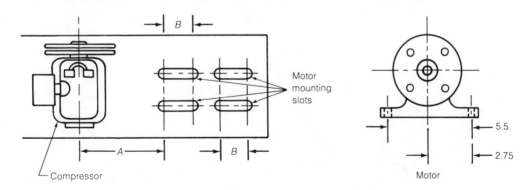

12.10 Design a belt drive for a bucket elevator for continuous service. Input shaft speed is 300 ± 10 rpm and the 75-hp dc series wound motor that drives it has a measured output speed of 690 rpm. Center distance should be 30 ± 5 in. Give sheave diameters, belt number, number of belts, and installation and take-up distances.

[*Ans.* Compare for both 5VX and 8V cross-section drives. Use of 8V1320 belts requires fewer belts.

12.11 A change in handling procedures requires that the bucket elevator described in problem 12.10 operate at 210 rpm and that the dc motor be slowed accordingly to maintain the same speed ratio. If the motor still delivers 75 hp at the slower speed will it be necessary to change belts and/or sheaves? Use standard sheaves. Would you recommend increasing the center distance to between 90 and 120 in.? Why?

12.12 Would you recommend a belt drive for a continuous service mine hoist in which the direct drive sheave on the hoist turns at 50 ± 2 rpm and the 100-hp dc compound wound driving motor turns at 575 rpm? Although the center distance is unrestricted, compactness is desirable. If possible, use at least one standard sheave. Specify the number of belts, the belt number, sheave diameters, center distance, and installation and take-up allowances. If you would not recommend a belt drive, explain why not.

12.13 It is necessary to redesign the drive in problem 12.9 for replacement of the single-phase motor with a squirrel cage motor of the same power. If it is necessary to change belts, give the new belt number, the number of belts, the sheave diameters, the center distance, and the installation and take-up allowances.

12.14 A split phase, 5-hp, 1750-rpm ac motor is to drive a woodworking bandsaw at 790 ± 10 rpm under normal service, with a speed at or near 790 preferred. Center distance may be from 49 to 5.6 in. with $\theta = 150°$. It is recommended that belt drives be designed using the largest sheaves that are compatible with available space and other design constraints. (The largest sheave pairs will not be used in the following examples.)

[*Ans.* 2 belts, center distance of 53.3 in.]

12.15 A $(25 \pm 2):1$ speed reduction for 9-h/day service is to be designed as shown in the figure to drive a line shaft for an antique woodworking shop. It is to be driven by a 50-hp, 3-cylinder gasoline engine geared down to turn at 600 rpm. The center distances may be between 50 and 70 in. and the two may be different. Can stock sheaves be used? Can both belts be the same? Specify belt numbers, number of belts, sheave diameters, and center distances.

[*Ans.* Stock sheaves may be used; same belts may be used.]

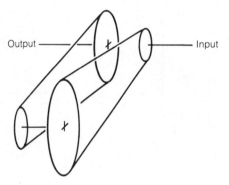

12.16 Calculate the static belt tension for the belt in problem 12.15.

12.17 Show that if the groove included angle is denoted by α that the belt tension formula for a V-belt may be written as

$$\frac{F_1}{F_2} = e^{k\theta}$$

where $k = \mu/\sin(\alpha/2)$. (*Hint:* Review the band brake derivation (Chapter 11) and observe that $\alpha/2$ is measured from the mid plane of the sheave, so that when $\alpha = \pi$ the above expression reduces to the band brake formula.)

12.18 Guided by the figure below, show that the resultant centrifugal force on a belt as it passes over a sheave may be written as

$$F_c = \frac{mv^2}{r} \int_{-\theta/2}^{\theta/2} \cos \psi \, d\psi = 2 \, \frac{mv^2}{r} \sin \frac{\theta}{2} = (F_1 + F_2) \sin \frac{\theta}{2}$$

Hence,

$$\frac{F_1 + F_2}{2} = F_s = \frac{mv^2}{r}$$

where the units of n are revolutions/second, F_c acts along the line between shaft centers, and m denotes mass/unit length.

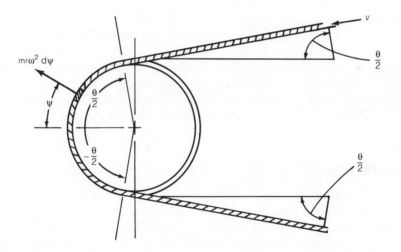

Section 12.7

Note: The following 5 problems form a sequence leading to to the estimation of the life of a V-belt.

12.19 Find the belt tensions F_1 and F_2 during high- and low-power demand for a compressor driven by a 20-hp internal combustion engine at 4000 rpm. The service factor is 1.4, the center distance is 20.4 in., and three 0.380 Plain HD belts, 71 in. long, are used. Large and small sheave outside diameters are 14.00 and 4.50 in. The machine operates at an input power of 20 hp for 10% of the time and at an input power of 14 hp for 90% of the time, corresponding to design horsepowers of 28 and 19.6, respectively.

[*Ans.* $F_1 = 85.809$ lb, $F_2 = 19.715$ lb.]

12.20 Find tensions F_{1e}, F_{2e}, F_{be}, F_{ce}, and the belt velocity for the V-belt drive described in problem 12.19 for both high- and low-power demand.

[*Ans.* $F_{1e} = 11.291$ lb, $F_{2e} = 2.594$ lb at high power, $F_{be} = 5.071$ for large sheave, $F_{ce} = 1.256$ lb, and $v = 4660.029$ ft/min.]

12.21 Find stresses σ_m and σ_0 for the belt drive described in problem 12.19 using the values found in problem 12.20.

[*Ans.* $\sigma_m = 10\,846$ psi, $\sigma_0 = 6\,953$ psi for large sheave, high power.]

12.22 Find the expected life of the belt described in problem 12.19 using the results of problem 12.21 and the values of the coefficients given in Table 12.7.1.

[*Ans.* 23 045 h (2 yr., 7.5 mo. continuous service).]

12.23 Show that the belt life found in problem 12.22 may be increased by increasing the belt length to 100 in. What is the expected life of the longer belts?

[*Ans.* 59 104 h.]

Note: Longer life may be achieved by replacing the belt drive with a gear drive, which may significantly increase the initial cost of the unit. Belt drives should not be used for intermittent service that includes appreciable idle time because the elastomers tend to deteriorate over time. Chain drives do not deteriorate over time, but their lubrication must be maintained.

12.24 Design two belt drives for a dough mixer in continuous service whose input shaft turns at 400 rpm as driven by a high-torque ac motor at 875 rpm and delivering 10 hp. The center distance for the first drive should be that closest to 17 in, and for the second drive it should be that closest to 23 in. The largest standard sheaves should be used for both drives for a speed ratio within 10% of the driven speed. Compare with problem 12.5.

12.25 Repeat problem 12.24 using the smallest standard sheaves for both drives. Do not repeat if only one pair of standard sheaves is available for the required speed range.

Sections 12.9–12.12

12.26 Explain in one sentence why equation 12.10.6*a* was used for chains instead of an approximate formula similar to that used in the design of V-belt drives.

12.27 Design a chain drive for a paper mill mixer, or beater, driven by a 30 hp-high torque ac motor that operates at 1160 rpm. The mixer operates at 135 ± 5 rpm for 8 h/day. The motor shaft diameter is 2.0 in. and the beater shaft diameter is 4.0 in., with the center distance between them to be 90 ± 12 in. Give the chain number, the pitch, the number of strands, the number of links, and the number of teeth on the small and large sprockets. Reduce polygonal action by using more than one strand but no more than three strands. Also calculate the maximum rpm if galling is to be prevented and recommend a lubrication type.

[*Ans.* 3 strands, number 60 chain.]

12.28 Design a single-strand chain drive to replace the belt drive in problem 12.9. Motor shaft diameter is $\frac{7}{8}$ in. and the mixer shaft diameter is 1 in. The small sprocket must have 12 or more teeth. Give the chain number, its length in pitches, the number of teeth in the large and small sprockets, the center distance, and the lubrication type.

12.29 Use the geometry shown in Figure 12.10.2 to show that the radial load on a sprocket is given by

$$F_r = \frac{396\,000}{\pi n_1 p}\, \text{hp}\, \csc \frac{\pi}{N_1}$$

if the tension in the slack side of the drive is neglected.

12.30 A wood chipper used in municipal park maintenance is driven by a 4-cylinder, 20-hp internal combustion engine with an output shaft 2.50 in. in diameter which turns at 1500 rpm. The input shaft to the chipper is 3.0 in. in diameter and turns at 600 rpm. The cutting heads are mechanically connected to the input shaft—no hydraulic couplings are employed. The center distance should be from 42 to 48 in. Specify the chain number, the chain length in pitches, the center distance, the sprocket pitch diameters, and the lateral loading on the motor output shaft.
   [*Ans.* 2 strands, number 60 chain.]

12.31 Compare the drive designed in problem 12.30 with those obtained by using either 80 or 100 chain. Specify the chain lengths, the center distances, the sprocket pitch diameters, and the lubrication type.

12.32 Design a two-strand chain drive for a bucket lift on a garbage truck if the sprocket speed is 80 rpm as driven as driven by a 5-hp dc motor with a 1.5-in.-diameter output shaft. There is no large sprocket because the chain is connected directly to the bucket. Do not use the chain multiplication factor; in this application two chains provide twice the hp/chain of one chain. If the service factor is 1.5, specify the chain number, the number of teeth in the sprocket, and find the radial load on the bearing at the sprocket when the motor is at its rated power.
   [*Ans.* 1 strand, 80 chain.]

12.33 Design a chain drive for problem 12.10. Motor shaft diameter is 3.0 in. A bucket elevator is also known as a bucket conveyor. Recommend a chain number, center distance, chain length in pitches, the number of teeth in the large and small sprocket, and the lubrication type.

12.34 Design a chain drive for a rotary screen for segregating gravel according to size in a testing laboratory. The screen has a 1.25-in.-diameter shaft, is uniformly fed, and is driven at 120 ± 2 rpm by a normal torque ac motor rated at 2 hp at 1160 rpm, with a $\frac{7}{8}$-in.-diameter shaft. Center distance should be 36 ± 2 in., and the small sprocket pitch diameter should not be greater than 3.0 in. Specify the chain number, the number of strands, the center distance, the pitch diameter of each sprocket, the lubrication, and the number of teeth on each sprocket.

12.35 The gathering head of a coal mining machine (service factor 1.4) is to rotate at 20 ± 2 rpm when driven by a 60-hp hydraulic motor at operating at 30 rpm. The center distance between the 4-in.-diameter head shaft and the 3.5-in.-diameter motor shaft should be 45 ± 2 in. Design a chain drive to satisfy these requirements and the condition that the small sprocket must have a minimum of 12 teeth. Give the chain number, the number of strands, the chain length in pitches, the number of teeth on each sprocket, the lubrication type, and the final center distance in inches. See the comment following problem 12.36.
   [*Ans.* 4 strands, number 240 chain.]

12.36 Recommend a chain drive for an oven auger driven by a 1.5-hp capacitor start ac motor operating at 1725 rpm. The auger center distance is to be 48 ± 4 in., it is to rotate at 140 ± 10 rpm, and its shaft diameter is to be $\frac{7}{8}$ in. The motor shaft is $\frac{5}{8}$ in. in diameter and the lateral force on its bearing should not exceed 60 lb. Give the chain number, the chain length in pitches and inches, the sprocket pitch diameters, the center distance, and the lubrication type. (Since $\frac{1}{4}$-in. and 3-in. pitches are missing from Table 12.10.6, use the values given for $\frac{3}{8}$-in. and $2\frac{1}{2}$-in. pitches, respectively.)

12.37 Demonstrate that the numerical solution for C from equation 12.10.6 differs from that given by equation 12.10.6a by less than 0.0001 pitches when $N_1 = 20$ teeth, $N_2 = 120$ teeth , and the initial center distance is 47 pitches if the chain length is 170 pitches. Also show that if $N_1 = 20$ teeth, $N_2 = 200$ teeth, and the initial center distances is 37 pitches the difference is of the order of 0.036 pitches if the chain length is 206 pitches. In both calculations ϕ in equation 12.10.6a is calculated using the initial center distance. Thus, we may deduce that equation 12.10.6 must be used whenever the speed ratio is large and the center distance short enough that the sprockets are nearly in contact and whenever there is any question as to the required accuracy of the center distance calculation. Equation 12.10.6a if often satisfactory whenever the speed ratio is small, the sprockets are separated by more than the diameter of the large sprocket, and the final chain length is that even number of pitches nearest to the initial length calculated from equation 12.10.5.

CHAPTER THIRTEEN

SPUR GEARS

NOTATION

a	addendum (l) or tooth parameter (1)	p_N	gear ratio parameter
a_n	coefficient	p_n	gear ratio parameter
a_1^*	standard long addendum, pinion (l)	q_N	gear ratio parameter
a_2^*	standard short addendum, gear (l)	q_n	gear ratio parameter
b	dedendum (l)	P_d	diametral pitch (not pitch diameter) $(1/l)$
B	backlash, in units of length (l)	p_b	base pitch (l)
c	center distance, clearance (l)	p_c	circular pitch (l)
C_a	application factor for pitting	Q_m	dynamic load capacity (ml/t^2)
C_f	surface condition factor	R	gear ratio (1) or radius (l)
C_H	hardness factor	r	radius (l)
C_L	life factor	r_i	pitch radius, gear number i (l)
C_m	load distribution factor	r_a	addendum radius (l)
C_o	overload factor	r_b	base radius (l)
C_p	elastic coefficient	r_d	dedendum radius (l)
C_R	safety factor	r_p	pitch radius (l)
C_s	size factor	T	torque (ml^2/t^2)
C_T	temperature factor	T_{10}	tooth life for 90% survival (1)
C_v	dynamic factor	t	tooth thickness (l)
d_b	base circle diameter (l)	u	load cycles (1)
d_p	pitch circle diameter (l)	v	velocity (l/t)
E	error factor	w	tooth face width (l)
e	Weibull exponent (1)	y	Lewis factor (1) or tooth ratio parameter (1)
F	force (ml/t^2)	y_n	coefficient in gear ratio calculation
F_d	AGMA dynamic force (ml/t^3)		
F_t	tangential component of tooth force (ml/t^2)	γ	generation angle, involute tooth (1)
G_{10}	expected gear life for 90% survival	γ_a	generation angle at addendum circle (1)
h	working depth (l)	γ_d	generation angle at dedendum circle (1)
I	geometry factor	δ	distance, internal contact ratio (l)
J	AGMA gear strength factor	η	angle of rotation (1)
K	gear life coefficient	θ	coordinate angle (1)
K_a	AGMA coefficient	Λ	angular backlash (1)
K_m	AGMA coefficient	λ	constant (1)
K_s	AGMA coefficient	μ	$1/\rho_1 + 1/\rho_2$ $(1/l)$
K_v	AGMA coefficient	v	Poisson's ratio (1)
L_{10}	life of gear and pinion in mesh in cycles (1)	ρ	radius of curvature of the involute profile (l)
l	length of involute curve	σ	stress (m/lt^2)
l_1	contact length, external gear 1	σ_c	allowable contact stress number
l_2	contact length, external gear 2	ϕ	pressure angle or angle of rotation (1)
l_0	contact length, internal gear	ψ	involute angle or angle of rotation (1)
l^*	contact interval along involute profile	ψ_a	value of ψ on the addendum circle (1)
m	module (l)	ω	angular rotation $(1/t)$
N	number of teeth (1)	ζ	angle measured from tooth rooth on base circle (1)
N_c	contact ratio (number of teeth in contact) (1)		
n	rotational speed, rpm (l/t) or an integer (1)	Ω	units conversion factor (t^2/ml)

The discussion of gearing is divided between two chapters, this and the next, to distinguish gearing that is limited to parallel shafts from the two following types: one that may used for either parallel or nonparallel shafts and one that is limited to nonparallel shafts. Many of the gears types to be discussed are shown in Figure 13.0.1.

General properties of involute gears are discussed before the introduction of standard tooth proportions. Methods of gear production will not be discussed because the introduction of numerically controlled machines offers the prospect of more flexible methods of gear manufacturer which may make obsolete the older gear forming procedures, known as gear generation. Nonstandard tooth profiles will be considered because of their performance advantages and the nascent trend toward their greater use.

Although tooth stresses due to a specified steady tooth load may be most accurately calculated by either the finite element method, partially available on personal computers,[6] or the modified Heywood method,[7] both are too long to be considered here. We shall, instead, consider the Lewis formula with the Barth correction, a recent NASA method, and the AGMA method. The latter will be described only briefly

Figure 13.0.1 (a) Spur gear, (b) worm gear, (c) helical gear, (d) rack, (e) internal gear, and (f) straight bevel gear. *Source: Boston Gear, Quincy, MA.*

because it involves multiplicative coefficients that are to be evaluated based upon the design engineer's personal experience.

A more important practical problem is that of estimating the life of two gears in mesh as they transmit power from one to the other and, thus, subject the teeth to cyclic loading. This problem is addressed by a method recently published by the National Aeronautics and Space Administration as a result of extensive gear studies at the Lewis Research Center. That method, which provides an estimate of gear life based upon the life of a tooth under cyclic loading, is described in greater detail, outlined with a flow-chart, and demonstrated by example.

13.1   SPUR GEAR TEETH: CONJUGATE ACTION

The basic problem solved by gearing is that of assuring that two thick disk in contact will roll against one another without slipping. The circular edge of each disk is termed the *pitch circle* and its diameter is termed the *pitch diameter*, denoted by d_p. The point of contact of the pitch circles is known as the *pitch point*.

The requirement that the teeth added to these disks not interfere with the uniform rotation that one disk would induce in the other is known as the requirement of *conjugate action*. The formal statement of the *law of conjugate action* is that to transmit uniform rotary motion from one shaft to another by means of gearing, the perpendicular to a tooth profile (which coincides with its line of force) at its point of contact with a tooth from the other gear must always pass through a fixed point on the line of centers between the two shafts.[2]

To duplicate the torque transfer from one disk to another it is also necessary that the line of force from each gear tooth pass through the pitch point, since this is the point where disks that roll without slipping transfer torque from one to the other. Thus, the pitch point is that fixed point on the line of centers through which the perpendicular to a tooth profile at its point of contact must pass if the law of conjugate action is to be satisfied.

Because of the manufacturing tolerances it is not possible to produce gears in which the teeth mesh perfectly. There will, therefore, always be some space between the trailing face of one tooth and the leading face of the next tooth, as shown in Figure 13.1.1. This distance is known as *backlash*, and it is the amount the driving gear will rotate without moving the driven gear when the direction of rotation is reversed. It may be largely eliminated in small gears that transmit little torque by replacing the gear by two gears, each of half the thickness of the pinion, with a slight angular displacement between them so that one gear bears against one flank of the pinion and the other gear bears against the other flank of the pinion.

Spur gearing is characterized by teeth cut perpendicular to the plane of the gear, as in Figure 13.0.1a, with the tooth profile designed for maximum strength consistent with the requirement for conjugate action. Although various profiles have been found to produce conjugate action, the *involute profile* not only satisfies this requirement but satisfies it even if the distance between the shafts differs slightly from that assumed in the gear design. It has, therefore, become the most commonly used profile.

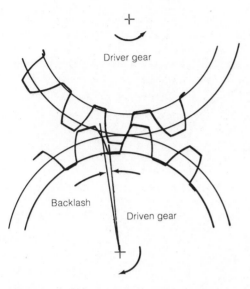

Figure 13.1.1 Backlash between two gears in mesh. Backlash may be expressed as the angle shown here, or as the equivalent distance along the pitch circle.

13.2 TERMINOLOGY RELATED TO INVOLUTE TEETH

An involute curve is defined as that curve traced by the end of a cord as it is held taut and unwrapped from a circular cylinder, or disk, as shown in Figure 13.2.1a. Hence, any force perpendicular to the involute must lie along a line tangent to the disk used to generate it. It is this characteristic that makes the involute a useful gear tooth profile. It means that the direction of the force is constant between two contracting involute teeth and that a constant torque may be transferred from one gear to another by means of involute teeth. It also follows from the involute geometry that a constant rotational speed in the driving gear provides a constant rotational speed in the driven gear.

To use an involute curve as the profile for teeth in mesh on both sides of the pitch circle, we must unwrap the cord from an imginary disk whose radius $r_b(r_b = d_b/2)$ is smaller than that of the pitch circle, r_p ($r_p = d_p/2$), as shown in Figure 13.2.1b. This smaller circle is known as the *base circle*.

The common tangent to the base circles of the two gears in mesh is known as the *pressure line* because it coincides with the line of action of forces between teeth in mesh. The angle between the pressure line and the common tangent to the pitch circles at the pitch point is known as the *pressure angle* and will be denoted by ϕ.

From the geometry shown in Figure 13.2.1 we see that the base circle radius r_b and the pitch circle radius r_p are related through the pressure angle according to

$$r_b = r_p \cos \phi \qquad (13.2.1)$$

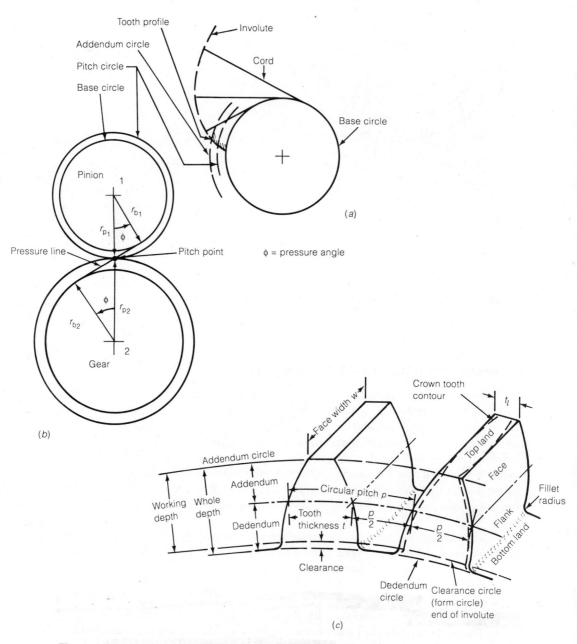

Figure 13.2.1 Illustration of gear terminology for involute tooth profile.

Thus, the pressure angle determines the tooth contour because it determines the ratio of the base radius to the pitch radius.

After the tooth contour is fixed by the selection of the pressure angle, three additional parameters are required to determine the size of the tooth: the *circular pitch*, which is the distance along the pitch circle between corresponding points on adjacent teeth; the *addedum*, which is the radial distance outward from the pitch circle to the tip of the tooth; and the *dedendum*, which is the radial distance inward from the pitch circle to the root (bottom) of the tooth. Hidden in the dedendum is the choice of the radius r_c of the *clearance circle*, which is the circle where the involute profile terminates. It is hidden in the sense that the dedendum has been defined to be the radial distance from the pitch circle to the clearance circle plus the clearance necessary to allow free movement of teeth from the mating gear. Consequently $r_a > r_c \geq r_b$ where r_a is the addendum circle radius.

Most of the remaining design considerations depend either directly or indirectly upon the choices made for these parameters.

According to the above definitions, the circular pitch for a gear with N teeth is given by

$$p_c = \pi \frac{d_p}{N} \qquad\qquad (13.2.2)$$

Since the involute profile starts on the base circle, it is convenient to define the *base pitch* in a similar fashion, namely

$$p_b = \pi \frac{d_b}{N} = \pi \frac{d_p \cos \phi}{N} = p_c \cos \phi \qquad\qquad (13.2.3)$$

In practice, however, tooth size is not given in terms of either the base pitch or the circular pitch, but in terms of either the *diametral pitch* P_d, defined by

$$P_d = \frac{N}{d_p} \qquad\qquad (13.2.4)$$

or in terms of the *module m* defined by

$$m = \frac{d_p}{N} \qquad\qquad (13.2.5)$$

Traditionally the diametral pitch has been associated with the Old English system of units and the module with SI units. For these definitions it follows that

$$P_d = \frac{1}{m} \qquad\qquad (13.2.6)$$

This relation may not be used for conversion from diametral pitch to module sizes, however, because it does not account for the different units of length used in the old english and SI systems. To account for units we must write that

$$2.54/P_d = m \qquad (13.2.7)$$

In other words, interchange of P_d and m in formulas involving these quantities requires use of (13.2.6), but conversion between tooth sizes in P_d and m requires use of (13.2.7).

Three regions may be defined for the location of the addendum circle, as illustrated in Figure 13.2.2. If it lies between its base circle and the base circle of the mating gear it is in region 1, for which the addendum circle radius r_{a_1} satisfies

$$r_{b_1} \leqq r_{a_1} \leqq c - r_{b_2} \qquad (13.2.8)$$

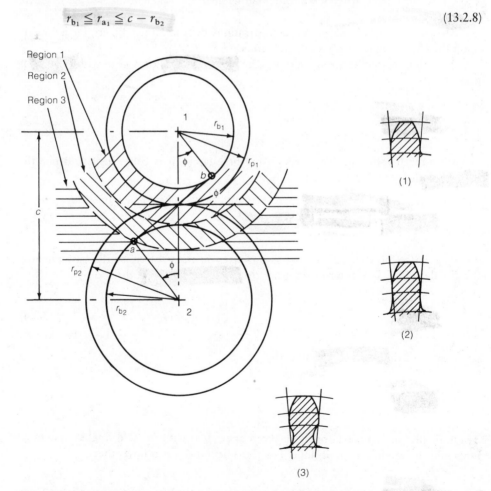

Figure 13.2.2 Illustration of regions 1, 2, and 3 for pinion (gear 1). Tooth profiles of teeth on gear 2 when pinion addendum is in region 1 (1), in region 2 (2), and in region 3 (3).

where r_{b_1} is the base circle radius for gear 1, c is the center distance given by

$$c = r_{p_1} + r_{p_2} \qquad\qquad (13.2.9)$$

and r_{b_2} is the base circle radius for gear 2. If r_{a_1} satisfies (13.2.8) then the involute profile of gear 1 does not extend beyond the involute portion of the teeth in gear 2 and the dedendum for gear 2 must be equal or greater than to the addendum plus clearance of gear 1. Hence, the selection of the addendum for one gear determines the minimum dedendum for the teeth on any gears meshing with it.

In region 2 the addendum circle radius is greater than $c - r_{b_2}$ but does not extend beyond point a in Figure 13.2.2, known as the *interference point*. In region 2 a typical tooth on gear 2 consists of an involute portion extending from its addendum circle its base circle and a noninvolute portion termed a *trochoid* which extends from its base circle to its dedendum circle. The trochoid on one gear is calculated to permit passage, plus a specified clearance, of the edges of the top lands of the teeth on gears meshing with it.

Region 3 is that region for which the addendum circle of gear 1 encloses a, the interference point on gear 2. In this region the tip of a tooth from a rack, Figure 13.6.1, will interfere with radial cuts inward from the ends of the involute tooth faces at the base circle. This interference may be eliminated by *undercutting* the teeth on gear 2, which obviously reduces their thickness and thereby reduces their strength.

In principle, there is no justification for designing undercut teeth since conjugate action requires that involute portions on the teeth on one gear must engage the involute portions of the teeth on any other gears. In practice, it may be justified in some situations; for example, if gear 1 meshes with several other gears, say gears 2 and 3, proper mesh between gears 1 and 2 may require undercutting of the teeth on gear 3, provided, of course, that the undercut teeth are still strong enough to operate without early failure.

In this section we have referred to two gears in mesh as gears 1 and 2. Whenever two gears of different diameter are in mesh the smaller is termed the *pinion* and the larger is termed the *gear*. Terminology has not been standardized for a gear that meshes with several smaller gears all of different sizes.

13.3 THE INVOLUTE PROFILE

As mentioned in Section 13.1, the involute is now the dominant gear tooth profile because it not only satisfies the law of conjugate action but it continues to do so even if the center distances are incorrect. The governing equations for the profile of an individual tooth may be derived from the geometry shown in Figure 13.3.1, where r_p denotes the pitch circle radius, r_b the base circle radius, and r the radius to a generic point on the tooth profile. Angle ϕ is the pressure angle and point C the interference point. Angle θ is the angle between the radius to the beginning point of the involute

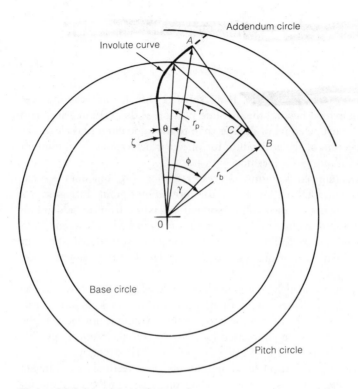

Figure 13.3.1 Involute tooth geometry.

on the base circle and the radius to point where the involute crosses the pitch circle. It will appear in a subsequent equation.

We begin our derivation by observing that the length of the imaginary cord from A to B is $r_b\gamma$ because that length of cord was originally wrapped about the base circle from B to the beginning of the involute. Since angle OBA is a right angle we have that

$$r^2 = r_b{}^2 + (r_b\gamma)^2 \tag{13.3.1}$$

which may be written as

$$r = r_b(1 + \gamma^2)^{1/2} \tag{13.3.2}$$

From this same triangle and the definition of the involute we also find that

$$r_b\tan(\gamma - \zeta) = r_b\gamma \tag{13.3.3}$$

from which we may calculate ζ from the expression

$$\zeta = \gamma - \tan^{-1}\gamma \tag{13.3.4}$$

Thus, the involute curve is defined in terms of the parametric equations 13.3.2 and 13.3.4 where the parameter γ may be incremented from 0 to γ_a, to generate a tooth profile from the base circle to the addendum circle.

Standard and nonstandard tooth profiles, however, often begin from a circle outside of the base circle. This circle where the involute profile of each tooth actually begins has been termed either the *clearance circle*, because the clearance begins at this circle, or the *form circle*, because the involute form ends at this circle. Hence, the clearance circle may lie anywhere between the pitch and base circles.

If we let r_a and r_e represent the radii of the addendum clearance circles, respectively, we may then calculate the γ values for points on the clearance and addendum circles from

$$\gamma_c = \left[\left(\frac{r_c}{r_b} \right)^2 - 1 \right]^{1/2} \qquad \text{for } r_p \geq r_c \geq r_b \qquad (13.3.5)$$

$$\gamma_a = \left[\left(\frac{r_a}{r_b} \right)^2 - 1 \right]^{1/2} \qquad\qquad (13.3.6)$$

Angle θ in Figure 13.3.1 may be found by recalling that for an involute

$$r_b(\theta + \phi) = r_b \tan \phi \qquad (13.3.7)$$

so that after canceling r_b in (13.3.7) we have

$$\theta = \tan \phi - \phi \qquad (13.3.8)$$

Describing both sides of an involute tooth may be simplified by expressing the involute profile in terms of an angle ψ from the center line of each tooth to the radius r from the center of the pitch circle. Since the width of the tooth at the pitch circle is given by π/N, it is evident from Figure 13.3.2 that

$$\psi = \theta + \frac{\pi}{2N} - \zeta \qquad (13.3.9)$$

One side of a tooth about its center line may be drawn or cut by incrementing γ from γ_c to γ_a, equations 13.3.5 and 13.3.6, and for each γ calculating the corresponding r from equation 13.3.2 and the corresponding ψ from equations 13.3.4 and 13.3.9. The other side of the tooth may be formed from the same data by replacing ψ by $-\psi$. Succeeding teeth may be generated from these data by rotating the center line through a multiple of $2\pi/N$. The space between teeth may consist of segments of the dedendum circle joined to the involute profiles of the teeth by the fillets or by any other curve that provides adequate strength along with clearance for teeth of the mating gear.

A flowchart for a program entitled INPRO to calculate data for the generation of a single tooth is given in Figure 13.3.3. When the program asks for the addendum

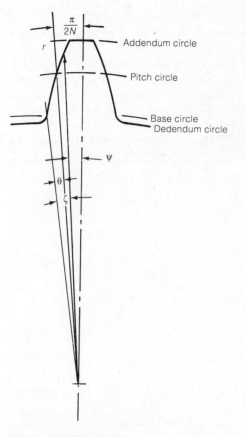

Figure 13.3.2 Parametric angles used in the description and plotting (or cutting) of an involute tooth profile.

radius the user is to enter either the magnitude of the radius or 0. Entering 0 causes the program to then prompt for the angle substended by the top land. It calculates $\pm\psi$ to give coordinates of both sides of the tooth.

Formulas for fillet contours and for the dedendum circle have not been included in the flowchart because of the variety of choices.

EXAMPLE 13.3.1

Write a program to plot a module 5 involute tooth with a 20° pressure angle for a gear with a pitch diameter of 40 mm, for a gear with a pitch diameter of 100 mm, and for a gear with a pitch diameter of 500 mm. Let the addendum be 5 mm and the dedendum minus the clearance be 3.0154 mm. (This

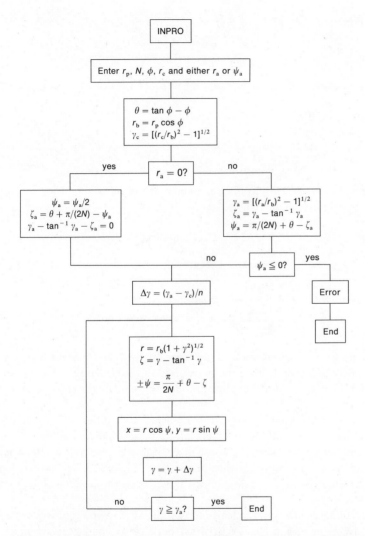

Figure 13.3.3 Flowchart for program INPRO to calculate the involute profile for an involute tooth. Top land, fillet radii, and the dedendum circle are not included.

is a nonstandard tooth, as we shall see in Section 13.5.) Explain any difference in the tooth contours. Neglect radii at the root, i.e., below the clearance circle.

First calculate the base circle radii from equation 13.2.1 for the pitch circle diameters of 40, 100, and 500 mm and find them to be 18.7939, 46.9846, and 234.9232 mm, respectively. Thus, the clearance circle diameters will be equal to or less than the base circle diameters in the first two gears, and greater than the base circle diameters in last. For all three the angle θ is found from equation 13.3.8 to be 0.854°.

From equation 13.3.6 we find that the values of γ at the addendum and the clearance circles are

$$\gamma_a = \left[\left(\frac{255}{234.9232}\right)^2 - 1\right]^{1/2} = 0.4222 \text{ rad.} = 24.1885°$$

and

$$\gamma_d = \left[\left(\frac{246.9846}{234.9232}\right)^2 - 1\right]^{1/2} = 0.1592 \text{ rad.} = 9.1224°$$

respectively, for teeth on the gear having a 500-mm pitch diameter. Upon combining these values with those found for the 40- and 100-mm pitch diameters we have that

d_p (mm)	γ_a (degrees)	γ_c (degrees)
40	50.2599	0.0000
100	34.8655	0.0000
500	24.1885	9.1224

The profile for the gear with a pitch diameter of 40 mm is formed by an involute from the base circle to the addendum circle and by a trochoid from the base circle to the dedendum circle. The three profiles generated are displayed in Figure 13.3.4. Radial lines and fillets complete the tooth profile for the gear with a 40 mm pitch diameter.

The profile in Figure 13.3.4a shows the combined effects of using a tooth size that is abnormally large for the pitch diameter of the gear and of using a dedendum circle smaller that the base circle. Figure 13.3.4b displays the profile characteristics of a tooth that begins at the base circle where the profile at the base circle is tangent to the radius to that point. Last, Figure 13.3.4c shows a profile characteristic of a gear in which the dedendum circle is much larger than the base circle, causing the profile to appear trapazoidal.

Comparison of tooth profile 13.3.4a with b and c clearly shows one of the two reasons why undercut pinion teeth tend to fail more frequently than gear teeth when both pinion and gear are made from the same material: the root thickness is less, so that the stresses are greater. The second reason, perhaps not evident from the tooth profile, is that pinion teeth are loaded more frequently than gear teeth because the pinion turns faster. For example, if the pinion pitch diameter is half that of the gear the pinion will rotate twice as fast as the gear so that a pinion tooth will experience two cyclic loads for every one cyclic load experienced by a gear tooth. It is for these reasons that pinions are often made from a stronger material (and/or with a different heat treatment) than that used for the mating gear.

At this point we have the relations necessary to calculate the backlash due to an incorrect center distance between two gears that mesh as shown in Figure 13.2.2. Angular separation between two teeth that were in contact at the pitch point before the center distance was increased by amount Δc may be found by observing that increasing the center distance will cause the actual pressure angle to increase to ϕ' and cause the gears to contact along new pitch circles with radii $r'_{p_1} = r_{b_1}/\cos \phi'$ and $r'_{p_2} = r_{b_2}/\cos \phi'$.

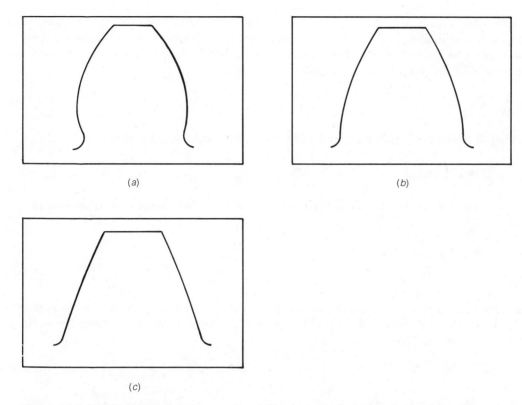

Figure 13.3.4 Effect of pitch diameter upon the profile of a tooth of constant size (module, addendum, and dedendum remain unchanged). (a) $d_p = 40$ mm, (b) $d_p = 100$ mm, (c) $d_p = 500$ mm.

From (13.2.1) and (13.2.9) we have by analogy that

$$(c + \Delta c) \cos \phi' = r_{b_1} + r_{b_2} = c \cos \phi \qquad (13.3.10)$$

so that

$$\cos \phi' = \frac{1}{1 + \dfrac{\Delta c}{c}} \cos \phi \qquad (13.3.11)$$

Let θ' denote the value of ζ on the new pitch circles, also known as the operating circles, so that by analogy with equation 13.3.8 we have

$$\theta' = \tan \phi' - \phi' \qquad (13.3.12)$$

Since $\zeta = \theta$ on the pitch circles of each gear before increasing the center distance and $\zeta = \theta'$ after increasing the center distance, the total angular separation is the sum of the angles each gear must rotate to again make contact along the line of centers, i.e., $2(\theta' - \theta)$. If the driving gear were to reverse direction after the center distance was increased, it would move through twice this angle before making contact with a tooth on the mating gear. Hence, the total angular backlash Λ involved in changing direction sometime after the center distance was increased becomes

$$\Lambda = 4(\theta' - \theta) \tag{13.3.13}$$

Backlash B in units of length along each of the new pitch circles may be written as

$$B = 2(r'_{p_1} + r'_{p_2})(\theta' - \theta) = 2c'(\theta' - \theta) \tag{13.3.14}$$

Differentiation of (13.3.12), simplification by means of a trigonometric identity, and use of (13.3.10), produces

$$\frac{d\theta}{dc} = \frac{\sin \phi}{r_{b_1} + r_{b_2}} \tag{13.3.15}$$

which implies that the effect of center distance errors may be decreased by decreasing the pressure angle. Thus, $14.5°$ teeth are often used for position control, $25°$ teeth are used for power transmission, and $20°$ teeth for general purpose.

13.4 CONTACT RATIO FOR INVOLUTE TEETH

Perhaps a more descriptive phrase than contact ratio would be contact number, since the contact ratio, also known as the contact duration, is defined as the average number of teeth in contact for two gears in mesh. It is important because it affects both the power that may be transferred from one gear to the other and the smoothness of the transferred motion in the presence of manufacturing defects. Increasing the contact ratio for greater power and more uniform speed is, perhaps, the major incentive for the increased use of nonstandard gears.

Since the involute profile can be constructed only for teeth extending outward from the base circle, it follows that involute teeth can be in contact only along the pressure line from one base circle to the other. From Figure 13.4.1 and the definition of the involute it follows that the length of that portion of the pressure line from the pitch point to point C_2 on gear 2 is equal to the distance along the base circle from C_2 to the starting point B_3 of the involute profile of that tooth of gear 2 making contact at the pitch point with a tooth of the mating gear. The remaining portion of the pressure line from the pitch point to C_1 on gear 1 is equal to the length along the base circle of gear 1 from C_1 to point B_2 at the starting point of the involute profile for the tooth of gear 1 which also contacts the pitch point. Thus, we see that

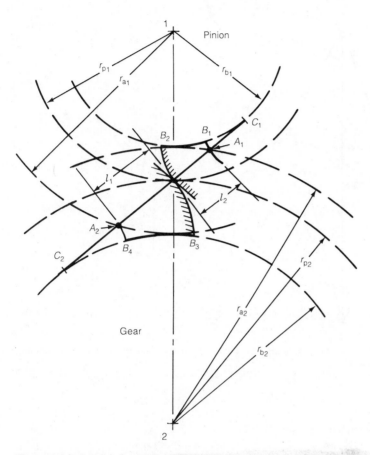

Figure 13.4.1 Contact length along the pressure line and the equivalent arc lengths along the gear and pinion base circles.

the length of the pressure line divided by the base pitch is the maximum number of teeth that can be in contact when the two gears are in mesh.

The number of teeth that may be in contact depends upon where the addendum of the teeth from each gear intersects the pressure line, as illustrated in Figure 13.4.1. In that figure let the addendum circle of gear 2 intersect the pressure line at A_1 and let the addendum circle from gear 1 (the pinion) intersect the pressure line at A_2. Hence, any contact between other teeth on the pinion and gear must lie between points A_1 and A_2.

The pressure line, the gear centers, and the line between centers in Figure 13.4.1 has been reproduced in Figure 13.4.2 to reduce line clutter. From the geometry shown in that figure we find that l_1 and l_2 may be expressed by

$$l_2 = (r_{a_2}^2 - r_{b_2}^2)^{1/2} - r_{p_2} \sin \phi \tag{13.4.1}$$

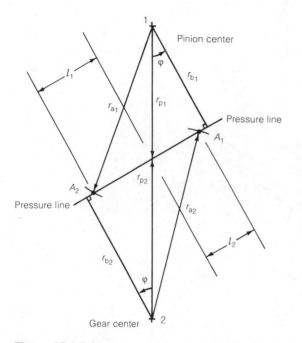

Figure 13.4.2 Geometry used in calculating the lengths l_1 and l_2 and the contact ratio for involute gears.

and

$$l_1 = (r_{a_1}{}^2 - r_{b_1}{}^2)^{1/2} - r_{p_1} \sin \phi \tag{13.4.2}$$

so that the contact ratio for these addendums may be written as

$$N_c = \frac{l_1 + l_2}{p_b} \tag{13.4.3}$$

It is also apparent from the geometry of Figure 13.4.2 that the maximum contact ratio may be written as

$$N_{c_{max}} = \frac{c \sin \phi}{p_c \cos \phi} = \frac{c}{p_c}(\theta + \phi) \tag{13.4.4}$$

when the addendum radius of each gear just contacts the interference point of the other. Increasing the center distance increases N_c because of the greater radii of curvature for the pitch circles, decreasing p_c increases N_c by narrowing the teeth, but it also reduces the addendum circle diameter. Increasing the sum of θ and ϕ by increasing θ will increase N_c by extending the dedendum radius to permit teeth with longer addendums on each gear.

Substitution from (13.4.1) and (13.4.2) into (13.4.3) yields the following expression for contact ratio N_c when the addendum radius is in either regions 1 or 2, Figure 13.2.2:

$$N_c = \frac{1}{p_b}[(r_{a_1}^2 - r_{b_1}^2)^{1/2} + (r_{a_2}^2 - r_{b_2}^2)^{1/2} - c\sin\phi] \tag{13.4.5}$$

The same reasoning is used in calculating the contact ratio for internal gearing: contact occurs along the pressure line. In this case, however, contact occurs outside of the interval between the points of tangency of the pressure line with the base circles of the planet and ring gears, as drawn in Figure 13.4.3.

The distance c between centers of an external and internal gear in mesh is the difference of the gear radii rather than the sum. The internal gear tooth profile is still an involute constructed from a base circle except that contact with teeth on the planet gears is on the concave side of the involute of the internal teeth. Just from the geometry alone we may correctly deduce that the contact ratio for mesh between external and internal gears is greater than between external gears having teeth of the same diametral pitch. A larger contact ratio, use of multiple gears to transfer power between the sun and ring gear, and compactness all make planetary gears an attractive means for changing power and speed.

Tooth contact lies within the length of line l_0 in Figure 13.4.4 so that the contact ratio is given by the length of this line divided by the base pitch. From the geometry shown we see that l_0 may be written as

$$l_0 = (r_{a_1}^2 - r_{b_1}^2)^{1/2} - \delta \tag{13.4.6}$$

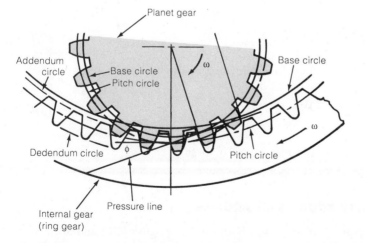

Figure 13.4.3 Contact between an internal gear (ring gear) and a planet gear. Note that the diameter of the addendum circle is smaller than that of the dedendum circle in the case of an internal gear.

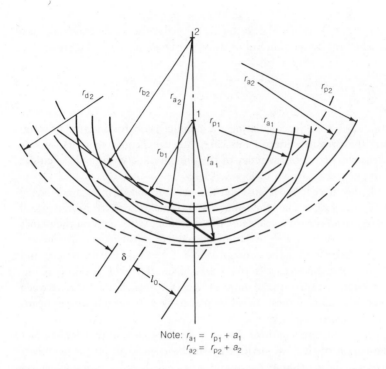

Note: $r_{a_1} = r_{p_1} + a_1$
$r_{a_2} = r_{p_2} + a_2$

Figure 13.4.4 Geometry involved in the calculation of the contact ratio (number) for an external involute pinion and an internal involute gear. *Note:* $c = r_{p_2} - r_{p_1}$

where length δ may be written as

$$\delta = (r_{a_2}{}^2 - r_{b_2}{}^2)^{1/2} - c \sin \phi \tag{13.4.7}$$

Upon combining (13.4.6) and (13.4.7) and dividing by the base pitch we have

$$N_c = \frac{(r_{a_1}{}^2 - r_{b_1}{}^2)^{1/2} - (r_{a_2}{}^2 - r_{b_2}{}^2)^{1/2} + c \sin \phi}{p_b} \tag{13.4.8}$$

as the contact ratio for a ring and planet gear in mesh.

13.5 STANDARD INVOLUTE SPUR GEAR PROFILES

Standard involute spur gear profiles in both the Old English (OE) and SI (metric) units have been established by the American Gear Manufacturers Association (AGMA) and some have been adopted by the American National Standards Institute (ANSI). The spur gear nomenclature and profiles are specified in standards ANSI B6.1, ANSI

B6.14, and AGMA 207.06. The proportions shown in Figure 13.5.1 are abstracted from them for the so-called *full depth* gears for which $a + b = 2.0/P_d$ for all pressure angles, for 20° stub teeth, for coarse ($P_d < 19.99$), and for fine ($20 \leq P_d \leq 120$) pitch gears. Note that the term full depth does not mean that the addendum extends to the interference point of the mating gear, but only that it extends the full amount allowed by the standard tooth proportions.

Rack proportions and gear tooth proportions for standard teeth of the same size (same P_d or m) are obviously identical because the two must mesh. Involute profiles on a gear become straight lines on a rack.

Undercut pinion teeth of standard proportion may be avoided by increasing the addendum and decreasing the dedendum to increase tooth thickness on the pitch circle and reduce the space between teeth. Thinner teeth and larger spaces on the gear pitch circle are made by reducing the addendum and increasing the dedendum. This *long and short addendum* gearing is made with standard cutters.

Representative standard tooth sizes are shown in full scale in Figure 13.5.2. Recall from Figure 13.3.4 that the thickness of these teeth at the pitch circle and their addendums and dedendums are the only characteristics of the tooth sizes shown in Figure 13.5.2 that are the same for teeth of the same size (the same P_d) on different diameter gears. Thickness at the top land and the tooth contour above and below the pitch circle depend upon the diameter of the base circle upon which they are generated.

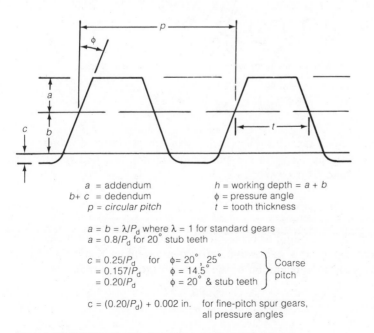

a = addendum

$b + c$ = dedendum

p = circular pitch

h = working depth = $a + b$

ϕ = pressure angle

t = tooth thickness

$a = b = \lambda/P_d$ where $\lambda = 1$ for standard gears

$a = 0.8/P_d$ for 20° stub teeth

$c = 0.25/P_d$ for $\phi = 20°, 25°$

  $= 0.157/P_d$ $\phi = 14.5°$ } Coarse pitch

  $= 0.20/P_d$ $\phi = 20°$ & stub teeth

$c = (0.20/P_d) + 0.002$ in.   for fine-pitch spur gears, all pressure angles

Figure 13.5.1 Standard tooth profiles, rack configuration.

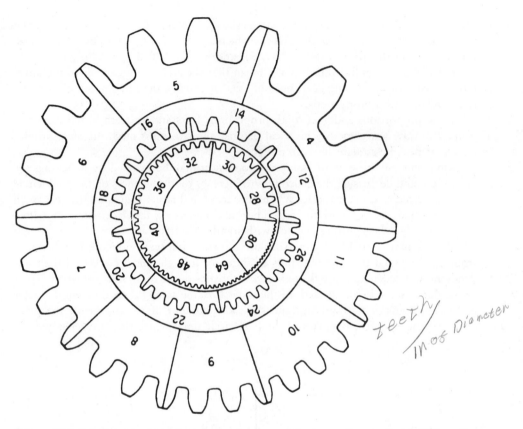

Figure 13.5.2 Actual size of standard teeth for values of P_d shown. *Source: Barber-Colman Co., Loves Park, IL.*

To specialize the contact ratio formula (13.4.5) to teeth of standard proportion, rewrite r_{a_1}, r_{a_2}, r_{b_1}, and a_{b_2} in terms of the pitch radii r_{p_1} and r_{p_2} and rewrite the addendums in term of these radii as well, to get

$$a_1 = a_2 = \lambda m = 2\lambda r_{p_1}/(N_1) = 2\lambda r_{p_2}/(N_2)$$

for standard tooth profiles according to equation 13.2.6 and the relations included in Figure 13.5.1 Substitution of this expression into equations 13.4.1 and 13.4.2 for lengths l_1 and l_2 involved in the definition of the contact ratio shows that

$$l_1 = r_{p_1}\left\{\left[\left(1 + \frac{2\lambda_1}{N_1}\right)^2 - \cos^2 \phi\right]^{1/2} - \sin \phi\right\} \tag{13.5.1}$$

and

$$l_2 = r_{\text{p}_2} \left\{ \left[\left(1 + \frac{2\lambda_1}{N_2} \right)^2 - \cos^2 \phi \right]^{1/2} - \sin \phi \right\}$$ (13.5.2)

may be smaller for standard profile teeth than for nonstandard teeth because λ is limited to either 1.0 or 0.8. This is one motivation for considering nonstandard tooth profiles.

13.6 MINIMUM NUMBER OF TEETH TO PREVENT UNDERCUTTING

Undercutting will occur between a gear and a rack, as pictured in Figure 13.6.1, whenever the addendum of the rack extends beyond the interference point shown in the figure. From this geometry we find that the maximum addendum that may be used without undercutting is given by

$$a_1 \leqq r_{\text{p}_2} - r_{\text{b}_2} \cos \phi = r_{\text{p}_2}(1 - \cos^2 \phi) = r_{\text{p}_2} \sin^2 \phi$$ (13.6.1)

Since the addendum is related to the tooth size in standard gears according to

$$a_1 = \frac{\lambda}{P_{\text{d}}} = \lambda m$$ (13.6.2)

as listed in Figure 13.5.1, equation 13.6.1 sets a limit on the minimum number of standard teeth that may be generated on a gear if undercutting is to be prevented.

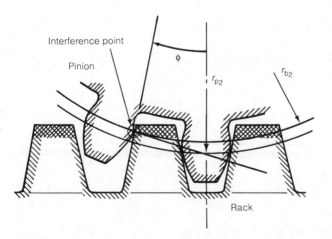

Figure 13.6.1 Pinion and rack in mesh. Crosshatched section of rack teeth causes undercutting.

Upon substituting for a_2 from (13.6.2) into (13.6.1) we find

$$N \gtrless \frac{2\lambda}{\sin^2 \phi} \tag{13.6.3}$$

By setting ϕ successively equal to 14.5°, 20°, and 25° this relation yields the following minimum number of standard teeth that a gear may have if undercutting of its teeth by a rack is to be avoided.

N_{min} (standard)	N_{min} (stub)	ϕ (degrees)
32	26	14.5
18*	14	20.0
12	9	25.0

* Relation 13.6.3 yields 17.09 for standard teeth with a 20° pressure angle. Rounding to 17 implies slight under cutting.

EXAMPLE 13.6.1

Plot the profiles for nonstandard proportion $m = 5$ and $m = 10$ teeth for pressure angles of 14.5° and 20° for a pitch circle diameter of 240 mm and a top land thickness that subtends an angle of $0.25\psi/(2N)$ relative to the center of the pitch circle.

   We begin with $\phi = 14.5° = 0.2531$ radians and $m = 5$, corresponding to 48 teeth. First use equation 13.3.8 to calculate θ from

$$\theta = \tan \phi - \phi = \tan 0.2531 - 0.2531 = 0.005545 \text{ rad}$$

to solve equation 13.3.9 for $\zeta = \zeta_a$ at the addendum circle, where

$$\psi_a = \frac{0.25\pi}{2N} = \frac{\pi}{2N} + \theta - \zeta_a$$

Thus,

$$\zeta_a = \frac{3\pi}{8N} + \theta = \frac{3\pi}{384} + 0.005545 = 0.030087 \text{ rad}$$

Next, solve for γ_a by substituting into equation 13.3.4 to obtain

$$\zeta_a = 0.030087 = \gamma_a - \tan^{-1} \gamma_a$$

01 LBL "GAM"	13 ATAN	25 LBL 02
02 RAD	14 RCL 02	26 RCL 03
03 "ZETA=?"	15 X<>Y	27 ST+02
04 PROMPT	16 –	28 GTO 01
05 STO 01	17 RCL 01	
06 "ERROR ?"	18 X<>Y	29 LBL 03
07 PROMPT	19 –	30 DEG
08 STO 04	20 STO 03	31 "GAM"
09 RCL 01	21 RCL 04	32 ARCL 02
10 STO 02	22 X<=Y?	33 AVIEW
11 LBL 01	23 GTO 02	34 STOP
12 RCL 02	24 GTO 03	35 END

Figure 13.6.2 Program listing for GAM which solves equation 13.3.4 for γ. Required input data are the value of ζ and the acceptable error for γ. *Note*: THE ERROR? prompt requests the maximum acceptable error in the value of γ that is returned. Errors less than 0.000 01 radians correspond to errors in angle less than 0.0018° or approximately 0.65 seconds of arc.

This equation may be solved by repeated iteration. A short program to do this, entitled GAM, is shown in Figure 13.6.2 when written for the HP-41CX. From it we find that $\gamma_a = 0.466943$ rad. We may calculate the base circle radius from equation 13.2.1 to find

$$r_b = r_p \cos \phi = 116.178 \text{ mm}$$

and then substitute r_b into equation 13.3.2 to find the addendum radius to be

$$r_a = r_b(1 + \gamma^2)^{1/2} = 116.178(1 + 0.466892^2)^{1/2} = 218.217 \text{ mm}$$

A program entitled TOOTH that was written for these calculations in Microsoft Quick Basic is shown in Figure 13.6.3. Lines 20 through 25 inclusive implement an iteration procedure to solve (13.3.4) for γ_a. The remainder of it is written to find the x and y coordinates at 11 points, 10 intervals, along the involute for one side of a tooth from $x = r \sin \psi$ and $y = r \cos \psi - r_c$ where r_c denotes the clearance circle. These values are printed on the screen and also stores in file PROFILE.DAT for plotting.

The profiles of all four teeth are displayed in Figure 13.6.4.

EXAMPLE 13.6.2

Compare the addendum circle diameters and the top land thicknesses of $m = 5$ and $m = 10$ teeth of standard proportion for 14.5° and 20° pressure angles for teeth of a gear having a 240-mm-diameter pitch circle with those found in Example 13.6.1.

```
'Program TOOTH.BAS to calculate the involute profile for given N,
'pressure angle, pitch radius, clearance radius, and top land angle.

DEFSNG A-H, P-Z
OPEN "PROFILE.DAT" FOR OUTPUT AS 1

INPUT "Enter the number of teeth and the pressure angle ", N, PHI
PI=3.141592654 : PHI=PHI*PI/180
INPUT "Enter the pitch radius. ", RP : RB=RP*COS(PHI)
PRINT "The base circle radius is " RB
INPUT "Enter the clearance radius.", RC : PHRP=180/N
PRINT "Angle at pitch radius is " PHRP "degrees. "
INPUT "Enter the top land angle in degrees.", PHIM : PHIM=PHIM*PI/180
THETA=TAN(PHI)-PHI
ZETA=PI/(2*N)+THETA-PHIM/2 : GAMA=0 : EP=0 : GAMO=SQR(((RC/RB)^2)-1)
INPUT "To convert from mm to in., enter 1. ", CN

20    DO
25      GAMA=GAMA+EP
30      EP=ATN(GAMA)-GAMA+ZETA
35    LOOP UNTIL ABS(EP)<0.00001

40    PRINT "Gamma is " GAMA : I=O
45    DELGA=(GAMA-GAMO)/10 : GAM=GAMO

50    DO UNTIL I=11
55      I=1+1 : PSII=PI/(2*N)+THETA
60      R=RB*SQR(1+GAM^2) : ZET=GAM-ATN(GAM) : PSI=PSII-ZET
65      X=R*SIN(PSI) : Y=R*COS(PSI)-RC*COS(PSII) : GAM=GAM+DELGA
70      PRINT #1, X, Y
75      IF CN=1 THEN
80        R=R/15.4 : X=X/25.4 : Y=Y/25.4 : PRINT "I, RA, X, Y are " I R X Y
85      ELSE
90        PRINT "I, RA, X, Y, are " I R X Y
95      END IF
100     LOOP : PRINT " "
105   PRINT "Addendum radius is " R
110   CLOSE #1

115   END
```

Figure 13.6.3 Listing for program TOOTH which calculates points on an involute profile.

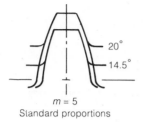

m = 5
Standard proportions

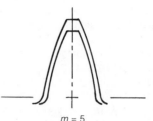

m = 5
Nonstandard proportions

Standard proportion involute profiles end at fillets labeled with the pressure angle

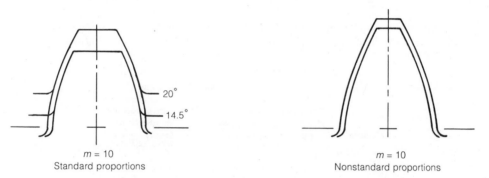

m = 10
Standard proportions

m = 10
Nonstandard proportions

Figure 13.6.4 Tooth profiles for $m = 5$ and $m = 10$ teeth of standard and nonstandard proportions (height and thickness are twice actual dimensions).

These values may be found from the requirement that the addendum for teeth of standard profile is given by $a = m$, so that the addendum radii for $m = 5$ and $m = 10$ teeth on a pitch circle of 120 mm are 125 and 130 mm. Angle ψ_a may be found by first recalling that $\theta = 0.005545$ rad for $\phi = 14.5°$ from Example 13.6.1 and from

$$\theta = \tan \phi - \phi = \tan 0.3491 - 0.3491 = 0.014904 \text{ rad}$$

for $\phi = 20°$. After using equation 13.3.6 to find γ_a we turn to equation 13.3.4 to solve for ζ_a. With these values known we substitute into equation 13.3.9 to find the corresponding values of ψ_a. Top land thicknessess are given by

$$t = 2r_a \sin \psi_a$$

where the factor of 2 enters because ψ_a was the angle from the centerline of the tooth.

```
'Program entitled CTLAND to calculate the top land for a given
'addendum radius

INPUT ''Enter pitch circle radius and pressure angle. '' , RP, PHI
PI = 3.141592654 : PHI = PHI*PI/180
INPUT ''Enter the number of teeth and the addendum. '' , N, AD
RB = RP*COS(PHI) : RA = RP + AD : GAMA = SQR((RA/RB)^2 - 1
ZET = GAMA - ATN(GAMA) : THET = TAN(PHI) - PHI : PSI = THET + PI/(2*N) - ZET
W = 2*RA*SIN(PSI) : PSI = 2*PSI*180/PI
PRINT ''TOP land angle and top land width are '' PSI '' and '' W

END
```

Figure 13.6.5 Program listing for CTLAND which calculate the top land thickness and the angle it subtends at the center of the pitch circle.

A short computer program to find the top land thickness and the angle subtended at the center of the pitch circle by the top land is shown in Figure 13.6.5. The comparison is shown below.

| | | **Tooth Thickness** | |
Module	Pressure Angle (degrees)	Standard Proportions (mm)	Nonstandard Proportions (mm)
5	14.5	4.795	2.100
10	14.5	8.896	4.399
5	20.0	3.864	2.078
10	20.0	7.154	4.337

EXAMPLE 13.6.3

Design a pinion and gear having pitch diameters of 240 and 400 mm to drive a radar dish. Tooth size of $m = 5$ or greater is preferred. Both power and smooth motion are required: power to allow the dish to rotate against high wind loads and smooth rotation to allow accurate fixes on enroute aircraft. Either standard or nonstandard gears will be acceptable. Consider 14.5° and 20° pressure angles and accept the gear set with the largest contact ratio. Tooth width will be adjusted to provide the required strength.

Use the programs displayed in Examples 13.6.1 and 13.6.2 to obtain the results shown below. The base circle radii for 14.5° may be calculated from equation 13.2.1 to obtain

$$r_{b_1} = 120 \cos 14.5° = 116.178 \text{ mm} \qquad r_{b_2} = 200 \cos 14.5° = 193.630 \text{ mm}$$

Use the addendum radii calculated for the $m = 5$ tooth as listed at the top of Table 13.6.1 to calculate

TABLE 13.6.1 CONTACT RATIOS FOR $m = 5$ AND $m = 10$ TEETH

Module	Pressure Angle (degrees)	Pitch circle Diameter (mm)	Number of Teeth	Addendum circle Diameter (mm)	Contact Ratio
5	14.5	240	48	128.217	
5	14.5	400	80	208.944	3.461
5	14.5	240	48	125.000 s	
5	14.5	400	80	205.000 s	2.192
10	14.5	240	24	134.353	
10	14.5	400	40	215.890	2.723
10	14.5	240	24	130.000 s	
10	14.5	400	40	208.000† s	1.781
5	20.0	240	48	126.842	
5	20.0	400	80	207.209	2.131
5	20.0	240	48	125.000 s	
5	20.0	400	80	205.000 s	1.465
10	20.0	240	24	132.459	
10	20.0	400	40	213.384	1.941
10	20.0	240	24	130.000 s	
10	20.0	400	40	210.00 s	1.518

s, standard proportions; † stub tooth required to avoid undercutting.

the contact ratio from equation 13.4.5. Thus, with $p_b = \pi m \cos 14.5° = 15.208$ mm,

$$N_c = [(128.217^2 - 116.178^2)^{1/2} + (208.944^2 - 193.603^2)^{1/2} - 320 \sin 14.5°]/15.208 = 3.461$$

Select the $m = 5$ tooth with a 14.5° pressure angle because the improved contact ratio may be obtained with a tooth extension of less than 4 mm compared to the addendum radius of a $m = 5$ tooth of standard proportions.

This selection is simplified in this example in that the tooth width is to be chosen to provide the necessary tooth strength. More detailed analysis (experimental or finite element) may be necessary if the tooth width is limited. This is because the greater thickness of the $m = 10$ tooth may more than compensate for its decreased contact ratio as compared to the load distribution over three teeth that is obtained from the $m = 5$ tooth, both for a 14.5° pressure angle.

The small contact ratio for teeth with a pressure angle of 20° disqualified them from further consideration. They are more commonly employed where power transfer is of primary importance.

Comparative profiles for standard and nonstandard addendums for $m = 5$ and $m = 10$ teeth are shown in Figure 13.6.4 for 14.5° and 20° pressure angles. The increased contact ratio obtained from $m = 5$ teeth having a 14.5° pressure angle and nonstandard proportions does not seem unreasonable when we see the extent of unused profile when standard proportions are imposed.

The apparent increase in tooth size for one with a 20° pressure angle as compared to one with a 14.5° pressure angle is caused by drawing both profiles from the base circle. In the actual gear the top lands would coincide and the tooth with the 20° pressure angle would extend deeper into the gear because its base circle is smaller. The teeth in Figure 13.6.4 were drawn from the base circle to avoid overlapping profiles since they both have the same thickness at the pitch circle.

13.7 TOOTH STRENGTH

Two of the many formulas and techniques that have been suggested for estimating tooth strength will be described: the Lewis formula with the Barth correction and the AGMA method. Infinite gear life is assumed for gears whose stresses are within the limits set by these formulas.

Improved estimates require more detailed procedures which better account for the details of the gear profile and the applied load. One improved method is a finite element program developed by the MacNeal-Schwendler Corporation.[6] Another is a modified Heywood formula which was found to be in good agreement with experimental results collected as part of an extensive gear study funded by the U.S. Army Aviation Research and Development Command.[7] This second method is most easily used with a personal computer because it involves lengthy formulas and auxiliary calculations.

Lewis Formula with the Barth Correction

The Lewis formula with the Barth correction for dynamic forces is given in relation 13.7.1. The first of the two forms is for the usual commercial quality gears operating at pitch line velocities of less than 3000 ft/min, while the second is for high quality commercial gears operating at pitch line velocities less than 6000 ft/min. Both have been used for many years in industry[8] as a simple means of estimating the necessary tooth strength. The group of quantities on the left side of equations 13.7.1 is known as the stress sensitivity; it is convenient to use equations in these terms because the expression on either side of the equal sign is dimensionless and is, therefore, independent of the units used. Once the stress intensity factor has been calculated it is easy to find the effect of changing tooth dimensions.

In these empirical formulas σ denotes the safe working stress for the gear, where

$$\frac{\sigma w p_c}{F} = \frac{1}{y}\left(1 + \frac{v}{600}\right) \qquad \text{for general commercial quality,} \\ v < 3000 \text{ ft/min}$$

$$\frac{\sigma w p_c}{F} = \frac{1}{y}\left(1 + \frac{v}{1200}\right) \qquad \text{for high commercial quality,} \\ v < 6000 \text{ ft/min}$$

$$(13.7.1)$$

in which w represents the face width in inches, y the tooth form factor in Table 13.7.1, v the pitch line velocity in ft/min, and F the force in pounds tangential to the pitch circle. Typical stress levels for common gear materials are listed in Table 13.7.2.

TABLE 13.7.1   VALUES OF THE LEWIS FORM FACTOR *y* FOR PRESSURE ANGLES INDICATED

Number of Teeth	Pressure Angle *y*		
	14.50	20.00	25.00
10	0.056	0.064	. . .
12	0.067	0.078	0.077
13	0.071	0.083	0.082
14	0.075	0.088	0.086
15	0.081	0.092	0.091
16	0.081	0.094	0.095
17	0.084	0.096	0.099
18	0.086	0.098	0.103
19	0.088	0.100	0.107
20	0.090	0.102	0.111
21	0.092	0.104	0.116
22	0.093	0.105	0.120
23	0.094	0.106	0.124
24	0.095	0.107	0.128
25	0.097	0.108	0.132
26	0.098	0.110	0.136
27	0.099	0.111	0.133
28	0.100	0.112	0.130
30	0.101	0.114	0.133
34	0.104	0.118	0.138
38	0.106	0.122	0.142
43	0.108	0.126	0.147
50	0.110	0.130	0.152
60	0.113	0.134	0.156
75	0.115	0.138	0.161
100	0.117	0.142	0.166
150	0.119	0.146	0.169
300	0.122	0.150	0.175
Rack	0.124	0.154	0.180

TABLE 13.7.2   SAFE VALUES OF STATIC STRESS FOR SEVERAL METALS

Material		(s) Lb. per Sq. in.
Bronze		8 000
Cast Iron		10 000
Steel	.20 Carbon (Untreated)	12 000
	.20 Carbon (Case-hardened)	25 000
	.30 Carbon (Untreated)	15 000
	.40 Carbon (Untreated)	20 000
	.40 Carbon (Heat-treated)	30 000
	.40 C. Alloy (Heat-treated)	40 000

Source: Boston Gear, Quincy, MA.

Formulas 13.7.1 are replaced by a single formula for phenolic laminated gears, given in equation 13.7.2, for $v < 5000$ ft/min. and $\sigma \leqq 6000$ psi

$$\frac{\sigma w p_c}{F} = \frac{1}{y} \left(\frac{250 + v}{150} + 0.25 \right) \tag{13.7.2}$$

It also holds for gears of compressed cotton with steel shrouds for $\sigma \leqq 8000$ psi.[8]

Values of y for pressure angles and numbers of teeth not given in Table 13.7.1 may be found from a graphical method outlined in AGMA Standard 218.01, December 1982, or from an algorithm given in AGMA Technical Paper P139.03, October 1981. A comment on page 65 of AGMA Standard 218.01 implies that some contributors to it believed the graphical method to be more accurate.

AGMA Method for Estimating Tooth Stress

AGMA Std. 218.01 replaces earlier standards 210.02, 211.02, 220.02 and 221.02 for spur, helical, and herringbone gears. It does not produce unique answers because the relation for the tooth bending stress contains four factors that are not objectively defined: Their values are to be assigned by the engineer based upon previous experience. In fact, on p. 1 of ref. 9 it states:

> The knowledge and judgement required to evaluate the various rating factors come from years of accumulated experience in designing, manufacturing, and operating gear units. Empirical factors given in this Standard are general in nature. AGMA Application Standards may use other empirical factors that are more closely suited to the particular use. This Standard is intended for use by the experienced gear designer, capable of selecting reasonable values for the factors. It is not intended for use by the engineering public at large.

The so-called fundamental formula used for evaluation of gear stress by the experienced gear designer is

$$\frac{\sigma w p_c}{F_t} = \frac{K_a K_s K_m \pi}{J K_v} \tag{13.7.3}$$

in which F_t is the force component tangential to the pitch circle, K_a is the application factor for bending stress, K_s is the size factor for bending, K_m is the load distribution factor, K_v is the dynamic factor, and J is the geometry factor for bending.

Taking these factors in order, factor K_a is not defined in either graphical or tabular form. Section 9.1 of the standard deals with this factor and reads in part:

> The application factor makes allowance for any externally applied loads in excess of the nominal tangential load, W_t [replaced by F_t in equation 13.7.13]. Application factors can only be established after considerable field experience is gained in a particular application.[9]

Factor K_s is now taken as 1.0, pending later recommendations for other values. Factor K_m is said to include such effects as manufacturing accuracy, alignment of gears axes, elastic deflection, and so on, and should be "evaluated by appropriate analysis." It apparently should be less than 1.0.

The formula

$$K_v = \frac{F_t}{F_t + F_d} \tag{13.7.4}$$

defines factor K_v in terms of the tangential force F_t and dynamic force F_d, which is the "incremental dynamic tooth load due to the dynamic response of the gear pair to the transmission error excitation, not including the transmitted tangential load."[9]

Factor J may be calculated by a lengthy graphical procedure described in AGMA 218.01 or, for standard gears, may be read from a graph, such as that reproduced in Figure 13.7.1 for teeth with a 20° pressure angle.

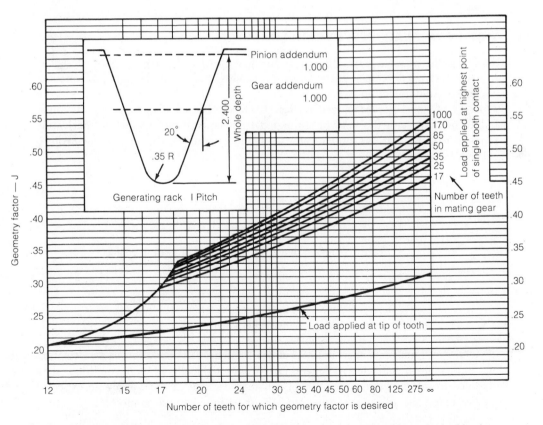

Figure 13.7.1 AGMA J factor for involute teeth with 20° pressure angle and standard addendum. *Source: American Gear Manufacturers Association, Alexandria, VA. Reprinted with permission.*

In summary, these two methods represent two different approaches to the problem of selecting a tooth size to transmit torque from gear to gear. The Lewis formula with the Barth correction for fatigue is an attempt to find an approximate value of tooth stress with a minimum of mathematics. The AGMA method is directed to the design problem of relating tooth stress and tooth life to a variety of factors whose specific effect may be found by appropriate theoretical methods in some cases and from the designer's experience in many other cases for which no generally accepted theoretical method has yet been developed.

In the majority of gear applications where the expected loads fall within a well-defined range and where adequate lubrication is maintained we may use the analytical methods represented by refs. 6 and 7, which have been found to be in reasonable agreement with a number of experimental results. Although the programs derived from these methods are generally easy to use, a thorough discussion of the methods themselves exceeds the page limitations of this text. We shall, therefore, continue to rely upon the Lewis-Barth approximation to estimate tooth strength.

13.8   TOOTH AND GEAR LIFE: CYCLIC LOADS

The relations between tooth life and tooth loads used as the basis of gear design in this section were developed at the Lewis Research Center of the National Aeronautics and Space Administration (NASA). Tooth life testing at the Lewis Research Center for a period of time, as summarized in Ref. 10, has indicated that, based upon similar failure modes, tooth life may be estimated from equation 13.8.1, whose derivation stems from considerations similar to those used in determining the fatigue life of rolling element bearings.

$$T_{10} = K \, F^{-4.3} \, w^{3.9} \, \mu^{-5} \, l^{*-0.4} \tag{13.8.1}$$

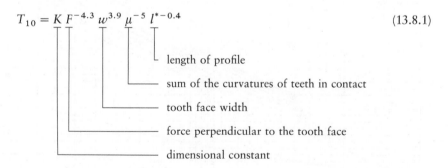

- length of profile
- sum of the curvatures of teeth in contact
- tooth face width
- force perpendicular to the tooth face
- dimensional constant

Once programmed, equation 13.8.1 and its auxiliary relations allow easy calculation of a more realistic gear life as a function of its expected load and mesh conditions than obtained from simpler relations for an isolated gear that neither supply an expected life nor reflect the influence of the specific gear with which it is to mesh.

In equation 13.8.1

$$K = 9.18 \times 10^{18} \qquad \text{for force in pounds, length in inches}$$
$$= 6.44 \times 10^{9} \qquad \text{for force in newtons, length in millimeters}$$

where T_{10} is the life of a tooth in millions of load cycles for which no more than 10% of a large number of similar teeth will fail, F is the force component perpendicular to the tooth profile, w is the face width in inches, μ is the sum of the curvatures of the teeth in contact as calculated from (13.8.2), and l is the length of the involute profile.

If we assume that the contact ratio is less than 2.0 the length of the loaded region may be modeled by the distribution shown in Figure 13.8.1, denoted by l^*, and calculated from equation 13.8.3.[10]

The quantities in these formulas are calculated from the relations

$$\mu = \frac{1}{\rho_1} + \frac{1}{\rho_2} \tag{13.8.2}$$

$$l^* = 2r_{b_1}\theta_1(\theta_1 + \theta_2 + \theta_3) \tag{13.8.3}$$

in which θ_1, θ_2, and θ_3 are defined by

$$\theta_1 = \frac{1}{2r_{b_1}}(2p_b - l_1 - l_2) \tag{13.8.4}$$

$$\theta_2 = \frac{1}{r_{b_1}}(l_1 + l_2 - p_b) \tag{13.8.5}$$

$$\theta_3 = \frac{1}{r_{b_1}}[c\sin\phi - (r_{a_2}^2 - r_{b_2}^2)^{1/2}] \tag{13.8.6}$$

where the radii of curvature appearing in equation 13.8.2 are given by

$$\rho_1 = r_{b_1}(\theta_2 + \theta_3) \tag{13.8.7}$$

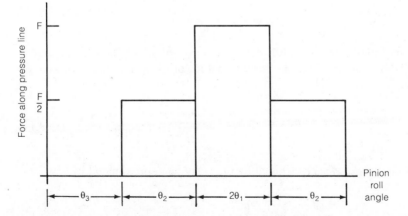

Figure 13.8.1 Force distribution along tooth profile. Two pairs of teeth are in contact along intervals θ_2 and only one pair is in contact along interval $2\theta_1$.

and

$$\rho_2 = c \sin \phi - \rho_1 \tag{13.8.8}$$

Equations 13.8.1 may be written in terms of hours by multiplying K by $10^6/60n$ for n in rpm. Thus,

$$K = 153 \times 10^{21}/n \qquad \text{force in pounds, length in inches}$$

$$K = 107 \times 10^{12}/n \qquad \text{force in newtons, length in millimeters}$$

These relations may be used to derive an expression of the life of the gear itself if the gear is considered to have failed when any one of its teeth has failed. Using elementary probability theory it may be shown[10] that 90% of a number of similar gears should have a life equal to, or greater than, the G_{10} life given by equation 13.8.9, which is

$$G_{10} = \frac{T_{10}}{u(N_i)^{1/e}} \qquad i = 1,2 \tag{13.8.9}$$

In this equation e is the Weibull exponent, which is 2.5 gears made from UNS G93106 steel, N_i is the number of teeth on gear i, and u denotes the number of load cycles per revolution imposed upon a representative tooth; i.e., the number of times it engaged a load bearing tooth from another gear. Thus u is 1.0 for each gear of a two-gear set, but is 2.0 for a gear driven by two pinions, as is customary for heavily loaded gears.

When applied to an idler gear, which transfers power from one gear to another in which each tooth is loaded first in one direction and then in the opposite direction, equation 13.8.9 is replaced by

$$G_{10} = \frac{T_{10}}{u(2N_i)^{1/e}} \tag{13.8.10}$$

The L_{10} life of two gears in mesh is defined in terms of a large number of similar gear pairs and is that number of cycles for which 10% of them have failed. It may be calculated from[10]

$$L_{10} = \left(\frac{Q_m}{F}\right)^{4.3} \tag{13.8.11}$$

where the *dynamic load capacity*, denoted by Q_m, may be approximated by

$$Q_m = \left\{\frac{K^{2.5}w^{9.75}}{\mu^{12.5}l*N_1[1 + (N_1/N_2)^{1.5}]}\right\}^{0.093} \tag{13.8.12}$$

in which N_1 and N_2 are the numbers of teeth on the pinion and gear, respectively, and the remaining terms are those that appeared in relation 13.8.1.

The order of calculation for programming these relations is displayed in Figure 13.8.2.

EXAMPLE 13.8.1

Estimate the life of a gear set in which the pinion has 36 teeth, the gear has 54 teeth, and the force along the pressure line is 3275 lb. The pressure angle is 20°, both pinion and gear are made from UNS G93106 steel, the face width is 0.75 in., and the diametral pitch of the teeth is 3.

All of the calculations required will be displayed so that this example may be used to check the operation of a program written according to the flowchart displayed in Figure 13.8.2. The input data are

$$\phi = 20° \qquad\qquad P_d = 3$$

$$N_1 = 36 \qquad\qquad N_2 = 54$$

$$w = 0.75 \text{ in.} \qquad\qquad u = 1 \text{ (only 1 gear in contact)}$$

$$a_1 = 0.333 \text{ in.} = 1/P_d \qquad a_2 = 0.333 \text{ in.} = 1/P_d$$

The calculated values are

$$d_1 = N_1/P_d = 12.00 \text{ in.} \qquad d_2 = N_2/P_d = 18.00 \text{ in.}$$

$$r_{p_1} = 6.00 \text{ in} \qquad r_{p_2} = 9.00 \text{ in.}$$

$$c = r_{p_1} + r_{p_2} = 15.000$$

$$r_{b_1} = r_{p_1} \cos \phi = 5.638 \text{ in.} \qquad r_{b_2} = r_{p_2} \cos \phi = 8.457 \text{ in.}$$

$$r_{a_1} = 6.333 \text{ in} \qquad r_{a_2} = 9.333 \text{ in.}$$

$$l_1 = (6.333^2 - 5.638^2)^{1/2} - 6.00 \sin 20° = 0.8323 \text{ in.}$$

$$l_2 = (9.333^2 - 8.457^2)^{1/2} - 9.00 \sin 20° = 0.8695 \text{ in.}$$

$$p_c = \pi/3 = 1.047 \text{ in.} \qquad p_b = 1.047 \cos 20° = 0.984 \text{ in.}$$

$$\theta_1 = [2(0.984) - 0.8695 - 0.8323]/[2(5.638)] = 0.0236 \text{ rad.}$$

$$\theta_2 = (0.8695 + 0.8323 - 0.984)/5.638 = 0.1273 \text{ rad.}$$

$$\theta_3 = [15 \sin 20° - (9.333^2 - 8.457^2)^{1/2}]/5.638 = 0.2098 \text{ rad.}$$

$$\rho_1 = (0.1273 + 0.2098)5.638 = 1.9006 \text{ in.}$$

$$\rho_2 = 15 \sin 20° - 1.9006 = 3.2297 \text{ in.}$$

$$\mu = 1/1.9006 + 1/3.2297 = 0.8358$$

$$l^* = 2(5.638)0.0236(0.0239 + 0.1273 + 0.2098) = 0.0960 \text{ in.}$$

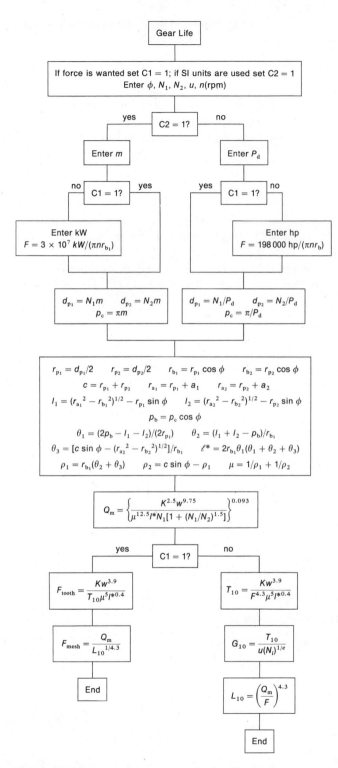

The C1 = 1 choice is used to find the force that may be expected from an existing gear. This question may arise in the selection of equipment from a vendor, in reexamining gear trains that have performed well, or in comparing gear trains from a competitor. The C1 = 0 choice may generally be used in the design of new gear trains, and both may be used in the re-design of those that have failed prematurely.

Figure 13.8.2 Flowchart for program entitled GEAR LIFE.

Upon using these values we find that

$$T_{10} = \frac{9.18 \times 10^{18}(0.75^{3.9})}{3275^{4.3}(0.8358^5)0.0960} = 14\,347 \text{ million cycles}$$

as the probable life of any one tooth, and

$$G_{10} = \frac{14\,347}{1(36^{1/2.5})} = 3422 \text{ million pinion cycles}$$

as the probable life of the pinion and

$$G_{10} = \frac{14\,347}{1(54^{1/2.5})} = 2909 \text{ million gear cycles}$$

as the probable life of the gear.

The L_{10} life of the pair is found from the dynamic load capacity, which is

$$Q_m = \left\{ \frac{(9.18 \times 10^{18})^{2.5}0.75^{9.75}}{0.8358^{12.5}(0.0960)36(1 + 0.6667^{1.5})} \right\}^{0.093} = 20\,820 \text{ lb}$$

and

$$L_{10} = \left(\frac{20\,820}{3275} \right)^{4.3} = 2845 \text{ million pinion cycles}$$

If the pinion turns at 1725 rpm and if the machine is used for one 8-h shift/day at 5 days/week for 50 weeks/year it will accumulate

$$1725(60)8(5)50 = 207 \times 10^6 \text{ cycles/year}$$

for an estimated life of $2845/207 = 13.7$ years. Since uncertainty in the calculation of the L_{10} life of the pair may be of the order of 20% because of possible variations in the physical constants used, the results should be interpreted as indicating a life of about 11.0 to 16.4 years.

AGMA Life Estimates

AGMA standards 218.01 and 411.02 assume different failure mechanisms. The first[11] assumes the existence of a surface endurance limit, with the implication that stresses below the elastic limit will not cause pitting. The second[12] indicates no such limit and implies decreased life with increased load. We shall consider only the first of these because of its more recent publication date and of data published in 1985 by SKF that indicates a similar limit exists for rolling element bearings.[†]

No explicit AGMA formulas exist at this time for the L_{10} life of gears. An indirect relation does exist: the pitting resistance power rating that is given by

$$P = \frac{n_p w I C_v}{\Omega C_s C_m C_f C_a} \left(\frac{d_p \sigma C_L C_H}{C_p C_T C_R}\right)^2 \tag{13.8.13}$$

in which

$$\Omega = 126\,000 \qquad \text{for power in units of } hp$$
$$= 1.91 \times 10^7 \qquad \text{for power in units of } kW$$

Pinion rpm is denoted by n_p and the face width, in millimeters or inches, is denoted by w. A sequence of formulas, graphical procedures, and geometrical calculations on pages 10–20 of the AGMA 218.01 standard are involved in the determination of I for gears of standard proportions and common loads. They are not reproduced here because factors C_s, C_m, C_f, and C_a in equation 13.8.13 are to be selected by the gear designer on the basis of years of experience. These factors are the size factor, the load distribution factor, the surface condition factor, and the application factor, respectively. Factor C_v, the dynamic factor for pitting, is equal to K_v as given by equation 13.7.4; factor σ_c, the allowable contact stress number, may be read from Table 13.8.1; and factor C_L, the life factor for pitting, may be read from Figure 13.8.3. Factor C_H, the hardness ratio factor, may be calculated from

$$C_H = 1 + A(N_g/N_p - 1) \tag{13.8.14}$$

for $\mathrm{BHN_p}/\mathrm{BHN_g} \leq 1.7$ and by

$$CH = 1 + B(450 + \mathrm{BHN_g}) \tag{13.8.15}$$

for a pinion surface hardness equal to or greater than 49 HRC (Rockwell hardness on the C scale), and for $180 \leq \mathrm{BHN_g} \leq 400$, where $\mathrm{BHN_g}$ and $\mathrm{BHN_p}$ denote the Brinell hardness of the gear and pinion, respectively, when measured with a 10 mm ball supporting 3 000 kg. (The load is specified in kg, not N, in the standard.) In these relations

$$A = 0.00898(\mathrm{BHN_p}/\mathrm{BHN_p}) - 0.00829$$

and

$$B = 0.00075e^{-0.0112f}$$

[†] Ioannides, E., and Harris, T.A., *A New Fatigue Life Model for Rolling Bearings*, Ball Bearings Journal, No. 224, SKF Industries, King of Prussia, PA, October 1985.

TABLE 13.8.1 STANDARD VALUES FOR (a) σ_c, (b) C_p, (c) C_o, AND (d) C_R WHICH APPEAR IN EQUATION 13.8.13

(a) Allowable Contact Stress Number, σ_{ac}

Material	AGMA Class	Commercial Designation	Heat Treatment	Minimum Hardness at Surface	s_{ac}, lb/in.^{2}	(MPa)
Steel	A-1		Through-hardened and tempered	180 BHN and less	85–95 000	(590–660)
	thru			240 BHN	105–115 000	(720–790)
				300 BHN	120–135 000	(830–930)
	A-5			360 BHN	145–160 000	(1 000–1 100)
				400 BHN	155–170 000	(1 100–1 200)
			Flame or induction hardened	50 HRC	170–190 000	(1 200–1 300)
				54 HRC	175–195 000	(1 200–1 300)
			Carburized and case-hardened	55 HRC	180–200 000	(1 250–1 400)
				60 HRC	200–225 000	(1 400–1 550)
		AISI 4140	Nitrided	48 HRC	155–180 000	(1 100–1 250)
		AISI 4340	Nitrided	46 HRC	150–175 000	(1 050–1 200)
		Nitralloy 135M	Nitrided	60 HRC	170–195 000	(1 170–1 350)
		$2\frac{1}{2}$% Chrome	Nitrided	54 HRC	155–172 000	(1 100–1 200)
		$2\frac{1}{2}$% Chrome	Nitrided	60 HRC	192–216 000	(1 300–1 500)
Cast Iron	20		As-cast	—	50–60 000	(340–410)
	30		As-cast	175 BHN	65–75 000	(450–520)
	40		As-cast	200 BHN	75–85 000	(520–590)
Nodular (Ductile)	A-7-a	60-14-18	Annealed	140 BHN	90–100% of	
	A-7-c	80-55-06	Quenched and tempered		s_{ac} value	
				180 BHN	of steel	
	A-7-d	100-70-03	''	230 BHN	with same	
	A-7-e	120-90-02	''	270 BHN	hardness	
Malleable Iron (Pearlitic)	A-8-c	45007	—	165 BHN	72 000	(500)
	A-8-e	50005	—	180 BHN	78 000	(540)
	A-8-f	53007	—	195 BHN	83 000	(570)
	A-8-i	80002	—	240 BHN	94 000	(650)
Bronze	Bronze 2	AGMA 2C	Sand cast	Tensile strength minimum 40 000 lb/in^2 (275 MPa)	30 000	(205)
	Al/Br 3	ASTM B-148-52 Alloy 9C	Heat treated	Tensile strength minimum 90 000 lb/in^2 (620 MPa)	65 000	(450)

(continued)

TABLE 13.8.1   (Continued)

(b) Elastic Coefficient, C_p

Pinion Material and Modulus of Elasticity, E^a		Gear Material and Modulus of Elasticity, E^a			
		Steel	**Cast Iron**	**Aluminum Bronze**	**Tin Bronze**
		30×10^6	19×10^6	17.5×10^6	16×10^6
Steel	30×10^6	2300	2000	1950	1900
Cast Iron	19×10^6	2000	1800	1800	1750
Aluminum Bronze	17.5×10^6	1950	1800	1750	1700
Tin Bronze	16×10^6	1900	1750	1700	1650

Poisson's Ratio $= 0.30$
[a]   When more exact values of E are obtained from roller contact tests, they can be used.

(c) Overload Factors, c_o[b]

	Character of Load On Driven Machine		
Power Source	**Uniform**	**Moderate Shock**	**Heavy Shock**
Uniform	1.00	1.25	1.75 or higher
Light Shock	1.25	1.50	2.00 or higher
Medium Shock	1.50	1.75	2.25 or higher

(d) Factor of Safety, c_R

Requirements of Application	c_R
High Reliability	1.25 or higher
Fewer than One Failure in 100	1.00
Fewer than One Failure in Three	0.80[c]

[c]   At this value plastic profile deformation might occur rather than pitting.

[b]   This table is for speed decreasing drives only. For speed increasing drives add $.01 \left[\dfrac{N_G}{N_P} \right]^2$ to the factors.

Source: AGMA 218.01, Dec., 1982, American Gear Manufacturers Association, Alexandria, VA. Reprinted with permission.

where f denotes the surface finish of the pinion in microinches. No SI equivalent is listed.

Factor C_p is known as the elastic coefficient. It is given by

$$C_p{}^2 = \pi \left[\left(\frac{1 - v_p}{E_p} \right)^2 + \left(\frac{1 - v_g}{E_g} \right)^2 \right]$$

in terms of Poisson's ratio v_p and Young's modulus E_p for the pinion and the corresponding quantities v_g and E_g for the gear. Factor C_T is known as the temperature factor, which is set equal to 1.0 for temperatures no greater than 120 °C (250 °F). The reliability factor for pitting resistance, C_R, is given by Table 13.8.2, made up of factors

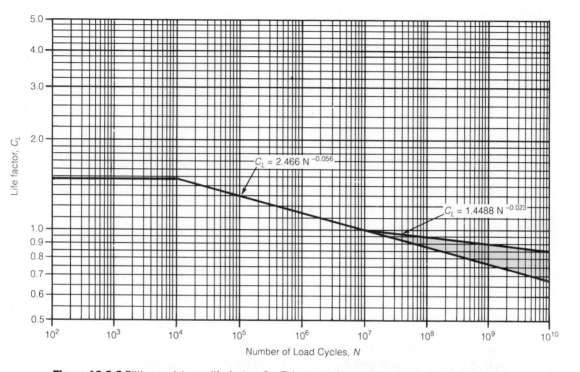

Figure 13.8.3 Pitting resistance life factor, C_L. This curve does not apply where a service factor, C_{SF}, is used. The choice of C_L above 10^7 cycles is influenced by: lubrication regime, failure criteria, smoothness of operation required, pitchline velocity, gear material cleanliness, material ductility and fracture toughness, and residual stress. *Source: AGMA 218.01 Dec. 1982, American Gear Manufacturers Assn., Alexandria, VA. Reprinted with permission.*

to modify the allowable stresses that appear in Table 13.8.1, which is based upon the probability of one failure in 1×10^9 cycles. The right-hand column in Table 13.8.2 lists the changed probabilities associated with the C_R factors listed in the left-hand column. See the standard for additional details.

TABLE 13.8.2

Application Requirements for Fewer than One Failure in	C_R
10 000 cycles	1.50
1 000	1.25
100	1.00
10	0.85*

* Plastic flow may occur rather than pitting.
Abstracted from Table 8, p. 41, AGMA 218.01, Dec. 1982, American Gear Manufacturers Association, Arlington, VA.

13.9 SPEED RATIO AND THE NUMBER OF TEETH

Shaft speed ratios are often of considerable importance in machine operation, especially in equipment that is driven by ac motors which are most economically produced to operate at certain speeds standardized by the the National Electrical Manufacturers Association (typically 575, 690, 870, 1160, 1750, 3450 rpm or values fairly close to these; i.e., 1725 rpm instead of 1750 rpm). For simplicity the shaft ratios are usually given in decimal form, such as 2.68:1 or as 2.68, which means that the pinion shaft makes 2.68 rotations for each rotation of the gear shaft. The engineer is left with the problem of selecting the number of teeth on each gear, N_1 and N_2, say, such that their ratio has the desired value.

By observing that the tangential velocity of the pitch circles of the two gears must be equal at their common point of contact, it follows that

$$v = r_{p_1}\omega_1 = r_{p_2}\omega_2 \tag{13.9.1}$$

so that

$$\frac{\omega_1}{\omega_2} = \frac{r_{p_2}}{r_{p_1}} = \frac{d_{p_2}}{d_{p_1}}\frac{P_d}{P_d} = \frac{N_2}{N_1} \tag{13.9.2}$$

If the ratio mentioned above is to be obtained the engineer is required to find integer values of N_1 and N_2 such that $N_2/N_1 = 2.68$, where N_1 is the number of teeth on the pinion and N_2 is the number of teeth on the gear. Since it is expensive to make a large number of teeth and since the small teeth resulting from using 100 standard teeth on the pinion and 268 standard teeth on the gear may be too weak for the loads to be transmitted, the engineer must find some other less obvious integer pair.

Perhaps the simplest means of finding a suitable ratio is to use the program outlined in Ref. 13, which is based upon the theory of continuing fractions and Euclid's algorithm for the greatest common denominator. From that theory the desired ratio R may be written as

$$R = \frac{p_N}{q_N} \tag{13.9.3}$$

to within some specified error E, defined by

$$E = \left| R - \frac{p_N}{q_N} \right| \tag{13.9.4}$$

Quantities p_N and q_N are acceptable values of p_n and q_n, which are defined by

$$p_n = a_n p_{n-1} + p_{n-2}$$
$$q_n = a_n q_{n-1} + q_{n-2} \tag{13.9.5}$$

where

$$a_n = \text{INT}\left(\frac{y_{n-2}}{y_{n-1}}\right) \tag{13.9.6}$$

$$y_n = -a_n y_{n-1} + y_{n-2}$$

for $n > 2$. Here notation INT signifies that a_n is the largest integer that is equal to or less than the ratio (y_{n-2}/y_{n-1}). For $n = 1,2$ these quantities are defined by

$$
\begin{aligned}
p_1 &= a_1 & a_1 &= \text{INT}(R) \\
q_1 &= 1 & y_1 &= \text{FRC}(R)
\end{aligned} \tag{13.9.7}
$$

$$
\begin{aligned}
p_2 &= a_1 a_2 + 1 & a_2 &= \text{INT}(1/y_1) \\
q_2 &= a_2 & y_2 &= 1 - a_2 y_1
\end{aligned} \tag{13.9.8}
$$

where $\text{FRC}(R) = R - \text{INT}(R)$.

Returning to the problem of finding integers smaller than 100 and 268 to obtain the ratio 2.68, use of relations 13.8.5 through 13.8.8 yields

$$
\begin{aligned}
a_1 &= \text{INT}(2.68) = 2 & a_2 &= \text{INT}(1/0.68) = 1 \\
p_1 &= 2 & p_2 &= 3 \\
q_1 &= 1 & q_2 &= 1 \\
y_1 &= 0.68 & y_2 &= 1 - 0.68 = 0.32
\end{aligned}
$$

$$
\begin{aligned}
a_3 &= \text{INT}(0.68/0.32) = 2 & a_4 &= 8 \\
p_3 &= 8 & p_4 &= 67 \\
q_3 &= 3 & q_4 &= 25 \\
y_3 &= 0.04 & y_4 &= 0
\end{aligned}
$$

The process terminates after the fourth step for whatever value of E is selected because $R - q_4/p_4 = 0$; i.e.,

$$R = \tfrac{67}{25} = 2.6800 = N_2/N_1$$

or $N_1 = 25$ and $N_2 = 67$. It is no accident that the rational number found by this method is in its lowest terms, it has been proven in the study of continued fractions that q_N/p_N will always be in its lowest terms (the numerator and denominator have no common factor other than 1.0).[13]

The flowchart for a program to calculate gear ratios according to these relations is shown in Figure 13.9.1 and a flowchart for a factoring routine to find the factors of p_N and q_N is shown in Figure 13.9.2.

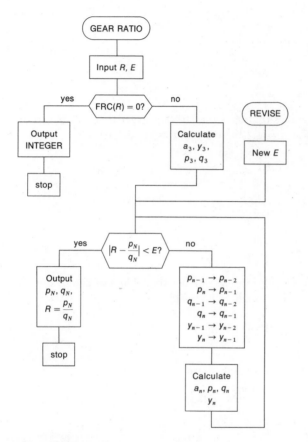

Figure 13.9.1 Flowchart for the GEAR RATIO program.

EXAMPLE 13.9.1

Design a gear train to provide an angular velocity ratio of $\sqrt{5}$. Note that since this is an irrational number it is essential that an acceptable error value E, be specified.

To demonstrate the effects of a range of E values, use the program flowcharted in Figure 13.9.1 to find that

E	p_N	q_N
1×10^{-2}	38	17
1×10^{-3}	38	17
1×10^{-4}	161	72
1×10^{-5}	682	305
1×10^{-6}	2 889	1 292
1×10^{-7}	12 238	5 473
1×10^{-8}	12 238	5 473
1×10^{-9}	115 920	51 841

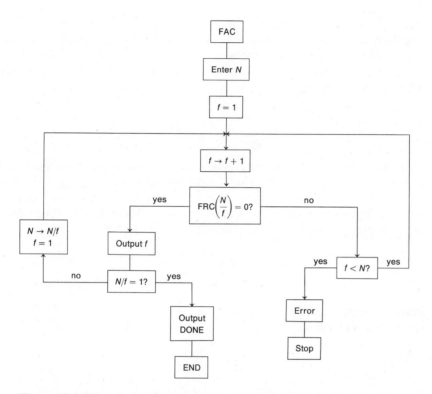

Figure 13.9.2 Flowchart of the factoring program FAC to find all of the prime factors in N.

If the error associated with $N_1 = 2889$ and $N_2 = 1292$ were acceptable, use of the factoring routine shown in Figure 13.9.2 could be used to find that $2889 = 27 \times 107$ and that $1291 = 19 \times 68$ so that a gear train as shown in Figure 13.9.3 could be assembled, where

$$R = (N_4/N_3)(N_2/N_1) = \tfrac{27}{19} \times \left(\tfrac{107}{68}\right)$$

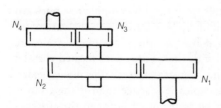

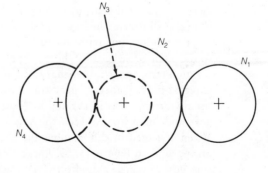

Figure 13.9.3 Gear train with four gears.

Whenever the above ratios are unsuitable because the ratio between two gears in mesh is too large, the individual pair ratios may be altered by adding more gears or shafts. For example, suppose the ratio is $\frac{37}{13}$ but that the pair ratio should be less than 2.00 and that teeth with a pressure angle of $20°$ are to be used. Because use of a pinion with 13 teeth and a standard profile will cause undercutting, this ratio must be replaced by $\frac{74}{26}$ and then written as

$$\frac{37}{13} = \frac{74}{26} = \left(\frac{74}{40}\right)\left(\frac{40}{26}\right)$$

which requires four gears on three shafts.

In the event both integers in the rational approximation to a gear ratio are large and prime, a small change in the ratio within the given acceptable error will usually eliminate the difficulty. This technique may also be used to find a number of tooth ratios for a specified gear ratio and acceptable error. For example, suppose we are to find tooth numbers for a gear ratio of 2.9464 ± 0.0001. When $2.94643 \pm 1 \times 10^{-5}$ is entered into the program to find a ratio it outputs

$$\frac{165}{56} = \frac{45}{24}\ \frac{33}{21}$$

If we next enter a different number with the given error, such as $2.946387 \pm 1 \times 10^{-6}$, which differs from the specified ratio by 0.00013, the numbers

$$\frac{1264}{429} = \frac{32}{22}\ \frac{79}{39}$$

are produced. This procedure may be programmed to produce a variety of tooth ratios for a particular input ratio and permissible error as described in Ref. 14.

13.10 PLANETARY GEAR TRAINS

Planetary gear trains usually consists of one or more internal gears and several external gears. An internal gear, or ring gear, is one in which the teeth are cut on the inside of a ring, as illustrated in Figure 13.4.3, with the base and pitch circles unchanged relative to one another, but with the addendum, dedendum, and root circles reversed relative to the pitch circle because the gear material extends from the addendum circle outward for an internal gear rather than from the addendum circle inward, as in the case of an external gear. Planetary gear trains are so named because the gears in the train are usually arranged as illustrated in Figure 13.10.1 with the central gear termed the sun gear and the gears between the sun gear and the ring gear termed the planet gears. When arranged in this manner the gear train occupies less space than if the train is arranged as in Figure 13.9.3. Large torques may be transmitted between the sun and ring gears because the effective contact ratio between the sun and planet gears and

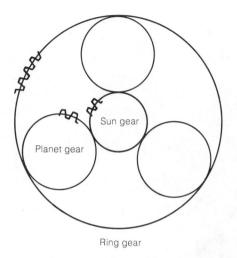

Figure 13.10.1 Gear arrangement in a planetary gear train.

between the planet and ring gears is equal to the contact ratio for one planet gear multiplied by the number of planet gears. An example of their use is pictured in Figure 13.10.2 where four-wheel drive may be obtained without multiple long drive shafts by using one hydraulic pump, four hydraulic motors, and four compact planetary gear trains, one of which is shown in Figure 13.10.3

Figure 13.10.2 Industrial crane using hydraulic motors and planetary gears on each wheel instead of a transmission, drive shaft, and differential. *Source: Auburn Gear Inc., Auburn, IN.*

Figure 13.10.3 Planetary gear used in a wheel. *Source: Auburn Gear Inc., Auburn, IN.*

Examples of the variety of planetary gear trains that have been designed are shown in Figures 13.10.4 and 13.10.5 along with schematics to aid in the derivation of their speed ratios.

13.11 SPEED RATIO CALCULATIONS FOR PLANETARY GEAR TRAINS

Four methods are generally used for deriving the ratio of input to output speed or its inverse. The first is a kinematic method based upon the rotation transferred from one gear to another, the second is known as the tabular method and is a modification of the first, the third is a dynamic method based upon the velocities transferred from gear to gear, and the fourth is a power transfer method which assumes no loss of power within the gear train.

The first, third, and fourth methods will be demonstrated for the planetary gear train shown schematically in Figure 13.11.1. The second method will not be considered because of its similarity to the first.

The planetary gear train shown in Figure 13.11.1 is widely used in automatic transmissions because the range of output speed ratios that can be obtained by controlling the speed of one or more of its components.

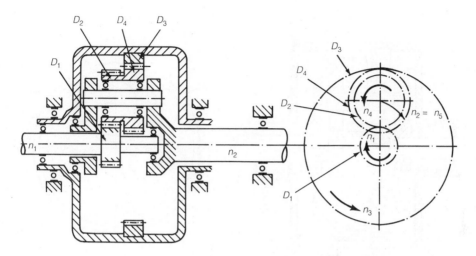

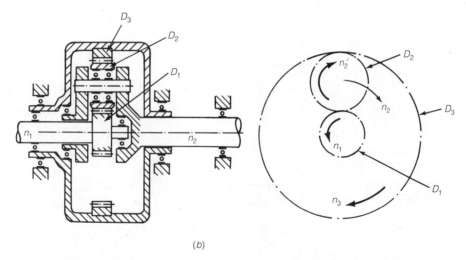

(a)

(b)

Figure 13.10.4 (a) Planetary gear train with double planet gears; (b) planetary gear train with single planetary gears. *Source: SKF Industries, Inc., Bearings Group, King of Prussia, PA.*

Kinematic and force analysis of the gear train may be had by considering the sun gear, the carrier, one of the planetary gears (since its motion is identical to that of the others), and the ring gear. After numbering the gears and introducing the angles of rotation as shown in Figure 13.11.2 it follows that if the gear train starts from the configuration shown in Figure 13.11.2a, where ϕ and ψ are the driving rotations, the planet and ring gears can pass through the positions shown in (b) for a particular choice of ϕ, ψ, and η.

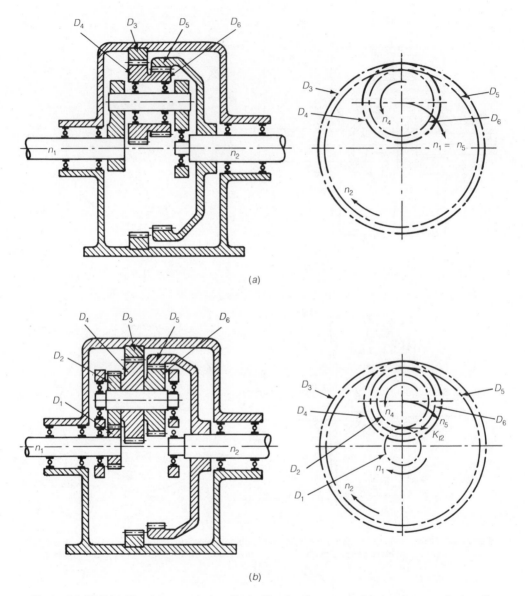

(a)

(b)

Figure 13.10.5 (a) Planetary gear train with double planetary gears; (b) planetary gear train with triple planetary gears. *Source: SKF Industries, Inc., Bearings Group, King of Prussia, PA.*

    Although the motion is continuous, we may analyze its kinematics by first imagining that the planet, carrier, and ring gears rotate as a unit with the rotation of the sun gear through angle ψ and then that the carrier is held stationary and the sun gear continues to rotate to angle ϕ. This continued rotation with the carrier stationary

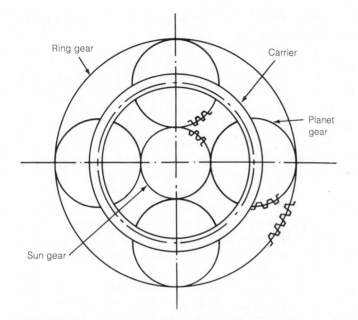

Figure 13.11.1 Schematic of a planetary gear train consisting of a sun gear, four planet gears, a carrier, and a ring gear.

1 Sun gear
2 Planet gear
3 Ring gear

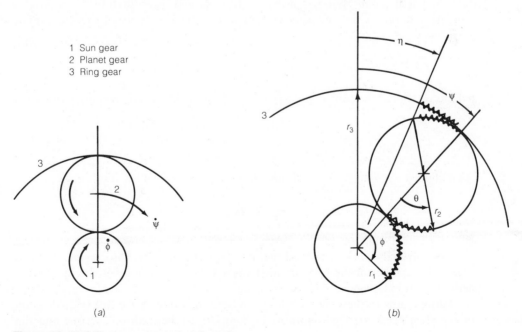

(a)   *(b)*

Figure 13.11.2 Planetary gear train schematic.

causes the planet to rotate through angle θ, which in turn causes the ring gear to rotate in the opposite direction by an amount $\psi - \eta$, where

$$r_1(\phi - \psi) = r_2\theta$$
$$r_2\theta = r_3(\psi - \eta) \tag{13.11.1}$$

so that upon solving for η we have that

$$\eta = \frac{r_3\psi - r_2\theta}{r_3} = \psi - \frac{r_2}{r_3}\theta \tag{13.11.2}$$

Substitution for θ from (13.11.1) into (13.11.2) yields

$$\dot{\eta} = \left(1 + \frac{N_1}{N_3}\right)\dot{\psi} - \frac{N_1}{N_3}\dot{\phi} \tag{13.11.3}$$

after differentiating with respect to time and replacing the pitch radii ratios with the corresponding tooth ratios. Returning to Figure 13.11.2 we find that positive angular velocities correspond to clockwise rotation for the sun gear, the carrier, and the ring gear.

Analysis of the planetary gear train by means of velocities may be accomplished from the initial geometry shown in Figure 13.11.2a and reproduced in Figure 13.11.3 with the addition of the necessary velocities. The linear velocity of the ring gear is given by

$$r_0\dot{\psi} - r_2\dot{\theta} = r_3\dot{\eta} \tag{13.11.4}$$

and the velocity of the contact point between the sun and planed gears may be written as

$$r_1\dot{\phi} = r_0\dot{\psi} + r_2\dot{\theta} \tag{13.11.5}$$

Upon eliminating θ between equations 13.11.4 and 13.11.5 and writing $2r_0$ as

$$2r_0 = (r_1 + r_2) + (r_3 - r_2) = r_1 + r_3$$

we again obtain equation 13.11.3.

Not only is the velocity method the simplest of the three in this example, but it is usually the simplest of the three in other cases as well. It is often used with another method as an independent check of results.

Turning now to the power method, we begin by drawing the free body diagram for the planet gear as acted upon by the forces from the teeth on the ring and sun gears and by the stub shaft attached to the carrier. Choosing the direction of each force

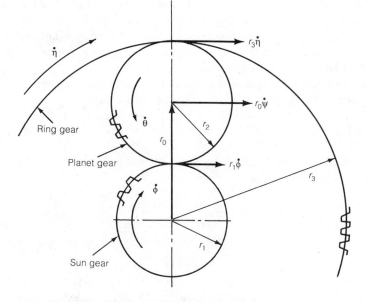

Figure 13.11.3 Schematic of the planetary gear train in Figure 13.11.1 showing peripheral velocities of each gear and the velocity of the center of a representative planet gear.

may be the most difficult part of this analysis because they may not be in the directions initially expected. Perhaps the most evident force is F_{12}, which is due to the sun gear (gear 1) acting upon the planet gear (gear 2), as shown in Figure 13.11.4, if the sun gear is to rotate in the clockwise direction.

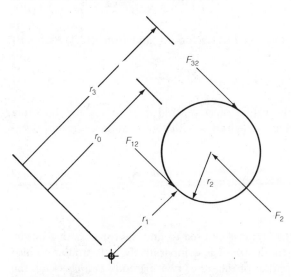

Figure 13.11.4 Forces acting on the planet gear.

By considering the moment equilibrium about the center of the ring gear we find that F_{32} from the ring gear on the planet gear must be as illustrated. Finally, by taking moments about the contact point between gears 1 and 2 (between sun and planet gear) we find that the force from the carrier on the planet gear, F_2, must also be as shown in Figure 13.11.4.

If we take clockwise rotation as positive in Figure 13.11.4 and if we assume no frictional losses in the gear train the power balance may be written as

$$T_1\dot{\phi} - T_2\dot{\psi} + T_3\dot{\eta} = 0 \tag{13.11.6}$$

where the torques involved may be related to the forces shown by taking moments about the center of the sun gear to obtain

$$T_1 = F_{12}r_1 \qquad T_2 = F_2r_0 \qquad T_3 = F_{32}r_3 \tag{13.11.7}$$

From the equations of static equilibrium about the center of the sun gear we have that

$$F_2r_0 = F_{32}r_3 + F_{12}r_1 \tag{13.11.8}$$

and from static equilibrium about the center of the planet gear we have

$$F_{32} = F_{12} \tag{13.11.9}$$

Substitution for T_1, T_2, and T_3 in (13.11.6) from (13.11.7) yields

$$F_{12}r_1n_1 - F_2r_0n_2 + F_{32}r_3n_3 = 0 \tag{13.11.10}$$

so that further substitution from equation 13.11.9 for F_{12} and from (13.11.8) for F_2 yields

$$F_{32}r_1n_1 - (F_{32}r_3 + F_{32}r_1)n_2 + F_{32}r_3n_3 = 0 \tag{13.11.11}$$

in which n_1, n_2, and n_3 denote the values of $\dot{\phi}$, $\dot{\psi}$, and $\dot{\eta}$ in rpm. After factoring out F_{32}, dividing by r_3 and replacing radius ratios with tooth ratios the resulting expression becomes

$$\dot{\eta} = \left(1 + \frac{N_1}{N_3}\right)\dot{\psi} - \frac{N_1}{N_3}\dot{\phi} \tag{13.11.12}$$

which is identical to (13.11.3).

Since equations 13.11.7 and 13.11.8 must be used to find the forces acting on the teeth in order to select the proper tooth size, it is comparatively little trouble to also use the torque method along with either of the other two methods as a check on the speed ratio relations.

We shall now examine the various speed ratios that may be delivered by a planetary gear set in an automotive automatic transmission. In first gear the speed ratio is obtained by holding the ring gear stationary with a band brake, driving the sun gear from the torque converter, and connecting the carrier by means of a disk clutch to the drive shaft. In this case $\dot{\eta} = 0$ so from (13.11.12) we have

$$\dot{\phi} = \left(1 + \frac{N_3}{N_1}\right)\dot{\psi} \qquad (13.11.13)$$

where $\dot{\phi}$ represents the torque converter speed and $\dot{\psi}$ the drive shaft speed, which may or may not be the wheel speed, depending upon the gear ratios in the differential. If we select $N_1 = 90$ and $N_3 = 210$ then $\dot{\psi} = 0.300\dot{\phi}$; i.e., the drive shaft speed is less than that of the torque converter.

Second gear is often obtained by using a second planetary gear train of the same type, known as the reaction gear set, in which the ring gear from the input gear set drives the carrier of the reaction set through a clutch which is engaged when the transmission is shifted into second. The sun gear in the reaction set is held stationary and the ring gear of the reaction set drives the carrier of the input set through another clutch that is also automatically engaged. If $\dot{\eta}'$ denotes the velocity of the ring gear in the reaction set and $\dot{\psi}'$ the velocity of the carrier in the reaction set, then

$$\dot{\eta}' = \left(1 + \frac{N_1'}{N_3'}\right)\dot{\psi}' \qquad (13.11.14)$$

Since the ring gear of the reaction set is driven by the carrier of the input set $\dot{\eta}'$ in (13.11.14) is replaced by $\dot{\psi}$ and since the ring gear of the input set is driven by the carrier of the reaction set, $\dot{\psi}'$ in (13.11.14) is replaced by $\dot{\eta}$. After these substitutions have been made $\dot{\eta}$ from (13.11.14) may be substituted into (13.11.12) and the resulting expression solved for $\dot{\phi}$ to yield

$$\dot{\phi} = \frac{N_3}{N_1}\left(1 + \frac{N_1}{N_3} - \frac{1}{1 + \frac{N_1'}{N_3'}}\right)\dot{\psi} \qquad (13.11.15)$$

If we select $N_1' = 74$ and $N_3' = 200$ and use the previously selected values for N_1 and N_3 we find that $\dot{\phi} = 1.6302\,\dot{\psi}$ or $\dot{\psi} = 0.613\,\dot{\phi}$.

The third gear speed ratio is had by driving the sun and ring gears at the same speed from the torque converter and driving the output shaft from the carrier. Since $\dot{\eta} = \dot{\phi}$ it follows from (13.11.12) that $\dot{\phi} = \dot{\psi}$; the torque converter and drive shaft rotate at the same speed.

In fourth gear, or overdrive, the sun gear is held stationary by a band brake, the ring gear is driven through a clutch by the torque converter, and the carrier drives the output shaft. With $\dot{\phi} = 0$ equation 13.11.12 yields

$$\dot{\eta} = \left(1 + \frac{N_1}{N_3}\right)\dot{\psi} \qquad (13.11.16)$$

so that the previous values of N_1 and N_3 lead to $\dot{\eta} = 1.43\dot{\psi}$. In other terms, the converter rpm is 0.7 that of the drive shaft.

Finally, reverse may be had by driving the sun gear, holding the carrier stationary, $(\dot{\psi} = 0)$ and driving the output shaft from the ring gear. From equation (13.11.12) we have that

$$\dot{\eta} = -\frac{N_1}{N_3}\dot{\phi} \tag{13.11.17}$$

for a speed ratio of $\dot{\eta} = -0.429\dot{\phi}$.

Automotive Differential

An automotive differential, pictured in Figure 13.11.5a, may be considered as a modified planetary gear train with two sun gears, two planet gears, and a carrier which has become the case. The case is driven by a ring gear, Figure 13.11.5b, which is driven by a hypoid pinion gear attached to the drive shaft, as shown in Figure 13.11.6d.

A schematic of the sun and planet gears is shown in Figure 13.11.6a. The sun gears have been renamed *wheel gears* and the planet gears have been renamed *differential pinion gears* to correspond to common terminology.

Their velocities may be calculated using the geometry shown in Figure 13.11.6b where r_1 denotes the pitch radius of wheel gear 1, which rotates with angular velocity ω_1, and r_3 denotes the pitch radius of one of the differential pinions, which rotates with angular velocity ω_3. The angular velocity of the carrier, or case, is represented by ω_2. Thus, the axis of the differential pinions rotates about the axis of the wheel gears with velocity ω_2 as shown in Figure 13.11.6b. If the angular velocity of the other wheel gear is denoted by ω_4, we find from summation of the velocity vectors at the pitch line of the differential pinion gear, Figure 13.11.6c, that

$$r_1\omega_1 = r_1\omega_2 - r_3\omega_3 \tag{13.11.18}$$

and that

$$r_1\omega_4 = r_1\omega_2 + r_3\omega_3 \tag{13.11.19}$$

Upon adding equations 13.11.18 and 13.11.19 we have

$$\omega_4 + \omega_1 = 2\omega_2 \tag{13.11.20}$$

and upon subtracting equation 13.11.18 from equation 13.11.19 we have

$$\omega_4 - \omega_1 = 2\omega_3 \frac{r_3}{r_1} \tag{13.11.21}$$

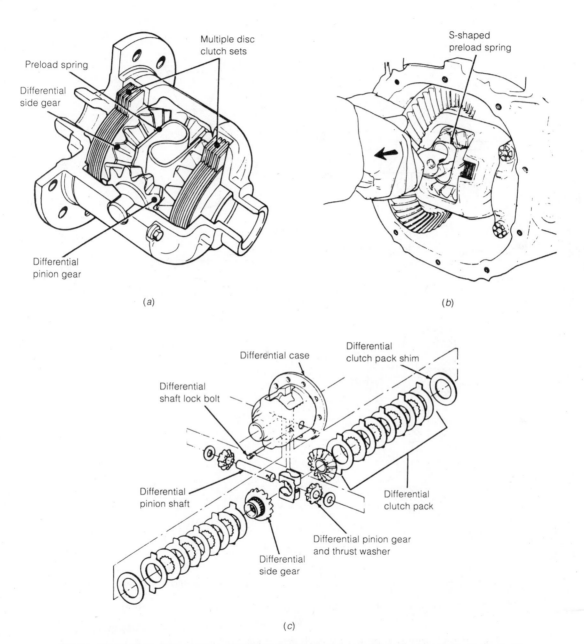

Figure 13.11.5 (a) Differential assembly; (b) differential assembly with ring gear attached; (c) exploded view of differential assembly showing the clutch plates for limited slip differential. *Source: Ford Motor Co., Dearborn, MI.*

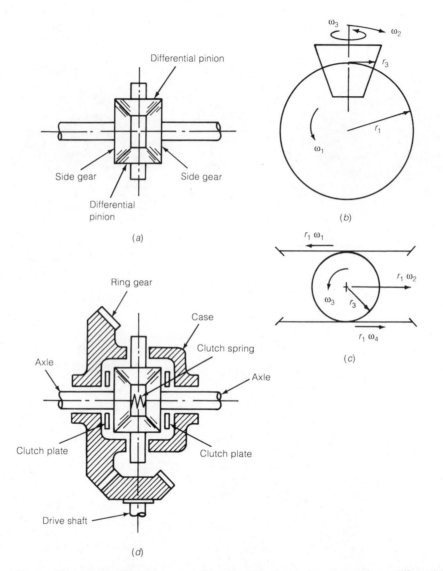

Figure 13.11.6 (a) Differential gears; (b) schematic of one side gear and one differential pinion, side view; (c) top view schematic of one pinion and side gear; (d) cross section of differential with ring gear, clutch plates, and drive shaft pinion shown.

To find the torques supplied by each wheel gear, let F_1 and F_4 denote the forces acting on the wheel gears and let F_2 denote the force acting on the axis of the differential pinion gear at radius r_2, which is the pitch radius of the ring gear, Figure 13.11.6d. Upon taking moments about the axis of the wheel gears, Figure 13.11.7a, we find

$$F_2 r_2 = (F_1 + F_4) r_1 \tag{13.11.22}$$

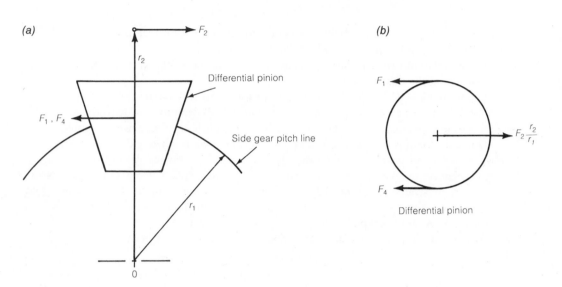

Figure 13.11.7 Forces acting on a differential pinion gear: (a) side view; (b) top view.

Upon taking moments about the center of the differential pinion when it is not accelerating we find that

$$F_1 = F_4$$

so that equation 13.11.22 becomes

$$2F_1 = 2F_4 = F_2 \frac{r_2}{r_1}$$

or

$$T_1 = T_4 = T_2/2 \qquad (13.11.23)$$

Thus, each wheel receives equal torque whenever ω_3 is constant.

Use of only one differential pinion gear instead of two does not affect the above derivation because if we had considered both of them the force on each would have been $F_2/2$ from the ring gear. The forces on the wheel gears would have been $F_1/2$ and $F_4/2$, for no net change in equation 13.11.23.

From these relations we see that during straight line motion each wheel receives the same torque and, since $\omega_1 = \omega_4$ due to road friction, $\omega_3 = 0$ according to equation 13.11.21. Thus, from equation 13.11.20, $\omega_1 = \omega_4 = \omega_2$ and the speed ratio of the wheels relative to the drive shaft is ω_2/ω_5, where ω_5 is the angular velocity of the drive shaft pinion, Figure 13.11.6d.

The automotive differential is used because it allows the vehicle to turn without forcing one wheel to slide. When ω_1 is not equal to ω_4 the differential pinion gears rotate to accomodate the speed differences according to equation 13.11.21. From equation 13.11.20 we see that as one wheel slows down the other speeds up accordingly, so that the relation between them is linear, as it should be for no slip, and the speed of the driveshaft is unaffected by the turn.

Equation 13.11.23 reveals a distinct disadvantage to the differential as described thus far; on slick surfaces the torque on both wheels will drop to that delivered to the wheel on the slipperier surface. Hence, it acts as though both wheels were slipping.

One cure for this is to add a slip clutch outside of each wheel gear and a spring between the wheel gears to engage the clutches and thus provide greater minimum torque whenever the wheels slip. Components of a multiple plate clutch added for this purpose are illustrated in Figure 13.11.5c. (Metal to metal cone clutches have also been used.) This is only a partial cure, however, because it contributes some frictional loss on every turn. A more expensive cure, not yet commonly available, is to engage a clutch only when a wheel slips, as detected by one of the sensors used in antilock braking systems.

13.12   GEAR LUBRICATION

Lubrication recommendations by AGMA depend upon whether the gear train is open or closed. Lubrication recommendations for open gear trains are limited to gears operating at speeds below 3600 rpm or pitch line velocities below 5000 rpm. Moreover, they do not apply to applications in the food and drug industries where the product may come in contact with the gear lubricants.

Four types of lubricants are considered: R & O gear oils, extreme pressure (EP) gear oils, residual compounds–diluent materials, and special compounds.[11] The R & O gear oils are petroleum-based oils which have Saybolt viscosities (defined in Chapter 9) ranging from 626 to 34670 SSU at 100 °F (corresponding to AGMA Lubricant Numbers 4 to 12 inclusive) and from 850 to 1000 SSU at 210 °F (corresponding to AGMA Lubricant Number 13). The low end of this range includes freely flowing oils and the high end of the range includes oils that flow very slowly. Extreme pressure (EP) oils are petroleum-based lubricants similar to the R & O lubricants except they contain special additives (graphite, molybdenum disulfide, lithium, etc.) to increase the film strength of the oil and/or to deposit a film with a low coefficient of friction on the tooth surfaces. Residual compounds–diluent-type lubricants are mineral oils or EP oils in the more viscous grades which have been mixed with a volatile solvent for easy application. After the solvent has evaporated the surface is coated with the original viscous oil. The special compound category contains the remaining lubricants which have properties not yet classified.

These oils are applied in a variety of ways, including immersion of the gears in an oil bath, operating the gears in a stream of oil or under a slowly dripping feed, and intermittent manual oiling with either an oil can or a brush. Extremely viscous oils used to lubricate large, heavily loaded gears, as in drag lines, are usually manually applied with a brush.

Lubricants for enclosed industrial drives are to be used at temperatures below 95 °C (200 °F) and are classified as R & O (rust and oxidation inhibited gear oils), extreme pressure (EP) lubricants, compound gear oils, and synthetic gear lubricants.[21] These R & O oils bear AGMA Lubricant Numbers 1 through 6, corresponding to viscosities from 41.4 to 352 centistokes at 40 °C. The compound gear oils are a mixture of petroleum base 3 to 10% of fatty or synthetic oils and have viscosity ranges from AGMA Lubricant Number 7 to 8A, corresponding to a range of from 414 to 1100 centistokes. Extreme pressure oils in AGMA Lubrication Numbers from 1 to 8A are formed from either R & O or compound gear oils by the inclusion of additives. Synthetic gear lubricants include diesters, polyglycol, and numerous synthetic hydrocarbons, some of which are more stable, have a greater tolerance for impurities, and may operate over a greater temperature range. Methods of application are similar to those for open gearing, dependent upon the viscosity of the oil at the application temperature.

13.13 GEAR MATERIALS

Gears have been made from structural nylon, and similar plastics for relatively light loads and from bronze, cast iron, and steel. Typical steels used for gears include AISI 1045, 4140, 4340, 8620, and 9310 steel. The first of these, the UNS G10450 steel, is fairly soft and is used where the hard surface steels may suffer from case crushing and attendant cracking, Figure 13.13.1, under heavy loads. The softer steel response to heavy loads is to exhibit cold flow and tooth deformation, Figure 13.13.2, rather than

Figure 13.13.1 An example of case crushing; large longitudinal cracks. *Source: ANSI/AGMA 110.04, American Gear Manufacturers Association, Alexandria, VA. Reprinted with permission.*

Figure 13.13.2 Cold flow of a gear made from UNS G10450 (ANSI 1045) steel. *Source: ANSI/AGMA 110.04, American Gear Manufacturers Association, Alexandria, VA. Reprinted with permission.*

TABLE 13.13.1   GENERAL RECOMMENDATIONS ON CHOICE OF STEEL FOR DIFFERENT GEAR SIZES

Pitch of Teeth	Wall Thickness,[a] in.	Hardness[b]	Gear Steels
		Through-hardening	
10–30	$\frac{1}{2}$	200 BHN	1045, 1137, 1335, 4047
		300 BHN	1045, 1060, 3140, 4047
5–15	1	200 BHN	1045, 1060, 3140, 4047
		300 BHN	2340, 3140, 3250, 4140, 4340, 4640
$2\frac{1}{2}$–8	2	200 BHN	1060, 2340, 3250, 4340, 5145, E52100
		300 BHN	2340, 3250, 4340, 4640, 8640, 9840
$1\frac{1}{4}$–4	4	200 BHN	2340, 3250, 4140, 4340, 4640, 9840
		300 BHN	3250, 4340
		Carburizing	
10–30	$\frac{1}{2}$	58 Rockwell C	1015, 1025, 1118, 1320, 4023, E9310
5–15	1	58 Rockwell C	2137, 4620, 6120, 8260, E9310
$2\frac{1}{2}$–8	2	58 Rockwell C	4620, E9310

[a] "Wall thickness" is based on thickest section of rim, web, or shaft that must develop the minimum hardness throughout.
[b] Hardnesses are minimum; 300 BHN would normally be a range of 300 to 350 BHN.
Source: D.W. Dudley, ed., *Gear Handbook* (New York: McGraw-Hill, 1962).

TABLE 13.13.2 GENERAL CHARACTERISTICS OF NONFERROUS GEAR MATERIALS*

Material	Modulus of Elasticity psi × 10^6	Density, lb per cu in.	Ultimate Strength, psi	Yield Strength, psi
Phosphor bronze	15	0.32	40,000	18,000
Manganese bronze	14–18	0.31	65,000	30,000
Aluminum bronze, heat-treated	19.4	0.27	100,000	50,000
Silicon bronze	15	0.31	45,000	20,000
Aluminum alloy	10.6	0.10	68,000	48,000
Zinc alloy	14	0.24	40,000	26,000
Magnesium alloy	6.5	0.065	44,000	32,000
Brass alloy	15	0.306	76,000	45,000
Nylon (73° F)	0.4	0.036	11,200	8,500
Phenolic laminates:				
Paper	1.5 (length) 1.0 (crosswise)	0.042	13,500	10,000
Fabric	1.0 (length) 0.8 (crosswise)	0.042	13,500	10,000

* The values shown in this table are typical. Individual compositions may vary appreciably.
Source: D.W. Dudley, *Gear Handbook* (New York: McGraw-Hill, 1962).

cracking. Hardenable steels, such as the UNS G86200 and G93100, on the other hand, are less subject to scoring and wear, so that the choice of steel is strongly dependent upon the expected service conditions of the gear. Gear size also is a factor in the selection of materials, as indicated in Table 13.13.1.

Typical nonferrous gear materials and some of their properties are shown in Table 13.13.2.

REFERENCES

1. Dudley, D.W. *Evolution of the Gear Art*, American Gear Manufacturers Association, Washington, DC, 1969.

2. Buckingham, E., *Analytical Mechanics of Gears*, McGraw-Hill, New York, 1949, p. 2.

3. *Gear Nomenclature (Geometry)*, ANSI B6. 14-1976 (AGMA 112.05, June 1976), American Gear Manufacturers Association, Washington, DC.

4. *USA Standard System Tooth Proportions for Coarse-Pitch Involute Spur Gears,* ANSI B6. 1-1968 (AGMA 201.02, Aug. 1968), American Gear Manufacturers Association, Washington, DC.

5. *AGMA Standard System Tooth Proportions for Fine-Pitch Involute Spur Gears* (AGMA 207.06, Nov. 1974), American Gear Manufacturers Association, Washington, DC.

6. Blakely, K., Lahey, R., and McLean, D., MSC/Pal: An FE companion for the PC, *Computers in Mechanical Engineering*, 3 (4), 32–41 (1985).

7. Cornell, R.W., Compliance and stress sensitivity of spug gear teeth, *Journal of Mechanical Design, Trans. ASME*, 103, 447–459 (1981).

8.   *Boston Gear Mechanical Products Catalog—No. MP76*, Boston Gear, Quincy, MA, 1976.

9.   *AGMA Standard for Rating the Pitting Resistance and Bending Strength of Spur and Helical Involute Gear Teeth* (AGMA 218.0-Dec. 1982), American Gear Manufacturers Association, Washington, DC.

10.   Coy, J.J., Townsend, D.P., and Zaretsky, E.V., *An Update on the Life Analysis of Spur Gears*, NASA CP-2210, June 9–11, 1981. (Also see Coy, J.J., Townsend, D.P., and Zaretsky, E.V., *Gearing*, NASA RP-1152 AVSCOM TR-84-C-15, Dec. 1985.)

11.   *AGMA Standard for Rating the Pitting Resistance and Bending Strength of Spur and Helical Involute Gear Teeth*, American Gear Manufacturers Association, Arlington, VA, Dec. 1982.

12.   *AGMA Standard Design Procedure for Aircraft Engine and Power Take-off Spur and Helical Gears*, AGMA 411.02, American Gear Manufacturers Association, Washington, DC, Sept. 1966, reaffirmed July 1974.

13.   Orthwein, W.C., Determination of gear ratios, *Journal of Mechanical Design, Trans. ASME*, 104, 775–777 (1982).

14.   Orthwein, W.C., Finding gear teeth ratios with a personal computer, *Computers in Mechanical Engineering*, 4 (1), 36–40 (1985).

15.   *Lubrication of Industrial Open Gearing*, AGMA 251.02, American Gear Manufacturers Association, Washington, DC, 1981.

16.   *Lubrication of Industrial Enclosed Gear Drives*, AGMA 250.04, American Gear Manufacturers Association, Washington, DC, 1981.

PROBLEMS

Section 13.2

13.1   If gears 1 and 2 are properly in mesh, if the pitch diameter of gear 1 is 20 mm, if the center distance is 25 mm, and if gear 1 turns at 575 rpm, what is the rpm of gear 2?
   [Ans. 383.33 rpm.]

Supplementary Problems

13.1   **a**   A gear with a pitch diameter of 1 in. has 32 teeth. Find its diametral pitch, circular pitch, and module. Compare m calculated from $P_d = 1/m$ and $m = d_p/N$. Reconcile the different values.

b   Are any gears having integer values of m (i.e., 1,2,3, . . .) interchangeable with gears having integer values of P_d?

c   What are the outside diameters of two gears in mesh with a speed ratio of 2.5:1 if their center distance is 9.8 in. and if the pinion has 28 teeth? The addendum is $1/P_d$.

13.2   To show that conjugate action is not realized for primative gearing, show that for pin-and-peg gears

$$\frac{\dot{\Omega}}{\dot{\omega}} = \frac{\cos \omega - \dfrac{c}{r} \sin^2 \Omega}{1 + \left(\dfrac{c}{r}\right)^2 - 2\dfrac{c}{r} \cos \omega} \cdot \frac{\sin \omega}{\sin \Omega \cos \Omega}$$

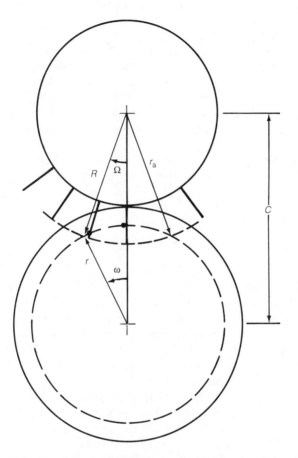

Primative pin-and-peg gears in mesh with pins on gear 1 and pegs on gear 2

using the geometry and notation in the figure in which r is the radius of the peg circle on the upper gear and R is the radius from the center of the lower gear to the contact point between a pin on the lower gear with a peg in the peg circle of the upper gear. Thus, r is constant and R varies with Ω.

13.3 Show that the outside diameter of a gear whose involute teeth have a top land thickness, shown in Figure 13.2.1c, which subtends 20% of the angle subtended by the tooth thickness at the pitch circle may be found from the two equations

$$\frac{2\pi}{5N} + \theta = \gamma - \tan^{-1}\gamma \qquad r_a = r_b(1 + \gamma^2)^{1/2}$$

for given values of N and ϕ.

Section 13.3

13.4   If computer graphics capability is available, use the relations that define an involute curve to construct a single involute tooth to be used on a gear with 35 teeth and having a pitch diameter of 5.0 in., a pressure angle of 20°, a fillet radius of 0.01 in., and an addendum of 0.11 in. Working depth of the tooth is to extend from the base circle to the addendum circle.

13.5   Use the program written for problem 13.4 with a routine to increment the polar angle to construct the entire gear. Locate the pitch circle, the addendum circle, the dedendum circle, and the root circle.

13.6   Calculate the center distance if the gear in problem 13.4 is to mesh with a gear having a pitch diameter of 20 in. Will undercutting occur if the addendum is allowed to increase until the teeth become pointed (until the top land is zero)? *Note:* Pointed teeth should never be produced even for very small loads.

13.7   Repeat problem 13.4 for a gear having a pitch diameter of 125 mm, a fillet radius of 1.0 mm, and an addendum of 10 mm. Compare this tooth depth with that of a tooth of standard addendum and dedundum.

13.8   Repeat problem 13.6, but for the gears described in problem 13.7.

13.9   Calculate the maximum addendum for module 5 gear teeth if they are to not extend into region 3, Figure 13.2.2. The pinion pitch diameter is 95 mm, the gear has 30 teeth, and the pressure angle is 14.5°. Can teeth of standard proportion be used for this pressure angle? Why?

$m = 5mn$

13.10   Locate a pair of small spur gears with involute teeth and mark the face of several teeth of one gear with chalk lines (or lines with any material that will smear when rubbed and may be later cleaned from the tooth face). Place the two gears on shafts in a frame so they are in mesh and then rotate them and notice that the lines are blurred in a direction toward a strip where no smuding is found. The region of no blurring indicates that sliding on each tooth face is from the dedendum to the pitch circle and from the addendum to the pitch circle.

13.11   Calculate the backlash for two similar gears in mesh if the center distance is incorrectly adjusted to 181 mm instead of 180 mm. Each gear has 30 teeth, a 90-mm pitch diameter, and a pressure angle of 14.5°. Assume no backlash when the center distance is 180 mm.

13.12   Show that if the pressure angle changes by

$$\Delta\phi = \frac{\Delta c}{c \tan \phi}$$

then the center distance is changed by an amount Δc.

Sections 13.4 and 13.5

13.13   Calculate the number of teeth of standard proportion in contact for a pinion having a pitch diameter of 120 mm and a gear having a pitch diameter of 300 mm for an $m = 6$ tooth and compare with the contact ratio for $m = 3$ and $m = 4$ teeth for a 20° pressure angle. Also find the thickness of the top land on the pinion for each of these tooth sizes.

[*Ans.* $N_c = 1.656$ for $m = 6$, 1.784 for $m = 3$, 1.737 for $m = 4$.]

13.14 Calculate the contact ratio for the gears and tooth sizes listed in problem 13.13 when nonstandard tooth proportions are used where the addendum extends either to the base circle of the other gear or to the distance where the top land is $0.05p_c$. Explain why the contact ratio for the $m = 6$ tooth for the nonstandard gear is less than the contact ratio for the standard tooth.

[*Ans.* $N_c = 1.586$ for $m = 6$, 1.998 for $m = 4$, 2.372 for $m = 3$.]

13.15 Show that the contact ratio for a rack and pinion is given by

$$p_b N_c = (r_{a_1}{}^2 - r_{b_1}{}^2)^{1/2} - r_{p_1} \sin \phi + \frac{a_r}{\sin \phi}$$

where

$$N_c \le \frac{(r_{a_1}{}^2 - r_{b_1}{}^2)^{1/2}}{p_b}$$

and where $a_r \le r_{p_1} \sin \phi$ must hold if undercutting is to be avoided.

13.16 Select a 20° pressure angle, standard tooth proportions, and a tooth size in SI units for a pinion having a pitch diameter of 95 mm and a gear having a pitch diameter of 150 mm. Calculate the contact ratio. Compare with problem 13.9. (*Hint:* Calculate the maximum addendum to the interference point, compare with addendum for teeth of standard proportion, select module, and calculate the contact ratio.)

13.17 Repeat problem 13.16 using nonstandard tooth proportions in which the maximum length addendums are used subject to the condition that the top land width be no smaller than $0.4p_c$. (*Hint:* After calculating the maximum addendum to the interference point for each gear calculate the minimum top land thickness in terms of subtended angle and solve for ζ_a, use a bisection program to solve (13.3.4) for γ_a with $\zeta = \zeta_a$, calculate r_a, and compare with maximum addendum found above. Use admissible addendums to find the contact ratio.)

13.18 List and explain two effects of increasing the pressure angle on the teeth selected in problem 13.17.

13.19 Compare the contact ratio for a gear with a 50-mm pitch diameter when in mesh with an external gear having a 150-mm pitch diameter and module 2 teeth with that when the gear with a 50-mm pitch diameter is in mesh with an internal gear that also has a 150-mm pitch diameter and module 2 teeth. Use a 20° pressure angle and standard tooth proportions. Recall that for internal gears $r_a = r_p - a$, where a denotes the addendum.

Section 13.6

13.20 Find the minimum number of teeth for special 15°, 22°, and 28° pressure angle teeth if the addendum is to be given by 1.1 m or $1.1/P_d$.

13.21 Just as undercutting for an external gear has been referenced to its meshing with a rack, so undercutting for an internal gear with N teeth may be referenced to its mesh with an external gear with $N - 1$ teeth. Examination of Figure 13.4.4 will show that undercutting may be avoided if δ in equation 13.4.7 is equal to or greater than zero. By setting δ equal to zero show that the minimum number of teeth is given by the relation

$$N = \frac{2a}{m \sin^2 \phi} \left[1 + \left(1 + \sin^2 \phi - \frac{m^2}{4a^2} \sin^4 \phi \right)^{1/2} \right]$$

Also show that the minimum number of teeth of standard proportion for an internal gear to prevent undercutting for pressure angles of 14.5°, 20°, and 25° are 64, 35, and 23, respectively. In this expression a denotes the rack addendum. Also write the above equation in the Old English system; i.e., in terms of P_d. (*Hint:* Justify setting δ to zero in relation 13.4.7, let $d_{p_1} = (N-1)m$, $d_{p_2} = Nm$, and solve for N.

Section 13.7

13.22 Show that the radial force on a gear transmitting hp horsepower at n rpm is given by

$$F_r = (396\,000\ \mathrm{hp}\ \tan \phi)/(\pi n d_p)\ \mathrm{lb} \qquad (d_p \text{ in inches})$$

using the notation of this chapter.

13.23 Show that the radial force of a gear transmitting kW kilowatts at n rpm is given by

$$F_r = (6 \times 10^7\ \mathrm{kW}\ \tan \phi)/(\pi n d_p)N \qquad (d_p \text{ in mm})$$

using the notation of this chapter.

13.24 Find the radial forces (see problem 13.22), the center distances, and the rotational speeds for the gear train shown in the figure. Input power is 30 hp at 600 rpm, the input pinion is to have 26 teeth, the output gear is to have 48 teeth, and each gear pair is to reduce the speed by a factor of 1.50. All gear teeth have a standard profile, have $P_d = 2$ and a 20° pressure angle. Assume no power losses.

[*Ans.* $d_{p_1} = 13$ in., $d_{p_2} = 19.5$ in. $d_{p_3} = 16$ in., $d_{p_4} = 24$ in., $F_{r_1} = F_{r_2} = 176.46$ lb, $F_{r_3} = F_{r_4} = 215.06$ lb.]

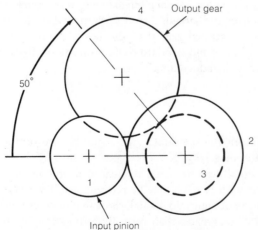

Input pinion
(26 teeth, 30 hp, 600 rpm)

13.25 Use the Lewis criterion with the Barth correction to determine the face width for the gear train described in problem 13.24 to transmit 30 hp if the working stress is limited to 6000 psi. Assume only one tooth from each gear is in contact.

13.26 a Show that the maximum addendum without undercutting for gears 1 and 2 when they are in mesh is given by (problem 13.24)

$$
\left.
\begin{aligned}
a_{\max_1} &= (c^2 \sin^2 \phi + r_{b_1}{}^2)^{1/2} - r_{p_1} \\
a_{\max_2} &= (c^2 \sin^2 \phi + r_{b_3}{}^2)^{1/2} - r_{p_2}
\end{aligned}
\right\}
\quad \text{for } c = r_{p_1} + r_{p_2}
$$

b Show that if the top lands of gears 2 and 4 in problem 13.24 subtend an angle of $0.3\pi/N$ no undercutting will occur and that the addendum for gear 2 is 0.6279 in. and for gear 4 is 0.6433 in.

c Calculate the contact ratios for gears 1 and 2 when they have the same addendum and for gears 3 and 4 when they have the same addendum. The addendums were calculated in problem 13.26b.

d Calculate the tooth width for the high contact ratio gears in problem 13.26 when loaded as in problem 13.24.

13.27 Find the ratio of the contact length (a number) for a pinion with 26 teeth, $P_d = 8$, when in contact with an external gear having 72 teeth and in contact with an internal gear having 72 teeth for a pressure angle of $20°$ and teeth of standard proportion.

Section 13.8

13.28 Find T_{10}, G_{10}, and L_{10} for two gears where $N_1 = N_2 = 28$, pitch diameters are 89.0 mm, addendum diameters are 95.2 mm, pressure angle is $20°$, face width is 2.79 mm, and the force tangent to the pitch circle is 1970.88 N.

[*Ans.* F on pressure line is 2097.37 N, $T_{10} = 38.45 \times 10^6$ cycles, $G_{10} = 10.14 \times 10^6$ cycles, and $L_{10} = 7.62 \times 10^6$ cycles.]

13.29 Estimate the L_{10} life of a pinion and gear with standard teeth where the pinion has 21 teeth, the gear has 36 teeth, and the pitch diameter of the pinion is 3.00 in. The face width is 0.75 in., the pressure angle is $20°$, and the gears are to transmit 18.5 hp at a pinion speed of 1050 rpm.

[*Ans.* $L_{10} = 1863.0$ million cycles.]

13.30 Repeat problem 13.29 with a pressure angle of $14.5°$

[*Ans.* $L_{10} = 1112.5$ million cycles. This explains decreased use of $14.5°$, teeth for power transmission.]

13.31 Repeat problem 13.29 with an addendum of 0.15 in. on each tooth.

[*Ans.* $L_{10} = 2214.4$ million cycles. This justifies the interest in nonstandard teeth.]

13.32 Design a high commercial quality gear and pinion with standard teeth to transmit 50 kW at a pinion speed of 950 rpm and a speed ratio of 1.5 revolutions of the pinion to 1.0 revolution of the gear. The pinion diameter should be no larger than 200 mm and the L_{10} life of the gear set should be 15 years at 6000 h/yr. Use the Lewis-Barth equation to select the initial tooth size and face width, which should be no more than 6 times the tooth thickness at the pitch circle. The largest integer in the contact ratio is the minimum number of teeth in contact. Finally, calculate the L_{10} life based upon this width and increase w if necessary to achieve the required life. A 20° pressure angle is preferred and the maximum safe static stress for the steel used is 132 MPa.

Section 13.9

13.33 Calculate the number of teeth on a gear and pinion to achieve a speed ratio of 2.156 to within 0.001.

13.34 Calculate the number of teeth on each of four gears to obtain a gear ratio of 4.3874 to within 0.0002. The preferred gear train has no ratio between gears in mesh greater than 2.5.

13.35 Find a gear train to provide a speed ratio of 1.486 700 62. Factor the numerator and denominator to obtain meshing gear pairs having tooth ratios less than 1.25 if possible. Sketch the gear train and show the numbers of teeth on each gear. An accuracy of 0.000 000 003 or better is required. *Note:* Do not truncate the ratio. Use of 1.4867 will produce gears with more teeth. Try both to demonstrate this.

13.36 Find a gear train to provide a speed ratio of π to 5 places to within an accuracy of 1×10^{-5}.

13.37 Design a compact gear train to produce a speed ratio of $2.095\,860\,566 \pm 1.0 \times 10^{-9}$ using module 4 teeth with a 25° pressure angle and standard proportions. Assume all shafts are 50 mm in diameter, that they have support bearings on each side of the housing, and that a 10-mm clearance is required between shaft and gear outside diameters.

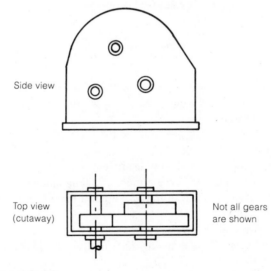

Side view

Top view
(cutaway)

Not all gears
are shown

Section 13.10

13.38 Calculate the speed ratio for the planetary gear train shown in Figure 13.10.3a.

13.39 Calculate the speed ratio for the planetary gear train shown in Figure 13.10.1. This is the planetary gear train used in the power wheel. Its planet gears are attached to the nonrotating base plate so that the planet gears rotate but do not translate.

13.40 Calculate the speed ratio for the planetary gear train shown in Figure 13.10.4a.

13.41 Calculate the speed ratio for the planetary gear train shown in Figure 13.10.5b using the rotational, or kinematic, method described with the aid of Figure 13.11.2. Number the gears as in the figure where

gear 1 = sun gear   gear 4 = largest planet gear

gear 2 = smallest planet gear   gear 5 = smaller ring gear

gear 3 = larger ring gear   gear 6 = intermediate planet gear

13.42 Repeat problem 13.41 using the velocity method as demonstrated in obtaining equations 13.11.4 and 13.11.5. Show that this method is simpler and that the results agree with those obtained in problem 13.41 by writing the speed ratio in terms of the gear radii and using the relations that

$$r_3 = r_1 + r_2 + r_4$$

$$r_5 = r_1 + r_2 + r_6$$

13.43 Calculate the speed ratio for the planetary gear train shown in the figure, with no ring gear but two sun gears and two sets of planet gears.

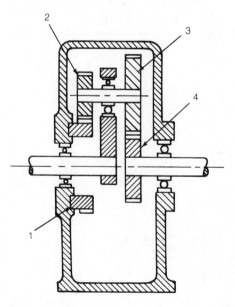

Planetary gear train with no ring gear but two sun gears and two sets of planet gears

13.44 Show that for a planetary gear train as illustrated in Figure 13.10.1 the ratio of the pitch diameter of the planetary gear to the sun gear is related to the speed ratio SR according to

$$\frac{d_1}{d_0} = \frac{1}{2}(SR - 1)$$

when the shafts of the planet gears are held motionless. (As when they are attached to a rigid frame or when the carrier is not allowed to rotate.)

13.45 Design a planetary gear train of the configuration used in Figure 13.11.1 to transmit 30 hp at an input shaft speed of 1000 rpm and an output shaft speed of 200 rpm. Use the minimum diameter ring gear consistent with the requirement that its tooth face width not exceed 2 in. and that the material working stress not exceed 12 000 psi for any of the gear teeth. Use as many planets gears as may be contained between the ring and sun gears. Use standard proportion teeth with a 20° pressure angle and a minimum of 18 teeth in the sun gear and in the planet gears. Use the Lewis-Barth relation to calculate the tooth strength of the sun and planet gears and assume that the corresponding ring gear teeth will also have adequate strength. When using the Lewis-Barth formula first assume that only one tooth from each planet gear will transmit its torque to the ring gear. Will the tooth width change if we calculate the contact ratio and assume that the load will be evenly divided among the integer number of teeth equal to, or less than, the contact ratio (i.e., if the contact ratio is 2.85, use 2 teeth)?

13.46 Find the speed ratio ψ/ϕ for the planetary gear train shown in the figure. Internal gear number 4 is stationary, gear 1 delivers power from the input shaft to the gear train, and ring (internal) gear 5 delivers power from the gear train to the output shaft. All teeth are the same size. Planet gears 2 and 3 are keyed to the same shaft which is free to rotate in the bearing in the carrier, which is not connected to either the input or the output shafts. Use of the velocity method is recommended.

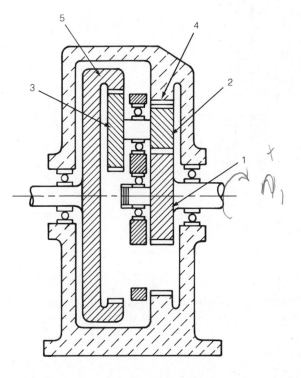

CHAPTER FOURTEEN

STRAIGHT BEVEL, HELICAL, AND WORM GEARS

NOTATION

A	area (l^2)		p_{t_g}	transverse pitch, worm gear (l)
C_Γ	thermal cooling constant $(m/t^2\,^\circ F)$		p_{t_w}	transverse pitch, worm (l)
c	constant (1)		r	shaft radius, generic radius (l)
c	center distance (l)		r_c	average collar radius $[r_c = (OD + ID)/2]$ (l)
d_a	addendum diameter (l)		r_f	radius of the front cone of a bevel gear (l)
d_r	diameter of rear cone (l)		r_o	mean radius, bevel gear (l)
d_p	pitch diameter (l)		r_p	pitch radius (l)
d_{p_e}	equivalent diametral pitch (l)		r_r	radius of the back cone, bevel gear (l)
d_{p_g}	pitch diameter of worm gear (l)		T	torque (ml^2/t^2)
d_{p_w}	pitch diameter of worm (l)		T_g	torque available at the gear (ml^2/t^2)
E	modulus of elasticity (Young's modulus) (m/lt^2)		T_w	torque applied to the worm (ml^2/t^2)
F	force (ml/t^2)		v	pitch line velocity (l/t)
F_a	axial force (ml/t^2)		v_n	velocity component normal to a tooth (l/t)
F_n	force normal (perpendicular) to a tooth face (ml/t^2)		w	tooth width (l)
			x	coordinate (l)
F_r	radial force (ml/t^2)		y	Lewis factor or a coordinate (1), (l)
F_{st}	tangential force to equivalent spur gear (ml/t^2)		z	coordinate (l)
F_t	tangential force (ml/t^2)		β	angle of lateral force reaction on a worm (l)
f	friction force (ml/t^2)		δ	deflection (l)
H_d	heat dissipated (ml^2/t^2)		Γ	temperature $(^\circ)$
I	moment of area in bending (l^4)		γ	pitch angle (1)
$\hat{i}, \hat{j}, \hat{k}$	unit vectors in directions x, y, z, respectively		κ	worm friction angle (1)
l	length (l)		λ	lead angle (1)
M	moment (ml^2/t^2)		μ	friction coefficient (1)
m	module (l)		μ_c	friction coefficient at worm thrust collar (1)
N	number of teeth (1)		ρ	radius of curvature (l)
N_g	number of teeth on a gear (worm gear) (1)		σ	direct (tensile or compressive) stress (m/lt^2)
N_w	number of teeth (threads) on a worm (1)		τ	shear stress (m/lt^2)
N_v	virtual number of teeth (1)		ϕ	pressure angle (1)
n	rotational speed in rpm $(1/t)$		ψ	helix angle (1)
P	power rating (ml^2/t^3)		Ψ	angle between shafts (1)
P_d	diametral pitch (l)		ω	angular velocity $(1/t)$
p_a	axial pitch (l)			
p_c	circular pitch (l)		C_m	ratio correction factor $(-)$
p_n	normal pitch (l)		C_s	materials factor $(-)$
p_s	pseudo-pitch, worm gear (l)		C_v	velocity factor $(-)$
p_t	transverse pitch (l)		u	units conversion factor $(C_s C_m C_v/u \sim mt^{-2}l^{-0.8})$

Chapter 13 was concerned with spur gears and power/motion transfer between parallel shafts. In this chapter we consider other tooth types and power/motion transfer between nonparallel shafts. Since many more gear designs have been developed for smooth power transfer between parallel and nonparallel shafts than can be discussed within the page limits allotted, only three designs will be considered in detail, based upon their analytical simplicity and wide usage: straight bevel, helical, and worm. Several of the other more common designs will be mentioned, however, to give the reader some idea of the range of choices available.

14.1 BEVEL GEAR TERMINOLOGY AND VARIATIONS

Bevel gears are those designed to induce nonslip motion between two cones, termed pitch cones, which are in contact along a generator of each, as shown in Figure 14.1.1, so that the axes of rotation intersect at the common vertex of each cone. (A generator is defined as a line from the apex of the cone to its base. It defines the surface of the cone as it makes a complete circuit about the base.)

The diameters of the base of the pitch cones, whose radii are denoted by r_{p_1} and r_{p_2} in Figure 14.1.1, are defined to be the pitch diameters of the bevel gears. Radii associated with the generators of the front and back cones and selected AGMA recommended terminology for bevel gears are also shown in Figure 14.1.1. The front and back cone surfaces are each perpendicular to the surface of the pitch cone. The teeth are perpendicular to these cones and extend from the front to the back cone, as shown in Figure 14.1.1.

Following the convention introduced in the discussion of spur gears, the smaller of two bevel gears is again known as the pinion and the larger as the gear. If the two pitch cones for a pair of bevel gears in mesh have the same included angle and if the teeth have the same width the bevel gears are identical and are known as *miter* gears.

Bevel gears with straight involute teeth, which if extended would intersect at the vertex of each pitch cone, are known as straight bevel gears, Figures 14.1.2 and 14.1.3, teeth are crowned in a manner similar to that employed for spur gears and for the same reason: to account for manufacturing and misalignment errors.

Skew bevel gears employ straight teeth whose extensions are tangent to a circle on the mating bevel gear as shown; i.e., they are angled relative to the axis of rotation of the gear blank on which they are cut. *Spiral* bevel gears differ from skewed bevel gears in that the teeth are both skewed and curved. *Zerol* bevel gears have curved teeth with a radial chord, Coniflex bevel gears have slightly convex tooth faces, and Formate and Revacycle bevel gears have special tooth profiles that will not be described here.

If the axes of rotation are offset as pictured at the bottom of Figure 14.1.3, if a conical pinion is used, and if the teeth on the pitch cone are curved as shown in Figure 14.1.4 the resulting gears are termed *hypoid* gears. These gears were first widely used by the automotive industry in their effort to lower the driveshaft below the floor of

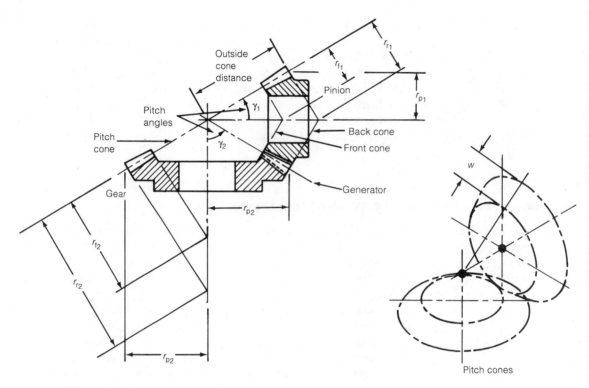

Figure 14.1.1 Bevel gear terminology. *Source: AGMA 112.05 June 1976, American Gear Manufacturers Assn., Alexandria, VA. Reprinted with permission*

the passenger compartment in rear drive vehicles. See Ref. 2 for additional details on the design of hypoid gears.

Crown gears are bevel gears in which one of the pitch cones has been deformed into a plane, with the crown gear teeth extensions meeting at a point as illustrated in

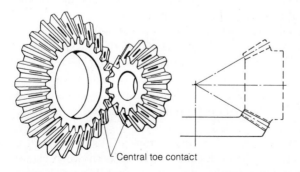

Central toe contact

Figure 14.1.2 Straight bevel gears showing correct tooth contact pattern. Contact areas are indicative of the extent of crowning. *Source: AGMA 331.01 Nov. 1969, American Gear Manufacturers Assn., Alexandria, VA. Reprinted with permission.*

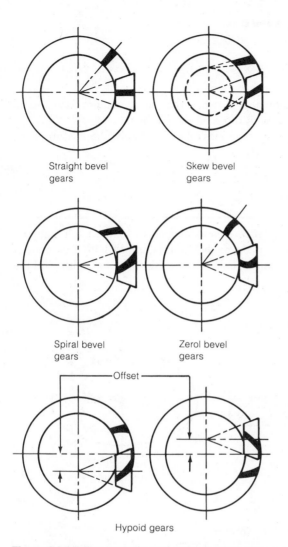

Straight bevel
gears

Skew bevel
gears

Spiral bevel
gears

Zerol bevel
gears

Offset

Hypoid gears

Figure 14.1.3 Five varieties of bevel-type gears. Note the change in tooth curvature of the hypoid gears for positive and negative offset. *Source: AGMA 112.05 June. 1976, American Gear Manufacturers Assn., Alexandria, VA. Reprinted with permission.*

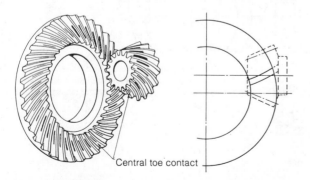

Central toe contact

Figure 14.1.4 Detail of hypoid gears with a right-hand pinion and showing the correct tooth contact pattern. *Source: AGMA 331.01 Nov. 1969, American Gear Manufacturers Assn., Alexandria, VA. Reprinted with permission.*

Figure 14.1.5. Use of a spur gear pinion also shown in that figure, requires that special teeth be cut on the gear disk to mesh with those on the pinion.

The speed ratio for straight bevel gears in mesh may be simply obtained from the geometry of the pitch cones, Figure 14.1.1, as

$$\frac{n_1}{n_2} = \frac{d_{p_2}}{d_{p_1}} = \tan \gamma_2 = \text{ctn} \, \gamma_1 \qquad (14.1.1)$$

where the diameter of the base of the *pitch cone* is defined to be the pitch diameter of the gear and where γ is the cone half angle, as shown in Figure 14.1.1. Gear catalogs often list bevel gears according to pressure angle of the teeth, the face width, and the

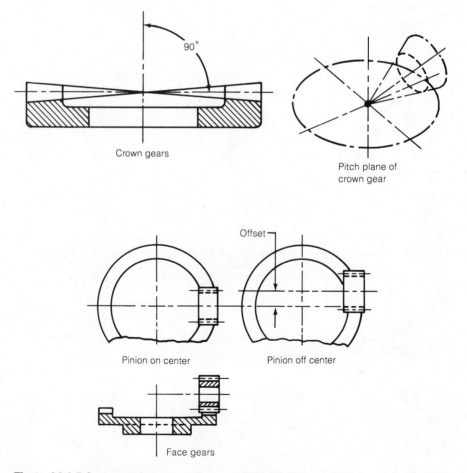

Figure 14.1.5 Crown and face gears. *Source: AGMA 112.05 June. 1976, American Gear Manufacturers Assn., Alexandria, VA. Reprinted with permission.*

pitch diameter. Face width w is given by

$$w = \frac{r_{r_1} - r_{f_1}}{\tan \gamma_1} = \frac{r_{r_2} - r_{f_2}}{\tan \gamma_2} \qquad (14.1.2)$$

in terms of the radii of the front and rear cones. The *rear cone* radius is related to the pitch diameter according to

$$r_{r_1} = \frac{d_{p_1}}{2 \cos \gamma_1} \qquad r_{r_2} = \frac{d_{p_2}}{2 \cos \gamma_2} \qquad (14.1.3)$$

The term rear cone is not standard. It has been used in place of *back cone* to designate its distance by r_r rather than r_b, which may be confused with the base radius.

14.2 VIRTUAL TEETH FOR STRAIGHT BEVEL GEARS

The linear reduction of tooth size from the back cone to the front cone of a bevel gear increases the difficulty of presenting a closed formula for the stresses within a tooth of one gear as induced by a tooth from the other bevel gear. Lewis attempted to account for the decreasing tooth size by associating the number of teeth listed in the table of y values with the virtual number of teeth of a straight bevel gear and by introducing a multiplicative term involving the pitch diameters at the front and back (rear) pitch cones.[3] We shall use this analysis for estimating tooth strength in the next section, with the understanding that a finite element analysis should be employed for a more realistic estimate of tooth stresses.

The relation for the virtual number of teeth may be derived by first noting that the circular pitch is defined to lie in the base plane of the back cone. If we temporarily omit the 1 and 2 subscripts, since these relations hold for all straight bevel gears, we have

$$p_c = \frac{\pi d_p}{N} = \frac{\pi d_r \cos \gamma}{N} \qquad (14.2.1)$$

If this circular pitch were associated with a spur gear having pitch radius r_r, which is the length of the back cone generator that is perpendicular to the axis of the tooth, the relation would become

$$p_c = \frac{\pi d_r}{N_v} \qquad d_r = 2r_r \qquad (14.2.2)$$

where N_v is defined as the *virtual* number of teeth for the bevel gear. Upon equating (14.2.1) and (14.2.2) and solving for N_v we find that

$$N_v = \frac{N}{\cos \gamma} \qquad (14.2.3)$$

From another point of view, N_v is the number of teeth in a fictitious spur gear made by imagining the rear cone of the bevel gear to be cut along a generator, the cone opened up and spread flat to form a sector of a spur gear, and sufficient teeth added to complete the pitch circle of the spur gear, as in Figure 14.2.1

Customarily the tooth size for bevel gears is described just in terms of the diametral pitch of the tooth profile at the base of the pitch cone which is given by

$$P_d = \frac{N_v}{d_r} = \frac{N/\cos\gamma}{d_p/\cos\gamma} = \frac{N}{d_p} \qquad m = \frac{d_p}{N} \qquad (14.2.4)$$

which holds for any straight bevel gear.

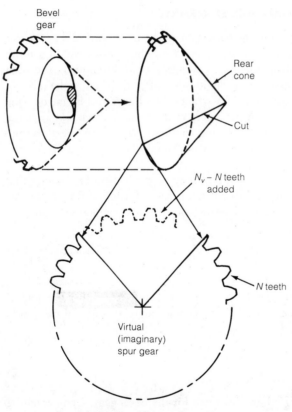

Figure 14.2.1 Development of the rear cone of a bevel gear to produce the associated virtual spur gear.

14.3 TOOTH STRESS AND INDUCED FORCES AND MOMENTS FOR STRAIGHT BEVEL GEARS

Since the virtual number of teeth found from (14.2.6) is generally not an integer, the corresponding y value to be substituted into the Lewis equation for the beam strength of a bevel gear tooth is usually obtained from Table 13.7.1 by interpolation and then substituted into the relation[3]

$$\frac{\sigma w p_c}{F_t} = 3 \frac{d_p{}^2(d_p - d_p')}{y(d_p{}^3 - d_p'{}^3)}$$

(14.3.1)

where d_p' is the pitch radius at the small end. It is given by

$$d_p' = d_p - 2w \sin \gamma$$

(14.3.2)

in terms of face width w. Note that since the expression on the right side of equation 14.3.1 is independent of the units used its form is the same for either Old English or SI units.

Although not explicitly stated by Buckingham in his review of the Lewis equation for bevel gears, it would appear that to be cautious the force F_t in this calculation of the tooth strength should be the load acting at the midpoint of the tooth face rather than at the base of the pitch cone.[3] Reduced tooth rigidity at the inner end of the tooth was cited by Buckingham as the reason for recommending that the face width of the bevel gear teeth be held to a minimum and that under no circumstances should it be greater than one third of the cone distance.

The AGMA formula for tooth strength will not be discussed because it is similar in form to equation 13.7.3 and involves at least one coefficient that is to be selected on the basis of the designer's experience.

The force F_n perpendicular to the load-bearing face of a tooth on a bevel gear has components as shown in Figure 14.3.1b. Since this normal (i.e., perpendicular) force lies in a plane perpendicular to the tooth face, it follows that its component in a plane containing the axis of rotation of the gear must be as shown in Figure 14.3.1a and that its mutually perpendicular components must be as drawn. Thus, the axial, radial, and normal forces are given by

$$F_a = F_t \tan \phi \sin \gamma$$

(14.3.3)

$$F_r = F_t \tan \phi \cos \gamma$$

(14.3.4)

$$F_n = \frac{F_t}{\cos \phi}$$

(14.3.5)

where F_a is the axial force, F_r the radial force, and F_t is the tangential force. Since F_a was assumed to act at the midpoint of the tooth face, it also contributes to a bending

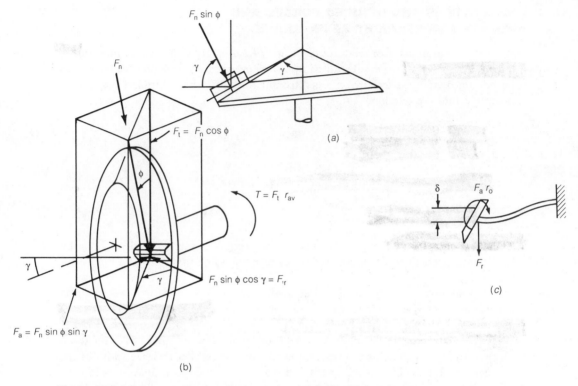

Figure 14.3.1 Force components acting on an involute tooth of a bevel gear.

moment in the supporting shaft, given by

$$M = F_a r_o \tag{14.3.6}$$

where r_o is given by

$$r_o = r_p - \frac{w}{2} \sin \gamma \tag{14.3.7}$$

Observe that the bending moment tends to reduce the shaft deflection because it acts to oppose the bending moment induced by the lateral force F_r, as shown in Figure 14.3.1c. Consequently the lateral deflection of a cantilevered shaft attached to a bevel gear may be approximated by

$$\delta = \frac{4F_t}{\pi E r^4} \left[\frac{l^3}{3} \cos \gamma - \frac{l^2}{4} (d_p - w \sin \gamma) \sin \gamma \right] \tan \phi \tag{14.3.8}$$

The forces and moments given above are those to be considered in selecting the shaft size and in choosing the bearings to be used with bevel gears.

EXAMPLE 14.3.1

Design a bevel gear set with a $20°$ pressure angle teeth and a speed ratio of 2.1:1 in which the pinion rotates at 690 rpm at an input of 150 hp. The gear material has a yield stress of 90 000 psi and the safety factor is to be 2.0. Tooth width should be approximately one-fourth the outside cone distance. See Figure 14.1.1.

From equation 14.1.1. $\tan \gamma_2 = 2.1$ so that

$$\gamma_1 = 25.4633°$$

$$\gamma_2 = \frac{\pi}{2} - \gamma_1 = 64.5367°$$

so that from (14.3.1) it follows that with $w \sin \gamma = d_p/8 \quad (d_p' = 0.75 \, d_p)$

$$\frac{\sigma w p_c}{F_t} = \frac{3(0.25)}{y(1 - 0.75^3)} = 1.2973/y$$

which may be used to find wp_c. Stress is given by $90\,000/2 = 45\,000$ psi and the tangential force component may be found from

$$F_t = \frac{198\,000}{\pi r_m n} \, \text{hp} = \frac{2(150)198\,000}{\pi 690 d_m} = \frac{27\,402}{d_m} \, \text{lb} \qquad d_m = d_{p1} - w \sin \gamma$$

so

$$wp_c = \frac{1.2973}{y} \frac{F_t}{\sigma} = \frac{1.2973(27\,402)}{45\,000 y d_m} = \frac{0.7900}{y d_m}$$

We also require that $N_1 = N_{v_1} \cos \gamma_1 > 18 \cos 25.4633° = 16.25$ to prevent undercutting. Because $n_1/n_2 = 2.1$, however, we must select $N_1 = 20$ to obtain $N_2 = 42$, and integer. Thus, from (14.2.1)

$$w d_{p1}(d_{p1} - w \sin \gamma) = \frac{N_1}{\pi} \frac{0.7900}{y}$$

From the tooth restriction that $w \sin \gamma = r_p/4$

$$w = \frac{d_p}{8 \sin \gamma} = \frac{d_p}{8 \sin 25.4633°}$$

so that

$$(d_p)^3 = \frac{20}{\pi} \frac{0.7900}{0.2544} \frac{1}{y} = \frac{19.769}{y}$$

Since y depends upon the virtual number of teeth, calculate that

$$N_v = \frac{N}{\cos \gamma} = \frac{20}{\cos 25.4633°} = 22.1518$$

and use this value and linear interpolation from Table 13.7.1 to obtain $y = 0.1052$. Thus, $d_{p1} = 5.7278$ and

$$P_d = \frac{N}{d_{p1}} = \frac{20}{5.7278} = 3.492 \text{ in.}^{-1}$$

If standard, integer value teeth are to be used, select a slightly larger tooth for which $P_d = 3.0$. Thus, $d_{p1} = 6.6667$ in., $d_{p2} = 14.000$ in., and $w = 1.938$ in. Substitution of the y value for 20 teeth into the expression for the stress sensitivity yields

$$\frac{\sigma w p_c}{F_t} = \frac{1.2973}{0.1052} = 12.332$$

The tangential force F_t may be calculated after finding d_m from

$$d_m = d_p - w \sin \gamma = 6.6667 - 0.8333 = 5.8333 \text{ in.}$$

Upon solving the stress sensitivity for σ after substitution of the above value for w along with $p_c = \pi/P_d = 1.0472$ in. and $F_t = 27\,402/5.8333 = 4697.5$ lb we find that $\sigma = 28\,544$ psi.

14.4 INVOLUTE HELICAL GEARS

As the name implies, the teeth on helical gears are involutes formed by the diagonal end of a sheet of paper as it is unwound from its base circle, as shown in Figure 14.1.1a. Thus, either end of the sheet of paper plays the role of the taut cord used to describe the involute for a spur gear. Right- and left-hand teeth are defined to lie along helices that correspond to right- and left-hand threads of a screw coincident with the axis of the cylinder. A right-angle gear train that uses both helical and bevel gears is shown in Figure 14.4.2. In it the large gear at the far end is right-handed while the small gear in mesh with it is left-handed.

Although helical gears are more expensive to produce than spur gears, they have the advantage of having more teeth in contact for the same pitch diameter and the same tooth size (same diametral pitch or module) when they are on parallel shafts than do

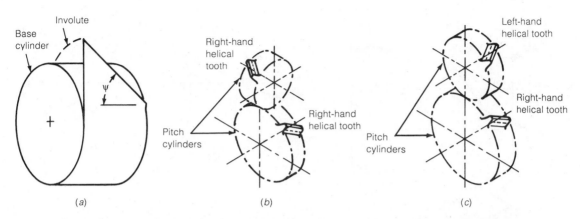

Figure 14.4.1 (a) Involute surface formed by unwrapping a taut paper strip from the base cylinder; (b) crossed helical gears; (c) parallel helical gears. *Source: AGMA 112.05 June 1976, American Gear Manufacturers Assn., Alexandria, VA. Reprinted with permission.*

spur gears. Consequently, they can transmit more power for a given pitch diameter and can do it more smoothly and quietly. These are the reasons for their use in better quality gear trains, such as that shown in Figure 14.4.2.

Another advantage of helical gears is that they may be used to transfer power between nonparallel shafts that do not intersect, as illustrated in Figure 14.4.1c. However,

Figure 14.4.2 Gear train using helical and bevel gears. *Source: PT Components, Inc., Link-Belt Drive Division, Indianapolis, IN.*

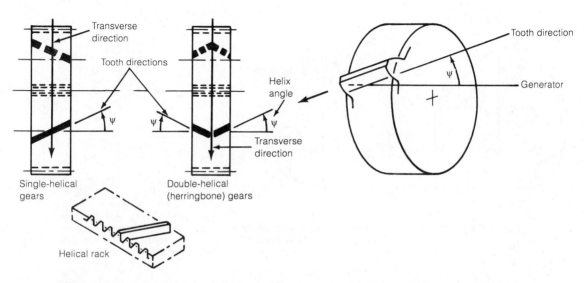

Figure 14.4.3 Helix angle, tooth direction, and transverse direction for helical and herringbone gears.

as will be shown in the next section, helical gears gain these benefits at the expense of inducing an axial force on their shafts, so that each must be fitted with bearings that can resist thrust. This disadvantage may be overcome with still more expensive machining to produce herringbone (or double-helical) gears, shown in Figure 14.4.3.

The angle between the direction of the tooth and a generator of the pitch cylinder that passes through the tooth as shown in Figure 14.4.3 is known as the helix angle. Because of the helix angle the distance between similar points on adjacent teeth is commonly described by three different pitches: the normal circular pitch p_n, the transverse circular pitch p_t, and the axial pitch p_a, as illustrated in Figure 14.4.4. They are defined and related to one another as

$$p_t = \frac{\pi d_p}{N} = \frac{\pi}{P_{d_t}} = \pi m_t \qquad (14.4.1)$$

$$p_n = p_t \cos \psi \qquad (14.4.2)$$

$$p_a = \frac{p_t}{\tan \psi} = \frac{p_n}{\sin \psi} \qquad (14.4.3)$$

where N is the actual number of teeth on the gear and d_p is the diameter of the pitch cylinder. It is the normal pitch p_n that determines the tooth size required for two helical gears to properly mesh. The relation between p_n and either the normal diametral pitch or the normal module is

$$P_{d_n} = \frac{\pi}{p_n} \qquad m_n = \frac{p_n}{\pi} \qquad (14.4.4)$$

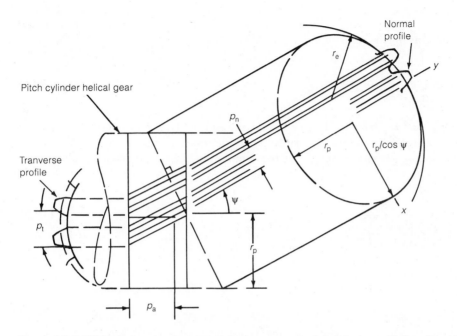

Figure 14.4.4 Helical gear showing normal, transverse, and axial pitch along with the pitch cylinder and projected ellipse whose maximum radius of curvature is the equivalent radius.

Thus, two helical gears can properly mesh only if they both have the same values of normal diametral pitch P_{d_n} or normal module m_n.

Note that if the helix angle goes to zero the normal pitch equals the transverse pitch, the axial pitch goes to infinity, and the diameter of the pitch cylinder becomes the pitch diameter of the pitch circle of the resulting spur gear.

To obtain a relation between the diametral pitch and the number of teeth, substitute for p_n in equations 14.4.2 and 14.4.4 and replace p_t from equation 14.4.1 to obtain the appropriate relation

$$P_{d_n} = \frac{N}{d_p \cos \psi} \quad \text{or} \quad m_n = \frac{d_p \cos \psi}{N} \tag{14.4.5}$$

where the first relation holds in the OE system and the second in the SI system of units.

Upcoming calculation of tooth strength in Section 14.6 using the Lewis formula will again require that we calculate the virtual number of teeth N_v to select the y factor. Calculation of N_v is based upon replacing the helical gear with a spur gear having teeth whose circular pitch is equal to the normal pitch of the helical gear teeth, as pictured in Figure 14.4.3, and whose pitch radius equals the largest radius of curvature of the ellipse formed by cutting the pitch cylinder with a plane perpendicular to the tooth direction on the helical gear.

We may solve either of the following spur gear equations for N_v once we know the equivalent pitch diameter d_{p_e}:

$$P_{d_n} = \frac{N_v}{d_{\rho_e}} \quad \text{or} \quad m_n = \frac{d_{p_e}}{N_v} \tag{14.4.6}$$

where p_{d_n} and m_n refer to the pitch diameter and module, respectively, of a typical helical gear tooth in a plane perpendicular to the tooth.

To find d_{p_e} we first observe that the equation of the ellipse described above may be written as

$$1 = \frac{x^2}{\left(\dfrac{r_p}{\cos\psi}\right)^2} + \frac{y^2}{r_p^{\;2}} \tag{14.4.7}$$

which can be written in the form

$$y^2 = r_p^{\;2} - x^2 \cos^2\psi \tag{14.4.8}$$

which is more convenient for differentiation for substitution into the formula for the radius of curvature from analytical geometry; namely,

$$\rho = \frac{\left[1 + \left(\dfrac{dy}{dx}\right)^2\right]^{3/2}}{\dfrac{d^2 x}{dx^2}} \tag{14.4.9}$$

Differentiation of (14.4.8) to obtain its first and second derivatives leads to

$$\frac{dy}{dx} = -\frac{x}{y}\cos^2\psi$$
$$\frac{d^2 y}{dx^2} = -\frac{\cos^2\psi}{y} - \frac{x^2}{y^3}\cos^4\psi \tag{14.4.10}$$

so that substitution of these expressions into (14.4.9) yields

$$\rho = \frac{\left[1 + \left(\dfrac{x}{y}\right)^2\cos\psi\right]^{3/2}}{\dfrac{\cos^2\psi}{y} + \dfrac{x^2}{y^2}\cos^4\psi} \tag{14.4.11}$$

There is no need to substitute for y because the largest radius of curvature is found at $x = 0$, $y = r_p$, at which point equation 14.4.11 yields

$$\rho = \frac{r_p}{\cos^2 \psi} \tag{14.4.12}$$

so that the equivalent diameter becomes

$$d_{p_e} = \frac{d_p}{\cos \psi} \tag{14.4.13}$$

Upon solving equation 14.4.6 for N_v and substituting for P_d or m_n from equation 14.4.5 and for d_{p_e} from equation 14.4.13 we have

$$N_v = \frac{d_{p_e}}{p_n} = \frac{d_{p_e}}{p_t \cos \psi} = \frac{N}{\cos^3 \psi} \tag{14.4.14}$$

as the desired relation between the virtual number N_v and the actual number of teeth N on the helical gear.

To maintain continuous tooth contact and smooth operation it is necessary that the projection of a tooth onto the plane of rotation be greater than the transverse circular pitch, i.e., that the face contact ratio N_f given by

$$N_f = \frac{w \tan \psi}{p_t} = \frac{w \sin \psi}{p_n} \tag{14.4.15}$$

be greater than 1.0. It is assumed that contact between teeth is across the entire width w of the gear blank. Actual contact width should be used for w whenever contact is over less than the entire width of the gear.

It is common practice, as in equation 14.4.5, to use the width of the gear blank for helical gears in place of the tooth width as measured along the tooth axis because the width of the gear is easier to measure.

14.5 INDUCED AXIAL AND RADIAL FORCES ON THE GEAR SHAFT

Figure 14.5.1 shows the forces acting on a tooth of a helical gear that is well enough lubricated for the frictional forces to be negligible. These forces are, therefore, components of a force vector normal to the tooth face. Since this normal force lies in a plane perpendicular to the tooth direction and at an angle ϕ relative to a plane tangent to the pitch cylinder, (its average location over time is assumed to be at the intersection

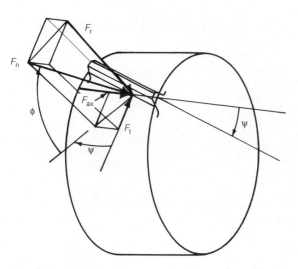

Figure 14.5.1 Force components acting upon a tooth of a helical gear. Negligible friction forces assumed.

of the tooth face and the pitch cylinder) its components are given by

$$F_t = F_n \cos \phi \cos \psi \qquad (14.5.1)$$

$$F_a = F_n \cos \phi \sin \psi \qquad (14.5.2)$$

$$F_r = F_n \sin \phi \qquad (14.5.3)$$

It is often more convenient, however, to express the components in terms of the tangential force because this component is directly related to the power transmitted. Hence,

$$F_n = \frac{F_t}{\cos \phi \cos \psi} \qquad (14.5.4)$$

$$F_a = \frac{\cos \phi \sin \psi}{\cos \phi \cos \psi} = F_t \tan \psi \qquad (14.5.5)$$

$$F_r = \frac{\tan \phi}{\cos \psi} F_t \qquad (14.5.6)$$

Single helix gears also induce a bending moment in the shaft as given by

$$M = r_p F_t \tan \psi \qquad (14.5.7)$$

In some designs it may be useful to notice that by choosing the appropriate direction of the tooth (right or left hand) the moment (14.5.7) may be used to reduce the lateral deflection at the gear of the supporting shaft by having it act in opposition to the moment induced by the lateral load, as in Figure 14.5.2. Shaft deflection at the gear, according to the Bernoulli-Euler beam theory, is given by

$$\delta = \frac{F_t l^2}{EI}\left(\frac{l}{3}\frac{\tan\phi}{\cos\psi} - \frac{d_p}{4}\tan\psi\right) \qquad (14.5.8)$$

Similarly, the bending stress at the shaft support (considered as a rigid support in bending) may be reduced by selecting the shaft length such that the term in parentheses in (14.5.9) is either small or zero.

$$\sigma = \frac{4F_t}{\pi r^3}\left(l\frac{\tan\phi}{\cos\psi} - r_p\tan\psi\right) \qquad (14.5.9)$$

Maximum shear stress at the support is given by

$$\tau_{max} = \frac{2F_t}{\pi r^2}\left[\left(\frac{2}{3}\frac{\tan\phi}{\cos\psi}\right)^2 + \left(\frac{r_p}{r}\right)^2\right]^{1/2} \qquad (14.5.10)$$

The width of the gear itself has been neglected in the derivation of equations 14.5.8 through 14.5.10, although it is often not negligible. Thus, these equations predict shaft stresses that may be larger than the actual values.

Double helical, or herringbone, gears induce no axial forces and no associated bending moment because the forces from the right and left-hand teeth cancel one another. Thus, force F_a and bending moment M vanish while equations 14.5.4 and 14.5.6 remain unchanged for herringbone gears. In either case the forces and moments calculated from the pertinent equations are used in the selection of bearings for the gear shaft according to the methods explained in Chapters 8 and 9.

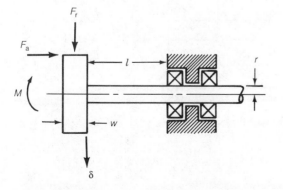

Figure 14.5.2 Bending force and moment induced by a helical gear.

14.6 SPEED RATIO AND TOOTH STRESS FOR HELICAL GEARS

Calculation of the speed ratio for helical gears whose axes cross as illustrated in Figure 14.6.1*a* is based upon the observation that although the teeth in contact slide relative to one another, they have equal velocities v_n in the direction of the normal to the line of contact between engaged teeth. If the gears were transparent and were viewed from the direction shown in Figure 14.6.1*a* a pair of teeth in contact would appear as in

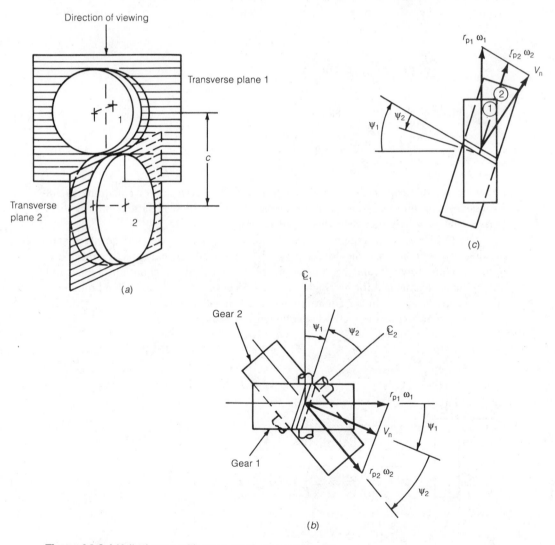

Figure 14.6.1 Helical gears with cross axes—two cases.

Figure 14.6.1b, where the solid lines denote a tooth from gear 1 and the middle solid line and the dashed line denote the contacting tooth from gear 2. The transverse velocity vector for the tooth on gear 1 is labeled $r_1 \omega_1$ and the transverse velocity vector for the tooth on gear 2 is labeled $r_2 \omega_2$. Since the common tooth direction for both teeth divides the angle between shafts it follows from the geometry that v_n is given by

$$v_n = r_{p_1} \omega_1 \cos \psi_1 = r_{p_2} \omega_2 \cos \psi_2 \tag{14.6.1}$$

so that

$$\frac{\omega_1}{\omega_2} = \frac{r_{p_2} \cos \psi_2}{r_{p_1} \cos \psi_1} \tag{14.6.2}$$

is the ratio of their angular velocities. Recall from equations 14.4.1 and 14.4.2 that

$$p_n = \frac{\pi d_p}{N_i} \cos \psi_i \tag{14.6.3}$$

for $i = 1,2$, so that upon replacing r_{p_i} in equation 14.6.2 by $d_{p_i}/2$ we get

$$\frac{n_1}{n_2} = \frac{\omega_1}{\omega_2} = \frac{d_{p_2} \cos \psi_2}{d_{p_1} \cos \psi_1} = \frac{N_2}{N_1} \tag{14.6.4}$$

since both gears must have teeth with the same p_n if they are to mesh. Here n_1 and n_2 represent the respective rpm of shafts 1 and 2.

The angle between shafts is the sum of the helix angles when both gears are of the same hand (both right hand or both left hand) as shown in Figure 14.6.1b; it is the difference between helix angles when both gears have teeth of opposite hands (one right hand and the other left hand). Thus, when helical gears are used to transfer power between parallel shafts the teeth on one gear must be left hand and those on the other right hand. Conversely, two right-hand or two left-hand helical gears may be used to transfer power between shafts at right angles to one another.

It also is evident from the geometry shown in Figure 14.6.1a that the center distance c is given by

$$2c = d_{p_1} + d_{p_2} \tag{14.6.5}$$

for all mesh configurations of helical gears. Substitution for d_{p_1} and d_{p_2} from equation 14.4.5 yields

$$2c = \frac{1}{P_{d_n}} \left(\frac{N_1}{\cos \psi_1} + \frac{N_2}{\cos \psi_2} \right) \quad \text{or} \quad 2c = m_n \left(\frac{N_1}{\cos \psi_1} + \frac{N_2}{\cos \psi_2} \right) \tag{14.6.6}$$

where the first holds for OE units and the second for SI units. In these expressions the angle between shafts, denoted by Ψ, is given by

$$\Psi = \psi_1 + \psi_2 \tag{14.6.7}$$

in which ψ_1 and ψ_2 are positive in the direction in which Ψ is positive.

Since the center distance c and the speed ratio N_1/N_2 are often determined by a machine's performance requirements, the design engineer usually must solve these equations for ψ_1 and ψ_2. This may be accomplished using the program flowcharted in Figure 14.6.2 and described in more detail in Ref. 7. Briefly, the procedure is to replace ψ_2 by $\Psi - \psi_2$ and solve the resulting equation using the bisection routine given in Chapter 1.

A modified form of the Lewis equation with a correction factor for velocity and manufacturing tolerances has been used to estimate the tooth stress intensity factor for helical gears. Thus,

$$\frac{\sigma w p_{\mathrm{n}}}{F_{\mathrm{t}}} = \frac{K}{y} \tag{14.6.8}$$

where[8]

$$K = 1 + \frac{v^{1/2}}{50} \qquad \text{for gears that are formed by hobbing or shaping}$$

and

$$K = \left(1 + \frac{v^{1/2}}{78}\right)^{1/2} \qquad \text{for gears that are finished by shaving or grinding}$$

and where y is the Lewis factor, w is the width of the gear blank, and the tangential velocity v in the plane rotation is in feet/minute. It is common practice to let F_{t} denote the tangential force component in the plane of rotation of the gears in equation 14.6.8 and to let w represent the width of the blank in on which the teeth are cut. This convention may be justified by recalling that to be consistent with the use of the Lewis relation the force used should be that component perpendicular to the tooth face, and the width should be the face width measured along the axis of the tooth itself. If we let these quantities be represented by F_{n} and w' respectively, the stress intensity may be given by the ratio on the right side of equation 14.6.9:

$$\frac{\sigma w' p_{\mathrm{n}}}{F_{\mathrm{nt}}} = \frac{\sigma p_{\mathrm{n}} w/\cos \lambda}{F_{\mathrm{t}}/\cos \lambda} = \frac{\sigma p_{\mathrm{n}} w}{F_{\mathrm{t}}} \tag{14.6.9}$$

According to Figure 14.5.1, the tangential component F_{t} is related to the tangential component of F_{n} by $F_{\mathrm{n}} = F_{\mathrm{t}}/\cos \lambda$ and the face width w' is related to the width w of

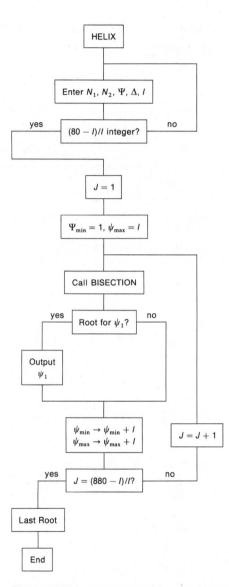

Figure 14.6.2 Flowchart of program HELIX to find values of ψ_1 satisfying equation 14.6.6.

the gear blank by $w = w'/\cos \gamma$. Upon making these substitutions we obtain the second of the relations in display 14.6.9, which is clearly equal to the ratio on the right.

Finally, the more realistic stresses that may be obtained from a finite element analysis are recommended whenever extreme service conditions may induce stresses which are near the yield or fracture stresses in the tooth. In making such an analysis is extremely important to consider gear dynamics and possible shaft misalignment in order to reasonably simulate forces acting upon the tooth face under expected service conditions.

EXAMPLE 14.6.1

Select the helix angles for a helical pinion and gear to be mounted on shafts with an included angle of 48° to provide a speed ratio 2.6296 ± 0.0001 and to transmit 200 kW at a pinion speed of 960 rpm. The center distance is to be 327 mm and teeth with a 20° pressure angle are preferred. The speed ratio may be satisfied by $N_1 = 27$ teeth and $N_2 = 71$ teeth. The teeth are to be formed by hobbing and the steel from which they are to be cut has a working stress of 414 MPa after a safety factor and a nominal stress concentration factor have been included.

To solve equation 14.6.6 for the helix angles we need to supply the tooth size. We may arrive at an initial estimate of the required tooth size by turning to Figure 14.6.3, which is a plot of

$$\frac{2c}{N_1 m_{\mathrm{n}}} = \frac{1}{\cos \psi_1} + \frac{N_2/N_1}{\cos(\Psi - \psi_1)} \tag{14.6.6}$$

and calculating the right side for $N_2/N_1 = 2.6296$, $\Psi = 48°$, and $\psi_1 = 20°$ and 30° to obtain an estimate of representative values of the left side for a desirable range of helix angles. From this estimation we

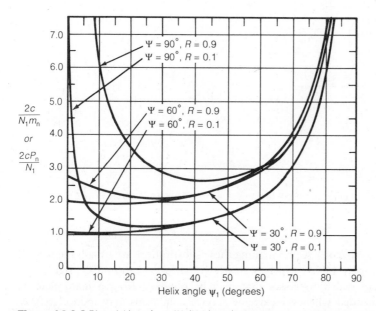

Figure 14.6.3 Plot of $1/\cos \psi_1 + (N_2/N_1)/\cos \psi_2$ for the values shown for $R = N_2/N_1$.

find that (using $N_1/N_2 = 0.38$. Either gear may be N_1.)

$$2c/N_1 m_n \simeq 4$$

Hence, our first estimate for m_n is

$$m_n = c/2N_1 = \frac{327}{54} = 6.056 \rightarrow 6$$

To find the helix angles with the aid of the program shown in Figure 14.6.2, enter $N_1 = 27$, $N_2 = 71$, $c = 327$ mm, $m_n = 6$ and select a search interval of $5°$ and a maximum error in ψ_1 of $0.001°$ to find that $\psi_1 = 20.2606°$ and $\psi_2 = 27.7394°$. (The other solution, $\psi_1 = 44.52°$ and $\psi_2 = 3.48°$, is ignored because of the large axial force and large Hertzian stresses on the teeth at an angle of $44.52°$.)

To calculate the tooth load on the pinion, which is more heavily loaded than the gear because of its smaller diameter and more rapid rotation, we must first find the virtual number of teeth on the pinion by turning to equation 14.4.14 to find that

$$N_v = N/\cos^3 \psi = 27/\cos^3 20.2606 = 32.70$$

With this information we turn to Table 13.7.1 in Chapter 13 and find by linear interpolation that

$$y = 0.1162$$

for teeth with a $20°$ pressure angle.

The tangential velocity of the pinion on the pitch cylinder is given by

$$v = \pi d_p n = \pi 172.68(960) = 520\,791 \text{ mm/min}$$

where the pitch diameter of the pinion was calculated from equation 14.4.4, $p_n = \pi m_n = 6\pi$, along with equation 14.4.2,

$$p_t = p_n/\cos \psi = 6\pi/\cos 20.2602° = 20.0927 \text{ mm}$$

and equation 14.4.1,

$$d_p = N p_t/\pi = 27(6)/\cos 20.2602° = 172.6840 \text{ mm}$$

Substitution into the expression for K for hobbed teeth may be simplified for future use with SI units by observing that for K to be dimensionless the constant 50 must have units of $v^{1/2}$ in $(\text{ft/min})^{1/2}$ so that we may convert it to mm/min by multiplying by $[(12)25.4]^{1/2}$ to obtain

$$K = (1 + v^{1/2}/873) = (1 + 721.65/873) = 1.827$$

From the formula for power in kW we find that the tangential component of force on the gear is

$$F_t = 3 \times 10^7 \text{ kW}/\pi r_p n = 3 \times 10^7 (200)/[\pi 86.342(960)] = 23\,041 \text{ N}$$

Upon rewriting equation 14.6.8 in the form

$$w = KF_t/y\sigma p_n$$

we have that the width of the gear blank must be equal to, or greater than,

$$w = 1.827(23\,041)/[0.1162(414)6] = 145.84 \text{ mm}$$

Finally, turn to equation 14.4.15 for the face contact angle to find that for the minimum width

$$N_f = (w \sin \psi)/p_n = (146 \sin 20.2606)/6\pi = 2.68$$

which indicates that we also have taken advantage of the ability of helical gears to transfer power and motion smoothly.

14.7 WORM GEARS

A worm and gear may be considered a variation of crossed helical gears in which the pinion pitch cylinder is extended and the axial pitch is decreased, with the result shown in Figure 14.7.1a. By convention, the inclination of the teeth will be described in terms

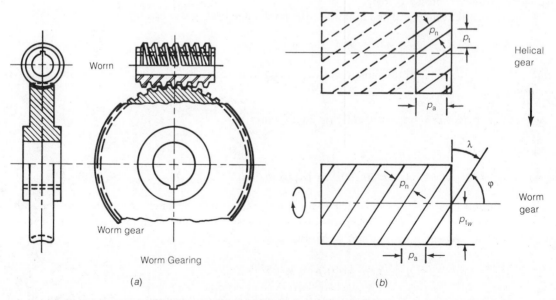

Worm

Worm gear

Worm Gearing

(a)

p_n
p_t
p_a

Helical gear

λ
φ
p_n
p_{tw}
p_a

Worm gear

(b)

Figure 14.7.1 Worm gearing and the related helical gears. The lower sketch in (b) represents the developed surface of a two-tooth worm. *Source: AGMA 112.05 June 1976, American Gear Manufacturers Assn., Alexandria, VA. Reprinted with permission.*

of their lead angle, a term borrowed from screw notation, where the lead angle λ is related to the helix angle as shown in Figure 14.7.1b, namely

$$\psi = \pi/2 - \lambda \tag{14.7.1}$$

Thus, from either equation 14.4.3 or Figure 14.7.1 it is evident that the axial, normal, and transverse pitches are related according to

$$p_n = p_a \cos \lambda = p_t \sin \lambda \tag{14.7.2}$$

Similar relations hold for the double-enveloping, or Conedrive, worm, shown in Figure 14.7.2, which has the advantage of a larger number of teeth in contact. It is used in applications where large torques are required along with small gear diameters, so that the increased number of teeth in contact is worth the additional machining costs associated with double-enveloping worm gears.

Consistent with the similarity between worm and helical gears, the speed ratio may be found from equation 14.6.4 by setting $\psi_1 = \psi_2$ and identifying the first with the worm and the second with the gear to obtain

$$\frac{n_w}{n_g} = \frac{N_g}{N_w} \tag{14.7.3}$$

Figure 14.7.2 Photograph of a double-enveloping worm and gear showing thrust shoulders on the worm. *Source: Ex-Cell-0 Corporation, Cone Drive Operations, Traverse City, MI.*

where N_g and n_g denote the number of teeth (equals number of starts) and the speed of the gear and N_w and n_w denote similar quantities for the worm. The transverse pitch of the worm it related to its diameter according to

$$\frac{\pi d_{pw}}{N_w} = p_{tw} \tag{14.7.4}$$

so that the pitch diameter may be written as

$$d_{pw} = \frac{N_w}{\pi} \frac{p_n}{\sin \lambda} = \frac{N_w}{P_{d_n} \sin \lambda} \tag{14.7.5}$$

Similarly, the transverse pitch of the gear may be written as

$$\frac{\pi d_{pg}}{N_g} = p_{tg} = p_{aw} \tag{14.7.6}$$

where the equality between p_{tg} for the gear and p_{aw} for the worm follows from Figure 14.7.3. It is because of this equality that the pitch diameter of the gear may be written

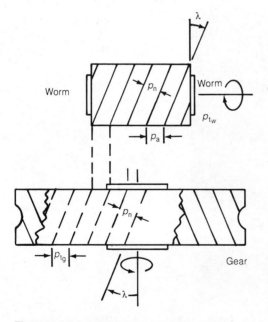

Figure 14.7.3 Worm and gear displaced from the mesh condition. Broken lines on the gear represent the teeth at the bottom of the gear where its teeth engage the worm.

in a mixed notation involving both the axial pitch and the normal pitch of the worm as

$$d_{p_g} = \frac{N_g}{\pi} p_a = \frac{N_g p_n}{\pi \cos \lambda} = \frac{N_g}{P_{d_n} \cos \lambda} \qquad (14.7.7)$$

Upon division of equation 14.7.7 by 14.7.5 we find that

$$\tan \lambda = \frac{N_w d_{p_g}}{N_g d_{p_w}} = \frac{n_g d_{p_g}}{n_w d_{p_w}} \qquad (14.7.8)$$

where the second term of 14.7.8 was found with the aid of 14.7.3. With it we may select a lead angle from the shaft speeds and the center distance, given in terms of $P_{d_n} = \pi/p_n$ or $m_n = p_n/\pi$, by

$$2c = d_{p_w} + d_{p_g} = \frac{1}{P_{d_n}} \left(\frac{N_w}{\sin \lambda} + \frac{N_g}{\cos \lambda} \right) \quad \text{or} \quad 2c = m_n \left(\frac{N_w}{\sin \lambda} + \frac{N_g}{\cos \lambda} \right)$$

$$(14.7.9)$$

At this stage of the analysis the choice of the lead angle clearly is not unique. This allows the design engineer some room for choice, based upon the desired ratio between input and output torque desired and the operating conditions expected. As we shall learn in the next section, the lead angle and the friction between worm and gear teeth can, for example, be chosen such that the worm cannot be driven by a torque applied to the gear. A worm and gear pair designed with this characteristic is known as a self-locking.

14.8   FORCES BETWEEN WORM AND GEAR

Because the force components acting on the worm are paired with equal and opposite components acting on the gear, primary attention will be directed to the forces acting on the worm. The resultant of the forces distributed over the several teeth in contact will be approximated by a single force vector acting on the face of a worm tooth (similar to a power screw thread) at a point on the pitch cylinder where it is in contact with the pitch circle of the gear. Thus, the forces, moments and reactions acting on the worm to maintain equilibrium when the worm is not accelerating or decelerating may be depicted as in Figure 14.8.1.

   This figure shows the forces that exist when the worm drives the gear in opposition to an external torque applied to the gear. An example of an external torque applied to the gear would be the weight of a load being raised by a hoist whose drum is driven by a worm and gear as shown in Figure 14.8.2a.

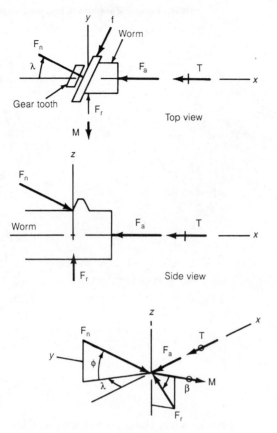

Top view

Side view

Vectors only

Figure 14.8.1 Force and moment vectors acting upon the worm in a worm gear drive. Equal and opposite forces act on the gear.

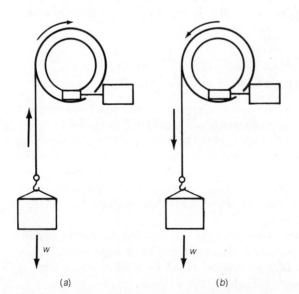

Figure 14.8.2 (a) Worm gear driven hoist raising a weight (worm opposing gear torque). (b) Worm gear driven hoist lowering a weight (worm driven with torque applied to the gear).

Vectors describing these forces and moments may be written relative to the coordinates shown as

$$\mathbf{F} = F_n(-\hat{\mathbf{i}} \cos \phi \cos \lambda - \hat{\mathbf{j}} \cos \phi \sin \lambda - \hat{\mathbf{k}} \sin \phi) \tag{14.8.1}$$

$$\mathbf{f} = \mu F_n(\hat{\mathbf{i}} \sin \lambda - \hat{\mathbf{j}} \cos \lambda) \tag{14.8.2}$$

$$\mathbf{F}_r = F_r(\hat{\mathbf{j}} \cos \beta + \hat{\mathbf{k}} \sin \beta) \tag{14.8.3}$$

$$\mathbf{T} = -\hat{\mathbf{i}}T \qquad \mathbf{M} = \hat{\mathbf{j}}M \qquad \mathbf{F}_a = \hat{\mathbf{i}}F_a \tag{14.8.4}$$

where F_n represents the force normal the face of the worm tooth, f denotes the friction force acting upon the worm, $\mathbf{F}_r$ and $\mathbf{F}_a$ denote radial and axial forces required to maintain equilibrium, T signifies the applied torque when the worm is driving and the resisting torque when the worm is driven, and M depicts the moment that prevents the worm from overturning. The $\hat{\mathbf{i}},\hat{\mathbf{j}},\hat{\mathbf{k}}$ unit vectors that appear in these relations are directed along the positive x,y,z axes, respectively.

Force and moment equilibrium equations involving these quantities may be written as

$$\mathbf{F}_n + \mathbf{f} + \mathbf{F}_a + \mathbf{F}_r = 0 \tag{14.8.5}$$

$$\mathbf{r}_p \times (\mathbf{F}_n + \mathbf{f}) + \mathbf{T} + \mathbf{M} = 0 \tag{14.8.6}$$

where $\mathbf{r}_p = \hat{\mathbf{k}}r_p$. In writing the equations that follow from equating each of the components of these equations to zero we shall eliminate F_n by means of the substitution

$$F_n = \frac{F_t}{\cos \phi \cos \lambda} \tag{14.8.7}$$

where F_t is the force component tangential to the gear.

Beginning with the x component of (14.8.5), we find that it yields the axial force that must be supported by the thrust bearing on the worm shaft:

$$F_a = F_t(1 - \tan \kappa \tan \lambda) \tag{14.8.8}$$

where

$$\tan \kappa = \frac{\mu}{\cos \phi} \tag{14.8.9}$$

From the y component of (14.8.5) we find that

$$F_r \cos \beta = F_t(\tan \lambda + \tan \kappa) \tag{14.8.10}$$

and from the z component we have

$$F_r \sin \beta = \frac{F_t}{\cos \lambda} \tan \phi \qquad (14.8.11)$$

Force $F_r \cos \beta$ given by (14.8.10) is the axial force that must be resisted by a bearing on the gear shaft and the force $F_r \sin \beta$ given by (14.8.11) is the radial force that must be resisted by bearings on the gear shaft. Upon squaring and summing (14.8.10) and (14.8.11) we have

$$F_r = F_t \left[\frac{\tan^2 \phi}{\cos^2 \lambda} + (\tan \lambda + \tan \kappa)^2 \right]^{1/2} \qquad (14.8.12)$$

as one of two contributors to the radial load applied to the bearings on the worm shaft. Angle β at which this radial force acts may be found by dividing (14.8.11) by (14.8.10) to get

$$\tan \beta = \frac{\tan \phi}{\sin \lambda + \tan \kappa \cos \lambda} \qquad (14.8.13)$$

Turning next to the moment equilibrium equation (14.8.6) we find from the x component that the required torque to drive the worm is

$$T = r_{pw} F_t (\tan \lambda + \tan \kappa) \qquad (14.8.14)$$

and from the y component we find that the overturning moment applied to the worm is, relative to a point on the centerline of the worm,

$$M = r_{pw} F_t (1 - \tan \kappa \tan \lambda) \qquad (14.8.15)$$

This equation provides the other contributor to the load applied to the radial bearings on the worm gear shaft.

The thrust-bearing friction was neglected in equation 14.8.14 because it usually is negligible compared to the other torques. Whenever it cannot be neglected equation 14.8.14 should be replaced by

$$T = r_{pw} F_t (\tan \lambda + \tan \kappa) + \mu_c r_c F_a \qquad (14.8.16)$$

where μ_c denotes the bearing friction and r_c is the effective radius at which it acts. Replacement of F_t in (14.8.16) by substitution from (14.8.8) yields the screw formula used in Chapter 6. (Force F_r and moment M vanish for a power screw because of the symmetry of the nut.)

One or more of the algebraic signs in equations 14.8.15 and 14.8.16 change if the direction of rotation changes. In particular, if the worm turns to move with the torque applied to the gear, as in the case of lowering the load in Figure 14.8.2*b*, the direction of rotation of the worm, the friction force, and the torque vector will all reverse. Since reversal of directions is indicated by a change in sign of the corresponding vector, equation 14.8.16 will take the form

$$\frac{T}{F_t} = r_{pw}(\tan \kappa - \tan \lambda) + \mu_c r_c (1 + \tan \kappa \tan \lambda) \tag{14.8.17}$$

after F_a is replaced by F_t according to (14.8.8).

If the direction of the friction force is reversed in Figure 14.8.1 the resulting diagram may be used to describe the situation where the lead angle and the friction between the teeth of the worm and gear are such that a torque applied to the gear can drive the worm against a resisting torque T. This is known as *backdriving*, and the torque required for equilibrium is given by

$$\frac{T}{F_t} = r_{pw}(\tan \lambda - \tan \kappa) + \mu_c r_c (1 + \tan \kappa \tan \lambda) \tag{14.8.18}$$

It follows from this equation that if

$$\tan \kappa \geqq \tan \lambda \tag{14.8.19}$$

holds, no force will be required to prevent backdriving; in other words, the worm and gear combination will be self-locking.

To find the torque multiplication obtained from a worm gear combination with friction included, let T_g represent the torque acting on the gear, where $T_g = F_t r_{pg}$ in terms of the force tangent to the gear at its pitch radius. The torque ratio for the worm and gear combination thus becomes

$$\frac{T_g}{T_w} \frac{r_{pw}}{r_{pg}} = \frac{1}{\tan \lambda + \tan \kappa + \mu_c \dfrac{r_c}{r_{pw}} (1 - \tan \lambda \tan \kappa)} \tag{14.8.20}$$

Note that the torque ratio is improved by using a small pressure angle, i.e., 14.5° instead of 20° or 25°.

By placing the ratio of pitch diameters on the left side of this relation the right side may be more easily plotted to show the effects of the lead angle, the pressure angle, and the coefficients of friction between the worm and gear teeth and at the thrust collar, as has been done in Figure 14.8.3. If the friction coefficient between the teeth of the worm and gear is 0.02 and the thrust collar friction is negligible we find from Figure 14.8.2*a* that the torque ratio may be as large as 13.7 for a lead angle of 3.0° and that it is strongly dependent upon the lead angle of the worm. We also find that when the friction coefficient is this small the effect of changing the pressure angle

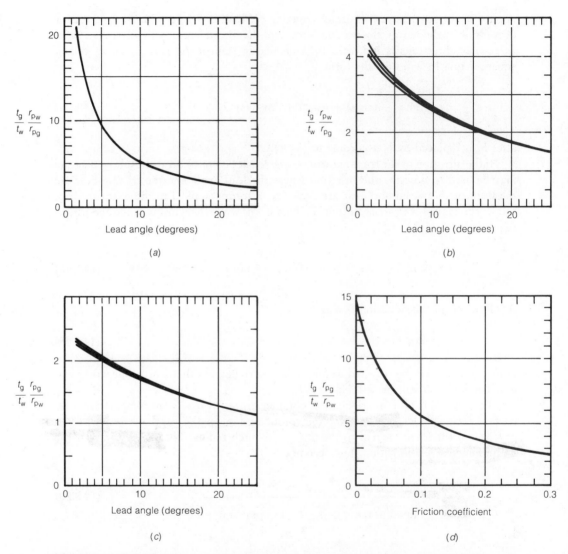

Figure 14.8.3 Modified torque ratio $T_g r_{pw}/T_w r_{pg}$ as a function of (a) the lead angle for $\mu = 0.02$, $\mu_c = 0$; (b) the lead angle for $\mu = 0.20$, $\mu_c = 0$; (c) the lead angle for $\mu = \mu_c = 0.2$; and (d) the friction coefficient μ at the threads for $\mu_c = 0$ and a lead angle λ of 3°.

14.5 to 25° is less than the line width of the curve. As the friction coefficient between the worm and gear increases to 0.20, however, the torque ratio for a lead angle of 3.0° drops to about 3.9, the dependence upon the lead angle becomes less dramatic, and the effect of the pressure angle becomes detectable, as is evident from Figure 14.8.2b. If the friction coefficient at the thrust collar is also allowed to increase, as shown in Figure 14.8.3c, the torque ratio decreases to a maximum of 2.2 or less for a lead angle of 3.0°, the differences between pressure angles has less influence, and the variation of the torque ratio with the lead angle becomes more linear.

The influence of the friction coefficient between the worm and gear teeth upon the torque ratio is displayed in Figure 14.8.3d for a lead angle of 3° and negligible friction at the thrust collar. Since the torque ratio T_g/T_w is obtained by multiplying the right side or (14.8.20) by the ratio of the gear to worm pitch diameters, the effects of the parameters on the right side of (14.8.20) upon the torque ratio may be multiplied by a factor of five or more. Clearly, reduction in the torque ratio that may be caused by sliding friction between worm and gear teeth and by friction at the thrust bearing, as shown in Figure 14.8.3, along with the attendant power losses, emphasize the importance of using of low-friction lubricants between the worm and the gear and of installing rolling element bearings at the thrust collar.

14.9 TOOTH STRENGTH

The Lewis formula for spur gears may be applied to worm gears by replacing the circular pitch by the normal pitch, as was done in the application of the Lewis formula to helical gears. It may also be written in terms of the axial pitch by appealing to equation 14.7.2, so that the stress sensitivity becomes

$$\frac{\sigma w p_n}{F_t} = \frac{1}{y} \qquad \frac{\sigma w p_a}{F_t} = \frac{1}{y \cos \lambda} \tag{14.9.1}$$

This relation determines tooth size only in the case of slow speeds and/or heavily loaded drives, and then only if the stress sensitivity found from the Lewis relation is greater than the stress sensitivity in shear, which is

$$\frac{\tau w p_a}{F_t} = \frac{1.5}{c} \tag{14.9.2}$$

where

$$
\begin{aligned}
c &= 0.60 && \text{for } 14.5° \text{ pressure angle} \\
  &= 0.70 && \text{for } 20.0° \text{ pressure angle} \\
  &= 0.75 && \text{for } 25° \text{ and } 30° \text{ pressure angles}
\end{aligned}
$$

and where σ and τ represent direct and shear stresses. As gear speeds increase the dynamic loads become more demanding than the static loads and, therefore, determine tooth dimensions. The reader is referred to Ref. 3 for the details of including dynamic effects which involve mass and acceleration considerations beyond the accepted scope of a machine design text.

14.10 WORM GEAR POWER RATING (AGMA)

ANSI/AGMA Std. 6034-A87, March 1988, is prefaced by the caveat that

> The knowledge and judgment required to evaluate the various rating factors come from years of accumulated experience in designing, manufacturing, and operating gear units. Empirical factors given in this Standard are suited to the particular use shown. This Standard is intended for use by the experienced gear designer, capable of selecting reasonable values for the factors. It is not intended for use by the "engineering public at large."

The required input power in horsepower or kilowatts is given by

$$\text{hp} = \frac{Z}{396\,000} \quad \text{or} \quad \text{kW} = \frac{Z}{3 \times 10^7} \tag{14.10.1}$$

where Z is given by

$$Z = \left(n_g d_{pg} + \frac{\mu n_w d_{pw}}{\cos \phi \, \cos^2 \lambda} \right) \pi F_t \tag{14.10.2}$$

The first term in Z represents power output and the second represents the losses due to friction. Mean worm and gear diameters d_{pg} and d_{pw} are defined to be the diameters in mm or in. of the midpoints of the working depths of the teeth on the worm and gear respectively.

Tangential force on the gear tooth, denoted by F_t, is given by

$$F_t = d_{pg} 0.8 w K_s C_m C_v / u \tag{14.10.3}$$

where $w \leq 0.67 d_{pg}$ is required. Coefficient K_s, the material factor, may be read from curves in Figure 14.10.1, coefficient C_m, the ratio correction factor, may be found from Table 14.10.1, and coefficient C_v, the velocity factor, may be obtained from Table 14.10.2 for OE units or from Table 14.10.3 for SI units. Denominator u is a conversion factor that takes on the values

$$u = 1.0 \qquad \text{for OE units}$$

$$u = 75.948 \qquad \text{for SI units}$$

The units of these coefficients and of the conversion factor are not given in the standard. Coefficients of friction as a function of the sliding velocity at the worm diameter d_{pw} are given in Table 14.10.4. Lubricants should conform to recommendations in ANSI/AGMA 6034-A87.

Diameter d_{pw} that has been termed pitch diameter for the worm is also defined to be the diameters at the midpoint of the working depth of the teeth in the ANSI/AGMA

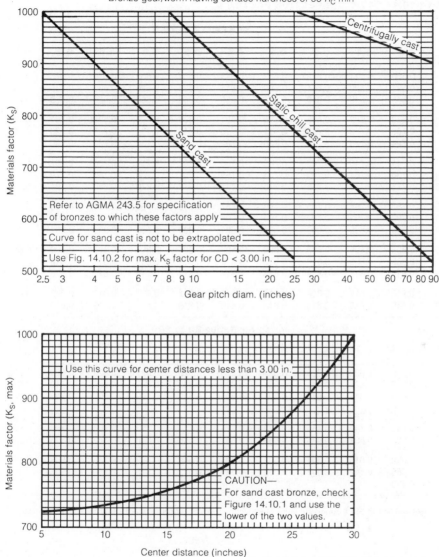

Figure 14.10.1 Materials factor K_s for worm and gear. *Source: AGMA 440.04 Oct. 1971, American Gear Manufacturers Assn., Alexandria, VA. Reprinted with permission.*

6034-A87, March 1988, standard. Diameters d_{p_g} and d_{p_w} are designated as the mean diameters of the gear and worm in the standard and are denoted by D_m and d_m. The gear is referred to either as the gear or as the wormgear, one word, in the above standard and elsewhere.

TABLE 14.10.1   RATIO CORRECTION FACTOR C_m

Ratio Range (m_G)	C_m for Ratios 3 to 19.9									
	0.0	0.1	0.2	0.3	0.4	0.5	0.6	0.7	0.8	0.9
3–3.9	0.500	0.511	0.522	0.532	0.543	0.554	0.562	0.570	0.577	0.585
4–4.9	0.593	0.598	0.604	0.609	0.615	0.620	0.625	0.630	0.635	0.640
5–5.9	0.645	0.649	0.652	0.656	0.659	0.663	0.666	0.669	0.673	0.676
6–6.9	0.679	0.682	0.685	0.688	0.691	0.694	0.696	0.699	0.701	0.704
7–7.9	0.706	0.708	0.710	0.711	0.713	0.715	0.717	0.719	0.720	0.722
8–8.9	0.724	0.726	0.728	0.730	0.732	0.734	0.736	0.738	0.740	0.742
9–9.9	0.744	0.746	0.747	0.749	0.750	0.752	0.754	0.755	0.757	0.758
10–10.9	0.760	0.761	0.763	0.764	0.765	0.767	0.768	0.769	0.770	0.782
11–11.9	0.773	0.774	0.775	0.776	0.777	0.778	0.779	0.780	0.781	0.782
12–12.9	0.783	0.784	0.785	0.786	0.787	0.788	0.788	0.789	0.790	0.791
13–13.9	0.792	0.793	0.794	0.795	0.795	0.796	0.796	0.797	0.798	0.798
14–14.9	0.799	0.800	0.800	0.801	0.801	0.802	0.803	0.803	0.804	0.804
15–15.9	0.805	0.805	0.806	0.806	0.807	0.807	0.807	0.807	0.808	0.808
16–16.9	0.809	0.809	0.810	0.810	0.811	0.811	0.811	0.811	0.812	0.812
17–17.9	0.813	0.813	0.814	0.814	0.814	0.814	0.815	0.815	0.815	0.815
18–18.9	0.816	0.816	0.816	0.817	0.817	0.817	0.817	0.817	0.818	0.818
19–19.9	0.818	0.818	0.818	0.818	0.819	0.819	0.819	0.819	0.820	0.820

Ratio Range (m_G)	C_m for Ratios 20 to 100									
	0	1	2	3	4	5	6	7	8	9
20–29	0.820	0.822	0.823	0.824	0.825	0.825	0.826	0.826	0.826	0.826
30–39	0.825	0.825	0.825	0.824	0.823	0.822	0.821	0.820	0.818	0.816
40–49	0.815	0.812	0.810	0.807	0.804	0.802	0.799	0.796	0.792	0.789
50–59	0.785	0.782	0.799	0.775	0.771	0.767	0.763	0.759	0.754	0.750
60–69	0.745	0.740	0.735	0.729	0.724	0.718	0.712	0.706	0.700	0.694
70–79	0.687	0.681	0.675	0.669	0.662	0.655	0.648	0.642	0.635	0.629
80–89	0.622	0.615	0.609	0.602	0.595	0.589	0.582	0.575	0.568	0.562
90–99	0.555	0.549	0.542	0.536	0.529	0.523	0.516	0.510	0.503	0.497
100	0.490									

Source: AGMA 440.04 Oct. 1971, American Gear Manufacturers Assn., Alexandria, VA. Reprinted with permission.

14.11   HEAT DISSIPATION FOR WORM GEARS

The power dissipation due to friction between the worm and the gear and the turbulence induced in the oil-filled gear housing both contribute to the generation of heat that must be dissipated.

A common textbook approximation for the rate of heat dissipation H_d is that proposed by Walker in 1944[5] wherein

$$H_d = C_\Gamma A(\Gamma_o - \Gamma_a) \tag{14.11.1}$$

TABLE 14.10.2 VELOCITY FACTOR C_v, OE UNITS

Velocity Range (fpm)	C_v for Velocities 0 to 359 fpm									
	0	1	2	3	4	5	6	7	8	9
0–9	—	0.649	0.649	0.648	0.648	0.647	0.646	0.646	0.645	0.644
10–19	0.644	0.643	0.643	0.642	0.641	0.641	0.640	0.639	0.639	0.648
20–29	0.638	0.637	0.636	0.635	0.634	0.634	0.633	0.633	0.633	0.632
30–39	0.631	0.631	0.630	0.629	0.629	0.628	0.628	0.627	0.626	0.626
40–49	0.625	0.624	0.624	0.623	0.623	0.622	0.621	0.621	0.620	0.619
50–59	0.619	0.618	0.618	0.617	0.616	0.616	0.615	0.614	0.614	0.613
60–69	0.613	0.612	0.611	0.611	0.610	0.609	0.609	0.608	0.608	0.607
70–79	0.606	0.606	0.605	0.604	0.604	0.603	0.603	0.602	0.601	0.601
80–89	0.600	0.599	0.599	0.598	0.598	0.597	0.596	0.596	0.595	0.595
90–99	0.594	0.593	0.592	0.592	0.592	0.591	0.590	0.590	0.589	0.589
100–109	0.588	0.587	0.575	0.574	0.573	0.572	0.572	0.572	0.571	0.571
110–119	0.582	0.581	0.581	0.581	0.580	0.580	0.579	0.578	0.577	0.577
120–129	0.576	0.575	0.575	0.574	0.574	0.573	0.572	0.572	0.571	0.571
130–139	0.570	0.569	0.569	0.568	0.568	0.567	0.566	0.566	0.565	0.565
140–149	0.564	0.563	0.563	0.562	0.561	0.561	0.561	0.560	0.559	0.558
150–159	0.558	0.557	0.556	0.556	0.555	0.554	0.554	0.553	0.553	0.552
160–169	0.551	0.551	0.550	0.549	0.549	0.548	0.548	0.547	0.546	0.546
170–179	0.545	0.544	0.544	0.543	0.543	0.542	0.542	0.541	0.540	0.540
180–189	0.539	0.539	0.538	0.538	0.537	0.536	0.536	0.535	0.535	0.534
190–199	0.534	0.533	0.533	0.532	0.531	0.531	0.530	0.530	0.529	0.529
200–209	0.528	0.527	0.527	0.526	0.526	0.525	0.525	0.524	0.524	0.523
210–219	0.522	0.522	0.521	0.521	0.520	0.520	0.519	0.518	0.518	0.517
220–229	0.517	0.516	0.516	0.515	0.515	0.514	0.513	0.513	0.512	0.512
230–239	0.511	0.511	0.510	0.510	0.509	0.508	0.508	0.507	0.507	0.506
240–249	0.506	0.505	0.504	0.504	0.503	0.503	0.502	0.502	0.501	0.501
250–259	0.500	0.499	0.499	0.498	0.498	0.497	0.497	0.496	0.496	0.495
260–269	0.494	0.494	0.493	0.493	0.492	0.492	0.491	0.490	0.490	0.489
270–279	0.489	0.488	0.488	0.487	0.487	0.486	0.485	0.485	0.484	0.484
280–289	0.483	0.483	0.482	0.482	0.481	0.480	0.480	0.479	0.479	0.478
290–299	0.478	0.477	0.476	0.476	0.475	0.475	0.474	0.474	0.473	0.473
300–309	0.472	0.471	0.471	0.470	0.470	0.469	0.469	0.468	0.468	0.467
310–319	0.467	0.466	0.466	0.465	0.465	0.464	0.464	0.463	0.463	0.462
320–329	0.462	0.461	0.461	0.460	0.460	0.460	0.458	0.458	0.458	0.457
330–339	0.456	0.456	0.455	0.455	0.454	0.454	0.453	0.453	0.452	0.452
340–349	0.451	0.451	0.450	0.450	0.499	0.499	0.448	0.488	0.477	0.477
350–359	0.446	0.445	0.445	0.444	0.444	0.443	0.443	0.442	0.442	0.441

Source: AGMA 440.04 Oct. 1971, American Gear Manufacturers Assn., Alexandria, VA. Reprinted with permission.

TABLE 14.10.3   VELOCITY FACTOR C_v, SI UNITS

Velocity Range (m/s)	C_v for Velocities 0 – 30 m/s									
	0.0	0.1	0.2	0.3	0.4	0.5	0.6	0.7	0.8	0.9
0–0.9	0.6500	0.6449	0.6311	0.6176	0.6043	0.5914	0.5787	0.5663	0.5542	0.5423
1.0–1.9	0.5307	0.5193	0.5082	0.4973	0.4867	0.4762	0.4660	0.4561	0.4463	0.4367
2.0–2.9	0.4274	0.4182	0.4093	0.4005	0.3919	0.3835	0.3753	0.3672	0.3594	0.3517
3.0–3.9	0.3442	0.3368	0.3296	0.3225	0.3156	0.3088	0.3137	0.3089	0.3042	0.2997
4.0–4.9	0.2954	0.2913	0.2873	0.2835	0.2798	0.2762	0.2728	0.2694	0.2662	0.2631
5.0–5.9	0.2601	0.2572	0.2543	0.2516	0.2489	0.2463	0.2438	0.2413	0.2390	0.2366
6.0–6.9	0.2344	0.2322	0.2300	0.2279	0.2259	0.2239	0.2220	0.2201	0.2182	0.2164
7.0–7.9	0.2146	0.2129	0.2112	0.2095	0.2079	0.2063	0.2048	0.2033	0.2018	0.1871
8.0–8.9	0.1989	0.1975	0.1961	0.1947	0.1934	0.1921	0.1908	0.1896	0.1883	0.1871
9.0–9.9	0.1859	0.1848	0.1836	0.1825	0.1814	0.1803	0.1892	0.1781	0.1771	0.1761
10.0–10.9	0.1751	0.1741	0.1731	0.1721	0.1712	0.1703	0.1693	0.1684	0.1675	0.1667
11.0–11.9	0.1658	0.1649	0.1641	0.1633	0.1625	0.1616	0.1609	0.1601	0.1593	0.1585
12.0–12.9	0.1578	0.1570	0.1563	0.1556	0.1548	0.1541	0.1534	0.1527	0.1521	0.1514
13.0–13.9	0.1507	0.1501	0.1494	0.1488	0.1481	0.1475	0.1469	0.1463	0.1457	0.1451
14.0–14.9	0.1445	0.1439	0.1433	0.1427	0.1422	0.1416	0.1411	0.1405	0.1400	0.1394
15.0–15.9	0.1389	0.1384	0.1378	0.1330	0.1323	0.1316	0.1310	0.1303	0.1297	0.1291
16.0–16.9	0.1284	0.1278	0.1272	0.1266	0.1260	0.1254	0.1248	0.1243	0.1237	0.1231
17.0–17.9	0.1226	0.1220	0.1215	0.1209	0.1204	0.1198	0.1193	0.1188	0.1183	0.1178
18.0–18.9	0.1173	0.1168	0.1163	0.1158	0.1153	0.1148	0.1143	0.1138	0.1134	0.1129
19.0–19.9	0.1124	0.1120	0.1115	0.1111	0.1107	0.1102	0.1098	0.1093	0.1089	0.1085
20.0–20.9	0.1081	0.1077	0.1072	0.1068	0.1064	0.1060	0.1056	0.1052	0.1048	0.1045
21.0–21.9	0.1041	0.1037	0.1033	0.1029	0.1026	0.1022	0.1018	0.1015	0.1011	0.1007
22.0–22.9	0.1004	0.1000	0.0997	0.0993	0.0990	0.0987	0.0983	0.0980	0.0976	0.0973
23.0–23.9	0.0970	0.0967	0.0963	0.0960	0.0957	0.0954	0.0951	0.0948	0.0945	0.0942
24.0–24.9	0.0938	0.0935	0.0932	0.0930	0.0927	0.0924	0.0921	0.0918	0.0915	0.0912
25.0–25.9	0.0909	0.0906	0.0904	0.0901	0.0898	0.0895	0.0893	0.0890	0.0887	0.0885
26.0–26.9	0.0882	0.0879	0.0877	0.0874	0.0873	0.0869	0.0867	0.0864	0.862	0.0859
27.0–27.9	0.0857	0.0854	0.0852	0.0849	0.0847	0.0845	0.0842	0.0840	0.0838	0.0835
28.0–28.9	0.0833	0.0831	0.0828	0.0826	0.0824	0.0822	0.0819	0.0817	0.0815	0.0813
29.0–29.9	0.0811	0.0808	0.0806	0.0804	0.0802	0.0800	0.0798	0.0796	0.0794	0.0792
30.0	0.0790	—	—	—	—	—	—	—	—	—

Source: AGMA 440.04 Oct. 1971, American Gear Manufacturers Assn., Alexandria, VA. Reprinted with permission.

where A represents the external surface area of the housing in square feet, Γ_o the oil temperature, and Γ_a the ambient air temperature in °F. Heat transfer coefficient C_Γ is expressed in units of foot-pounds/(minute square feet °F) so that H_d is in units of foot-pounds/minute rather than in either Btu or calories. For purposes of calculation the area may be approximated as a function of center distance c as

$$A = 0.3c^{1.7} \tag{14.11.2}$$

for A in square feet and c in inches. Obviously 0.3 is a dimensional constant. Also for the purpose of textbook calculations the heat transfer coefficient C_Γ may be related

TABLE 14.10.4 COEFFICIENT OF FRICTION μ

Velocity range (fpm)	μ for Velocities 0–359 fpm									
	0	1	2	3	4	5	6	7	8	9
0–9	0.150	0.115	0.111	0.107	0.103	0.099	0.097	0.095	0.094	0.092
10–19	0.090	0.089	0.088	0.087	0.086	0.085	0.084	0.083	0.082	0.081
20–29	0.080	0.0792	0.0786	0.0779	0.0772	0.0765	0.0758	0.0751	0.0744	0.0737
30–39	0.0730	0.0726	0.0722	0.0718	0.0714	0.0711	0.0707	0.0703	0.0699	0.0695
40–49	0.0691	0.0687	0.0684	0.0680	0.0676	0.0673	0.0669	0.0665	0.0661	0.0658
50–59	0.0654	0.0651	0.0647	0.0644	0.0640	0.0637	0.0634	0.0630	0.0627	0.0623
60–69	0.0620	0.0618	0.0616	0.0614	0.0612	0.0610	0.0608	0.0606	0.0604	0.0602
70–79	0.0600	0.0598	0.0596	0.0594	0.0592	0.0590	0.0588	0.0586	0.0584	0.0582
80–89	0.0580	0.0578	0.0576	0.0574	0.0572	0.0570	0.0568	0.0566	0.0564	0.0562
90–99	0.0560	0.0558	0.0556	0.0554	0.0552	0.0550	0.0548	0.0546	0.0544	0.0542
100–109	0.0540	0.0539	0.0538	0.0537	0.0536	0.0535	0.0534	0.0533	0.0532	0.0531
110–119	0.0530	0.0528	0.0527	0.0526	0.0525	0.0524	0.0523	0.0522	0.0521	0.0520
120–129	0.0519	0.0518	0.0517	0.0516	0.0515	0.0514	0.0513	0.0512	0.0511	0.0510
130–139	0.0509	0.0507	0.0506	0.0505	0.0504	0.0503	0.0502	0.0501	0.0500	0.0499
140–149	0.0498	0.0497	0.0496	0.0495	0.0494	0.0493	0.0492	0.0491	0.0490	0.0489
150–159	0.0488	0.0486	0.0485	0.0484	0.0483	0.0482	0.0481	0.0480	0.0479	0.0478
160–169	0.0477	0.0476	0.0475	0.0474	0.0473	0.0472	0.0471	0.0470	0.0469	0.0468
170–179	0.0467	0.0465	0.0464	0.0463	0.0462	0.0461	0.0460	0.0459	0.0458	0.0457
180–189	0.0456	0.0455	0.0454	0.0453	0.0452	0.0451	0.0450	0.0449	0.0448	0.0447
190–199	0.0446	0.0444	0.0443	0.0443	0.0441	0.0440	0.0439	0.0438	0.0437	0.0436
200–209	0.0435	0.0434	0.0434	0.0433	0.0432	0.0432	0.0431	0.0430	0.0429	0.0429
210–219	0.0428	0.0427	0.0427	0.0426	0.0425	0.0425	0.0424	0.0423	0.0422	0.0422
220–229	0.0421	0.0420	0.0420	0.0419	0.0418	0.0418	0.0417	0.0416	0.0415	0.0415
230–239	0.0414	0.0413	0.0413	0.0412	0.0411	0.0411	0.0410	0.0409	0.0408	0.0408
240–249	0.0407	0.0406	0.0406	0.0405	0.0404	0.0404	0.0403	0.0402	0.0401	0.0401
250–259	0.0400	0.0399	0.0399	0.0398	0.0397	0.0397	0.0396	0.0395	0.0394	0.0394
260–269	0.0393	0.0392	0.0392	0.0391	0.0390	0.0390	0.0389	0.0388	0.0387	0.0387
270–279	0.0386	0.0385	0.0385	0.0384	0.0383	0.0383	0.0382	0.0381	0.0380	0.0380
280–289	0.0379	0.0378	0.0377	0.0376	0.0376	0.0375	0.0374	0.0374	0.0373	0.0373
290–299	0.0372	0.0371	0.0370	0.0369	0.0369	0.0369	0.0368	0.0367	0.0366	0.0366
300–309	0.0365	0.0364	0.0364	0.0363	0.0363	0.0363	0.0362	0.0362	0.0361	0.0361
310–319	0.0361	0.0360	0.0360	0.0359	0.0359	0.0359	0.0358	0.0358	0.0357	0.0357
320–329	0.0357	0.0356	0.0356	0.0356	0.0356	0.0356	0.0355	0.0355	0.0354	0.0354
330–339	0.0354	0.0353	0.0353	0.0352	0.0352	0.0352	0.0351	0.0351	0.0350	0.0350
340–349	0.0350	0.0349	0.0349	0.0348	0.0348	0.0348	0.0347	0.0347	0.0346	0.0346
350–359	0.0346	0.0345	0.0344	0.0344	0.0344	0.0344	0.0343	0.0343	0.0342	0.0342

Source: AGMA 440.04 Oct. 1971, American Gear Manufacturers Assn., Alexandria, VA. Reprinted with permission.

to the rpm of the worm, denoted by n_w, according to

$$C_\Gamma = 19.0 + 0.02250n_w \qquad (14.11.3)$$

if the air near the housing is relatively still and by

$$C_\Gamma = 19.0 + 0.03656n_w \qquad (14.11.4)$$

If a fan is to move air by the housing at some undetermined velocity. In an actual design the effective heat transfer coefficient should be found from prototype measurements in order to account for actual surface conditions and for the effect of the location of the housing on the machine. Its location affects the air flow it receives and the amount of heat to be gained or lost through the housing mount.

EXAMPLE 14.11.1

As the plant engineer for a small job shop which is to rebuild several take-up roll drives for a plastics factory, you have been asked to recommend the AGMA power rating for the worm gearing on some of the original equipment, which appears to have been custom made. Essential data are

- Worm OD = 1.030 in.
- $\phi = 14.5°$
- $w = 0.450$ in.
- $c = 2.010$ in.
- $n_w = 1725$ rpm
- Tooth depth = 0.302 in.

- Gear OD = 3.571 in.
- $N_w = 1$
- $N_g = 23$
- Hardened worm (60 Rc). Gear appears to have been sand cast

Assume a standard tooth profile so that the pitch diameter of the gear should be close to

$$d_{p_g} = \text{outside diameter} - \text{whole depth} = 3.751 - 0.302$$
$$= 3.269 \text{ in.}$$

Worm gear manufacturers may describe worm gear teeth in terms of a pseudo-pitch P_s, termed "pitch" in their catalogs. It is defined as $P_s = \pi/p_{t_g}$ so that $d_{p_g} = N_g/P_s$, similar to the spur gear relation. Worm gear teeth are often sized in integer values of P_s rather than P_d. According to the above definition, the pseudo-pitch (not a term shown in the AGMA standards) may be found from

$$P_s = \frac{23}{3.269} = 7.036 \text{ in.}^{-1}$$

Since this is quite close to 7.0, assume that $P_s = 7.00$. Thus, the corrected pitch diameter becomes

$$d_{p_g} = \frac{N}{P_s} = \frac{23}{7} = 3.2857 \text{ in.}$$

and from equation 14.7.11 the pitch diameter of the worm should be

$$d_{p_w} = 2c - d_{p_g} = 4.030 - 3.2857 = 0.7443 \text{ in.}$$

If this is the correct pitch diameter for the worm, and if the worm tooth profile were designed according to the addendum recommended by AGMA 341.02 for coarse pitch cylindrical worm gears,[6] which is $a = 1/P_s = p_{a_w}/\pi$ (given in the standard as $a = 0.3183 \, p_a$ instead of $a = p_a/\pi$) then the outside diameter of the worm should be

$$d_{a_w} = 0.7443 + 2/7.00 = 1.030 \text{ in.}$$

Since this is the diameter measured in the shop, our assumptions appear to be correct.

The lead angle may be found from equation 14.7.8 to be

$$\tan \lambda = \frac{3.2857}{23(0.7443)} = 0.1919 \qquad \lambda = 10.8649°$$

Determination of tangential force F_t according to equation 14.10.3 requires that we first find coefficient K_s for a sand cast gear and a hardened worm (Rockwell hardness Rc $\geq$ 58) from Figure 14.10.1 and find coefficients C_m and C_v from Tables 14.10.1 and 14.10.2. Hence, we have that $K_s = 943$, that $C_m = 0.824$, and that $C_v = 0.450$. This last coefficient was found by entering Table 14.10.2 at the sliding velocity already calculated.

Substitution of these values along with the gear tooth width and the worm pitch radius, which is taken to be its mean (average) radius, into equation 14.10.3 yields

$$F_t = 3.2857^{0.8}0.45(943)0.824(0.450) = 407.536 \text{ lb}$$

From Table 14.10.4 we see that μ changes so slowly that the jump between 342 and 343 rpm is due to rounding μ to four places. Since μ is given as 0.0349 for both 341 and 342 rpm, we shall assume that this value best approximates μ at 341.5 rpm. Likewise, since it is 0.0348 from 343 to 345 rpm, we shall assume that it best approximates μ at 344 rpm. Linear interpolation between these values yields $\mu = 0.03487$ at $v = 342.264$ rpm.

Substitution of these values into equation 14.10.2 produces

$$Z = \left(\frac{1725}{23} 3.2857 + \frac{0.03487(1725)0.7443}{\cos 14.5° \cos^2 10.8649°} \right) \pi 407.436$$

$$= 315\,428.065 + 61\,371.677 \text{ in-lb/min}$$

so that substitution into 14.10.1 yields

$$\text{hp} = 0.7965 + 0.1550 = 0.9469 \simeq 0.95$$

EXAMPLE 14.11.2

Estimate the oil temperature rise above the ambient temperature of 70 °F for the worm gear described in Example 14.11.1 when no fan is used.

We must calculate power loss H_d in ft.-lb./min., thermal capacity C_Γ in units of lb./(ft.-min.-°F), and area A in ft.^{2} before we can evaluate equation 14.11.1 for oil temperature Γ_o.

Since 1 hp = 33 000 ft.-lb./min. we immediately find that

$$H_d = 0.154(33\,000) = 5082.0 \text{ ft.-lb/min.}$$

From equation 14.11.3 we have that

$$C_\Gamma = 19 + 0.02250(1725) = 57.813 \text{ lb./(ft.-min.-°F)}$$

The center distance may be calculated from equation 14.7.9 after we calculate P_n from 14.7.5, which yields

$$P_{d_n} = 1/(0.7443 \sin 10.8649°) = 7.1278 \text{ in.}^{-1}$$

Thus,

$$c = \frac{1}{2(7.1278)} \left(\frac{1}{\sin 10.8649°} + \frac{23}{\cos 10.8649°} \right)$$
$$= 2.015 \text{ in.}$$

With the center distance known we may calculate area A from equation 14.11.2 to obtain

$$A = 0.3(2.015)^{1.7} = 0.987 \text{ ft.}^2$$

Upon solving equation 14.11.1 for Γ_o we have that

$$\Gamma_o = \Gamma_a + \frac{H_d}{C_\Gamma A} = 70 + \frac{5082.0}{57.813(0.987)} = 159.06 \,°F$$

Hence, the estimated oil temperature is approximately 160 °F.

14.12 MATERIALS FOR BEVEL, HYPOID, AND WORM GEARS

It has been the practice of gear manufacturers to use a more restricted list of materials for bevel and hypoid gears than that for spur gears, as given in Table 14.12.1. Turning next to worm gears, we find that although the worms themselves may be made from

TABLE 14.12.1   OUTLINE OF BEVEL AND HYPOID MATERIALS AND HEAT-TREAT PROCESSING FOR PRECISION-TYPE GEARS

Type	Precision Fine-Pitch (light load)	Precision, Load-Carrying		
		Fine Pitch	Aircraft Accessory Drives — Medium Pitch	
			Medium Stress	High Stress
Typical application	Instruments	Instruments, High-Speed Tools	Medium Stress	High Stress
Materials	1. .40% C steel (AISI 1040 or 1141) 2. Bronze (SAE 65) 3. Stainless steel (type 416)	1. Nitralloy 2. AISI 4340 / Medium-carbon alloy steel (AISI 4140, 4340)	Low-carbon medium-alloy steel (AISI 4620, 8617)	Low-carbon high-alloy steel (AISI 3310, 9310)
Heat-treatment and associated processing		1. Quench and draw rough blank 2. Finish machine 3. Cut teeth 4. Nitride 1. Heat-treat blank (34 Rockwell C+) 2. Finish blank 3. Grind teeth from solid 1. Heat-treat blank (34 Rockwell C max.) 2. Finish blank 3. Grind teeth from solid 1. Heat-treat blank (34 Rockwell C max.) 2. Finish blank 3. Cut teeth 4. Flame-harden teeth	Carburize and harden 1. Carburize and harden 2. Grind teeth	1. Carburize and harden 2. Grind teeth
Machine for hardening operation		Gear flame hardener (Gleason No. 1)	Quenching press (Gleason No. 16 and No. 529) Roller-quenching machine (Gleason No. 140)	

TABLE 14.12.2   TYPICAL GEAR BRONZES AND THEIR APPLICATION

Alloy	Composition (%)	Ultimate (psi)	Yield (psi)	Application
Bronze	Cu 88; Sn 10 Zn 2	46 000	19 000	Low-speed worm gears
Phosphor Bronze (chill cast)	Cu 89; Sn 10 Pb 0.25	50 000	22 000	Medium-speed worm gears
Nickel Phosphor Bronze (chill cast)	Cu 88; Sn 10.5 Ni 1.5 Pb 0.2	55 000	28 000	Medium-speed worm gears
Leaded Phosphor Bronze (sand cast)	Cu 87.5; Sn 11 Pb 1.5	50 000	22 000	High-speed worm gears
Aluminum Bronze	Cu 89; Al 10 Fe 1	65 000	27 000	Low-speed worm gears

one of the steels with good surface hardening properties for low wear, the gear is often made from phosphor bronze although bronze and alloys with nickel, lead, and aluminum are also used (see Table 14.12.2).

REFERENCES

1.   *Gear Nomenclature, Terms, Definitions, Symbols, and Abbreviations*, AGMA/ANSI 112.05, American Gear Manufacturers Association, Alexandria, Va, June 1976.

2.   Shtipelman, B.A., *Design and Manufacture of Hypoid Gears*, Wiley, New York, 1978.

3.   Buckingham, E., *Analytical Mechanics of Gears*, McGraw-Hill, New York, 1949.

4.   *Practice for Enclosed Cylindrical Wormgear Speed Reducers and Gearmotors*, ANSI/AGMA 6034-A87, American Gear Manufacturers Association, Alexandria, VA, March 1988.

5.   Walker, H., Thermal rating of worm gear boxes, *Proceedings of the Institution of Mechanical Engineers*, 151, 326–335 (1944) [British].

6.   *Design of General Industrial Coarse-Pitch Cylindrical Wormgearing*, AGMA 341.02, American Gear Manufacturers Association, Alexandria, VA., May 1970.

7.   Orthwein, W.C., Helical and worm gear design, *Computers in Mechanical Engineering*, 6 (4), 38–43 (1988).

8.   Burr, A.H., *Mechanical Analysis and Design*, Elsevier, New York, 1981.

PROBLEMS

Sections 14.1 and 14.2

14.1 Find the pitch angle for a bevel pinion and gear set that provides a 2:1 speed ratio. Also find the cone distances, the virtual numbers of teeth, and the diametral pitch if the pitch diameter of the pinion is 2.0 in. and it has 24 teeth.

14.2 Compare the number of virtual teeth for straight bevel gears having 60 teeth and pitch angles of 20°, 40°, and 60°.

14.3 Find the axial and radial forces and the bending moment transferred from the bevel and pinion gears described in problem 14.2 to their shafts.

Sections 14.2 and 14.3

14.4 Estimate the axial and radial forces and the bending moment induced at the pinion and at the gear of a straight bevel gear set having a 3:1 speed ratio which is rated at 1.10 hp at a pinion speed of 450 rpm. Diametral pitch is 8, the pressure angle is 20°, and the pinion has 16 teeth, with a face width equal to one-tenth of the outside cone distance.

   [*Ans. w* = 0.316 in., F_t = 162.2 lb.]

14.5 To maintain the backlash within acceptable bounds in a rather unusual design it is necessary that the deflection at the pinion of the bevel gear and pinion in problem 14.4 not exceed $\frac{1}{50}$th of the addendum at the rear cone when standard full depth teeth are used. Can this condition be satisfied if the unsupported length of the cantilever shaft is 5.50 in. and if the shaft diameter is 0.50 in.? $E = 3 \times 10^7$ psi.

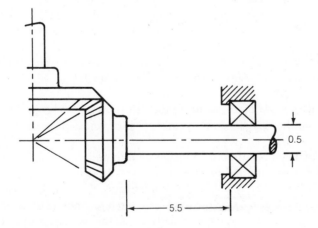

Cantilevered straight bevel gear.

14.6 Design a bevel gear set to provide a 2:1 ratio and to be driven by a 10-kW motor operating at 1725 rpm. Use a safety factor of 2.0 relative to the yield stress (415 MPa) and let

$$(w/d_p) \sin \gamma \leq 0.125.$$

Use $N > 18$ to be sure of no undercutting for a 20° pressure angle. Specify the pitch diameters, the pitch angles, the number of teeth, and the module. Solution of this problem may require several trial values for m. Note the variation of F_t, σ, and the stress sensitivity factor with changing m and d_p.

14.7 Design a bevel gear set to provide a 4.4 to 1 speed ratio using a safety factor of 2.20 relative to a yield stress of 70 000 psi. Let

$$(w/d_p) \sin \gamma \leqq 0.15$$

and use a 20° pressure angle. Specify the pitch angles, numbers of teeth, diametral pitch, and the tooth thickness at the pitch circle. Pinion speed is 1750 rpm at 2 hp and its pitch diameter may be no longer than 2.00 in. Use standard teeth and select $N > 18$ to prevent undercutting. Could a smaller number of actual teeth be used without undercutting? If so, what is that number? If not, why not?

14.8 Estimate the axial and radial forces and the bending moment induced by the pinion and bevel gear described in problem 14.7.

Section 14.4

14.9 Compare the torque of a spur gear and a helical gear when the teeth have the same normal force F_n, when both gears have been made from gear blanks of the same thickness, and when the normal diametral pitch of the teeth in the helical gear is equal to the diametral pitch of the spur gear. $N = 20$ for the spur gear and $N = 18$ for the helical gear. What is the helix angle for the helical gear? Helical gears with a face contact ratio greater than 1.0 may transmit more torque than corresponding spur gears because there is always more than one tooth in contact. See problem 14.12.

14.10 What is the minimum face width that a helical gear with a normal diametral pitch of 12 and a helix angle of 22° may have to insure continuous tooth contact?
   [*Ans.* 0.6989 in.]

14.11 A helical gear with a helix angle of 26° is to be used to transmit 1.5 hp at 690 rpm. Find the axial and radial forces that the bearings on the gear shaft must withstand when $N = 32$, $P_{d_n} = 10$, and $\phi = 20°$.

14.12 Since the profile in the plane of rotation of an involute tooth on a helical gear is a true involute (the profile in the plane of the normal to the tooth face is not; it is the projection of a true involute at an angle of ψ), we may calculate the contact ratio N_c from equation 13.4.5 for the tooth profile on the face of the gear using the transverse pitch p_t. Show with the aid of a sketch of the contact area that the number of teeth actually in contact will be greater than N_c if the face contact ratio is greater than 1.0. Demonstrate this for $N_c = 1.70$ and $N_f = 1.2$. The contact area may be defined as a rectangular region whose edge dimensions are $N_c p_t$ and w, where w is the width of the gear blank.

14.13 Show that

$$P_{d_t} = \pi/p_t = P_{d_n} \cos \psi$$

and that

$$P_{d_a} = \pi/p_a = P_{d_n} \sin \psi$$

Sections 14.4–14.6

14.14 Show that the velocity correction, or Barth corrections, may be written as

$$\frac{\sigma w p_n}{F_t} = \frac{1}{y}\left[1 + \left(\frac{v}{762\,000}\right)^{1/2}\right]^{1/2}$$

for velocity in mm/min for gears formed by hobbing and by

$$\frac{\sigma w p_n}{F_t} = \frac{1}{y}\left[1 + \left(\frac{v}{1\,854\,403}\right)^{1/2}\right]^{1/2}$$

for gears formed by shaping and grinding.

14.15 Design a helical gear set for teeth having standard proportions and with a 20° pressure angle for a speed ratio of 2.3:1 with an input power rating of 3 hp at 1160 rpm. The angle between shaft is 30° and the center distance is to be 3.80 in. Choose $\psi_1 = \psi_2 = 15°$ as shown in the figure and specify the pitch diameter of the pinion and gear in the normal direction, the number of teeth on each, the diametral pitch of the tooth profile, the tooth thickness at the pitch circle, the thickness of the gear blands, and whether the teeth are right- or left-handed on the pinion and gear. Use the Lewi-Barth equation for gears formed by shaving or grinding to select the tooth size. The working stress must not exceed 18 000 psi.

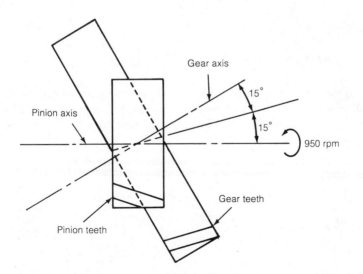

Helical gears on non co-planar shafts at an angle of 30°.

14.16 Redesign the gear set described in problem 14.15 using angles of $\psi_1 = 45°$ and $\psi_2 = 20°$ and $N_1 = 20$ teeth. Provide all of the information required in that problem.

14.17 Design a helical gear set using teeth of standard proportion with a 20° pressure angle and a speed ratio of 3:1 with an input power of 5 kW at 950 rpm and a center distance of 105 mm with shafts at an angle of 60° between them. Choose $\psi_1 = \psi_2 = 30°$ as shown and specify the pitch diameter of the pinion and gear, the number of teeth on each, the module, whether the teeth on the pinion and gear are right- or left-handed, and the thickness of the gear blank. Use the Lewis-Barth equation for gears formed by grinding to select the tooth size. The working stress is 83 MPa and the fact contact ratio should be 1.1 or greater.

14.18 Select the helix angle for two mating helical gears to transmit 360 kW between shafts at an angle of 70° using teeth with a 25° pressure angle. The center distance should be 642.0 mm and the steel used for the gears should not support a stress greater than 85 MPa. Pinon speed is 960 rpm and the gear speed should be 320 rpm. Experience with a somewhat similar gear set, but with the shafts at a different angle, indicates that a standard proportion module 10 tooth with $N_1 = 25$ teeth and $N_2 = 75$ teeth may provide adequate strength. Tooth surfaces are to be ground.

14.19 Redesign the gears described in problem 14.17 using $\psi_1 = 35°$ and $\psi_2 = 25°$ and supply the specifications requested in that problem. Use the same number of teeth as selected for problem 14.17 if possible. If not, modify N_1 and N_2 to satisfy $w \leq 6p_n$.

14.20 Can the gears designed in problem 14.17 be used for shafts that are parallel? If so, what is the speed ratio, the center distance, and the maximum input power rating at this angle? If not, why not?

14.21 Show that the sliding velocity between contacting teeth of two helical gears in mesh may be written as

$$v_s = r_{p_1}\omega_1 \sin \psi_1 + r_{p_2}\omega_2 \sin \psi_2$$

or as

$$v_s = r_{p_1}\omega_1 \sin \psi_1 - r_{p_2}\omega_2 \sin \psi_2$$

corresponding to the orientation of their axes, as shown in Figure 14.6.1b or c.

14.22 Compare the relative sliding velocity of the gears as arranged in problem 14.17 with that when they are arranged as in problem 14.19. Which arrangement would dissipate less energy?

Section 14.7

14.23 Show that

$$2C = \frac{1}{P_d}\left(\frac{N_w}{\sin \lambda} + \frac{N_g}{\cos \lambda}\right) \quad \text{or} \quad 2C = m\left(\frac{N_w}{\sin \lambda} + \frac{N_g}{\cos \lambda}\right)$$

for worm gears.

Section 14.8

14.24 Find the axial and radial forces on the worm and gear when the tangential force is 2200 N, the pressure angle is 25°, and the friction coefficient between worm and gear is 0.04. The lead angle is 10.5°.

14.25 Find the bearing loads on the worm described in problem 14.24 if they are located as shown in the figure 14.25. $d_{pw} = 60$ mm.

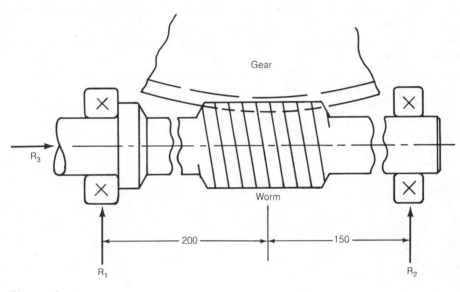

Worm and gear.

14.26 Find the torque ratio for a gear set in which the speed ratio is 48 : 1, the center distance is 10 in., the worm has a single tooth thread with a pitch diameter of 4.00 in., the lead angle is 4.76364°, and $\phi = 14.5°$. The gear pitch diameter is 16.00 in., the collar mean radius r_c is 3.00 in., $\mu_c = 0.08$, and $\mu = \mu_t = 0.10$. Compare this torque ratio with that which could be had if $\mu_t = 0$ and $\mu_c = 0$. This emphasizes the importance of using rolling element bearings in worm gear drives.

14.27 Calculate the torque ratio between input and output shafts of a double-reduction worm gear. (A double-reduction worm gear consists of a worm that drives a gear that is on the same shaft as a second worm that drives a second gear. Input and ouput shafts are parallel, but offset.) Each worm has two teeth and each gear has 60 teeth. The friction coefficient between worm and gear teeth for each pair is 0.04 and the friction coefficient at each thrust collar is 0.003. The thrust collar mean diameter is 3.60 in., the lead angle is 4.50, and the center distance for each worm and gear is 6.1198 in.

14.28 Calculate the efficiency of the double-reduction worm gear described in problem 14.27. (Efficiency is as 100% multiplied by the ratio of the torque that is actually transferred to that which could be transferred if there were no friction.

Section 14.10

14.29 Find the AGMA power rating of a worm and gear set described in problem 14.26 if the worm speed is 2180 rpm and if the gear was sand cast with $w = 1.60$ in.

Section 14.11

14.30 Calculate the power loss and the temperature rise of the oil above the ambient temperature for example 14.11.2 when the friction coefficient between worm and gear teeth is 0.04. All other quantities remain as in the example.

14.31 Calculate the temperature rise if a fan is used to cool the housing in problem 14.30.

14.32 The reduction gear described in problems 14.26 and 14.29 is to be used on a small excavator in a position where in some circumstances it may come in contact with the operator. Should the operator be protected by a cooling fan in the unit, should a heavy wire guard cage be placed around the reduction gear, or should both be used? The expected ambient temperature could rise to 115 °F in some parts of the world where it may be used.

CHAPTER FIFTEEN

MOTOR CHARACTERISTICS

NOTATION

c	damping constant	R_{RH}	rheostat resistance
I	effective moment of inertia of the load as measured at the motor	T	torque
		t	time
I_A	armature current	V	voltage
k	motor constant	ϕ	magnetic flux
n	revolutions per minute	ω	angular rotation
R_A	rotor (armature) resistance		

This chapter differs from the previous chapters in that it is not concerned with the design of a particular mechanical component or a particular class of components, but with the information needed to select a motor (electric) or an engine (internal combustion) to drive a mechanical system. The system may be as simple as a fan or as complex as a printing press or a textile machine. Selection consists of determining both the starting torque and the torque required as a function of the operating speed of the driven machine and then selecting a motor or engine that can deliver these required starting and running torques.

Hence, typical torque–speed curves for electric motors and internal combustion engines will be briefly considered along with the torque–speed curves for various driven machines.

Several books from the 1940s and 1950s are included in the references because they were written for users of electric motors. No recent books of this sort have been found. The newer textbooks on electric motors have been written for electrical engineers and consequently emphasize principles of motor design and operation. Many may not include speed–torque curves for all of the motor types discussed, and speed–torque curves for driven machines are omitted. Recent development of permanent magnet motors, reluctance motors, hysteresis motors, and solid state controls are noted in two recent texts listed at the end of the references. Their inclusion of speed–torque curves may be better than average.

The author thanks James Adams of the Applications Department of Baldor Electric Co., Fort Smith, Arkansas, and Vernold Feiste of Southern Illinois University for reviewing this chapter and providing valuable comments. Any errors or omissions are the fault of the author.

15.1 ELECTRIC MOTOR CLASSIFICATION

Electric motors will be divided into two categories: those operating on alternating current (AC), wherein the voltage varies sinusoidally, and those operating on direct current (DC), wherein the voltage is constant for a given operating condition. AC motors are also characterized as polyphase or single-phase (two wires and ground). Polyphase motors most commonly occur as three-phase motors (three wires and a ground wire), although some two-phase motors and some specialized motors with more than three phases have been manufactured.

All motors manufactured in the United States according to the NEMA (National Electrical Manufacturers Association) standards have frames and mounts that are completely interchangeable between manufacturers.

The classification of the motors to be discussed is summarized in Table 15.1.1.

TABLE 15.1.1   ELECTRIC MOTOR CLASSIFICATION

Motor Designation	Category			Comments[a]
	AC		DC	
	1-phase	3-phase		
Capacitor-start	X			Induction
Capacitor-start, capacitor-run	X			Induction
Split-capacitor	X			Induction
Shaded-pole	X			Induction
Repulsion	X			Induction
Repulsion-start induction	X			Induction
Repulsion-induction	X			Induction
Types A, B, C, D, & F[b]		X		Induction or Commutator
Synchronous (wound rotor)	X	X		Commutator
Reluctance (contoured rotor)	X	X		Synchronous
Hysteresis (no rotor winding)	X	X		Synchronous
Universal[c]	X		X	Commutator
Shunt wound			X	Commutator
Series wound[c]	X		X	Commutator
Compound wound			X	Commutator
Permanent Magnet			X	Commutator

[a]   Low-maintenance motors; induction motors have no electrical connection to the rotor, which is also known as the armature. Commutated motors have a segmented commutator and use carbon (or other soft conducting material) to conduct external electrical power to the rotor through the commutator.

[b]   Most motors of this type are induction motors in rating above 50 hp.

[c]   Small power series wound motors are known as universal motors.

15.2 ALTERNATING CURRENT MOTORS

Alternating current motors are classified according to four electrical types by NEMA: (1) induction motors, (2) synchronous motors, (3) part-winding-start motors, and (4) series-wound motors. Induction motors have a number of subclassifications according to designations suggested by NEMA. Part-winding motors have reduced initial starting torque because only part of the armature winding is energized then the motor is started. The remainder is energized as the motor speed increases. Series-wound motors have their rotor and field circuits in series.

Many of the motor descriptions will not be accompanied by a figure because their external appearances are very similar. In fact, if their nameplates, which describe their electrical characteristics, were removed it would be difficult, if not impossible, to distinguish between some of them without electrical and performance measurements.

Induction motors operate at fixed speeds determined by the number of poles arranged in the winding and the frequency of the supplied power. They are relatively inexpensive, rugged, and required little maintenance because they have no electrical connections (no brushes, commutators, or slip rings) on the rotor. When an induction motor is overloaded, however, heating will increase until the breakdown torque is reached, at which point the motor will quickly stop. Unless power is shut off immediately after the rotor stops the motor will be ruined from overheating.

Induction motors may have a variety of speed–torque curves depending upon their starting circuitry and upon the rotor resistance. In broad terms they may be subdivided into squirrel-cage induction motors and wound-rotor induction motors. Squirrel-cage motors are so named because of the arrangement of the conductors in their rotors, as shown in Figure 15.2.1. The space between conductors is filled nonconducting material to give the final appearance shown in Figure 15.2.2.

A wound-rotor induction motor is one in which the squirrel-cage rotor is replaced by one having winding or coils that are connected to slip rings so that the rotor circuit may be connected to an external circuit through brushes. The reason for this construction is that the torque–speed curve of the motor may be changed by changing the resistance in the rotor circuit as shown in Section 15.5, Figure 15.5.8.

An induction motor will run slightly less than synchronous speed (by 7% or less) whenever it delivers more than zero torque because it is the difference in the rotor's rotational speed and the rotational speed of the magnetic field produced by the field coils (or stator) that induces current in the rotor and thereby produces torque between the rotor and the field.

Truly synchronous motors have rotor windings that are connected to external circuitry through brushes and a commutator (see Figure 15.4.1) to control the magnetic field induced by current in the rotor winding. They may use an induction winding for starting and to add frequency stability.

Single-phase squirrel-cage induction motors are classified by NEMA as split-phase, resistance-start, or capacitor motors. A split-phase motor is equipped with an auxiliary

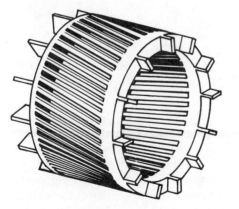

Figure 15.2.1 Squirrel-cage rotor without laminations. *Source: T.C. Lloyd*, Electric Motors and Their Applications, *Wiley-Interscience, New York, 1969. Reprinted by permission of John Wiley & Sons, Inc.*

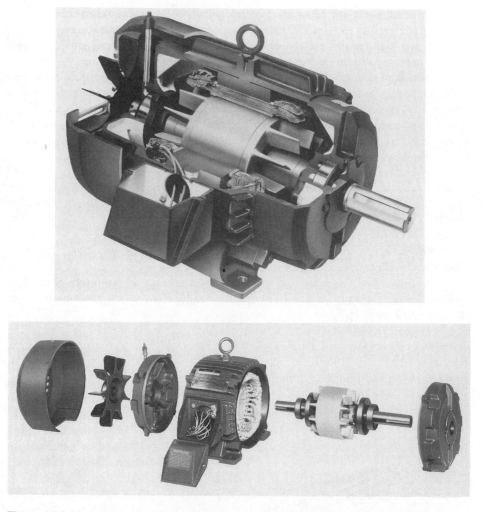

Figure 15.2.2 Induction motor with squirrel-cage rotor: (a) cutaway view, (b) exploded view. *Source: Reliance Electric Co., Cleveland, OH.*

winding displaced 90° in its magnetic position from the main winding. A split-phase motor with a resistance in series with the auxiliary winding is termed a resistance-start motor. Capacitor motors are further subdivided as capacitor-start, permanent-split capacitor, or two-value capacitor. As the names imply, the capacitor is in the circuit only during starting in a capacitor-start motor but remains in the circuit in a permanent capacitor motor. The two-value capacitor motor uses different capacitors for starting and for running.

Some capacitor motors, as shown in Figure 15.2.3, are easily identified by the externally mounted capacitor. This external appearance, however, does not indicate

Figure 15.2.3 The distinctive appearance of a capacitor motor with the external capacitor housing.

whether it is a capacitor-start, permanent-split capacitor, or a two-value capacitor motor. In applications where the motor and associated equipment are sold as a unit in a single housing, as in air conditioners, the capacitor for the capacitor motor may not be fastened to the motor itself.

A shaded-pole motor is one with an auxiliary short-circuited coil enclosing a part of each stator pole.

Single-phase wound-rotor motors are classified by NEMA as repulsion, repulsion-start induction motors, and repulsion-induction motors. The stator is connected directly to the input power while the rotor winding is connected to the input power through a commutator (see Figure 15.4.1) in the repulsion motor. If the winding is short-circuited at a predetermined speed it becomes equivalent to a squirrel-cage motor, and is termed a repulsion-start induction motor. If the rotor contains both a repulsion motor and a squirrel-cage winding it is termed a repulsion-induction motor.

15.3 POLYPHASE MOTORS

Most industrial motors of one horsepower or larger are three-phase induction motors because they are smaller and cheaper than single-phase motors of equal power. They range in capacity from less than one horsepower to 10 000 hp. In spite of their size advantage, fractional horsepower three-phase motors are not used in appliances and home workshop equipment because of the added expense of changing from single-phase to three-phase house wiring.

Three-phase motors are classified as designs A, B, C, D, and F on the basis of their speed–torque characteristics, which will be shown in Section 15.5.

Figure 15.4.1 Disassembled universal motor showing stator, rotor, and attached commutator and cooling fan.

15.4 UNIVERSAL MOTORS

Universal motors are classified separately by NEMA. They are series-wound motors which operate at about the same speed and power on direct current or single-phase alternating current at approximately the same root mean square voltage. They have commutators, as shown in Figure 15.4.1, and may be either series-wound (field and rotor circuits in series) or compensated series-wound (with a compensated field winding).

15.5 MOTOR SELECTION: SPEED–TORQUE CURVES FOR AC MOTORS

Speed–torque curves for both single-phase and polyphase AC motors are of the general form shown in Figure 15.5.1. Some of the common terminology used in describing the AC motor characteristics is included.

Understanding motor curves

Determining the power required to drive the machine is only the first step in sizing the motor. The next step is to examine the motor performance curves to see if the motor has enough starting torque to overcome machine static friction, to accelerate the load to full running speed, and to handle maximum overload.

Basic motor performance information is contained in the speed vs. torque curve. The curve shown here is for a general-purpose, squirrel-cage induction motor. It shows how output torque varies as speed increases from zero to synchronous speed with rated voltage and frequency applied to the motor.

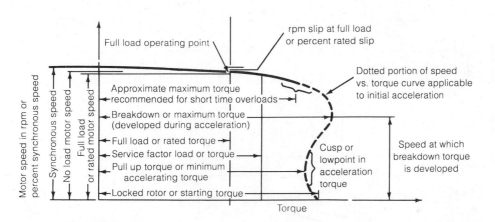

Figure 15.5.1 Representative speed–torque curve and associated terminology. *Source: Reprinted from* Machine Design, *May 13, 1982. Copyright 1982 by Penton/IPC, Cleveland, OH.*

Capacitor-start induction-run motors give a high starting torque which continues until the operating speed is attained. Permanent-split capacitor motors may have slightly higher starting torque, equal full-load torque, and slightly higher efficiency. The starting torque for a repulsion-start induction motor is similar to that of the capacitor-start motor except that the torque drops off faster as the design speed is reached. Although the split-phase motors have relatively smaller starting torque, they still have adequate starting torque for many applications. Shaded-pole motors represent the least expensive of the motors listed and are at the low end of the starting torque range. They are, therefore, used for fans, blowers, and similar easily started equipment.

The five NEMA design classifications, A, B, C, D, and F, for three-phase motors have different speed–torque curves, as illustrated in Figure 15.5.2. Designs A and B are known as normal starting torque motors.

Design A motors differ from design B motors in that their starting current and breakdown (or pull-up) torque is larger, so they are capable of greater load acceleration. Design B motors are standard, general purpose, three-phase motors with low starting current, normal torque, and normal slip. Design C motors have larger breakaway, (starting) torque than either A or B motors, low starting current, and low slip. They

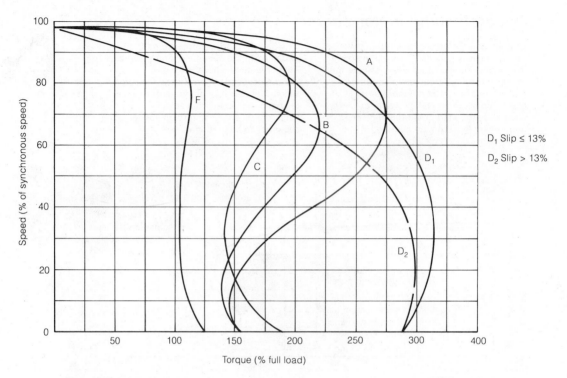

Figure 15.5.2 Speed–torque curves for three phase motors; designs A, B, C, D, and F. *Source: Reprinted from* Machine Design, *May 13, 1982. Copyright 1982 by Penton/IPC, Cleveland, OH.*

are, therefore, used for hard to start applications such as plunger pumps, piston compressors, and conveyors.

Design D motors have large breakaway torque and high slip: 4, 6, and 8 pole design D motors may have a starting torque that is 275% or more of their full load torque. They are used for punch presses, shears, oil well pumps, and similar high-inertia machinery where the rotor speed fluctuates, or slips, during the work cycle as mechanical energy stored at low torque and high rpm, usually in a flywheel, is transformed into work in another part of the cycle at high torque and low rpm. In these applications the motor must be chosen to have a torque that is large enough to accelerate the flywheel, or similar inertia, up to operating speed before the next heavy load is encountered. They are also used in multimotor drives that operate in parallel mechanically, as on conveyors and draglines.

Design F motors are rarely used because they have low starting torque. It is their relatively low starting current that justifies their use where the supply current is limited, as in remote locations or on small ships. Due to their low slip and low pull-up torque, they are limited to centrifugal pumps, blowers, and similar low inertia, easily started machines when used without a clutch. Machines with slightly greater inertia may be driven using a clutch; the clutch is engaged after the motor is first brought up to speed.

All of the speed–torque curves discussed have been displayed as percentages of full load and of operating speed in order to be independent of any particular power rating and line frequency.

Single-phase shaded-pole motors have a speed–torque curve that is a composite of the curves for each winding, and appear as in Figure 15.5.3 with typical speed and torque values shown. Because of relatively low efficiency and low starting torque, these

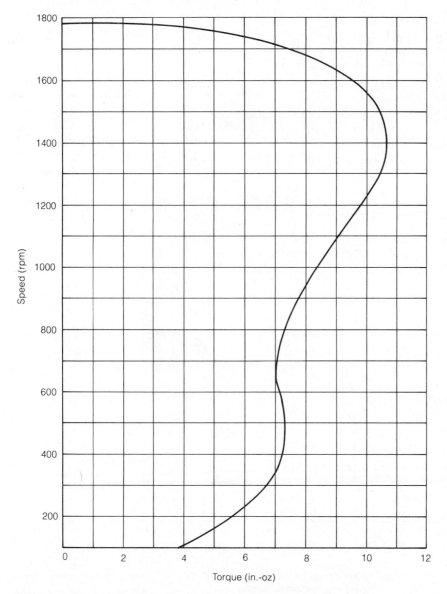

Figure 15.5.3 Representative speed–torque curve for a shaded-pole motor.

motors are used only for light duty appliances such as direct drive fans, phonograph turn-tables, and blowers. Their advantage is low manufacturing cost. Capacitor-start motors have greater starting torque than shaded pole motors, but require a centrifugal switch to remove the capacitor when 65 to 70% of the synchronous speed is approached. Typical speed–torque characteristics are shown by curve A in Figure 15.5.4. Permanent split capacitor motors dispense with the centrifugal switch. Although the efficiency may exceed that of other single-phase motors, the locked rotor torque is low, as illus-trated by curve C in Figure 15.5.4. Two-value capacitor motors retain the centrifugal switch and add another capacitor during the run portion. Hence, they are known as capacitor-start capacitor-run motors. Speed–torque curves for these motors are similar to that shown by curve B in Figure 15.5.4.

Split-phase motors are generally considered to be low to moderate starting torque motors, as illustrated in Figure 15.5.5. Starting torque would be zero if it were not for the auxiliary winding which cancels a phase shift and thereby provides the starting torque. Because this winding may seriously overheat if used for more than a few seconds, these motors are limited to about $\frac{1}{3}$ hp and to small appliances and office machines that require only short acceleration times.

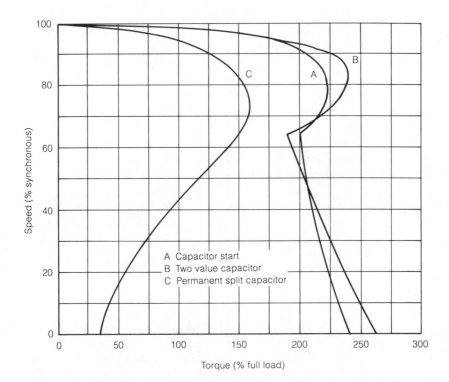

Figure 15.5.4 Speed–torque curves for (A) capacitor start, (B) two value capacitor, and (C) per-manent split capacitor induction motors.

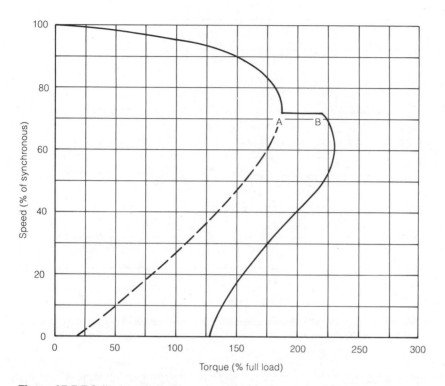

Figure 15.5.5 Split-phase induction motor speed–torque curve. Jump A-B occurs at the opening of the centrifugal switch.

Repulsion-start and the so-called universal motors represent different construction and performance. They are characterized by having brushes, commutator, and segmented rotor windings, as shown in Figure 15.4.1.

Repulsion-start and repulsion-induction motors are characterized by moderate to large starting torque and differ in that the repulsion-start motor, whose speed–torque curve is shown in Figure 15.5.6, uses a centrifugal switch to short-circuit the commutator at about 75% of the synchronous speed. This converts the motor to a squirrel-cage motor over the last 25% of the range. Repulsion-induction motors have a somewhat higher starting torque, as shown in Figure 15.5.7, than do repulsion-start motors and, therefore, may be preferred for applications that require starting at full load, such as conveyors and stokers.

As demonstrated by the speed–torque curves in Figure 15.5.8, an advantage of wound rotor construction is that adding different values of external resistance results in different characteristics. By switching from one to another it is possible to obtain a variety of speed–torque curves.

Small universal motors, which can operate on either AC or DC power, are widely used in small machines, such as vacuum cleaners, sewing machines, food mixers, electric

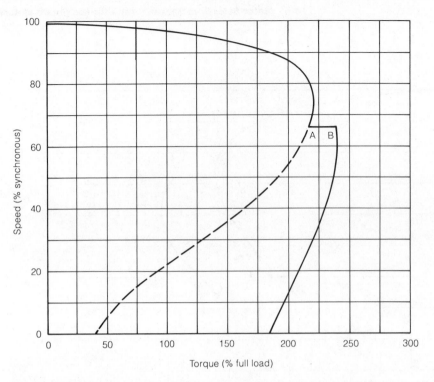

Figure 15.5.6 Repulsion-start induction motor speed–torque curve. Jump A-B occurs at the opening of the centrifugal switch.

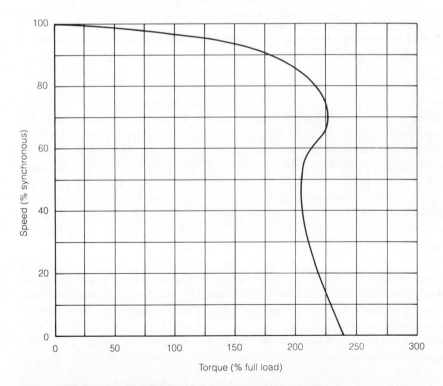

Figure 15.5.7 Speed–torque curve for a repulsion-induction motor.

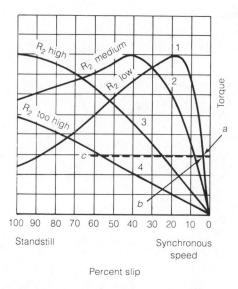

Figure 15.5.8 Effect of rotor resistance upon the speed–torque curves for wound-rotor induction motors. *Source: T.C. Lloyd*, Electric Motors and Their Applications, *Wiley-Interscience, New York 1969. Reprinted by permission of John Wiley & Sons, Inc.*

drills, and hedge trimmers. As is evident from Figure 15.5.9, these are not constant speed motors because the speed increases as the torque decreases. Consequently, large universal motors should not be used where part failure, such as breakage of a belt, could cause the motor to lose its load and overspeed. This is not a problem in small ($\frac{1}{16}$ hp or less) motors because wind and bearing resistance provides enough of a load to prevent dangerous speeds.

Synchronous motors with zero slip (i.e., always in synchronism with the driving frequency) are available for either polyphase or single-phase supply, with single-phase type including spit-phase, capacitor-start, permanent-split capacitor, and shaded pole. All of them operate at precisely the frequency of the supply voltage as long as their torque limitations are not exceeded.

Synchronous motors require more maintenance than induction motors because of commutator and brush wear.

Other characteristics depend upon the power range of the motor. Fractional horsepower synchronous motors require only an AC supply and self-starting circuitry; i.e., shaded-pole or capacitor-start. Synchronous motors of 1 hp and greater usually require DC excitation of the rotor and AC exitation of the stator. In addition, these motors must be started by having the rotor connected to a starting circuit, by having a squirrel-cage starter winding on the rotor in addition to the synchronous winding, or by a starter motor connected to the motor shaft through a overrunning clutch (described in the chapter on clutches and brakes). As noted in a previous paragraph, addition of an induction winding provides additional frequency stability.

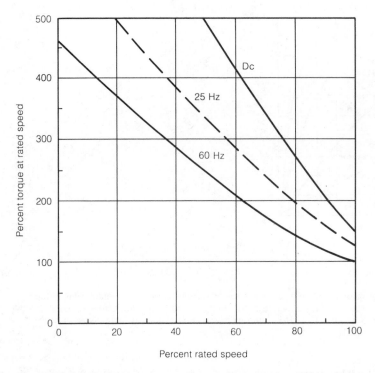

Figure 15.5.9 Torque–speed curves for a universal motor at 60 Hz, 25 Hz, and DC.

The constant speed range from 0 to *A* or to *B* in the speed–torque curve for a synchronous motor in Figure 15.5.10 merely indicates that changing the torque has no steady-state effect. It does have a transient effect. When the torsional load is changed the motor will hesitate momentarily until the phase relations between rotor and stator change to accomodate the new power requirement and then it will again rotate in synchronism with the supply frequency.

As a synchronous motor is started and accelerated up to speed by either the starter winding or the starter motor a point will be reached slightly below its synchronous speed where it jumps into synchronization. The torque developed to make the final pickup in speed is termed the *pull-in torque*. It is indicated by point *A* in Figure 15.5.10.

If the driven load is increased until the motor is overloaded another point will be reached at which the motor can no longer remain in synchronization and still provide the larger torque. The point at which it drops out of synchronization is termed the *pull-out torque*. It is indicated by point *B* in Figure 15.5.10. Neither of these terms should be confused with the pull-up torque, which is the minimum torque developed in accelerating from rest to the speed at which breakdown occurs, both illustrated in Figure 15.5.10.

Two other types of synchronous motors are known as reluctance motors and as hysteresis motors. The reluctance motor has magnetic poles on the rotor produced by

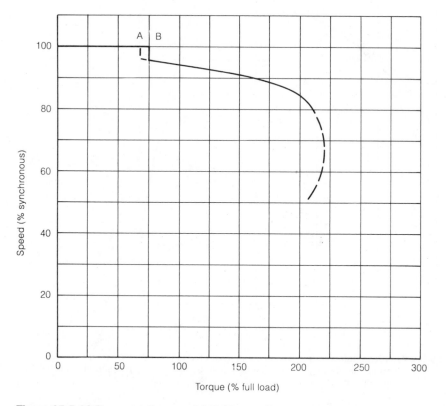

Figure 15.5.10 Representative speed–torque curve for a synchronous motor. Point *A* denotes the pull in torque and point *B* denotes the pull out torque.

either a noncylindrical rotor or by auxiliary windings. It may be made self-starting by adding a squirrel-cage winding. The torque–speed characteristics of a reluctance motor are shown in Figure 15.5.11. The rotor of a hysteresis motor is simply a cylinder of magnetic material: no winding, or protrusions, no laminations. It operates from the magnetic poles induced in the rotor by the rotating magnetic field of the stator. Torque is induced because hysteresis losses in the rotor prevent its magnetic field from exactly following the field changes in the stator. Perhaps their most common application is that of electric clock motors. Torque–speed curves for hysteresis motors this size and larger are similar to that shown in Figure 15.5.12.

Although the speed–torque curves discussed here may be obtained from motor manufacturers, they are not the curves usually supplied to the buyer. The more common curves are the so-called performance curves displayed in Figure 15.5.13, which show the speed in rpm, the efficiency, the current in amperes, the power factor, the kilowatts required, and the torque as a function of horsepower. Rather than extending the torque curve to 0 rpm, the starting torque is simply indicated by a notation listing the locked-rotor (or starting) torque. Normal operating speed is in the region of maximum effi-

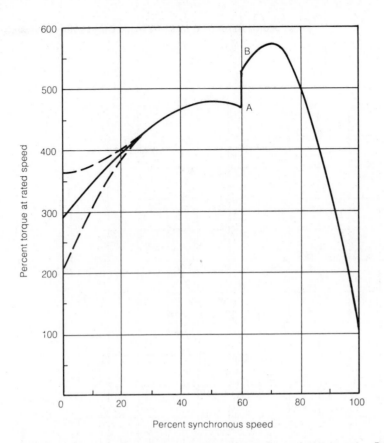

Figure 15.5.11 Representative torque–speed curve for a reluctance motor. Dashed curves indicate variation in starting torque as determined by the starting position of the rotor. Discontinuity at *AB* occurs at switching.

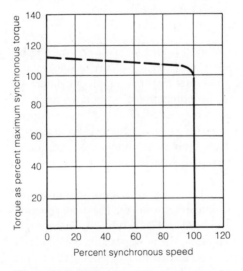

Figure 15.5.12 Representative torque-speed curve for a hysteresis motor. Dashed curve is estimated, depending upon rotor material.

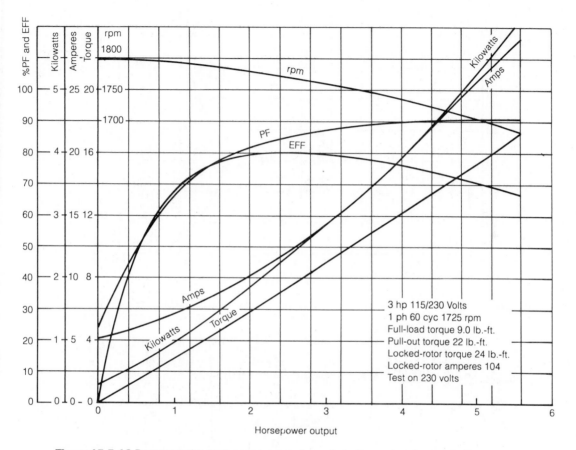

Horsepower output

Figure 15.5.13 Representative performance curve for an induction motor. *Source: Baldor Electric Company, Fort Smith, AR. Reprinted with permission.*

ciency, which for the motor shown is approximately 83% for an output between 2 and 4 hp and centered about 1740 rpm. A small portion of the speed–torque curve may be plotted from the performance curve by plotting the torque against speed as calculated from the corresponding horsepower. The result is shown in Figure 15.5.14 (page 908).

15.6 MOTOR SELECTION: SPEED–TORQUE CURVES FOR DC MOTORS

Direct current motors are classified according to their rotor windings relative to their field windings. The four standard classifications are series, shunt, compound, and permanent magnet. Their respective circuits are shown symbolically in Figure 15.6.1 and their torque–speed curves are shown in Figure 15.6.2.

Shunt-wound motors have good speed control characteristics because, as may be seen from these curves, they have the smallest change in torque for a given change in

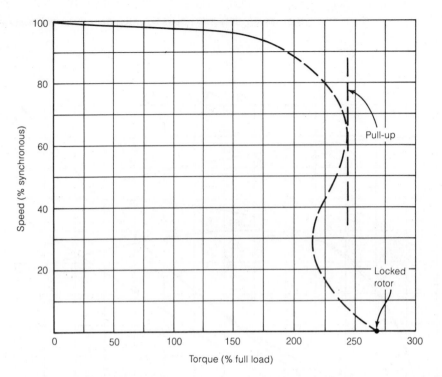

Figure 15.5.14 Portion of the speed–torque curve included in the performance curve (solid curve) along with the pull-up boundary and locked-rotor value. Dashed curve represents estimated values.

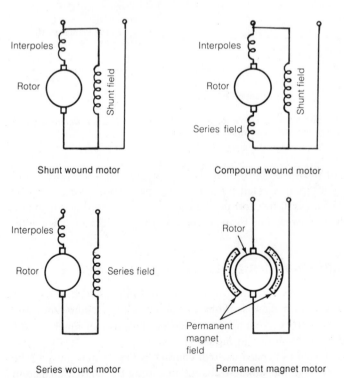

Figure 15.6.1 Schematic showing winding differences between shunt, compound, series, and permanent magnet wound DC motors. (Interpoles are stator windings designed to compensate for field distortion due to the rotor.)

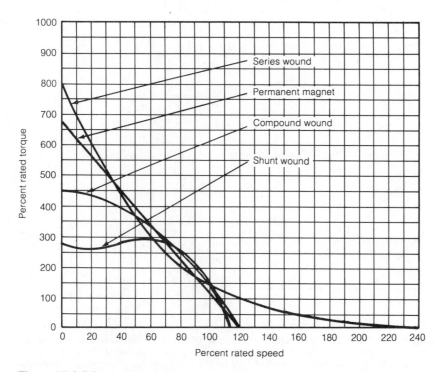

Figure 15.6.2 Torque–speed curves for shunt-wound, compound-wound, series-wound, and permanent magnet DC motors.

speed of all of the DC motor types. Since their no-load speed is generally no more than about 15% above the rated speed, they are safe if the load is lost. They are also simple to control in applications where reversing direction is important, as in rolling mills for steel production.

Compound-wound motors provide a compromise between the speed stability of shunt wound motors and the greater starting torque associated with series-wound motors. Increasing the strength of the series field relative to the shunt field is said to increase the compounding and provide characteristics more like that of a series motor. Although standard compounding is about 12%, it may approach 50% for hard starting applications, such as hoists.

Series-wound motors offer large starting torques and high torque output per ampere at the expense of good speed control. They are ideally suited for traction motors, which are direct drive motors used in electric and diesel–electric railroad locomotives and underground mine cars, because they can deliver large torque at low speeds, as in starting or in pulling a heavy load up an incline. (Direct drive means that the wheels are mounted on opposite ends of the rotor shaft.) Once on level, straight, track they may run at high speed using only that torque required to overcome wind and bearing resistance.

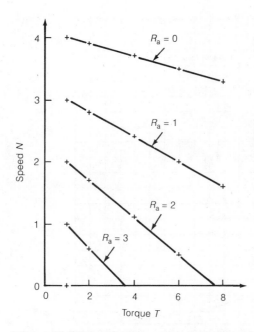

Figure 15.6.3 Speed–torque curves for saturated (constant) field series (and universal) motors with rotor resistance control. *Source: A. Kusko*, Solid State D-C Motor Drives, *MIT Press, Cambridge, MA, 1969. Copyright 1969 by the Massachusetts Institute of Technology Press. Reprinted with permission.*

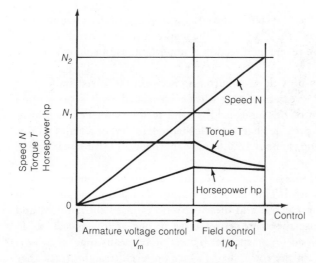

Figure 15.6.4 Speed–torque curve for a DC motor with solid state control of rotor (armature) voltage of a shunt motor and with field control. *Source: A. Kusho*, Solid State D-C Motor Drive, *MIT Press, Cambridge, MA, 1969. Copyright 1969 by the Massachusetts Institute of Technology Press. Reprinted with permission.*

Direct drives are desirable for series wound motors because they can rapidly accelerate to dangerous speeds under no-load conditions. (Dangerous speeds may be those in excess of 5000 rpm, which may cause motor explosion due to large centrifugal forces on the rotor.)

Permanent magnet DC motors have permanent magnets in place of electrically excited stator coils. They have good starting torque capability, are energy efficient, but with poorer speed regulation compared to compound-wound motors. Although they can be dynamically braked and reversed, it should be done at low rotor voltages—generally less than about 10% of rated voltage. Typical speed–torque curves for these motors are shown in Figures 15.6.2 and 15.6.7.

Solid state control of the rotor voltage in shunt motors leaves the magnetic field at full amplitude and thus provides constant torque over a range of rotational speeds, as indicated in Figure 15.6.4.

Series DC motors may be controlled by a combination of resistance control and solid state voltage control. The effect of series resistance control is exhibited in Figure 15.6.5, where we find that the speed–torque curves are lowered as the resistance is in-

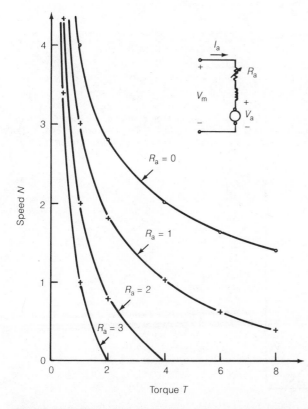

Figure 15.6.5 Speed–torque curves obtained from resistance control of a series DC motor. *Source: A. Kusko, Solid State D-C Motor Drives, MIT Press, Cambridge, MA, 1969. Copyright 1969 by the Massachusetts Institute of Technology Press. Reprinted with permission.*

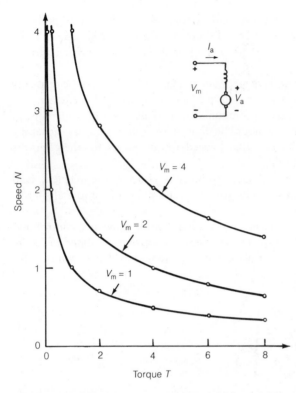

Figure 15.6.6 Speed–torque curves obtained from solid state voltage control of a series DC motor
Source: A. Kusko, Solid State D-C Motor Drives, *MIT Press, Cambridge, MA, 1969. Copyright 1969 by the Massachusetts Institute of Technology Press. Reprinted with permission.*

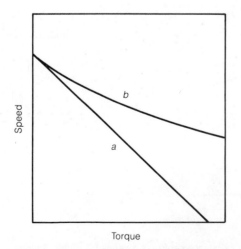

Figure 15.6.7 Representative speed–torque for (*a*) a permanent magnet motor with constant DC voltage supply, and (*b*) with a rectified AC voltage supply.

creased. Increased voltage, on the other hand, raises the speed–torque curves, as is evident from Figure 15.6.6.

15.7 SPEED CONTROL OF ELECTRIC MOTORS

Speed control of DC motors is accomplished by adjusting one or more of the parameters in the equation

$$n = \frac{V - I_A(R_A + R_{RH})}{k\phi} \tag{15.7.1}$$

where n designates the rotational speed and V denotes the voltage supplied to the motor. Rotor current is denoted by I_A, the rotor resistance by R_A, the rheostat resistance by R_{RH}, and the magnetic flux by ϕ, where k is a constant. Although early DC motors were often controlled with a rheostat (a variable resistor) which adjusted the magnetic flux ϕ by adjusting the field current, newer motors employ solid state voltage control to regulate V because they dissipate less power and may be capable of control over a greater speed range.

Speed control of AC motors using voltage control is usually more difficult because it depends upon the speed–torque curves of both the motor and the load. For example, consider speed control by means of voltage control for a fan and for a buffer, both driven by a shaded-pole motor. By superimposing the speed–torque curves for the fan and the motor and adding the speed–torque curves of the motor at various voltages as in Figure 15.7.1 we see that as the voltage is reduced the speed of the fan moves

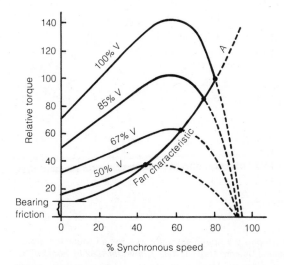

Figure 15.7.1 Speed regulation of a wound-rotor induction motor and fan combination using voltage control.

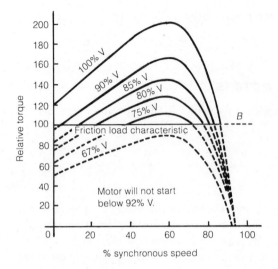

Figure 15.7.2 Speed–torque curve for a shaded-pole motor and resistance load. *Source: E.K. Bottle, Fractional Horse-power Electric Motors, Griffin, London, 1948. Reprinted with permission.*

to the point where the respective speed–torque curves intersect. It is evident, therefore, that speed control by means of voltage control is effective over about 40% of the synchronous speed range of the fan, from about 40 to 80% of the synchronous speed. Superposition of these speed–torque curves for the motor upon the speed–torque curve

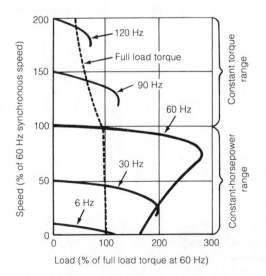

Figure 15.7.3 Effect of frequency variation upon the speed–torque curve for an induction motor. *Source: Reprinted from Machine Design, May 13, 1982. Copyright 1982 by Penton/IPC, Cleveland, OH.*

for the buffer, Figure 15.7.2, indicates that speed control by means of voltage control becomes less effective in this application. The voltage cannot be reduced by much more than 25%, which in this particular instance also provides about a 25% variation in synchronous motor speed; i.e., from about 90 to 65%. Speed control for AC motors also may be obtained using a variable frequency output from an electronic inverter. Frequency adjustment to control AC motor speed typically produces speed–torque curves similar to those shown in Figure 15.7.3 for an induction motor.

15.8 DRIVEN MACHINE SPEED–TORQUE CURVES

Proper matching of a motor to a driven machine can be accomplished only after the speed–torque curve of the driven machine is known. Unfortunately, many manufacturers have not conducted experiments to find the speed–torque curves for their product for a variety of reasons, including the difficulty of devising techniques to obtain the data for a full range of speeds and torques. When these curves are at hand, however, it becomes a relatively straightforward matter to select the appropriate motor for a particular machine.

In the absence of this information, an estimate of the speed–torque curve may be made using data at one point, such as the rated speed at full torque, and placing a typical curve for machines of this type through that point.

A typical curve for a friction device, such as a belt or disk sander, a buffer, and so on, is shown in Figure 15.8.1. The difference between static and dynamic friction accounts for the initial drop in torque as the speed increases from zero for the solid line. Since material on the surface may increase the friction coefficient as it is heated

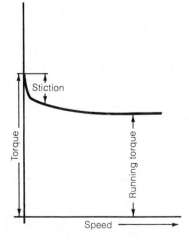

Figure 15.8.1 Torque-speed curve of a friction machine (sander, buffer, etc.). *Source: E.K. Bottle, Fractional Horse-power Electric Motors, Griffin, London, 1948. Reprinted with permission.*

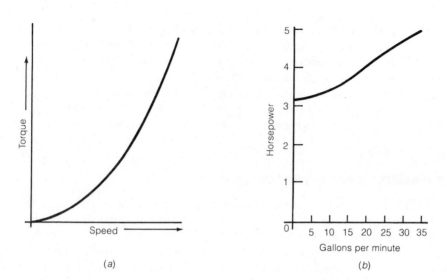

(a)

(b)

Figure 15.8.2 Torque–speed curves for (a) a fan, and (b) a centrifugal pump. *Source: E.K. Bottle, Fractional Horse-Power Electric Motors, Griffin, London, 1948. Reprinted with permission.*

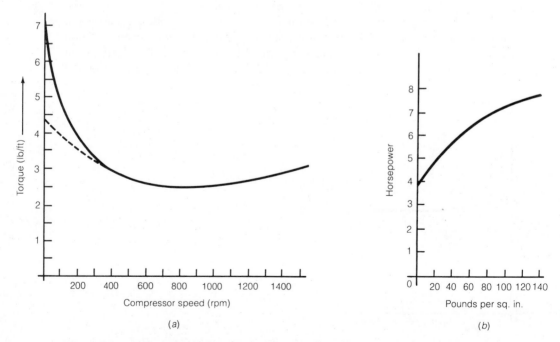

(a)

(b)

Figure 15.8.3 Torque-speed curves for (a) small piston compressor with no relief valve, and (b) piston compressor with relief valve. *Source: E.K. Bottle, Fractional Horse-Power Electric Motors, Griffin, London, 1948. Reprinted with permission.*

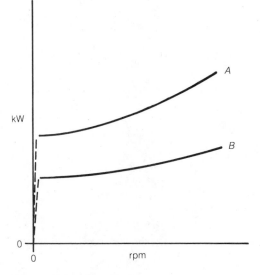

Figure 15.8.4 Power vs. depth of cut curves for a milling machine. Curves *A* and *B* represent two different advance rates.

due to buffing, the dashed curve must also be considered. Fan and blower curves appear as in Figure 15.8.2*a*. Centrifugal pumps show a similarly rising curve, but with the origin displaced vertically because of the inertia of the fluid in the pump, as sketched in Figure 15.8.2*b*. Small piston type compressors usually display a curve similar to that shown in Figure 15.8.3*a*. Larger piston type compressors which supply a reservoir with pressure regulation tend to follow a curve similar to that in Figure 15.8.3*b*. Finally, a typical speed–torque curve for a cylinder-boring machine is drawn in Figure 15.8.4 in terms of depth of cut for two different advance rates. Transformation of data in this form to that required for motor selection will be discussed in the next section.

15.9   MOTOR SELECTION: MATCHING THE MOTOR TO THE DRIVEN MACHINE

As noted earlier, choosing an electric motor to drive a particular machine is largely a matter of comparing their speed–torque curves to assure that adequate torque is available over the operating range of the driven machine. Two factors are of particular importance: (1) the capability of the motor to handle changes in the load without danger to the motor itself, and (2) its ability to accelerate the load up to its operating speed, or speeds, in a reasonable amount of time. We shall consider the first of these factors in this section and the second in the next section.

As an example of selecting a direct drive motor for a particular machine, suppose that a motor is to be selected for an industrial floor buffer. Since the buffer's speed–

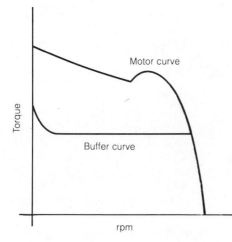

Figure 15.9.1 Superposition of the torque–speed curves for a capacitor-start induction motor and a buffer.

torque curve is similar to that shown in Figure 15.8.1, it will be necessary to use a motor with a large starting torque, such as that represented by curve A for a capacitor-start motor in Figure 15.5.4. Superposition of the curves as in Figure 15.9.1 indicates that the torque available for accelerating to operating speed during start-up is the difference between the two curves and that the power required is proportional to the area between them.

If an induction motor with a speed–torque curve as illustrated by curve C in Figure 15.5.4 had been selected the superimposed curves may have appeared as in Figure 15.9.2. This combination would clearly result in a buffer that would never start unless lifted from the floor because the initial torque provided by the motor is less than the torque required to rotate the unit, even without considering that the static friction is less than the dynamic friction.

Although the motor represented by curve C in Figure 15.5.4 was unsuited for the floor buffer, it is better suited to driving a fan than either of the possible choices for the buffer. Comparison of its speed–torque curve with that of a fan, Figure 15.8.2a indicates that it may give a smoother start to the fan because of lower initial torque and will definitely give better performance at operating speed because of its greater torque at the higher speed range. In Figure 15.9.3 curve A represents the speed–torque for the induction motor and curve B represents the speed–torque curve for the fan.

Turning to the boring machine whose performance was given in terms of power as a function of depth for a particular speed, we must relate the power to the torque, as shown in Figure 15.9.4, which must then be related to the torque–speed curves as illustrated. In this particular example curve A may hold for 1725 rpm and curve B for 1450 rpm and the torque–speed curve used is determined by the applied voltage, such as voltages, V_1, V_2, and V_3.

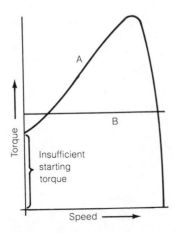

Figure 15.9.2 Superposition of the torque–speed curves for a wound-rotor induction motor and a buffer.

To demonstrate the use of these curves, suppose we wish to cut at a depth of 0.5 mm, at a motor speed of 1750 rpm, and at an advance rate corresponding to curve A. If we enter chart 1 at the lower end of curve A, read over to the 1750 rpm line in chart 2, and then down to the torque axis to find what torque is required of a minimum cut of a few hundredths of a millimeter. If we repeat the process for a point at the deepest cut for which the machine is designed we find the range of the required torque. Upon transferring this torque to the 1750 line in chart 3 we have the torque range at 1750 rpm for the motor we select. If an induction motor is used, the curves

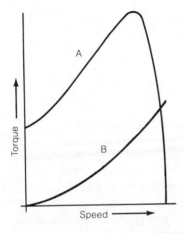

Figure 15.9.3 Superposition of the torque–speed curve for a low-impedance wound-rotor motor and a fan.

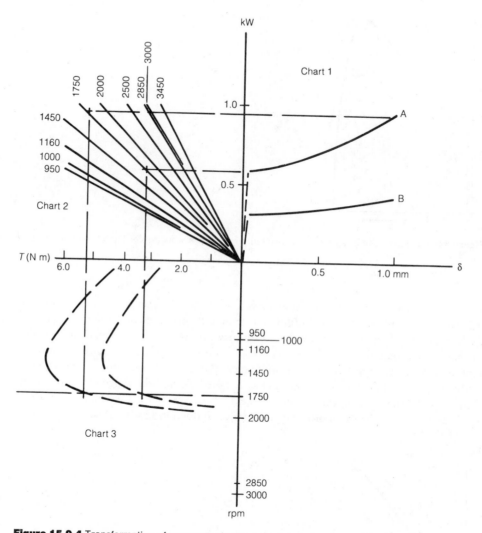

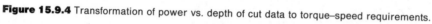

Figure 15.9.4 Transformation of power vs. depth of cut data to torque–speed requirements.

for the motor selected may appear as shown by the broken curve in chart 3 for a series of voltages. These curves also indicate the range of voltage control required.

The discussion in this section has been based upon an implied direct drive between the motor and the driven unit. A mismatch between motor and driven unit speed torque curves may, in some cases, be overcome by use of an appropriate clutch between the motor and the driven unit.

Other considerations that may arise in special applications are the cleanliness of the motor in food and drug processing facilities, its isolation from any internal arcing in explosive environments, such as petroleum refineries or grain elevators, and its ability to operate at very hot or very cold temperatures. Most problems of this type may be solved using special enclosures or special motors, such as explosion-proof motors, which are stock items in low horsepowers.

15.10 TIME TO ACCELERATE TO OPERATING SPEED

Whenever the time required for the motor to accelerate its load from one speed to another is an important aspect of motor selection, the acceleration time may be estimated from the speed–torque curves and the relation

$$I\frac{d\omega}{dt} + c\omega = T \tag{15.10.1}$$

Since the speed–torque curves for most motors are usually found experimentally, numerical integration using the trapazoidal rule is often the most practical means of arriving at an estimate of the acceleration time. In equation 15.10.1 time is represented by t, the net torque (the difference between the motor torque and any resisting torque provided by the load, such as a weight on a hoist) by T, the effective moment of inertia (see Section 11.8) by I, the damping constant by c, and the angular velocity in radians/second by ω. Integration of (15.10.1) yields the acceleration time as (x substituted for ω in the integrand)

$$t = I\int_0^\omega \frac{dx}{T - cx} \qquad T = T(\omega) \tag{15.10.2}$$

in which the damping constant is denoted by c.

EXAMPLE 15.10.1

Recommend an AC motor to drive a conveyor belt having a constant torque–speed curve. The equivalent moment of inertia of the conveyor system at the motor ouput shaft is 0.53 slugs and the equivalent moment of inertia of the material on the belt is expected to be 0.34 slugs. Curve B in Figure 15.10.1 represents the torque–speed curve for the conveyor itself and curve A represents the curve for conveyor plus product. Estimate the start-up time to accelerate to its operating speed of 1730 rpm when the damping is negligible. The starting torque of the motor should not be so large as to accelerate the conveyor to the point where material on it is shifted or upset during start-up.

The requirement for a soft start motivated consideration of a permanent split capacitor induction motor having the torque–speed curve shown in Figure 15.10.1. Table 15.10.1 lists values read from the motor torque–speed curve and from the larger of the two conveyor curves. It is the difference between these two torques, listed as T, that is used in equation 15.10.2. Numerical integration with the trapazoidal rule between 0 and 180.64 radians/s yields $t = 1.29$ min. as the longest starting time.

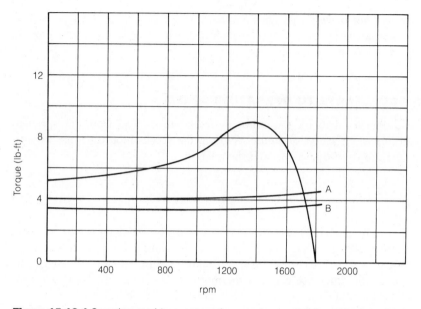

Figure 15.10.1 Superimposed torque–speed curves for an electric motor and a friction load.

TABLE 15.10.1   TORQUE–SPEED DATA FOR MOTOR AND CONVEYOR

n	ω	T_1	T_2	ΔT	1/ΔT
0	0	5.20	4.00	1.20	0.775
100	10.47	5.30	4.01	1.29	0.775
200	20.94	5.35	4.03	1.32	0.758
300	31.42	5.43	4.04	1.39	0.719
400	41.89	5.50	4.05	1.49	0.671
500	52.36	5.65	4.06	1.59	0.629
600	62.83	5.80	4.07	1.73	0.578
700	73.30	6.00	4.09	1.91	0.524
800	83.78	6.25	4.10	2.15	0.465
900	93.25	6.60	4.13	2.47	0.405
1000	104.72	7.00	4.15	2.85	0.351
1100	115.19	7.70	4.18	3.52	0.284
1200	125.66	8.45	4.20	4.25	0.235
1300	136.14	8.95	4.23	4.72	0.212
1400	146.60	8.97	4.30	4.67	0.214
1500	157.08	8.55	4.33	4.22	0.237
1600	167.55	7.50	4.38	3.12	0.321
1700	178.02	5.50	4.42	1.08	0.962
1720	180.12	—	—	0.36*	2.778
1725	180.64	—	—	0.18*	5.556
1730	181.17	4.45	4.45	0.00	—

*   Obtained by linear interpolation between 1.08 and 0.00.

15.11 GASOLINE AND DIESEL ENGINES

Typical speed–torque curves for gasoline engines are as shown in Figure 15.11.1 when plotted in terms of specific torque, which is the total torque divided by the cubic inch displacement of the engine. Inclusion of the specific horsepower is to emphasize that the power output increases with speed to a point beyond the peak torque, but also reduces its slope as it approaches its maximum. It is because of the relatively narrow range of speed at which the engine delivers 80% power or better that speed change transmissions are used. By changing the gear ratio over a set of ranges the engine itself may operate at near 80% of maximum power or better while for a broad range of transmission output speeds.

Speed control is obtained by controlling the amount of fuel metered to each cylinder during the combustion stroke. Increasing the fuel increases the force released during combustion and hence increases the acceleration of the piston during that stroke.

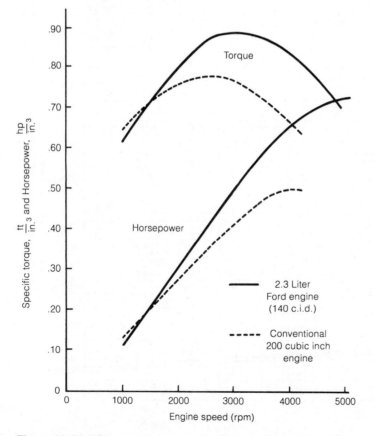

Figure 15.11.1 Torque-speed curves for two gasoline engines. *Source: Ford Motor Company, Dearborn, MI.*

Smooth speed control is obtained if the increase in fuel is matched to a proper increase of oxygen, a change in the ignition time, and control of the temperature of the combustion chamber and the mixing of the air and fuel in the chamber prior to and during combustion.

Engine performance curves for a diesel engine with two different fuel pump calibrations are shown in Figures 15.11.2 and 15.11.3. Although both torque–speed curves are qualitatively similar their different negative slopes with increasing speed are impor-

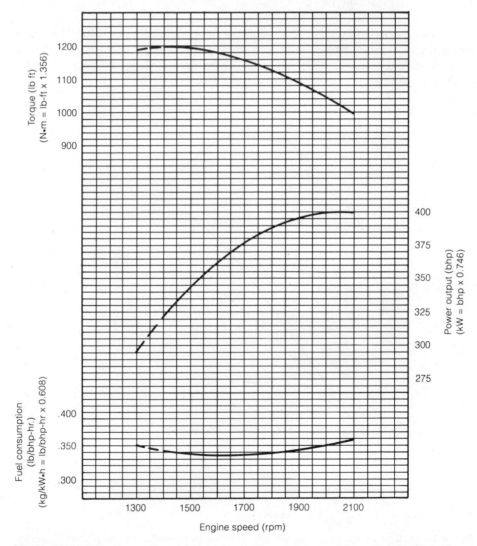

Figure 15.11.2 Torque, power, and fuel consumption versus speed for a diesel engine. *Source: Cummins Engine Company, Inc., Columbus, IN.*

tant because of the different power speed curves that result. The power speed curve shown in Figure 15.11.2 is desirable for driving an electric generator where the engine is to run at one speed and to deliver maximum power at that speed. The power speed curve in Figure 15.11.3, on the other hand, is desirable for driving a farm tractor so that when the load increases, as in plowing through soils of different consistency, the reduced speed will be associated with increased power output. This type of power–speed curve is also desirable for crushers and grinders, which may be subject to fluctuating loads.

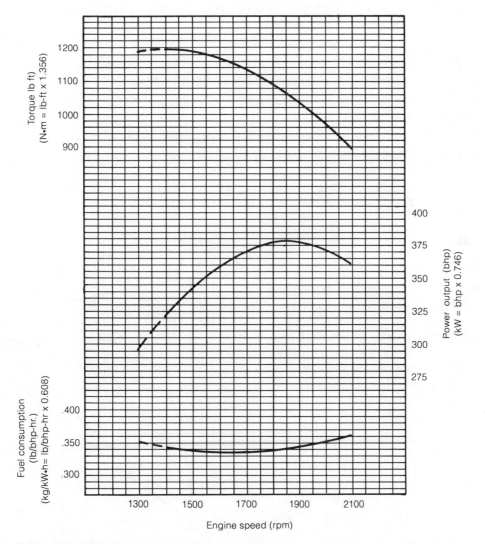

Figure 15.11.3 Torque, power, and fuel consumption versus speed for a diesel engine. *Source: Cummins Engine Company, Inc., Columbus, In.*

Matching these engines to a particular driven unit begins with selection of an engine to supply the required power and is followed by selection of the fuel pump calibration to provide the desired power–speed curve. A transmission is then selected to match the engine output speed, usually at 80% power or greater, to the required input speed, or speeds, of the driven unit.

REFERENCES

1. *Motors and Generators*, ANSI/NEMA Standards Publication MG 1-1978, National Electrical Manufacturers Association, Washington, DC.

2. Lloyd, T.C., *Electric Motors and Their Applications*, Wiley-Interscience, New York, 1969.

3. Bottle, E.K., *Fractional Horse-power Electric Motors*, Griffin, London, 1948.

4. Siskind, C.S., *Direct Current Machinery*, McGraw-Hill, New York, 1952.

5. 1984 Electrical and Electronics Reference Issue, *Machine Design*, 56 (12), Penton/IPC, Cleveland, OH (31 May 1984).

6. Kusko, A., *Solid-State D-C Motor Drives*, MIT Press, Cambridge, MA, 1969.

7. Matsch, L.W., and Morgan, J.D., *Electromagnetic and Electromechanical Machines*, 3d ed., Harper & Row, New York, 1986.

8. Chapman, S.J., *Electric Machinery Fundamentals*, McGraw-Hill, New York, 1985.

PROBLEMS

Section 15.5

15.1 Plot the speed–torque curve for the 623M motor whose performance curves are shown in the figure. Note that the full-load torque is listed as 4.5 lb-ft and that the locked-rotor torque is listed as 28 lb-ft, so that the torque scale as a percentage of full load torque tends well beyond 350%. Locked-rotor torque is the starting torque.

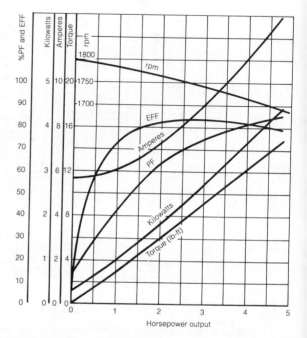

Source: Baldor Electric Company, Fort Smith, AR. Reprinted with permission.

Section 15.8

15.2 Sketch the torque–speed curve for the chuck drive on metal-working lathe and for the needle drive for a sewing machine. Justify your choices for starting conditions and for operating conditions.

15.3 Sketch the expected speed–torque curve for a ball mill operating at 6 rpm with a motor speed of 690 rpm. Assume the weight of the drum is one-third the weight of the material to be pulvarized and that the material rises to 40° before falling from the baffles. Recommend a three-phase motor to drive the ball mill. The motor should be able to start the mill when it is fully loaded, but with the center of gravity of the load below the axis of drum rotation, as shown in the figure.

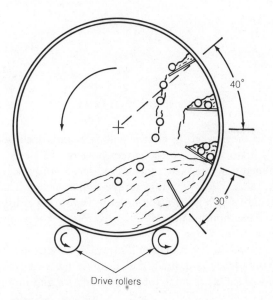

Ball mill schematic

Section 15.9

15.4 Describe the starting and running characteristics of a disk sander when driven by a motor with the speed–torque curve *a* in the figure as compared with one driven by a motor with curve *b*

[*Ans.* When motor *a* is used the sander will start quickly (and tend to rotate in the hands of the operator) and will not easily stall if the operator pushes on it or if it encounters somewhat adhesive materials. Motor *b* will not jerk as it starts, but may stall if pressed too hard against a surface or if it used against somewhat adhesive materials.]

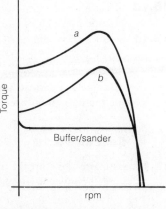

15.5   Select a single-phase motor to drive the mixer for a thixotropic material (such as dripless paints) for which the torque–power curve is that shown in the figure. Justify your choice on the basis of running torque and start-up time. Large initial torque is desired.

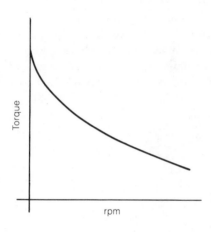

15.6   Your company has decided to compete in the market for electric golf carts and forklifts. What electric motor would you recommend for traction for their battery powered golf cart? What type of speed control would you recommend? Why? Sketch the desired speed–torque curve for the golf cart, with careful attention to all expected starting torque requirements. The vice-president for sales would like you to avoid using a gearshift in order to have a lower list price.

15.7   What type of electric motor would you recommend for the lift drive for a battery operated forklift? What controls would you recommend? Why?

15.8   Select an electric motor, either AC or DC, to drive a rolling mill in which the direction of the rotation must change. Although constant torque is desired, the customer has been told that constant torque may not always be possible. Is constant torque possible for the motor you selected to power the rollers in the mill?

15.9   If the rolling mill is to be produced for a location where skilled electronics maintenance may be difficult to find, would you change your recommendation for problem 15.8?

15.10   Select an electric motor for a residental garage door opener to be produced by your company. Include a sketch of the speed–torque curve you expect and use it to justify your selection.

15.11   Select a three-phase AC motor to drive the hydraulic pump to supply hydraulic power for the die cylinders that force the dies together and for the ram that forces the molten metal between the dies in a die-casting machine. Small starting torque is to be expected because the die casting machine is operated only after the motor is at operating speed. The hydraulic pump is sized for the maximum pressure.

Section 15.10

15.12   Estimate the start-up time for the motor and conveyor belt in example 15.10.1 when the un-loaded conveyor torque–power curve, curve B, is used.

15.13 An economy version of the conveyor belt system described in example 15.10.1 is to be produced with the rolling element bearings replaced by grease-lubricated journal bearings. This introduces a damping factor $c = 0.0035$ lb-s/rad. Find the starting time for the empty conveyor.

15.14 Compare the start-up time for the motor and conveyor in example 15.10.1 with that when the capacitor-start motor is replaced with a split-phase motor whose speed–torque relation is shown in the table.

[*Ans.* 0.62 min., 48% of time in Example 15.10.1.]

NET TORQUE VS. SPEED FOR A SPLIT PHASE INDUCTION MOTOR DRIVING THE BUFFER IN EXAMPLE 15.10.1

rpm	ω	ΔT	$1/\Delta T$
0	0	2.519	0.397
86.25	9.032	2.667	0.375
172.50	18.064	2.994	0.334
258.75	27.096	3.367	0.297
345.00	36.120	4.630	0.216
431.25	45.160	5.464	0.183
517.50	54.192	6.410	0.156
603.75	63.225	7.299	0.137
690.00	72.257	8.264	0.121
776.25	81.289	9.259	0.108
862.50	90.321	10.000	0.100
948.75	99.353	10.638	0.094
1035.00	108.385	10.989	0.091
1121.15	117.825	10.989	0.091
1207.50	126.449	10.526	0.095
1242.00	130.062	10.101	0.099
1242.00	130.062	10.101	0.099
1293.75	135.481	7.246	0.138
1380.00	144.513	6.711	0.149
1466.25	153.545	6.061	0.165
1552.50	162.577	4.202	0.238
1638.75	171.609	0.840	1.190
1690.50	177.029	0.420	2.381
1707.75	178.835	0.000	—

Section 15.11

15.15 Using a gasoline engine with a displacement of 642 in.^{3} and a speed–torque curve identical to that in Figure 15.11.1, recommend a gear ratio, or a set of gear ratios, to be used to power a small dredge for mine reclamation pits which will allow the engine to operate between 85 and 100% of specific torque when the output shaft of the gear train is to operate from 10 and 230 rpm. Assuming the gear train to be 93% efficient, plot the horsepower (not specific horsepower) delivered to the gear train output shaft over the speed range. Plot in increments of 20 rpm.

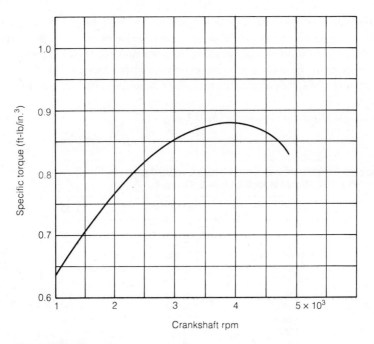

Figure P15.15

15.16 A motor manufacturer has just developed a gasoline engine with a speed–torque curve as shown in the figure. Its price is comparable with that of the engine described in problem 15.15. Would you recommend changing to the new engine? Why?

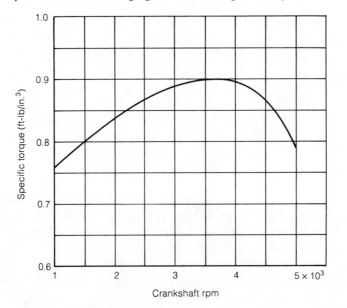

CHAPTER SIXTEEN

HYDRAULICS

NOTATION

A	area (l^2)	Pe	pressure energy (ml^2/t^2)
a	constant $(\ )$	p	pressure (m/lt^2)
b	constant $(\ )$	Q	flow volume (l^3)
E	energy (ml^2/t^2)	r	radius (l)
F	force (ml/t^2)	t	time (t)
g	acceleration due to gravity (l/t^2)	V	volume (l^3)
hp	horsepower (ml^2/t^3)	v	velocity (l/t)
K	fluid correction factor $(\ )$	x	displacement or a variable $(l), (\ )$
k	spring constant (spring rate) (m/t^2)	y	variable $(\ )$
KE	kinetic energy (ml^2/t^2)	z	vertical displacement, height (l)
kW	kilowatts $(ml^2/t^2\_\_)$	β	accumulator factor, or beta factor (1)
L, l	length (l)	ρ	mass/unit volume (m/l^3)
m	mass (m)	μ	dynamic (absolute) viscosity (m/lt)
n	polytropic gas exponent (1)	ξ	reference pressure drop per unit length (m/l^2t^2)
PE	potential energy (ml^2/t^2)		

This chapter is intended to be only an introduction to hydraulics. In that context, the following sections are intended to acquaint the reader with some of the more common hydraulic components, with the notation accepted by the American National Standards Institute (ANSI) for drawing hydraulic circuits, with typical data provided by manufacturers of hydraulic components, and with the design of hydraulic circuits.

16.1 HYDRAULIC EQUIPMENT

Perhaps the most common machines with an extensive and obvious use of hydraulics are backholes and end loaders, examples of which are shown in Figures 16.1.1 and 16.1.2, although hydraulic brakes in automobiles may account for the largest number of hydraulic systems. In all of these examples the hydraulic systems were installed because they enabled the designer to significantly magnify and/or transfer forces.

Inclusion of hydraulic components in the earth-moving machines enables relatively large forces to be applied at locations remote from the engine with comparatively little additional weight and complexity. In these and other machines, such as presses and

Figure 16.1.1 Hydraulically activated backhoe, or shovel. Hydraulic cylinders control the articulation of the shovel and of the boom and arm. *Photo by the author.*

Figure 16.1.2 Small end loader, front, with a small backhoe attached at the rear of the tractor. Both are hydraulically controlled. *Photo by the author*.

mining machines, an engine-driven pump can power thrust cylinders and/or torque motors at remote locations under the control of a single operator at a central location. Although the pump, thrust cylinder, and controls are the most evident of the hydraulic components involved, the system requires inclusion of various other components to be described in the following paragraphs.

16.2   ABSOLUTE AND GAGE PRESSURE

In working with hydraulic components it is necessary to distinguish between absolute pressure and gage pressure when selecting pumps and reservoirs. *Absolute pressure* is that pressure measured relative to an absolute vacuum; *gage pressure* is the pressure difference between the pressure being measured and the ambient pressure. Unless otherwise specified, the ambient pressure is assumed to be atmospheric pressure. Even though the atmospheric pressure at a particular location depends upon the altitude and the weather, it is assumed in most design calculations to be 14.5 psi in the Old English system of units and 1 *bar* ($=10^5$ Pa) in the SI system. In those formulas that hold for either reference system (i.e., involving only pressure differences) no distinction will be made. Otherwise psia will denote absolute pressure in pounds per square inch, psig will denote gage pressure in the same units, and bar will denote gage pressure in SI units. In practice, other units of pressure are also used, such as the height of a water column or the height of a column of mercury that will produce the specified pressure at its base. In this text the height of a mercury column will refer to absolute pressure only.

16.3 BERNOULLI'S THEOREM

Bernoulli's theorem in its simplest form follows from the conservation of energy in a nonviscous, incompressible fluid. The kinetic energy of a volume of fluid of mass m moving with velocity v is given by

$$\text{KE} = \tfrac{1}{2}mv^2 \qquad (16.3.1)$$

and the potential energy is given by

$$\text{PE} = mgz \qquad (16.3.2)$$

where z is the elevation above a reference elevation and g is the acceleration of gravity. Upon writing the pressure energy at pressure p as

$$P_e = pV \qquad (16.3.3)$$

where V is the reference volume, we may write the total energy of the reference volume of fluid as

$$\tfrac{1}{2}mv^2 + mgz + pV = EV \qquad (16.3.4)$$

Upon dividing by the reference volume this may be written as

$$\tfrac{1}{2}\rho v^2 + \rho gz + p = E \qquad (16.3.5)$$

Since (16.3.5) holds for any points 1 and 2 in the circuit, we may write

$$\tfrac{1}{2}\rho(v_1{}^2 - v_2{}^2) + \rho g(z_1 - z_2) + p_1 - p_2 = 0 \qquad (16.3.6)$$

We shall consider this as the standard form for the Bernoulli theorem. Note that in the form given in equation 16.3.6 the pressure may be either gage pressure or absolute pressure. The units, however, must be consistent: If g is in ft/s^2 then v must be in ft/s, p must be in lb/ft^2, ρ must be in lb/ft^3, and z must be in ft. In the SI system if g is in m/s^2 then v must be in m/s, p must be in N/m^2, ρ must be in kg/m^3, and z must be in m.

This formula is particularly important in the design of hydraulic systems because it clearly shows the relation between pressure and the flow velocity in a hydraulic line. If one is increased the other is decreased; it is an important relation which forces the hydraulic system designer to always think of the effect of one upon the other.

EXAMPLE 16.3.1

A pressure gage mounted at station 1 in a fluid line with an internal diameter of 1.0 in. reads 644.0 psig for a fluid flow of 18.0 gpm (gal/min) of hydraulic oil which weighs 47.4 lb/ft^3. It passes through a reducer to a pressure hose with an internal diameter of 0.75 in., to a gear motor on a robot arm that moves 29 in. above and below the level of station 2 as shown in Figure 16.3.1. Find the gage pressure at stations 2, 3, and 4. Energy losses at the reducer are negligible [1 gal = 231 in.^{3}].

From equation 16.3.6 we may write

$$\rho \frac{v_1^2}{2} + p_1 + \rho g z_1 = \rho \frac{v_2^2}{2} + p_2 + \rho g z_2$$

where $p_1 = 664.0 \times 12^2 = 92\,736$ psfg. The velocities may be found from

$$v_1 = \frac{\dot{Q}}{A_1} \qquad v_2 = \frac{\dot{Q}}{A_2} \quad \text{and} \quad v_2 = v_1 \frac{A_1}{A_2}$$

along with

$$v_1 = \frac{18(231)}{60(12)^3} \frac{(12)^2}{\pi(0.5)^2} = 7.353 \text{ ft/s}$$

and

$$v_2 = 7.353 \frac{\pi(0.5)^2}{\pi(0.375)^2} = 13.072 \text{ ft/s}$$

Since there is no change in elevation between stations 1 and 2 the z terms cancel in (16.3.6) so that upon solving for p_2 at station 2 we find that

$$p_2 = \frac{\rho}{2}(v_1^2 - v_2^2) + p_1$$

$$= \frac{47.4}{2(32.2)}(7.353^2 - 13.072^2) + 92\,736.0 = 92\,650.025 \text{ psfg}$$

$$= 643.4 \text{ psig}$$

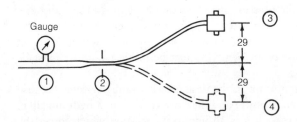

Figure 16.3.1 Pressure line (1), reduced section (2), and torque motor at positions (3) and (4).

Because there is a change in elevation at station 3

$$p_3 = p_2 + \rho g(z_2 - z_3)$$
$$= 92\,650.025 + 47.4\left(0 - \frac{29}{12}\right) = 92\,535.475 \text{ psfg} = 642.607 \text{ psig}$$

At station 4 we find

$$p_4 = 92\,650.025 + 47.4\left(\frac{29}{12}\right) = 92\,764.575 \text{ psfg} = 644.2 \text{ psig}$$

16.4 HYDRAULIC CYLINDERS

Some of the more common types of hydraulic cylinders are displayed in Figure 16.4.1. Styles *a* and *b* are the most common. Since both cylinders show two ports, one at the head end and one at the cap end, the distinction between single- and double-acting cylinders is whether fluid pressure is delivered under external control to both ends or to just one. The piston and rod in the single-acting cylinder may be extended by forcing fluid into the port at the cap end. It is allowed to drain from that port as the rod is retracted by an external force. The port at the head end may be used to admit air or fluid as the rod is retracted. Obviously single-acting cylinders may also be retracted hydraulically and extended mechanically.

Double-acting cylinders, as represented by Figure 16.4.1*a*, have ports on each end of the cylinder so that the piston and rod may be moved hydraulically in either direction. Terminology for the major parts of a hydraulic cylinder of either type is shown in Figure 16.4.2.

If inertial forces are high the cap end of the cylinder will undergo a shock when the piston and rod contact the cap. This shock may be reduced by installing a hydraulic cushion as shown in Figure 16.4.3. At the cap end it consists of a cushion spear, a needle valve, and a secondary drain line from the needle valve as shown in the figure. As the piston approaches the end cap the tapered portion of the cushion spear partially closes the larger port at the cap end and slows the flow from the cylinder, thus decelerating the piston. When the cushion spear finally closes the larger port the draining of the remaining fluid is slowed still further as it is diverted through the small needle valve orifice. The cushion sleeve performs the same function at the head end.

Hydraulic cylinders are usually used to transform work into a particular combination of force and distance. Consider two cylinders connected by a hydraulic line as shown in Figure 16.4.4 and let force F_1 act upon cylinder 1. If the cross-sectional area of the piston is denoted by A_1 the pressure required in the fluid to hold piston 1 in

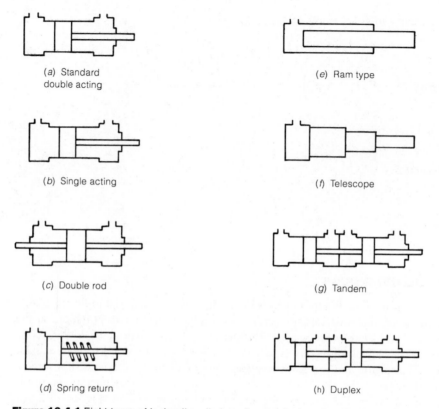

(a) Standard
double acting

(e) Ram type

(b) Single acting

(f) Telescope

(c) Double rod

(g) Tandem

(d) Spring return

(h) Duplex

Figure 16.4.1 Eight types of hydraulic cylinders. *Source: Parker-Hannifin Corp., Cleveland, OH.*

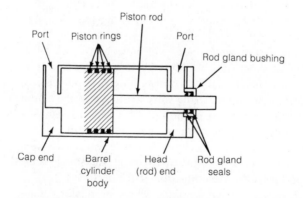

Figure 16.4.2 Terminology for major components of a hydraulic cylinder. *Source: Parker-Hannifin Corp., Cleveland, OH.*

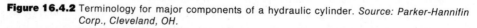

Cushioning action

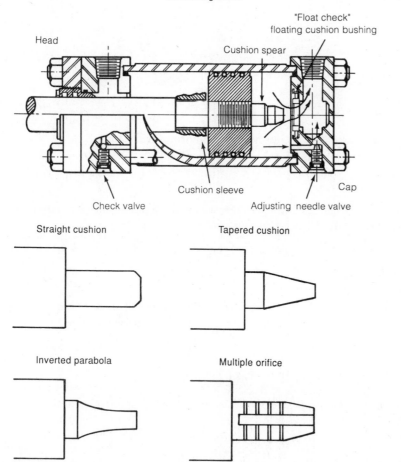

Straight cushion

Tapered cushion

Inverted parabola

Multiple orifice

Figure 16.4.3 Cushion sleeve to cushion motion toward the head end, cusion spear for motion to-ward the cap end, and four varieties of cusion spears. *Source: Parker-Hannifin Corp., Cleveland, OH.*

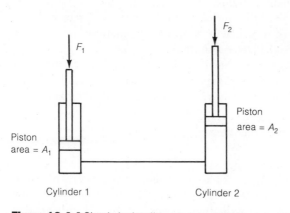

Figure 16.4.4 Simple hydraulic system to show force, area, and cylinder displacement relations. The hydraulic fluid is considered to be incompressible.

equilibrium will be given by

$$p_1 A_1 = F_1 \tag{16.4.1}$$

Hence, the force necessary to hold piston 2 in equilibrium will be given by

$$p_1 A_2 = F_2 \tag{16.4.2}$$

since the pressure is unchanged throughout a stationary fluid. Upon elimination of p_1 between (16.4.1) and (16.4.2) we find that

$$F_2 = \frac{A_2}{A_1} F_1 \tag{16.4.3}$$

If force F_2 moves piston 2 then piston 1 must also move if none of the incompressible fluid leaves or enters the system. If no energy is lost the work done by piston 2 must be equal that done by piston 1. Thus,

$$F_1 x_1 = F_2 x_2 \tag{16.4.4}$$

where x_1 and x_2 denote the displacements of pistons 1 and 2, respectively. Substitution for F_2 in (16.4.4) from (16.4.3) yields

$$x_1 = \frac{A_2}{A_1} = x_2 \tag{16.4.5}$$

Hence, the increased force on piston 1 is obtained at the expense of increased motion of piston 2. This last relation could have been obtained directly by equating volumes if we had not wanted the intermediate relations.

Although equation 16.4.3 will not hold during motion because of pressure losses in the hydraulic lines and cylinders, which will be discussed in more detail in the following sections, it will hold once the pistons and fluid come to rest.

Equation 16.4.4 is an approximation to the actual motion and force relations because of energy losses due to viscosity and turbulence. These losses are usually negligible, however, when compared to the energy being transmitted from cylinder to cylinder.

16.5 PRESSURE INTENSIFIERS

A pressure intensifier, usually known simply as an intensifier, may be constructed as shown in Figure 16.5.1, where the pistons in two cylinders of different diameters are connected mechanically. Since the forces on the two pistons are equal, it follows that if they are held in equilibrium by pressurized fluid in each cylinder the pressures must

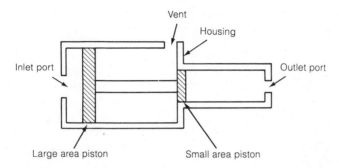

Figure 16.5.1 Schematic of a pressure intensifier. *Source: Parker-Hannifin Corp., Cleveland, OH.*

be related according to

$$p_1 A_1 = F_1 = F_2 = p_2 A_2 \tag{16.5.1}$$

so that

$$p_2 = \frac{A_1}{A_2} p_1 \tag{16.5.2}$$

holds in the steady state condition; i.e., when the pistons are not accelerating or decelerating.

16.6 PRESSURE GAGES

Most pressure gages on machines at this time are either mechanical or electrical. Hydraulic gages, manometers, are found on some wind tunnels.

Perhaps the most common of the many varieties of mechanical gages that have been devised are the Bourdon tube and the plunger types. The Bourdon tube is a curved, flattened, hollow tube which may be formed in a variety of shapes, as shown in Figure 16.6.1. Regardless of the particular configuration used, the tube tends to straighten when pressurized, thus increasing its radius of curvature. By attaching one end of the tube to a pressure source and the other to a linkage as shown in Figure 16.6.2 the motion may be displayed on a dial indicator. The linkage is designed both to produce a linear relation between pressure and angular displacement on the gage face and to compensate for the effect of temperature changes upon the position of Bourdon tube.

In the plunger type of gage the Bourdon tube is replaced by a spring-loaded plunger that is moved in and out by pressure changes on the free end of the plunger. Pressure readings may be inscribed upon the plunger itself or its motion may be translated into a dial reading by a rack on the plunger and a gear segment on the dial arm.

Electric pressure indication may be obtained by means of a piezoelectric sensor which transforms a pressure-induced force into an electric charge that is electronically

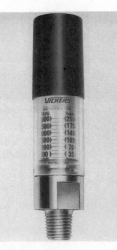

Figure 16.6.1 One type of Bourdon tube. *Source: Vickers, Inc., Troy, MI.*

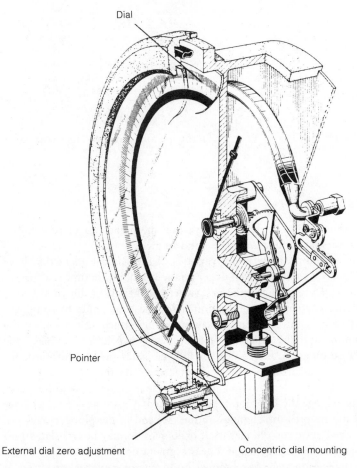

Dial

Pointer

External dial zero adjustment

Concentric dial mounting

Figure 16.6.2 Sectional view, Bourdon tube gage. *Source: Instrument Division, Dresser Industries, Stratford, CT.*

Figure 16.6.3 A piezoelectric pressure sensors. *Source: Kistler Instrument Corporation, Amherst, NY.*

transformed by a coupler into a voltage that is proportional to the pressure. A typical piezoelectric sensor is pictured in Figure 16.6.3. The active element in a piezoelectric sensor is the piezoelectric crystal whose structure causes it to produce opposite electrical charges on its upper and lower surfaces when it is subjected to a compressive force perpendicular to those surfaces.

Other pressure transducers may replace the piezoelectric element with a parallel plate capacitor in which one plate deforms with pressure or with a detection plate, or diaphragm, upon which one or more strain gages are mounted. All of these pressure detectors are particularly useful for automatic control of hydraulic machines.

16.7   PRESSURE CONTROL VALVES

The simplest pressure control valve may seem to be a spring-loaded plug in a fluid line. The operation of such a valve is unsatisfactory, however, because its behavior depends not only upon the static pressure, but also upon the stagnation pressure of the moving fluid and the large spring rate (spring constant) that is necessary to hold the valve closed. The effects may be reduced by designing a pressure control valve as shown in Figure 16.7.1 where a *spool* (two pistons connected by a rod) is spring loaded and where the valve body contains a *primary port* (where the fluid enters), a *secondary port* (where the fluid exits) and an internal passage to either the primary port, as in Figure 16.7.1*a* or to the secondary port, as shown in Figure 16.7.1*c*. The first of these is known as a *normally closed* or *simple relief* valve and the second as a *normally open* or *pressure reducing* valve. In both valve styles the adjustable spring force acts to hold the spool against the bottom of the valve body and the fluid pressure entering through the primary port has no effect on spool motion because it acts against equal spool areas.

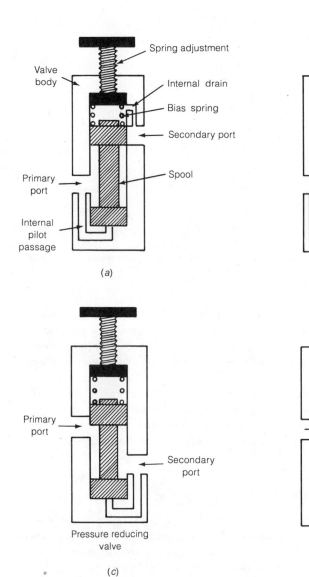

Spring adjustment

Valve body

Internal drain

Bias spring

Secondary port

Primary port

Spool

Internal pilot passage

(a)

(b)

Primary port

Secondary port

Pressure reducing valve

(c)

(d)

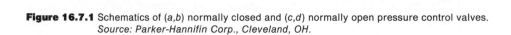

Figure 16.7.1 Schematics of (a,b) normally closed and (c,d) normally open pressure control valves. *Source: Parker-Hannifin Corp., Cleveland, OH.*

In the normally closed valve the spool will move upward to open the valve if primary pressure acting on the bottom of the spool over the area of the internal passage increases enough to produce a force sufficient to push the spool upward against the spring force. This is the *relief pressure* at which the valve opens to permit fluid flow to reduce the pressure on the primary side of the valve.

In the normally open valve, Figure 16.7.1*c*, fluid flows through the valve until the pressure on the secondary side transmitted to the bottom of the spool valve increases to the set pressure where sufficient force acts to move the spool upward and begins to close the valve. As the secondary port closes turbulence increases and the pressure drop across the valve increases, thus diminishing the pressure on the secondary side of the valve. Thus, it acts as a pressure reducing valve.

Although the pilot pressure may be affected by the flow velocity according to the Bernoulli relation, there is negligible stagnation pressure to affect it because of the low velocity in the pilot line.

16.8   PILOT-OPERATED RELIEF VALVES

A pilot-operated relief valve may also be known as a pilot-operated pressure control valve. Its advantage over the pressure control valve described in the previous section is that it operates over a smaller pressure range, as illustrated in Figure 16.8.1.

It consists of two valves, a pilot valve and a main valve. The main valve is capable of a large flow rate and is biased with a light spring (having a small spring rate, or spring constant) while the pilot valve is capable of a small flow and is biased with a heavy spring (having a large spring rate, or spring constant). As shown in Figure 16.8.2, the main valve consists of a light bias spring, a spool with a small hole drilled through it, and the valve body containing the primary and secondary ports. The valve body may also contain the pilot valve, which consists of a heavy bias spring, an adjusting screw, and a plug, known as a *dart*, which opens and closes the pilot line to the pilot valve. (Terms primary port and secondary port are reserved for the main valve.) The hole in the spool is to permit a low-velocity flow to the spring side of the spool to equalize the pressure across the spool so that it is held in the closed position by the force of the bias spring only. The pilot line from the spring side of the spool to the dart also supplies pressure from the primary port to the pilot valve. As long as the fluid pressure is below that pressure required to unseat the dart no fluid will flow through

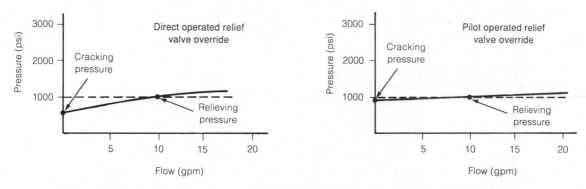

Figure 16.8.1 Pressure-flow curves for direct-operated and pilot-operated relief valves. *Source: Parker-Hannifin Corp., Cleveland, OH.*

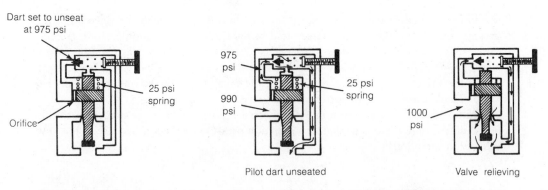

Figure 16.8.2 Schematic of a pilot-operated relief valve. *Source: Parker-Hannifin Corp., Cleveland, OH.*

the valve except for a small amount that may leak past the spool valve that plugs the secondary port. Fluid that does leak from the secondary port is returned to the tank through a drain line.

When the pressure exceeds the relief pressure set by means of the screw adjustment, the pilot pressure to the dart unseats the dart and fluid pressure from the primary port acts on the top of the spool valve, as shown in Figure 16.8.2. The forces on the spool are now unbalanced and the spool opens to permit flow from the secondary port. The improved sensitivity of the valve is due to the small spring force biasing the spool valve as compared to the large spring force use to close the dart. The ratio of these spring forces may be of the order of 50:1 or greater.

Remote Pilot Adjustment

Relief pressure of a pilot-operated relief valve may be adjusted downward at a remote location by providing the pilot valve portion of the relief valve with a port and hydraulic line as shown in Figure 16.8.3. In this arrangement the unseating of the pilot

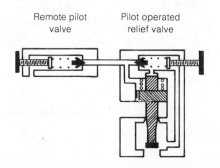

Figure 16.8.3 Remote pilot valve used to vent a pilot-operated relief valve. *Source: Parker-Hannifin Corp., Cleveland, OH.*

dart may be controlled at the remote location. The relief pressure cannot be adjusted upward at the remote location.

Venting

To *vent* a relief valve is to release the pilot pressure that acts on the the dart and on the spring end of the spool in a pilot operated relief valve. This is generally a feature of a hydraulic system in which the pump continues to run while the hydraulics are idle during part of the parent machine's duty cycle. To save power and wear during this time a relief valve in the high-pressure portion of the system may be vented to allow fluid under low pressure to open the secondary port on the relief valve and return to the reservoir. Thus, the pump merely circulates fluid back to the reservoir under low pressure while waiting for the idle time to end.

16.9 CHECK VALVES

A check valve is a valve with a spring-loaded plug which blocks a fluid line, as shown in Figure 16.9.1, and its is to open or close depending upon the direction of the flow.

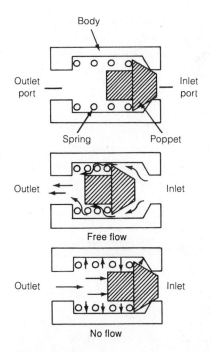

Figure 16.9.1 Schematic of a check valve. *Source: Parker-Hannifin Corp., Cleveland, OH.*

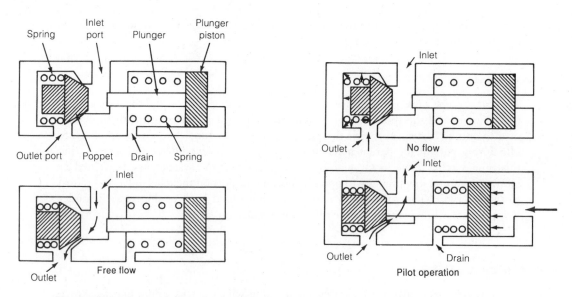

Figure 16.9.2 Schematic of a pilot-operated check valve. *Source: Parker-Hannifin Corp., Cleveland, OH.*

Pilot-operated check valves, Figure 16.9.2, may open either if the direction of flow reverses or if the pilot pressure acting on the plunger piston forces it open. Its primary application is in the latter situation where the valve remains closed until opened by the pilot pressure which is controlled elsewhere in the hydraulic circuit.

16.10 HYDRAULIC PUMPS

Although many hydraulic pumps and motors appear to be interchangeable in that they operate on the same principles and have similar parts, they often have design differences that make their performances better as either motors or pumps. Moreover, some motors have no pump counterparts.

In this chapter we shall consider only positive displacement pumps: those that deliver a particular volume of fluid with each revolution of the input drive shaft. This terminology is used to distinguish them from centrifugal pumps and turbines.

Gear Pump

Perhaps the simplest of these pumps is the gear pump, shown in Figure 16.10.1, in which the fluid is captured in the spaces between the gear teeth and the housing as the gears rotate. Flow volume is controlled by controlling the speed of the driver. Although these pumps may be noisy unless well designed, they are simple and compact.

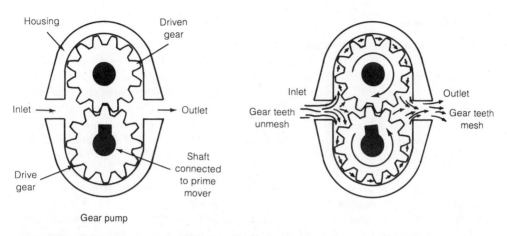

Figure 16.10.1 Gear pump schematic. *Source: Parker-Hannifin Corp., Cleveland, OH.*

Gerotor Pump

Another version of the gear pump is the gerotor whose cross section is shown schematically in Figure 16.10.2. The internal gear has one tooth fewer than the external gear, which causes its axis to rotate about the axis of the external gear. The geometry is such that on one side of the internal gear the space between the inner and outer gears increases for one-half of each rotation and on the other side it decreases for the remaining half of the rotation.

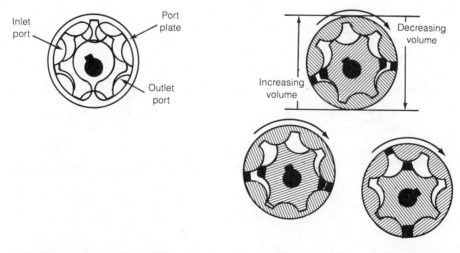

Figure 16.10.2 Schematic of the elements of a gerotor pump. *Source: Parker-Hannifin Corp., Cleveland, OH.*

Vane Pumps

The vane pump shown schematically in cross section in Figure 16.10.3a is similar to the gear pump in that it traps fluid between the vanes. Inlet and outlet ports are as shown in Figure 16.10.3b. The drive shaft center line is displaced from the housing center line, Figure 16.20.3a, and has uniformly spaced vanes mounted in radial slots so that the vanes can move radially inward and outward to always maintain contact with the housing. A so-called intravane device may be incorporated into the design which uses hydraulic pressure to reduce the radial force between the vanes and the housing in order to reduce wear on both of them. Pumps with this feature are generally considered to have a long life.

The unbalanced pressure between the inlet and outlet side of the pump may cause large bearing loads on the rotor shaft. This bearing load may be largely eliminated by rearranging the inlet and outlet ports as shown in Figure 16.10.4 and by machining the housing interior to an eliptical cross-section to make what is known as balanced vane pump. In this configuration the forces on the rotor from the pressure acting on the vanes cancel because of the symmetry of the pressure distribution about the rotor when the inlet and outlet ports are arranged as shown. Fluid enters and exists through port plates at each end of the housing.

Advantages of vane pumps over gear pumps are that they may give higher pressures and they may be designed to provide variable output without having to control the speed of the prime mover—an electric motor, diesel engine, etc.

The design modification required for a variable volume output from a pump having a circular interior cross section is that of mounting the housing between end plates so that the axis of the cylinder in which the vane rotates may be shifted relative to the axis of the rotor, as shown in Figure 16.10.5. Zero flow is obtained when the axis of

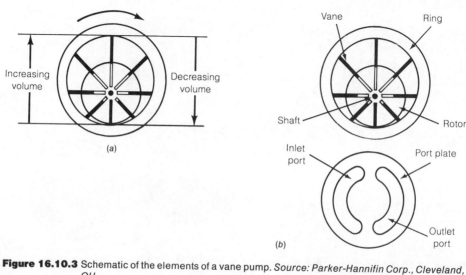

Figure 16.10.3 Schematic of the elements of a vane pump. *Source: Parker-Hannifin Corp., Cleveland, OH.*

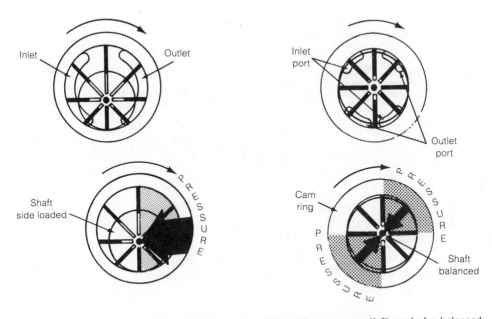

Figure 16.10.4 Cross-sectional schematic of an unbalanced vane pump (*left*), and of a balanced vane pump (*right*). *Source: Parker-Hannifin Corp., Cleveland, OH.*

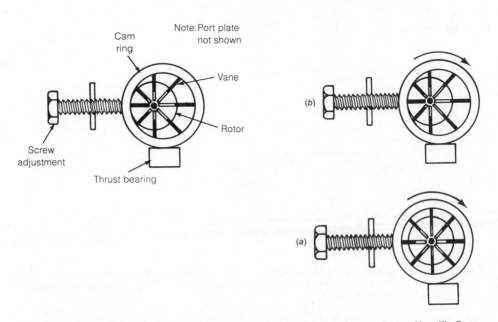

Figure 16.10.5 Cross-sectional schematic of a variable vane pump. *Source: Parker-Hannifin Corp., Cleveland, OH.*

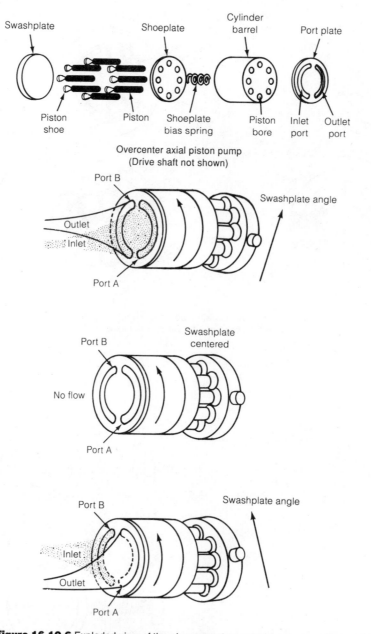

Figure 16.10.6 Exploded view of the elements of an axial piston pump (top) and assembled (bottom) to show operation of the pistons as the cylinder barrel rotates. *Source: Parker-Hannifin Corp., Cleveland, OH.*

the rotor and the housing are coincident as in Figure 16.10.5*a* and maximum flow is obtained when they are displaced as in Figure 16.10.5*b*. The housing and rotor may be split into two cartridges to produce a balanced vane pump as described in Ref. 1.

Axial Piston Pump

One- and two-cylinder reciprocating piston pumps as used for air compressors are not used for hydraulic power because of the oscillatory nature of the output volume. (This irregularity is largely eliminated in pneumatic systems by having the compressor supply an air reservoir which in turn supplies the pneumatic system.)

Instead, an axial piston compressor with eight or more cylinders, as illustrated in Figure 16.10.6, is employed to provide a flow of fluid which may be represented as a ripple superimposed upon a constant flow. Its major components are the *swashplate*, the *axial pistons* with *shoes*, the *cylinder barrel*, the *shoeplate*, the *shoeplate bias spring*,

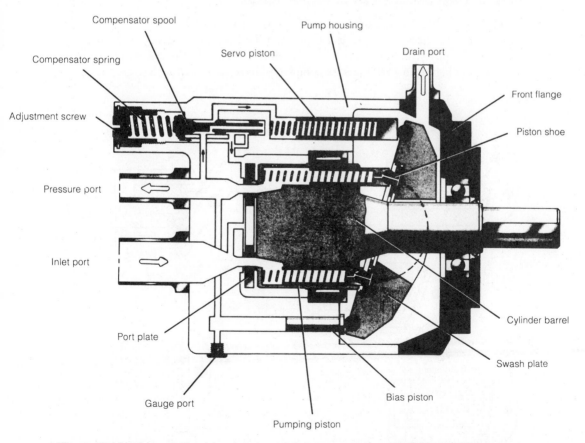

Figure 16.10.7 Cross-sectional drawing of an axial piston pump with compensation. The servo piston controls the angle of the swash plate. *Source: Parker-Hannifin Corp., Cleveland, OH.*

and the *port plate*. The shoeplate and the shoeplate bias spring hold the pistons against the swashplate, which is held stationary while the cylinder barrel is rotated by the prime mover. The cylinder, the shoeplate, and the bias spring all rotate with the input shaft, thus forcing the pistons to move back and forth in their respective cylinders in the cylinder barrel. Input and output flows are separated by the stationary port plate with its kidney-shaped ports.

Output volume may be controlled by changing the angle of the swashplate. As angle α between the normal to the swashplate and the axis of the drive shaft in Figure 16.10.7 goes to zero the flow volume decreases. If angle α becomes negative (if the normal to the swashplate moves above the axis of the drive shaft) the flow reverses; i.e., what was the input port now becomes the output port and the output port becomes the input port. Axial piston pumps with this feature are known as *overcenter* axial piston pumps.

Pressure Compensated Axial Piston Pumps

In these pumps the angle α of the swashplate is controlled by a spring-loaded piston which senses the pressure at a selected point in the hydraulic system. As the pressure increases the piston may decrease α in an effort to decrease the system pressure, as illustrated in Figure 16.10.8. Pressure compensation is often used with overcenter axial

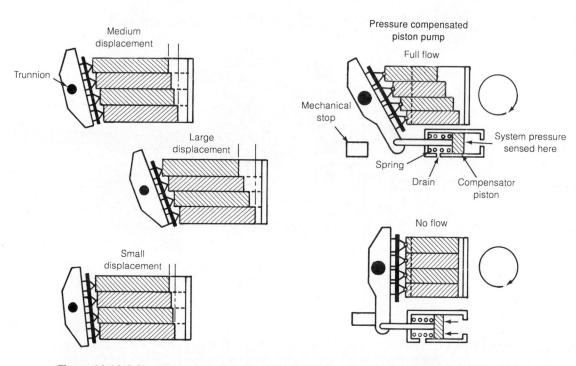

Figure 16.10.8 Simplified schematic of the operation of the compensator piston in controlling the angle of the swashplate to control output flow rate. *Source: Parker-Hannifin Corp., Cleveland, OH.*

piston pumps in hydrostatic transmissions to control the rotational speed and direction of hydraulic motors.

16.11   HYDRAULIC MOTORS

Hydraulic motors include gear, vane, gerotor, and axial vane types whose principles of operation are similar to those of the corresponding motors described in the last section. They differ from pumps in that they may be designed to rotate in either direction, they may have different seals to sustain high pressure at low rpm, or they may have different bearings to withstand large transverse loads so they may drive sprockets, gears, or road wheels on vehicles.

Other radial piston and crankshaft motors, as represented in Figure 16.11.1, have no pump counterpart. In these motors a rotating valve which distributes the pressure to the pistons in sequence will cause the output shaft to rotate in the desired direction. The pistons are mounted in a block that surrounds the rotor and holds the pistons perpendicular to the rotor. Orientation of the block relative to the housing is maintained by means of an Oldham coupling. Each piston slides laterally on a flat surface inside the housing as it applies a force between the flat portion of the housing and the eccentric rotor, as shown in Figure 16.11.2.

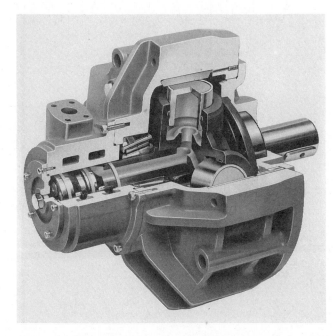

Figure 16.11.1 Cutaway view of a radial piston hydraulic motor showing pistons, housing, and eccentric shaft. Details of the Oldham coupling not shown. *Source: Rotary Power, Inc., Columbus, OH.*

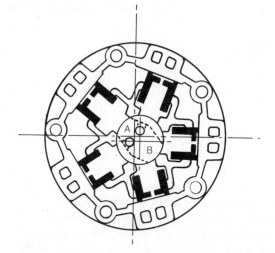

Figure 16.11.2 Cross-section of a radial piston motor showing the pentagonal cylinder block which mounts on the crankshaft eccentric which supplies pressurized fluid. *Source: Copyright 1985 Society of Automotive Engineers, Inc., Warrendale, PA. Reprinted with permission.*

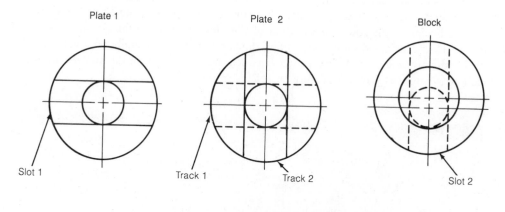

Figure 16.11.3 Front and side views of plates 1 and 2 and the block, which provides an Oldham coupling relation.

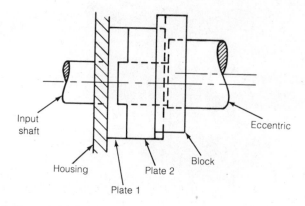

A simplified schematic to display the principle of operation of an Oldham coupling in this application is given in Figure 16.11.3. It may consist of end plate 1, coupling plate 2, and block 3, which contains the pistons and the eccentric portion of the shaft. Slot 1 is cut into plate 1 and accepts track 1 which is part of plate 2. Track 2 is perpendicular to track 1 and is located on the opposite side of plate 2 from track 1. Slot 2 is cut into the block and accepts track 2. Thus, any displacement of the block relative to plate 1, which is attached to the housing, may be decomposed into components parallel to tracks 1 and 2. As the shaft turns the center of the eccentric and of the block will describe a circle about the center of the housing but the block itself will not rotate. Pistons, block, and housing, therefore, will always maintain their proper orientation relative to one another.

16.12 DIRECTIONAL CONTROL VALVES

Directional control valves are means of providing external control to hydraulic cylinders and motors. They may be either manually, hydraulically, or electrically activated. Their external appearance is represented by the valves shown in Figure 16.12.1 and their internal construction is schematically represented in Figure 16.12.2. The single

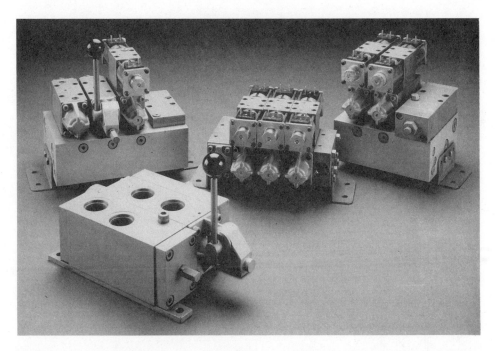

Figure 16.12.1 External appearance of solenoid and manually controlled directional valves. *Source: Parker-Hannifin Corp., Cleveland, OH.*

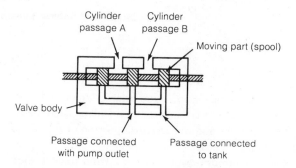

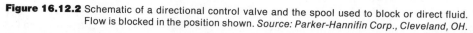

Figure 16.12.2 Schematic of a directional control valve and the spool used to block or direct fluid. Flow is blocked in the position shown. *Source: Parker-Hannifin Corp., Cleveland, OH.*

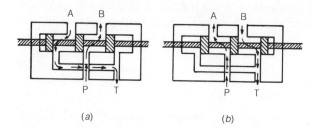

(a)   (b)

Figure 16.12.3 Schematic of a direction valve positioned in (a) to direct fluid from pump P to fluid line B and from fluid line A to the tank (reservoir). In position (b) the direction of flow is reversed. *Source: Parker-Hannifin Corp., Cleveland, OH.*

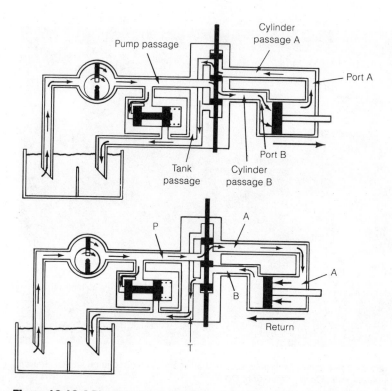

Figure 16.12.4 Directional control valve connected to a pump and cylinder to illustrate its use in a circuit. The spool is vertical in this schematic. A pressure relief valve is shown below and to the right of the pump. *Source: Parker-Hannifin Corp., Cleveland, OH.*

internally moving part is cylindrical and is known as the *spool*. Its function is to divert flow from the incomming port to the desired output port as depicted in Figure 16.12.3. When connected to a hydraulic cylinder as illustrated in Figure 16.12.4, it enables the operator to manually extend or retract the piston rod in a two-directional hydraulic cylinder. Replacement of the lever by a solenoid allows the control to be exercised by a microprocessor.

16.13 ACCUMULATORS

An acumulator is a tank that will accumulate and hold fluid under pressure. They are available in four main styles: the piston accumulator, the bladder accumulator, the diaphragm accumulator, and the pneumatic accumulator. Piston accumulator may be further classified as weight loaded, spring loaded, or pneumatic, depending upon the mechanism used to maintain pressure by forcing the piston against the fluid. The internal construction of the accumulators shown in Figure 16.13.1 is shown in Figure 16.13.2.

Accumulators are used to maintain pressure in the presence of fluctuating flow volume, to absorb shock when pistons are abruptly loaded or stopped, as in the case

Figure 16.13.1 External appearance of a bladder type and piston type accumulators. *Source: Parker-Hannifin Corp., Cleveland, OH.*

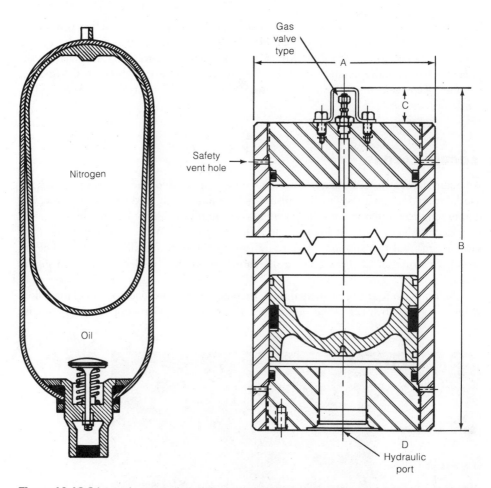

Figure 16.13.2 Internal construction of the bladder and piston type accumulators shown in Figure 16.13.1. *Source: Parker-Hannifin Corp., Cleveland, OH.*

of planers, rock crushers, pressure rollers, etc., or to supplement pump delivery in circuits where fluid can be stored during other parts of the cycle.

16.14 STANDARD HYDRAULIC SYMBOLS

Drawing Figure 16.14.1 at the end of this section was a time-consuming task even though it was drawn schematically by omitting many construction details. The time and effort to draw and modify design drawings for hydraulic systems may be greatly reduced by employing a set of standard design symbols to denote hydraulic components. Standard symbols accepted by the American National Standards Institute (ANSI) in cooperation with the American Society of Mechanical Engineers are listed in ANSI

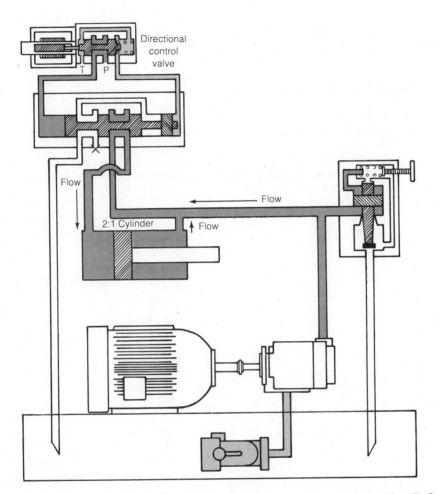

Figure 16.14.1 Schematic for a regenerative cylinder circuit. *Source: Parker-Hannifin Corp., Cleveland, OH.*

Y 32.10-1967 (Revised 1979). Selected symbols from that standard for some of the hydraulic components described in previous sections will be displayed in the remainder of this section.

Two different conventions have been accepted for joining and crossing hydraulic lines. They are

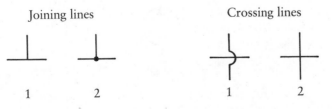

The remaining conventions and symbols used in the remainder of this chapter follow.

- Main hydraulic lines are drawn as solid lines,—
- Pilot lines are drawn as long dashes, – – –
- Exhaust and drain line are drawn as short dashes, - - -
- Check valves are drawn as   B ——————⟨O⟩———— A

where flow is permitted from A to B, but not from B to A.

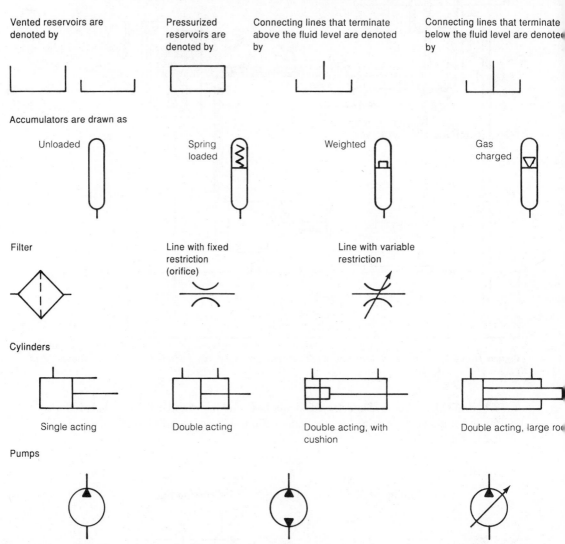

Vented reservoirs are denoted by

Pressurized reservoirs are denoted by

Connecting lines that terminate above the fluid level are denoted by

Connecting lines that terminate below the fluid level are denoted by

Accumulators are drawn as

Unloaded   Spring loaded   Weighted   Gas charged

Filter   Line with fixed restriction (orifice)   Line with variable restriction

Cylinders

Single acting   Double acting   Double acting, with cushion   Double acting, large rod

Pumps

Single directional   Bi-directional   Variable displacement (simplified)

Other activation symbols (continued)

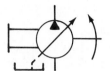

Variable displacement
(complete)

Variable displacement,
pressure compensated
(simplified)

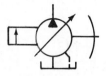

Variable displacement,
pressure compensated
(complete)

Hydraulic motor

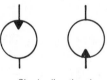

Single directional

Bidirectional

Electric motor

Internal combustion engine

On-off valve

Manual operation

Directional valves

Two-way, two-position valve,
normally closed

Two-way, two-position valve,
normally open

Three-way (three-passage),
two-position valve

Normal position

Four-way (four-passage),
two-position valve

Normal position

Four-way, three-position valve,
normal position

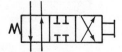

Manually operated
spring return

Four-way, three-position valve,
normal position

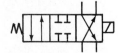

Solenoid operated
spring return

Other activation symbols (continued)

Other passages for the center block in three-position valves

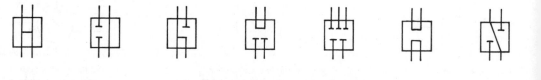

Other activation symbols

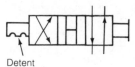

Detent

Manual

Spring return

Pushbutton

Spring return

Push-pull lever

Spring return

Pedal or treadle

Mechanical

Pressure compensated

Pilot pressure, remote supply

Pilot pressure, internal supply

2:1 Cylinder

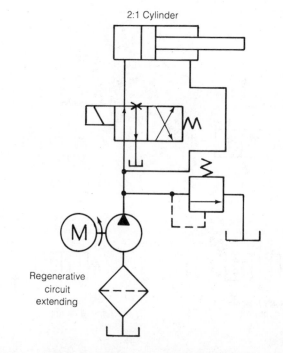

Regenerative
circuit
extending

Figure 16.14.2 Symbolic circuit for the regenerative cylinder shown
in Figure 16.14.1. *Source:
Parker-Hannifin   Corp.,
Cleveland, OH.*

EXAMPLE 16.14.1

Figure 16.14.1 is a schematic of a hydraulic system to extend a piston at relatively high speed at low pressure and retract it at slow speed under full pressure. Note that the pump pressure is applied to both sides of the piston as it is extended. Since the area of the rod side of the piston is less than the area on the other side by the amount of the cross-sectional area of the rod, the extension force is the product of the pressure and the rod cross section. Fluid from the head end of the cylinder flows to the cap end to provide the flow rate required for the rapid advance of the piston. Cylinders connected in this manner are said to be in a *regenerative* circuit.

The symbolic representation of this same circuit is shown in Figure 16.14.2. One of the disadvantages of symbolic representation is that the difference in effective area on opposite sides of the piston is not explicitly shown. It must be deduced from the pressure in the lines for the two positions of the directional control valve. Clearly these symbolic diagrams must be carefully checked to separate inferences from mistakes.

16.15 FLUID POWER TRANSMITTED

To derive formulas for calculating the power transmitted to a particular unit we may return to the fundamental formula for power, which is

$$P = Fv \tag{16.15.1}$$

where P denotes the power, F the force, and v the velocity. Write force $\dot{F}$ as

$$F = pA \tag{16.15.2}$$

in terms of pressure p and cross-sectional area A over which the pressure acts. If L is the distance traveled in time t by a point that moves with the fluid that flows through the hose or cylinder of cross-sectional area A, then the power required to move that fluid is given by

$$P = pA \, \partial L/\partial t = pAv = p \, \partial V/\partial t \tag{16.15.3}$$

Let the rate of change of volume $\partial V/\partial t$ be denoted by $\dot{Q}$ in units of gallons per minute or liters per minute.

Recall that 1 horsepower (hp) = (ft-lb/min)/33 000 and that one gallon = 231 in.^{3}, so that ($\dot{Q} = \partial Q/\partial t$)

$$\text{hp} = \frac{\text{ft-lb}}{\text{min}} = \frac{\text{ft}^3}{\text{min}} \frac{\text{lb}}{\text{ft}^2} = \frac{231 \text{ gal/min}}{(12)^3} \frac{144 \text{ (lb)}}{\text{in}^2} = \frac{(\text{psi})(\text{gal/min})}{1714.3} = \frac{p\dot{Q}}{1714.3}$$

$$\tag{16.15.4}$$

in OE units, where p is in units of lb/in.^{2} and Q is in gal/min. In SI units 1 kilowatt (kW) = (N m/s) and 1 bar = 10^5 N/m^2 = 0.1 MPa. Thus, we have

$$kW = \frac{N}{m^2}\frac{m^3}{s} = 10^{-2}\,(bar)\,\frac{10^3\,(L)}{60\,(min)} = \frac{bar(L/min)}{600} = \frac{p\dot{Q}}{600} \qquad (16.15.5)$$

where p is in units of bars and Q is in liters/minute. (The FAA reports atmospheric pressure to aircraft pilots in millibars.)

16.16 ACCUMULATOR SIZING

Accumulator size depends upon the amount of fluid to be stored and the means used to supply pressure to the fluid stored in the accumulator. If a weight above a piston is used, the accumulator must be large enough to hold the fluid and the volume of the weight and piston. Likewise, if a spring is used provision must be made for the piston and the compressed length of the spring. Since the design of compression springs was outlined in Chapter 5, we shall not repeat the procedures for calculating the length of a compression spring designed to provide the required force on the piston.

When gas pressure is used, either in a bladder or above a piston, the sizing of the accumulator requires that we consider the behavior of the gas as it is being compressed by the incomming fluid.* Its behavior is usually considered to be that of a polytropic gas, which includes isothermal and reversible adiabatic changes as special cases by selecting the appropriate value of the exponent. (Recall that an isothermal process is one in which the compression is slow enough for the temperature of gas to remain constant. An adiabatic process, on the other hand, is one which is so rapid that no heat is lost and the temperature rises accordingly.)

According to the polytropic gas equation

$$pV^n = p_0 V_0{}^n \qquad (16.16.1)$$

Let the accumulator volume and pressure be denoted by V_f and p_f when it is filled with the desired amount of fluid and by V_e and p_e when it is empty. Let V_s denote the volume of stored fluid, so that

$$V_s = V_e - V_f \qquad (16.16.2)$$

From (16.16.1) it follows that

$$p_f V_f{}^n = p_e V_e{}^n \qquad (16.16.3)$$

---

*   An inert gas, such as nitrogen, should be used in an accumulator to prevent an explosion or reaction with the hydraulic oil.

so that upon solving (16.16.3) for V_f and substituting into (16.16.2) we find the required volume V_e of the accumulator to be given by

$$V_e = \frac{V_s}{1 - \left(\dfrac{p_e}{p_f}\right)^{1/n}} \beta \qquad\qquad (16.16.4)$$

where the value of n is read from Figure 16.16.1 and β is an experimentally measured factor which is given by[2]

$$\beta = 1.24 \text{ for bladder type accumulators}$$
$$= 1.11 \text{ for piston type accumulators}$$

A different relation must be used if an accumulator is to be used as a shock absorber. We may begin the derivation by observing that the kinetic energy of the moving fluid will equal the work done on the resisting mechanism in the accumulator.

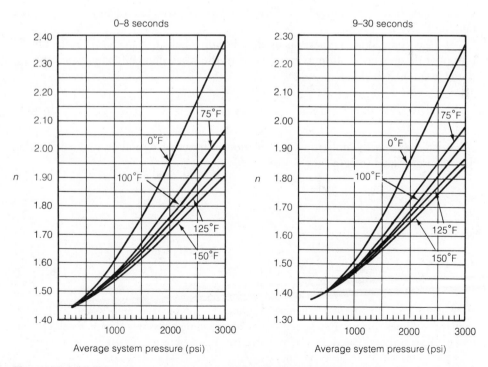

Figure 16.16.1 Polytropic exponent n as a function of the average system pressure for five temperatures. Linear interpolation for intermediate temperatures is usually acceptable. *Source: Parker-Hannifin Corp., Cleveland, OH.*

If it is spring loaded the relation becomes

$$\tfrac{1}{2}mv^2 = \tfrac{1}{2}ky^2 \qquad k = \frac{\rho}{A_p l}\left(\dot{Q}\frac{A_a}{A_p}\right)^2 \tag{16.16.5}$$

where m denotes the mass of the fluid, v its velocity, k the spring rate (spring constant), and y the amount the spring is compressed. If the length of hose or pipe from the pump to the valve causing the shock is l and if ρ represents the mass/volume of the fluid, then solving (16.16.5) for k gives

$$k = \frac{\rho}{A_p l}\left(Q0\frac{A_a}{A_p}\right)^2 \tag{16.16.6}$$

where A_p and A_a denote the cross-sectional area of the pipe (hose) and accumulator, respectively. Since $A_p l = A_p y$, the necessary accumulator volume V_a may be written as

$$V_a = A_p l + V_k \tag{16.16.7}$$

where V_k indicates the volume required by the spring and piston. In this derivation it was assumed that the volume above the piston was vented so that there was no pressure buildup above the piston.

If a gas-charged accumulator is used, either the bladder or piston type, the kinetic energy of the moving fluid must be equated to the work done in compressing the gas in the accumulator. Thus,

$$\tfrac{1}{2}\rho V_o v^2 = -\int_{V_f}^{V_e} F\,dx = -\int_{V_f}^{V_e} p\,dV \tag{16.16.8}$$

Solving for p from the polytropic gas relation in terms of p_e and V_e and substituting into the right-hand integral in (16.16.8) yields

$$\tfrac{1}{2}\rho V_o v^2 = -p_e V_e^{\,n}\int_{V_f}^{V_e} V^{-n}\,dV \tag{16.16.9}$$

which may be integrated to give

$$\tfrac{1}{2}\rho V_o v^2 = \frac{p_e V_e}{n-1}\left[\left(\frac{p_f}{p_e}\right)^{\frac{n-1}{n}}-1\right] \tag{16.16.10}$$

after using $V_e/V_f = (p_f/p_e)^{1/n}$, where V_e and V_f denote the empty and full accumulator volumes and p_e and p_f are the corresponding pressures. Upon recalling that $v = \dot{Q}/A_p$ and that $V_e = A_p l$ we may solve for V_e to obtain

$$V_e = \frac{(n-1)\rho\dot{Q}^2 l}{2A_p p_e\left[\left(\dfrac{p_f}{p_e}\right)^{(n-1)/n}-1\right]} \tag{16.16.11}$$

EXAMPLE 16.16.1 (modified from Ref. 2)

Select a piston accumulator to supplement pump flow to a cylinder that has a 10.0-in. stroke, a 4.0-in.-diameter rod, and a 10.0-in. bore. The extension-retraction time is to be 6.0 s with an idle time of 1.5 min for a total cycle time of 96.0 s. The minimum pressure to activate the cylinder is 1000 psi, the maximum system pressure is 2000 psi, and the gas temperature in the accumulator is to be 100 °F. Also find the minimum pump capacity that may be used.

Extension of the piston will require a fluid volume of

$$V_{ext} = \pi r_c^2 l = \pi 25.0(10) = 785.398 \text{ in.}^3$$

while retraction of the piston will require a volume of

$$V_{ret} = \pi(r_c^2 - r_r^2)l = \pi(25 - 4)10 = 659.735 \text{ in.}^3$$

where r_c and r_r represent the radii of the cylinder and rod, respectively. Since the pump must supply the sum of these volumes in 96 s, required capacity must be no less than

$$\dot{Q} = (785.398 + 659.734)/96 = 15.053 \text{ in.}^3/\text{s} = 903.207 \text{ in.}^3/\text{min}$$
$$= 3.910 \text{ gal/min}$$

If the accumulator is to be filled during the 1.5 min when the cylinder is idle, it must be able to store a volume V_s given by

$$V_s = 90(15.053) = 1354.770 \text{ in.}^3$$

Before we can evaluate equation 16.15.4 we must select an appropriate value of n from Figure 16.16.1 for a draining time of 6 s. Reading from an average pressure of 1500 psi we find $n = 1.65$. Hence, from equation 16.16.4 we find

$$V_e = \frac{1354.77}{1 - (0.5)^{1/1.65}} 1.11 = 4384.09 \text{ in.}^3 = 18.979 \text{ gal}$$

and we shall recommend a standard 19.0-gal piston-type accumulator.

Equation (16.16.4) was used because the accumulator was not sized to act as a shock absorber. We shall consider the shock absorber application next.

EXAMPLE 16.16.2 (from Ref. 2)

Select a bladder-type accumulator to absorb the shock from closing a directional valve to a line 80 ft long having a cross-sectional area of 1.40 in.² and supplying fluid at 100 gal/min at a pressure of 2500 psi. The fluid pressure should not rise more than 10%. The accumulator is designed to operate at 100 °F and the fluid's specific weight per unit volume is 54.3 lb/ft³.

After calculating the average pressure p from $p = (p_{max} + p_{min})/2$ we may turn to Figure 16.16.1 to find $n = 1.92$ from the chart for times from 0 to 8 s. If we decide to work in units of feet we must convert the weight per unit volume to mass per unit volume to find $\rho = 1.686$ sl/ft^3. The flow rate is 0.2228 ft^3/s through an area of 1.4 in.^{2}. By looking at its location in equation 16.16.11 we find that it is not necessary to change the pressures to units of square feet if we do not change the pipe cross section to square feet. This is because the pressures enter either as a ratio or as a product with the pipe cross section. Thus, the minimum volume necessary is

$$V_e = \frac{0.92(1.686)(0.2228^2)80}{2(1.4)2500\left[\left(\dfrac{2750}{2500}\right)^{0.92/1.92} - 1\right]} = \frac{6.160}{327.099} = 0.0188 \text{ ft}^3$$

$$= 32.5 \text{ in.}^3 = 0.141 \text{ gal}$$

Recommend a standard one-quart bladder-type accumulator.

16.17 PISTON ACCELERATION AND DECELERATION

In manufacturing applications it is essential to be able to estimate the time required for each operation in order to be able to satisfy time requirements based upon intended production schedules. Whenever hydraulic cylinders are involved it is, therefore, necessary to be able to calculate the piston extention and retraction times.

We shall begin our analysis of piston behavior by considering piston velocity and acceleration as a function of the system parameters.

Calculation of the steady state velocity of the piston is quite simple because the fluid is essentially incompressible. It is the volumetric flow rate divided by the cross-sectional area A_c of the cylinder. Thus,

$$v_r = \frac{\dot{Q}}{A_c} \tag{16.17.1}$$

where v_r denotes the velocity of the piston and rod.

During the time the rod and piston are accelerating the equilibrium equation for the rod and any load moving with it is given by

$$m\frac{d^2x}{dt^2} = pA_c - (F_r + f) \tag{16.17.2}$$

If we assume the pressure is constant, which is a reasonable approximation if the lines and fittings are large enough to produce only negligible pressure losses, we may integrate (16.17.2) to get

$$\frac{dx}{dt} = \frac{pA_c - (F_r + f)}{m}t \tag{16.17.3}$$

when the piston starts from rest. When set equal to the piston's maximum steady state velocity v_r we find that

$$t_a = \frac{mv_r}{pA_c - (F_r + f)} \tag{16.17.4}$$

is the time to accelerate to velocity v_r. The distance required for the piston to reach this velocity may be calculated by integrating (16.17.3) with respect to time and using the condition that the motion started from $x = 0$ to obtain

$$x_a = \frac{mv_r^2}{2[pA_c - (F_r + f)]} = \frac{t_a v_r}{2} \tag{16.17.5}$$

where F_r denotes the force opposing the motion of the piston and f denotes the friction and fluid losses at the exhaust ports. Here m represents the total accelerating mass: the piston, rod, and any mass being accelerated by the piston and rod.

If the stroke of the cylinder is less than x_a the piston will accelerate over the entire stroke.

Expressions for deceleration time and distance may be obtained from equations 16.17.4 and 16.17.5 by replacing p by $(p - p_c)$ and where p_c is the pressure in the cushion; i.e., after the spear has begun to plug the primary exit port. Pressure p_c usually depends upon x.

Finally, the time for the piston to accelerate, move at constant velocity, and decelerate may be estimated from

$$t_t = t_a + t_d + \frac{s - (x_a + x_d)}{v_r} \tag{16.17.6}$$

where the stroke is of length s, and where t_d and x_d denote the deceleration time and deceleration distance, respectively.

Hydraulic pistons usually move relatively slowly because the ratio between hose and cylinder cross-sectional areas is generally small and because large line losses are associated with large velocities. Consequently, acceleration times and distances are usually negligible. This will be demonstrated in the problems associated with this section at the end of the chapter.

16.18 FILTERS

It is widely believed that most of the premature failure of hydraulic systems is due to dirt in the hydraulic oil that degrades its lubricating properties. Large particles of solid material can interfere with the operation of pilot valves and block the flow through small orifices. Filters, such as that shown in Figure 16.18.1, are added to hydraulic circuits to remove foreign matter without adding appreciably to pressure loss in the

Figure 16.18.1 A typical hydraulic fluid filter. *Source: Schroeder Brothers Corp., McKees Rocks, PA.*

circuit. Usually only one filter is added to most hydraulic systems on machines unless a particular component is especially sensitive to dirt and must have extra protection.

Filters are usually added in the reservoir, in the pump intake from the reservoir, or in the return line to the reservoir to remove solid particles that may enter the fluid from wear of components in the system, from careless handling of hydraulic fluid, or other sources.

Motivation for the first two choices is that the pump is usually the most expensive single component in the system and that foreign matter tends to collect in the reservoir because the flow velocity is low—it acts as a settling tank. The disadvantage of this location is that if the filter becomes clogged it can starve the pump and cause extensive damage. This possibility may be largely eliminated by placing the filter in the return line, but with the risk of damaging the filter by forcing large particles through it. A pilot-operated check valve may be used to bypass a clogged filter, but at the expense of circulating foreign matter that should have been filtered out. Another alternative is to stop the system when the pressure across a filter exceeds a limiting value.

Particulate matter is described in terms of its largest dimension in micrometers (microns), where a micrometer is 1×10^{-6} meters, and filters are classified in terms of their ability to entrap these particles by means of a β value. Symbol β is immediately followed by a number that denotes the diameter of the particles involved according to the relation

$$\beta_d = \frac{\text{number of particles of diameter } d \text{ upstream from the filter}}{\text{number of particles of diameter } d \text{ downstream from the filter}}$$

The β ratio is usually plotted as a straight line on semilog paper, as shown in Figure 16.18.2. Data for this graph are obtained according to the procedures outlined in standard ANSI B 93.31-1973, which describes the hydraulic circuit to be used and the measurements to be made.

Most fluid filters are not rated for particles less than 3 μm; β_d may be taken as 1.0 to $d < 3$. Human hair diameter is about 58 ± 5 μm.

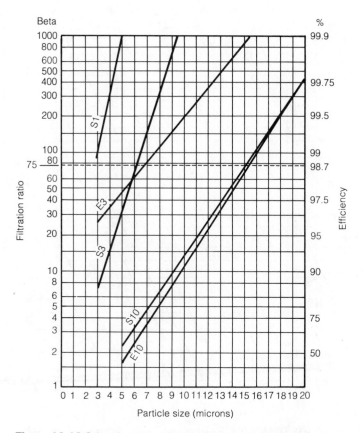

Figure 16.18.2 An element performance graph which gives β as a function of the particle size in microns (m^{-6}). *Source: Schroeder Brothers Corp., McKees Rocks, PA.*

16.19 PIPE AND HOSE PRESSURE LOSSES

Pressure losses as a consequence of flow through pipes has been investigated for several hundred years, with no uniformly accepted formulas yet obtained for flow through pipes and hoses that are connected with a variety of connectors, elbows, and fittings. The theory as it is understood at his time is well explained in Ref. 3, which also includes a satisfactory, but lengthy, method for using it in pipe flow calculations.

None of the computer programs that are available for flow evaluation have been listed in the references because the accuracy of their output often depends upon the problem to which they are applied.

In this chapter we shall use simple, experimentally obtained, pressure loss formulas provided by Currie, Sullivan, and Rasmussen[4] to demonstrate design methods with a minimum of calculation. Other manufactures may use a Moody chart,[5] Manning

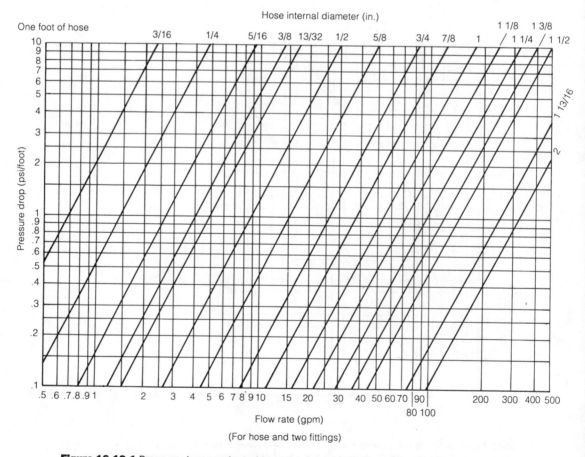

Figure 16.19.1 Pressure drop per foot of hose for the internal diameters shown. *Source: Parker-Hannifin Corp., Cleveland, OH.*

tables,[6,7] or similar well-established approximations to pressure losses. In most applications the errors introduced by these approximations appear to be less than the variations introduced by viscosity changes due to temperature fluctuations and variations between hydraulic oil suppliers.

Pressure loss data are often presented in graphical form on log–log scales as shown in Figure 16.19.1 through 16.19.4.

Rather than read pressure loss per foot from these curves with the attendant difficulty of interpolating between grid lines on a log–log scale, we shall use the methods of Chapter 1 and replace the straight lines (or nearly straight lines) by their equations. Since these lines may be described by relations of the form

$$\log y = a \log x + \log b \tag{16.19.1}$$

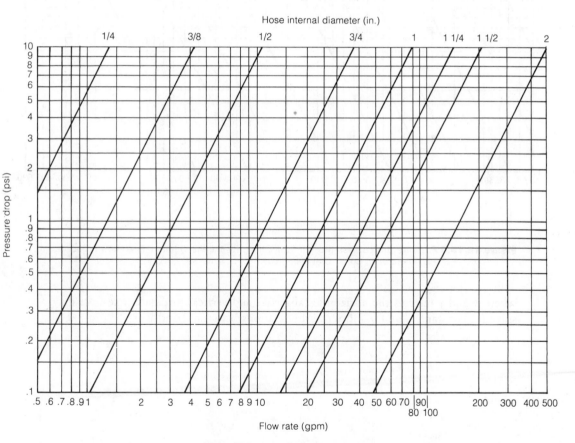

(Fitting ID is determined by the hose ID)

Figure 16.19.2 Pressure drop across two standard fittings for the inside diameters shown. *Source: Parker-Hannifin Corp., Cleveland, OH.*

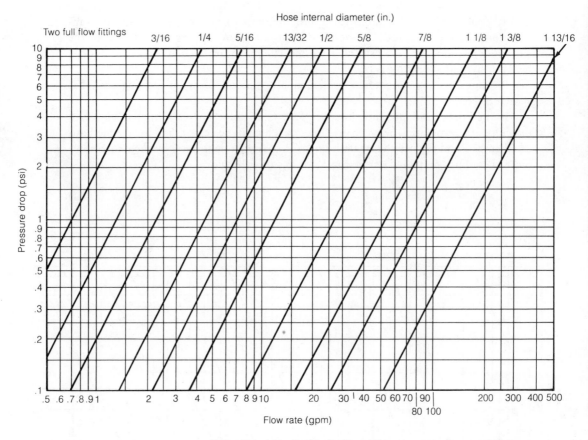

Hose internal diameter (in.)

(Fitting ID is determined by the hose ID)

Figure 16.19.3 Pressure drop across two full flow fittings for the inside diameters shown. *Source: Parker-Hannifin Corp., Cleveland, OH.*

where x and y refer to the horizontal (abscissa) and vertical (ordinate) axes, we may solve for constants a and b and write the relations between x and y as

$$\frac{y}{b} = x^a \tag{16.19.2}$$

where

$$a = \frac{\log \dfrac{y_1}{y_0}}{\log \dfrac{x_1}{x_0}} \tag{16.19.3}$$

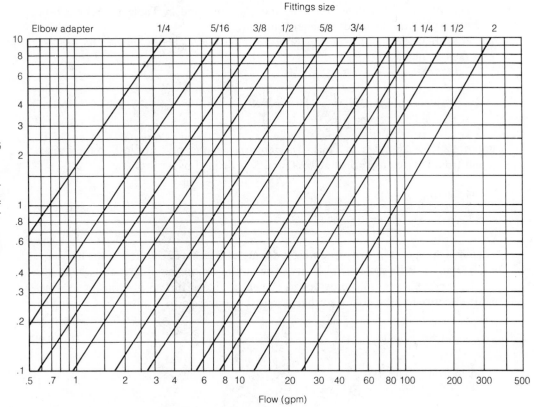

Figure 16.19.4 Pressure drop across one elbow adapter for the inside diameters shown. *Source: Parker-Hannifin Corp., Cleveland, OH.*

and where

$$\log b = \log y_0 - a \log x_0 \tag{16.19.4}$$

Thus, for the curves in the figures above, with x replaced by $\dot{Q}$ and y by $\Delta p/\Delta L$, we have

$$\frac{\Delta p}{\Delta L} = \left(\frac{\dot{Q}}{\dot{Q}_0}\right)^a \xi \qquad \xi = \left(\frac{\Delta p}{\Delta L}\right)_0 \tag{16.19.5}$$

where

$$a = \frac{\log \dfrac{\Delta p_1/\Delta L}{\Delta p_0/\Delta L}}{\log \dfrac{\dot{Q}_1}{\dot{Q}_0}} \tag{16.19.6}$$

Typical values calculated for a, ξ, and $\dot{Q}_0$, are listed in Table 16.19.1 for hoses and fittings having the inside diameters indicated.

Pressure loss for a hose or pipe of length L having N_b bends between 45 and 180° may then be obtained from

$$\Delta p = \left[(L + 0.55 N_b) \left(\frac{\Delta p}{\Delta L} \right) + \Delta p_f \right] K \tag{16.19.7}$$

in terms of $\Delta p / \Delta L$ given by (16.19.5). Here Δp_f denotes pressure loss at the fittings, found from Figures 16.19.3 or 16.19.4, and K is defined by

$$K = \left(\frac{\mu}{6.7} \right)^{0.25} + \left(\frac{\gamma}{0.86} \right)^{0.75} \tag{16.19.8}$$

where γ is the specific gravity (the weight density of water is 62.4 lb/ft³) of the hydraulic oil used and μ is its dynamic viscosity of the oil in centipoise. Regrettably, hydraulic oil viscosity data are often provided as shown in Figure 16.19.5 in terms of its kinematic viscosity in centistokes, so that it must be converted to centipoise according to

$$\rho(\mathrm{cSt}) = (\mathrm{cP}) \tag{16.19.9}$$

as taken from Chapter 10.

TABLE 16.19.1 PIPE/HOSE PRESSURE LOSS CONSTANTS

Component	Constant	Hose/Pipe Internal Diameter (ID)			
		$\frac{1}{4}$	$\frac{1}{2}$	1	$1\frac{1}{2}$
Hose & 2 fittings	a	1.872	1.907	1.950	1.921
	ξ	0.140	0.100	0.100	0.100
	$\dot{Q}_0$	0.500	2.680	16.50	45.50
Standard fittings	a	1.916	1.847	1.949	1.959
	ξ	1.430	0.100	0.100	0.100
	$\dot{Q}_0$	0.500	0.950	7.900	20.000
Full flow fittings	a	1.932	1.985	—	—
	ξ	0.160	0.100	—	—
	$\dot{Q}_0$	0.500	2.210	—	—
Elbow adapter	a	1.366	1.506	1.619	1.647
	ξ	0.650	0.100	0.100	0.100
	$\dot{Q}_0$	0.500	0.940	5.320	11.40

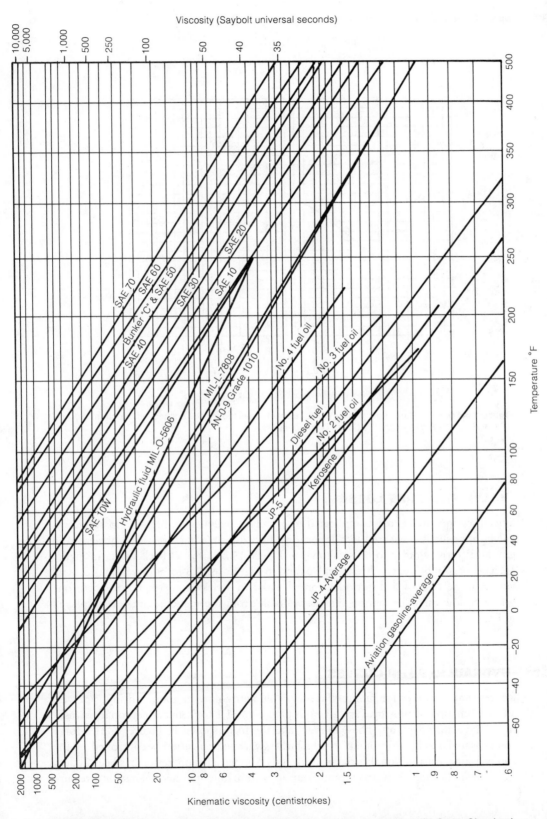

Viscosity (Saybolt universal seconds)

Kinematic viscosity (centistrokes)

Temperature °F

Figure 16.19.5 Viscosity–Temperature for selected oils. *Source: Parker-Hannifin Corp., Cleveland,*

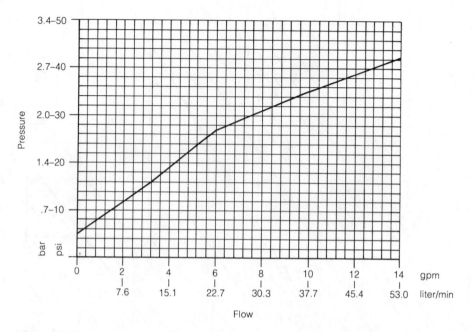

Figure 16.20.1 Pressure–flow rate curves for a check valve (flow in the unchecked direction). *Source: Parker-Hannifin Corp., Cleveland, OH.*

16.20   VALVE PRESSURE LOSSES

Since flow in valves is too complicated to calculate analytically, pressure losses across valves are usually measured in a laboratory by valve manufacturers and presented in graphical form, similar to those shown in Figure 16.20.1 for a check valve, in Figure 16.20.2 for a pilot-operated relief valve, and in Figure 16.20.3 for a directional control valve. In the case of a directional control valve, a series of curves is provided, one for each flow direction option that is available from the manufacturer, as displayed in Figure 16.20.4.

16.21   HYDRAULIC PUMP CURVES

Representative curves for pump efficiency, volumetric flow in gpm, and input power, all as a function of pressure, are given in Figure 16.21.1 for a fixed (stationary swash-plate) axial piston pump. These curves take the form shown in Figure 16.21.2 for a compensated axial piston pump in which the swashplate position is controlled by the compensator. An additional curve, that for fully compensated operation has been added. Recall from Section 16.10 that compensation is to maintain either a constant pressure

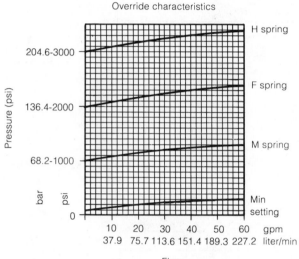

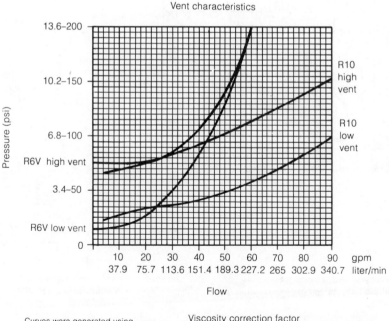

Viscosity correction factor							
Viscosity (SSU)	75	150	200	250	300	350	400
Percentage of Δp (approx)	93	111	119	126	132	137	141

Curves were generated using 100 SSU hydraulic oil. For any other viscosity, pressure drop will change as per chart.

Figure 16.20.2 Pressure–flow curves for a pilot-operated relief valve (top) and vent curves (bottom) along with viscosity correction data. *Source: Parker-Hannifin Corp., Cleveland, OH.*

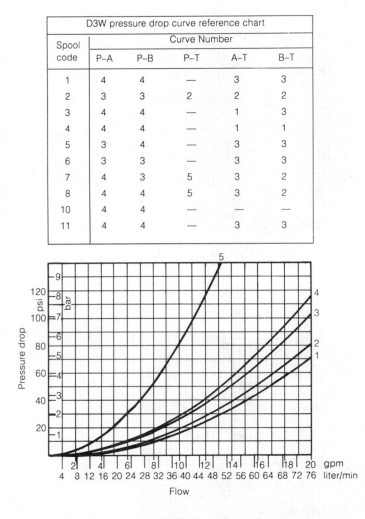

D3W pressure drop curve reference chart					
Spool code	Curve Number				
	P–A	P–B	P–T	A–T	B–T
1	4	4	—	3	3
2	3	3	2	2	2
3	4	4	—	1	3
4	4	4	—	1	1
5	3	4	—	3	3
6	3	3	—	3	3
7	4	3	5	3	2
8	4	4	5	3	2
10	4	4	—	—	—
11	4	4	—	3	3

Curves were generated using 100 SSU hydraulic oil. For any other viscosity, pressure drop will change as per chart.

Viscosity correction factor							
Viscosity (SSU)	75	150	200	250	300	350	400
Percentage of Δp (approx)	93	111	119	126	132	137	141

Figure 16.20.3 Pressure drop curves for the directional valve and spool codes shown in Figure 16.20.4 along with its viscosity correction factor. *Source: Parker-Hannifin Corp., Cleveland, OH.*

or a constant flow rate, and it is achieved by pressure from the output port being supplied to a control valve which in turn controls a piston which adjusts the angle of the swashplate to achieve either the intended maximum pressure or the intended maximum flow. The input power in the fully compensated mode is comparatively small

Model	Spool Symbol	Maximum Flow (gpm) 3000 psi	Model	Spool Symbol	Maximum Flow (gpm) 3000 psi
D3W1		20	D3W6		20
D3W2		20	D3W7		8
D3W3		20	D3W8		12
D3W4		20	D3W10		4
D3W5		20	D3W11		20

Figure 16.20.4 External appearance and spool code for a directional valve, solenoid controlled. *Source: Parker-Hannifin Corp., Cleveland, OH.*

because in this mode the pressure is at its preset maximum value and, consistent with (16.15.4) and (16.15.5), comparatively little fluid is flowing.

When motion begins, as in operating a cylinder, the pressure will fall as the fluid flow rate increases and the pump will go to the full flow mode to bring the pressure up to its set value.

The values of volumetric efficiency and overall efficiency for the outlet pressures increasing from left to right along the bottom horizontal scale are read from the efficieny scale at the top of the vertical scale on the left. Flow rate in gpm as a function of output pressure is read from the delivery gpm scale on the lower part of the vertical scale on the left. The dashed curves dropping down from the flow curve represent the flow when the limiting pressure is set at 2000 or 4500 psi. Other curves, having similar slope, may be added between these two for pressure settings between 2000 and 4500 psi. Input horsepower at full flow as a function of outlet pressure is read from

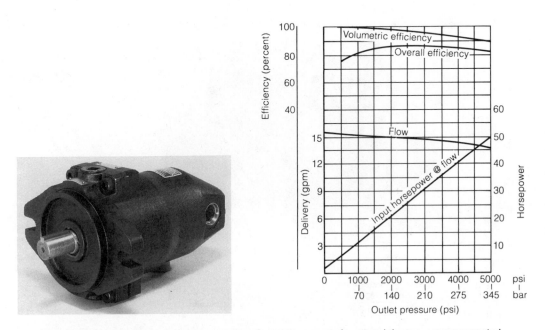

Figure 16.21.1 External appearance and performance curves for an axial vane, uncompensated, constant displacement, hydraulic pump (flow rate is nearly constant). *Source: Parker-Hannifin Corp., Cleveland, OH.*

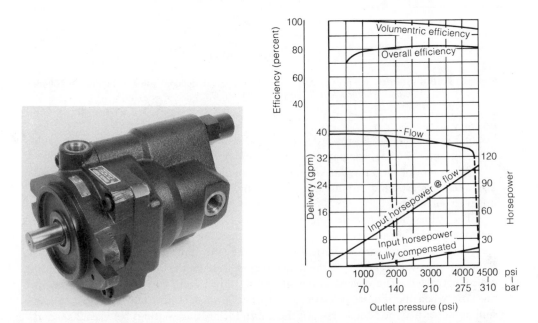

Figure 16.21.2 Performance curves for an axial piston, variable volume, compensated hydraulic pump. *Source: Parker-Hannifin Corp., Cleveland, OH.*

the horsepower scale on the vertical scale on the right, as is the input horsepower in the fully compensated mode.

Similar comments hold for a variable vane pump, whose performance curves are similar, as shown in Figure 16.21.3. As indicated, the manufacturer may also include a curve of the input torque. This is an aid to the designer who may plan to drive the pump with a belt drive because it may be used to specify belt tension.

Finally, Figure 16.21.4 displays representative input power and flow rate as a function of gear speed for a series of a series of output pressures for two pump models.

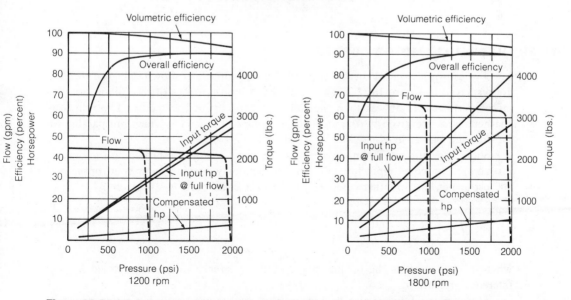

Figure 16.21.3 External appearance and performance curves for a variable volume, pressure compensated, vane pump. *Source: Parker-Hannifin Corp., Cleveland. OH.*

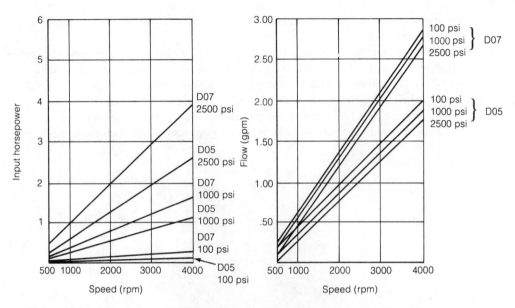

Figure 16.21.4 Input power and output flow rate as functions of gear speed for different pressure settings for a gear pump. *Source: Parker-Hannifin Corp., Cleveland, OH.*

Since both input power and flow rate are clearly nonlinear functions of pressure, orthogonal interpolation is recommended for intermediate pressures.

16.22   HYDRAULIC MOTOR CURVES

Typical torque–speed curves for a radial piston motor, as shown in Figure 16.22.1, are displayed in Figure 16.22.2. The nearly horizontal portions of the curves describe motor performance at less than the full flow capacity of the motor. The full flow characteristics are given by the inclined portions of the curves which are parallel to the constant power lines.

Some hydraulic motor manufacturers rate their motors by providing the output torque for a specified input pressure, while others may provide the output torque per 100 psi or per bar. In either case, the torque T at other pressures is given by

$$T = \left(\frac{T_o}{p_o}\right)p \tag{16.22.1}$$

where (T_o/p_o) denotes the torque rating or the reference torque T_o divided by the reference pressure p_o.

Motor manufacturers also provide a displacement per revolution rating which is the fluid volume V_r required to turn the rotor (or housing) one revolution. Thus, motor

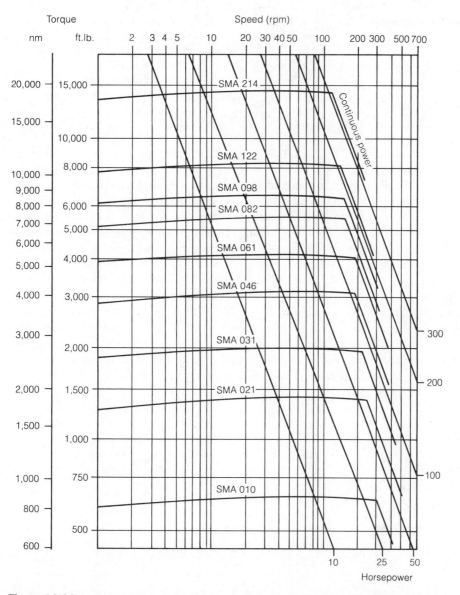

Figure 16.22.1 Hydraulic motor performance curves. Nearly horizontal curves are at less than rated power. Constant HP curves are at rated power. *Source: Rotary Power, Inc., Columbus, OH.*

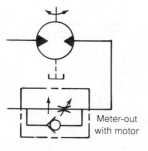

Meter-out
with motor

Figure 16.22.2 Motor with a pressure compensated flow control valve in output line to control speed. Since the motor is bidirectional the check valve is to bypass the speed control when the direction is reversed. The reservoir line is to drain motor leakage. *Source: Parker-Hannifin Corp., Cleveland, OH.*

speed may be found from

$$n = \frac{\dot{Q}}{V_r} \tag{16.22.2}$$

where n is the shaft speed in rpm and Q is the flow rate in the units of V_r.

From (16.22.2) we find that motor speed may be controlled by controlling the flow rate. Because of Bernoulli's theorem it is easier to obtain an accurate control of motor speed by controlling the flow leaving the motor than to regulate the incomming flow. That part of the circuit containing the motor and its speed control may be represented as in Figure 16.22.2.

Braking of a hydraulic motor is accomplished by controlling the pressure drop across the motor and thus regulating power according to equation 16.15.3. One of several possible circuits is shown in Figure 16.22.3. Braking is accomplished by positioning the directional control valve as shown to block flow to and from the motor. The pilot-operated relief valve shown in the circuit is to prevent overpressure if the motor continues to turn due to the moment of inertia of its load. Check valves 1 and 2 are set at higher than normal operating pressure but open at less than the pilot

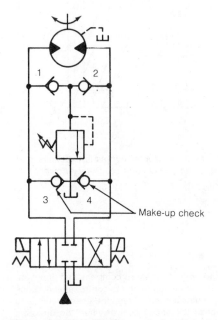

Figure 16.22.3 Circuit to control and brake a bidirectional motor under the control of a three-position solenoid controlled directional valve. *Source: Parker-Hannifin Corp., Cleveland, OH.*

pressure for the relief valve. They are to isolate the relief valve during powered operation. Check valves 3 and 4 are to permit backflow from the reservoir during braking
to prevent *cavitation*. Cavitation occurs when a solid surface moves away from a
contacting liquid so rapidly that cavities are formed in the liquid. This may cause damage to a motor by reducing lubrication and by causing surface deterioration as cav-
aties collapse and fluid impacts the surface of the motor's pistons, vanes, or gears.

Motor performance in a particular circuit depends upon both the motor and the
pump characteristics. Four different combinations with four different performance
characteristics are shown in Figure 16.22.4.

A constant displacement pump with a constant displacement motor, Figure 16.22.4*a*
provides a system in which the torque and speed of the motor are determined by the
flow rate and speed of the pump.

Replacement of the constant displacement motor in (*a*) with a variable displacement
motor, as shown in Figure 16.22.4*b*, results in a system in which the power to the
motor is determined by the input power to the pump, but motor speed and torque
may be controlled at the motor itself.

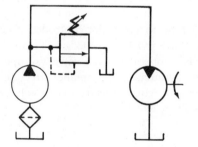

(*a*) Fixed hp, torque, and speed

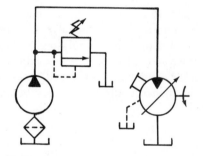

(*b*) Fixed hp, variable speed and torque

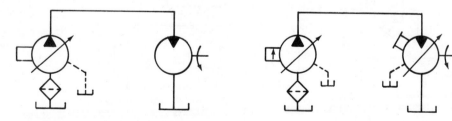

(*c*) Variable hp and speed, constant torque

(*d*) Variable hp, speed, and torque

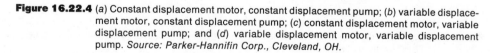

Figure 16.22.4 (*a*) Constant displacement motor, constant displacement pump; (*b*) variable displacement motor, constant displacement pump; (*c*) constant displacement motor, variable
displacement pump; and (*d*) variable displacement motor, variable displacement
pump. *Source: Parker-Hannifin Corp., Cleveland, OH.*

Replacement of the constant displacement pump in (*a*) with a variable displacement pump produces the system shown in (*c*) in which the motor torque is constant because the pressure delivered to it is constant. Motor speed and power are variable and may be controlled at the pump by controlling pump displacement, i.e., its output flow rate.

If both the constant displacement pump and the constant displacement motor in (*a*) are replaced by a variable displacement pump and a variable displacement motor we have the system shown in Figure 16.22.4*d* in which the power output of the motor may be controlled at the pump and the torque and speed of the motor may be controlled at the motor.

16.23 REPRESENTATIVE HYDRAULIC SYSTEMS

A simple hydraulic circuit to provide bidirectional control of a hydraulic cylinder is shown in Figure 16.23.1. It includes a motor with a clutch between the motor and pump, a filter in the motor intake line from the reservoir, and a manually operated directional control valve.

Figure 16.23.2 shows a regenerative hydraulic circuit which differs from that shown in Figure 16.14.2 in that the cylinder control is indirect; that is, the solenoid-operated, spring-return, directional-valve V-1 controls the flow to two pilot-controlled check valves that directly control flow to the cylinder. Valve PO-1 is opened to extend the cylinder and valve PO-2 is opened to retract it. Pilot-operated relief valve V-2 opens after the accumulator is charged. One advantage of this circuit is that smaller directional valves and solenoids may be used because they are not required to pass large volumes of fluid.

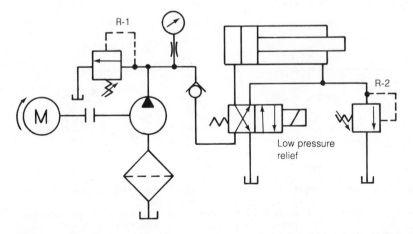

Figure 16.23.1 Hydraulic circuit for solenoid of a thrust cylinder. *Source: Parker-Hannifin Corp., Cleveland, OH.*

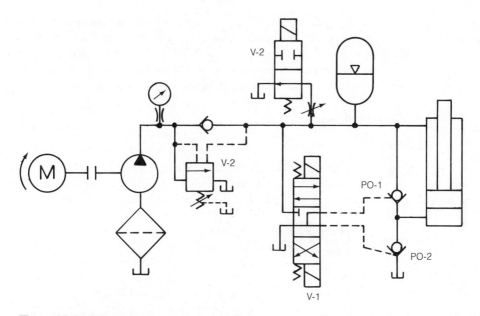

Figure 16.23.2 Regenerative hydraulic circuit with indirect control of a thrust cylinder. *Source: Parker-Hannifin Corp., Cleveland, OH.*

EXAMPLE 16.23.1

Select the components to construct the circuit shown in Figure 16.23.1. The cylinder must provide a force of 14 500 lb, it should have a 19-in. stroke, and its extension time should be no longer than 6 s.

Since we must know the operating pressure to specify the piston diameter and must know the flow rate to calculate the extension time, our piston selection and pump selection are coupled together. We may, therefore, begin our search for a suitable combination by first choosing a standard cylinder diameter, calculating the necessary pressure, and then searching for a pump with the necessary volumetric capacity and pressure. Cylinder lengths, incidentially, are not standard because it is easy to cut them to whatever length is required.

Begin with a $2\frac{1}{2}$-in. bore, which has an area of 4.909 in.^{2}, so that the required pressure is

$$p = F/A = 14\,500/4.909 = 2953.9 \text{ psi}$$

Since this is within the operating range of axial piston pumps, we shall next calculate the required pumping capacity from

$$\dot{Q} = vA = (19/6)4.909 = 15.545 \text{ in.}^3/\text{s}$$
$$= 932.710 \text{ in.}^3/\text{min} = 932.71/231 = 4.038 \text{ gpm}$$

Upon searching through catalogs from several manufacturers of hydraulic pumps we find one with the characteristics shown in Figure 16.23.3 that has a capacity of approximately 4.30 gpm at a compensated pressure setting of 3500 psi, based upon a curve for 3500 psi similar to the dashed curves shown for 2000 and 4500 psi. From the intersection of this curve with that for the input horsepower at full flow we find that it requires approximately 8.9 hp for an output power, according to (16.15.3), of

$$\text{hp} = \frac{p\dot{Q}}{1714.3} = \frac{3000(4.038)}{1714.3} = 7.525$$

These values lead to a calculated efficiency of 84.5%, as compared to 84% read from the overall efficiency curve. The estimated values used are, therefore, within acceptable error. Overall efficiency is defined as the product of the mechanical efficiency and the volumetric efficiency, where

$$\text{Mechanical efficiency} = \frac{\text{output work per revolution}}{\text{input work per revolution}}$$

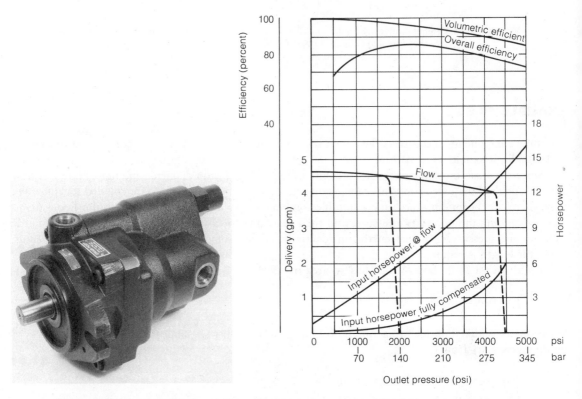

Figure 16.23.3 Axial piston, variable volume, pressure compensated hydraulic pump. *Source: Parker-Hannifin Corp., Cleveland, OH.*

and

$$\text{Volumetric efficiency} = \frac{\text{actual delivery per unit time}}{\text{calculated delivery per unit time}}$$

Hose and fitting losses may be estimated from equations 16.19.7 and 16.19.8. If the hydraulic fluid satisfies MIL-0-5606 and has a specific mass of 0.821 g/cm^3 then its viscosity at 80 °F in cP is

$$\mu = 0.821(20) = 16.42 \text{ cP}$$

after converting from the kinematic viscosity in cSt. With these values we now can evaluate K as given by relation 16.19.8. Thus,

$$K = \left(\frac{16.42}{6.7}\right)^{0.25} + \left(\frac{0.821}{0.860}\right)^{0.75} = 2.217$$

If the total hose length for the entire system is 49.8 ft, if there are 9 bends between 45 and 180° in the $\frac{1}{2}$-in. line, if there are four standard fittings, and if the other fittings are equivalent to 11 elbow adapters, then the fitting losses may be found from equation 16.19.5 and Table 16.19.1 to be

$$2 \, \Delta p_f = 2 \left(\frac{4.038}{0.95}\right)^{1.847} 0.10 = 2.896 \text{ psi}$$

for the standard fittings not included with the hose and

$$11 \, \Delta p_f = 11 \left(\frac{4.038}{0.94}\right)^{1.506} 0.10 = 9.880 \text{ psi}$$

for the elbow adapters. Inclusion of these results into equation 16.19.7 yields

$$\Delta p = \left[\left(49.8 + 0.55 \times 9\right) \left(\frac{4.038}{2.68}\right)^{1.907} 0.10 + 12.776 \right] 2.217 = 54.849 \rightarrow 54.8 \text{ psi}$$

as the total pressure loss due to hose and fittings. Directional valve losses may be estimated from Figures 16.20.3 and 16.20.4. The D3W1 configuration holds in this case; from P to A to pressurize the cap end and B to T to drain the head end of the cylinder as the piston is extended, then P to B to pressurize the head end and A to T to drain the cap end as the piston is retracted. In the center position the passages are blocked and the piston is held steady, except for leakage. From curve 4 in Figure 16.20.3 we read a pressure loss at the control valve of 5 psi for the passage from the pump to the cylinder and for the passage from the cylinder to the reservoir, for a total of 10 psi. Thus, the pressure loss around the circuit is 64.8 psi.

EXAMPLE 16.23.2

Replace the cylinder in the circuit for Example 1 with one having a 5-in. bore and a 23-in. stroke. Replace the pump with an axial piston pump set for a maximum pressure of 2500 psi and having a flow rate of 15 gpm at 2000 psi pressure. Find the maximum piston velocity, the static force, and the dynamic force the piston can exert at its maximum velocity.

Piston velocity may again be found from $v = \dot{Q}/A$, where

$$A = \pi(2.5)^2 = 19.635 \text{ in.}^2$$

So with

$$\dot{Q} = 15(231) = 3465.0 \text{ in.}^3/\text{min}$$

we have

$$v = 3456.0/19.635 = 176.471 \text{ in./min}$$
$$= 2.941 \text{ in./s}$$

After searching through catalogs for pump manufacturers we find that the one with the curve shown in Figure 16.23.4 may be used. Since the curve for a pressure setting for 2000 psi is shown explicitly, we may easily read the full flow power required as about 21 hp and the fully compensated power as about 2.5 hp.

Upon following the same steps as in Example 16.23.1 we find that the losses due to hoses and fittings are approximately

$$\Delta p = \left[(49.8 + 0.55 \times 9)\left(\frac{15.0}{2.68}\right)^{1.907} 0.10 + 2\left(\frac{15}{0.95}\right)^{1.847} 0.10 \right.$$
$$\left. + 11\left(\frac{15}{0.94}\right)^{1.506} 0.10 \right] 2.217 = 554.4 \text{ psi}$$

Upon reading the same curves as in the previous example, but for the larger flow rates, we find the valve losses to total 125.0 for a total of 679.5 psi. Consequently, the dynamic load capability of the piston at full speed has fallen by 34% to

$$F = 19.635(1320.5) = 25928 \text{ lb}$$

as compared to its static capacity of

$$F = 19.635(2000) = 39\,270 \text{ lb}$$

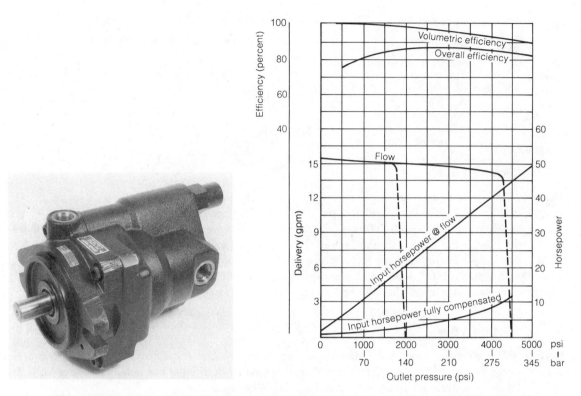

Figure 16.23.4 External appearance and performance curves for an axial piston variable displacement pump. *Source: Parker-Hannifin Corp., Cleveland, OH.*

As these numbers show, this larger power loss is because we have increased the flow through these small lines and fittings. Where space is limited, as in aircraft and some mining machines, it may be necessary to use small diameter lines. Otherwise, the design engineer should note that small lines reduce initial cost at the expense of increased operating costs.

REFERENCES

1. Lambeck, R.P., *Hydraulic Pumps and Motors*, Dekker, New York, 1983.

2. *Hydraulic Cylinders, Accumulators and Air-Oil Tanks,* Cat. 1100 H, Cylinder Div., Parker-Hannifin Corp., Des Plaines, IL (undated).

3. Benedict, R.P., *Fundamentals of Pipe Flow*, Wiley, New York, 1977, pp. 270–280.

4. Currie, W.E., Sullivan, H.L., and Rasmussen, R.G., *Pressure Drop through Hose*, Internal Report, Hose Products Div., Parker-Hannifin Corp., Wickliffe, OH (undated).

5. Moody, L.F., Friction factors for pipe flow, *Transactions of the ASME*, 66 (8), 671–678 (1944).

6. Manning, R., Flow of water in open channels and pipes, *Transactions of the Institute of Civil Engineers (Ireland)*, 20 (1890).

7. Brater, E.F., and King, H.W., *Handbook of Hydraulics*, 6th ed., McGraw-Hill, New York, 1980 pp. 6-39–6-62.

PROBLEMS

Section 16.3

16.1 A hydraulic system has a pressure gage on the supply line ($A = 1.0$ in.^{2}) to a hydraulic lift cylinder ($A = 17.0$ in.^{2}). When raising a load the gage reads 1440 psi. When it holds the load stationary the gage reads 2000 psi. If the hydraulic fluid weighs 50 lb/ft^3, what is the velocity of the fluid at the gage an instant before the flow was stopped?

16.2 The supply line in problem 16.1 has a section of pipe with a large inside diameter and a section with a small inside diameter. If the pressure differences when raising and holding the load are to be used to estimate the flow rate, where should the gage, or gages, be mounted? Why?

Section 16.4

16.3 What advantage does the tandem cylinder have over the standard double-acting cylinder? Compare the force exerted by the rod for each when both have the same supply pressure. Is there a disadvantage in using such a cylinder? What is it?

16.4 What would be an advantage of a duplex cylinder as compared to a tandem cylinder if each piston in the duplex can be individually controlled?

Section 16.14

16.5 Identify the components represented by symbols 1 through 7 in the figure. (Although the system is to have only one reservoir, it is conventional to show a reservoir near the return port to avoid multiple lines to one reservoir.)

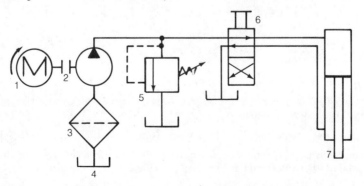

16.6 Identify symbols 1 through 5 in the figure.

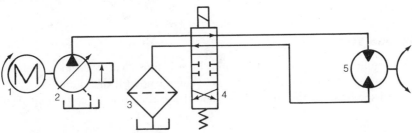

Section 16.14

16.7 Sketch the hydraulic circuit for an electric motor-driven pump to supply a bidirectional tandem thrust cylinder with the flow to be augmented by a gas-charged accumulator. Place the filter in the pump supply line. The manually operated directional control valve should allow the cylinder to be extended, held, and retracted.

16.8 Sketch the hydraulic circuit for an internal combustion engine-driven pump to drive either a bidirectional thrust cylinder or a bidirectional hydraulic motor. Use a pedal-operated directional valve to select cylinder or motor operation and a manually operated directional valve to control cylinder or motor direction, with a stop position in between. Include an accumulator to absorb the shock of any change in motion. Place the filter in the drain line.

16.9 Redraw the circuit in problem 16.8 to include a pilot-operated relief valve in the high-pressure line.

16.10 Redraw the circuit in problem 16.9 to include a pilot-operated check valve to bypass the filter if it becomes blocked.

16.11 Sketch a hydraulic that uses a pressure-intensified and a standard cylinder to obtain a large thrust in place of a tandem piston. What is the relation of the size of the pressure intensifier to the size and stroke of the thrust cylinder? (*Hint*: Consider the volume of fluid from the pressure intensifier and the volume required by the cylinder.)

Section 16.16

16.12 Recommend the size of a gas-charged bladder-type accumulator used to augment the flow to a double-acting tandem cylinder whose bore is 1.50 mm and whose stroke is 3.50 mm. Idle time is 70 s and the cycle time (extension and retraction) is 8 s. The gas temperature is 51.5 °C and the maximum and minimum system pressures are 175 and 65 bar (1 bar = 14.5 psi). Rod diameter is 40 mm.

16.13 If the space for the accumulator is expensive, would you recommend increasing the pump pressure or the pump flow rate? Why?

16.14 Select a spring-loaded accumulator to replace that chosen in problem 16.12 in which the diameter is 25% of the height and where the spring at full compression requires 20% of the fluid volume.

16.15 Select a bladder-type accumulator to cushion the shock of temporarily blocking the fluid to a bidirectional motor as its direction of rotation is changed. Pump capacity is 160 gpm the line pressure is 2000 psi, the maximum rated pressure is 2300 psi, and the gas temperature is 75 °F. The pipe length is 65 ft and its internal diameter is 1.375 in. The specific gravity of the oil used is 0.834.

Section 16.17

16.16 A hydraulic cylinder having a 300-mm bore is fitted with a cushion and a spear that reduces the flow velocity according to the curve shown below. If the pump flow rate is 960 L/min, find the rod speed before the cushion acts and the deceleration force at point A in the figure, where

$$\frac{d\dot{Q}}{dx} = -400 \text{ L/m-s}$$

(*Hint*: Recall that the retarding force f satisfies

$$f = -m\frac{dv}{dt} = -m\frac{dv}{dx}\frac{dx}{dt} = -mv\frac{dv}{dx}$$

and that $v = \dot{Q}/A$.) The effective mass of the piston and its load is 800 kg.

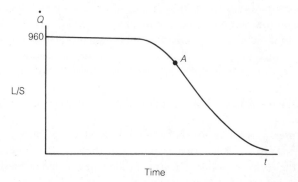

16.17 Pressure at the inlet port at the cap end of a hydraulic lift cylinder to elevate a 900 kg load is 14.0 bar. The cylinder bore is 140 mm, the stroke is 300 mm, and the friction between the rod and the head gland and the piston and the cylinder adds to 22.0 N. Find the extension time if the deceleration is 80 mm/s^2 and if the pump supplies 245 L/min.

16.18 Select a hydraulic motor to drive a rotary rock crusher at 10 rpm at an average torque of 1441 Nm. The motor should provide sufficient torque to accelerate the crusher to its operating speed in 1.5 seconds. The moment of inertia of the crusher is 2111 kg-m^2. Curves in Figure 16.22.1 indicate that it is reasonable to assume that the starting torque is constant during the acceleration time.

Sections 16.18–16.23

16.19 Why are jackhammers pneumatically (air) operated instead of hydraulically operated?

16.20 Select a pump to provide at least 2000 psi at 14 gpm to the hydraulic motor shown in the circuit in the figure. Pipe length is 21.3 ft, it has four bends between 45° and 180°, eight elbow adapters, two pairs of fittings, and two check valves. The check valves do not function simultaneously. Assume the relief valve does not function during normal operation and use the motor curves in Section 16.22.

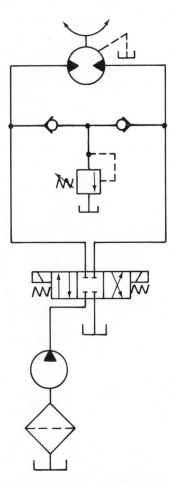

16.21 Sketch the circuit for a bidirectional variable displacement pump driving a bidirectional variable displacement motor. Include braking capability, a filter (the standard filter symbol is used), and return flow capability to prevent cavitation.

16.22 Would you recommend use of the circuit in the figure for problem 16.21 for general use? (*Hint*: Would it be satisfactory for high-inertia loads which may not stop quickly?)

APPENDIX C

DESIGN EXAMPLE

Design a custom, inexpensive, three-wheel repair vehicle to carry the operator and up to a 200-lb load for a customer's factory. Its most demanding performance requirement is that it be able to drive up a 20° slope at no less than 7.6 mph. An L_{10} life of 5 years at 50 weeks/year, 5 days/week, and 3 hours/day is desired for all components except for maintenance items such as tires, air filters, and spark plugs. Price is an important consideration, along with relatively low maintenance .

ESTIMATE DESIGN LOADS

Our first task is that of estimating the design load and the design power. If we use a canvas or fiberglass shell we estimate that we can hold the empty weight of the cart to about 600 lb, so that with a 200-lb payload and a 250-lb driver we have a design load of 1050 lb. We shall initially select 15-in.-diameter wheels because they are commonly used on golf carts and riding lawn mowers and are, therefore, readily available and relatively inexpensive, and we shall calculate the required torque from the geometry shown in Figure C1. We shall assume that the rolling friction for a vehicle of weight W for these wheels on a factory floor is 0.02W, based upon a rolling friction coefficient for dirt roads and automotive tires.[1] Thus,

$$T = Wr_{\mathbf{w}} \sin \theta + 0.02\, Wr_{\mathbf{w}} = 1050(7.5)(\sin 20° + 0.02) = 2850.9$$

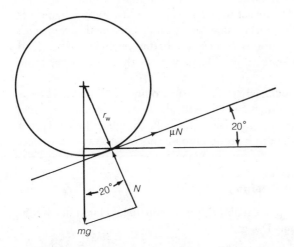

Figure C1 Torque acting on the rear wheels due to gravity.

The required horsepower may be found from

$$\text{hp} = Tn\pi/198\,000 = 2850.9(170.309)\pi/198\,000 = 7.70$$

using velocity $v = 7.6(5280)12/60 = 8025.6$ in./min from which we find that $n = v/\pi d = 8025.6/15\pi = 170.309$ rpm

ENGINE OR MOTOR SELECTION

Because of our assumption of a 600-lb total cart weight, we cannot consider electric power because of the battery weight. Hence, we must select a single-cylinder gasoline engine to drive the cart.

We now must decide whether we shall adhere strictly to the customer's specifications and provide the minimum machine to meet these requirements or whether we shall build in added assurance that the vehicle will satisfy the customer. In particular, shall we use a 7.5-hp engine, which almost meets the power requirement just estimated, and is made by few manufacturers, or shall we use an 8-hp engine, which provides some extra performance and is made by many manufacturers? After searching a supplier's catalog, we decide to use an 8-hp engine (shipping weight 55 to 78 lb) to have a selection of engines from more suppliers and to assure customer satisfaction since the 7.5-hp engine is almost the same weight (shipping weight 64 to 73 lb) and price.[2]

DRIVE TRAIN DESIGN

Clutch

Our next decision is that of transferring power from the motor output shaft to the wheels. We cannot use a direct drive because the starter would have to move the cart to start the engine. We must, therefore, elect to use a clutch to disconnect the power train during starting. (We shall let the marketing department decide whether they wish to have manual and electric starting options because the battery, ignition circuit, and starter motor—shipping weight 5 lb—probably will add no more than 25 lb to the gross weight.[2])

Accordingly, we shall next consider what clutch to recommend. Possible choices are (1) a manually operated plate clutch, which will allow us to use a transmission with more than one speed range, or (2) a centrifugal clutch, which limits us to a single-speed transmission. Since we are to hold down cost and weight, we shall reject the first choice.

Centrifugal Clutch Preliminary Design

According to the formula for the torque capability for a centrifugal clutch derived in Section 11.5 of Chapter 11 the torque T is given by

$$T = 4\mu r N \left[\left(\frac{\pi n}{30}\right)^2 mr_c - k_s \delta \right] \frac{\sin(\phi_0/2)}{\phi_0 + \sin \phi_0} \tag{C1}$$

Since the torque is zero at initial contact, we also have that

$$k\delta = mr_c\omega_1{}^2 \qquad \omega_1 = \pi n_1/30 \tag{C2}$$

where ω_1 is the speed at which the clutch initially engages. Substitution for $k\delta$ from equation (C2) into equation (C1) yields

$$mr_c(\omega^2 - \omega_1{}^2) = \frac{T}{4\mu rN}\frac{\phi_0 + \sin\phi_0}{\sin(\phi_0/2)} \tag{C3}$$

Thus, we may select any combination mass m and radius r_c to the center of gravity (cg) that satisfies equation (C3). Before we can reasonably select r_c for a centrifugal clutch we must select an input shaft diameter in order to find the space it will occupy. We may estimate the shaft diameter by solving hp $= Tn\pi/198\,000$ for the maximum torque and then substituting into the shear stress formula $\sigma_{z\theta} = Tr/J = 2T/(\pi r^3)$ for a round shaft.

From Figure C2 we find that the maximum torque is delivered at 2800 rpm when the engine is correctly tuned. Thus

$$T = 198\,000(8)/(2800\pi) = 180.072 \text{ in-lb}$$

Recall, incidentally, that the clutch should be placed on the motor output shaft because it operates at the highest rpm and, therefore, requires the lowest torque for the power specified.

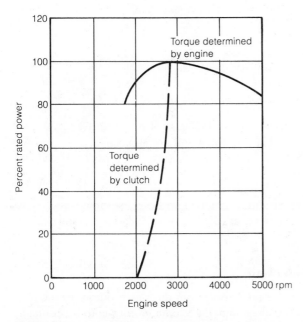

Figure C2 Torque–speed curve for two-cycle engine and centrifugal clutch.

We must next select the steel to be used in the shaft. From the steels listed in Chapter 2 we shall select UNS G10950, which has a yield stress in tension of 55 000 psi. According to the HMH criterion its yield stress in shear may be estimated from $\sigma_y = 55\,000/\sqrt{3} = 31754$ psi. If we use a safety factor of 2.5 and a stress concentration factor of 5 at the key, we find that

$$r = [2TK\zeta/(\pi\sigma_y)]^{1/3} = [2(180.072)2.5(5)/(31754\pi)]^{1/3} = 0.356 \text{ in.}$$

for a diameter of 0.712 in. We shall increase that 0.75 in. in order to use a standard bar diameter and eliminate machining.

If we assume a nominal 1.0 in. diameter hub to attach the weights, springs and guides it appears that a drum diameter of 3.5 in. may be a reasonable first approximation. We shall initially assume that each of the two weights (based upon the chain saw centrifugal clutch configuration) occupies a 90° arc and substitute these values into (C3) to obtain

$$mr_c = \frac{181}{\omega^2 - \omega_1{}^2} \frac{(\pi/2 + 1)2}{4(0.4)3.5(2)} = \frac{58.75}{\omega^2 - \omega_1{}^2}$$

The difference in the square of the angular velocities in the denominator implies that ω for full torque delivery must be close to ω_1 if the mass is to be large enough to have the structural strength to transmit the necessary torque. Since most small gasoline engines start at about 1750 rpm, we shall design for initial engagement at 2000 rpm (so the clutch does not engage as the engine is started) and shall consider full torque at 2050 rpm. Thus, $\omega_1 = 2000\pi/30 = 204.440$ rad/s and $\omega = 2050\pi/30 = 214.675$ rad/s. Let $r_c = 3.0$ in., so that from

$$mr_c = \frac{58.75}{\left(\dfrac{2050}{30}\right)^2 - \left(\dfrac{2000}{30}\right)^2} = 0.0265$$

we have that $m = 0.00882$ slugs $= 0.284$ lb $= 4.54$ oz. Use of full torque at 2050 rpm, rather than the rated torque, assures that the clutch will not slip.

Examination of product information from manufacturers of centrifugal clutches shows that at least one manufacturer offers a clutch with an 8 hp capacity that engages at 2000 rpm. Its diameter is 3.5 in., its length (the axial dimension) is 2.06 in. and its bore is 0.75 in. Although it is offered with a different sprocket, it should be easy to change sprockets. Larger power ratings are available in 4.5 in. diameters and 2.25 in. lengths.

Gear, Belt, or Chain Drive Selection

We shall next turn our attention to the drive train from the clutch output to the rear wheels. We shall select a speed ratio that permits the maximum torque to be available at a ground speed of 7.6 mph. Since maximum torque will be available if the engine turns at 2800 rpm, the speed ratio for the drive train must be

$$n_e/n_w = 2800/170.309 = 16.441.$$

in terms of the engine speed n_e. Unless the engine is carefully tuned at all times, the maximum torque may not always occur at 2800 rpm. We shall assume maximum torque may occur from 2750 to 2850 rpm, corresponding in a speed ratio variation from 16.147 to 16.734.

All gear drives will not be considered because of the expense of bevel gears to connect the clutch to the rear axle and of the weight of the gear housings at each end of the drive shaft. We shall, therefore, consider a belt or a chain drive with a spur gear reducer following the clutch. The speed ratio available from either a belt or chain drive will be limited by the diameter of the large sheave or sprocket, which must be small enough relative to the tire diameter to assure reasonable ground clearance. We shall assume reasonable ground clearance to be 3.5 in., so that the sheave/sprocket diameter must be no more than 8.0 in. Since the chain formulas are based upon a life of 15 000 h and the belt formulas on 20 000 h, we can easily satisfy the required life of 3(5)50(5) = 3750 h.

Belt Drive Selection

First consider a belt drive. Turn to Table 12.5.2 to find that the smallest commercial sheave diameter is 2.2 in. Consequently, the small sheave rpm must be 170.309(8)/2.2 = 619.306 rpm for a speed ratio of 3.64. If we consider the cart service loads to be similar to a sand or gravel conveyor belt, we may find the service factor from Table 12.5.1 to be 1.3. Thus, the design horsepower, Dhp, is given by Dhp = 1:3(8) = 10.4. Use Figure 12.5.1 to find that either a 3VX or a 5VX belt is recommended. From Table 12.5.3 we find that the pitch diameter d_p of the small sheave is given by $d_p = 2.2 - 0.05 = 2.15$ in., so the large sheave diameter may be found from $D_p = 2.15(3.64) = 7.826$. This corresponds to an outside diameter of 7.87 in. The closest standard sheave diameter is 8.00 in., for a pitch diameter of 7.95 in. Use of these two sheaves provides a speed ratio of 7.95/2.15 = 3.70, so that the gear train between the engine and the small sheave must have a speed ratio between 4.364 and 4.523 to obtain the design speed for the cart.

If we use a trial center distance of 12.00 in. in equation 12.5.10 we find a belt length of 40.72 in., which is near the commercial length of 40.00 in. for the 3VX400 belt from Table 12.5.6. Substitution of $L = 40.00$ in. into Equation 12.5.11 yields a center distance of 11.617 in., which is satisfactory. Upon turning to equations 12.5.12 and 12.5.13 and the appropriate tables for the constants for the horsepower per belt we find that it can supply no more than 0.185 hp/belt.

Since the smallest sheave diameter for the 5VX belts is 4.40 in., it will not provide a speed ratio near 3.64 without using too large a sheave on the rear axle. Belt drives are, therefore, eliminated from consideration.

Chain Drive Selection

To select a chain drive we turn to Table 12.11.1 to find that a service factor of 1.2 is recommended for a smooth load driven by an internal combustion engine with a mechanical drive. Thus, the design horsepower is Dhp = 8(1.2) = 9.6.

Next, read up to 9.6 hp in the single strand column in Figure 12.11.1 and then over to 700 rpm (arbitrarily chosen for a 4:1 ratio relative to 2800 rpm) to find that a number 50 chain is the suggested first choice. This chain has a pitch of 0.625 in., so from equation 12.10.2 we find the number of teeth in the 8.0 in. sprocket on the rear axle to be

$$N_2 = \pi/\sin^{-1}(1/D) = \pi/\sin^{-1}(1/12.8) \simeq 40$$

where $D = 12.8$ is the diameter of the large sprocket in units of pitch. From equations 12.10.8 and 12.10.9 we find that the small sprocket must have at least 21 teeth in order to deliver 9.6 hp at 700 rpm, corresponding to a speed ratio of 1.9. Use of three strands will reduce the small sprocket to one with 9 teeth, for a speed ratio of 4.44.

In an attempt to use fewer strands, we shall consider a number 60 chain and a pitch of 0.750 in. The large sprocket diameter is 10.667 pitches for this chain, so that according to equation 12.10.2

$$N_2 = \pi/\sin^{-1}(1/D) = \pi/\sin^{-1}(1/10.667) = 33.46$$

Set $N_2 = 33$ teeth for the large sprocket and set $N_1 = 9$ teeth for the small sprocket, giving a speed ratio of 3.667. From equations (12.10.8) and (12.10.9) we find that two strands of number 60 chain are required.

Before proceeding we shall turn to Table 12.11.3 to find the maximum bore diameter for a 9 tooth sprocket with a 3/4 in. pitch. Since the table stops at 11 teeth we may extrapolate to 9 teeth by observing that the maximum bore decreases approximately 1/4 in. for a decrease of one tooth in the range between 14 and 11 teeth. Thus, we estimate that the maximum bore for a 9 tooth sprocket to be 3/4 in.

To check the adequacy of this shaft diameter we shall first turn to the horsepower formula, written as

$$T = \frac{198\,000 \text{ hp}}{n\pi} = \frac{198\,000(8)}{624.528\pi} = 807.341 \text{ in-lb}$$

where $n = 3.667(170.309) = 624.523$ rpm. If we use a output shaft from the gear train between the motor and the small sprocket made from normalized UNS G10950 steel we have $\sigma_{y_s} = \sigma_{y_t}/\sqrt{3} = 55\,000/\sqrt{3} = 31\,754$ psi, so that the shaft radius for a safety factor of 2 is given by

$$r = \left(\frac{2T\zeta}{\pi\sigma_\theta}\right)^{1/3} = \left[\frac{2(807.341)^2}{31754\pi}\right]^{1/3} = 0.319 \text{ in.}$$

for a minimum shaft diameter of 0.638 in. Increasing the diameter to 0.750 in. will increase the safety factor to 3.249. Hence, we can accept a 3/4 in. output shaft from the engine driven gear train and use a 9 tooth sprocket for a 60 chain.

Chain length formula 12.10.5 in Chapter 12 may be used here to find the number of pitches in the drive chain for a given center distance in units of pitch. Upon entering an initial center distance of $c = 12$ in. $= 16$ pitches along with $N_1 = 9$ and $N_2 = 33$ we find that a chain length of 53.916 pitches is required. Change this length to 54 pitches (the nearest even number of pitches) and substitute into equation 12.10.6a to find a center distance of 16.0431 pitches or 12.032 in. This value differs from the length found from the exact relation, equation 12.10.6, by less than 0.0001 in.

Reduction Gear Train Design

The last step in designing the drive train is that of designing the reduction gear box between the centrifugal clutch and the small sprocket of the chain drive. Since the overall speed ratio may vary from 16.15 to 16.73 and since the chain drive speed ratio is $\frac{33}{9} = 3.67$ the speed ratio between the input and output shafts of the reduction gear box may lie anywhere between $16.15/3.67 = 4.40$ and $16.73/3.67 = 4.55$

Although we could use the automatic search program described in Ref. 4, we will find that it is not necessary in this instance if we observe that $4.5 = (2/1)(2.25/1)$ lies within the acceptable speed range. These ratios may be written as

$$\frac{N_2}{N_1}\frac{N_4}{N_3} = \frac{36}{18}\frac{45}{20}$$

for a 4-gear train which may be arranged as in Figure C3. Numbers N_1 and N_2 have been chosen so that no undercutting will occur for either 14.5 or 20° pressure angles.

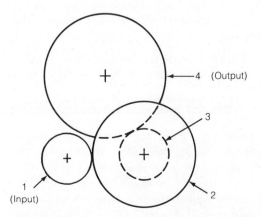

Figure C3 Gear train arrangement (reversing mechanism not shown).

REAR AXLE DESIGN

The minimum rear axle diameter as determined on the basis of shear stress may be found by assuming that 30% of the vehicle weight (180 lb), all of the load (200 lb), and the weight of the operator (250 lb, when standing in the truck bed to reach higher, for example) act on one shear plane at the edge of the wheel bearing. Thus,

$$r = [\zeta F/(\pi\sigma_y]^{1/2} = [2(630)/(\pi 31\,754]^{1/2} = 0.1124 \text{ in.}$$

$$d = 0.2245 \text{ in.}$$

for UNS G10950 steel and a safety factor of 2.0. The minimum radius on the basis of torsional loading may be found by assuming that 8 hp is delivered entirely to one wheel when the vehicle moves at 7.6 mph. When this occurs

$$T = 198\,000(8)/(170.309\pi) = 2960.52 \text{ in.-lb}$$

is the transmitted torque, so that for a safety factor of 2

$$r = [2T\zeta/(\pi\sigma_{y_s})]^{1/3} = [2(2960.52)2/(31\,754\pi)]^{1/3} = 0.4915 \text{ in.}$$

$$d = 0.9830 \simeq 1.000 \text{ in.}$$

ALTERNATIVE TO AN AUTOMOTIVE DIFFERENTIAL

Our next problem is that of either avoiding or reducing skidding of one of the driven rear wheels during a turn. An automotive differential will be our last choice because of its expense and its weight. In an attempt to find a cheap and lightweight alternative we may consider the following two mechanisms.

1. Drive only one wheel of two rear wheels. This has been used in some inexpensive carts. It causes uneven rear tire wear and may cause the cart to spin if two wheels are on a slippery surface and the drive wheel is not. We shall not use this alternative because of the frequency of oil spills in many factories and the attendant possibility of an accident.

2. A permanently engaged slip clutch with a single friction surface to provide continuous power to one wheel and to allow the other to slip when the ground friction exceeds a predetermined value.

The maximum torque delivered to the wheel that is allowed to slip in the second alternative may be chosen by assuming a load of 500 lb, and a maximum friction coefficient between tire and ground that we will accept before slip. This is a compromise between a one-wheel drive system (no slip clutch) that will reduce tire wear on the opposite wheel (by allowing it to roll freely) and supply full torque to the driving wheel during turns and a slip clutch that can deliver more power to the clutched wheel but at the expense of greater tire wear and reduced torque to the drive wheel during turns. If we take ice and loose snow as equivalent to spilled water, oil, or grease, whose friction coefficients for tires range from 0.02 to 0.2, we shall select that torque that would allow both wheels to drive when $\mu = 0.2$. Thus

$$T = \mu W r = 0.2(500)7.5 = 750 \text{ in-lb}$$

Recall the results of problem 11.21 to select the clutch diameter and to select the springs needed to supply the necessary contact force. Solve the torque expression for the outside radius r_o using a lining pressure of 200 psi to find that

$$r_o = \left(\frac{3\sqrt{3}T}{2\pi\mu p_{max}}\right)^{1/3} = \left[\frac{3\sqrt{3}\,750}{2\pi 0.4(200)}\right]^{1/3} = 1.979 \text{ in.}$$

We shall accept these values because the clutch lining will wear very slowly for this low contact pressure. The associated engagement force for a lining friction coefficient of 0.4 is given by

$$F = 2\pi p_{max}\left(1 - \frac{1}{\sqrt{3}}\right)\frac{r_o^2}{\sqrt{3}} = 2\pi 200.0(0.4226)\frac{1.98^2}{\sqrt{3}} = 507.18 \text{ lb}$$

As noted above, this is a compromise between having enough power to move the vehicle on level ground whenever the other wheel is on a slippery surface and having a slip clutch with small enough maximum torque to allow it to be overridden by the wheel during a turn.

The clutch and large sprocket will be enclosed in the same welded housing, with one clutch plate attached to the sprocket. The sprocket itself may be cantilevered from the housing and the mating clutch pressure plate and lining splined to a shaft cantilevered from the opposite side of the housing, and counterbored to fit over a stub shaft extension of the sprocket shaft so that when assembled the two shafts will be equivalent to a shaft extending from one side of the housing to the other. A Belleville spring between a reaction plate pinned to the shaft and the clutch pressure plate may apply force F.

Axial force may be applied by enclosing the clutch in a cylindrical housing that is welded to the sprocket at one end and threaded at the other. A threaded cap with a hole for the shaft is tightened against the Belleville spring and the cylinder length is selected to give the correct spring compression when the cap is fully tightened.

A clutch of similar construction, but for lower torque and located near the middle of the rear axle, has been used by at least one riding lawnmower manufacturer.

BEARING SELECTION

The last design problem we shall consider is that of selecting bearing for the three wheels. If we use journal bearings we will also have to provide thrust bearings to prevent the axial load from either side from adding to the engagement force on the slip clutch which we have used in place of a differential. Addition of structure to resist axial thrust in either direction will require either additional rigidity of the frame or a cast housing around the slip clutch and rear axle. (Farm tractors use cast steel housings and shafts that have adjacent radial and thrust bearings on the front wheels. Riding lawn mowers may use journal bearings and thrust washers.)

Bearing selection will be based upon a vertical load of 400 lb on the front wheel and of 500 lb on each of the rear wheels. The sum of these loads exceeds 1050 lb because we have attempted to partially account for unknown load distributions. Since the loads are not significantly different in terms of bearing selection, we shall use the same bearings on all three wheels.

An axial load of 342 lb on each wheel will be assumed. This corresponds to a vertical load of 1000 lb on one wheel when the cart is on a 20° incline, and is justified by the assumption that each wheel will only have inboard wheel bearings. Hence $F_a/F_r = \tan 20° = 0.354$. The L_{10} life in cycles for the prescribed five year bearing life becomes

$$170.309(60)3(5)50(5) = 38\,319\,525 \text{ cycles}$$

We may first search for an angular contact radial ball bearing because of the combined axial and radial loading. Upon calculating the life of the angular contact radial ball bearing 7305B, which has a 25 mm bore, listed in Table 10.4.5 (p. 587), we find that $F_a/F_r < e = 1.14$ so $X = 1.0$, $Y = 0.0$, and $L_{10} = 75.15$ milllion cycles. Since this is almost twice the required life, we shall consider radial contact ball bearings from Table 10.4.4 (p. 586), inasmuch as they can resist some axial loading. Bearing 6205RS has a suitable 25 mm bore but has only a L_{10} life of 16.53 million cycles under these load conditions. If we press fit a shoulder on the shaft, however, we can use bearing 6206RS, which has a 30 mm bore, $C_o = 2250$ lb, and $C = 3360$. From $F_a/C_o = 0.152$ we find by interpolation that $e = 0.328$ so that F_a/F_r is larger than e and $X = 0.56$, $Y = 1.352$, and $P = 988.64$ lb. Thus, $L_{10} = 39\,255\,752$ cycles, which is only 2.4% longer than the required life.

In this example we have considered a cart with two rear wheels because it required analysis of more components than would a cart with two wheels in front and one driving wheel in the rear. Now that we have completed the design, it is clear that the one rear wheel has at least two advantages and one disadvantage. The advantages are (1) that no differential or slip clutch is required and (2) that the cart has greater stability when turning. The major disadvantage is that the vehicle stops when the single drive wheel loses traction. Three-wheel carts with a single rear drive wheel have been common in Switzerland.

COMMENT

This example has been concerned with a design that compromised performance for price. The balance between design quality and price differs from one industry to another and may change with time. At this time the least compromise, or highest quality, appears to be in the design of military aircraft, with commercial aircraft a close second. Industrial and military equipment, commerical ground transportation, private aircraft, automobiles, and home appliances appear to emphasize performance over price in the order listed. The engineering in farm equipment seems to range from very good to abominable; in private aircraft and automobiles it appears to range from good to poor.

PROBLEMS

C.1   Select the tooth size for the reduction gear train, calculate the gear diameters, find the center distances, recommend either ball, roller, or journal bearings. If journal bearings are selected, give the recommended oil level. Give reasons for your choices.

C.2   No reversing mechanism was designed for the cart. Suggest a modification to the reduction gear to allow the operator to shift to reverse when the clutch is not engaged.

C.3   Select the bearings for the reduction gear train with the reversing mechanism included.

C.4   Complete the design of the clutch used as an alternative to an automotive differential. Show the housing, indicate whether it is a casting or welded assembly , and indicate whether sheet metal may be used for any part of it. Briefly describe its assembly procedure.

REFERENCES

1.   *Automotive Handbook*, 18th German ed., 1st English ed., Robert Bosch GmbH, Stuttgart, 1976

2.   *Grainger's* Catalog No. 371, Spring 1987, W.W. Grainger, Inc. Chicago, IL.

3.   Orthwein, W.C., *Clutches and Brakes,* Marcel Dekker, New York, 1986.

4.   Orthwein, W.C., Finding gear ratios with a personal computer, *Computers in Mechanical Engineering*, 4 (1), 36–40 (July 1985).

5.   *Charts and Tables for Stopping Distances of Motor Vechicles*, PN 22, Traffic Institute, Northwestern University, Evanston, IL, 1978, p. 12.

INDEX